DATE DUE

VOLUME FIVE HUNDRED AND SIXTY FIVE

METHODS IN
ENZYMOLOGY

Isotope Labeling of Biomolecules - Labeling Methods

METHODS IN ENZYMOLOGY

Editors-in-Chief

JOHN N. ABELSON and MELVIN I. SIMON
Division of Biology
California Institute of Technology
Pasadena, California

ANNA MARIE PYLE
Departments of Molecular, Cellular and Developmental
Biology and Department of Chemistry Investigator
Howard Hughes Medical Institute
Yale University

DAVID W. CHRISTIANSON
Roy and Diana Vagelos Laboratories
Department of Chemistry
University of Pennsylvania
Philadelphia, PA

Founding Editors

SIDNEY P. COLOWICK and NATHAN O. KAPLAN

VOLUME FIVE HUNDRED AND SIXTY FIVE

METHODS IN ENZYMOLOGY

Isotope Labeling of Biomolecules - Labeling Methods

Edited by

ZVI KELMAN

Biomolecular Labeling Laboratory, and Biomolecular Structure and Function Group, Institute for Bioscience and Biotechnology Research, National Institute of Standards and Technology and the University of Maryland, Rockville, Maryland, USA

AMSTERDAM • BOSTON • HEIDELBERG • LONDON
NEW YORK • OXFORD • PARIS • SAN DIEGO
SAN FRANCISCO • SINGAPORE • SYDNEY • TOKYO

Academic Press is an imprint of Elsevier

Academic Press is an imprint of Elsevier
225 Wyman Street, Waltham, MA 02451, USA
525 B Street, Suite 1800, San Diego, CA 92101-4495, USA
The Boulevard, Langford Lane, Kidlington, Oxford OX5 1GB, UK
125 London Wall, London, EC2Y 5AS, UK

First edition 2015

Copyright © 2015 Elsevier Inc. All rights reserved.

No part of this publication may be reproduced or transmitted in any form or by any means, electronic or mechanical, including photocopying, recording, or any information storage and retrieval system, without permission in writing from the publisher. Details on how to seek permission, further information about the Publisher's permissions policies and our arrangements with organizations such as the Copyright Clearance Center and the Copyright Licensing Agency, can be found at our website: www.elsevier.com/permissions.

This book and the individual contributions contained in it are protected under copyright by the Publisher (other than as may be noted herein).

Notices
Knowledge and best practice in this field are constantly changing. As new research and experience broaden our understanding, changes in research methods, professional practices, or medical treatment may become necessary.

Practitioners and researchers must always rely on their own experience and knowledge in evaluating and using any information, methods, compounds, or experiments described herein. In using such information or methods they should be mindful of their own safety and the safety of others, including parties for whom they have a professional responsibility.

To the fullest extent of the law, neither the Publisher nor the authors, contributors, or editors, assume any liability for any injury and/or damage to persons or property as a matter of products liability, negligence or otherwise, or from any use or operation of any methods, products, instructions, or ideas contained in the material herein.

ISBN: 978-0-12-803048-6
ISSN: 0076-6879

For information on all Academic Press publications
visit our website at http://store.elsevier.com/

CONTENTS

Contributors   xiii
Preface   xix

Section I
Labeling in Prokarya

1. Robust High-Yield Methodologies for ^{2}H and ^{2}H/^{15}N/^{13}C Labeling of Proteins for Structural Investigations Using Neutron Scattering and NMR   3

Anthony P. Duff, Karyn L. Wilde, Agata Rekas, Vanessa Lake, and Peter J. Holden

1. Introduction   4
2. Media Preparation   7
3. Unlabeled Protein Production   9
4. Deuterated Protein Production   14
5. Multiple Labeling of Proteins for NMR   17
6. Comments on the Method   18
7. Typical Deuteration Levels   22
Acknowledgments   23
References   23

2. Protein Labeling in *Escherichia coli* with ^{2}H, ^{13}C, and ^{15}N   27

J. Todd Hoopes, Margaret A. Elberson, Renae J. Preston, Prasad T. Reddy, and Zvi Kelman

1. Introduction   28
2. Selection of Induction Method   29
3. Special Considerations When Labeling with Deuterium   30
4. Plasmid and *E. coli* Strain Selection   30
5. Determination of Isotope Incorporation   32
6. Acclimation of *E. coli* to Growth in D_2O 35
7. Media Preparation   36
8. Protein Expression   40
Acknowledgments   43
References   43

3. *Escherichia coli* Auxotroph Host Strains for Amino Acid-Selective Isotope Labeling of Recombinant Proteins   45

Myat T. Lin, Risako Fukazawa, Yoshiharu Miyajima-Nakano, Shinichi Matsushita, Sylvia K. Choi, Toshio Iwasaki, and Robert B. Gennis

1. Introduction   46
2. *E. coli* Auxotrophs for Amino Acid-Selective Isotope Labeling   47
3. Methods   53
4. Conclusions   64
Acknowledgments   65
References   65

4. ^{19}F-Modified Proteins and ^{19}F-Containing Ligands as Tools in Solution NMR Studies of Protein Interactions   67

Naima G. Sharaf and Angela M. Gronenborn

1. Introduction   68
2. Protocol 1: Biosynthetic Amino Acid Type-Specific Incorporation of ^{19}F-Modified Aromatic Amino Acids   73
3. Protocol 2: Site-Specific Incorporation of Fluorinated Amino Acids Using a Recombinantly Expressed Orthogonal Amber tRNA/tRNA Synthetase Pair in *E. coli*   79
4. General Considerations for ^{19}F-Observe NMR Experiments   84
5. ^{19}F-Modified Protein-Observe NMR Experiments   86
6. NMR Experiments with ^{19}F-Containing Ligands   89
Acknowledgments   91
References   92

5. Biopolymer Deuteration for Neutron Scattering and Other Isotope-Sensitive Techniques   97

Robert A. Russell, Christopher J. Garvey, Tamim A. Darwish, L. John R. Foster, and Peter J. Holden

1. Introduction   98
2. Deuterated Biopolyesters   100
3. Deuterated Chitosan   106
4. Deuterated Cellulose   111
Acknowledgments   117
References   118

6. Production of Bacterial Cellulose with Controlled Deuterium–Hydrogen Substitution for Neutron Scattering Studies   123

Hugh O'Neill, Riddhi Shah, Barbara R. Evans, Junhong He, Sai Venkatesh Pingali, Shishir P. S. Chundawat, A. Daniel Jones, Paul Langan, Brian H. Davison, and Volker Urban

1. Introduction   124
2. The Occurrence and Properties of Cellulose   127
3. Deuteration of Bacterial Cellulose   129
4. Characterization of Deuterated Cellulose   133
Acknowledgments   142
References   142

7. Isotopic Labeling of Proteins in *Halobacterium salinarum*   147

Thomas E. Cleveland IV and Zvi Kelman

1. Introduction   148
2. Growth and Maintenance of *Halobacterium salinarum*   149
3. Purification of Proteins from *Halobacterium salinarum*   153
4. Summary   163
Acknowledgments   164
References   164

8. Amino Acid Selective Unlabeling in Protein NMR Spectroscopy   167

Chinmayi Prasanna, Abhinav Dubey, and Hanudatta S. Atreya

1. Introduction   168
2. Method Description   171
3. Applications of Selective Unlabeling   182
4. Conclusions   186
Acknowledgments   187
References   187

Section II
Labeling in Eukarya

9. Isotope Labeling of Eukaryotic Membrane Proteins in Yeast for Solid-State NMR   193

Ying Fan, Sanaz Emami, Rachel Munro, Vladimir Ladizhansky, and Leonid S. Brown

1. Introduction   194
2. Expression and Isotope Labeling in *P. pastoris*: Background   196

3. Isotope Labeling of Membrane Proteins in *P. pastoris*	197
4. Outlook	207
Acknowledgments	207
References	207

10. Development of Approaches for Deuterium Incorporation in Plants — 213
Barbara R. Evans and Riddhi Shah

1. Introduction	214
2. Challenges of Plant Cultivation in D_2O	215
3. Analysis of Deuterium-Labeled Plant Biomass	220
4. Deuterium Labeling of Plants for Metabolic Studies	223
5. Production of Deuterated Plants for Structural Studies	225
Acknowledgments	239
References	239

11. Isotope Labeling of Proteins in Insect Cells — 245
Lukasz Skora, Binesh Shrestha, and Alvar D. Gossert

1. Insect Cells as Expression System	246
2. General Considerations for Isotope Labeling in Insect Cells	250
3. Amino Acid Type-Specific Isotope Labeling in Insect Cells	252
4. Uniform Isotope Labeling in Insect Cells	256
5. Applications	268
6. Protocols	274
References	284

12. Effective Isotope Labeling of Proteins in a Mammalian Expression System — 289
Mallika Sastry, Carole A. Bewley, and Peter D. Kwong

1. Introduction	290
2. Overview of Mammalian Expression	291
3. Protein Expression	299
4. NMR Characterization of Expressed Protein	300
5. Conclusions	302
6. Materials	303
Acknowledgments	303
References	303

Section III
In Vitro Labeling

13. *Escherichia coli* Cell-Free Protein Synthesis and Isotope Labeling of Mammalian Proteins — 311
Takaho Terada and Shigeyuki Yokoyama

1. Introduction — 312
2. The *E. coli* Cell-Free Protein Synthesis Method — 316
3. Stable Isotope Labeling of Proteins — 323
Acknowledgments — 339
References — 340

14. Rapid Biosynthesis of Stable Isotope-Labeled Peptides from a Reconstituted *In Vitro* Translation System for Targeted Proteomics — 347
Feng Xian, Shuwei Li, and Siqi Liu

1. Introduction — 348
2. Equipment, Materials, and Buffers — 350
3. Section 1: DNA Template Preparation — 351
4. Section 2: Peptide Synthesis with PURE System — 355
5. Section 3: Enrichment and Digestion of Synthesized Peptide — 358
6. Section 4: Quantification of PURE-Synthesized Peptide — 360
7. An Example — 363
8. Summary and Discussion — 365
Acknowledgment — 365
References — 365

15. Labeling of Membrane Proteins by Cell-Free Expression — 367
Aisha LaGuerre, Frank Löhr, Frank Bernhard, and Volker Dötsch

1. Introduction — 368
2. Core Considerations for the Cell-Free Generation of MP Samples — 369
3. Specific Challenges of NMR Studies with MPs — 372
4. An Emerging Perspective: NMR with NDs — 374
5. Labeling of Cell-Free Synthesized MPs with Stable Isotopes — 375
6. Reducing Scrambling Problems — 376
7. Perdeuteration of Cell-Free Synthesized MPs — 380
8. Conclusion — 382
Acknowledgments — 382
References — 382

16. Selective Amino Acid Segmental Labeling of Multi-Domain Proteins   389
Erich Michel and Frédéric H.-T. Allain

1. Introduction   390
2. Methods   396
3. Conclusion   417
Acknowledgments   419
References   419

17. Labeling Monosaccharides With Stable Isotopes   423
Wenhui Zhang, Shikai Zhao, and Anthony S. Serianni

1. Introduction   424
2. Terminology to Describe Different Monosaccharide Isotopomers   425
3. Introducing ^{13}C into Monosaccharides   427
4. Multiple Labeling of Aldoses Via Chain Inversion   438
5. Labeling at the Internal Carbons of Aldoses   439
6. Extension to Biologically Important Aldoses   442
7. Relative Carbonyl Reactivities in Osones—Synthesis of Labeled 2-Ketoses   444
8. Manipulation of Three-Carbon Building Blocks in Enzyme-Mediated Aldol Condensation   445
9. Manipulation of Isotopically Labeled D-Fructose 17 and L-Sorbose 25   452
10. Concluding Remarks   452
References   455

Section IV
RNA Labeling

18. Stable Isotope-Labeled RNA Phosphoramidites to Facilitate Dynamics by NMR   461
Christoph H. Wunderlich, Michael A. Juen, Regan M. LeBlanc, Andrew P. Longhini, T. Kwaku Dayie, and Christoph Kreutz

1. Theory   462
2. Equipment   463
3. Materials   464
4. Protocol   467
5. Step 1: Synthesis of 6-^{13}C-Uridine TOM Phosphoramidite   468
6. Step 2: Synthesis of 6-^{13}C-Cytidine TOM Phosphoramidite   473
7. Step 3: Chemical RNA Synthesis   477
8. Step 4: Applications   484
9. Conclusions   490
Acknowledgments   491
References   492

19. *In Vivo*, Large-Scale Preparation of Uniformly ^{15}N- and Site-Specifically ^{13}C-Labeled Homogeneous, Recombinant RNA for NMR Studies — 495

My T. Le, Rachel E. Brown, Anne E. Simon, and T. Kwaku Dayie

1. Theory — 497
2. Equipment — 499
3. Materials — 502
4. Protocol — 510
5. Step 1: Pilot of the Expression of the Recombinant tRNA-Scaffold Plasmid in Wild-Type K12 *E. coli* — 512
6. Step 2: Double Selection of High-Expressing *E. coli* Clones — 514
7. Step 3: Large-Scale Expression in Labeled SPG Minimal Media — 517
8. Step 4: Total Cellular RNA Extraction — 518
9. Step 5a: Purification of the Recombinant tRNA-Scaffold Using Anion-Exchange Chromatography — 520
10. Step 5b: Purification of the Recombinant tRNA-Scaffold Using Affinity Chromatography — 523
11. Step 6: Excision and Purification of the RNA of Interest — 524
12. Step 7: NMR Applications — 527
13. Conclusion — 530
Acknowledgments — 532
References — 532

20. Cut and Paste RNA for Nuclear Magnetic Resonance, Paramagnetic Resonance Enhancement, and Electron Paramagnetic Resonance Structural Studies — 537

Olivier Duss, Nana Diarra dit Konté, and Frédéric H.-T. Allain

1. Introduction — 538
2. Cut and Paste RNA Approach — 540
3. Production of Small (<10 nts) Isotopically Labeled RNAs — 546
4. Protocol A: Production of Small Spin-Labeled RNA Fragments — 552
5. Protocol B: Production of Unlabeled and Isotopically Labeled RNA Fragments — 554
6. Protocol C: Ligation — 557
7. Summary and Outlook — 559
Acknowledgments — 560
References — 561

Author Index — *563*
Subject Index — *595*

CONTRIBUTORS

Frédéric H.-T. Allain
Institute for Molecular Biology and Biophysics, ETH Zürich, Zürich, Switzerland

Hanudatta S. Atreya
NMR Research Centre, Indian Institute of Science, Bangalore, Karnataka, India

Frank Bernhard
Institute of Biophysical Chemistry, Centre for Biomolecular Magnetic Resonance, J.W. Goethe-University, Frankfurt-am-Main, Germany

Carole A. Bewley
Laboratory of Bioorganic Chemistry, National Institute of Diabetes and Digestive and Kidney Diseases, National Institutes of Health, Bethesda, Maryland, USA

Leonid S. Brown
Department of Physics, and Biophysics Interdepartmental Group, University of Guelph, Guelph, Ontario, Canada

Rachel E. Brown
Department of Chemistry and Biochemistry, Department of Cellular Biology and Molecular Genetics, Center for Biomolecular Structure and Organization, University of Maryland, College Park, Maryland, USA

Sylvia K. Choi
Center for Biophysics and Computational Biology, University of Illinois at Urbana-Champaign, Urbana, Illinois, USA

Shishir P.S. Chundawat
Department of Chemical and Biochemical Engineering, Rutgers University, Piscataway, New Jersey, USA

Thomas E. Cleveland IV
NIST Center for Neutron Research, Gaithersburg, and Biomolecular Structure and Function Group, Institute for Bioscience and Biotechnology Research, National Institute of Standards and Technology and the University of Maryland, Rockville, Maryland, USA

Tamim A. Darwish
National Deuteration Facility, Bragg Institute, Australian Nuclear Science and Technology Organisation, New South Wales, Australia

Brian H. Davison
Biosciences Division, Oak Ridge National Laboratory, Oak Ridge, Tennessee, USA

T. Kwaku Dayie
Department of Chemistry and Biochemistry, Center for Biomolecular Structure and Organization, University of Maryland, College Park, Maryland, USA

Nana Diarra dit Konté
Institute for Molecular Biology and Biophysics, ETH Zürich, Zürich, Switzerland

Volker Dötsch
Institute of Biophysical Chemistry, Centre for Biomolecular Magnetic Resonance, J.W. Goethe-University, Frankfurt-am-Main, Germany

Abhinav Dubey
NMR Research Centre, and Institute Mathematics Initiative, Indian Institute of Science, Bangalore, Karnataka, India

Anthony P. Duff
National Deuteration Facility, Bragg Institute, ANSTO, Lucas Heights, New South Wales, Australia

Olivier Duss
Institute for Molecular Biology and Biophysics, ETH Zürich, Zürich, Switzerland

Margaret A. Elberson
Biomolecular Labeling Laboratory, Institute for Bioscience and Biotechnology Research, National Institute of Standards and Technology and the University of Maryland, Rockville, Maryland, USA

Sanaz Emami
Department of Physics, and Biophysics Interdepartmental Group, University of Guelph, Guelph, Ontario, Canada

Barbara R. Evans
Chemical Sciences Division, Oak Ridge National Laboratory, Oak Ridge, Tennessee, USA

Ying Fan
Department of Physics, and Biophysics Interdepartmental Group, University of Guelph, Guelph, Ontario, Canada

L. John R. Foster
Bio/Polymer Research Group, Centre for Advanced Macromolecular Design, School of Biotechnology & Biomolecular Science, University of New South Wales, Sydney, New South Wales, Australia

Risako Fukazawa
Department of Biochemistry and Molecular Biology, Nippon Medical School, Bunkyo-ku, Tokyo, Japan

Christopher J. Garvey
National Deuteration Facility, Bragg Institute, Australian Nuclear Science and Technology Organisation, New South Wales, Australia

Robert B. Gennis
Department of Biochemistry, University of Illinois at Urbana-Champaign, Urbana, Illinois, USA

Alvar D. Gossert
Novartis Institutes of BioMedical Research, Novartis Campus, Basel, Switzerland

Angela M. Gronenborn
Department of Structural Biology, University of Pittsburgh School of Medicine, Pittsburgh, Pennsylvania, USA

Junhong He
Biology and Soft Matter Division, Oak Ridge National Laboratory, Oak Ridge, Tennessee, USA

Peter J. Holden
National Deuteration Facility, Bragg Institute, Australian Nuclear Science and Technology Organisation, and Bio/Polymer Research Group, Centre for Advanced Macromolecular Design, School of Biotechnology & Biomolecular Science, University of New South Wales, Sydney, New South Wales, Australia

J. Todd Hoopes
Biomolecular Labeling Laboratory, Institute for Bioscience and Biotechnology Research, National Institute of Standards and Technology and the University of Maryland, Rockville, Maryland, USA

Toshio Iwasaki
Department of Biochemistry and Molecular Biology, Nippon Medical School, Bunkyo-ku, Tokyo, Japan

A. Daniel Jones
Department of Biochemistry and Molecular Biology, and Department of Chemistry, Michigan State University, East Lansing, Michigan, USA

Michael A. Juen
Institute of Organic Chemistry and Center for Biomolecular Sciences Innsbruck, University of Innsbruck, Innsbruck, Austria

Zvi Kelman
Biomolecular Labeling Laboratory, and Biomolecular Structure and Function Group, Institute for Bioscience and Biotechnology Research, National Institute of Standards and Technology and the University of Maryland, Rockville, Maryland, USA

Christoph Kreutz
Institute of Organic Chemistry and Center for Biomolecular Sciences Innsbruck, University of Innsbruck, Innsbruck, Austria

Peter D. Kwong
Vaccine Research Center, National Institute of Allergy and Infectious Diseases, National Institutes of Health, Bethesda, Maryland, USA

Vladimir Ladizhansky
Department of Physics, and Biophysics Interdepartmental Group, University of Guelph, Guelph, Ontario, Canada

Aisha LaGuerre
Institute of Biophysical Chemistry, Centre for Biomolecular Magnetic Resonance, J.W. Goethe-University, Frankfurt-am-Main, Germany

Vanessa Lake
National Deuteration Facility, Bragg Institute, ANSTO, Lucas Heights, New South Wales, Australia

Paul Langan
Biology and Soft Matter Division, Oak Ridge National Laboratory, Oak Ridge, Tennessee, USA

My T. Le
Department of Chemistry and Biochemistry, Center for Biomolecular Structure and Organization, University of Maryland, College Park, Maryland, USA

Regan M. LeBlanc
Department of Chemistry and Biochemistry, Center for Biomolecular Structure and Organization, University of Maryland, College Park, Maryland, USA

Shuwei Li
Institute for Bioscience and Biotechnology Research, University of Maryland, Rockville, Maryland, USA

Myat T. Lin
Department of Biochemistry, University of Illinois at Urbana-Champaign, Urbana, Illinois, USA

Siqi Liu
CAS Key Laboratory of Genome Sciences and Information, Beijing Institute of Genomics, Chinese Academy of Sciences; BGI-Shenzhen, Shenzhen, and Sino-Danish Center/Sino-Danish College, University of Chinese Academy of Sciences, Beijing, PR China

Frank Löhr
Institute of Biophysical Chemistry, Centre for Biomolecular Magnetic Resonance, J.W. Goethe-University, Frankfurt-am-Main, Germany

Andrew P. Longhini
Department of Chemistry and Biochemistry, Center for Biomolecular Structure and Organization, University of Maryland, College Park, Maryland, USA

Shinichi Matsushita
Department of Biochemistry and Molecular Biology, Nippon Medical School, Bunkyo-ku, Tokyo, Japan

Erich Michel
Institute of Molecular Biology and Biophysics, ETH Zürich, Zürich, Switzerland

Yoshiharu Miyajima-Nakano
Department of Biochemistry and Molecular Biology, Nippon Medical School, Bunkyo-ku, Tokyo, Japan

Rachel Munro
Department of Physics, and Biophysics Interdepartmental Group, University of Guelph, Guelph, Ontario, Canada

Hugh O'Neill
Biology and Soft Matter Division, Oak Ridge National Laboratory, Oak Ridge, Tennessee, USA

Sai Venkatesh Pingali
Biology and Soft Matter Division, Oak Ridge National Laboratory, Oak Ridge, Tennessee, USA

Chinmayi Prasanna
NMR Research Centre, Indian Institute of Science, Bangalore, Karnataka, India

Renae J. Preston
Biomolecular Labeling Laboratory, Institute for Bioscience and Biotechnology Research, National Institute of Standards and Technology and the University of Maryland, Rockville, Maryland, USA

Prasad T. Reddy
Biomolecular Labeling Laboratory, and Biomolecular Structure and Function Group, Institute for Bioscience and Biotechnology Research, National Institute of Standards and Technology and the University of Maryland, Rockville, Maryland, USA

Agata Rekas
National Deuteration Facility, Bragg Institute, ANSTO, Lucas Heights, New South Wales, Australia

Robert A. Russell
National Deuteration Facility, Bragg Institute, Australian Nuclear Science and Technology Organisation, and Bio/Polymer Research Group, Centre for Advanced Macromolecular Design, School of Biotechnology & Biomolecular Science, University of New South Wales, Sydney, New South Wales, Australia

Mallika Sastry
Vaccine Research Center, National Institute of Allergy and Infectious Diseases, National Institutes of Health, Bethesda, Maryland, USA

Anthony S. Serianni
Department of Chemistry and Biochemistry, University of Notre Dame, Notre Dame, Indiana, USA

Riddhi Shah
Biology and Soft Matter Division, Oak Ridge National Laboratory, Oak Ridge, and Bredesen Center for Interdisciplinary Research and Graduate Education, University of Tennessee, Knoxville, Tennessee, USA

Naima G. Sharaf
Department of Structural Biology, University of Pittsburgh School of Medicine, Pittsburgh, Pennsylvania, USA

Binesh Shrestha
Novartis Institutes of BioMedical Research, Novartis Campus, Basel, Switzerland

Anne E. Simon
Department of Chemistry and Biochemistry, Department of Cellular Biology and Molecular Genetics, Center for Biomolecular Structure and Organization, University of Maryland, College Park, Maryland, USA

Lukasz Skora
Novartis Institutes of BioMedical Research, Novartis Campus, Basel, Switzerland

Takaho Terada
RIKEN Structural Biology Laboratory, Yokohama, Japan

Volker Urban
Biology and Soft Matter Division, Oak Ridge National Laboratory, Oak Ridge, Tennessee, USA

Karyn L. Wilde
National Deuteration Facility, Bragg Institute, ANSTO, Lucas Heights, New South Wales, Australia

Christoph H. Wunderlich
Institute of Organic Chemistry and Center for Biomolecular Sciences Innsbruck, University of Innsbruck, Innsbruck, Austria

Feng Xian
CAS Key Laboratory of Genome Sciences and Information, Beijing Institute of Genomics, Chinese Academy of Sciences; BGI-Shenzhen, Shenzhen, and Sino-Danish Center/Sino-Danish College, University of Chinese Academy of Sciences, Beijing, PR China

Shigeyuki Yokoyama
RIKEN Structural Biology Laboratory, Yokohama, Japan

Wenhui Zhang
Department of Chemistry and Biochemistry, University of Notre Dame, Notre Dame, Indiana, USA

Shikai Zhao
Omicron Biochemicals, Inc., South Bend, Indiana, USA

PREFACE

Biomolecules labeled with stable isotopes (mainly ^{2}H, ^{13}C, ^{15}N) are often used to obtain structural information using techniques such as small-angle neutron scattering, neutron reflectometry, and nuclear magnetic resonance (NMR). Stable isotope labeling of biomolecules can be classified into five categories: uniform labeling, in which all atoms in the molecules are labeled; fractional labeling, in which not all atoms are labeled but the labeling is distributive along the biomolecule; site-specific labeling, in which a specific amino acid or specific atom (e.g., a specific carbon on the sugar ring) is labeled; selective labeling, in which one type of amino acid or nucleotide is labeled throughout the protein, peptide, or nucleic acid; and segmental labeling, in which only a part of the molecule (e.g., a domain within a protein) is labeled.

The aim of this volume of *Methods in Enzymology* is to provide comprehensive methodologies for the production and purification of labeled biomolecules of various types and from different sources. The accompanying volume describes different experimental techniques that use labeled biomolecules.

The first step is to produce the biomolecule in the presence of a labeled chemical, precursor, or amino acid. The most commonly used system to label protein is heterologous expression in *Escherichia coli* in the presence of media containing ^{2}H, ^{13}C, and/or ^{15}N. Two chapters provide detailed descriptions of uniform and fractional protein labeling in bacteria using fermenters (Chapter 1 by Duff and coworkers) or shaking flasks (Chapter 2 by Hoopes et al.). In Chapter 3, Lin et al. describe the use of *E. coli* auxotroph host strains for amino acid-selective labeling of protein. Sharaf and Gronenborn describe two methods for ^{19}F-labeling of protein in *E. coli*: one for selective amino acid incorporation and a second for site-specific incorporation of ^{19}F-modified amino acids (Chapter 4).

In addition to proteins, bacteria are also used to label other biomolecules. In Chapter 5, Russell and coworkers describe the use of bacteria to produce deuterated polyhydroxyalkanoate biopolyesters and cellulose, and in Chapter 6 O'Neill et al. describe a method for the deuteration of bacterial cellulose. Proteins can also be labeled in archaea as described by Cleveland and Kelman in Chapter 7 on the labeling of proteins in *Halobacterium salinarum*. One approach to achieve selective amino acid labeling is via amino acid–selective unlabeling, as described by the laboratory of Hanudatta Atreya in Chapter 8.

Many proteins, in particular eukaryotic proteins, cannot be expressed in bacteria in an active, folded, form. In addition, when posttranslational modifications (such as glycosylation) are required for proper folding or activity, bacteria cannot be used for expression. Therefore, methods have been developed to express proteins in eukaryotic cells. Fan et al. describe the use of yeast to label membrane proteins (Chapter 9), and Peter Holden's laboratory describes the use of yeast to make the polysaccharide chitosan (Chapter 5). Evans and Shah describe approaches for deuterium incorporation in plants (Chapter 10). Skora et al. describe the use of insect cells for uniform labeling of proteins or selective labeled amino acid incorporation (Chapter 11). In Chapter 12, Sastry et al. describe the use of an adenovirus-based system for two types of protein labeling in mammalian cells: uniform labeling with ^{13}C and/or ^{15}N, or selective labeling using an amino acid.

Biomolecule labeling can also be performed *in vitro*, and several chapters in the volume provide protocols. Terada and Yokoyama (Chapter 13) and Xian and coworkers (Chapter 14) describe two approaches for the labeling of mammalian proteins using an *E. coli* cell-free system, while LaGuerre et al. provide a protocol for the labeling of membrane proteins with a cell-free system (Chapter 15). A procedure for segmental labeling of an individual domain in a multidomain protein is described by Michel and Allain (Chapter 16). Zhang et al. describe *in vitro* chemical methods to introduce stable isotopes into monosaccharides (Chapter 17).

Labeled nucleotides aid in structural analysis of nucleic acids by NMR. The contribution by Wunderlich et al. describes chemoenzymatic methods for site-specific labeling of RNA (Chapter 18). The contribution from the laboratory of Kwaku Dayie describes methods for uniform ^{15}N and site-specific ^{13}C labeling of RNA in *E. coli* (Chapter 19). Duss et al. describe protocols for making very short labeled RNAs as well as RNA fragments of any desired size (Chapter 20).

I hope that this and the accompanying volume prove useful for the scientific community by providing descriptions of techniques for biomolecule labeling and their use in structural analysis.

I would like to acknowledge the help of Dr. Lori M. Kelman during the course of editing these volumes.

Edited by ZVI KELMAN

SECTION I

Labeling in Prokarya

CHAPTER ONE

Robust High-Yield Methodologies for ^{2}H and ^{2}H/^{15}N/^{13}C Labeling of Proteins for Structural Investigations Using Neutron Scattering and NMR

Anthony P. Duff*,[1], Karyn L. Wilde*, Agata Rekas*, Vanessa Lake*, Peter J. Holden[†]

*National Deuteration Facility, Bragg Institute, ANSTO, Lucas Heights, New South Wales, Australia
[†]National Deuteration Facility, Bragg Institute, Australian Nuclear Science and Technology Organisation, New South Wales, Australia
[1]Corresponding author: e-mail address: anthony.duff@ansto.gov.au

Contents

1. Introduction — 4
2. Media Preparation — 7
3. Unlabeled Protein Production — 9
   3.1 Transformation of Expression Cells and Staged Culturing — 9
   3.2 1 L Bioreactor Culture — 10
4. Deuterated Protein Production — 14
   4.1 Deuteration in 90% D_2O for SANS — 14
   4.2 Deuteration in 100% D_2O with Unlabeled Glycerol — 15
   4.3 Perdeuteration: 100% D_2O and Deuterated Carbon Source — 16
   4.4 Deuteration Level Quantification — 16
5. Multiple Labeling of Proteins for NMR — 17
   5.1 ^{15}N Labeling — 17
   5.2 ^{13}C Labeling — 18
6. Comments on the Method — 18
   6.1 Plasmid Choice — 18
   6.2 Precipitating Media — 19
   6.3 Use of Commercial Supercompetent Expression Cells — 19
   6.4 Staged Starter Cultures — 21
   6.5 Gentle Handling of the Cultures — 21
   6.6 ^{15}N Ammonium Hydroxide or Sodium Hydroxide as Base Feed for pH Control — 21
7. Typical Deuteration Levels — 22
Acknowledgments — 23
References — 23

Methods in Enzymology, Volume 565
ISSN 0076-6879
http://dx.doi.org/10.1016/bs.mie.2015.06.014

Abstract

We have developed a method that has proven highly reliable for the deuteration and triple labeling (^{2}H/^{15}N/^{13}C) of a broad range of proteins by recombinant expression in *Escherichia coli* BL21. Typical biomass yields are 40–80 g/L wet weight, yielding 50–500 mg/L purified protein. This method uses a simple, relatively inexpensive defined medium, and routinely results in a high-yield expression without need for optimization. The key elements are very tight control of expression, careful starter culture adaptation steps, media composition, and strict maintenance of aerobic conditions ensuring exponential growth. Temperature is reduced as required to prevent biological oxygen demand exceeding maximum aeration capacity. Glycerol is the sole carbon source. We have not encountered an upper limit for the size of proteins that can be expressed, achieving excellent expression for proteins from 11 to 154 kDa and the quantity produced at 1 L scale ensures that no small-angle neutron scattering, nuclear magnetic resonance, or neutron crystallography experiment is limited by the amount of deuterated material. Where difficulties remain, these tend to be cases of altered protein solubility due to high protein concentration and a D$_2$O-based environment.

ABBREVIATIONS AND SYMBOLS

D ^{2}H or deuterium
D$_2$O heavy water, 100% deuterated water, deuterium oxide
DOT dissolved oxygen tension
Deuterated deuterium labeled, deuterium atoms in place of hydrogen (^{1}H) atoms
glycerol-d glycerol-d8, glycerol with all hydrogen atom positions occupied by deuterium atoms
IPTG isopropyl-β-D-thiogalactopyranoside
NMR nuclear magnetic resonance
NR neutron reflectometry
SANS small-angle neutron scattering

1. INTRODUCTION

The production of deuterated proteins in *Escherichia coli* has been a practice pursued for several decades in order to produce labeled proteins for structural studies using nuclear magnetic resonance (NMR), small-angle neutron scattering (SANS), and other techniques. It is required in SANS and neutron reflectometry (NR) studies of protein complexes for the labeling of one protein with respect to another, taking advantage of the strong contrast between ^{1}H and ^{2}H in neutron scattering. In NMR, it is often required to reduce the prevalence of the highly NMR-active nuclei ^{1}H that when overabundant dampens and broadens the signal. In neutron crystallography, it is required to achieve increased signal through reduction of incoherent neutron scattering noise and neutron absorbance, and to avoid the problem of negative ^{1}H density smearing with positive carbon density, as observed by neutron diffraction.

When Moore (1979) reviewed their production of deuterated ribosomal materials for SANS, it was known that *E. coli*, like many other bacteria, can grow in almost fully deuterated media. They used a minimal medium, designated "M9," which was adapted from Anderson's work (Anderson, 1946). Initially, they selected deuterated acetic acid, as the cheapest fully deuterated carbon source readily metabolized by *E. coli*. Subsequently, deuterated succinic acid was utilized due to its ease of synthesis and better growth rates. Although deuterated glucose was available, at that time, it was prohibitively expensive and thus seldom used. Others at this time similarly achieved deuteration using deuterated acetate and succinate (LeMaster & Richards, 1982). These early protocols were difficult to follow in practice, requiring considerable attention and care to achieve moderate yields of biomass at best (Vanatalu et al., 1993). Supplementation of minimal media with deuterated complex additives to reduce the synthesis burden has also been utilized for production of deuterated proteins. Deuterated algal hydrolysate was used, for example, to supplement M9 medium in partially deuterated solvent to produce uniformly partially deuterated troponin C (Olah, Rokop, Wang, Blechner, & Trewhella, 1994). Labeling of proteins for NMR structural analysis has been routinely achieved by using labeled amino acids in minimal medium, enabling amino acid-specific labeling (Torchia, Sparks, & Bax, 1989), and perdeuteration, achieving 96% deuteration (Gamble, Clauser, & Kossiakoff, 1994).

The use of hydrogenated carbon sources in D_2O to achieve high protein deuteration levels has been most successful with acetate (Sosa-Peinado, Mustafi, & Makinen, 2000; Venters et al., 1995). The metabolism of acetate involves significant scrambling of the three carbon-bonded hydrogen atoms in the metabolic biosynthesis of amino acids. In contrast, metabolic biosynthesis of amino acids from glucose demonstrates less scrambling, achieving a maximum average protein deuteration level of 86% (LeMaster, 1994).

In reviewing protein isotopic labeling strategies used in the NMR community, Gardner and Kay (1998) described predictable patterns of carbon source proton scrambling. In 2007, a study of options for the carbon source, indicated that among small-molecule carbon sources, glycerol outperforms other molecules—pyruvate, fumarate, succinate, and acetate, in terms of lag-free growth rate and maximum final density (Paliy & Gunasekera, 2007).

The classic method of adaptation of a culture from H_2O to D_2O typically involves three to four steps of subculturing (Paliy, Bloor, Brockwell, Gilbert, & Barber, 2003). Moore (1979), for example, reported the production of 0.4 g biomass wet weight per gram of acetate as achieved by stepwise subculturing from glucose to acetate in H_2O, then to 80% D_2O, and then to 100% D_2O. One-step adaptation to 100% D_2O is noted to be possible but unreliable, producing long, variable, unbounded lag times (Venters et al., 1995). Gardner and Kay (1998) noted, in the context of adaptation methodology, the importance of minimized lag time associated with inoculation. After dismissing acetate as unreliable, Leiting, Marsilio, and O'Connell (1998) reported a time-efficient method of predictable deuteration using a brief adaptation protocol. They began with a fresh transformation and observed decreasing growth rates and biomass yields with increasing deuteration, and quantified deuteration level results as a function of media deuteration and choice of hydrogenated or deuterated glucose as the carbon source.

The preparation and use of preadapted *E. coli* transformation cells, requiring regular plating and colony selection for strain maintenance, was reported by Paliy et al. (2003). High-yield expression for specific proteins has been reported a number of times (Sivashanmugam et al., 2009; Sosa-Peinado et al., 2000). Cost-efficient methods of growing biomass in unlabeled media and introducing or transferring to labeled media for induction of expression of the labeled protein have been described (Cai et al., 1998; Marley, Lu, & Bracken, 2001). This method is of particular advantage when expensive ^{13}C-labeled precursor molecules are used.

Protein perdeuteration is more challenging than partial deuteration because of the increased stress on microbial metabolism from the use of a deuterated carbon source in addition to D_2O. Moreover, it requires a compromise between the cost of labeled carbon sources and their efficiency as substrates, for example, acetate and succinate are cheaper but produce lower yields than glucose. The perdeuteration of myoglobin using succinate-*d6* achieved high protein yields with a deuteration level exceeding 98% (Shu, Ramakrishnan, & Schoenborn, 2000).

In the 2000s, the establishment of "D-lab" at Grenoble (http://www.ill.eu/sites/deuteration/) resulted in a resurgence in protein deuteration for neutron crystallography and SANS. Examples of their work, specifically noting their choice of carbon source, are: the perdeuteration of pyrophosphatase using CELTONE (Tuominen et al., 2004); the perdeuteration of human aldose reductase (Hazemann et al., 2005) using succinate-*d*; perdeuteration of cytochrome P450cam using glycerol-*d* achieving 98% deuteration (Meilleur, Contzen, Myles, & Jung, 2004), later optimized to achieve 99.5% (Meilleur, Dauvergne, Schlichting, & Myles, 2005). Advice from Michael Haertlein of the D-lab was seminal in our choice of glycerol as a carbon source and the use of kanamycin resistance-based vectors.

The use of bioreactors is critical for high yield, reliable, controlled culture production, especially for partial deuteration in D_2O using a hydrogenated carbon source, where the dominant consumable cost is the D_2O. However, bioreactor culturing in minimal medium to high density introduces some issues not normally encountered in low cell density, rich medium flask cultures. Glucose, when used as the carbon and energy source, is known to generate significant end-product inhibition, due to accumulation of by-products that if not removed, limit the growth rate (Paalme, Tiisma, Kahru, Vanatalu, & Vilu, 1990; Roe, O'Byrne, McLaggan, & Booth, 2002). Acetate accumulation is particularly problematic as it affects several physiological functions of *E. coli* including growth rate and oxygen uptake (Xu, Jahic, & Enfors, 1999). Similarly, in batch culturing, the bulk addition of ammonium can lead to inhibition of ammonium transport (Jayakumar, Hong, & Barnes, 1987). In high-density cultures, oxygen is frequently the limiting factor for growth rate and biomass density (Stanbury, Whitaker, & Hall, 1995).

We now report our development of a method that has proven highly reliable for the deuteration of a broad range of proteins. Typical yields are 40–80 g/L wet weight, yielding 50–500 mg purified protein per liter of media. Published results using this method are: the deuteration of human galectin-2 for SANS studies of the PEGylated molecule (Chen et al., 2012); the deuteration of the surfactant peptide DAMP4 for NR (Dimitrijev-Dwyer et al., 2012), in a study of its surfactant properties; the deuteration of syntaxin in a SANS study of the structure of protein complexes involved in vesicle capture (Christie et al., 2012); the deuteration of calmodulin in a SANS study of the interaction of calmodulin with an HIV protein (Taylor et al., 2012); and the triple labeling of the fungal hydrophin, EAS, enabling solid-state NMR in the first structural study of a functional amyloid (Morris et al., 2012). Here, we describe the production protocol in detail using a variety of examples highlighting factors critical to reproducible success.

2. MEDIA PREPARATION

To establish that our method is applicable to a particular protein, we first produce nonlabeled protein using a method as close as possible to the method to be used for deuterated expression. This approach provides unlabeled protein for trial purification at a scale produced by this method.

Table 1 "ModC1" Medium Composition

Solution	Component	g/L	mM
Bulk solution	NH_4Cl	2.58	48.23
	KH_2PO_4	2.54	18.66
	Na_2HPO_4	4.16	29.30
	K_2SO_4	1.94	11.13
	Glycerol	40	434
Additive A (1000 × stock)	$FeSO_4 \cdot 7H_2O$	0.02	0.719
	Trisodium citrate	0.088	0.3410
Additive B (1000 × stock)	$MnSO_4 \cdot H_2O$	0.005	0.0348
	$ZnSO_4 \cdot 7H_2O$	0.0086	0.0299
	$CuSO_4 \cdot 5H_2O$	0.00076	0.00304
Additive C (1000 × stock)	Thiamine	0.048	0.1596
Additive D (1000 × stock)	$MgSO_4 \cdot 7H_2O$	0.67	2.72

Kanamycin is added to 40 μg/L for kanamycin-resistant vectors. The pH is controlled by automated addition of ammonium hydroxide (28%, w/v).
Modified, with glycerol replacing dextrose, from Middelberg, O'Neill, Bogle, and Snoswell (1991).

For the first-time user, it also allows familiarization with the method and equipment before the use of expensive media components, such as D_2O and deuterated carbon sources.

Prepare separate solutions of hydrogenated (^1H) and deuterated (^2H) minimal media (ModC1). Media of varying D_2O percentage is prepared by mixing appropriate volumes of these solutions. The composition of ModC1 minimal medium is given in Table 1. Due to a propensity for ModC1 to precipitate, additive solutions A, B, C, and D are prepared separately and added to the bulk solution at the time of use, along with the antibiotic kanamycin for plasmid-transformed cell selection (based on use of kanamycin-resistant vectors).

Solution Preparation

1. Dissolve the bulk solution components (salts and carbon source[1]) in H_2O or D_2O. Autoclave or filter-sterilize (0.22 μm) solutions with H_2O. D_2O-containing solutions are filter-sterilized (0.22 μm).

---

[1] For deuteration levels less than perdeuteration, deuterated medium includes unlabelled glycerol as the sole carbon source. For perdeuteration, glycerol-*d* (Sigma-Aldrich 447498) is the sole carbon source. Refer to Section 4 for further details.

2. Prepare the four additive solutions separately (A, B, C, and D) at 1000× concentration with both H_2O and D_2O solvent and sterilize solutions by filtration (0.22 μm). Store at room temperature.
3. Dissolve kanamycin sulfate salt at 40 mg/mL (1000×) and filter-sterilize (0.22 μm). Store at −20 °C.

Prepare ModC1 by adding the bulk solution to a culture flask or bioreactor, followed by additives A, B, C, D, and kanamycin. Some precipitate will form. Shake or stir from this point onward to prevent settling of precipitate.

3. UNLABELED PROTEIN PRODUCTION

Protein production, whether labeled or unlabeled, proceeds via nearly identical methods, beginning with a standard transformation of commercial competent cells, through three flask cultures of increasing volume, to inoculation of a bioreactor, as illustrated in Scheme 1.

3.1 Transformation of Expression Cells and Staged Culturing

3.1.1 Transformation

Use 1 μL of miniprep plasmid supercoiled DNA (100–300 ng) of a kanamycin-resistant construct to transform one 50 μL reaction tube of *E. coli* BL21Star™(DE3) commercial chemically competent cells (Invitrogen™ C6010-03, transformation efficiency of 1×10^8 cfu/μg). We transform, largely according to the manufacturer's instructions, as follows:
- Thaw the reaction tube on ice for 30 min, then add the plasmid and gently swirl with the pipette tip, and return the tube to ice for 30 min.

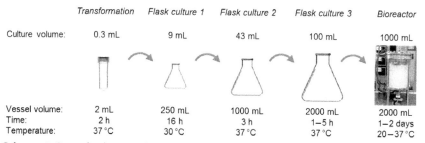

	Transformation	Flask culture 1	Flask culture 2	Flask culture 3	Bioreactor
Culture volume:	0.3 mL	9 mL	43 mL	100 mL	1000 mL
Vessel volume:	2 mL	250 mL	1000 mL	2000 mL	2000 mL
Time:	2 h	16 h	3 h	1–5 h	1–2 days
Temperature:	37 °C	30 °C	37 °C	37 °C	20–37 °C

Scheme 1 Staged culture scale-up from transformation to the bioreactor. The transformation reaction is used to inoculate three progressively larger flask cultures before inoculation of the bioreactor. The flasks are sealed and optical density (OD_{600}) of the culture in the flasks remains between 0.1 and 1.0. The times indicated correspond to hydrogenated media. These are longer for cooler temperatures and deuterated media.

- Heat shock by incubating the tube in a 42 °C water bath for 30 s, and then return the tube to ice for 2 min.
- Add 250 μL of the rich SOC medium (Invitrogen 15544-034: 2% tryptone, 0.5% yeast extract, 10 mM NaCl, 2.5 mM KCl, 10 mM MgCl$_2$, 10 mM MgSO$_4$, and 20 mM glucose).
- Incubate at 37 °C for 2 h (without shaking).

As a quality check on the competent cells and plasmid, take 1 μL of the reaction mixture, mix with 20 μL SOC medium, and spread onto an LB-kanamycin (40 μg/L) agar plate. Incubate at 37 °C overnight and record the number of colonies.

3.1.2 Flask Culture 1

Add the transformation reaction (299 μL) to 10 mL freshly prepared ModC1 medium and shake overnight at 220 rpm at 30 °C, in a sealed 250 mL flask. After 16 h, take a 1 mL sample to measure OD$_{600}$. If OD$_{600}$ measurement is between 0.5 and 1.0, use the remaining 9 mL to inoculate Flask Culture 2. If OD$_{600}$ is less than 0.5, transfer the culture to a 37-°C shaking incubator to continue growing to an OD$_{600}$ between 0.5 and 1.0. If the OD$_{600}$ measures greater than 1.0, discard and start again.

3.1.3 Flask Culture 2

Add 9 mL of Flask Culture 1 to a sealed 1 L flask containing 36 mL freshly prepared medium. Shake at 220 rpm at 37 °C for two generations. In H$_2$O, two generations will be approximately 3 h. Measure OD$_{600}$, using 1 mL samples, at the beginning and end of this incubation for growth rate verification and retain for SDS–PAGE analysis for verification of expression control.

3.1.4 Flask Culture 3

Use the remaining 43 mL from Flask Culture 2 to inoculate 59 mL of freshly made medium in a sealed 2 L flask. Shake at 220 rpm at 37 °C until OD$_{600}$ is 1.0. Again, take two 1 mL samples, at the beginning and end of this incubation, for growth rate verification and SDS–PAGE analysis. Use the remaining 100 mL to inoculate the 1 L bioreactor culture.

3.2 1 L Bioreactor Culture

3.2.1 Bioreactor Set-Up and Inoculation

One hour before use, add 900 mL fresh medium to a calibrated and sterilized 2 L bioreactor (Real Time Engineering). Activate the airflow (1 L/min), impeller (600 rpm) and temperature control (37 °C), allowing parameter measurements to reach steady baselines (temperature, pH, and dissolved oxygen tension (DOT)). Calibrate the DOT probe to 100%. Set any proportional–integral–derivative feedback controls to firstly increase impeller speed to maximum (1200 rpm), and secondly to decrease vessel temperature (to a minimum of 15 °C), to maintain a DOT set-point of 75%. Fix a pH set-point of 6.6. Shortly before inoculation, add the additives A, B, C, D, and kanamycin at 1 mL/L. Activate recording of bioreactor run data.

When Flask Culture 3 has achieved an OD_{600} of 1.0, use the entire volume to inoculate the bioreactor, producing an expected initial OD_{600} measurement of 0.1.

3.2.2 Running the Bioreactor to Induction and Harvest

After inoculation, add 100 μL of Antifoam 204 (Sigma-Aldrich A8311). Prepare the base feed and connect the base feed line to the bioreactor. Typically the base used is 28% (w/v) ammonia which fumes ammonia. Prime the base feed tubing to the peristaltic pump, so as to not cause input of ammonia gas. Ammonium hydroxide should ideally be added below the culture medium surface to avoid purging of gaseous ammonia from dripping ammonium hydroxide. The base feed will be required when the culture OD_{600} reaches approximately 5.

The growth rate in the flasks and bioreactor should be steady and predictable. Growth rate predictive equations, outlined in the following section, allow prediction of the time for protein expression induction (addition of isopropyl-β-D-thiogalactopyranoside (IPTG)). Time until induction can be increased by decreasing the culture temperature. Measure OD_{600} values regularly and when OD_{600} approaches 16, lower the culture temperature to the desired protein-specific expression temperature. Take the preinduction sample for later SDS–PAGE analysis and add IPTG to 1 mM. If the vessel temperature is to be lowered below 20 °C, also add 100 μL of the antifoam Silicone Oil DC 200 (Sigma-Aldrich 85412) in addition to the Antifoam 204 added previously; both are required in concert below 20 °C. Regular postinduction samples should also be taken for OD_{600} measurements and subsequent SDS–PAGE analysis. Depending on the stability of the protein, continue expression for a fixed time or until exhaustion of the carbon source which will be apparent from a sudden rise in DOT and pH, and cessation of the automated base feed. Upon induction time completion, stop the bioreactor and take a final sample for OD_{600} measurement and SDS–PAGE analysis. Harvest the culture by centrifugation at

$8000 \times g$, 4 °C, for 30 min and collect a sample 20 μL of the supernatant for SDS–PAGE analysis. Freeze the biomass in liquid nitrogen and store at −80 °C.

3.2.3 Growth Rate Expectations and Prediction

An excellent real-time indicator of health of the culture is given by the doubling or generation time (g) for OD_{600} (a proxy for cell biomass) during exponential growth and whether it matches expectations of a typical healthy culture.

Generation time can be calculated from the reciprocal of the gradient of the plot of

$$\log_2\{OD_{600}\} \text{ versus time.}$$

Two-point estimates of generation time can be obtained from the formula:

$$\text{Generation time}, g = \frac{(t_2 - t_1) \times \log\{2\}}{\log\left\{\frac{OD_2}{OD_1} \times \frac{V_2}{V_1}\right\}}$$

where at time point t_i, the culture volume is V_i, and optical density is OD_i. OD is the optical density $OD_{600\,nm}$. V = volume of the culture. The term $\frac{V_2}{V_1}$ allows for generation time measurement between points where a culture of volume V_1 has been diluted to become a culture of volume V_2.

We observe lag-free inoculations and consistent generation times with well-handled cultures, for all uninduced cultures from Flask Culture 1 to the bioreactor. Notable examples of generation times are:

Medium	Temperature (°C)	Generation time (g) (h)
ModC1 H_2O	37	1.5
ModC1 90% D_2O	37	3

Generation time should be continually monitored and compared with expected values. Significant deviation from expected values indicates probable errors in preparations and handling, the inability to predict future growth characteristics, and a loss in confidence of a strong induced expression level.

If a culture is growing with a generation time (g) corresponding to expected values for the medium type and temperature, then the time until the culture reaches a specific optical density can be estimated. For example, if intending to induce expression, as we prefer, at $OD_{600nm} = 16$, the time until induction is given by the following formula:

$$\text{Time until induction } (OD = 16)$$

$$t_{Induction} = \frac{g \times \log\left\{\frac{16}{OD}\right\}}{\log\{2\}}.$$

It may be desirable for practical reasons to alter the temperature to obtain a different generation time so that the time of induction is at a suitable time convenient to the scientist. We have observed that the general Arrhenius relationship between chemical/biochemical activity and temperature is accurately held for uninduced exponentially growing *E. coli* in these media. A decrease in temperature of 10 °C will decrease the growth rate approximately twofold.

3.2.4 SDS–PAGE Analysis of Expression Control and Expression Level

To verify expression control and satisfactory postinduction expression, samples pre- and postinduction should be analyzed by reducing SDS–PAGE. To ensure that all samples are visualized at similar loading level, SDS–PAGE samples should contain similar biomass (biomass = volume × OD). This can be achieved, for example, by choosing a sample volume according to the following formula:

$$\text{Sample volume} = \frac{1 \text{ mL}}{3 \times OD}.$$

Spin the sample just enough to make a soft pellet, discard the supernatant, and freeze at −20 °C for later analysis. Run the samples as per standard SDS–PAGE, stain, and destain. Examine the resulting SDS–PAGE gel for the absence of preinduction expression, strong induced expression postinduction, and retention of the overexpressed protein to the preferred time of harvest. Only after verifying satisfactory expression in nondeuterated protein production, should deuterated production be commenced.

4. DEUTERATED PROTEIN PRODUCTION

Deuterated protein production follows the same method as unlabeled expression, except for the use of deuterated water and/or carbon source and base, and increased duration of the protocol due to longer generation times. Production with different deuteration level targets involves different choices of media deuteration or carbon source within the same generic method. The four most common production conditions—unlabeled, labeled for a neutron scattering match point equivalent to 100% D_2O, maximum deuteration without use of a labeled carbon source, and perdeuteration using deuterated glycerol throughout, are illustrated in Scheme 2.

In deuterated production, use a bioreactor pH set-point of 6.2, which corresponds to a pD of 6.6 (Covington, Paabo, Robinson, & Bates, 1968), equivalent to the effective acidity in the hydrogenated production. Use deuterated base, 25% (w/v) ammonium-*d4* deuteroxide (Sigma-Aldrich 176702) in D_2O, as the base feed for perdeuteration.

4.1 Deuteration in 90% D_2O for SANS

Deuteration in 90% D_2O, using hydrogenated glycerol, is commonly undertaken for deuterium labeling of proteins for SANS, achieving a

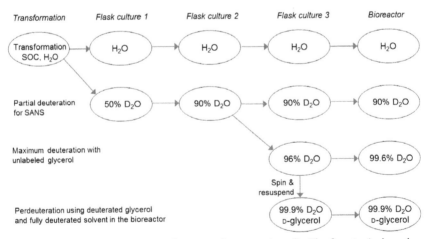

Scheme 2 Stepped adaptation pathways to deuterated media. The four typical production conditions are indicated. Unlabeled production (top row) provides a test that the protocols utilized will produce induced protein expression by the method and scale intended for labeled production. Varying levels of protein deuteration are achieved by following the branches (arrows) to lower rows.

nonexchangeable hydrogen-site deuteration level of approximately 75% giving a neutron scattering contrast match point at approximately 100% D_2O.

4.1.1 Culture Adaptation to 50% D_2O then 90% D_2O

To produce protein of 75% D labeling, use the same method of transformation as described for hydrogenated expression. Broadly follow the Scheme 1 in Section 3.1 with the exception of variations listed below.

Flask Culture 1 should be prepared at 50% D_2O, achieved by mixing 5 mL of H_2O-based ModC1 with 5 mL of D_2O-based ModC1 (refer to Section 2). Note that this 50% D_2O Flask Culture 1 will grow more slowly than the hydrogenated version. Shake at 220 rpm, at 37 °C, in a sealed 250 mL flask for 20–28 h to achieve the required OD_{600} measurement of 0.5–1.0. Sample 1 mL for verification of OD_{600}, and use the remaining 9 mL for inoculation of the 90% D_2O Flask Culture 2.

Flask Culture 2, at 90% D_2O, should be prepared by adding the remaining 9 mL of the 50% D_2O Flask Culture 1 to 36 mL of freshly prepared 100% D_2O ModC1 in a 1 L flask. Seal the flask and shake at 220 rpm at 37 °C for two generations (6 h). Take 1 mL samples, at the beginning and end of this incubation for growth rate verification and SDS–PAGE analysis, leaving 43 mL for inoculation of Flask Culture 3.

The final inoculum is prepared (Flask Culture 3) by adding 59 mL of 90% D_2O freshly prepared ModC1 medium to a 2 L flask and inoculating with the remaining volume from Flask Culture 2. Seal and shake at 220 rpm at 37 °C until the OD_{600} reaches 1.0. The generation time should be approximately 3 h. Allowing for two 1 mL samples to track OD_{600} and SDS–PAGE analysis, use the remaining 100 mL to inoculate the bioreactor, prepared as described in Section 3.2.

4.2 Deuteration in 100% D_2O with Unlabeled Glycerol

Deuterium-labeled protein using 100% D_2O and hydrogenated glycerol is a common request in our experience. Proceed similarly to the production of deuterated protein from 90% D_2O in Section 4.1 until the method changes at Flask Culture 3.

4.2.1 Culture Adaptation to 90% D_2O then 100% D_2O

Prepare Flask Culture 3 using 59 mL of freshly prepared 100% D_2O ModC1, inoculating with 43 mL 90% D_2O from Flask Culture 2. This results in a culture with a D_2O concentration of approximately 96%. Seal and shake at 37 °C until OD_{600} reaches 1.0. As this culture is continuing to adapt to declining H_2O, the generation time will be decreasing, with measurable values typically between 6 and 4 h. When an OD_{600} of 1.0 is achieved, use the 100 mL to inoculate the bioreactor containing 100% D_2O-based ModC1. Use deuterated ammonia, 25% (w/v) ND_4OD in D_2O (Sigma-Aldrich 176702), as the base for pH control.

4.3 Perdeuteration: 100% D_2O and Deuterated Carbon Source

Perdeuteration requires a perdeuterated carbon source, and removal of the H_2O necessarily present in the initial steps of adaptation. Usually, glycerol-*d8* (Sigma-Aldrich 447198) is considered the best choice for carbon source, and it is used from Flask Culture 1 onward.

Optionally, the final bioreactor culture D_2O concentration may be increased from 99.6% to 99.9% by resuspending settled cells of Flask Culture 3 in fresh perdeuterated medium. The cost of this option involves the disposal of an unused fraction of the medium, a loss of some of the culture, and risk of stress to the culture. A compromise is required between the yield of resuspended cells and viability of cells pelleted.

To settle and resuspend the cells, transfer Flask Culture 3 to an iced water bath to chill the cells while swirling. Transfer the culture to a sterile centrifuge tube, in a prechilled rotor. Spin at $2000 \times g$ for 5 min and remove the supernatant. Gently resuspend the pellet in ice-chilled, freshly prepared perdeuterated medium. Return the resuspended cells to a 2-L flask and resume shaking at 37 °C until OD_{600} is 1.0.

A resuspended perdeuterated culture has virtually no ^1H, with the longest generation times and the least precisely predictable growth rates. Typical generation times measured are approximately 8 h at the start of the bioreactor culture, decreasing to approximately 5 h at the time of induction.

4.4 Deuteration Level Quantification

Quantification of the nonexchangeable deuteration level can be determined by analysis of matching peptide MALDI-TOF MS peaks from non-deuterated and deuterated samples obtained and visualized by SDS–PAGE. Hydrogen atoms bonded to carbon atoms are assumed to be nonexchangeable whereas hydrogen atoms bonded to nitrogen, oxygen, or sulfur atoms are considered to be exchangeable. The protein samples are extracted from the SDS–PAGE gels, subjected to partial trypsin hydrolysis, and the mass spectroscopy spectra obtained.

To calculate deuteration percentage of a protein, first match the observed peaks from the nondeuterated sample to theoretical trypsin digest fragments of the protein sequence. Then, manually match these to the corresponding peaks from the deuterated protein. For each peak pair successfully matched to a peptide, the deuteration level can be determined by dividing mass difference by the number of nonexchangeable hydrogen atoms multiplied by the deuteron–proton weight difference:

Deuteration level of peptide fragment
$$= \frac{\text{mass}(\text{deuterated}) - \text{mass}(\text{hydrogenated})}{n(\text{nonexchangeable H sites}) \times (\text{mass}(D) - \text{mass}(H))}$$

Typically, 5–10 nonoverlapping peptides will match peaks unambiguously, representing 20–60% of the protein sequence. The statistics for mass(deuterated) − mass(hydrogenated) and n(nonexchangeable H sites) are summed to compute a final deuteration level. Uncertainty in the deuteration level is estimated as the standard deviation of individual peptide deuteration levels plus the fraction full width half maximum for the strongest deuterated sample fragment peak. This value is approximately 3–5% for proteins produced from 90% deuterated medium and is negligible for perdeuterated samples.

5. MULTIPLE LABELING OF PROTEINS FOR NMR

Deuteration is also advantageous for NMR-based structural studies when combined with ^{15}N labeling, and ^{13}C labeling. Although the method design that has been described in Sections 3 and 4 was motivated by the need for cost-efficient, high-yield deuterium labeling, the reliability of the method has proven of value for double (^{2}H/^{15}N or ^{15}N/^{13}C) and triple labeling (^{2}H/^{15}N/^{13}C). Partial ^{2}H labeling with uniform ^{15}N and/or ^{13}C double labeling is achieved by using ^{15}N-labeled ammonium chloride as the nitrogen source and/or ^{13}C-glycerol as the carbon source. Perdeuteration with ^{13}C labeling requires a compromise between level of perdeuteration and expense and purity of the double-labeled carbon source required. The method for double and triple labeling is essentially the same as described in Section 4 with minor modifications.

5.1 ^{15}N Labeling

Labeling with ^{15}N is achieved by substituting the ammonium chloride of ModC1 medium (in the bulk solution) with ^{15}N ammonium chloride

(Sigma-Aldrich 299251). ^{15}N can be added not only in the medium but also supplied in the base feed used to control pH in the bioreactor culture (e.g., ^{15}N ammonium hydroxide). Alternatively, avoiding the use of a labeled base feed, the following protocol is used:

- Increase the NH$_4$Cl component of ModC1 medium to 5.16 g/L ^{15}NH$_4$Cl throughout.
- Add an additional 2.58 g/L ^{15}NH$_4$Cl to the bioreactor culture when OD$_{600}$ reaches approximately 10, 20, and 30.
- Use a non-nitrogen base feed such as 5 M sodium hydroxide or sodium deuteroxide in D$_2$O (Sigma-Aldrich 176788) for pH control.

The increased ammonium concentration causes a small decrease in growth rate, but otherwise this modification has little impact on the method described in Sections 3 and 4. ^{15}N labeling may be performed with or without ^{2}H labeling, and with or without ^{13}C labeling, without further modifications beyond those described here.

5.2 ^{13}C Labeling

^{13}C labeling is achieved by replacing unlabeled glycerol with glycerol-^{13}C (Sigma-Aldrich 489476) as the carbon source of ModC1 medium throughout the protocols described in Sections 3 and 4. As described in Section 5.1, ^{13}C labeling may be performed in conjunction with ^{2}H labeling and/or ^{15}N labeling.

5.2.1 Perdeuteration with ^{13}C Labeling

Perdeuteration with ^{13}C requires using a double-labeled carbon source which is considerably more expensive and typically only available at a lower deuterium purity. Double-labeled glycerol is prohibitively expensive. The use of glucose in D$_2$O, resulted in unusual growth characteristics at OD$_{600}$ > 10, and so we reduced the density of production when using double-labeled glucose, to use 10 g/L glucose-^{2}H,^{13}C and inducing at an OD$_{600}$ of 4–6. This was the approach taken for triple-labeled expression of the hydrophobin EAS$_{\Delta 15}$ for solid-state NMR analyses.

6. COMMENTS ON THE METHOD

6.1 Plasmid Choice

The deuteration of proteins by recombinant expression described here involves *E. coli* (Baneyx, 1999) standard laboratory strain BL21, using kanamycin-resistant plasmids incorporating the T7 promoter system (Studier & Moffatt, 1986) and using the *lacI* repressor (Dubendorff &

Studier, 1991) to reduce expression before induction. Initial attempts at deuteration using ampicillin-resistant plasmids and not using a plasmid-encoded *lacI* repressor met with failures to scale to 1 L cultures, or to reach optical densities OD_{600} approaching 10. Changing to kanamycin is considered preferable, due to its superior bacteriocidal properties and persistence during cultivation.

Inclusion of the *lacI* operon on the plasmid increases the level of the lactose repressor present, creating tighter expression control. Expression control is important because, in its absence, we consistently observed strong leaky expression during adaption to minimal media (whether H_2O or D_2O), followed by decreasing expression and insensitivity to addition of IPTG. This was consistent with existing knowledge that leaky expression is a negative selective pressure, whether through harm to the host cells, or by causing plasmid or expression instability (Baneyx, 1999). As deuterated expression requires many generations of adaption to minimal deuterated media, we consider it is important to avoid leaky expression. SDS–PAGE of positive and a negative example is shown in Fig. 1.

6.2 Precipitating Media

A bane of a beginner using minimal media is the propensity of the media components to precipitate during preparation, especially before or during sterile filtration. This is a problem with ModC1, as with other minimal media including M9. The propensity of the ModC1 media to precipitate is worse with the substitution of glycerol for the original glucose, and worse again with the substitution of D_2O for H_2O. We identified the components that precipitated on mixing, and chose to mix them only immediately before use of the media. On mixing, a fine precipitate forms to some degree, but does not appear to influence the growth of the culture. Components that precipitate on mixing are primarily the copper (II) and citrate-chelated iron(II), which immediately form an orange precipitate.

The citrate-chelated iron(II) $1000 \times$ solution turns a deep brown with weeks of storage, but this is not associated with precipitation or any other kind of separation, and when used to prepare media, it appears to have no effect on culture growth. Minor variations in starting OD due to minor precipitation may be observed.

6.3 Use of Commercial Supercompetent Expression Cells

When testing the growth of a culture under different conditions, it is very helpful to have a reproducible inoculum. We found the use of commercial

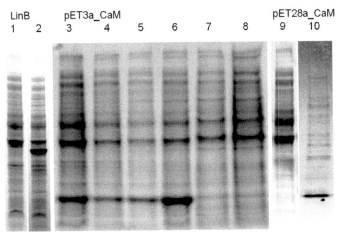

Figure 1 SDS–PAGE of deuterated expression—illustrative examples. Lanes 1 and 2 show preinduction and harvest samples (9.5 h postinduction at 25 °C) from a 1 L production, showing a very strong expression of perdeuterated LinB. Lanes 3–8 demonstrate a failed expression of calmodulin, from a production run previous to the application of the method described here, using a now-considered unsuitable plasmid, pET3a. Lanes 3 and 4: preinduction samples, showing strong uncontrolled expression of the 17 kDa protein. Lane 5: 2.1 h postinduction. Lane 6: 4.6 h postinduction displaying strong expression. Lane 7: 11 h postinduction sample, with the calmodulin no longer abundantly present. Lane 8: harvest sample. Calmodulin was unable to be purified by affinity chromatography from either the pellet or the supernatant. Lane 9 shows a preinduction sample of the same protein subcloned to the preferred vector, pET28a, showing no clear sign of leaky expression. Lane 10 shows the 11.3 h postinduction harvest sample, from which 400 mg (from 1 L medium) was purified. This protein was used in the SANS study of the complex between calmodulin and the HIV1 protein, MA (Taylor et al., 2012).

chemically competent expression cells, transformed with a saturating quantity of miniprep plasmid, to be time efficient and satisfactorily reproducible.

The use of the transformation reaction for the culture inoculation, without plating on agar, minimizes time and generations between transformation and harvest and was found to result in consistently strong induced protein expression. It is known that expression cells containing T7 expression plasmids are prone to lose the ability to be induced to overexpress at high level (Baneyx, 1999). Logically, it follows that this loss of inducible expression is reduced by decreasing the number of generations between transformation and induction. This is achieved by using nearly all of every transformed line from the initial transformation.

6.4 Staged Starter Cultures

The condition of the inoculum is well known to have a major effect on the performance of culture scale-up (Stanbury et al., 1995). We observed this holds true for subcultures being adapted to both minimal media and deuterated media. We assumed that the condition or health of a culture is reflected in its growth rate, relative to previous maximum growth rates in matching D_2O concentrations, corrected for temperature. We determined that inocula were successful when kept within the relatively narrow OD_{600} range of 0.1–1.0. Flask cultures with an initial OD_{600} below 0.1 had varied results, in both nondeuterated and deuterated media, sometimes failing to maintain previous growth rates or exhibiting a significant lag phase. Above an OD_{600} of 1.0, a subsequent decrease in growth rate is evident. We deduce that this is due to oxygen limitation, because the same decrease is not observed in an aerated bioreactor measuring an OD_{600} of 1.0, and at this cell density, the pH has not shifted significantly. In considering that oxygen limitation is stressful to bacteria using glycerol as the sole carbon source, we chose to stringently avoid conditions leading to this state.

Culture flasks were sealed to eliminate deuterium–hydrogen exchange with the moisture in air. They are sealed for hydrogenated cultures as well as with deuterated cultures because a major purpose of unlabeled production is to test expression and equipment in conditions as similar as possible to the intended deuterated growth. An advantage of sealed flasks is that cultures do not concentrate due to evaporation, especially if extended incubation time is required.

6.5 Gentle Handling of the Cultures

Again noting that plasmid loss by expression cells is highly variable and can be fast and significant (Baneyx, 1999), we adhere to gentle handling practices of the cultures. This includes minimizing the time that a flask is removed from a shaker to take a sample, avoiding settling of cells. If cells are spun, they are first chilled by swirling the flask in wet ice. It is likely that the effect of both of these precautions is the avoidance of oxygen limitation occurring due to temporary cessation of culture agitation.

6.6 ^{15}N Ammonium Hydroxide or Sodium Hydroxide as Base Feed for pH Control

The standard method, developed for ^2H labeling, uses 2.58 g/L ammonium chloride as the sole initial nitrogen source, with additional ammonium provided through the base feed controlling pH. This arrangement has the advantage of negligible inhibition by ammonia, and provision of further ammonia being only in response to biomass production. However, when producing ^{15}N-labeled protein, it may be preferred to not use a labeled ammonium base feed. We therefore explored the quantity of ammonia required, and the effect of increased initial ammonium chloride concentration in ModC1, in developing the method used for ^{15}N labeling.

A bioreactor culture prepared as per the standard method but with sodium hydroxide in place of ammonium hydroxide, resulted in the culture failing at an OD_{600} of approximately 12. In a set of flask cultures, initial ammonium chloride concentration in ModC1 bulk solution was tested from 2.58 to 25.8 g/L, and exponential growth rates were measured. Increasing to an initial concentration of 5.16 g/L ammonium chloride resulted in a decrease of growth rate of 6%. 7.74 g/L of ammonium chloride produced a decrease in growth rate of 23%. Larger amounts of ammonium chloride produced further decreasing growth rates, with the growth rate decreasing near linearly with increasing initial ammonium chloride. Therefore, we concluded that for ^{15}N labeling, an initial concentration of 5.16 g/L ^{15}N ammonium chloride was reasonable to use. Estimating a consumption of approximately 2.58 g/L ammonium chloride per 10 OD_{600} units, we chose to add 2.58 g/L ^{15}N ammonium chloride at OD_{600} points of approximately 10, 20, and 30 as described in Section 5.1.

7. TYPICAL DEUTERATION LEVELS

The method as described is the result of theory, development, and refinement, and the results described stringently adhered to that method, demonstrating its reliability and robustness. We have found that if the protein expresses well in hydrogenated ModC1, then it similarly expresses well in labeled media, with similar yield.

Deuteration levels obtained have been consistent and predictable. Production in 90% D_2O using 40 g/L unlabeled glycerol consistently achieves a nonexchangeable deuteration level of $74 \pm 2\%$. Deuteration in 100% D_2O, including triple and double labeling, using 40 g/L nondeuterated glycerol achieves a deuteration level $88 \pm 2\%$. Perdeuteration, using 100% D_2O and 40 g/L D-glycerol (98%) has repeatedly achieved 99% deuteration.

The method has been stringently applied in approximately 20 cases. Wet weight yields are typically in the range of 40–80 g/L of media, from which purification yields range from 50 to 500 mg. Variation in protein molecular weight, DNA source, and hydrophobicity were not observed to affect protein production using ModC1 minimal medium and our method. Purification was analogous to that obtained from protein produced from biomass grown on rich media such as Luria Broth. The reliability of our method gives confidence that a time- and cost-efficient approach to deuterating or multiple labeling a protein is readily available.

ACKNOWLEDGMENTS

The National Deuteration Facility (http://www.ansto.gov.au/ndf) is partially funded by the National Collaborative Research Infrastructure Strategy, an initiative of the Australian Federal Government. Technical support for operation of bioreactors and media preparation was provided by Robert A. Russell and Marie Gillon. We thank Anton P.J. Middelberg for advice on choice of defined medium for high biomass yield, and Scott Mimms and Orsola Regalia for helpful discussions.

REFERENCES

Anderson, E. H. (1946). Growth requirements of virus-resistant mutants of *Escherichia coli* strain "B". *Proceedings of the National Academy of Sciences of the United States of America, 32*(5), 120–128.

Baneyx, F. (1999). Recombinant protein expression in *Escherichia coli*. *Current Opinion in Biotechnology, 10*(5), 411–421.

Cai, M. L., Huang, Y., Sakaguchi, K., Clore, G. M., Gronenborn, A. M., & Craigie, R. (1998). An efficient and cost-effective isotope labeling protocol for proteins expressed in *Escherichia coli*. *Journal of Biomolecular NMR, 11*(1), 97–102.

Chen, X., Wilde, K. L., Wang, H., Lake, V., Holden, P. J., Middelberg, A. P. J., et al. (2012). High yield expression and efficient purification of deuterated human protein galectin-2. *Food and Bioproducts Processing, 90*(3), 563–572.

Christie, M. P., Whitten, A. E., King, G. J., Hu, S.-H., Jarrott, R. J., Chen, K.-E., et al. (2012). Low-resolution solution structures of Munc18:Syntaxin protein complexes indicate an open binding mode driven by the Syntaxin N-peptide. *Proceedings of the National Academy of Sciences of the United States of America, 109*(25), 9816–9821.

Covington, A. K., Paabo, M., Robinson, R. A., & Bates, R. G. (1968). Use of the glass electrode in deuterium oxide and the relation between the standardized pD (paD) scale and the operational pH in heavy water. *Analytical Chemistry, 40*(4), 700–706.

Dimitrijev-Dwyer, M., He, L., James, M., Nelson, A., Wang, L., & Middelberg, A. P. J. (2012). The effects of acid hydrolysis on protein biosurfactant molecular, interfacial, and foam properties: pH responsive protein hydrolysates. *Soft Matter, 8*(19), 5131–5139.

Dubendorff, J. W., & Studier, F. W. (1991). Controlling basal expression in an inducible T7 expression system by blocking the target T7 promoter with lac repressor. *Journal of Molecular Biology, 219*(1), 45–59.

Gamble, T. R., Clauser, K. R., & Kossiakoff, A. A. (1994). The production and X-ray structure determination of perdeuterated Staphylococcal nuclease. *Biophysical Chemistry, 53*(1–2), 15–25.

Gardner, K. H., & Kay, L. E. (1998). The use of H-2, C-13, N-15 multidimensional NMR to study the structure and dynamics of proteins. *Annual Review of Biophysics and Biomolecular Structure, 27*, 357–406.

Hazemann, I., Dauvergne, M. T., Blakeley, M. P., Meilleur, F., Haertlein, M., Van Dorsselaer, A., et al. (2005). High-resolution neutron protein crystallography with radically small crystal volumes: Application of perdeuteration to human aldose reductase. *Acta Crystallographica. Section D, Biological Crystallography, 61*(Pt. 10), 1413–1417.

Jayakumar, A., Hong, J. S., & Barnes, E. M. (1987). Feedback inhibition of ammonium (methylammonium) ion transport in *Escherichia coli* by glutamine and glutamine analogs. *Journal of Bacteriology, 169*(2), 553–557.

Leiting, B., Marsilio, F., & O'Connell, J. F. (1998). Predictable deuteration of recombinant proteins expressed in *Escherichia coli*. *Analytical Biochemistry, 265*(2), 351–355.

LeMaster, D. M. (1994). Isotope labeling in solution protein assignment and structural analysis. *Progress in Nuclear Magnetic Resonance Spectroscopy, 26*, 371–419, Part 4(0).

LeMaster, D. M., & Richards, F. M. (1982). Preparative-scale isolation of isotopically labeled amino acids. *Analytical Biochemistry, 122*(2), 238–247.

Marley, J., Lu, M., & Bracken, C. (2001). A method for efficient isotopic labeling of recombinant proteins. *Journal of Biomolecular NMR, 20*(1), 71–75.

Meilleur, F., Contzen, J., Myles, D. A. A., & Jung, C. (2004). Structural stability and dynamics of hydrogenated and perdeuterated cytochrome P450cam (CYP101). *Biochemistry, 43*(27), 8744–8753.

Meilleur, F., Dauvergne, M. T., Schlichting, I., & Myles, D. A. A. (2005). Production and X-ray crystallographic analysis of fully deuterated cytochrome P450cam. *Acta Crystallographica. Section D, Biological Crystallography, 61*, 539–544.

Middelberg, A. P. J., O'Neill, B. K., Bogle, I. D. L., & Snoswell, M. A. (1991). A novel technique for the measurement of disruption in high-pressure homogenization: Studies on *E. coli* containing recombinant inclusion bodies. *Biotechnology and Bioengineering, 38*(4), 363–370.

Moore, P. B. (1979). The preparation of deuterated ribosomal materials for neutron scattering. *Methods in Enzymology, 59*, 639–655.

Morris, V. K., Linser, R., Wilde, K. L., Duff, A. P., Sunde, M., & Kwan, A. H. (2012). Solid-state NMR spectroscopy of functional amyloid from a fungal hydrophobin: A well-ordered beta-sheet core amidst structural heterogeneity. *Angewandte Chemie (International Edition in English), 51*(50), 12621–12625.

Olah, G. A., Rokop, S. E., Wang, C. L., Blechner, S. L., & Trewhella, J. (1994). Troponin I encompasses an extended troponin C in the Ca(2+)-bound complex: A small-angle X-ray and neutron scattering study. *Biochemistry, 33*(27), 8233–8239.

Paalme, T., Tiisma, K., Kahru, A., Vanatalu, K., & Vilu, R. (1990). Glucose-limited fed-batch cultivation of *Escherichia coli* with computer-controlled fixed growth rate. *Biotechnology and Bioengineering, 35*(3), 312–319.

Paliy, O., Bloor, D., Brockwell, D., Gilbert, P., & Barber, J. (2003). Improved methods of cultivation and production of deuterated proteins from *E. coli* strains grown on fully deuterated minimal medium. *Journal of Applied Microbiology, 94*(4), 580–586.

Paliy, O., & Gunasekera, T. S. (2007). Growth of *E. coli* BL21 in minimal media with different gluconeogenic carbon sources and salt contents. *Applied Microbiology and Biotechnology, 73*(5), 1169–1172.

Roe, A. J., O'Byrne, C., McLaggan, D., & Booth, I. R. (2002). Inhibition of *Escherichia coli* growth by acetic acid: A problem with methionine biosynthesis and homocysteine toxicity. *Microbiology—SGM, 148*, 2215–2222.

Shu, F., Ramakrishnan, V., & Schoenborn, B. P. (2000). Enhanced visibility of hydrogen atoms by neutron crystallography on fully deuterated myoglobin. *Proceedings of the National Academy of Sciences of the United States of America, 97*(8), 3872–3877.

Sivashanmugam, A., Murray, V., Cui, C. X., Zhang, Y. H., Wang, J. J., & Li, Q. Q. (2009). Practical protocols for production of very high yields of recombinant proteins using *Escherichia coli*. *Protein Science, 18*(5), 936–948.

Sosa-Peinado, A., Mustafi, D., & Makinen, M. W. (2000). Overexpression and biosynthetic deuterium enrichment of TEM-1 beta-lactamase for structural characterization by magnetic resonance methods. *Protein Expression and Purification*, *19*(2), 235–245.

Stanbury, P. F., Whitaker, A., & Hall, S. J. (1995). *Principles of fermentation technology* (2nd ed.). Amsterdam: Pergamon.

Studier, F. W., & Moffatt, B. A. (1986). Use of bacteriophage T7 RNA polymerase to direct selective high-level expression of cloned genes. *Journal of Molecular Biology*, *189*(1), 113–130.

Taylor, J. E., Chow, J. Y. H., Jeffries, C. M., Kwan, A. H., Duff, A. P., Hamilton, W. A., et al. (2012). Calmodulin binds a highly extended HIV-1 MA protein that refolds upon its release. *Biophysical Journal*, *103*(3), 541–549.

Torchia, D. A., Sparks, S. W., & Bax, A. (1989). Staphylococcal nuclease: Sequential assignments and solution structure. *Biochemistry*, *28*(13), 5509–5524.

Tuominen, V. U., Myles, D. A. A., Dauvergne, M. T., Lahti, R., Heikinheimo, P., & Goldman, A. (2004). Production and preliminary analysis of perdeuterated yeast inorganic pyrophosphatase crystals suitable for neutron diffraction. *Acta Crystallographica. Section D, Biological Crystallography*, *60*, 606–609.

Vanatalu, K., Paalme, T., Vilu, R., Burkhardt, N., Junemann, R., May, R., et al. (1993). Large-scale preparation of fully deuterated cell components. Ribosomes from *Escherichia coli* with high biological activity. *European Journal of Biochemistry*, *216*(1), 315–321.

Venters, R. A., Huang, C. C., Farmer, B. T., Trolard, R., Spicer, L. D., & Fierke, C. A. (1995). High-level H-2/C-13/N-15 labeling of proteins for NMR studies. *Journal of Biomolecular NMR*, *5*(4), 339–344.

Xu, B., Jahic, M., & Enfors, S. O. (1999). Modeling of overflow metabolism in batch and fed-batch cultures of *Escherichia coli*. *Biotechnology Progress*, *15*(1), 81–90.

CHAPTER TWO

Protein Labeling in *Escherichia coli* with ^{2}H, ^{13}C, and ^{15}N

J. Todd Hoopes*,[1], Margaret A. Elberson*, Renae J. Preston*, Prasad T. Reddy*,†, Zvi Kelman†

*Biomolecular Labeling Laboratory, Institute for Bioscience and Biotechnology Research, National Institute of Standards and Technology and the University of Maryland, Rockville, Maryland, USA
†Biomolecular Labeling Laboratory, and Biomolecular Structure and Function Group, Institute for Bioscience and Biotechnology Research, National Institute of Standards and Technology and the University of Maryland, Rockville, Maryland, USA
[1]Corresponding author: e-mail address: hoopesj@umd.edu

Contents

1. Introduction 28
2. Selection of Induction Method 29
    2.1 Manual Induction 29
    2.2 Autoinduction 29
    2.3 Manual Versus Autoinduction 29
3. Special Considerations When Labeling with Deuterium 30
4. Plasmid and *E. coli* Strain Selection 30
5. Determination of Isotope Incorporation 32
6. Acclimation of *E. coli* to Growth in D_2O 35
7. Media Preparation 36
    7.1 Stock Solutions 37
    7.2 Media Composition 39
8. Protein Expression 40
    8.1 Manual Induction 41
    8.2 Autoinduction 42
Acknowledgments 43
References 43

Abstract

A number of structural biology techniques such as nuclear magnetic resonance spectroscopy and small-angle neutron scattering can be performed with proteins with nuclei at natural isotope abundance. However, the use of proteins labeled with stable isotopes (^{2}H, ^{13}C, and ^{15}N) enables greater experimental flexibility. In this chapter, several methods for uniform and fractional protein labeling with stable isotopes using *Escherichia coli* in a defined media are described. The methods described can be used for labeling with single or multiple isotopes.

1. INTRODUCTION

Stable isotopes are widely used in studies of biomolecules. The use of isotopic labeling in nuclear magnetic resonance (NMR) spectroscopy, small-angle neutron scattering (SANS), neutron reflectometry, and proteomic mass spectroscopy research has revolutionized the analysis of biomolecules (Atreya & Chary, 2001; Capel et al., 1987; Jacrot, 2001; Shu, Ramakrishnan, & Schoenborn, 2000; Stewart, Thomson, & Figeys, 2001; Woods & Darie, 2014). While these techniques can be performed with biomolecules at natural isotope abundance, the use of isotopically labeled proteins allows for additional experimental approaches.

Escherichia coli is the most commonly used platform for the labeling of protein with stable isotopes for several reasons. It is the most commonly used host for heterologous expression of proteins and therefore can be readily adapted to the production of labeled proteins. *E. coli* is easy and inexpensive to grow with well-defined media required for protein labeling. In addition, *E. coli* can be grown in any biochemical laboratory without specialized equipment (e.g., bioreactor, CO_2 incubator, special aseptic techniques, etc.). When perdeuterated proteins are needed, *E. coli* has an advantage over mammalian expression systems; it can grow in the presence of 100% deuterium oxide (D_2O), while mammalian cells cannot grow in the presence of more than 35% D_2O (Murphy, Desaive, Giaretti, Kendall, & Nicolini, 1977).

Laboratories that routinely express recombinant proteins in *E. coli* should be able to adapt their protocols, host cells, and plasmids to produce proteins labeled with ^2H, ^{13}C, or ^{15}N. Of course, some changes to protocols will be necessary; in particular, the cells will have to be grown in minimal media.

Stable isotope labeling of proteins can be divided into five types: uniform labeling in which all atoms in the protein are labeled, fractional labeling in which not all atoms are labeled but label is randomly distributed along the protein, amino acid-specific labeling in which one type of amino acid residue is labeled (e.g., all alanines), site-specific labeling in which a unique amino acid residue is labeled, and segmental labeling in which only a part (segment or domain) of the protein is labeled (Atreya, 2012).

Chemicals labeled with ^2H and ^{13}C are expensive. Therefore, before performing a labeling experiment one should optimize protein expression and purification protocols to get the highest yield possible per liter of culture.

Similarly, the type of labeling, cell type, and plasmid need to be considered carefully prior to starting the experiment.

Here, we describe methods for the uniform and fractional labeling of proteins expressed in *E. coli* using shaker flasks. The chapter starts with general issues to be considered when the expression of label protein is needed, followed by specific expression protocols used in our laboratory.

2. SELECTION OF INDUCTION METHOD

There are two basic approaches to protein expression in *E. coli*. Manual induction uses an inducer molecule (e.g., a sugar metabolite or a sugar analogue such as isopropyl β-D-1-thiogalactopyranoside (IPTG)) or a temperature shift (Reddy, Peterkofsy, & McKenney, 1989) to begin protein expression. The other approach, autoinduction, uses a combination of carbon sources to induce expression at an appropriate point. General considerations for each option are discussed below.

2.1 Manual Induction

Manual induction is the most common way to induce plasmid-based protein expression in *E. coli*. Manual induction provides tighter control of expression and can be used with any induction system. This system can be easily optimized for a specific protein because the induction point, inducer level, and length of induction can all be controlled.

2.2 Autoinduction

Autoinduction media were originally developed by Studier and were based on the function of the lac operon in a mixture of glucose, glycerol, and lactose under diauxic growth conditions (Studier, 2005). During the initial growth period, glucose is used preferentially as a carbon source, the lac operon is repressed, and the protein is not expressed. As glucose is depleted from the medium, the repression of the lac operon is relieved, leading to the activation of other sugar uptake systems and protein expression. In addition, the autoinduction method can be adapted for expression systems that use other carbon sources as inducers (e.g., arabinose or rhamnose).

2.3 Manual Versus Autoinduction

In manual induction, the culture growth needs to be frequently monitored because induction should occur at mid-log phase. One of the main advantages of the autoinduction method is its convenience. It requires very little monitoring of cultures from inoculation through harvest and cells can reach high density. Another advantage of the autoinduction system is that switching from glucose to an alternate carbon source (e.g., lactose, arabinose, or rhamnose) occurs gradually, resulting in a more gradual induction of the protein of interest. This gradual induction, in some cases, results in a larger fraction of soluble protein. On the other hand, autoinduction cannot be used for expression systems that do not use a carbon source that is a catabolite for *E. coli* (e.g., temperature shift). Another issue to consider is the cost associated with the different induction systems. While both manual and autoinduction can be used for labeling with ^{2}H, ^{13}C, and ^{15}N, labeled carbon sources (other than glucose) are expensive. Therefore, when labeling with ^{13}C, it is more economical to use manual induction.

3. SPECIAL CONSIDERATIONS WHEN LABELING WITH DEUTERIUM

Usually, the growth of *E. coli* is not affected in media containing ^{13}C and/or ^{15}N in lieu of the natural abundance isotopes. However, growth in the presence of D_2O is more complicated. While the presence of 70% D_2O has little effect on *E. coli* growth, media containing a higher concentration of D_2O requires cells to be "acclimated" or growth may stall (Moore & Engelman, 1976). In addition, cells acclimated to 100% D_2O may exhibit marked changes in growth rate and metabolic processes (Hochuli, Szyperski, & Wüthrich, 2000; Mann & Moses, 1971; Paliy & Gunasekera, 2007; van Horn & Ware, 1959).

For many applications, perdeuteration is not required and partial (fractional) labeling with ^{2}H is sufficient. Fractional deuteration is achieved by varying the ratio of $D_2O:H_2O$ in the media. Table 1 shows the percent deuteration of a sample protein when a specific percent D_2O is included in M9 media. When higher than 95% deuteration is needed, deuterated glucose should be used in the media.

4. PLASMID AND *E. COLI* STRAIN SELECTION

A large number of plasmids and *E. coli* strains are available for protein expression, each with unique features needed for specific expression and/or purification requirements (e.g., codon usage, disulfide bond formation, affinity tags, etc.). It is beyond the scope of this article to address all of these

Table 1 Percent Incorporation of ^{2}H into a Protein Expressed in *E. coli* Cells Grown in M9 Media Containing Different Percentages of D_2O

D_2O (%)	Mass of Protein (Da)	Increase in Mass (Da)	^{2}H Incorporation (%)
0	33,629	0	0
20	33,857	+228	17
40	34,074	+445	33
60	34,331	+702	52
80	34,611	+982	73
100	34,971	+1342	100

E. coli BL21(DE3) cells were transformed with a pET28a vector harboring the gene encoding for MSP 1E3D1 (Bayburt, Grinkova, & Sligar, 2002). Cells were grown in a 250-mL baffled shake flask containing 100 mL of media, and the protein was purified in H_2O-containing buffers by immobilized metal ion affinity chromatography.

The molecular weight of the samples was determined by MALDI mass spectroscopy using a Bruker microflex LRF instrument. Samples were prepared by layering 1 μL matrix, 1 μL sample, and 1 μL matrix, drying after each addition. The matrix was α-cyano-4-hydroxycinnamic acid (HCCA) dissolved to saturation in 30:70 (v/v) acetonitrile to 0.1% trifluoroacetic acid (TFA). The concentration of the protein samples was approximately 5 mg/mL in buffer containing no more that 10 m*M* NaCl.

The 100% ^{2}H incorporation into proteins when cells were grown in 100% D_2O is high. Usually, about 95–98% incorporation of ^{2}H can be observed when cells are grown under these conditions.

issues, but general information on basic aspects of protein expression has been reviewed (Baneyx, 1999; Rosano & Ceccarelli, 2014; Terpe, 2006). The protocols described in this chapter are applicable to most commonly used *E. coli* strains and plasmids.

The promoter that drives protein expression needs to be considered when choosing a plasmid for protein labeling. For example, while the araBAD and rhaPBAD promoters can be used for ^{2}H and ^{15}N labeling they are not recommended when uniform labeling with ^{13}C is required because *E. coli* catabolizes both arabinose and rhamnose used to induce these promoters (labeled arabinose and rhamnose are very expensive). The lac promoter, on the other hand, can be used for ^{13}C labeling because it can be induced by IPTG, which can not be catabolized by *E. coli*.

The choice of a host strain is usually dependent on the specific vector and promoter used for expression (e.g., the T7 promoter), or a specific requirement of the expressed protein (e.g., S–S bond formation). However, cost is important when labeling proteins. As labeled chemicals and D_2O are expensive, one should aim to get the highest possible biomass per liter. Therefore, when possible, avoid *E. coli* strains derived from K-12 and use B strain derivatives (e.g., BL21) because these tend to be more robust and result in higher biomass yield (Waegeman et al., 2013). Growth in D_2O places stress on the cells, as does growth in antibiotic. Therefore, when cells are grown in 100% D_2O only the antibiotic for the expression plasmid needs to be included (even if the strain requires additional antibiotics, for example, for the maintenance of a plasmid expressing rare tRNAs), although this should be experimentally determined for each protein and expression system.

5. DETERMINATION OF ISOTOPE INCORPORATION

Determining the level of ^{13}C or ^{15}N incorporation into a purified protein is relatively easy using matrix assisted laser desorption/ionization (MALDI) mass spectrometry. The observed mass increase is compared to the theoretical mass increase of the uniformly labeled protein. Table 2 summarizes the mass for each amino acid residue when labeled with ^{2}H, ^{13}C, or ^{15}N. The mass of the N-terminal and C-terminal residues (end effect) is accounted for by the addition of 18.02 Da to the total. By comparing the ratio between the observed mass and the theoretical mass, the percent labeling can be determined. The observed increase in mass divided by the theoretical increase in mass (assuming complete labeling), multiplied by 100 yields the percent isotope incorporated.

Most purification protocols for labeled proteins use H_2O and not D_2O. This complicates the calculation of percent ^{2}H incorporation, as some of the ^{2}H is readily exchangeable with the ^{1}H in the buffer. Table 2 shows the mass for each amino acid residue when it is labeled with ^{2}H and equilibrated in H_2O buffer (the exchangeable hydrogens are assumed to be ^{1}H). Table 2 also provides the number of nonexchangeable hydrogens for each residue. The mass of the N-terminal and C-terminal residues (end effect) is accounted for by the addition of 18.02 Da to the total.

Note: based on our experience when using modified Tyler media (MTM) (Tyler et al., 2005) and the protocols described here, if proteins are labeled with ^{13}C, ^{15}N, and ^{2}H, the calculation for the theoretical mass can assume that ^{13}C and ^{15}N are incorporated at 95% or more (see Table 3).

Note: for ^{13}C, ^{15}N, and $^{13}C/^{15}N$ labeling, there are a number of web sites that will calculate the theoretical protein mass if uniformly labeled.

Note: the size differences between the unlabeled and labeled proteins are such that they cannot be evaluated using sodium dodecyl sulfate polyacrylamide gel electrophoresis (SDS-PAGE) (for an example, see Fig. 1).

Table 2 Theoretical Molecular Mass of Amino Acid Residues

Residue Type	Unlabeled Residue Mass	Number of Nonexchangeable Hydrogens	Mass of ^{2}H-Labeled Residue	Number of Carbons	Mass of ^{13}C-Labeled Residue	Number of Nitrogens	Mass of ^{15}N-Labeled Residue
Ala	71.08	4	75.10	3	74.05	1	72.06
Arg	156.19	7	163.23	6	162.14	4	160.15
Asn	114.10	3	117.12	4	118.06	2	116.08
Asp	115.09	3	118.11	4	119.05	1	116.07
Cys	103.14	3	106.16	3	106.11	1	104.13
Glu	129.12	5	134.15	5	134.07	1	130.10
Gln	128.13	5	133.16	5	133.08	2	130.11
Gly	57.05	2	59.06	2	59.03	1	58.04
His	137.14	5	142.17	6	143.09	3	140.11
Ile	113.16	10	123.22	6	119.11	1	114.15
Leu	113.16	10	123.22	6	119.11	1	114.15
Lys	128.17	9	137.23	6	134.13	2	130.15
Met	131.20	8	139.25	5	136.15	1	132.18
Phe	147.18	8	155.23	9	156.10	1	148.16
Pro	97.12	7	104.16	5	102.07	1	98.10

Continued

Table 2 Theoretical Molecular Mass of Amino Acid Residues—cont'd

Residue Type	Unlabeled Residue Mass	Number of Nonexchangeable Hydrogens	Mass of ^{2}H-Labeled Residue	Number of Carbons	Mass of ^{13}C-Labeled Residue	Number of Nitrogens	Mass of ^{15}N-Labeled Residue
Ser	87.08	3	90.10	3	90.05	1	88.07
Thr	101.11	5	106.14	4	105.07	1	102.09
Trp	186.21	8	194.26	11	197.13	2	188.19
Tyr	163.18	7	170.22	9	172.11	1	164.16
Val	99.13	8	107.05	5	104.09	1	100.12

Within a polypeptide chain fully labeled with ^{2}H, ^{13}C, or ^{15}N (deuterated mass assumes the polypeptide is equilibrated in H$_2$O-containing buffer).

Table 3 Percent Labeling of an Expressed Protein in *E. coli* Cells Grown in the Presence of ^{2}H, ^{13}C, or ^{15}N

Condition	Mass of Protein (Da)	Labeling Efficiency
Unlabeled	29,064	Not applicable
100% ^{13}C-glucose	30,314	97%
100% ^{15}N-ammonium chloride	29,368	95%
100% D$_2$O	30,459	98%

E. coli BL21 (DE3) cells were transformed with a pET28a vector harboring the gene encoding for proliferating cell nuclear antigen (PCNA, TK0535; Ladner, Pan, Hurwitz, & Kelman, 2011). Cells were grown in MTM containing ^{13}C, ^{15}N, or ^{2}H and induced at an OD$_{600}$ of 1.25 at 37 °C for 4 h. The protein was purified as previously described (Ladner et al., 2011).
The molecular weight of the purified protein was determined by MALDI mass spectroscopy using a Bruker microflex LRF instrument. Samples were prepared by layering 1 μL matrix, 1 μL sample, and 1 μL matrix, drying after each addition. The matrix was HCCA dissolved to saturation in 30:70 (v/v) acetonitrile to 0.1% TFA. Samples were at a concentration of approximately 5 mg/mL in buffer containing no more than 10 m*M* NaCl.

6. ACCLIMATION OF *E. COLI* TO GROWTH IN D$_2$O

- Early in the morning select a fresh single colony from a LB agar plate and inoculate a culture tube containing 5 mL of modified Tyler media (MTM) (Tyler et al., 2005) (Section 7.2.2) made with 50% D$_2$O and the appropriate antibiotics.
- Incubate the culture at 37 °C with shaking at 250 rpm.
- After 6 h of growth, check for turbidity of the culture. If the culture is turbid, add 5 mL of 37 °C MTM made with 100% D$_2$O (1:2 dilution resulting in 75% D$_2$O).
- Divide the mixture into two culture tubes and place the tubes back in the shaker and allow the culture to continue to grow until the cells become turbid again.
- Use all 10 mL to inoculate a 500-mL baffled shake flask containing 120 mL of 100% D$_2$O in MTM that has been warmed to 37 °C.
- Allow the culture to grow overnight (at least 16 h) at 37 °C with shaking at 200 rpm.
- The next day centrifuge the culture at 3500 × *g* for 20 min using a rotor prewarmed to 37 °C.
- Resuspend the cells in 60 mL of MTM made with 100% D$_2$O and preheated to 37 °C.

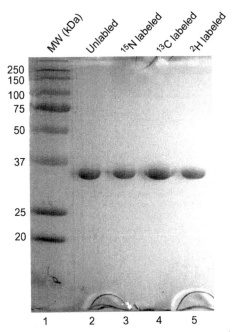

Figure 1 Protein labeled with stable isotopes cannot be distinguished using SDS-PAGE. E. coli BL21(DE3) cells were transformed with a pET28a vector harboring the gene encoding for proliferating cell nuclear antigen (PCNA, TK0535; Ladner et al., 2011). Cells were grown in MTM-containing ^{2}H, ^{13}C, or ^{15}N and induced at OD$_{600}$ of 1.25 at 37 °C for 4 h. The protein was purified as previously described (Ladner et al., 2011). One microgram of purified protein was fractionated on 15% SDS-PAGE and the gel was stained with Coomassie brilliant blue. Lane 1, molecular mass marker; Lane 2, unlabeled PCNA; Lane 3, PCNA labeled with ^{15}N; Lane 4, PCNA labeled with ^{13}C; and Lane 5, PCNA labeled with ^{2}H.

- Add 15 mL to each shake flask to inoculate the large cultures described below (see Day 2 in Sections 8.1 and 8.2).

Note: when labeled carbon and/or nitrogen is required in addition to deuterium, the cells are first acclimated to D_2O in the presence of unlabeled carbon and nitrogen. Only after the cells are acclimated to 100% D_2O is the ^{13}C-glucose and/or ^{15}NH$_4$Cl added.

7. MEDIA PREPARATION

M9 is one of the most commonly used minimal media and it is the base for many other minimal media in use (Marley, Lu, & Bracken, 2001). However, M9 lacks trace elements and vitamins, resulting in a relatively low cell mass per liter compared to rich media. Some laboratories add trace elements and vitamin to M9, resulting in increased cell mass per liter (Marley et al., 2001; Paliy & Gunasekera, 2007). Our laboratory routinely uses MTM in order to achieve cell mass approaching that of rich media. MTM works well for the production of most proteins.

Most minimal media use a phosphate buffer. Phosphate, however, may produce a precipitate in the presence of divalent cations. As some proteins require a high level of divalent cation for proper folding (e.g., C-reactive protein, which requires 5–10 mM calcium ion in the media), minimal media based on a phosphate buffer system is not suitable if a precipitate forms. For the expression of these proteins, our laboratory uses a modified Neidhardt media which includes 3-(N-morpholino)propanesulfonic acid (MOPS) as the buffering agent and therefore contains a lower level of phosphate (Neidhardt, Bloch, & Smith, 1974).

In the section below, protocols are provided for making media for manual and autoinduction. The stock solutions will be described first, followed by media preparation. The protocols are for the preparation of unlabeled media. When ^{15}N labeling is required, the NH$_4$Cl in the nitrogen phosphate stock (NPS) (Section 7.1.2) is replaced with ^{15}NH$_4$Cl, and when ^{13}C labeling is needed the carbon source stock (CSS) (Section 7.1.3) is replaced with 5 g/L ^{13}C-glucose. Unless perdeuterated labeling is required, stock solutions can be made with H$_2$O and not D$_2$O (to reduce the cost). Using H$_2$O-based stock solutions will result in media that is about 90% D$_2$O. For all media, the appropriate antibiotic(s) should also be included (not mentioned in the recipes).

Note: reducing the ^{13}C-glucose to 2.5 g/L will give a final OD$_{600}$ of about 4.0. This level of growth may be sufficient for most projects.

When making the stock solutions listed below, the components should be added in the order they are listed.

7.1 Stock Solutions
7.1.1 20× Modified Neidhardt Stock
600 mL H$_2$O
168 g MOPS
14.4 g Tricine
25 g NH$_4$Cl
4.8 g K$_2$SO$_4$
117 g NaCl

Stir the solution until fully dissolved; adjust pH to 6.7 with NaOH, add H_2O to 1 L and autoclave.

Store at ambient temperature.

7.1.2 20× Nitrogen Phosphate Stock

600 mL H_2O
50 g NH_4Cl
14.2 g Na_2SO_4
136 g KH_2PO_4
142 g Na_2HPO_4

Stir until fully dissolved, add H_2O to 1 L and autoclave.

Store at ambient temperature.

Note: if you store NPS at 4 °C, it will precipitate. It can be redissolved by heating in a 50 °C water bath with occasional shaking. It can then be used directly without cooling.

7.1.3 50× Carbon Source Stock

300 g glycerol
125 g D-glucose

Add H_2O to 1 L. Filter sterilize and store at 4 °C.

Note: when labeling with ^{13}C, the CSS is replaced with ^{13}C-glucose (no CSS stock is used). Due to the expense of ^{13}C-glucose, we add the ^{13}C-glucose as powder directly to the media at the point of inoculation (a total of 5 g/L ^{13}C-glucose).

7.1.4 50× Autoinduction Carbon Source Stock

300 g glycerol
35 g D-glucose
100 g lactose (or arabinose, rhamnose, etc., depending on the induction system)

Add H_2O to 1 L. Filter sterilize and store at 4 °C.

Note: upon standing a precipitate may form. Autoinduction carbon source stock (ACSS) can be warmed in a 35–45 °C water bath to dissolve the lactose completely into solution. However, shaking the bottle and immediately adding the ACSS to the media will work as well.

7.1.5 10,000× Trace Elements Stock

36 mL of H_2O
50 mL of 0.1 M $FeCl_2$ in 0.01 M HCl

1 mL of 1 M MnCl$_2$
1 mL of 1 M ZnCl$_2$
1 mL of 0.2 M CoCl$_2$
2 mL of 0.1 M CuCl$_2$
1 mL of 0.2 M NiCl$_2$
2 mL of 0.1 M Na$_2$MoO$_4$
2 mL of 0.1 M Na$_2$SeO$_3$
2 mL of 0.1 M H$_3$BO$_3$
2 mL of 1 M CaCl$_2$

Add stock minerals to the H$_2$O in the order listed to give a final volume of 100 mL. Filter sterilize and divide into 0.5 mL aliquots. Store at −20 °C.

Note: the FeCl$_2$ needs to be made fresh before use. All other solutions can be made in advance and stored at ambient temperature.

Note: make sure that each trace element is fully dissolved before adding to the trace elements stock (TES).

7.1.6 1000× Vitamin Stock

78 mL of H$_2$O
4 mL of 5 mM vitamin B12
2 mL of 10 mM nicotinic acid
2 mL of 10 mM pyridoxine HCl
2 mL of 10 mM thiamine HCl
2 mL of 10 mM 4-aminobenzoic acid
5 mL of 100 μM folic acid
5 mL of 100 μM riboflavin

Add stocks to the water to yield 100 mL final volume. Filter sterilize and divide into 1 mL aliquots. Store at −20 °C.

Note: individual vitamin stock (VS) solutions should be stored at −20 °C.

Note: there are commercially available VSs, such as minimum essential media (MEM) vitamin solution (Life Technologies), which can be substituted for the VS. However, we have not evaluated these products.

7.2 Media Composition

The recipes described below are for making 2 L of media. When making media containing ^{13}C or ^{15}N, it may be easier to autoclave only the water in the flask and store the flask until needed. For D$_2$O, the media should be mixed fresh and not stored.

Note: all stock solutions should be added once the flask has cooled to below 55 °C.

Note: as mentioned earlier, depending upon the labeling needed, for deuterium labeling replace the H_2O with D_2O, for carbon labeling replace the CSS with 5 g of ^{13}C-glucose per liter of media, or for nitrogen labeling replace the NPS with NPS containing $^{15}NH_4Cl$ instead of NH_4Cl.

Note: the carbon source, trace elements, and vitamins should be added just prior to use. If the carbon source is added to the media prior to autoclaving, the media will turn light brown. This should not markedly affect growth of the cultures. However, this browning does indicate a slight decomposition of the carbon source and is not recommended.

7.2.1 Modified Neidhardt Media

1.75 L H_2O
2 mL 1 M $MgSO_4$
200 μL TES
40 mL CSS
200 mL modified Neidhardt stock
2 mL VS

The pH will be below 6.5 and the conductivity will be ~11 mS. Adjust the pH to 7.2 with NaOH. Just prior to use, add 5 mL of NPS.

7.2.2 Modified Tyler Media

1.85 L H_2O
2 mL 1 M $MgSO_4$
200 μL TES
40 mL CSS
100 mL NPS
2 mL VS

The pH will be approximately 6.8 and the conductivity will be ~16 mS.

7.2.3 Modified Tyler Medium for Autoinduction

The recipe is the same as MTM, but 40 mL of ACSS should replace the 40 mL CSS.

The pH will be approximately 6.8 and the conductivity will be ~16 mS.

8. PROTEIN EXPRESSION

The protocols described below are to grow 6 L of culture in four 4-L baffled flasks (1.5 L per flask). We found that for well-expressed proteins, this yields a sufficient amount of labeled purified protein for routine analysis by

NMR or SANS. This section is divided into two protocols: one for MTM and the other for autoinduction (AMTM).

Note: when a large quantity of protein is needed, we would recommend growing *E. coli* using a fermenter.

Note: depending on the label(s) needed, the H_2O, glucose, and/or ammonium chloride need to be replaced with their labeled counterparts (Section 7).

Note: if labeling with ^2H is required, the cells need to be acclimated to growth in D_2O (Section 6) prior to the beginning of the protocols described below.

8.1 Manual Induction
8.1.1 Day 1
Make 6 L of media (Section 7.2) to be used the next day.
- Aliquot into 4×4-L baffled flasks (1.5 L per flask).
- Media can be stored at 4 °C for up to a week.

For ^{13}C and/or ^{15}N labeling
- Make a starter culture: add 100 mL of MTM (Section 7.2.2) to a 500-mL baffled flask.
- Add the appropriate antibiotic to the media.
- Inoculate the flask with a freshly transformed colony late in the day. Shake the cultures at 30 °C, 250 rpm overnight.

Note: the starter inoculum is grown at 30 °C to prevent the cells from reaching stationary phase, as this will increase lag time when added to the large volume flask and can cause stalled cultures. The temperature may need to be reduced to 25 °C if the inoculum will be incubated for longer than 20 h.

8.1.2 Day 2
- Raise the temperature of the shaker to 37 °C. Do not remove the starter culture.
- Add the appropriate antibiotic to the $4 \times$ 4-L baffled flasks made on Day 1.
- Warm the flasks in hot water or an incubator to 30–35 °C.
- Place the flasks into a 37 °C incubated shaker and shake at 185 rpm for 15 min.
- Inoculate each flask with 15 mL of the overnight culture. The remainder of the overnight culture can be centrifuged at $3500 \times g$ for 20 min and the pellet frozen for use as an uninduced control.

- If labeling with D_2O, inoculate each flask with 15 mL of the acclimated overnight culture as described in Section 6.

From here on the protocol is the same whether labeling with D_2O or not.
- Take an initial OD_{600} and monitor OD_{600} every hour until the culture reaches 0.8 OD_{600}. Then monitor the OD_{600} every 20 min. It will take 3–5 h total time to reach the induction cell density depending on the *E. coli* strain and plasmid used.
- Induce protein expression when the cultures reach OD_{600} 1.25–1.5.
- Allow cells to grow an additional 3–6 h.
- Harvest cells by centrifuging at 3500 × *g* for 20 min. Discard the supernatant and record the cell mass.
- Immediately lyse cells with a sonicator, cell disruptor, or French Press, or store the cell pellet at −80 °C for later use.

Note: 3–6 h is a standard induction time for cells growing at 37 °C. However, if protein expression requires lower temperature, induce for 16–24 h.

Note: if lower temperatures (30 °C or lower) are needed for appropriate expression, the temperature of the culture needs to be reduced before the cells reach induction density. The cells should be at the desired induction temperature for at least 20 min prior to induction. This can be achieved by cooling down the culture in ice water before placing it in a shaker at the desired temperature. However, rapid cooling stresses the cells. When D_2O is used the cells are already under stress and thus the rapid cooling may not be the desired technique. So, alternatively, the culture can be moved to a lower temperature incubator when the OD_{600} reach 0.8. It will take about an hour for the culture to reach the new, lower, temperature. The exact time and OD_{600} needed must be determined empirically.

Note: if an ultralow freezer is not available, −20 °C will suffice for short-term (several weeks) storage of the cell pellet, but should be avoided for longer periods.

8.2 Autoinduction

8.2.1 Day 1
Same procedure as Day 1 (Section 8.1).

8.2.2 Day 2
Same procedure as Day 2 (Section 8.1). However, there is no need to closely monitor cell density. Check the OD_{600} every few hours to make sure the culture is growing. If the culture has not reached an OD_{600} of 5.0 by the

end of the day, reduce temperature to 30 °C and allow cells to grow overnight.

8.2.3 Day 3

- Check the OD_{600} of the cultures. If the cultures have reached an OD_{600} of at least 5.0 harvest the cells as described for manual induction, above.
- If the cultures are not at an OD_{600} of 5.0, raise the temperature to 37 °C and monitor OD_{600} every hour until cultures reach an OD_{600} of 5.0.
- Harvest the cells as described for manual induction, above.
- Process the cell pellet or store it as described for manual induction, above.

Note: the culture may be allowed to grow until it reaches an OD_{600} of 7.0 which may increase the yield.

ACKNOWLEDGMENTS

We would like to thank Dr. Lori Kelman for critical reading of the manuscript.

Disclaimer. Certain commercial equipment, instruments, and materials are identified in this paper in order to specify the experimental procedure. Such identification does not imply recommendation or endorsement by the National Institute of Standards and Technology.

REFERENCES

Atreya, H. S. (Ed.), (2012). *Advances in experimental medicine and biology*: Vol. 992. Isotope labeling in biomolecular NMR. Netherlands: Springer.

Atreya, H. S., & Chary, K. V. (2001). Selective "unlabeling" of amino acids in fractionally ^{13}C labeled proteins: An approach for stereospecific NMR assignments of CH$_3$ groups in Val and Leu residues. *Journal of Biomolecular NMR, 19*, 267–272.

Baneyx, F. (1999). Recombinant protein expression in *Escherichia coli*. *Current Opinion in Biotechnology, 10*, 411–421.

Bayburt, T. H., Grinkova, Y. V., & Sligar, S. G. (2002). Self-assembly of discoidal phospholipid bilayer nanoparticles with membrane scaffold proteins. *Nano Letters, 2*, 853–856.

Capel, M. S., Engelman, D. M., Freeborn, B. R., Kjeldgaard, M., Langer, J. A., Ramakrishnan, V., et al. (1987). A complete mapping of the proteins in the small ribosomal subunit of *Escherichia coli*. *Science, 238*, 1403–1406.

Hochuli, M., Szyperski, T., & Wüthrich, K. (2000). Deuterium isotope effects on the central carbon metabolism of *Escherichia coli* cells grown on a D_2O-containing minimal medium. *Journal of Biomolecular NMR, 17*, 33–42.

Jacrot, B. (2001). The study of biological structures by neutron scattering from solution. *Reports on Progress in Physics, 39*, 911–953.

Ladner, J. E., Pan, M., Hurwitz, J., & Kelman, Z. (2011). Crystal structures of two active proliferating cell nuclear antigens (PCNAs) encoded by *Thermococcus kodakaraensis*. *Proceedings of the National Academy of Sciences of the United States of America, 108*, 2711–2716.

Mann, L. R., & Moses, V. (1971). Properties of *Escherichia coli* grown in deuterated media. *Folia Microbiologica, 16*, 267–284.

Marley, J., Lu, M., & Bracken, C. (2001). A method for efficient isotopic labeling of recombinant proteins. *Journal of Biomolecular NMR, 20*, 71–75.

Moore, P. B., & Engelman, D. M. (1976). The production of deuterated *E. coli*. *Brookhaven Symposia in Biology, 27,* V12–V23.

Murphy, J., Desaive, C., Giaretti, W., Kendall, F., & Nicolini, C. (1977). Experimental results on mammalian cells growing *in vitro* in deuterated medium for neutron-scattering studies. *Journal of Cell Science, 25,* 87–94.

Neidhardt, F. C., Bloch, P. L., & Smith, D. F. (1974). Culture medium for enterobacteria. *Journal of Bacteriology, 119,* 736–747.

Paliy, O., & Gunasekera, T. S. (2007). Growth of *E. coli* BL21 in minimal media with different gluconeogenic carbon sources and salt contents. *Applied Microbiology and Biotechnology, 73,* 1169–1172.

Reddy, P. T., Peterkofsy, A., & McKenney, K. (1989). Hyperexpression and purification of the *Escherichia coli* adenylate cyclase. *Nucleic Acids Research, 17,* 10473–10488.

Rosano, G. N. L., & Ceccarelli, E. A. (2014). Recombinant protein expression in *Escherichia coli*: Advances and challenges. *Frontiers in Microbiology, 5,* 1–17.

Shu, F., Ramakrishnan, V., & Schoenborn, B. P. (2000). Enhanced visibility of hydrogen atoms by neutron crystallography on fully deuterated myoglobin. *Proceedings of the National Academy of Sciences of the United States of America, 97,* 3872–3877.

Stewart, I. I., Thomson, T., & Figeys, D. (2001). ^{18}O labeling: A tool for proteomics. *Rapid Communications in Mass Spectrometry, 15,* 2456–2465.

Studier, F. W. (2005). Protein production by auto-induction in high density shaking cultures. *Protein Expression and Purification, 41,* 207–234.

Terpe, K. (2006). Overview of bacterial expression systems for heterologous protein production: From molecular and biochemical fundamentals to commercial systems. *Applied Microbiology and Biotechnology, 72,* 211–222.

Tyler, R. C., Sreenath, H. K., Singh, S., Aceti, D. J., Bingman, C. A., Markley, J. L., et al. (2005). Auto-induction medium for the production of [U-^{15}N]- and [U-^{13}C, U-^{15}N]-labeled proteins for NMR screening and structure determination. *Protein Expression and Purification, 40,* 268–278.

van Horn, E., & Ware, G. C. (1959). Growth of *Bacterium coli* and *Staphylococcus albus* in heavy water. *Nature, 184,* 833.

Waegeman, H., De Lausnay, S., Beauprez, J., Maertens, J., De Mey, M., & Soetaert, W. (2013). Increasing recombinant protein production in *Escherichia coli* K12 through metabolic engineering. *New Biotechnology, 30,* 255–261.

Woods, A. G., & Darie, C. C. (Eds.), (2014). *Advances in experimental medicine and biology: Vol. 806. Advancements of mass spectrometry in biomedical research.* Netherlands: Springer.

CHAPTER THREE

Escherichia coli Auxotroph Host Strains for Amino Acid-Selective Isotope Labeling of Recombinant Proteins

Myat T. Lin[*,1], Risako Fukazawa[†], Yoshiharu Miyajima-Nakano[†], Shinichi Matsushita[†], Sylvia K. Choi[‡], Toshio Iwasaki[†,2], Robert B. Gennis[*]

[*]Department of Biochemistry, University of Illinois at Urbana-Champaign, Urbana, Illinois, USA
[†]Department of Biochemistry and Molecular Biology, Nippon Medical School, Bunkyo-ku, Tokyo, Japan
[‡]Center for Biophysics and Computational Biology, University of Illinois at Urbana-Champaign, Urbana, Illinois, USA
[2]Corresponding author: e-mail: tiwasaki@nms.ac.jp

Contents

1. Introduction	46
2. *E. coli* Auxotrophs for Amino Acid-Selective Isotope Labeling	47
2.1 Selection of Appropriate *E. coli* Auxotroph Strains	50
2.2 Challenges for Amino Acid-Selective Isotope Labeling	53
3. Methods	53
3.1 Homologous Expression and Amino Acid-Selective Labeling of a Membrane Protein Complex Cyt bo_3 with *E. coli* Auxotrophs	55
3.2 Heterologous Expression and Preparation of the Selectively ^{15}Nε-Glutamine-Labeled FdxB of *P. putida* JCM 20004 (On the ^{14}N(Natural Abundance, N/A)-Protein Background)	57
3.3 Heterologous Expression and Preparation of ^{14}N(In Natural Abundance, N/A) Lysine-Labeled ARF on the ^{15}N-Protein Background by Reverse Labeling Technique Using ML40K1	61
3.4 Heterologous Expression and Preparation of ^{14}N(N/A) Tyrosine-Labeled ARF on the ^{15}N-Protein Background by Reverse Labeling Technique Using RF4RIL	63
4. Conclusions	64
Acknowledgments	65
References	65

---

[1] Current address: Department of Molecular Biology and Genetics, Cornell University, Ithaca, NY 14853, USA.

Abstract

Enrichment of proteins with isotopes such as ^{2}H, ^{15}N, and ^{13}C is commonly carried out in magnetic resonance and vibrational spectroscopic characterization of protein structures, mechanisms, and dynamics. Although uniform isotopic labeling of proteins is straightforward, efficient labeling of proteins with only a selected set of amino acid types is often challenging. A number of approaches have been described in the literature for amino acid-selective isotope labeling of proteins, each with its own limitations. Since *Escherichia coli* represents the most cost-effective and widely used host for heterologous production of foreign proteins, an efficient method to express proteins selectively labeled with isotopes would be highly valuable for these studies. However, an obvious drawback is misincorporation and dilution of input isotope labels to unwanted amino acid types due to metabolic scrambling *in vivo*. To overcome this problem, we have generated *E. coli* auxotroph strains that are compatible with the widely used T7 RNA polymerase overexpression systems and that minimize metabolic scrambling. We present several examples of selective amino acid isotope labeling of simple and complex proteins with bound cofactors, as an initial guide for practical applications of these *E. coli* strains.

1. INTRODUCTION

Amino acid-selective isotope labeling of proteins is a powerful approach in structural and functional studies of proteins. Enrichment of proteins at selected residue types with ^{15}N and/or ^{13}C isotopes greatly simplifies the chemical shift assignments in nuclear magnetic resonance (NMR) spectroscopic studies (Verardi, Traaseth, Masterson, Vostrikov, & Veglia, 2012). We have also employed similar approaches to characterize the interactions between enzymes and their cofactors such as a semiquinone and an iron–sulfur cluster by pulsed electron paramagnetic resonance (EPR) spectroscopy (Iwasaki et al., 2012; Iwasaki, Samoilova, Kounosu, Ohmori, & Dikanov, 2009; Lin et al., 2012; Lin, Samoilova, Gennis, & Dikanov, 2008). In addition, vibrational spectroscopies such as resonance Raman and Fourier transform infrared also greatly benefit from specific isotope labeling in determining protein–ligand and protein–protein interactions (Haris, 2010; Rotsaert, Pikus, Fox, Markley, & Sanders-Loehr, 2003). All these techniques provide deeper mechanistic insights into the functions of proteins that are often not readily accessible from their crystal structures.

Escherichia coli has been the most commonly used host for heterologous production of foreign proteins due to its straightforward and inexpensive cultivation, well-known genetics, and highly developed biotechnological tools (Zerbs, Frank, & Collart, 2009). If a protein of interest can be successfully expressed in *E. coli* and purified in its functional forms, using *E. coli* for selective amino acid isotope labeling is likely the most reasonable and cost-effective option. Similarly, selective unlabeling of proteins with individual amino acids or metabolic precursors may also be useful in NMR peak assignments (Bellstedt et al., 2013; Rasia, Brutscher, & Plevin, 2012). However, *E. coli* possesses biosynthetic and biodegradation networks for all 20 amino acids (Waugh, 1996), and these pathways intersect in a complex manner such that one amino acid can be readily converted into other types. The result is often the dilution and scrambling of the input isotope label to undesired amino acid types.

The extent of metabolic scrambling in *E. coli* depends on not only the amino acid types and the location of the isotopic label (Waugh, 1996) but also the bacterial growth conditions. For example, simply adding individual ^{15}N- and/or ^{13}C-labeled histidine, lysine, or methionine in the growth media will effectively enrich proteins labeled at the intended amino acids when cultured in the growth medium supplied with other unlabeled amino acid types (Iwasaki et al., 2009; O'Grady, Rempel, Sokaribo, Nokhrin, & Dmitriev, 2012). However, attempting this with most other amino acids under the same conditions results in significant dilution and scrambling of the isotope labels (O'Grady et al., 2012). ^{15}Nα labels, commonly used in NMR experiments for chemical shift assignments, are specifically readily transferred from one amino acid type to another by any of the four aminotransferases present in *E. coli* (encoded by *aspC*, *avtA*, *ilvE*, and *tyrB*) (Waugh, 1996). These problems can be minimized by genetic engineering of the host strains to optimize for the desired label distribution.

2. *E. COLI* AUXOTROPHS FOR AMINO ACID-SELECTIVE ISOTOPE LABELING

The approach is to produce proteins in *E. coli* auxotrophic strains in which enzymes that result in dilution and scrambling of the label (Waugh, 1996) have been eliminated, avoiding mutations or groups of mutations that would result in impeding cell growth. A set of isogenic auxotrophic host strains has been generated using *E. coli* parent strains, BL21 (DE3) or C43(DE3), which are designed for the widely used T7 RNA polymerase overexpression system to enhance their utility. These strains (Table 1) were generated using the λ-Red recombinase system (Iwasaki et al., 2012; Lin et al., 2011). Because of the complex and versatile nature of the bacterial metabolic network, these *E. coli* auxotrophic expression host strains can "minimize" possible metabolic scrambling of the input isotopic labels of many different amino acid types, but cannot "completely eliminate"

Table 1 *E. coli* Auxotroph Strains and Their Properties (http://www.nms.ac.jp/fesworld/EcoliStrains.html)

Strain	Parent Strain	Genotype	Amino Acid Requirement and/or Labeling
ML2	C43(DE3)	cyo::kan ilvE	Ile, Leu[a]
ML3	C43(DE3)	cyo::kan hisG	His
ML6	C43(DE3)	cyo::kan ilvE avtA	Ile, Leu[a], Val
ML8	C43(DE3)	cyo::kan argH	Arg
ML12	C43(DE3)	cyo::kan ilvE avtA aspC	Ile, Leu[a], Tyr[a], Val
ML14	C43(DE3)	tyrA	Tyr
ML17	C43(DE3)	glnA	Gln
ML21	C43(DE3)	tyrA hisG	His Tyr
ML24	C43(DE3)	cyo::cat ilvE avtA aspC hisG	His, Ile, Leu[a], Tyr[a], Val
ML25	C43(DE3)	cyo ilvE avtA aspC hisG asnA asnB	Asn, His, Ile, Leu[a], Tyr[a], Val
ML26	C43(DE3)	cyo ilvE avtA aspC hisG argH	Arg, His, Ile, Leu[a], Tyr[a], Val
ML31	C43(DE3)	cyo ilvE avtA aspC hisG argH metA	Arg, His, Ile, Leu[a], Met, Tyr[a], Val
ML36	C43(DE3)	cyo ilvE avtA aspC hisG metA	His, Ile, Leu[a], Met, Tyr[a], Val
ML40K1	C43(DE3)	cyo ilvE avtA aspC hisG argH metA lysA	Arg, His, Ile, Leu[a], Lys, Met, Tyr[a], Val
ML42	C43(DE3)	cyo ilvE avtA aspC hisG argH metA lysA thrC asnB	Arg, His, Ile, Leu[a], Lys, Met, Thr, Tyr[a], Val
ML43	C43(DE3)	cyo ilvE avtA aspC hisG argH metA lysA thrC asnA asnB	Arg, Asn, His, Ile, Leu[a], Lys, Met, Thr, Tyr[a], Val
YM138	C43(DE3)	cysE	Cys
YM154	C43(DE3)	cysE	Cys (pACYC-based plasmid harboring argU, ileY and leuW)
MS1	C43(DE3)	cysE hisG	Cys His

Table 1 E. coli Auxotroph Strains and Their Properties (http://www.nms.ac.jp/fesworld/EcoliStrains.html)—cont'd

Strain	Parent Strain	Genotype	Amino Acid Requirement and/or Labeling
RF1	BL21 CodonPlus (DE3)-RIL	glyA	Gly
RF2	BL21 CodonPlus (DE3)-RIL	thrC	Thr
RF4	BL21 CodonPlus (DE3)-RIL	aspC tyrB	Asp[b], (Phe), Tyr
RF5	BL21 CodonPlus (DE3)-RIL	aspC tyrB hisG	Asp[b], His, (Phe), Tyr
RF6	BL21 CodonPlus (DE3)-RIL	proC	Pro
RF8	BL21 CodonPlus (DE3)-RIL	asnA asnB	Asn
RF10	BL21 CodonPlus (DE3)-RIL	lysA	Lys
RF11	C43(DE3)	metA	Met
RF12	BL21 CodonPlus (DE3)-RIL	trpA trpB	Trp
RF13	BL21 CodonPlus (DE3)-RIL	aspC tyrB trpA trpB	Asp[b], (Phe), Trp, Tyr

[a]Although *tyrB* has not been knocked out, it may be still possible to use these strains for selective isotope labeling with L-leucine and L-tyrosine as indicated by adding L-tyrosine (0.4–1 mM) in the growth media (Lin et al., 2011; Yang, Camakaris, & Pittard, 2002). In our experience, this strategy could not prevent the metabolic scrambling in the case of long-term cultivation (Iwasaki et al., 2012).

[b]Although these strains display auxotroph for L-aspartate (Fig. 1), we have not tested the labeling efficiency of L-aspartate.

this mixing in many cases. In most cases, the metabolic scrambling is reduced so that the label distribution in the isolated protein is sufficient for the purposes of many protein structure–function studies. We have successfully applied some of these strains for preparing proteins used for magnetic resonance studies of selectively labeled metalloproteins (Iwasaki et al., 2012; Lin et al., 2012).

2.1 Selection of Appropriate *E. coli* Auxotroph Strains

Consideration of only the limited number of metabolic pathways that provide direct connections leading to scrambling of the input label to other amino acids (Waugh, 1996) works remarkably well in designing auxotrophs with acceptably low levels of label scrambling in many cases, but additional mutations can often decrease scrambling of the input label further. Auxotrophs for selective labeling of 10 of the amino acids each require only a single-gene knockout (Waugh, 1996): arginine, cysteine, glutamine, glycine, histidine, isoleucine, lysine, methionine, proline, and threonine (Table 2). Within this set, for example, the deletion of *thrC* (previously considered to be an ideal genotype for threonine labeling) eliminates most but not all of the scrambling of an input ^{15}N-threonine label (R. Fukazawa & T. Iwasaki, unpublished). Metabolic scrambling may be reduced to lower levels by also deleting the *ilvA* gene involved in the isoleucine biosynthesis pathway (see Waugh, 1996).

Auxotrophs for the remaining 10 amino acids each require two or more genes to be deleted, but most of them share the requirement for deletion of the genes encoding the four common aminotransferase genes, *aspC*, *avtA*, *ilvE*, and *tyrB* (Waugh, 1996). We have generated *E. coli* strains that are appropriate for selective labeling of seven other amino acids: alanine, asparagine, leucine, phenylalanine, tryptophan, tyrosine, and valine (Table 2). Although we currently do not have an ideal *E. coli* strain to selectively label L-serine, it should be straightforward to generate one since three of the four genes that need to be knocked out have already been deleted in some of our strains (Table 2).

Designing auxotrophs with acceptable levels of scrambling for each of the two remaining amino acids, L-aspartate and L-glutamate, remains challenging due to their central roles in metabolism, which means that they are involved in networks of additional pathways to those described by Waugh (1996). This limits the ability for selective labeling of these amino acids. Deletion of the *aspC* and *tyrB* genes gives rise to L-aspartate auxotrophs

Table 2 List of Ideal *E. coli* Genotypes (Based on Waugh, 1996) and Auxotrophic Strains Available

Amino Acid	Ideal *E. coli* Genotypes	*E. coli* Auxotroph Strains from Table 1
Ala	*aspC avtA ilvE tyrB*[a]	ML12, ML24, ML25, ML26, ML31, ML36, ML40K1, ML42, ML43
Arg	*argH*	ML8, ML26, ML31, ML40K1, ML42, ML43
Asn	*asnA asnB*	ML25, ML43, RF8
Asp	*asd asnA asnB aspC tyrB*	Not available (see RF4, RF5, and RF13 in Table 1)
Cys	*cysE*	YM138, YM154, MS1
Gln	*glnA*	ML17
Glu	*argH aspC gdh glnA gltB ilvE proC tyrB*	Not available
Gly	*glyA*	RF1
His	*hisG*	ML3, ML21, ML24, ML25, ML26, ML31, ML36, ML40K1, ML42, ML43, MS1, RF5
Ile	*ilvE*	ML2, ML6, ML12, ML24, ML25, ML26, ML31, ML36, ML40K1, ML43
Leu	*ilvE tyrB*[a]	ML2, ML6, ML12, ML25, ML26, ML31, ML36, ML40K1, ML42, ML43
Lys	*lysA*	ML40K1, ML42, ML43, RF10
Met	*metA*	ML31, ML36, ML40K1, ML42, ML43, RF11
Phe	*aspC ilvE tyrB*[a]	ML12, ML24, ML25, ML26, ML31, ML36, ML40K1, ML42, ML43, RF4[b], RF5[b], RF13[b]
Pro	*proC*	RF6
Ser	*cysE glyA serB trpB*	Not available
Thr	*thrC*	ML42, ML43, RF2
Trp	*trpB tyrB*	RF12, RF13
Tyr	*aspC tyrB*[a]	RF4, RF5, RF13, ML12, ML24, ML25, ML26, ML31, ML36, ML40K1, ML42, ML43
Val	*avtA ilvE*	ML4, ML6, ML12, ML24, ML25, ML26, ML31, ML36, ML40K1, ML42, ML43

[a]In ML strains, *tyrB* has not been knocked out. Addition of L-tyrosine (0.4–1 mM) may be used to suppress *tyrB* (Lin et al., 2011; Yang et al., 2002), but in our experience this strategy could not prevent the metabolic scrambling in the case of long-term cultivation (Iwasaki et al., 2012).
[b]We have not tested if RF4, RF5, and RF13 could be used for selective L-phenylalanine isotope labeling (i.e., *ilvE* has not been knocked out).

(strains RF4, RF5, and RF13; Table 1, Fig. 1), but this genotype is probably not sufficient to limit scrambling of a label within L-aspartate.

One complication is that we have been unable to knockout *tyrB* in C43 (DE3) and *ilvE* in BL21(DE3) for unknown reasons. Thus, for selective labeling of L-tyrosine, strains RF4 and RF5, both having the deletion of *tyrB* and *aspC* genes in BL21(DE3), can be used (Iwasaki et al., 2012). Alternatively, L-tyrosine is known to suppress expression of *tyrB* (Yang et al., 2002), and this can be useful in minimizing label scrambling using some of the C43 (DE3) auxotrophic strains (Table 1; applicable only for the short-term

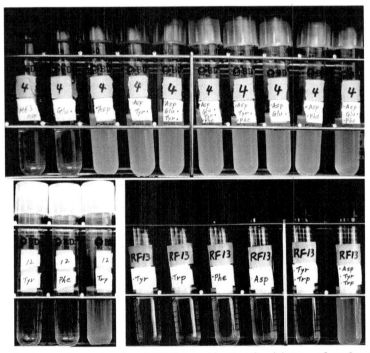

Figure 1 L-Aspartate auxotrophy of RF4 (#4, top) having the deletions of *aspC* and *tyrB* and representing an ideal genotype for tyrosine labeling (Table 1). For comparison, RF12 (#12, bottom) displays L-tryptophan auxotrophy, whereas RF13 (#13, bottom), having the knockouts of *aspC*, *tyrB*, *trpA*, and *trpB*, and representing ideal genotypes for tyrosine and tryptophan labeling, does not grow in M63 minimal medium in the presence of either L-tyrosine, L-phenylalanine, or L-tryptophan alone (see also Table 1). These examples illustrate that *E. coli* amino acid metabolic network is more complex than the simple sums of individual amino acid biosynthesis pathway catalogs (Waugh, 1996) (cf. Table 1). Concentration(s) of amino acid(s) added to M63 minimal medium (where exist): 0.4 g/L L-glutamate, 0.4 g/L L-aspartate, 0.17 g/L L-tyrosine, 0.13 g/L L-phenylalanine, and 0.1 g/L L-tryptophan. (See the color plate.)

cultivation, as in Sections 3.1 and 3.2). For example, by adding L-tyrosine (0.4–1 m*M*) in the medium, we were able to selectively label leucine using strain ML6 (Lin et al., 2011). In another example, ML14 and ML21 strains have *tyrA* deletions and display auxotrophy for L-tyrosine. However, these strains are suitable only for the selective labeling of the tyrosine side chain and should not be used for selective labeling of proteins with α-^{15}N-tyrosine due to the possible scrambling of ^{15}N isotope by the transaminases encoded by *aspC* and *tyrB*.

2.2 Challenges for Amino Acid-Selective Isotope Labeling

One remaining challenge is to generate the BL21(DE3)- or C43(DE3)-derived *E. coli* strains having the deletions of all four genes encoding the four cognate general aminotransferases, *aspC*, *avtA*, *ilvE*, and *tyrB* (Waugh, 1996). The extent of metabolic mixing of an input α-^{15}N label of some amino acid types would be then significantly minimized, which has clear merit in NMR experiments for chemical shift assignments.

Another challenge is to engineer L-aspartate auxotrophs. Several of our auxotrophs such as RF4, RF5, and RF13 are deficient in growth in the absence of L-aspartate (Fig. 1). Further limiting of the metabolic scrambling of an input L-aspartate isotope label beyond what is available with these auxotrophs will require additional genetic engineering, targeting other genes in pathways connecting L-aspartate to the metabolic pool.

L-Glutamate is perhaps the most challenging amino acid for selective isotope labeling as it plays a central role in nitrogen metabolism. A reverse labeling strategy may provide one possible option (Verardi et al., 2012). In this approach, all of the amino acids with the exceptions of the amino acid(s) of interest (such as L-glutamate) are added to the medium in their unlabeled forms along with ^{15}N-ammonium chloride, which leads to selective labeling of the missing amino acid(s). An alternative strategy, if it is practical for the target protein, is to use a cell-free expression system. This eliminates most problems associated with isotopic scrambling due to the absence of enzymes other than those involved in transcription and translation (Klammt et al., 2004; Sobhanifar et al., 2010; Takeda, Ikeya, Guntert, & Kainosho, 2007).

3. METHODS

In general, optimal conditions for the overexpression of proteins in *E. coli* can vary considerably (Zerbs et al., 2009). For high-level production of simple and complex proteins with bound cofactors required for biophysical studies, the following factors may be considered:

1. Coupling efficiency of the apoprotein translation and folding speeds with those of cofactor insertion and/or processing may be tuned by the growth rate of the *E. coli* expression host strains by temperature control, e.g., at 18, 22, 25, or 37 °C. In some cases, working at lower temperatures (e.g., at 22 or 25 °C) may give a better result for obtaining a holoprotein form (Iwasaki et al., 1999; Kounosu et al., 2004; Roman et al., 1995).

2. The translational coupling efficiency may also be adjusted by the aeration of the culture, which, if using a shaker/incubator, is influenced by the speed (e.g., at a slow rate (120–150 rpm) vs. fast rate (250 rpm)) and medium volume of the shaking culture (e.g., 0.5, 1, or 1.5 L of the medium in a 2-L flask).

3. For producing sufficient amount of recombinant holoproteins, the time after isopropyl-β-D-thiogalactopyranoside (IPTG) induction may need to be optimized. For example, in our experience, high-level heterologous productions of archaeal Rieske-type iron–sulfur proteins and mammalian nitric oxide synthase holoproteins in *E. coli* require a long-term cultivation, typically more than 48 h (Iwasaki et al., 2012, 1999; Roman et al., 1995). This could be a potential disadvantage for a possible enhancement of the metabolic scrambling of an input isotope label.

4. For many proteins with bound cofactors, addition of appropriate metabolic precursor compounds in the growth medium may be required (Iwasaki et al., 1999; Roman et al., 1995).

5. For heterologous expression, high-level production in *E. coli* may require extra copies of tRNA genes for the cognate rare codons (Iwasaki et al., 2012; Kounosu et al., 2004). In some cases, incorporation of extra plasmid coding for *E. coli* chaperonin genes (e.g., *groELS*) (Iwasaki et al., 1999; Roman et al., 1995) or specific genes required for cofactor biosynthesis into the expression host strains may also be necessary.

In addition, selective isotope labeling requires the use of a defined growth medium, which could result in the lower expression of recombinant proteins (Kounosu et al., 2004) compared to growth in the rich media (such as LB) that are usually employed. This may also require optimization of expression for each target protein.

The auxotrophic strains with one or two gene deletions generally grow at a similar rate as the wild-type strain when appropriate amino acids are supplemented, except for those having the *cysE* deletion (YM138, YM154, and MS1; Table 1), which usually displays a slower growth. Auxotrophs with multiple deletions, such as ML25, ML40, and ML43, grow significantly more slowly than the parent C43(DE3) strain. Therefore, pilot cultures of the auxotrophic strains with different minimal media, combinations of amino acids, and conditions for growth and induction of expression should be tested to establish an optimal protocol for expression of the target protein prior to proceeding with expensive isotope labels. To aid this, we describe below several examples of the protocols that we have applied to produce selectively isotope-labeled cytochrome bo_3 ubiquinol oxidase (Cyt bo_3) from *E. coli* (Section 3.1) and simple iron–sulfur proteins from *Pseudomonas putida* JCM 20004 (ISC-like [2Fe-2S] ferredoxin (FdxB)) (Section 3.2) and *Sulfolobus solfataricus* P1 (archaeal Rieske-type [2Fe-2S] ferredoxin (ARF)) (Sections 3.3 and 3.4).

3.1 Homologous Expression and Amino Acid-Selective Labeling of a Membrane Protein Complex Cyt bo_3 with *E. coli* Auxotrophs

We grow C43(DE3) auxotrophic strains containing pET17b vector to express the *E. coli* Cyt bo_3 in modified M63 minimal medium (Frericks, Zhou, Yap, Gennis, & Rienstra, 2006). The preparation of the growth medium and the homologous expression of Cyt bo_3 are described below.

1. Prepare the concentrated stock chemicals listed in Table 3 and filter-sterilize them. L-Tyrosine solution requires brief and moderate heating before it dissolves. Store the sterilized stock solutions of amino acids, thiamine–HCl, and antibiotics such as ampicillin and kanamycin at $-20\ °C$ and those of $MgSO_4$, $CuSO_4$, and glucose at $4\ °C$.

2. For every liter of culture, dissolve 3 g KH_2PO_4, 7 g K_2HPO_4, 2 g NH_4Cl, and 8.3 mg $FeSO_4 \cdot 7H_2O$ in 900 ml of double-distilled H_2O in appropriate culture flasks and autoclave them. Let the solutions cool to the room temperature.

3. Immediately before inoculation, add thiamine–HCl, $CuSO_4$, $MgSO_4$, glucose, and appropriate antibiotics as well as the necessary amino acids to their final concentrations according to Table 3. For example, if we want to produce a Cyt bo_3 sample selectively labeled with $^{15}N\alpha$-histidine using the ML3 strain in Table 1, we would add only the $^{15}N\alpha$-histidine without the other amino acids. Finally, add the autoclaved double-distilled H_2O as necessary to reach the final intended volume for the cultures.

Table 3 Amino Acid Stock Solutions

Amino Acids and Chemicals	Final Concentrations (mg/L or as Indicated)	Stock Solutions (g/L)	Times Concentrated
L-Alanine	42	8.4	200
L-Arginine	105	105.0	1000
L-Asparagine	42	8.4	200
L-Aspartic acid	40	40.0	1000
L-Cysteine	36	7.2	200
L-Glutamic acid	736	147.2	200
L-Glutamine	731	146.2	200
Glycine	10	2.0	200
L-Histidine	16	16.0	1000
L-Isoleucine	39	7.8	200
L-Leucine	39	7.8	200
L-Lysine	44	44.0	1000
L-Methionine	45	9.0	200
L-Phenylalanine	50	5.0	100
L-Proline	230	46.0	200
L-Serine	420	84.0	200
L-Threonine	36	18.0	500
L-Tryptophan	20	2.0	100
L-Tyrosine	18	1.8	100
L-Valine	35	7.0	200
Ampicillin	100	100	1000
Chloramphenicol (in ethanol)	20	20	1000
Kanamycin	50	50	1000
$CuSO_4$	10 μM	10 mM	1000
$MgSO_4$	1 mM	1 M	1000
Glucose	2	200	100
Thiamine-HCl	10	10	1000
IPTG	0.5 mM	1 M	2000

4. Inoculate 3 ml of the minimal medium prepared in the previous step in a 15-ml tube with a single colony of the *E. coli* auxotroph growing on an LB-agar plate. Incubate the tube overnight at 37 °C, 250 rpm.
5. Enlarge the overnight 3 ml culture to the final volume (200–500 ml) the next morning and incubate it at 37 °C, 250 rpm.
6. When the cell density reaches an OD_{600} of ~0.5–0.6, add IPTG from the stock solution (Table 3) to a final concentration of 0.5 mM and continue the incubation for 4–6 h.
7. Harvest the cells in a centrifuge tube at $5000 \times g$ for 10 min. Proceed to the extraction and purification of Cyt bo_3 as described previously (Lin et al., 2011) or store the cells at -80 °C for purification later.
8. Extract the His-tagged Cyt bo_3 from *E. coli* membrane with 1% *n*-dodecyl β-D-maltoside detergent and purify it with Ni^{2+} affinity chromatography as described previously (Frericks et al., 2006).

3.2 Heterologous Expression and Preparation of the Selectively ^{15}Nε-Glutamine-Labeled FdxB of *P. putida* JCM 20004 (On the ^{14}N(Natural Abundance, N/A)-Protein Background)

The *fdxB* gene coding for the ISC-like [2Fe–2S] ferredoxin (FdxB) of *P. putida* JCM 20004 (formerly *Pseudomonas ovalis* IAM 1002) has been cloned and sequenced as a part of its *isc* gene cluster (DDBJ/EMBL/GenBank code AB109467) and heterologously overexpressed in *E. coli* BL21-CodonPlus(DE3)-RIL strain (Stratagene) using a pET28aFDXB-SG vector (based on a pET28a His-tag expression vector; Novagen), purified, and crystallized (Iwasaki et al., 2011) (PDB ID code: 3AH7). The following protocols are based on the procedure originally used for heterologous overproduction of the uniformly ^{15}N-labeled FdxB using *E. coli* BL21-CodonPlus(DE3)-RIL strain grown in the CHL-^{15}N (<97 atm%) medium (Chlorella Industry Co. Ltd., Fukuoka, Japan) (Lin et al., 2011).

1. For preparation of ^{15}Nε-glutamine-labeled FdxB on the ^{14}N(N/A)-protein background, *E. coli* C43(DE3) auxotroph strain ML17 (Table 1) having the deletion of *glnA* gene was used as parent cells for heterologous overexpression of the *P. putida fdxB* gene. When required, verify the deletion of the *glnA* gene in ML17 by a set of the PCR primers in Table 4 prior to use. Note that ML17 strain lacking the *glnA* gene is useful for selective labeling of ^{15}Nε-glutamine but that minor diffusion and dilution of the input (not severe as compared to L-glutamate and

Table 4 List of PCR Primers for Verification of Each Target Gene in Auxotrophs

Target Gene	PCR Primer Name	Primer Sequence (5′ → 3′)	Position of PCR primer in E. coli BL21(DE3) genomic DNA[a,e]		Position of the target gene in E. coli BL21 (DE3) genomic DNA[a]		Size of the target gene (bp)	Estimated PCR product size in knockout strain (bp)[b]	Estimated PCR product size in parent E. coli BL21 (DE3) (bp)[a]
			Start	End	Start	End			
cysE	cysE.ext.F	GAAATTACGCAAGATTCGCTGGTGC	3,650,170	3,650,194	3,650,936	3,651,757	822	1491	2313
	cysE.ext.R	CGAGGCGTTAGGCGATCAAATTCC	3,652,459	3,652,482					
lysA[e]	lysA.ext.F	TACGGGAAAGGCTGATGTAGTTCTCAC	2,817,868	2,817,894	2,815,913	2,817,175	1263	1407	2670
	lysA.ext.R	TGCGTTGGTCGTCCATGCCAAAATG	2,815,225	2,815,249					
avtA	avtA.ext.F	ATACCCGCCTGTTCCGTGAAG	3,601,336	3,601,356	3,601,851	3,603,104	1254	1102	2356
	avtA.ext.R	TCGGTTGCCGTACCTGTGAAG	3,603,671	3,603,691					
ilvE	ilvE.ext.F	CCAACGGTTAGGGATGGTTCG	3,842,158	3,842,178	3,842,720	3,843,649	930	1133	2063
	ilvE.ext.R	GATCATCGCATCAACCAGATCG	3,844,199	3,844,220					
glyA	glyA.ext.F	TGATTACGGCATTGGCTCGTC	2,549,300	2,549,320	2,550,053	2,551,306	1254	1496	2750
	glyA.ext.R	TAGATTTCCGCCTCGCGATTG	2,552,029	2,552,049					
aspC	aspC.ext.F	TTGATGACAGCGGCCTGACACTGATGCAG	988,858	988,886	989,603	990,793	1191	1531	2722
	aspC.ext.R	ACTCCAACTTCTTTGGTCTCGGTTGATGG	991,552	991,579					
tyrB	tyrB.ext.F	GCGAAGGCAAACCTGGTCAACGTTCC	4,173,729	4,173,754	4,174,517	4,175,710	1194	1556	2750
	tyrB.ext.R	GATTGACCAGCCCCCTACCCTACAATGG	4,176,451	4,176,478					
hisG	hisG.ext.F	TTACTCCCGGTAACTTGCCAGCCTC	1,985,275	1,985,299	1,985,989	1,986,888	900	1484	2384
	hisG.ext.R	ATCAGAAGCCACGAAATCCGGCGTAG	1,987,633	1,987,658					

argH	argH.ext.F	AAACCCTGTTGGCATGGGCGAAG		4,064,564	4,064,586	4,065,145	4,066,518 1374 1003	2377
	argH.ext.R	GCTGGTCCGCTCCAGCAACATCAC		4,066,917	4,066,940			
asnA	asnA.ext.F	TGAACGCGTTGGGATCTACCTGTG		3,815,290	3,815,313	3,815,950	3,816,942 993 1378	2371
	asnA.ext.R	ACGGTTCCTGAGCAGGTTGATGGTC		3,817,636	3,817,660			
asnB	asnB.ext.F	TCGAACCTGTGACCCCATCATTATGAG		656,550	656,576	657,381	659,045 1665 1152	2817
	asnB.ext.R	ACACCCACGGGCGCGGTTTTTATC		659,735	659,758			
metA	metA.ext.F	CATCAAATAAAGCGAAAGGCCATCCGTC		4,121,432	4,121,459	4,122,159	4,123,088 930 1440	2370
	metA.ext.R	CATTAGTGTAACTGATGGTGCCGTTAACC		4,123,773	4,123,801			
metL	metL.ext.F	TGCGAAAATCTGCCATCTGGCAAGG		4,037,680	4,037,704	4,038,376	4,040,808 2433 1390	3823
	metL.ext.R	CCACGATATGACGAATACCGTTATTCCAG		4,041,474	4,041,502			
thrA	thrA.ext.F	CTGGTGTTTGGTCGCGAAGATTCC		4,558,564	4,558,587	336	2798 2463 1323	3786
	thrA.ext.R	AATGCAATCCTGGCGCGATACTG		3379	3402			
tyrA[e]	tyrA.ext.F	GGTCAGCAATTGGTGCTCGTACAACG		2,607,155	2,607,180	2,605,456	2,606,577 1122 1191	2313
	tyrA.ext.R	GTCTACAGCCATCCGCAGCCATTC		2,604,868	2,604,891			
glnA	glnA.ext.F	TCAGGTAACGCTTTGCTGAGCAGCTG		3,962,550	3,962,575	3,963,317	3,964,726 1410 1515	2925
	glnA.ext.R	CAACAATCGGCTTCAGGCCGTAAGC		3,965,450	3,965,474			
thrC	thrC.ext.F	TCAGTGCTGGGAGCGTTTTTG		2997	3017	3733	5019 1287 1511	2798
	thrC.ext.R	AATCGGCTGACCAAACCAGAGC		5773	5794			
proC	proC.ext.F	CTCCCTTCCCATTATTGTCATTTATCCTC		368,849	368,877	369,454	370,263 810 1323	2133
	proC.ext.R	GTCATGAGAATAGGCGATCATACCATCAAAC		370,952	370,981			

Continued

Table 4 List of PCR Primers for Verification of Each Target Gene in Auxotrophs—cont'd

Target Gene	PCR Primer Name	Primer Sequence (5′ → 3′)	Position of PCR primer in E. coli BL21(DE3) genomic DNA[a,e] Start	End	Position of the target gene in E. coli BL21 (DE3) genomic DNA Start	End	Size of the target gene (bp)	Estimated PCR product size in knockout strain (bp)	Estimated PCR product size in parent E. coli BL21 (DE3) (bp)
trpAB	trpAB.ext.F	CAAATTAAGCGCAACGAGAAGATAGAGG	1,301,710	1,301,737	1,302,002	1,304,001	2000	397	2397
	trpAB.ext.R	CGGACTTGATTTTAATTCTGCTGTAGAGTC	1,304,077	1,304,106					
hisG (internal)[c]	hisG internal F	GACAGACAACTCTCGTTTACGCATAGC	1,985,991	1,986,017	1,985,989	1,986,888	900	No amplification[c]	839
	hisG internal R	GTTTTTCCATCGTTTCCCAGAACAG	1,986,805	1,986,829					
aspC (internal)[c]	aspC internal F	CACTTGTTCTTTTGTCAGGCCACTG	989,720	989,744	989,603	990,793	1191	No amplification[c]	922
	aspC internal R	GGCTGAACAGTATCTGCTCGAAAATG	990,616	990,641					
metA (internal)[c]	metA internal F	TAACCTGATGCCGAAGAAGATTGAAAC	4,122,281	4,122,307	4,122,159	4,123,088	930	No amplification[c]	729
	metA internal R	AGTAAATTACCGTGACTACGCCAGCTC	4,122,983	4,123,009					
Resistance marker cassette (in pKD3)	pKD3verN3 (used in conjunction with PCR primers ext.F or ext.R)	CGCAAGATGTGGCGTGTTACGGTG	Position in pKD3[d] 388	Position in pKD3[d] 411	–	–	–	–	No amplification[c]

[a]Estimated by "In silico PCR amplification" Web site using the E. coli BL(DE3) genomic DNA sequence (http://insilico.ehu.es/PCR/index.php?mo=Escherichia). Note that the genomic DNA sequence of C43(DE3) strain has not been deposited on the NCBI Web site.
[b]"Estimated PCR product size in knockout strain," is based on the difference between "estimated PCR product size in parent E. coli BL21(DE3) strain" and "estimated size of the knocked-out region," without considering the FRT (FLP recognition target) regions (Iwasaki et al., 2012; Lin et al., 2011).
[c]These internal PCR primers have been used to confirm no amplification upon PCR, when external primer sets did not work in several knockout strains.
[d]Position refers to that in the pKD3 plasmid sequence (GenBank accession: AY048742).
[e]Position refers to that of the E. coli BL(DE3) genomic DNA sequence on the NCBI Web site. In most cases, directions of the F and R primer sequences (Iwasaki et al., 2012; Lin et al., 2011) are the same as those on the NCBI Web site, respectively, except for those of the lysA and tyrA primers which are in reverse directions, respectively.

L-aspartate, which are located at the center of the *E. coli* metabolic pathways) can occur when used in the absence of appropriate nonlabeled amino acids because Nα-glutamine preferentially diffuses into glutamate.

2. Transform the pET28aFDXB-SG expression vector (Iwasaki et al., 2011) into the host strain ML17 (Table 1) by electroporation.
3. Let the transformants grow overnight at 25 °C in 1 L culture (in a 2-L flask) of the nonlabeled CHL medium (Chlorella Industry Co. Ltd., Fukuoka, Japan) containing 50 mg/L kanamycin, 0.2 mM FeCl$_3$, 0.5 g/L ^{14}N(N/A) L-glutamine, and 0.5 g/L nonlabeled Algal Amino Acid Mix (Chlorella Industry Co. Ltd., Fukuoka, Japan). Note that the nonlabeled Algal Amino Acid Mix is the commercial amino acid mixture derived from the acid extract of *Chlorella* and therefore does not contain L-glutamine and L-asparagine.
4. Harvest the cells in a centrifuge tube at 6500 rpm for 10 min at 10 °C.
5. Subsequently, inoculate the resulting cell pellet into a total of 2 L culture (using two 2-L flasks, each containing 1 L culture medium) of the freshly prepared, nonlabeled CHL medium (Chlorella Industry Co. Ltd., Fukuoka, Japan) containing 50 mg/L kanamycin, 0.2 mM FeCl$_3$, 0.5 g/L nonlabeled Algal Amino Acid Mix (Chlorella Industry Co. Ltd., Fukuoka, Japan), and ~0.45 g/L L-glutamine labeled at the ^{15}Nε2 position (98%+) (Cambridge Isotope Laboratories, Inc., Andover, MA, USA) for 30 min at 37 °C.
6. Add 1 mM IPTG to this culture and continue the incubation for 6 h at 30 °C.
7. Harvest the cells in a centrifuge tube at 6500 rpm for 10 min at 4 °C and store the resulting cell pellet at −80 °C until use.
8. Purify the ^{15}Nε-glutamine-labeled FdxB holoprotein with Ni^{2+} affinity chromatography as described previously (Iwasaki et al., 2011).

3.3 Heterologous Expression and Preparation of ^{14}N(In Natural Abundance, N/A) Lysine-Labeled ARF on the ^{15}N-Protein Background by Reverse Labeling Technique Using ML40K1

One drawback of amino acid-selective labeling is the labor and the expense associated with some amino acid isotope compounds to be high. A relatively inexpensive alternative method is to use the reverse labeling technique, i.e., the auxotrophic strain is grown on the ^{15}N- or ^{13}C-labeled culture medium to which the desired unlabeled (e.g., ^{14}N in natural abundance) L-amino acid is added. The following two examples describe the application of this reverse labeling technique for selective ^{14}N(N/A) amino acid labeling of archaeal

Rieske-type [2Fe–2S] ferredoxin (ARF) from *S. solfataricus* P1 (DSM 1616) on the ^{15}N-protein background, using auxotrophs (Iwasaki et al., 2012). In our experience, all BL21 CodonPlus (DE3)-RIL-derived auxotrophs (RF strains) in Table 1 display leaky expression (i.e., expression in the absence of IPTG inducer) under the applied conditions, which is not the case with C43(DE3)-derived auxotrophs. Although this may result in a suboptimal level of expression, the heterologous expression levels of our water-soluble iron–sulfur proteins using RF strains are usually higher than those using the C43(DE3) derivatives in Table 1.

The *arf* gene coding for the hyperthermostable ARF from *S. solfataricus* P1 has been cloned, sequenced (DDBJ/EMBL/GenBank code AB047031), and heterologously overexpressed in *E. coli* BL21-CodonPlus(DE3)-RIL strain (Stratagene) using a pET28aARF vector (based on a pET28a His-tag expression vector; Novagen), purified, and crystallized (Iwasaki et al., 2012; Kounosu et al., 2004). The following protocols in Sections 3.3 and 3.4 are based on the procedure originally used for heterologous overproduction of the uniformly ^{15}N-labeled ARF using *E. coli* BL21-CodonPlus(DE3)-RIL strain grown in the CHL-^{15}N (<97 atm%) medium (Chlorella Industry Co. Ltd., Fukuoka, Japan) (Iwasaki et al., 2012).

Note that a high-level expression of the archaeal *arf* gene in *E. coli* requires extra copies of tRNA genes for the *E. coli* rare codons (Kounosu et al., 2004) (incorporated into auxotrophs as a pACYC-based plasmid harboring *argU*, *ileY*, and *leuW* (Agilent Technologies) by electroporation). Moreover, as described below, our heterologous expression strategy developed for site-specific labeling of bacterial FdxB (Section 3.2) did not work with this archaeal ARF, for which we have found the absolute requirement of a much longer-term cultivation for effective heterologous production of a *holoprotein* form in *E. coli* C43(DE3) and BL21(DE3) strains (Iwasaki et al., 2012). Such a long-term cultivation of auxotrophs can be a potential threat for enhanced metabolic scrambling and/or dilution of the input isotope labels of certain amino acid types.

1. For preparation of ^{14}N (in natural abundance, N/A) lysine-labeled ARF on the ^{15}N-protein background, *E. coli* C43(DE3) auxotroph strain ML40K1 (Table 1) was used as parent cells for heterologous overexpression of the archaeal *arf* gene. When required, verify the deletion of each target chromosomal gene in ML40K1 by sets of the PCR primers in Table 4 prior to use.
2. Transform the pET28aARF expression vector (Kounosu et al., 2004) into the host strain ML40K1 by electroporation.

3. Let the transformants grow overnight at 25 °C in 1 L culture (in a 2-L flask) of the CHL-^{15}N ($\sim$97 atm%) medium (Chlorella Industry Co. Ltd., Fukuoka, Japan) containing 25 mg/L kanamycin, 17 mg/L chloramphenicol, 0.2 mM FeCl$_3$, 0.25 g/L MgSO$_4$·7H$_2$O, and 0.5 g/L Algal ^{15}N(98.7–99.2%)-Amino Acid Mix (Chlorella Industry Co. Ltd., Fukuoka, Japan).
4. Harvest the cells in a centrifuge tube at 6500 rpm for 10 min at 10 °C.
5. Subsequently, inoculate the resulting cell pellet into a total of 2 L culture (using two 2-L flasks, each containing 1 L culture medium) of the freshly prepared, CHL-^{15}N ($\sim$97 atm%) medium (Chlorella Industry Co. Ltd., Fukuoka, Japan) containing 25 mg/L kanamycin, 17 mg/L chloramphenicol, 0.2 mM FeCl$_3$, 0.25 g/L MgSO$_4$·7 H$_2$O, and 0.5 g/L Algal ^{15}N(98.7–99.2%)-Amino Acid Mix (Chlorella Industry Co. Ltd., Fukuoka, Japan).
6. Add 1 mM IPTG to this culture and continue the incubation for 5 h at 25 °C.
7. Continue the incubation for another 48–50 h at 25 °C in the presence of 0.42 g/L unlabeled L-lysine (Sigma Chemicals). Note that C43(DE3)-derivative auxotrophs display a strict control of IPTG induction for heterologous expression.
8. Harvest the cells in a centrifuge tube at 6500 rpm for 10 min at 4 °C and store the resulting cell pellet at −80 °C until use.
9. Purify the ^{14}N(N/A) lysine-labeled ARF holoprotein (on the ^{15}N-protein background) with Ni^{2+} affinity chromatography as described previously (Kounosu et al., 2004) and add 1 M NaCl before storage at −80 °C.

3.4 Heterologous Expression and Preparation of ^{14}N(N/A) Tyrosine-Labeled ARF on the ^{15}N-Protein Background by Reverse Labeling Technique Using RF4RIL

1. For preparation of ^{14}N(N/A) tyrosine-labeled ARF on the ^{15}N-protein background, *E. coli* BL21(DE3) auxotroph strain RF4 (Table 1) carrying a pACYC-based plasmid harboring extra copies of tRNA genes for the *E. coli* rare codons (RF4RIL) (Iwasaki et al., 2012) was used as parent cells for heterologous overexpression of the archaeal *arf* gene. When required, verify the deletion of each target chromosomal gene in RF4RIL by sets of the PCR primers in Table 4 prior to use.
2. Transform the pET28aARF expression vector (Kounosu et al., 2004) into the host strain RF4RIL by electroporation.

3. Let the transformants grow overnight at 25 °C in 1 L culture (in a 2-L flask) of the CHL-^{15}N (~97 atm%) medium (Chlorella Industry Co. Ltd., Fukuoka, Japan) containing 25 mg/L kanamycin, 17 mg/L chloramphenicol, 0.2 mM FeCl$_3$, 0.25 g/L MgSO$_4$·7H$_2$O, and 0.5 g/L Algal ^{15}N(98.7–99.2%)-Amino Acid Mix (Chlorella Industry Co. Ltd., Fukuoka, Japan).
4. Harvest the cells in a centrifuge tube at 6500 rpm for 10 min at 10 °C.
5. Subsequently, inoculate the resulting cell pellet into a total of 2 L culture (using two 2-L flasks, each containing 1 L culture medium) of the freshly prepared, CHL-^{15}N (~97 atm%) medium (Chlorella Industry Co. Ltd., Fukuoka, Japan) containing 25 mg/L kanamycin, 17 mg/L chloramphenicol, 0.2 mM FeCl$_3$, 0.25 g/L MgSO$_4$·7H$_2$O, and 0.5 g/L Algal ^{15}N(98.7–99.2%)-Amino Acid Mix (Chlorella Industry Co. Ltd., Fukuoka, Japan).
6. Add 1 mM IPTG to this culture and continue the incubation for 5 h at 25 °C. Note that BL21(DE3)-derivative auxotrophs display a relatively leaky mode of heterologous expression (even in the absence and/or the limited amount of IPTG) under the applied conditions.
7. Continue the incubation for another 48–50 h at 25 °C in the presence of ~0.183 g/L unlabeled L-tyrosine (Nacalai Tesque, Japan).
8. Harvest the cells in a centrifuge tube at 6500 rpm for 10 min at 4 °C and store the resulting cell pellet at −80 °C until use.
9. Purify the ^{14}N(N/A) tyrosine-labeled ARF holoprotein (on the ^{15}N-protein background) with Ni^{2+} affinity chromatography as described previously (Kounosu et al., 2004), and add 1 M NaCl before storage at −80 °C.

4. CONCLUSIONS

Amino acid-selective isotope labeling is a powerful tool required to make full use of magnetic resonance and vibrational spectroscopies as applied to protein structure and dynamics. For this purpose, we have generated a set of BL21(DE3) or C43(DE3)-derived *E. coli* amino acid auxotrophic expression host strains (Table 1) (Iwasaki et al., 2012; Lin et al., 2011), which can help minimize possible metabolic scrambling of the input isotopic labels of many different amino acid types. Examples of the use of these strains are presented (Iwasaki et al., 2012; Lin et al., 2012) to assist others to selectively label proteins of their interest with minimal effort and cost. All of these *E. coli* auxotrophic expression host strains except for YM154 have been deposited to a public strain bank (Addgene, www.addgene.org/Toshio_Iwasaki or www.addgene.org/Robert_Gennis) and are readily available.

ACKNOWLEDGMENTS

This auxotroph strain bank project was supported in part by the JSPS-NSF International Collaborations in Chemistry (ICC) Grant from JSPS (T.I.), the JSPS Grant-in-aid 24659202 (T.I.), and the Nagase Science and Technology Foundation Research Grant (T.I.). Additional funding was provided by the DE-FG02-87ER13716 (R.B.G.) Grant from Chemical Sciences, Geosciences and Biosciences Division, Office of Basic Energy Sciences, Office of Sciences, U.S. Department of Energy, and from the National Institutes of Health, HL16101 (R.B.G.). We thank Dr. Sergei A. Dikanov (University of Illinois at Urbana-Champaign) for pulsed EPR analysis.

REFERENCES

Bellstedt, P., Seiboth, T., Hafner, S., Kutscha, H., Ramachandran, R., & Gorlach, M. (2013). Resonance assignment for a particularly challenging protein based on systematic unlabeling of amino acids to complement incomplete NMR data sets. *Journal of Biomolecular NMR*, 57(1), 65–72. http://dx.doi.org/10.1007/s10858-013-9768-0.

Fricks, H. L., Zhou, D. H., Yap, L. L., Gennis, R. B., & Rienstra, C. M. (2006). Magic-angle spinning solid-state NMR of a 144 kDa membrane protein complex: *E. coli* cytochrome bo_3 oxidase. *Journal of Biomolecular NMR*, 36(1), 55–71. http://dx.doi.org/10.1007/s10858-006-9070-5.

Haris, P. I. (2010). Can infrared spectroscopy provide information on protein-protein interactions? *Biochemical Society Transactions*, 38(4), 940–946. http://dx.doi.org/10.1042/bst0380940.

Iwasaki, T., Fukazawa, R., Miyajima-Nakano, Y., Baldansuren, A., Matsushita, S., Lin, M. T., et al. (2012). Dissection of hydrogen bond interaction network around an iron-sulfur cluster by site-specific isotope labeling of hyperthermophilic archaeal Rieske-type ferredoxin. *Journal of the American Chemical Society*, 134(48), 19731–19738. http://dx.doi.org/10.1021/ja308049u.

Iwasaki, T., Hori, H., Hayashi, Y., Nishino, T., Tamura, K., Oue, S., et al. (1999). Characterization of mouse nNOS2, a natural variant of neuronal nitric-oxide synthase produced in the central nervous system by selective alternative splicing. *Journal of Biological Chemistry*, 274(25), 17559–17566.

Iwasaki, T., Kappl, R., Bracic, G., Shimizu, N., Ohmori, D., & Kumasaka, T. (2011). ISC-like [2Fe-2S] ferredoxin (FdxB) dimer from *Pseudomonas putida* JCM 20004: Structural and electron-nuclear double resonance characterization. *Journal of Biological Inorganic Chemistry*, 16(6), 923–935. http://dx.doi.org/10.1007/s00775-011-0793-8.

Iwasaki, T., Samoilova, R. I., Kounosu, A., Ohmori, D., & Dikanov, S. A. (2009). Continuous-wave and pulsed EPR characterization of the [2Fe-2S](Cys)$_3$(His)$_1$ cluster in rat mitoNEET. *Journal of the American Chemical Society*, 131(38), 13659–13667. http://dx.doi.org/10.1021/ja903228w.

Klammt, C., Lohr, F., Schafer, B., Haase, W., Dotsch, V., Ruterjans, H., et al. (2004). High level cell-free expression and specific labeling of integral membrane proteins. *European Journal of Biochemistry*, 271(3), 568–580.

Kounosu, A., Li, Z., Cosper, N. J., Shokes, J. E., Scott, R. A., Imai, T., et al. (2004). Engineering a three-cysteine, one-histidine ligand environment into a new hyperthermophilic archaeal Rieske-type [2Fe-2S] ferredoxin from *Sulfolobus solfataricus*. *Journal of Biological Chemistry*, 279(13), 12519–12528. http://dx.doi.org/10.1074/jbc.M305923200.

Lin, M. T., Baldansuren, A., Hart, R., Samoilova, R. I., Narasimhulu, K. V., Yap, L. L., et al. (2012). Interactions of intermediate semiquinone with surrounding protein residues at the Q_H site of wild-type and D75H mutant cytochrome bo_3 from *Escherichia coli*. *Biochemistry*, *51*(18), 3827–3838. http://dx.doi.org/10.1021/bi300151q.

Lin, M. T., Samoilova, R. I., Gennis, R. B., & Dikanov, S. A. (2008). Identification of the nitrogen donor hydrogen bonded with the semiquinone at the Q_H site of the cytochrome bo_3 from *Escherichia coli*. *Journal of the American Chemical Society*, *130*(47), 15768–15769. http://dx.doi.org/10.1021/ja805906a.

Lin, M. T., Sperling, L. J., Frericks Schmidt, H. L., Tang, M., Samoilova, R. I., Kumasaka, T., et al. (2011). A rapid and robust method for selective isotope labeling of proteins. *Methods*, *55*(4), 370–378. http://dx.doi.org/10.1016/j.ymeth.2011.08.019.

O'Grady, C., Rempel, B. L., Sokaribo, A., Nokhrin, S., & Dmitriev, O. Y. (2012). One-step amino acid selective isotope labeling of proteins in prototrophic *Escherichia coli* strains. *Analytical Biochemistry*, *426*(2), 126–128. http://dx.doi.org/10.1016/j.ab.2012.04.019.

Rasia, R. M., Brutscher, B., & Plevin, M. J. (2012). Selective isotopic unlabeling of proteins using metabolic precursors: Application to NMR assignment of intrinsically disordered proteins. *Chembiochem*, *13*(5), 732–739. http://dx.doi.org/10.1002/cbic.201100678.

Roman, L. J., Sheta, E. A., Martasek, P., Gross, S. S., Liu, Q., & Masters, B. S. S. (1995). High-level expression of functional rat neuronal nitric oxide synthase in *Escherichia coli*. *Proceedings of the National Academy of Sciences of the United States of America*, *92*(18), 8428–8432.

Rotsaert, F. J., Pikus, J. D., Fox, B. G., Markley, J. L., & Sanders-Loehr, J. (2003). N-isotope effects on the Raman spectra of Fe_2S_2 ferredoxin and Rieske ferredoxin: Evidence for structural rigidity of metal sites. *Journal of Biological Inorganic Chemistry*, *8*(3), 318–326. http://dx.doi.org/10.1007/s00775-002-0417-4.

Sobhanifar, S., Reckel, S., Junge, F., Schwarz, D., Kai, L., Karbyshev, M., et al. (2010). Cell-free expression and stable isotope labelling strategies for membrane proteins. *Journal of Biomolecular NMR*, *46*(1), 33–43. http://dx.doi.org/10.1007/s10858-009-9364-5.

Takeda, M., Ikeya, T., Guntert, P., & Kainosho, M. (2007). Automated structure determination of proteins with the SAIL-FLYA NMR method. *Nature Protocols*, *2*(11), 2896–2902. http://dx.doi.org/10.1038/nprot.2007.423.

Verardi, R., Traaseth, N. J., Masterson, L. R., Vostrikov, V. V., & Veglia, G. (2012). Isotope labeling for solution and solid-state NMR spectroscopy of membrane proteins. *Advances in Experimental Medicine and Biology*, *992*, 35–62. http://dx.doi.org/10.1007/978-94-007-4954-2_3.

Waugh, D. S. (1996). Genetic tools for selective labeling of proteins with α-^{15}N-amino acids. *Journal of Biomolecular NMR*, *8*(2), 184–192.

Yang, J., Camakaris, H., & Pittard, J. (2002). Molecular analysis of tyrosine-and phenylalanine-mediated repression of the *tyrB* promoter by the TyrR protein of *Escherichia coli*. *Molecular Microbiology*, *45*(5), 1407–1419.

Zerbs, S., Frank, A. M., & Collart, F. R. (2009). Bacterial systems for production of heterologous proteins. *Methods in Enzymology*, *463*, 149–168. http://dx.doi.org/10.1016/S0076-6879(09)63012-3.

CHAPTER FOUR

^{19}F-Modified Proteins and ^{19}F-Containing Ligands as Tools in Solution NMR Studies of Protein Interactions

Naima G. Sharaf, Angela M. Gronenborn[1]

Department of Structural Biology, University of Pittsburgh School of Medicine, Pittsburgh, Pennsylvania, USA
[1]Corresponding author: e-mail address: amg100@pitt.edu

Contents

1. Introduction — 68
2. Protocol 1: Biosynthetic Amino Acid Type-Specific Incorporation of ^{19}F-Modified Aromatic Amino Acids — 73
   - 2.1 Protocol Overview — 74
   - 2.2 Step 1: Preparation of Constructs and Transformation — 75
   - 2.3 Tip — 75
   - 2.4 Step 2: Expression of m-Fluoro-L-Tyrosine-Containing Protein — 75
   - 2.5 Tip — 76
   - 2.6 Tip — 78
   - 2.7 Tip — 78
3. Protocol 2: Site-Specific Incorporation of Fluorinated Amino Acids Using a Recombinantly Expressed Orthogonal Amber tRNA/tRNA Synthetase Pair in E. coli — 79
   - 3.1 Protocol Overview — 80
   - 3.2 Step 1: Preparation of Constructs and Transformation — 81
   - 3.3 Tip — 82
   - 3.4 Tip — 82
   - 3.5 Tip — 82
   - 3.6 Step 2: Expression of 4-(Trifluoromethyl)-Phenylalanine-Containing Protein — 82
   - 3.7 Tip — 83
   - 3.8 Tip — 84
4. General Considerations for ^{19}F-Observe NMR Experiments — 84
   - 4.1 Solubility and Stability of the NMR Sample — 85
   - 4.2 Spectrometer Magnetic Field Strength — 85
5. ^{19}F-Modified Protein-Observe NMR Experiments — 86
   - 5.1 Protein–Ligand Interactions — 86
   - 5.2 Protein Un/Folding — 87
   - 5.3 Protein Aggregation — 88
   - 5.4 Protein–Lipid Interactions — 88

6. NMR Experiments with ^{19}F-Containing Ligands   89
   6.1 Monitoring Line Broadening   89
   6.2 Competition-Based Experiments   90
   6.3 ^{19}F NMR-Based Biochemical Screening   91
   6.4 Magnetization Transfer Experiments   91
Acknowledgments   91
References   92

Abstract

^{19}F solution NMR is a powerful and versatile tool to study protein structure and protein–ligand interactions due to the favorable NMR characteristics of the ^{19}F atom, its absence in naturally occurring biomolecules, and small size. Protocols to introduce ^{19}F atoms into both proteins and their ligands are readily available and offer the ability to conduct protein-observe (using ^{19}F-labeled proteins) or ligand-observe (using ^{19}F-containing ligands) NMR experiments. This chapter provides two protocols for the ^{19}F-labeling of proteins, using an *Escherichia coli* expression system: (i) amino acid type-specific incorporation of ^{19}F-modified amino acids and (ii) site-specific incorporation of ^{19}F-modified amino acids using recombinantly expressed orthogonal amber tRNA/tRNA synthetase pairs. In addition, we discuss several applications, involving ^{19}F-modified proteins and ^{19}F-containing ligands.

1. INTRODUCTION

Solution NMR is an invaluable tool to provide structure and dynamic information on proteins and their complexes. To date, most biomolecular NMR applications use proteins that are uniformly labeled with ^{15}N and ^{13}C isotopes. The reasons for this are manifold: (i) the most common nuclei in proteins are protons, carbons, and nitrogen, (ii) replacement of the natural ^{12}C and ^{14}N isotopes by those that possess a nuclear spin of 1/2 (^{15}N and ^{13}C) does not influence protein structure and function, (iii) incorporation of these isotopes into proteins through biosynthetic microbial protein expression is relatively easy and cost-effective, using ^{15}N- and/or ^{13}C-containing nutrients in growth media, and (iv) a large suite of one-, two-, and three-dimensional NMR experiments has been developed over the last 30 years that relies on and exploits ^{15}N and ^{13}C magnetization manipulations, geared at yielding structural and dynamic information on protein and protein–ligand complexes.

Although classical ^{1}H, ^{13}C, and ^{15}N spectroscopic approaches have been used extensively to study proteins in solution, ^{19}F NMR is gaining increasing popularity. As discussed previously (Danielson & Falke, 1996; Gerig,

Table 1 Properties of Spin ½ NMR Active Nuclei

Nuclei	Natural Abundance (%)	I	γ (rad×s^{-1}×gauss^{-1})	Sensitivity (% vs. ^{1}H)
^{1}H	99.980	1/2	26,753	100
^{13}C	1.108	1/2	6,728	1.59
^{15}N	0.37	1/2	−2,712	0.104
^{19}F	100.00	1/2	25,179	83.3
^{31}P	100.00	1/2	10,841	6.63

1994), the 100% naturally abundant ^{19}F atom displays several properties that render it ideal for NMR exploitation: it possesses a spin 1/2 nucleus and a high gyromagnetic ratio that results in excellent sensitivity (83% of ^{1}H) (see Table 1). In addition, the shielding of the ^{19}F nucleus is dominated by a large paramagnetic term and, as a result, fluorine chemical shifts are exquisitely sensitive to changes in local environment (the chemical shift range is ~100-fold larger than that of ^{1}H). Another great advantage of using ^{19}F as an NMR probe is its absence from virtually all naturally occurring biomolecules, small and large. For this reason, studies of fluorinated biopolymers can be carried out in any routinely used buffer system or environment without suffering from interference by background signals. Thus, no special precautions are needed to remove buffer and additive signal intensity from the spectra. The van der Waals radius of the ^{19}F atom (1.47 Å) lies between those of hydrogen (1.2 Å) and oxygen (1.52 Å), and strategic substitution of ^{19}F atoms for hydrogens, hydroxyl groups, or carbonyl oxygens in biological molecules is considered weakly perturbing and often has little effect on a protein's biological activity (Campos-Olivas, Aziz, Helms, Evans, & Gronenborn, 2002; Danielson & Falke, 1996; Gerig, 1994; Kitevski-Leblanc & Prosser, 2012).

For 40 years now, the ^{19}F atom has been exploited in biological NMR (Arseniev et al., 1986; Chaiken, Freedman, Lyerla, & Cohen, 1973), and several reviews have covered the field over the last two decades (Danielson & Falke, 1996; Didenko, Liu, Horst, Stevens, & Wüthrich, 2013; Gerig, 1994; Kitevski-Leblanc & Prosser, 2012; Marsh & Suzuki, 2014). Indeed, the ^{19}F atom has been used as a molecular probe to gain insight into protein and peptide structure (Danielson & Falke, 1996; Gerig, 1994; Kawahara et al., 2012; Liu, Horst, Katritch, Stevens, & Wuthrich, 2012), protein–ligand interactions (Campos-Olivas et al., 2002; Danielson & Falke, 1996; Gerig, 1994;

Kitevski-Leblanc & Prosser, 2012; Luck & Falke, 1991; Rydzik et al., 2014), protein unfolding (Arseniev et al., 1986; Chaiken et al., 1973; Khan, Kuprov, Craggs, Hore, & Jackson, 2006; Kitevski-Leblanc, Hoang, Thach, Larda, & Prosser, 2013), protein aggregation (Danielson & Falke, 1996; Didenko et al., 2013; Gerig, 1994; Kitevski-Leblanc & Prosser, 2012; Li et al., 2009; Marsh & Suzuki, 2014; Suzuki, Brender, Hartman, Ramamoorthy, & Marsh, 2012), and protein dynamics (Fischer et al., 2003), clearly demonstrating the power and versatility of the fluorine probe for NMR.

Several methods to prepare ^{19}F-modified proteins have been described (Cellitti et al., 2008; Danielson & Falke, 1996; Frieden, Hoeltzli, & Bann, 2004; Hortin & Boime, 1983; Kitevski-Leblanc & Prosser, 2012; Peeler & Mehl, 2011; Sykes & Hull, 1978). These methods fall into three main categories: (i) posttranslational covalent attachment of ^{19}F-containing moieties to the protein, (ii) biosynthetic amino acid type-specific incorporation of ^{19}F-modified amino acids, and (iii) site-specific incorporation of ^{19}F-modified amino acids using recombinantly expressed orthogonal amber tRNA/tRNA synthetase pairs.

In brief, posttranslational covalent modification introduces ^{19}F atoms into the protein of interest by conjugating a ^{19}F-containing moiety to a reactive group, such as an -SH group on a solvent accessible cysteine (Liu et al., 2012; Rydzik et al., 2014). One advantage of this technique is the ability to incorporate the label into proteins for which biosynthetic labeling is cost-prohibitive, such as proteins expressed in mammalian cells. For residue-specific incorporation of ^{19}F-modified amino acids into proteins, expression is carried out in defined growth media, supplemented with the ^{19}F-modified amino acid. This method relies on the endogenous aminoacyl-tRNA synthetases to charge the ^{19}F-modified amino acid onto their cognate tRNAs. As a result, all codons recognized by the amino acid-specific tRNA will carry the ^{19}F-modified amino acid, and global incorporation of the ^{19}F-modified amino acid into all proteins will occur. To maximize the efficiency of ^{19}F-labeled amino acid incorporation, auxotrophic bacterial strains that cannot synthesize the amino acid that is to be replaced are sometimes used. However, even with these strains, substitution efficiencies using this approach are always less than 100%, due to partial incorporation of the natural amino acid, generated during normal cellular metabolism (Kim, Perez, Ferguson, & Campbell, 1990; Kitevski-Leblanc, Al-Abdul-Wahid, & Prosser, 2009). Still, for some amino acids, >90% labeling can be achieved if conditions are worked out carefully. Potentially, the most powerful and versatile method to introduce ^{19}F atoms into proteins is site-specific incorporation of ^{19}F-modified amino acids using a recombinantly introduced orthogonal amber tRNA/tRNA synthetase pair. This method is based on an extension of the genetic code beyond the natural 20 amino acids, first described by Noren and colleagues in 1989 (Noren, Anthony-Cahill, Griffith, & Schultz, 1989). It uses nonsense stop codons and nonsense suppressor-tRNAs to overturn termination of protein biosynthesis and requires engineered aminoacyl-tRNA synthetases that specifically acylate the suppressor-tRNAs with the nonnatural ^{19}F-modified amino acid *in vivo*, without interfering with other tRNA/synthetase pairs. Application of this approach requires the introduction of the amber nonsense codon (TAG) at any desired location in the protein-coding sequence, such that it replaces the natural amino acid codon in that location, in conjunction with introduction of a tailored orthogonal amber tRNA/tRNA synthetase pair. More specifically, a vector containing the protein amber-mutant gene is co-introduced into an *Escherichia coli* (*E. coli*) host strain along with a vector encoding an *in vitro* evolved orthogonal amber tRNA/tRNA synthetase pair that recognizes the ^{19}F-modified amino acid. Expression is then carried out in growth media supplemented with the ^{19}F-modified amino acid. One drawback of this methodology is the fact that the orthogonal amber tRNA/tRNA synthetase pair does not incorporate the ^{19}F-modified amino acid with 100% efficiency at the amber codon and that translation termination occurs to varying degrees. As a result, two protein products are invariably generated, an undesired unlabeled truncated protein, and a 100% labeled full-length protein. These two proteins can be separated during purification if a C-terminal affinity tag is used. All three methods are schematically illustrated in Fig. 1.

In addition to using the ^{19}F-atom as a molecular probe on the protein to investigate protein–ligand interactions, the ^{19}F-atom may also be incorporated into the ligand. Many protein-binding molecules have been modified with ^{19}F atoms, including nucleic acids, peptides, carbohydrates, and a myriad of small molecules (Chen, Viel, Ziarelli, & Peng, 2013; Fielding, 2007; Ghitti, Musco, & Spitaleri, 2014; Kiviniemi & Virta, 2010; Matei et al., 2013; Tanabe, Sugiura, & Nishimoto, 2010; Yu, Hallac, Chiguru, & Mason, 2013). Among these, incorporation of ^{19}F atoms into small molecule drugs is of particular interest, since the presence of ^{19}F atoms in small molecules has been shown to improve their pharmacokinetic and physicochemical properties. As a result, ^{19}F-containing chemicals are now routinely synthesized, and fluorine atoms are present in many drug-like molecules, with ~20% of all drugs on the market containing at least one fluorine atom (Dalvit & Vulpetti, 2011;

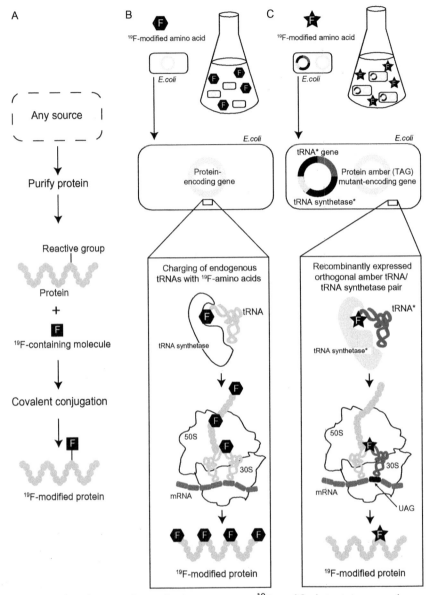

Figure 1 The three main methods to prepare ^{19}F-modified proteins are shown. (A) Posttranslational covalent conjugation of ^{19}F-containing moieties to the protein. (B) Biosynthetic amino acid type-specific incorporation of ^{19}F-modified amino acids. (C) Site-specific incorporation of ^{19}F-modified amino acids using recombinantly expressed orthogonal amber tRNA/tRNA synthetase pairs.

Muller, Faeh, & Diederich, 2007). The use of ^{19}F-containing ligands to probe protein–ligand interactions by NMR is also well established (Chen et al., 2013; Fielding, 2007; Ghitti et al., 2014; Hortin & Boime, 1983). ^{19}F-containing ligand-observe experiments have been used to provide information on ligand-binding affinity, specificity, and reversibility and to obtain ligand–protein structural constraints (Clore, Gronenborn, Birdsall, Feeney, & Roberts, 1984; Goudreau, Coulombe, & Faucher, 2013; Kim et al., 1990; Matei et al., 2013; Yu, Hajduk, Mack, & Olejniczak, 2006).

In this chapter, we provide protocols for the two biosynthetic methods that are used to label proteins with ^{19}F, employing an *E. coli* expression system: amino acid type-specific incorporation of ^{19}F-modified amino acids and site-specific incorporation of ^{19}F-modifed amino acids. We only discuss expression in *E. coli*, because of the ease, speed, and affordability of this system, compared to eukaryotic expression systems such as yeast, insect, or mammalian cells. Since relative large protein amounts are often required for ^{19}F NMR experiments of modified proteins, especially for ligand-library screening, bacterial expression is still the system of choice. We also include a section that describes basic ^{19}F NMR experiments and provide examples of the application of protein-observe and ligand-observe experiments using ^{19}F-modified proteins and ligands, respectively.

2. PROTOCOL 1: BIOSYNTHETIC AMINO ACID TYPE-SPECIFIC INCORPORATION OF ^{19}F-MODIFIED AROMATIC AMINO ACIDS

One popular method for incorporating ^{19}F-modified amino acids into proteins is through biosynthetic microbial protein expression (Evanics, Kitevski, Bezsonova, Forman-Kay, & Prosser, 2006; Frieden et al., 2004; Gerig, 1994; Luck & Falke, 1991). This method simply requires the addition of the ^{19}F-modified amino acid to the growth medium of the bacterial cells. Upon flooding the cells with a particular ^{19}F-modified amino acid, the appropriate aminoacyl-tRNA synthetase will mischarge its cognate tRNA with the ^{19}F-modifed amino acid, since a single ^{19}F atom does not substantially alter the properties of the amino acid. As a result, all proteins produced in the bacterial cell experience amino acid-specific incorporation of the ^{19}F-modified amino acid. The ability to mischarge the tRNA with the ^{19}F-modified amino acid relies on the permissivity, or relaxed substrate specificity, of most aminoacyl-tRNA synthetases. Since some aminoacyl-tRNA synthetases have limited permissivity, not all ^{19}F-modified amino acids can be incorporated into proteins using this method. Naturally, aminoacyl-tRNA synthetases bind to both, the unmodified and ^{19}F-modifed amino acids, and expression protocols, therefore, employ high concentrations of the ^{19}F-modified amino acid and/or depletion of any unmodified amino acid that is produced by normal cellular metabolism (Hortin & Boime, 1983). A common approach is to utilize auxotrophic bacterial strains. For the introduction of ^{19}F-modified aromatic amino acids, an alternative approach is also available: adding glyphosate, an inhibitor of a key step in aromatic amino acid biosynthesis, during cellular growth, suppresses the production of tryptophan, tyrosine, and phenylalanine (Dominguez, Thornton, Melendez, & Dupureur, 2001; Evanics et al., 2006; Hoeltzli & Frieden, 1994; Kim et al., 1990).

One drawback of using amino acid type-specific labeling of proteins is the subsequent need for specific resonance assignments, given that most proteins will contain several amino acids of the same type, and it is impossible to reliably predict the chemical shift of the ^{19}F resonance based on the position of the ^{19}F-modified amino acid in the protein structure. Resonance broadening or resonance overlap can further compound this impediment. Traditionally, resonance assignments of amino acid type-labeled proteins involves systematic, iterative substitution of all residues of the specific amino acid type in the protein with a structurally similar amino acid by site-directed mutagenesis (Aramini et al., 2014; Hyean-Woo et al., 2000; Li et al., 2009; Salopek-Sondi & Luck, 2002). Full spectral assignments can, in principle, be obtained by recording spectra of the complete set of single mutants and comparison of the remaining resonances in the mutants with those in the fully labeled native protein. Unfortunately, this method can fail, if the substitutions result in structural perturbations that alter the resonance frequencies of the remaining resonances. Two alternative approaches have been proposed and applied: (i) mutagenesis of a residue near the residue to be assigned (Drake, Bourret, Luck, Simon, & Falke, 1993) and (ii) NMR-based assignments that combine side-chain correlation experiments with traditional backbone assignments (Kitevski-Leblanc et al., 2009). The latter strategy relies on the availability of the protein structure or prior backbone ^{1}H, ^{13}C, ^{15}N resonance assignments.

Here, we provide a protocol for the global incorporation of ^{19}F-modified aromatic amino acids into a protein, using m-fluoro-tyrosine as an example.

2.1 Protocol Overview

The approaches that have been used to incorporate ^{19}F-modified aromatic amino acids into proteins using biosynthetic microbial expression vary in detail but share several key steps (Aramini et al., 2014; Kim et al., 1990; Li et al., 2009). First, since addition of the ^{19}F-modifed amino acid can cause toxicity and interfere with protein synthesis, as a precaution, bacteria usually are not grown from an inoculum that was grown in ^{19}F-modified amino acid-containing media. Second, as best as possible, the presence of any unmodified amino acid that is generated by normal cellular metabolism is suppressed. This is commonly accomplished by growing the cells in media that prevent or suppress generation of the unmodified amino acid, such as defined medium, and transferring the cells to defined medium supplemented with the ^{19}F-modified amino acid. The addition of glyphosate, which interferes with aromatic amino acid synthesis, is also used. Third, the concentration of the ^{19}F-modified aromatic amino acid in the medium at induction must be optimized to ensure maximal protein expression, without affecting cellular growth. Given the large set of possible variables in the different protocols, the method provided below should only be used as a starting point for further optimization.

2.2 Step 1: Preparation of Constructs and Transformation

Time: ~1 week

1.1. Insert the protein-encoding gene into a suitable pET vector, such as a pET21 (ampicillin resistant; EMD Millipore).

1.2. Transform competent BL21-Gold (DE3) cells (Agilent) with the vector (1.1), using the basic transformation procedure supplied by the vendor. Select for cells containing the vector by plating the transformation reaction onto LB-agar plates, containing 100 mg/L ampicillin.

1.3. To prepare a glycerol stock, inoculate a 5 mL LB culture (100 mg/L ampicillin) with a single colony of the transformed strain. Grow overnight at 37 °C, with shaking at 225 rpm.

1.4. Mix 750 μL of the cell culture with 500 μL of glycerol, flash freeze in liquid nitrogen, and store at −80 °C for use in Step 2.

2.3 Tip

Most expression vectors and bacterial strains should be compatible with the labeling method below. Therefore, we recommend the use of an expression vector and *E. coli* strain that give good expression of your particular protein.

See Fig. 2 for the flowchart of Step 1.

2.4 Step 2: Expression of *m*-Fluoro-L-Tyrosine-Containing Protein

Time: ~3 days

1.1. Inoculate 20 mL of LB culture medium with the bacterial glycerol stock.

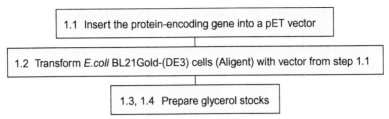

Figure 2 Flowchart for Protocol 1: Step 1.

1.2. Grow overnight at 37 °C, shaking at 225 rpm.
1.3. Centrifuge the 20 mL culture and remove the supernatant.
1.4. Inoculate 100 mL M9 medium (see Tables 2 and 3) to an OD_{600} of 0.2 using the LB cell pellet. This is your starter culture.
1.5. Grow overnight at 37 °C, shaking at 225 rpm.
1.6. Inoculate 1 L M9 minimal medium (see Tables 2 and 3) to an OD_{600} of 0.2 using the starter culture. This is your expression culture.
1.7. Grow at 37 °C with shaking at 225 rpm until the culture reaches an OD_{600} of 0.5, then reduce to the temperature designated for expression.
1.8. Wait 30 min.
1.9. Add the following (see Tip 1.5): 500 mg of m-fluoro-DL-tyrosine, 1 mL of 0.2 g/L glucose, 1 mL of 0.1 g/L NH_4Cl, 100 μL of 100 mM $ZnSO_4$, 0.5 mL of 1 M isopropyl β-D-1-thiogalactopyranoside (IPTG), and 1 g glyphosate.
1.10. Grow at a temperature designated for expression, shaking at 225 rpm for 5 h.
1.11. Harvest cells by centrifugation, resuspend the cell pellet in lysis buffer, flash freeze in liquid nitrogen, and store at −80 °C.

2.5 Tip

Protocols differ in the identity and quantity of the amino acids added before induction. Here are some common combinations of amino acids, added before induction: (i) >500 mg of the ^{19}F-modified aromatic amino acid, plus all unmodified amino acids at a final concentration of ~1.0 mg/L, (ii) combinations of the three aromatic amino acids at various concentrations, for example, tyrosine (50 mg/L), phenylalanine (50 mg/L), and tryptophan (35 mg/L, two aromatic amino acids would be unmodified while the third would be ^{19}F-modified), (iii) only the ^{19}F-modified aromatic amino acid at concentrations >500 mg/L. All these options require the absence of the unmodified amino acid analog for incorporation of the desired ^{19}F-modified aromatic amino acid.

Table 2 Stock Solutions

Media	Content	
10 × M9 salts[a]	Na_2HPO_4	66 g
	KH_2PO_4	30 g
	NH_4Cl	10 g
	NaCl	5 g
	Water	To 1 L
5000 × Trace metals[b]	$FeCl_3$ (powder)	486 mg
	2 M $CaCl_2$	500 μL
	1 M $ZnSO_4$	500 μL
	1 M $MnCl_2$	500 μL
	0.2 M Na_2MoO_4	500 μL
	0.2 M $CoCl_2$	500 μL
	0.2 M Na_2SeO_3	500 μL
	0.2 M $CuCl_2$	500 μL
	0.2 M $NiCl_2$	500 μL
	0.2 M H_2BO_3	500 μL
	ddH_2O	To 50 mL
ZY medium[a]	N-Z amine	10 g
	Yeast extract	5 g
	Water	To 1 L
25 × M[a]	Na_2HPO_4	88 g
	KH_2PO_4	85 g
	NH_4Cl	67 g
	Na_2SO_4	18 g
	Water	To 1 L
50 × 5052[b]	Glycerol	250 g
	Glucose	25 g
	α-Lactose	100 g
	Water	To 1 L
[19]F-modified amino acid solution: trifluoromethyl phenylalanine (tfmF)	tfmF	245 mg
	ddH_2O	8 mL
	8 M NaOH	1 mL

[a]Autoclave.
[b]Sterilize by passing through a 0.22-μm filter.

Table 3 Media for Protocol 1: Step 2

Media	Content	
Starter culture: 100 mL M9 medium	M9 salts (10×)	10 mL
	20% (w/v) glucose[b]	2 mL
	1 M MgSO$_4$[a]	100 µL
	0.2 M CaCl$_2$[a]	5 µL
	2.5 mg/mL biotin (pH 11)[b]	4 µL
	5000× trace metals[b]	2 µL
	5 mg/mL thiamine[b]	1 µL
	100 mg/mL ampicillin	100 µL
	Water[a]	To 100 mL
Expression culture: 1 L M9 medium	M9 salts (10×)	100 mL
	20% (w/v) glucose[b]	20 mL
	1 M MgSO$_4$[a]	1 mL
	0.2 M CaCl$_2$[a]	0.5 mL
	2.5 mg/mL biotin (pH 11)[b]	0.4 mL
	5000× trace metals[b]	200 µL
	5 mg/mL thiamine[b]	100 µL
	100 mg/mL ampicillin	1 mL
	Water[a]	To 1 L

[a]Autoclave.
[b]Sterilize by passing through a 0.22-µm filter.

2.6 Tip

The tryptophan precursor fluoroindole can also be used to incorporate ^{19}F-modified tryptophan into proteins (Crowley, Kyne, & Monteith, 2012).

2.7 Tip

Frequently, ^{19}F-modified aromatic amino acids can be incorporated into proteins without using glyphosate.

See Fig. 3 for the flowchart of Step 2.

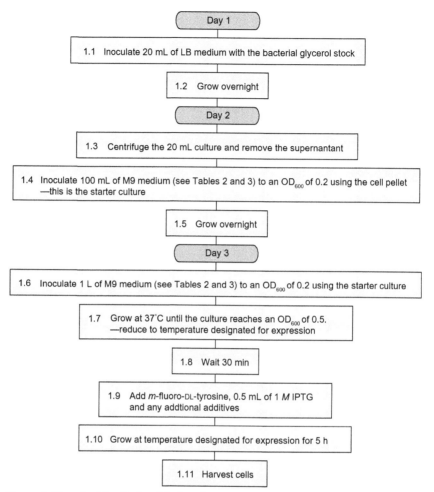

Figure 3 Flowchart for Protocol 1: Step 2.

3. PROTOCOL 2: SITE-SPECIFIC INCORPORATION OF FLUORINATED AMINO ACIDS USING A RECOMBINANTLY EXPRESSED ORTHOGONAL AMBER tRNA/tRNA SYNTHETASE PAIR IN *E. COLI*

Several protocols are available to site-specifically label proteins with a recombinantly expressed orthogonal amber tRNA/tRNA synthetase pair (Cellitti et al., 2008; Hammill, Miyake-Stoner, Hazen, Jackson, & Mehl, 2007). They all rely on several basic principles or steps: (i) in *E. coli*, three codons are signals for translation termination, UAG (amber), UAA (ochre), and UGA (opal), and they do not possess cognate tRNAs; (ii) an amber tRNA/tRNA synthetase pair from another organism can be evolved to uniquely charge the amber tRNA with the desired ^{19}F-modfied amino acid, and genes encoding the evolved amber tRNA and tRNA synthetase pair are then placed into a vector; (iii) in the expression vector that encodes the protein of interest, an amber stop codon is introduced at the sequence position where the ^{19}F-modified amino acid is to be incorporated; and (iv) the amber tRNA/tRNA synthetase pair and the protein of interest are co-expressed in *E. coli*. As a result of co-expression, the amber tRNA is charged by is cognate tRNA synthetase with the ^{19}F-modified amino acid and delivers the ^{19}F-modified amino acid at the position of the amber codon to the growing polypeptide chain during protein synthesis. Since this process is not 100% efficient, translation of the protein will sometimes terminate at the amber codon, resulting in two protein products, an undesired, unlabeled, truncated protein and the 100% singly labeled, full-length protein. The truncation product can be separated from the modified, full-length protein using a C-terminal affinity tag for purification.

Unfortunately, vectors encoding the amber tRNA/tRNA synthetase pairs for individual modified amino acids are not commercially available. Therefore, each tRNA/tRNA synthetase pair for any ^{19}F-modified amino acid has to be evolved *in vitro*, which takes time and resources. However, over the last few years, the first Unnatural Protein Facility has been established by Dr. Ryan Mehl at Oregon State University, providing researchers full access to current noncanonical amino acid protein production capability for academic studies. The facility will aid in testing and optimizing unnatural protein expression. Sometimes, a vector with the appropriate evolved orthogonal tRNA/tRNA synthetase pair may already be available and can be requested.

Here, we provide a protocol developed by the Mehl laboratory (Hammill et al., 2007; Peeler & Mehl, 2011) that has proven robust and successful in our hands for site-specific incorporation of fluorinated amino acids. The approach is compatible with the pET vector series (Novagen), which contains an IPTG-inducible T7 promoter, and is commonly employed for protein expression. In addition, the labeling protocol is compatible with auto-induction medium (Studier, 2005), which improves protein yields.

3.1 Protocol Overview

Evolution of the amber tRNA is guided by two important principles: (i) it must deliver the ^{19}F-modified amino acid into the growing polypeptide chain in response to the UAG amber codon and (ii) it should neither be acylated nor deacylated by any of the endogenous tRNA synthetases. The ^{19}F-modified amino acid is chosen such, that any endogenous aminoacyl-tRNA synthetases do not charge their cognate endogenous tRNAs with this particular ^{19}F-modified amino acid (Noren et al., 1989). This ensures (i) that the recombinantly expressed amber tRNA/tRNA synthetase pair only incorporates the chosen ^{19}F-modified amino acid and not any of the other 20 natural amino acids, and (ii) that endogenous tRNAs cannot deliver the ^{19}F-modified amino acid to the ribosome for incorporation into the growing polypeptide chain.

Since the endogenous tRNA/tRNA synthetases do not globally cause incorporation of the ^{19}F-modified amino acid into all proteins in the cell, the presence of the ^{19}F-modified amino acid in the growth medium is less toxic, and it can be added early in the protocol. In addition, the amber tRNA synthetase only binds to the ^{19}F-modified amino acids, and any natural amino acids created during cellular metabolism or added to the medium will not be bound. Therefore, this protocol can be carried out in rich medium. We illustrate the methodology with a protocol for incorporating 4-(trifluoromethyl)-L-phenylalanine into a protein.

3.2 Step 1: Preparation of Constructs and Transformation

Time: 1–3 weeks

- **1.1.** Use the pDule2 RS (streptomycin resistant) vector encoding the tailored orthogonal amber tRNA/tRNA synthetase pair; available from Dr. Ryan Mehl (http://mehl.science.oregonstate.edu).
- **1.2.** Insert the protein-encoding gene into a suitable expression vector, such as a pET21 (ampicillin resistant; EMD Millipore).
- **1.3.** Change the protein-coding sequence in the expression vector to the amber codon (TAG) at the location desired for amino acid substitution, using QuikChange (Stratagene).
- **1.4a.** Co-transform competent BL21-AI cells (Invitrogen) with the protein wild-type construct (1.2) and pDule2 RS (this strain will serve as a positive control), using the basic transformation procedure supplied by the vendor. Use at least 100 ng of each vector for the co-transformation, but do not add more than 2 μL of each plasmid stock solution. Select for cells carrying both vectors by plating the transformation reaction on LB-agar plates containing both 100 mg/L ampicillin and 50 mg/L streptomycin.

1.4b. Co-transform competent BL21-AI cells (Invitrogen) with the protein amber-mutant construct (1.3) and pDule2 RS, using the procedure described earlier (this strain will be used to express the ^{19}F-modified protein).

1.5. To prepare glycerol stocks, inoculate 5 mL LB (100 mg/L ampicillin and 50 mg/L streptomycin) with a single colony of the co-transformed strain (pDule2 and wild-type protein-encoding gene). Inoculate another 5 mL LB (100 mg/L ampicillin and 50 mg/L streptomycin) with a single colony of the transformant that will be used to express the ^{19}F-modified protein (pDule2/protein TAG mutant). Grow overnight at 37 °C with shaking at 225 rpm.

1.6. Mix 750 µL of each culture with 500 µL of glycerol, flash freeze in liquid nitrogen, and store at −80 °C for use in Step 3.

3.3 Tip

Make sure the protein construct possesses a C-terminal purification tag (preferably cleavable). This will ensure efficient purification away from the undesired, unlabeled, truncated protein.

3.4 Tip

If no transformants are found using the transformation protocol supplied by the vendor, try adding 250 µL Recovery Medium (Lucigen) instead of SOC medium and shake for 4 h at 225 rpm.

3.5 Tip

Not all protein sequence locations are amenable to incorporation of a ^{19}F-labeled amino acid. It is advisable to first test several different protein amber mutants (where the TAG mutation is substituted at different residue positions) in small-scale expressions (5 mL) to determine which residue positions can be labeled. A good trouble-shooting guide is provided by Hammill et al. (2007).

See Fig. 4 for the flowchart of Step 1.

3.6 Step 2: Expression of 4-(Trifluoromethyl)-Phenylalanine-Containing Protein

Time: 3 days

1.1. Inoculate 10 mL of noninducing medium (see Tables 2 and 4) with the bacterial glycerol stock made in Step 1.

1.2. Grow overnight at 37 °C, with shaking at 225 rpm.

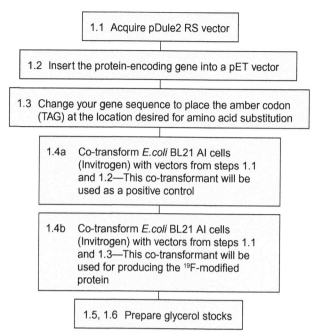

Figure 4 Flowchart for Protocol 2: Step 1.

1.3. Inoculate 1 L inducing medium (see Tables 2 and 4) with 10 mL of starter culture.
1.4. Immediately (~30 s) add freshly prepared 4-(trifluoromethyl)-L-phenylalanine (Peptech) solution to the 1 L culture, to a final concentration of 1 mM (see Table 2).
1.5. Grow for 24 h at the temperature designated for expression, shaking at 225 rpm.
1.6. Harvest cells by centrifugation, resuspend the cell pellet in lysis buffer, flash freeze in liquid nitrogen, and store at −80 °C.

3.7 Tip

It is important to perform a negative and positive control for each expression run. For the positive control, use the pDule2 RS/wild-type protein-encoding gene transformant without adding 4-(trifluoromethyl)-L-phenylalanine to the growth medium. For the negative control, use pDule2 RS/ protein amber-mutant gene without adding 4-(trifluoromethyl)-L-phenylalanine to the growth medium. SDS–PAGE analysis of the samples prepared from these two growths should show a band for the size of the

Table 4 Media for Protocol 2: Step 2

Media	Content	
Starter culture: 10 mL noninducing medium	ZY medium[a]	9 mL
	25 × M[a]	400 μL
	20% Glucose[b]	250 μL
	1 M MgSO$_4$[a]	20 μL
	100 mg/mL ampicillin	10 μL
	50 mg/mL streptomycin	10 μL
	5000 × trace metals[b]	2 μL
Expression culture: 1 L inducing medium	ZY medium[a]	1 L
	25 × M[a]	40 mL
	50 × 5052[b]	20 mL
	Starter culture	10 mL
	10% Arabinose[b]	5 mL
	1 M MgSO$_4$[a]	2 mL
	100 mg/mL ampicillin	1 mL
	50 mg/mL streptomycin	1 mL
	5000 × trace metals	200 μL

[a]Autoclave.
[b]Sterilize by passing through a 0.22-μm filter.

full-length protein (positive control) and a band at the size of the undesired truncated protein (negative control).

3.8 Tip

We found that the pDule2 RS vector can also be used to incorporate 4-(trifluoromethoxy)-L-phenylalanine into proteins (JRD fluorochemicals).
See Fig. 5 for the flowchart of Step 2.

4. GENERAL CONSIDERATIONS FOR ^{19}F-OBSERVE NMR EXPERIMENTS

One of the most critical points, if one wants to use ^{19}F NMR to study ^{19}F-modified proteins, is to obtain a high-quality 1D ^{19}F spectrum with good signal-to-noise. Here, we briefly discuss some factors that can

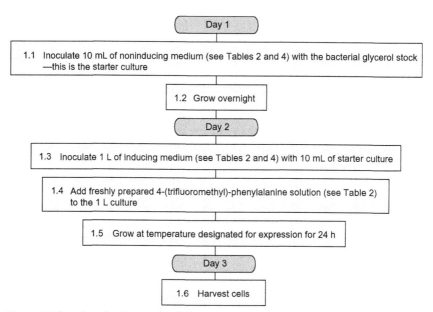

Figure 5 Flowchart for Protocol 2: Step 2.

influence the quality of such a NMR spectrum. All these issues have been covered previously in greater detail, and readers are referred to the references for further in-depth coverage.

4.1 Solubility and Stability of the NMR Sample

One of the most critical requirements for using NMR to study proteins is a sample of high quality and integrity. Therefore, sample preparation requires careful attention. In particular, solution conditions need to be optimized to ensure that the protein is soluble and stable for the length of the NMR experiment. Several factors that can influence solubility and stability include pH, temperature, buffer type, salt concentration, and additives (Cavanagh, Fairbrother, Palmer, Rance, & Skelton, 2007; Oppenheimer, 1989; Rule & Hitchenns, 2006).

4.2 Spectrometer Magnetic Field Strength

In ^{1}H NMR, any increase in magnetic field strength generally increases the experimental sensitivity and resolution. However, this is not necessarily the case for the ^{19}F nucleus, which can have a large and asymmetric chemical shift anisotropy (CSA). The CSA contribution to T_2 relaxation increases with the square of the magnetic field strength and leads to broad lines

(Yamamoto, Nagao, Hirai, Tai, & Suzuki, 2012). Therefore, the optimal magnetic field strength depends on the desired resolution and the line shape (Gerig, 1994; Kitevski-Leblanc & Prosser, 2012). In general, a fluorine probe with a small CSA is desirable. Therefore, trifluoromethyl probes are commonly used in covalent protein-labeling approaches; these probes exhibit relatively short T_1s, possess axially symmetric tensors, and have smaller chemical shift anisotropies compared to fluorinated aromatics.

5. ^{19}F-MODIFIED PROTEIN-OBSERVE NMR EXPERIMENTS

Protein-observe NMR experiments monitor the resonance of a ^{19}F-modified amino acid in the protein to elucidate various biological phenomena, such as protein–ligand binding, protein un/folding, protein aggregation, and protein dynamics. As with traditional ^{1}H, ^{13}C, and ^{15}N NMR experiments, ^{19}F NMR experiments provide complimentary information to that obtained from X-ray crystallography. In particular, NMR solution experiments provide insights into protein conformational dynamics and protein motions. ^{19}F-protein-observe NMR experiments have several advantages over traditional NMR experiments due to the favorable NMR properties of the ^{19}F nucleus. The increased sensitivity and limited number of resonances, resulting from the sparse placement of the ^{19}F-atom in a small number of amino acids, allows for the study of proteins that are challenging by traditional NMR, such as proteins prone to aggregation, those with a large molecular mass (>30 kDa), and membrane proteins. Since several excellent reviews are available that cover the field (Chen et al., 2013; Danielson & Falke, 1996; Gerig, 1994; Kitevski-Leblanc & Prosser, 2012; Marsh & Suzuki, 2014), we only discuss some applications that involve ^{19}F-modifed proteins.

5.1 Protein–Ligand Interactions

^{19}F NMR experiments have been used to monitor protein conformational changes upon ligand binding for proteins of various sizes, including soluble and membrane proteins. Notably, studies of the β_2-adrenergic receptor, which was modified by posttranslational covalent attachment of 2,2,2 trifluoroethanethiol (Liu et al., 2012), revealed a correlation between the physiological effects of ligands and the ^{19}F resonance line widths and intensities. In a follow-up study, the thermodynamic and kinetic parameters associated with ligand binding were determined (Horst, Liu, Stevens, & Wüthrich, 2013). ^{19}F NMR experiments have also been used to obtain quantitative data on protein–ligand binding. Labeling the enzyme New Delhi metallo-β-lactamase by covalent addition of 3-bromo-1,1,1-trifluoroacetone, Rydzik and colleagues were able to monitor the ^{19}F resonance on a loop that participates in ligand binding (Rydzik et al., 2014). For the two known inhibitors, D-captopril and L-captopril, the ^{19}F NMR-extracted K_D values were in good agreement with previously measured IC_{50} values. In addition, the authors were able to monitor the binding of weak inhibitors for which no structural information was available. In another study, the SPRY domain-containing SOCS box protein 2 was labeled by amino acid type-specific incorporation of 5-fluoro-tryptophan. ^{19}F NMR experiments were carried out, and the response of a ^{19}F-resonance located near a target-binding site for peptides and small molecules was monitored (Leung et al., 2014). Using the ^{19}F resonance as a reporter for protein–ligand binding, the authors were able to identify several compounds that bind near this site in fragment-based screening.

Although few, the above examples provide a vivid illustration that fluorine-labeled proteins and ^{19}F NMR can be exploited for effective and powerful screening of ligand libraries in the characterization and development of pharmacologically active compounds.

5.2 Protein Un/Folding

^{19}F NMR experiments are well suited to monitor protein un/folding since the large chemical shift dispersion of the ^{19}F nucleus allows for easy detection of resonances in different chemical environments. Unfolding studies of GFP, labeled by amino acid type-specific incorporation of *m*-fluoro-tyrosine, investigated the differences between pH-induced and guanidine hydrochloride-induced denatured states. The acid denatured state (pH 2.9) was noted to contain heterogeneous subensembles with different rotational correlation times (Khan et al., 2006), compared to the chemically denatured state and a fully acid denatured state (pH 1.5). ^{19}F NMR has also been applied to obtain information on folding intermediates. For calmodulin, labeled by amino acid type-specific incorporation of 3-fluorophenylalanine, solvent isotope shifts between H_2O with D_2O were used to characterize the native state and a folding intermediate along the heat-denaturation pathway. The folding intermediate was found to possess less water within the hydrophobic core and an increased hydrophobicity, compared to the protein in its native folded state (Kitevski-Leblanc et al., 2013).

5.3 Protein Aggregation

The high sensitivity and large chemical shift dispersion of the ^{19}F nucleus has also been exploited to monitor protein conformational changes during aggregation of proteins. In a study of α-synuclein, a protein associated with Parkinson's disease, amino acid type-specific incorporation of m-fluoro tyrosine and ^{19}F NMR measurements were used to examine the conformational changes of the protein in the presence of urea, spermine, or sodium dodecyl sulfate (SDS) (Li et al., 2009). SDS and spermine were found to accelerate protein aggregation in comparison to buffer alone, and no soluble low-molecular-weight intermediates were detected. ^{19}F NMR experiments have also been used in real-time measurements of amyloid formation of the islet amyloid polypeptide. This peptide was prepared by solid-phase synthesis with incorporation of 4-(trifluoromethyl)-phenylalanine at position 23 (Suzuki et al., 2012). The ^{19}F resonance of the ^{19}F-modified residue was followed from the solvent-exposed environment in the monomer to the buried environment in the aggregated state. Soluble intermediates were not detected. This is in contrast to a similar study conducted with the Alzheimer's Aβ$_{1-40}$ peptide, for which several oligomeric species were detected in real time during the lag phase of fibril formation (Suzuki et al., 2013).

5.4 Protein–Lipid Interactions

^{19}F NMR spectroscopy has also been used to investigate protein–lipid interactions. The binding of α-synuclein to small unilamellar vesicles (SUVs) was probed by ^{19}F NMR. In this study, two labeling strategies were used: (i) amino acid type-specific incorporation of 3-fluoro tyrosine and (ii) site-specific incorporation of 4-(trifluoromethyl)-phenylalanine. By exploiting the large differences in resonance line width and signal intensity in the presence and absence of SUVs, the authors were able to quantify the fraction of membrane-bound protein. In addition, they were able to monitor the effect of lipid composition on the interaction between α-synuclein and the membrane, and found that both, the head group and the acyl chain of the SUV, affect the affinity (Wang, Li, & Pielak, 2010). In follow-up studies by ^{19}F NMR and fluorescence, interactions with large unilamellar vesicles, whose composition is similar to that of the inner and outer mitochondrial membrane, were probed. From these studies, the strength of α-synuclein binding was found to depend on the presence of cardiolipin, a component of the inner mitochondrial membrane (Zigoneanu, Yang, Krois, Haque, &

Pielak, 2012). ^{19}F NMR has also been used to examine the emersion depth, secondary structure, and topology of the first transmembrane segment of the membrane protein diacyl glycerol kinase (DAGK). In this case, DAGK was modified posttranslationally by covalent addition of 3-bromo-1,1,1-trifluoropropanone, and oxygen, at a partial pressure of 100 atm, was used to obtain position-dependent, oxygen-induced, paramagnetic shifts of the fluorinated probes (Luchette, Prosser, & Sanders, 2002).

6. NMR EXPERIMENTS WITH ^{19}F-CONTAINING LIGANDS

Ligand-observe NMR experiments monitor resonance perturbations of a ^{19}F-containing ligand upon protein binding. Induced spectral changes are caused by differences in the apparent molecular mass of the ligand in the free state versus the bound state and by the chemical environment. In general, ligand-observe experiments are faster than protein-observe experiments and require less protein. These factors make ^{19}F-containing ligand-observe solution NMR experiments well suited to monitor protein–ligand interactions. Here, we discuss some experiments that use ^{19}F-containing ligands to monitor protein–ligand binding.

6.1 Monitoring Line Broadening

Line broadening in the presence of the protein is often an indicator of protein–ligand interactions. The degree of line broadening depends on many factors, including the transverse relaxation rate, the exchange rate, and the fraction of the ligand in the free and bound state. The ^{19}F nucleus provides distinct advantages over the ^{1}H nucleus for binding experiments. In ^{19}F NMR experiments, transverse relaxation, which is dominated by the large ^{19}F CSA, produces a large difference in resonance line width between the free ligand and the protein-bound ligand. As a result, ^{19}F-ligand-observe NMR experiments become more sensitive as the molecular mass of the protein increases. Moreover, in ^{19}F NMR, essentially no background signals interfere with resonance detection, therefore, no solvent suppression is needed. In addition, the fast- and slow-exchange regime on the NMR chemical shift scale is heavily biased toward slow exchange. For equivalent binding affinities, ^{1}H NMR experiments, may run into difficulties since resonance broadening beyond detection in the intermediate exchange regime can ensue. This is less problematic on the ^{19}F chemical shift scale since the overall shift range is significantly larger.

In an elegant study, Goudreau and colleagues characterized a series of benzodiazepine inhibitors that target the viral capsid protein (CA) of HIV-1, using a multipronged approach employing X-ray crystallography, and NMR (Goudreau et al., 2013). In particular, they used ^{19}F NMR to assess binding of inhibitors to HIV-1 Gag and subfragments such as, CA-Nucleocapsid (CA-NC). Two resonances for the free and bound inhibitor were observed in slow exchange on the ^{19}F chemical shift scale, consistent with the submicromolar activity ($IC_{50} = 0.89$ μM), measured using an *in vitro* capsid assembly assay. Inhibitor specificity was investigated in two ways: (i) by monitoring the resonance line shape of the ^{19}F-containing benzodiazepine inhibitor upon addition of the RSV Gag protein (a protein not interacting with the benzodiazepine inhibitor series). As expected, the ^{19}F resonance did not experience any noticeable chemical shift perturbation, and (ii) by investigating the effects of HIV-1 CA-NC binding on the ^{19}F resonance of a less potent inhibitor. The ^{19}F spectrum of this inhibitor contained only one single average broad peak, which indicated that binding was in fast exchange, consistent with the inhibitor's weaker potency ($IC_{50} = 33$ μM). ^{19}F experiments were also used to evaluate the reversibility of inhibitor binding, in order to exclude the possibility that covalent protein adducts are formed. This was achieved using competition experiments (described in more detail in the Section 6.2 later). In these experiments, an unlabeled inhibitor was titrated into a sample containing CA-NC and the ^{19}F benzodiazepine inhibitor. The single ^{19}F resonance of the CA-NC/^{19}F benzodiazepine complex disappeared concomitantly with the appearance of the free ^{19}F inhibitor signal.

6.2 Competition-Based Experiments

In competition-based experiments, two ligands that bind competitively to the protein are used. The first ligand that is added to the protein is a "spy" ^{19}F-containing molecule with a known weak-binding affinity and binding site. The "spy" ^{19}F resonance is then monitored during the addition of a second unlabeled ligand (with a higher affinity for the protein), which displaces the "spy" ^{19}F-containing molecule. The analysis of changes in the monitored resonance can be used to extract affinity constants for the unlabeled target-ligand. In addition, the binding site of the unlabeled target-ligand is confirmed, since nonspecific binders do not displace the "spy" ^{19}F-containing molecule. These experiments have been successfully carried out on ^{19}F-containing small molecules and ^{19}F-containing carbohydrates (Dalvit et al., 2005; Matei et al., 2013).

6.3 ^{19}F NMR-Based Biochemical Screening

This method employs a ^{19}F-modified enzyme substrate and its cognate unlabeled enzyme with the ^{19}F-modifed substrate resonance being monitored as it transitions from its initial state to its enzyme-modified state. In these experiments, titration with an inhibitor of the enzyme permits one to determine the IC$_{50}$ value (the inhibitor concentration at which 50% inhibition of the reaction is observed), based on the difference in the ^{19}F resonance intensity in the initial state and the enzyme-modified state. Such experiments can be modified for other purposes, including simultaneous screening against multiple proteins, determining the function of a newly sequenced protein, and for screening in living cells or cell extracts (Dalvit, Ardini, Fogliatto, Mongelli, & Veronesi, 2004; Dalvit et al., 2003; Forino et al., 2005; Frutos, Tarrago, & Giralt, 2006; Lambruschini et al., 2013; Veronesi et al., 2014).

6.4 Magnetization Transfer Experiments

Structural information on protein–ligand complexes can be obtained from nuclear Overhauser effects (NOEs), such as heteronuclear ^{19}F–^{1}H NOEs between a ^{19}F-containing ligand and protein hydrogens, and homonuclear ^{19}F–^{19}F NOEs between a ^{19}F-containing ligand and ^{19}F-modified protein. The ability to obtain protein–ligand constraints was demonstrated in a study of the antiapoptotic protein Bcl-xL, a drug target for cancer therapy. In this study, Yu and colleagues (Yu et al., 2006) were able to obtain heteronuclear NOEs between uniformly ^{13}C-labeled Bcl-xL and a ^{19}F-containing small molecule using ^{1}H–^{13}C HMQC experiments with and without ^{19}F-resonance saturation. In addition, structural constrains were derived, monitoring magnetization exchange between a ^{19}F-containing ligand and ^{19}F-modified Bcl-xL. Using these data, the researchers were able to identify intermolecular contacts between the ligand and the protein that were consistent with known high-resolution crystal structures of Bcl-xL in complex with similar small molecules.

ACKNOWLEDGMENTS

We would like to thank Dr. Ryan Mehl for providing the pDule2 RS vector and for his support, help, and guidance when setting up the methodology for site-specifically labeling proteins with the orthogonal amber tRNA/tRNA synthetase pair in our laboratory. We would also like to acknowledge Mike Delk for NMR technical support and the members of the Gronenborn laboratory for their invaluable insight and advice. We would like to specifically thank Drs. Rieko Ishima and Elena Matei for many useful discussions and help related to ^{19}F NMR. In addition, we thank Drs. Matthew Whitley, Teresa Brosentisch, and Christopher Barnes for critical reading of the chapter.

REFERENCES

Aramini, J. M., Hamilton, K., Ma, L.-C., Swapna, G. V. T., Leonard, P. G., Ladbury, J. E., et al. (2014). ^{19}F NMR reveals multiple conformations at the dimer interface of the non-structural protein 1 effector domain from influenza A virus. *Structure*, *22*(4), 515–525. http://dx.doi.org/10.1016/j.str.2014.01.010.

Arseniev, A. S., Kuryatov, A. B., Tsetlin, V. I., Bystrov, V. F., Ivanov, V. T., & Ovchinnikov, Y. A. (1986). ^{19}F NMR study of 5-fluorotryptophan-labeled bacteriorhodopsin. *FEBS Letters*, *213*(2), 283–288.

Campos-Olivas, R., Aziz, R., Helms, G. L., Evans, J. N. S., & Gronenborn, A. M. (2002). Placement of ^{19}F into the center of GB1: Effects on structure and stability. *FEBS Letters*, *517*(1–3), 55–60.

Cavanagh, J., Fairbrother, W. J., Palmer, A. G., III, Rance, M., & Skelton, N. J. (2007). *Protein NMR spectroscopy: Principles and practice* (pp. 1–28). Burlington, MA: Elsevier Academic Press.

Cellitti, S. E., Jones, D. H., Lagpacan, L., Hao, X., Zhang, Q., Hu, H., et al. (2008). In vivo incorporation of unnatural amino acids to probe structure, dynamics, and ligand binding in a large protein by nuclear magnetic resonance spectroscopy. *Journal of the American Chemical Society*, *130*(29), 9268–9281. http://dx.doi.org/10.1021/ja801602q.

Chaiken, I. M., Freedman, M. H., Lyerla, J. R. J., & Cohen, J. S. (1973). Preparation and studies of ^{19}F-labeled and enriched ^{13}C-labeled semisynthetic ribonuclease-S' analogues. *Journal of Biological Chemistry*, *248*(3), 884–891.

Chen, H., Viel, S., Ziarelli, F., & Peng, L. (2013). ^{19}F NMR: A valuable tool for studying biological events. *Chemical Society Reviews*, *42*(20), 7971–7982. http://dx.doi.org/10.1039/c3cs60129c.

Clore, G. M., Gronenborn, A. M., Birdsall, B., Feeney, J., & Roberts, G. C. (1984). ^{19}F-NMR studies of 3′,5′-difluoromethotrexate binding to Lactobacillus casei dihydrofolate reductase. Molecular motion and coenzyme-induced conformational changes. *Biochemical Journal*, *217*(3), 659–666.

Crowley, P. B., Kyne, C., & Monteith, W. B. (2012). Simple and inexpensive incorporation of ^{19}F-tryptophan for protein NMR spectroscopy. *Chemical Communications*, *48*(86), 10681–10683. http://dx.doi.org/10.1039/c2cc35347d.

Dalvit, C., Ardini, E., Flocco, M., Fogliatto, G. P., Mongelli, N., & Veronesi, M. (2003). A general NMR method for rapid, efficient, and reliable biochemical screening. *Journal of the American Chemical Society*, *125*(47), 14620–14625. http://dx.doi.org/10.1021/ja038128e.

Dalvit, C., Ardini, E., Fogliatto, G. P., Mongelli, N., & Veronesi, M. (2004). Reliable high-throughput functional screening with 3-FABS. *Drug Discovery Today*, *9*(14), 595–602.

Dalvit, C., Mongelli, N., Papeo, G., Giordano, P., Veronesi, M., Moskau, D., et al. (2005). Sensitivity improvement in ^{19}F NMR-based screening experiments: Theoretical considerations and experimental applications. *Journal of the American Chemical Society*, *127*(38), 13380–13385. http://dx.doi.org/10.1021/ja0542385.

Dalvit, C., & Vulpetti, A. (2011). Fluorine–protein interactions and ^{19}F NMR isotropic chemical shifts: An empirical correlation with implications for drug design. *ChemMedChem*, *6*(1), 104–114. http://dx.doi.org/10.1002/cmdc.201000412.

Danielson, M. A., & Falke, J. J. (1996). Use of ^{19}F NMR to probe protein structure and conformational changes. *Annual Review of Biophysics and Biomolecular Structure*, *25*(1), 163–195. http://dx.doi.org/10.1146/annurev.bb.25.060196.001115.

Didenko, T., Liu, J. J., Horst, R., Stevens, R. C., & Wüthrich, K. (2013). Fluorine-19 NMR of integral membrane proteins illustrated with studies of GPCRs. *Current Opinion in Structural Biology*, *23*(5), 740–747. http://dx.doi.org/10.1016/j.sbi.2013.07.011.

Dominguez, M. A., Thornton, K. C., Melendez, M. G., & Dupureur, C. M. (2001). Differential effects of isomeric incorporation of fluorophenylalanines into PvuII endonuclease. *Proteins, 45*(1), 55–61.
Drake, S. K., Bourret, R. B., Luck, L. A., Simon, M. I., & Falke, J. J. (1993). Activation of the phosphosignaling protein CheY. I. Analysis of the phosphorylated conformation by ^{19}F NMR and protein engineering. *Journal of Biological Chemistry, 268*(18), 13081–13088.
Evanics, F., Kitevski, J. L., Bezsonova, I., Forman-Kay, J., & Prosser, R. S. (2006). ^{19}F NMR studies of solvent exposure and peptide binding to an SH3 domain. *Biochimica et Biophysica Acta, 1770*(2), 221–230. http://dx.doi.org/10.1016/j.bbagen.2006.10.017.
Fielding, L. (2007). NMR methods for the determination of protein–ligand dissociation constants. *Progress in Nuclear Magnetic Resonance Spectroscopy, 51*(4), 219–242. http://dx.doi.org/10.1016/j.pnmrs.2007.04.001.
Fischer, M., Schott, A.-K., Kemter, K., Feicht, R., Richter, G., Illarionov, B., et al. (2003). Riboflavin synthase of Schizosaccharomyces pombe. Protein dynamics revealed by ^{19}F NMR protein perturbation experiments. *BMC Biochemistry, 4*(1), 1–19. http://dx.doi.org/10.1186/1471-2091-4-18.
Forino, M., Johnson, S., Wong, T. Y., Rozanov, D. V., Savinov, A. Y., Li, W., et al. (2005). Efficient synthetic inhibitors of anthrax lethal factor. *Proceedings of the National Academy of Sciences, 102*(27), 9499–9504. http://dx.doi.org/10.1073/pnas.0502733102.
Frieden, C., Hoeltzli, S. D., & Bann, J. G. (2004). The preparation of 19F-labeled proteins for NMR studies. *Methods in Enzymology, 380*, 400–415. http://dx.doi.org/10.1016/S0076-6879(04)80018-1.
Frutos, S., Tarrago, T., & Giralt, E. (2006). A fast and robust ^{19}F NMR-based method for finding new HIV-1 protease inhibitors. *Bioorganic & Medicinal Chemistry Letters, 16*(10), 2677–2681. http://dx.doi.org/10.1016/j.bmcl.2006.02.031.
Gerig, J. T. (1994). Fluorine NMR of proteins. *Progress in Nuclear Magnetic Resonance Spectroscopy, 26*, 293–370.
Ghitti, M., Musco, G., & Spitaleri, A. (2014). NMR and computational methods in the structural and dynamic characterization of ligand-receptor interactions. *Advances in Experimental Medicine and Biology, 805*, 271–304. http://dx.doi.org/10.1007/978-3-319-02970-2_12.
Goudreau, N., Coulombe, R., & Faucher, A. M. (2013). Monitoring binding of HIV-1 capsid assembly inhibitors using ^{19}F ligand- and ^{15}N protein-based NMR and X-ray crystallography: Early hit validation of a benzodiazepine series. *ChemMedChem, 8*, 405–414.
Hammill, J. T., Miyake-Stoner, S., Hazen, J. L., Jackson, J. C., & Mehl, R. A. (2007). Preparation of site-specifically labeled fluorinated proteins for ^{19}F-NMR structural characterization. *Nature Protocols, 2*(10), 2601–2607. http://dx.doi.org/10.1038/nprot.2007.379.
Hoeltzli, S. D., & Frieden, C. (1994). ^{19}F NMR spectroscopy of [6-^{19}F]tryptophan-labeled *Escherichia coli* dihydrofolate reductase: Equilibrium folding and ligand binding studies. *Biochemistry, 33*(18), 5502–5509.
Horst, R., Liu, J. J., Stevens, R. C., & Wüthrich, K. (2013). β$_2$-adrenergic receptor activation by agonists studied with ^{19}F NMR spectroscopy. *Angewandte Chemie International Edition, 52*, 10762–10765. http://dx.doi.org/10.1002/anie.201305286.
Hortin, G., & Boime, I. (1983). Applications of amino acid analogs for studying co- and post-translational modifications of proteins. *Methods in Enzymology, 96*, 777–784. http://dx.doi.org/10.1016/S0076-6879(83)96065-2.
Hyean-Woo, L., Sohn, J. H., Byung, Y., Jong-Whan, C., Jung, S., & Hyun-Won, K. (2000). ^{19}F NMR investigation of F1-ATPase of *Escherichia coli* using fluorotryptophan labeling. *Journal of Biochemistry, 127*(6), 1053–1056.
Kawahara, K., Nemoto, N., Motooka, D., Nishi, Y., Doi, M., Uchiyama, S., et al. (2012). Polymorphism of collagen triple helix revealed by ^{19}F NMR of model peptide [Pro-4

(R)-hydroxyprolyl-Gly] 3-[Pro-4(R)-fluoroprolyl-Gly]-[Pro-4(R)-hydroxyprolyl-Gly] 3. *The Journal of Physical Chemistry B, 116*(23), 6908–6915. http://dx.doi.org/10.1021/jp212631q.

Khan, F., Kuprov, I., Craggs, T. D., Hore, P. J., & Jackson, S. E. (2006). ^{19}F NMR studies of the native and denatured states of green fluorescent Protein. *Journal of the American Chemical Society, 128*(33), 10729–10737. http://dx.doi.org/10.1021/ja060618u.

Kim, H. W., Perez, J. A., Ferguson, S. J., & Campbell, I. D. (1990). The specific incorporation of labelled aromatic amino acids into proteins through growth of bacteria in the presence of glyphosate. Application to fluorotryptophan labelling to the H^+-ATPase of *Escherichia coli* and NMR studies. *FEBS Letters, 272*(1-2), 34–36.

Kitevski-Leblanc, J. L., Al-Abdul-Wahid, M. S., & Prosser, R. S. (2009). A mutagenesis-free approach to assignment of ^{19}F NMR resonances in biosynthetically labeled proteins. *Journal of the American Chemical Society, 131*(6), 2054–2055. http://dx.doi.org/10.1021/ja8085752.

Kitevski-Leblanc, J. L., Hoang, J., Thach, W., Larda, S. T., & Prosser, R. S. (2013). ^{19}F NMR Studies of a desolvated near-native protein folding intermediate. *Biochemistry, 52*(34), 5780–5789. http://dx.doi.org/10.1021/bi4010057.

Kitevski-Leblanc, J. L., & Prosser, R. S. (2012). Current applications of ^{19}F NMR to studies of protein structure and dynamics. *Progress in Nuclear Magnetic Resonance Spectroscopy, 62*, 1–33. http://dx.doi.org/10.1016/j.pnmrs.2011.06.003.

Kiviniemi, A., & Virta, P. (2010). Characterization of RNA invasion by ^{19}F NMR spectroscopy. *Journal of the American Chemical Society, 132*(25), 8560–8562. http://dx.doi.org/10.1021/ja1014629.

Lambruschini, C., Veronesi, M., Romeo, E., Garau, G., Bandiera, T., Piomelli, D., et al. (2013). Development of fragment-based n-FABS NMR screening applied to the membrane enzyme FAAH. *ChemBioChem, 14*(13), 1611–1619. http://dx.doi.org/10.1002/cbic.201300347.

Leung, E. W. W., Yagi, H., Harjani, J. R., Mulcair, M. D., Scanlon, M. J., Baell, J. B., et al. (2014). ^{19}F NMR as a probe of ligand interactions with the iNOS binding site of SPRY domain-containing SOCS box protein 2. *Chemical Biology & Drug Design, 84*(5), 616–625. http://dx.doi.org/10.1111/cbdd.12355.

Li, C., Lutz, E. A., Slade, K. M., Ruf, R. A. S., Wang, G.-F., & Pielak, G. J. (2009). ^{19}F NMR studies of α-synuclein conformation and fibrillation. *Biochemistry, 48*(36), 8578–8584. http://dx.doi.org/10.1021/bi900872p.

Liu, J. J., Horst, R., Katritch, V., Stevens, R. C., & Wuthrich, K. (2012). Biased signaling pathways in β$_2$-adrenergic receptor characterized by ^{19}F-NMR. *Science, 335*(6072), 1106–1110. http://dx.doi.org/10.1126/science.1215802.

Luchette, P. A., Prosser, R. S., & Sanders, C. R. (2002). Oxygen as a paramagnetic probe of membrane protein structure by cysteine mutagenesis and ^{19}F NMR spectroscopy. *Journal of the American Chemical Society, 124*(8), 1778–1781.

Luck, L. A., & Falke, J. J. (1991). ^{19}F NMR studies of the D-galactose chemosensory receptor. 1. Sugar binding yields a global structural change. *Biochemistry, 30*(17), 4248–4256.

Marsh, E. N. G., & Suzuki, Y. (2014). Using ^{19}F NMR to probe biological interactions of proteins and peptides. *ACS Chemical Biology, 9*(6), 1242–1250. http://dx.doi.org/10.1021/cb500111u.

Matei, E., André, S., Glinschert, A., Infantino, A. S., Oscarson, S., Gabius, H.-J., et al. (2013). Fluorinated carbohydrates as lectin ligands: Dissecting glycan–cyanovirin interactions by using ^{19}F NMR spectroscopy. *Chemistry, 19*(17), 5364–5374. http://dx.doi.org/10.1002/chem.201204070.

Muller, K., Faeh, C., & Diederich, F. (2007). Fluorine in pharmaceuticals: Looking beyond intuition. *Science, 317*(5846), 1881–1886. http://dx.doi.org/10.1126/science.1131943.

Noren, C. J., Anthony-Cahill, S. J., Griffith, M. C., & Schultz, P. G. (1989). A general-method for site-specific incorporation of unnatural amino-acids into proteins. *Science*, *244*(4901), 182–188.
Oppenheimer, N. J. (1989). Sample preparation. *Methods in Enzymology*, *176*, 78–89.
Peeler, J. C., & Mehl, R. A. (2011). Site-specific incorporation of unnatural amino acids as probes for protein conformational changes. *Methods in Molecular Biology*, *794*, 125–134. http://dx.doi.org/10.1007/978-1-61779-331-8_8.
Rule, G. S., & Hitchenns, T. K. (2006). NMR spectroscopy. In R. Kaptein (Ed.), *Vol. 5. Fundamentals of protein NMR spectroscopy* (pp. 1–27). Dordrecht, Netherlands: Springer.
Rydzik, A. M., Brem, J., van Berkel, S. S., Pfeffer, I., Makena, A., Claridge, T. D. W., et al. (2014). Monitoring conformational changes in the NDM-1 metallo-β-lactamase by ^{19}F NMR spectroscopy. *Angewandte Chemie International Edition*, *53*(12), 3129–3133. http://dx.doi.org/10.1002/anie.201310866.
Salopek-Sondi, B., & Luck, L. A. (2002). ^{19}F NMR study of the leucine-specific binding protein of *Escherichia coli*: Mutagenesis and assignment of the 5-fluorotryptophan-labeled residues. *Protein Engineering*, *15*(11), 855–859.
Studier, F. W. (2005). Protein production by auto-induction in high density shaking cultures. *Protein Expression and Purification*, *41*(1), 207–234.
Suzuki, Y., Brender, J. R., Hartman, K., Ramamoorthy, A., & Marsh, E. N. G. (2012). Alternative pathways of human islet amyloid polypeptide aggregation distinguished by ^{19}F nuclear magnetic resonance-detected kinetics of monomer consumption. *Biochemistry*, *51*(41), 8154–8162. http://dx.doi.org/10.1021/bi3012548.
Suzuki, Y., Brender, J. R., Soper, M. T., Krishnamoorthy, J., Zhou, Y., Ruotolo, B. T., et al. (2013). Resolution of oligomeric species during the aggregation of Aβ 1–40 using ^{19}F NMR. *Biochemistry*, *52*(11), 1903–1912. http://dx.doi.org/10.1021/bi400027y.
Sykes, B. D., & Hull, W. E. (1978). Fluorine nuclear magnetic resonance studies of proteins. *Methods in Enzymology*, *49*, 270–295. http://dx.doi.org/10.1016/S0076-6879(78)49015-9.
Tanabe, K., Sugiura, M., & Nishimoto, S.-I. (2010). Monitoring of duplex and triplex formation by ^{19}F NMR using oligodeoxynucleotides possessing 5-fluorodeoxyuridine unit as ^{19}F signal transmitter. *Bioorganic & Medicinal Chemistry*, *18*(18), 6690–6694. http://dx.doi.org/10.1016/j.bmc.2010.07.066.
Veronesi, M., Romeo, E., Lambruschini, C., Piomelli, D., Bandiera, T., Scarpelli, R., et al. (2014). Fluorine NMR-based screening on cell membrane extracts. *ChemMedChem*, *9*(2), 286–289. http://dx.doi.org/10.1002/cmdc.201300438.
Wang, G.-F., Li, C., & Pielak, G. J. (2010). ^{19}F NMR studies of α-synuclein-membrane interactions. *Protein Science*, *19*(9), 1686–1691. http://dx.doi.org/10.1002/pro.449.
Yamamoto, Y., Nagao, S., Hirai, Y., Tai, H., & Suzuki, A. (2012). Field-dependent ^{19}F NMR study of sperm whale myoglobin reconstituted with a ring-fluorinated heme. *Polymer Journal*, *44*(8), 907–912. http://dx.doi.org/10.1038/pj.2012.60.
Yu, L., Hajduk, P. J., Mack, J., & Olejniczak, E. T. (2006). Structural studies of Bcl-xL/ligand complexes using ^{19}F NMR. *Journal of Biomolecular NMR*, *34*(4), 221–227. http://dx.doi.org/10.1007/s10858-006-0005-y.
Yu, J.-X., Hallac, R. R., Chiguru, S., & Mason, R. P. (2013). New frontiers and developing applications in ^{19}F NMR. *Progress in Nuclear Magnetic Resonance Spectroscopy*, *70*(C), 25–49. http://dx.doi.org/10.1016/j.pnmrs.2012.10.001.
Zigoneanu, I. G., Yang, Y. J., Krois, A. S., Haque, M. E., & Pielak, G. J. (2012). Interaction of α-synuclein with vesicles that mimic mitochondrial membranes. *Biochimica et Biophysica Acta—Biomembranes*, *1818*(3), 512–519.

CHAPTER FIVE

Biopolymer Deuteration for Neutron Scattering and Other Isotope-Sensitive Techniques

Robert A. Russell*,[†], Christopher J. Garvey*, Tamim A. Darwish*, L. John R. Foster[†], Peter J. Holden*,[†],[1]

*National Deuteration Facility, Bragg Institute, Australian Nuclear Science and Technology Organisation, New South Wales, Australia
[†]Bio/Polymer Research Group, Centre for Advanced Macromolecular Design, School of Biotechnology & Biomolecular Science, University of New South Wales, Sydney, New South Wales, Australia
[1]Corresponding author: e-mail address: phx@ansto.gov.au

Contents

1. Introduction 98
2. Deuterated Biopolyesters 100
   2.1 Poly-3-Hydroxybutyrate Properties and Characterization 100
   2.2 Production of Deuterated PHB 101
   2.3 Polymer Extraction and Analysis 102
   2.4 Poly-3-Hydroxyoctanoate Biodeuteration for Characterization Studies 103
   2.5 PHO Production and Deuteration 103
   2.6 Selective Deuteration to Probe Metabolic Pathways 104
3. Deuterated Chitosan 106
   3.1 Chitosan from *Pichia pastoris* 106
   3.2 Deuteration of *P. pastoris* and Chitosan 107
   3.3 Biomass Production and Deuterium Adaptation 107
   3.4 Chitosan Extraction and Purification 109
   3.5 Characterization of Chitosan Biodeuteration 110
   3.6 Final Remarks 111
4. Deuterated Cellulose 111
   4.1 Organization and Biosynthesis of Bacterial Cellulose 111
   4.2 Production and Characterization of BC 113
   4.3 Final Remarks 117
Acknowledgments 117
References 118

Abstract

The use of microbial biosynthesis to produced deuterated recombinant proteins is a well-established practice in investigations of the relationship between molecular structure and function using neutron scattering and nuclear magnetic resonance spectroscopy. However, there have been few reports of using microbial synthetic capacity to produce labeled native biopolymers. Here, we describe methods for the production of deuterated polyhydroxyalkanoate biopolyesters in bacteria, the polysaccharide chitosan in the yeast *Pichia pastoris,* and cellulose in the bacterium *Gluconacetobacter xylinus*. The resulting molecules offer not only multiple options in creating structural contrast in polymer blends and composites in structural studies but also insight into the biosynthetic pathways themselves.

1. INTRODUCTION

Biopolymers have attracted considerable interest focussed on developing applications of native or composite materials across a range of outcomes including tissue engineering, drug delivery, food additives, aerogels and thin films for sensors and surface coatings, environmentally friendly replacements for petrochemical plastics in consumer packaging, and composite materials with desirable properties for a broad range of nanotech applications.

The potential of such materials, however, generates a strong need for physical characterization and structural elucidation. Investigation of composite materials frequently requires substantial quantities of labeled molecules in order to obtain structural information using techniques such as neutron scattering, nuclear magnetic resonance (NMR) spectroscopy, and vibrational spectroscopy, which are sensitive to labeling with the stable isotope deuterium (^{2}H). Neutron scattering techniques such as small-angle neutron scattering (SANS) and neutron reflectometry (NR) are particularly suited to providing physical and structural information at a nanoscale on polymer or composite samples. Data analysis gives averaged (statistical) information on parameters related to the spatial distribution of nuclei in a sample (scattering length density (SLD) correlation function) to yield simple parameters such as density, roughness, depth of thin-film layers, and spatial relationship between different sample components. The high penetration of neutrons enables real-time experiments on the effects of heating, humidity, pressure, and other environmental parameters. As SANS and NR rely on the probability of the coherent interaction of neutrons with the nuclei in a sample, the ability to replace light hydrogen (^{1}H) with deuterium (^{2}H) enables provision of contrast in samples with similar distributions of nuclei (chemistry) and lowers the incoherent background signal (Sears, 1992). These techniques are inherently low resolution and do not have requirements for highly site-specific labeling of nuclei.

By contrast, ^{2}H spectroscopy NMR provides very local information on the environment around nuclei via the nuclear quadrupolar interaction. The approach has long been used with solid-state synthetic polymers as a probe of orientation and dynamics, where site-specific deuteration and line shape simulation elucidate very local information on the motions and average orientations (Schmidt-Rohr & Spiess, 1994). The advantages include the wide range of dynamics (from Hz to MHz), which are not accessible by other techniques. The challenge is to avoid complex line shapes from multiple sites of deuteration which will need to be deconvolved.

By virtue of the extra mass of ^{2}H and the lack of significant perturbation to the chemical bond, substitution with ^{2}H will have a well-defined effect on the characteristic frequency (Vasko, Blackwell, & Koenig, 1972). This is useful for assignment of unknown spectroscopic peaks and coupling between vibrational states. FT-IR microscopy of polymer blends, combined with deuterium labeling of otherwise chemically similar components, has been used to investigate phase separation, facilitated by nonoverlapping IR stretching vibrations of their carbon–hydrogen and carbon–deuterium alkyl chains (C–D stretch, 2000–2250 cm^{-1}; C–H stretch, 2800–3050 cm^{-1}). This avoids the problems associated with methods that historically rely on identifying peaks from different components that are overlapping in nature, as well as problems related to the unpredictable shifts in IR signals due to intermolecular interactions and crystalline properties of polymers (Russell et al., 2014).

The use of microbial synthesis adapted to production in D$_2$O-based media not only offers the opportunity to produce such labeled molecules but also may reveal additional information about the biosynthetic process itself. The use of stable isotopes in the tracing of even quite complicated metabolic pathways is well established. Both mass spectrometry and NMR analysis are particularly useful for localization of deuterium substitution and for accurate quantitation of overall deuteration of the product. Mass spectrometry analysis of polymer fragmentation patterns following separation by gas chromatography has also revealed information on deuterium incorporation into biopolymers based on composition of the growth medium, revealing the contribution of H from carbon sources versus water itself (Russell, Holden, Wilde, Hammerton, & Foster, 2007).

In order to maximize experimental outcomes, the pattern of molecular deuteration needs to be matched to the experimental techniques. For NR and SANS, the requirements are for control of net deuteration as a means to control the SLD during use in a contrast variation series. In the case of stable isotope tracing of metabolic pathways, a highly specific characterization is required. Such understanding may be used to produce the very specific deuteration required for ^{2}H solid-state NMR studies of polymers.

2. DEUTERATED BIOPOLYESTERS

2.1 Poly-3-Hydroxybutyrate Properties and Characterization

Polyhydroxyalkanoates (PHAs) are natural, linear, thermoplastic polyesters synthesized by microorganisms as intracellular carbon reserves in response to limited nutrient availability and stress conditions. The members of the PHA family encompass a wide range of mechanical properties, from hard and brittle poly(3-hydroxybutyrate) (PHB) to soft and elastomeric poly(3-hydroxyoctanoate) (PHO) (Fig. 1). To better understand their material properties, it is necessary to characterize not just chain length/molecular weight but also the polymer chain behavior. Elucidation of individual chain conformation can provide insight into the behavior of the polymer as a solid film. In SANS investigations of chain conformation in solution and in molten or solid states (Jouault et al., 2010), scattering patterns from dilute solutions can be fit to various models of polymer chains (Beaucage et al., 1997; Foster, Schwahn, Pipich, Holden, & Richter, 2008). The contrast between two polymers in solution can be manipulated by deuterium labeling, whereby the scattering from a labeled polymer can be matched to the solvent (H_2O, D_2O, or a mixture of both) effectively masking it to reveal the structure of the nonlabeled polymer.

Deuterated PHB (D-PHB) has been reported in a limited number of characterization studies. Beaucage et al. (1997) used the contrast provided by protonated versus deuterated isotactic PHB to characterize important polymer chain characteristics using SANS. Selective deuteration of

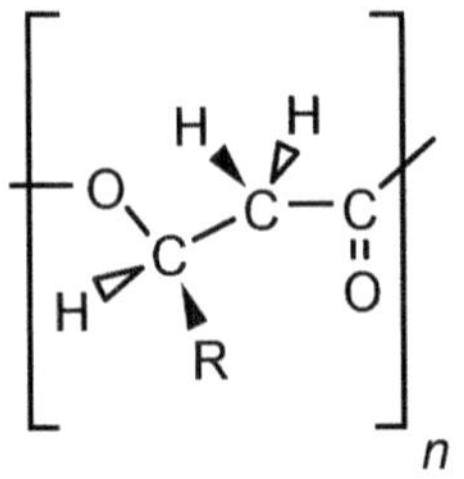

Figure 1 Structure of generic PHA monomer, where side chain R = C1–2 for short-chain length PHA and 4–9 for medium-chain length PHA.

exchangeable protons has also been reported for molecular dynamic analysis using NMR spectroscopy (Nozirov, Fojud, Szcześniak, & Jurga, 2000). *Cupriavidus necator* ATCC 17699 can be cultured to produce D-PHB from sodium d_3-acetate—an inexpensive deuterated carbon sources for bacterial growth. Partial deuteration, achieved using only deuterated carbon source in the growth medium, was used to determine deuterium distribution and isotope effect in isolated PHB by ^{1}H, ^{2}H, and ^{13}C NMR spectroscopy (Yoshie et al., 1992). Deuterated growth medium containing D_2O (up to 90%) and d_3-acetate produced D-PHB that was probed by ^{1}H NMR to determine deuterium distribution (Gross et al., 1992) and by SANS to determine polymer chain conformation in dilute solution (Yun, Russell, & Holden, 2008).

2.2 Production of Deuterated PHB

C. necator can be cultured in Luria-Bertani (LB) medium and streaked onto LB agar plates to obtain single colonies or stored as glycerol stocks at -80 °C. Biomass is cultured to sufficient cell density in LB medium, followed by polymer production in a minimal medium containing the desired carbon source.

LB medium: dissolve 10 g NaCl, 10 g peptone (pancreatic digest of casein), and 5 g yeast extract in water and autoclave at 121 °C for 20 min. To prepare agar plates or slopes, add 2% (w/v) agar prior to autoclaving.

Bring glycerol stock of *C. necator* to room temperature from -80 °C storage and inoculate 1 mL into 100 mL LB medium. Incubate overnight at 30 °C and 200 rpm to OD_{600nm} ~2. Confirm the absence of polymer inclusions by phase contrast microscopy (400× magnification or greater). Inoculate LB culture (1%, v/v) into a 2-L flask containing 1 L LB medium prepared with 90% D_2O:10% H_2O. Incubate until OD_{600nm} reaches 5–10 and confirm cells remain inclusion free. Oxygen limitation, which can induce PHB formation in the batch phase, can be alleviated by the use of baffled flasks to increase culture aeration, or culturing in bioreactors with dissolved oxygen (DO) controlled by stirrer speed. Harvest cells by centrifugation (4000 ×g at 4 °C for 30 min), discard supernatant, wash cells in water, and centrifuge to remove residual LB medium.

Prepare *nitrogen-free minimal medium* which promotes polymer formation from available carbon and energy source (Shi, Ashby, & Gross, 1996). Dissolve salts in 1 L D_2O and filter sterilize (0.2 μm membrane).

Na_2HPO_4	1.51 g
KH_2PO_4	26.5 g
$MgSO_4$	0.2 g
Sodium d_3-acetate (20 mM)	1.7 g

1 mL microelements solution (1000×):

$FeSO_4 \cdot 7H_2O$	2.78 g/L
$MnCl_2 \cdot 2H_2O$	1.16 g/L
$CoSO_4 \cdot 7H_2O$	2.81 g/L
$CaCl_2 \cdot 2H_2O$	1.67 g/L
$CuCl_2 \cdot 2H_2O$	0.17 g/L
$ZnSO_4 \cdot 7H_2O$	0.29 g/L

Resuspend and disperse the pellet in N-free medium and incubate (30 °C, 200 rpm). Formation of polymer inclusions may be observed after several hours' incubation, but may proceed for 1–2 days. Sample periodically and derivatize for PHB quantification by gas chromatography/mass spectrometry (GC–MS) (Russell et al., 2008). Harvest biomass-containing polymer inclusions by centrifugation and then lyophilize the frozen cell pellet prior to chloroform extraction of PHB.

2.3 Polymer Extraction and Analysis

Centrifuge samples (5–10 mL) taken during cell growth and polymer formation phases at 4000 × g for 20 min at 4 °C. Freeze the pellet prior to lyophilization for cell dry weight calculation and derivatization for GC–MS analysis. Methyl esters of the polymer are formed by refluxing the dry pellet with 3 mL chloroform and 3 mL acidified methanol (15% H_2SO_4 in methanol in a sealed tube at 110 °C for 3 h). Allow to cool and then mix 2 mL water into the solution resulting in separation of aqueous (upper) and organic (lower) fractions. Neutralize by washing with bicarbonate solution, remove aqueous phase, and then wash with water. The lower organic fraction containing the PHA methyl esters is transferred to a sample vial for quantitation and analysis by GC–MS (Russell et al., 2008).

2.4 Poly-3-Hydroxyoctanoate Biodeuteration for Characterization Studies

Deuteration has been shown to be a useful tool to probe the *in vivo* synthesis of biopolymers as well as the interaction between polymers when blended together. In the case of polymer biosynthesis, biodeuteration has been used for investigating the formation of membrane-bound polymer inclusions inside bacterial cells. Sequential feeding of chemically different polymer precursors has been used to infer core/shell arrangement of inclusion body formation from limited electron contrast under electron microscopy (Curley, Lenz, & Fuller, 1996). However, by adding hydrogenated followed by perdeuterated octanoic acid during batch–fed growth, ultra SANS conclusively demonstrated that inclusions with core/shell arrangement were obtained (Russell et al., 2008). This supported previous observations that polymer inclusions are built up in a structure from a central core to an outer shell.

2.5 PHO Production and Deuteration

PHO biopolyester is produced from octanoic acid via the β-oxidation pathway of various bacteria (Sudesh, Abe, & Doi, 2000). In *Pseudomonas oleovorans*, the PHO synthase family regulates biosynthesis of PHO which is promoted by unbalanced growth conditions, typically oxygen limitation, which can be induced by reducing stirrer speed in a bioreactor following batch growth.

P. oleovorans ATCC 29347 can be grown on modified E medium (Vogel & Bonner, 1956) with 20 mM octanoic acid as sole carbon source in the batch phase. Monitoring of DO bioreactor culture enables determination of carbon source depletion, typically seen by a rapid rise in DO. At this point, the stirrer speed is reduced, creating an oxygen-limited environment for PHO biosynthesis. Addition of carbon source (polymer precursor) during the fed stage will lead to a further drop in DO as the cell metabolism utilizes the available oxygen in the growth medium for polymer synthesis. Depletion of the carbon source produces a spike in DO as synthesis activity stops, and further precursor is added to accumulate polymer as DO is maintained below 20%.

Growth of *P. oleovorans* in a fed–batch fashion allows control of deuterium incorporation during cell and polymer production phases. Formation of partially or fully deuterated PHO can be achieved using deuterated carbon source and/or deuterated solvent (D_2O, 99.8% isotopic purity,

Sigma-Aldrich) in the growth medium. In order to produce the fully deuterated PHO monomer, the culture medium contained D_2O as the solvent, in addition to deuterated carbon source. Proton impurities from salts in the growth medium can be minimized by exchanging in D_2O prior to addition.

2.6 Selective Deuteration to Probe Metabolic Pathways

The contribution of solvent (H_2O/D_2O) versus precursor (fatty acid) deuterium to the formation of labeled biopolymer can be investigated by appropriate manipulation of growth and polymer production conditions followed by analysis using GC–MS or NMR spectroscopy. The GC–MS spectrum of the hydrogenated PHO monomer has an indicative major ion peak at m/z 103 (Russell et al., 2007). This ion fragment originates from the carbon backbone (rather than the side chain) and contains three hydrogen atoms that may be substituted by deuterium. Addition of d_{15}-octanoic acid as precursor produces a deuterated ion fragment m/z 105 indicative of deuterium substitution in all but one position. Addition of hydrogenated precursor to growth medium containing D_2O produces a fragment with a dominant m/z 104 peak corresponding to a single deuterium substitution in the main carbon chain, while perdeuterated PHO, indicated by the m/z 106 fragment, is obtained from growth in fully deuterated medium in both batch and fed phases (Russell et al., 2008).

NMR analysis does not rely on derivatization or fragmentation of the molecule and provides more accurate localization of deuterium substitution and quantitation of overall deuteration. PHO, as well as D-PHO extracted from biomass grown on d_{15}-octanoic acid and/or D_2O as described here, has been characterized using a combination of ^1H, ^2H (Russell et al., 2014), and ^{13}C NMR (reported here).

^{13}C NMR is able to identify whether H bonded to certain carbon atoms in a compound is completely or partially deuterated. In a standard $^{13}C\{^1H\}$ NMR (^{13}C with ^1H decoupled) experiment, ^{13}C NMR signals of carbons attached to protons appear as singlets, while carbon signals of carbons attached to deuteriums suffer from extensive splitting due to the C–D J coupling. To resolve the carbon signals where ^1H and/or ^2H atoms are attached, ^{13}C NMR $\{^1H, ^2H\}$ can be performed by decoupling both nuclei ^1H and ^2H. This shows carbon signals attached to protons and deuterium as singlets, where each is isotopically shifted (Fig. 2). ^{13}C signal shifts to lower frequency when more deuteriums are attached to it. Peaks around the 39 ppm region (circled) are expanded in Fig. 3 to highlight the characteristic NMR

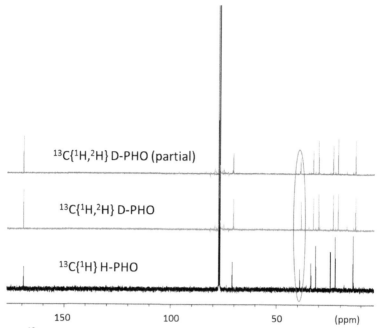

Figure 2 ^{13}C NMR in CDCl$_3$ from top to bottom: partially deuterated D-PHO (decoupling ^{1}H and ^{2}H), perdeuterated D-PHO (decoupling ^{1}H and ^{2}H), and completely protonated H-PHO (decoupling ^{1}H).

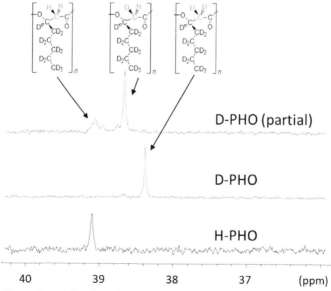

Figure 3 Expansion of Fig. 2 in the region of 39 ppm. (See the color plate.)

frequency region for the methylene carbon signal in PHO. The resonance shows an isotopic shift from 39.1 ppm in completely protonated PHO (CH_2) to 38.4 ppm in fully deuterated PHO (CD_2). Partially deuterated PHO (from bacteria grown in H_2O with deuterated carbon source) shows a signal at 38.8 ppm, which corresponds to partial deuteration of the methylene unit (CHD) and a small signal at 39.1 ppm, which is assigned to traces of completely protonated PHO. These results support earlier determination of deuterium substitution by GC–MS (Russell et al., 2008). The β-oxidation pathway is used by pseudomonads to convert fatty acids to PHAs (Sudesh et al., 2000), and the hydration of enoyl-CoA in this process is the step most likely responsible for deuterium substitution at the methylene carbon. This partial substitution at the methylene carbon atom is consistent with the findings of Yoshie et al. (1992), which suggests some stereoselectivity of deuterium substitution at this position.

3. DEUTERATED CHITOSAN

3.1 Chitosan from *Pichia pastoris*

Chitosan is a natural, widely abundant polysaccharide consisting of glucosamine (Glc) and *N*-acetylglucosamine (GlcNAc) subunits linked at β-1,4 positions into linear chains. Its structure, chemical properties, and biocompatibility have attracted enormous interest in exploring its biotechnological potential (Pillai, Paul, & Sharma, 2009). The proportion of Glc monomers gives the material its degree of deacetylation (DDA), and it is generally accepted that DDA greater than 60% is chitosan, being the approximate solubility limit in dilute acetic acid. DDA below this refers to the less soluble chitin, which forms the exoskeletons of crustaceans and insects and provides an abundant source material for the preparation of chitosan by deacetylation using routine alkaline treatment.

An alternative, naturally occurring source of chitosan is found in the cell walls of fungi, where it often coexists with chitin. Hyphal fungi such as the Mucorales typically yield ca. 7% chitosan from dry mycelia (Synowiecki & Al-Khateeb, 1997) and may be grown by solid-state or submerged (liquid) fermentation (Crestini, Kovac, & Giovannozzi-Sermanni, 1996). Nonhyphal yeasts such as *P. pastoris* have received little attention as a source of chitosan; however, a study of cell wall polysaccharides reported ca. 6% chitin content from dry cell mass (Chagas et al., 2009). While nonhyphal fungal cell walls may yield only 1–2% chitosan from dry biomass, the yield may be enhanced and the DDA tuned to a higher value by deacetylation of the chitin fraction (Trutnau, Suckale, Groeger, Bley, & Ondruschka, 2009).

3.2 Deuteration of *P. pastoris* and Chitosan

Deuteration of *P. pastoris* biomass has been reported for production of deuterated metabolites (Haon, Augé, Tropis, Milon, & Lindley, 1993; Massou et al., 1999), as well as for deuterated protein expression for NMR analysis (Morgan, Kragt, & Feeney, 2000). In 2014, biodeuteration of deuterated fatty acids including oleic acid was also reported (de Ghellinck et al., 2014).

The only reports of chitosan deuteration concern substitution of labile (solvent-exchangeable) protons using D_2O vapor (Smotrina & Smirnov, 2008; Valentin, Bonelli, Garrone, Di Renzo, & Quignard, 2007). Deuteration of the remaining, nonlabile, protons is desirable for the characterization of chitosan blended with other carbohydrate molecules. The structural characterization of such blends using FT-IR spectroscopy (Anicuta, Dobre, Stroescu, & Jipa, 2010) and neutron scattering techniques (Crompton, Forsythe, Horne, Finkelstein, & Knott, 2009; Wang, Qiu, Cosgrove, & Denbow, 2009) may be enhanced by contrasting deuterated chitosan against the other (protonated) molecules in the blend. Contrast can also be obtained using FT-IR spectroscopy, where the IR spectra show clear peak separation of the C–H and C–D stretch vibrations (Russell et al., 2014).

In order to produce deuterated chitosan, the methylotrophic yeast *P. pastoris* was selected for its ability to grow in liquid culture using methanol as sole carbon source. This organism has previously been grown on deuterated media for the production of deuterated sterols (Haon et al., 1993). In that study, a series of adaptation steps to increasing D_2O content were performed; however, the yield was a relatively low 16 g perdeuterated biomass/L using methanol as sole source of carbon. In the work reported here, high biomass yield of *P. pastoris* was achieved using a feed regime controlled by methanol concentration, yielding 101 g/L deuterated biomass which equates to 0.8 g deuterated biomass (wet weight)/g d_4-methanol.

3.3 Biomass Production and Deuterium Adaptation
3.3.1 Media Recipes
P. pastoris wild-type X-33 and recipes for yeast peptone dextrose (YPD) and yeast nitrogen base (YNB) media are sourced from the Invitrogen *Pichia*

Expression Kit User Manual (Invitrogen, Carlsbad, CA, USA). However, other recombinant strains used for expression of particular protein would likely have the same chitosan content in cell walls.

3.3.1.1 YPD Medium
Dissolve yeast extract 10 g, peptone 20 g, and dextrose (glucose) 20 g in 1 L of reverse osmosis water. Filter sterilize or autoclave at 121 °C for 20 min. For YPD agar plates, add 20 g/L agar to solution prior to autoclaving.

3.3.1.2 Buffered Minimal Methanol Medium
To 750 mL of water, add 100 mM potassium phosphate buffer–pH 6.0 (K_2HPO_4 and KH_2PO_4); 100 mL 10× YNB (13.4 g/100 mL of YNB [Invitrogen] filter sterilized); 2 mL 500× biotin (20 mg biotin in 100 mL water, and filter sterilized), and 5 mL of methanol. Adjust volume to 1 L and filter sterilize (0.22 μm).

For perdeuterated media, prepare stock solutions in D_2O and then flash evaporate to remove solvent-exchangeable protons. Redissolve solids in filter-sterilized D_2O at appropriate concentrations for stock solutions, purge with dry nitrogen and store in sealed containers.

3.3.2 *High Cell Density Growth of* P. pastoris *on Methanol*
Stock cultures of *P. pastoris* may be maintained on YPD agar plates or slopes stored at 4 °C, or as glycerol stocks stored at −80 °C. To prepare seed cultures, inoculate *P. pastoris* from glycerol stock or a single colony from YPD agar plate into 100 mL BMM (buffered minimal methanol) medium containing 0.5% (v/v) methanol. Incubate in a shaker–incubator at 30 °C and 220 rpm until OD_{600nm} ~1–5.

Inoculate seed culture into a bioreactor at 1% (v/v) containing BMM and grow at 30 °C to high cell density using a batch/fed regime. Maintain DO above 50% by controlling impeller speed, and pH above 5.0 by addition of ammonium hydroxide solution, which provides additional nitrogen to the culture medium. Methanol concentration can be maintained at 0.5% using a methanol probe (e.g., Raven Biotech Inc, USA) linked to a methanol dosing pump, to optimize the biomass yield.

3.3.3 *D_2O Adaptation and Production of Deuterated Biomass*
Adaptation to increasing D_2O in the culture medium is required, as inoculating directly into perdeuterated medium will produce inefficient or reduced

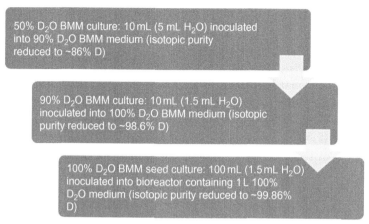

Figure 4 Carryover of water/proton depletion in D_2O adaptation subcultures (100 mL cultures, 10% inoculum transfer, theoretical 100% D_2O isotopic purity).

growth. The carryover of protonated culture medium in serial inoculations must also be considered when producing fully deuterated biomass (Fig. 4). Transfers of up to 10% (v/v) of an established culture (OD ∼1 in logarithmic growth phase) into fresh medium subcultures will be depleted in protons to less than 0.14%, following transfers described below. Proton transfer can be further reduced by performing additional subculture in 100% D_2O medium or growing to higher cell density and transferring 1% (v/v) inoculum.

Growth of *P. pastoris* in media with high D_2O concentration is much slower than in H_2O; however, a similar final cell concentration can be achieved (de Ghellinck et al., 2014). Doubling times of 8–10 h can be expected, compared with 2.5 h in H_2O. The biomass yield of *P. pastoris* grown on protonated BMM using the protocol above was 165 g/L wet cell pellet, OD_{600nm} 140. One gram biomass (wet weight) was obtained per gram methanol, whereas the yield in perdeuterated medium was slightly lower at 0.8 g wet wt/g d_4-methanol.

3.4 Chitosan Extraction and Purification

A modification of the widely used alkali/acid extraction of cell biomass was used for the extraction and purification of chitosan, based on methods of Trutnau et al. (2009) and Synowiecki and Al-Khateeb (1997):
- Freeze biomass and lyophilize, recording dry cell weight yield from wet cell pellet.

- *Deproteination*: Add dried biomass to 2 M NaOH at 1:30 (w/v), stirring at 70 °C for 3 h. Centrifuge (6000 × g, 15 min), wash pellet in water three times, freeze, and lyophilize.
- *Demineralization*: Add dried alkali-insoluble material to 1 M Tris–HCl (pH 7) +0.1 M EDTA at 1:10 (w/v), stirring at 70 °C for 3 h. Centrifuge (6000 × g, 15 min), wash pellet in water, freeze, and lyophilize.
- *Extraction and purification*: Add dried material to 10% acetic acid at 1:100 (w/v), stirring at 70 °C for 6 h. Centrifuge to collect acetic-soluble material, retaining the acetic-insoluble pellet for subsequent deacetylation. Filter (0.45 μm) the supernatant containing solubilized chitosan and then add 1 M NaOH until a cloudy precipitate appears around pH 6. Centrifuge (6000 × g for 15 min) to obtain the crude chitosan precipitate, wash in water, and repeat the acetic acid extraction to obtain purified chitosan.

The chitosan yield using this protocol is approximately 1% of cell dry weight. Yield optimization, as well as increased DDA of the final product, may be achieved by deacetylation of the alkali-insoluble material. Resuspend dry solids in 50% NaOH and reflux at 100–130 °C for 2 h. Wash solids in water to neutralize prior to acetic solubilization as described above (Trutnau et al., 2009).

3.5 Characterization of Chitosan Biodeuteration

Attenuated total reflection Fourier transform infrared spectroscopy (ATR FT-IR) experiments were performed using an FT-IR spectrometer (Nicolet 6700 Series) equipped with a single-reflection ATR and diamond crystal. The diamond ATR had a sampling area of approximately 0.5 mm^2, and the infrared spectra were collected at 4 cm^{-1} resolution over 128 scans. Samples were washed twice in D$_2$O and lyophilized prior to FT-IR analysis in order to reduce the dominant O–H/N–H peaks that occur between 2500 and 3500 cm^{-1} and allow the clear identification of C–H stretch (2800–3000 cm^{-1}) (Fig. 5). Deuteration of chitosan was determined by showing the absence of C–H stretch and appearance of C–D stretch at 2050–2250 cm^{-1} compared to protonated chitosan. The appearance of O–D/N–D peaks from 2300 to 2700 cm^{-1} in both protonated and deuterated samples was attributed to the exchange of the labile protons in deuterated solvents and did not interfere with identification of C–D stretching vibration (2000–2250 cm^{-1}).

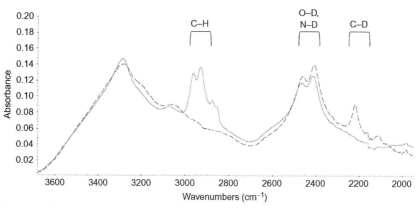

Figure 5 FT-IR spectra of protonated (solid red; gray in the print version) and deuterated (dashed blue; black in the print version) chitosan purified from *P. pastoris*.

3.6 Final Remarks

Biodeuteration of *P. pastoris* can provide a cost-effective source of deuterated chitosan, especially since it can be obtained as a by-product of deuterated protein expression. The chitosan yield is lower than reported for other, filamentous, fungi; however, the ability to grow in liquid culture using relatively cheap deuterated substrates makes this an attractive option when considering isotope-sensitive characterizations of this material.

4. DEUTERATED CELLULOSE

4.1 Organization and Biosynthesis of Bacterial Cellulose

Cellulose is a polymer of 2 β(1–4)-linked glucose units (cellobiose) assembled vectorially in a parallel fashion into nanoscale crystallites of well-defined dimensions. It is the most abundant natural biopolymer being the major structural component of the cell walls of higher plants and the basis of the important cellulosic fiber industry. Burgeoning interest in applications of natural polymers has included a focus on novel utilizations of cellulose nanofibres (Deepa et al., 2015; Hettrich, Pinnow, Volkert, Passauer, & Fischer, 2014; Jonoobi et al., 2015), which represent various chemical and physical alterations. Furthermore, advancing fundamental understanding of the starting material helps to expand the applications of this renewable material in new technologies (Eichhorn et al., 2010; Kim, Yun, & Ounaies, 2006). While the nanostructure of cellulose is preserved in many materials and applications, films of cellulose and its derivatives have proven useful as models for understanding more complex systems (Cheng et al., 2012; Su et al., 2015) and novel surfaces for technological applications (Eriksson, Notley, & Wagberg, 2007; Turner, Spear, Holbrey, & Rogers, 2004).

Cellulose chains are organized into nanoscale rectangular cross-sectioned crystallites called microfibrils, which are achieved in the rosette complex by simultaneous synthesis of nanostructure and polymer in higher plants (Doblin, Kurek, Jacob-Wilk, & Delmer, 2002) and some algae (Garvey, Keckes, Parker, Beilby, & Lee, 2006). Thus, both polymer chain and microfibrils are assembled in a metastable state with the cellulose chains arranged in a parallel manner (Nishiyama, Langan, & Chanzy, 2002; Nishiyama, Sugiyama, Chanzy, & Langan, 2003). Some limited deuteration of cellulose has been achieved in higher plant systems either through exchange of intracrystalline –OH groups at high pressure and temperature (Muller, Czihak, Schober, Nishiyama, & Vogl, 2000) or through growth of ryegrass seedlings in D_2O achieving 60% deuteration (Evans et al., 2014). The bacterium *Gluconacetobacter xylinus* is well known for its ability to produce cellulose (Hestrin & Schramm, 1954), and bacteria are more tolerant to high D_2O concentrations. Bacterial cellulose (BC) differs from higher plant celluloses in that the synthetic complex is linear (Ross, Mayer, & Benziman, 1991) rather than the rosette complex (Doblin et al., 2002). There are many similarities in the synthetic pathway, and for this reason, it has served as a model system for cellulose biosynthesis (Pear, Kawagoe, Schreckengost, Delmer, & Stalker, 1996). As a consequence of linear synthetic complex, BC is organized into sheets of hydrogen-bonded cellulose chains. The top and bottom of the sheet consist of axial C–H groups, and the edges are –OHs (Fig. 6). The different surfaces of the BC crystallite are dominated by the C–H group to a greater extent than is found for higher plant crystallites (Garvey, Parker, & Simon, 2005) and by implication the interactions between microfibrils are dominated by van der Waal rather than hydrogen-bonding interactions. Differences in chain packing include that BC is rich in the Iα (triclinic) lattice (Nishiyama et al., 2003), while higher plant celluloses tend toward mixtures (Atalla & Vanderhart, 1984) of the Iβ (monoclinic) (Nishiyama et al., 2002) and triclinic lattices. Despite these fundamental differences, BC composites are quite often considered as important model systems for higher plant cell wall mechanics (Astley, Chanliaud, Donald, & Gidley, 2003; Chanliaud, Burrows, Jeronimidis, & Gidley, 2002) and the dynamics of cellulose (Petridis et al., 2014). The exchange properties of –OH groups differ in the intra- and extracrystalline regions (Fig. 6). In the case of cellulose, it is well known that there are exchangeable (extracrystalline) and non-exchangeable (intracrystalline) –OH groups.

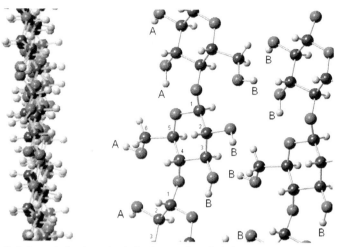

Figure 6 Cellulose chains in a single hydrogen-bonded sheet of the monoclinic lattice in bacterial cellulose. The left-hand side shows a number of cellulose chains edge on. The pink axial atoms are those nonexchangeable hydrogens (attached to carbon atoms), and the blue atoms are those equatorial hydrogens which participate in hydrogen bonding. The right-hand side shows a region of the hydrogen-bonded sheet with an edge occupying chain. The –OH groups at the edge the sheet, A, are exchangeable in the solvent. Those –OH groups labeled B are in the interior of the cellulose crystallite and do not normally exchange. (See the color plate.)

4.2 Production and Characterization of BC

Various carbon sources give high yields of cellulose (Mikkelsen, Flanagan, Dykes, & Gidley, 2009); however, glycerol is the cheapest commercially available suitable deuterated growth source. He et al. (2014) have produced incrementally deuterated BC produced from ^{2}H-glycerol (40 g/L) using a defined medium prepared in D_2O, which also contained vitamins (nine components), trace elements (six components), lactate, essential salts, and phosphate (four components). This is a quite laborious method and typical yields from 50 mL were 0.72 g of cellulose. In this study, we produced deuterated cellulose under the different conditions shown in Table 1. Our method involves production of deuterated cellulose from modified Hestrin and Schramm (1954) medium prepared in D_2O or H_2O in 70-mL sterile specimen jars under static incubation. The medium does contain undefined components (peptone and yeast extract) and 20 g/L glycerol but is far simpler to implement, and the yields per gram of ^{2}H-glycerol (the most expensive component of the media) were not greatly dissimilar to He et al. (2014). About 40 mL of static culture produced approximately 0.4 g of cellulose. The degree of nonexchangeable deuteration can be estimated from SLD of the materials obtained from SANS contrast variation. Control

Table 1 Different Deuteration Schemes

Carbon Source	Solvent	SLD × 10⁻⁶ (Å)	Deuteration Scheme
^{1}H-Glucose	H_2O	2.0	–
^{1}H-Glucose	D_2O	3.2	B* and some C–H
^{2}H-Glycerol	H_2O	3.8	Some C–H
^{2}H-Glycerol	D_2O	6.1	Large proportion
^{2}H-Glucose	H_2O	5.1	All nonexchangeable
^{2}H-Glucose	D_2O	6.2	Large proportion

B* refers to –OH groups in the interior of cellulose crystallites as labeled in Fig. 6.

of the SLD to create contrast between components is the key advantage conferred by molecular deuteration.

Both position and degree of labeling can be manipulated by production of BC using either H_2O or D_2O and deuterated or hydrogenated glycerol or glucose, offering a variety of labeling permutations the results of which are discussed below.

To prepare growth media, dissolve the following in 960 mL of H_2O or D_2O:

Peptone	5.0 g
Yeast extract	5.0 g
Na_2HPO_4	2.7 g
Citric acid	1.15 g

and adjust the pH to 5.0 with 2 M HCl or DCl, followed by autoclaving. About 40 mL of a filter-sterilized (0.22 μm) solution of deuterated or hydrogenated carbon substrate (50%, w/v glucose or glycerol) is added to give a final concentration of 20 g/L. Deuterated glucose-d_{12} and glycerol-d_8 were sourced from Sigma-Aldrich (cat. # 616338 and 447498, respectively). Modified Hestrin and Schramm agar may be prepared by adding 16 g/L agar prior to autoclaving.

To produce cellulose in media containing ^{1}H in either the carbon source or solvent (H_2O), a loopful of culture from an agar plate culture (7 days at 30 °C) is transferred into 5 mL of medium in a sterile centrifuge tube and incubated statically (without shaking) until a pellicle forms, after which 1 mL is transferred into 40 mL of medium in a 70-mL sterile medical sample jar and incubated until pellicle formation is complete. For totally deuterated media (glycerol-d_8 or glucose-d_{12} in D_2O-based Hestrin and Schramm medium), a preadapted inoculum (5 mL) is prepared in 90% D_2O, grown until pellicle formation and 1 mL is added to 40 mL of medium as above. Pellicles first appear after 14 days or more incubation and are usually complete after 20–24 days.

The resulting raw pellicles are tough, mechanically stable, and are largely composed of water. The mass fraction of cellulose can be determined gravimetrically determined after drying (~1%). The volume fraction of cellulose in the films is determined from this mass assuming isotropic shrinkage of the films during drying, a cellulose density of 1.5 g cm^{-3}, and characterization of the shrinkage by image analysis. The BC pellicles are washed well with water, boiled with 1 M NaOH for 2 h to remove cell debris and the NaOH washed from the pellicle to neutrality. FT-IR of the pellicles indicated typical spectra of deuterated cellulose, and the absence of the amide I (1.655 cm^{-1}) and II (1.545 cm^{-1}) bands from proteins (Giordano et al., 2001) was used to determine the removal of the cell materials.

To characterize the effects of deuteration on chain packing into the cellulose crystallites, we use wide angle X-ray scattering (WAXS) at an SAXS/WAXS beamline (Australia Synchrotron, Melbourne, Australia) (Kirby et al., 2013) shown in Fig. 7. To a good first approximation, the width of most peaks is dominated by the size effect in a direction normal to that particular crystallographic plane (Garvey et al., 2005). Importantly for this work, we do not find any resolvable difference in peak positions,

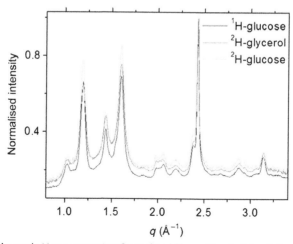

Figure 7 Wide-angle X-ray scattering from dried films of bacterial cellulose grown with different carbon sources in D_2O.

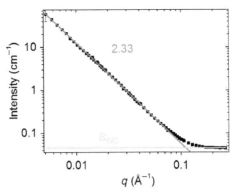

Figure 8 A typical SANS 1D curve from deuterated bacterial cellulose produced using deuterated glycerol in D_2O-based Hestrin and Schramm medium. There is linear region of slope approximately -2.3 at low q. The scattering intensity then adopts a constant value, the incoherent background, B_{INC}, which is a function of the volume density of various nuclei representative of the sample.

relative intensity, and the shape between samples under different deuteration schemes shown in Table 1.

Determination of the SLD can be made by the contrast variation SANS curves (Fig. 8) from samples in different mixtures of H_2O and D_2O from isotropic data on an absolute intensity scale (Lindner, 2004) as a function of the scattering vector, q, where q is defined as $4\pi \sin(\theta/2)/\lambda$, θ is the scattering angle, and λ is the wavelength of the scattered neutron. To calculate the SLD of the cellulose, we assume a simple two-phase behavior where, after subtraction of B_{INC}, the scattered intensity at low q is proportional to the square of the difference of the SLD:

$$I(q) = (\delta_{cellulose} - \delta_{solvent})^2 X - B_{INC}$$

where $\delta_{cellulose}$ is the SLD of cellulose, $\delta_{solvent}$ is the SLD of the water mixtures, and X is a factor that represents the structure of the material. We therefore determine the composition of the cellulose independent of any structural model of the cellulose phase.

The incorporation of different deuterated substrates into the polymer has been localized on hydrolyzed glucose units using deuterium NMR (Barnoud, Gagnaire, Odier, & Vincendo, 1971; Chumpitazi-Hermoza, Gagnaire, & Taravel, 1983). The results are consistent with knowledge of the biosynthetic pathways (Lin et al., 2013). These NMR results indicate a specific incorporation of the labeled glucose ring and site-specific incorporation of labeled glycerol into the glucose ring, which can be understood in terms of the biosynthetic pathways to cellulose synthesis. When D_2O is supplied to the growth media, it may be incorporated indirectly via sugar synthesis from glycerol and other molecules within the growth media (Bali et al., 2013). In the case of glucose with specific deuterium labeling and no other source of deuterium, this enables labeling of specific positions within the cellobiose unit. In the case of materials grown in D_2O with a hydrogenated carbon source, it is largely the inaccessible –OHs which are labeled with a small percentage being channeled metabolically into C–Hs on the cellobiose ring. Such materials may provide useful materials in the study of the role of intracrystalline dynamics in the properties of cellulose crystallites.

4.3 Final Remarks

The techniques described enable the production of BC crystallites of varying degrees of deuteration, and hence SLD, suitable for contrast variation for SANS studies of complex systems. X-ray diffraction indicates that there is no observable change in the morphology of the cellulose crystallites with deuteration. Using control of substrates and the potential to control precisely the site of deuteration, we are able to selectively deuterate various sites within the cellulose crystallite. Such materials may be suitable for the study of the internal dynamics of cellulose crystallite—an approach useful in studying outstanding questions about the role of disordered materials in the mechanical properties of cellulosic materials and the nature of phase transitions within the cellulose crystallite. When cellulosic materials are dissolved and reconstituted, for example, into films for reflectometry, then all exchangeable groups will be replaced in the solvent. In this case, it is not important to ensure that intracrystalline –OHs are deuterated. In both cases, it is important to include a measurement to calculate the scattering SLD of the cellulose.

ACKNOWLEDGMENTS

The technical support of Marie Gillon in production of bacterial cellulose is gratefully acknowledged. WAXS measurements on cellulose were made at the SAXS/WAXS beamline at the Australian Synchrotron, Melbourne, Australia. The National Deuteration Facility is partially funded by the National Collaborative Research Infrastructure Strategy, an initiative of the Australian Federal Government. We acknowledge the support of the Bragg Institute, Australian Nuclear Science and Technology Organization, and the Laboratoire Leon Brillouin (Gif-sur-Yvette, France) in providing the neutron research facilities used in this work.

REFERENCES

Anicuta, S.-G., Dobre, L., Stroescu, M., & Jipa, I. (2010). Fourier transform infrared (FTIR) spectroscopy for characterization of antimicrobial films containing chitosan. In *Analele Universitatii din Oradea Fascicula: Ecotoxicologie, Zootehniesitehnologii de Industrie Alimentara* (pp. 1234–1240). http://webhost.uoradea.ro/editura/reviste_en.html.

Astley, O. M., Chanliaud, E., Donald, A. M., & Gidley, M. J. (2003). Tensile deformation of bacterial cellulose composites. *International Journal of Biological Macromolecules, 32*(1–2), 28–35.

Atalla, R. H., & Vanderhart, D. L. (1984). Native cellulose: A composite of two distinct crystalline forms. *Science, 223*(4633), 283–285.

Bali, G., Foston, M. B., O'Neill, H. M., Evans, B. R., He, J., & Ragauskas, A. J. (2013). The effect of deuteration on the structure of bacterial cellulose. *Carbohydrate Research, 374*, 82–88.

Barnoud, F., Gagnaire, D., Odier, L., & Vincendo, M. (1971). Biosynthesis of deuterated bacterial cellulose: NMR study on levels of incorporation and localization of deuterium. *Biopolymers, 10*(11), 2269–2273.

Beaucage, G., Rane, S., Sukumaran, S., Satkowski, M. M., Schechtman, L. A., & Doi, Y. (1997). Persistence length of isotactic poly(hydroxy butyrate). *Macromolecules, 30*(14), 4158–4162.

Chagas, B., Cruz, L., João, C., Freitas, F., Oliveira, R., & Reis, M. A. (2009). Extraction and purification of cell wall polysaccharides from *Pichia pastoris* biomass. *New Biotechnology, 25*(Suppl. 1), S214.

Chanliaud, E., Burrows, K. M., Jeronimidis, G., & Gidley, M. J. (2002). Mechanical properties of primary plant cell wall analogues. *Planta, 215*(6), 989–996.

Cheng, G., Datta, S., Liu, Z. L., Wang, C., Murton, J. K., Brown, P. A., et al. (2012). Interactions of endoglucanases with amorphous cellulose films resolved by neutron reflectometry and quartz crystal microbalance with dissipation monitoring. *Langmuir, 28*(22), 8348–8358.

Chumpitazi-Hermoza, B. F., Gagnaire, D., & Taravel, F. R. (1983). Bacterial biosynthesis of cellulose from D-glucose or glycerol precursors labelled with deuterium. *Carbohydrate Polymers, 3*(1), 1–12.

Crestini, C., Kovac, B., & Giovannozzi-Sermanni, G. (1996). Production and isolation of chitosan by submerged and solid-state fermentation from *Lentinus edodes*. *Biotechnology and Bioengineering, 50*(2), 207–210.

Crompton, K. E., Forsythe, J. S., Horne, M. K., Finkelstein, D. I., & Knott, R. B. (2009). Molecular level and microstructural characterisation of thermally sensitive chitosan hydrogels. *Soft Matter, 5*, 4704–4711.

Curley, J. M., Lenz, R. W., & Fuller, R. C. (1996). Sequential production of two different polyesters in the inclusion bodies of *Pseudomonas oleovorans*. *International Journal of Biological Macromolecules, 19*, 29–34.

Deepa, B., Abraham, E., Cordeiro, N., Mozetic, M., Mathew, A., Oksman, K., et al. (2015). Utilization of various lignocellulosic biomass for the production of nanocellulose: A comparative study. *Cellulose, 22*(2), 1075–1090.

de Ghellinck, A., Schaller, H., Laux, V., Haertlein, M., Sferrazza, M., Maréchal, E., et al. (2014). Production and analysis of perdeuterated lipids from *Pichia pastoris* cells. *PLoS One, 9*(4), e92999.

Doblin, M. S., Kurek, I., Jacob-Wilk, D., & Delmer, D. P. (2002). Cellulose biosynthesis in plants: From genes to rosettes. *Plant and Cell Physiology, 43*(12), 1407–1420.

Eichhorn, S. J., Dufresne, A., Aranguren, M., Marcovich, N. E., Capadona, J. R., Rowan, S. J., et al. (2010). Review: Current international research into cellulose nanofibres and nanocomposites. *Journal of Materials Science, 45*(1), 1–33.

Eriksson, M., Notley, S. M., & Wagberg, L. (2007). Cellulose thin films: Degree of cellulose ordering and its influence on adhesion. *Biomacromolecules, 8*(3), 912–919.

Evans, B. R., Bali, G., Reeves, D. T., O'Neill, H. M., Sun, Q. N., Shah, R., et al. (2014). Effect of D_2O on growth properties and chemical structure of annual ryegrass (Lolium multiflorum). *Journal of Agricultural and Food Chemistry, 62*(12), 2595–2604.

Foster, L. J. R., Schwahn, D., Pipich, V., Holden, P. J., & Richter, D. (2008). Small-angle neutron scattering characterization of polyhydroxyalkanoates and their BioPEGylated hybrids in solution. *Biomacromolecules, 9*(1), 314–320.

Garvey, C., Keckes, J., Parker, I., Beilby, M., & Lee, G. (2006). Polymer nanoscale morphology in Chara australis Brown cell walls studied by advanced solid state techniques. *Cryptogamie Algologie, 27*(4), 391–401.

Garvey, C. J., Parker, I. H., & Simon, G. P. (2005). On the interpretation of X-ray diffraction powder patterns in terms of the nanostructure of cellulose I fibres. *Macromolecular Chemistry and Physics, 206*(15), 1568–1575.

Giordano, M., Kansiz, M., Heraud, P., Beardall, J., Wood, B., & McNaughton, D. (2001). Fourier transform infrared spectroscopy as a novel tool to investigate changes in intracellular macromolecular pools in the marine microalga Chaetoceros muellerii (Bacillariophyceae). *Journal of Phycology, 37*(2), 271–279.

Gross, R. A., Ulmer, H. W., Lenz, R. W., Tshudy, D. W., Uden, P. C., Brandt, H., et al. (1992). Biodeuteration of poly(β-hydroxybutyrate). *International Journal of Biological Macromolecules, 14*, 33–40.

Haon, S., Augé, S., Tropis, M., Milon, A., & Lindley, N. D. (1993). Low cost production of perdeuterated biomass using methylotrophic yeasts. *Journal of Labelled Compounds and Radiopharmaceuticals, 33*(11), 1053–1063.

He, J. H., Pingali, S. V., Chundawat, S. P. S., Pack, A., Jones, A. D., Langan, P., et al. (2014). Controlled incorporation of deuterium into bacterial cellulose. *Cellulose, 21*(2), 927–936.

Hestrin, S., & Schramm, M. (1954). Synthesis of cellulose by Acetobacter xylinum. II. Preparation of freeze-dried cells capable of polymerizing glucose to cellulose. *Biochemical Journal, 58*(2), 345–352.

Hettrich, K., Pinnow, M., Volkert, B., Passauer, L., & Fischer, S. (2014). Novel aspects of nanocellulose. *Cellulose, 21*(4), 2479–2488.

Jonoobi, M., Oladi, R., Davoudpour, Y., Oksman, K., Dufresne, A., Hamzeh, Y., et al. (2015). Different preparation methods and properties of nanostructured cellulose from various natural resources and residues: A review. *Cellulose, 22*(2), 935–969.

Jouault, N., Dalmas, F., Said, S., Di Cola, E., Schweins, R., Jestin, J., et al. (2010). Direct measurement of polymer chain conformation in well-controlled model nanocomposites by combining SANS and SAXS. *Macromolecules, 43*(23), 9881–9891.

Kim, J., Yun, S., & Ounaies, Z. (2006). Discovery of cellulose as a smart material. *Macromolecules, 39*(12), 4202–4206.

Kirby, N. M., Mudie, S. T., Hawley, A. M., Cookson, D. J., Mertens, H. D. T., Cowieson, N., et al. (2013). A low-background-intensity focusing small-angle X-ray scattering undulator beamline. *Journal of Applied Crystallography, 46*, 1670–1680.

Lin, S.-P., Loira Calvar, I., Catchmark, J., Liu, J.-R., Demirci, A., & Cheng, K.-C. (2013). Biosynthesis, production and applications of bacterial cellulose. *Cellulose, 20*(5), 2191–2219.

Lindner, P. (2004). Scattering experiments: Experimental aspects, initial data reduction and absolute calibration. In P. Lindner, & T. Zemb (Eds.), *Neutrons, X-rays, and light: Scattering methods applied to soft condensed matter* (pp. 23–48): Amsterdam: Elsevier.

Massou, S., Puech, V., Talmont, F., Demange, P., Lindley, N. D., Tropis, M., et al. (1999). Heterologous expression of a deuterated membrane-integrated receptor and partial deuteration in methylotrophic yeasts. *Journal of Biomolecular NMR, 14*(3), 231–239.

Mikkelsen, D., Flanagan, B. M., Dykes, G. A., & Gidley, M. J. (2009). Influence of different carbon sources on bacterial cellulose production by *Gluconacetobacter xylinus* strain ATCC 53524. *Journal of Applied Microbiology, 107*(2), 576–583.

Morgan, W. D., Kragt, A., & Feeney, J. (2000). Expression of deuterium-isotope-labelled protein in the yeast *Pichia pastoris* for NMR studies. *Journal of Biomolecular NMR, 17*(4), 337–347.

Muller, M., Czihak, C., Schober, H., Nishiyama, Y., & Vogl, G. (2000). All disordered regions of native cellulose show common low-frequency dynamics. *Macromolecules, 33*(5), 1834–1840.

Nishiyama, Y., Langan, P., & Chanzy, H. (2002). Crystal structure and hydrogen-bonding system in cellulose 1 beta from synchrotron X-ray and neutron fiber diffraction. *Journal of the American Chemical Society, 124*(31), 9074–9082.

Nishiyama, Y., Sugiyama, J., Chanzy, H., & Langan, P. (2003). Crystal structure and hydrogen bonding system in cellulose 1(alpha), from synchrotron X-ray and neutron fiber diffraction. *Journal of the American Chemical Society, 125*(47), 14300–14306.

Nozirov, F., Fojud, Z., Szcześniak, E., & Jurga, S. (2000). Molecular dynamics in poly[(R)-3-hydroxybutyric acid] biopolymer as studied by NMR. *Applied Magnetic Resonance, 18*(1), 37–45.

Pear, J. R., Kawagoe, Y., Schreckengost, W. E., Delmer, D. P., & Stalker, D. M. (1996). Higher plants contain homologs of the bacterial celA genes encoding the catalytic subunit of cellulose synthase. *Proceedings of the National Academy of Sciences of the United States of America, 93*(22), 12637–12642.

Petridis, L., O'Neill, H. M., Johnsen, M., Fan, B. X., Schulz, R., Mamontov, E., et al. (2014). Hydration control of the mechanical and dynamical properties of cellulose. *Biomacromolecules, 15*(11), 4152–4159.

Pillai, C. K. S., Paul, W., & Sharma, C. P. (2009). Chitin and chitosan polymers: Chemistry, solubility and fiber formation. *Progress in Polymer Science, 34*(7), 641–678.

Ross, P., Mayer, R., & Benziman, M. (1991). Cellulose biosynthesis and function in bacteria. *Microbiological Reviews, 55*(1), 35–58.

Russell, R. A., Darwish, T. A., Puskar, L., Martin, D. E., Holden, P. J., & Foster, L. J. R. (2014). Deuterated polymers for probing phase separation using infrared microspectroscopy. *Biomacromolecules, 15*(2), 644–649.

Russell, R. A., Holden, P. J., Wilde, K. L., Garvey, C. J., Hammerton, K. M., & Foster, L. J. R. (2008). *In vivo* deuteration strategies for neutron scattering analysis of bacterial polyhydroxyoctanoate. *European Biophysics Journal, 37*(5), 711–715.

Russell, R. A., Holden, P. J., Wilde, K. L., Hammerton, K. M., & Foster, L. J. R. (2007). Production and use of deuterated polyhydroxyoctanoate in structural studies of PHO inclusions. *Journal of Biotechnology, 132*(3), 303–305.

Schmidt-Rohr, K., & Spiess, H. W. (1994). *Multidimensional solid-state NMR and polymers*. London: Academic Press.

Sears, V. F. (1992). Neutron scattering lengths and cross sections. *Neutron News, 3*(3), 26–37.

Shi, F., Ashby, R., & Gross, R. A. (1996). Use of poly(ethylene glycol)s to regulate poly (3-hydroxybutyrate) molecular weight during *Alcaligenes eutrophus* cultivations. *Macromolecules, 29*(24), 7753–7758.

Smotrina, T., & Smirnov, A. (2008). Effect of water on relaxation processes in biopolymer sorbents. *Colloid Journal, 70*(3), 337–340.

Su, J., Garvey, C. J., Holt, S., Tabor, R. F., Winther-Jensen, B., Batchelor, W., et al. (2015). Adsorption of cationic polyacrylamide at the cellulose–liquid interface: A neutron reflectometry study. *Journal of Colloid and Interface Science, 448*, 88–89.

Sudesh, K., Abe, H., & Doi, Y. (2000). Synthesis, structure and properties of polyhydroxyalkanoates: Biological polyesters. *Progress in Polymer Science, 25*(10), 1503–1555.

Synowiecki, J., & Al-Khateeb, N. (1997). Mycelia of *Mucor rouxii* as a source of chitin and chitosan. *Food Chemistry*, *60*(4), 605–610.

Trutnau, M., Suckale, N., Groeger, G., Bley, T., & Ondruschka, J. (2009). Enhanced chitosan production and modeling hyphal growth of *Mucor rouxii* interpreting the dependence of chitosan yields on processing and cultivation time. *Engineering in Life Sciences*, *9*(6), 437–443.

Turner, M. B., Spear, S. K., Holbrey, J. D., & Rogers, R. D. (2004). Production of bioactive cellulose films reconstituted from ionic liquids. *Biomacromolecules*, *5*(4), 1379–1384.

Valentin, R., Bonelli, B., Garrone, E., Di Renzo, F., & Quignard, F. (2007). Accessibility of the functional groups of chitosan aerogel probed by FT-IR-monitored deuteration. *Biomacromolecules*, *8*(11), 3646–3650.

Vasko, P. D., Blackwell, J., & Koenig, J. L. (1972). Infrared and Raman spectroscopy of carbohydrates. Part II: Normal coordinate analysis of α-D-glucose. *Carbohydrate Research*, *23*(3), 407–416.

Vogel, H. J., & Bonner, D. M. (1956). Acetylornithinase of *Escherichia coli*: Partial purification and some properties. *Journal of Biological Chemistry*, *218*(1), 97–106.

Wang, Y., Qiu, D., Cosgrove, T., & Denbow, M. L. (2009). A small-angle neutron scattering and rheology study of the composite of chitosan and gelatin. *Colloids and Surfaces. B, Biointerfaces*, *70*, 254–258.

Yoshie, N., Goto, Y., Sakurai, M., Inoue, Y., Chujo, R., & Doi, Y. (1992). Biosynthesis and n.m.r. studies of deuterated poly(3-hydroxybutyrate) produced by *Alcaligenes eutrophus* H16. *International Journal of Biological Macromolecules*, *14*(2), 81–86.

Yun, S. I., Russell, R. A., & Holden, P. J. (2008). Determination of solution properties of poly(3-hydroxybutyrate) and deuterated poly(3-hydroxybutyrate) by SANS. *Polymeric Materials: Science and Engineering*, *98*, 877–878.

CHAPTER SIX

Production of Bacterial Cellulose with Controlled Deuterium–Hydrogen Substitution for Neutron Scattering Studies ☆

Hugh O'Neill[*,1], Riddhi Shah[*,†], Barbara R. Evans[‡], Junhong He[*], Sai Venkatesh Pingali[*], Shishir P.S. Chundawat[§], A. Daniel Jones[¶,‖], Paul Langan[*], Brian H. Davison[#], Volker Urban[*]

[*]Biology and Soft Matter Division, Oak Ridge National Laboratory, Oak Ridge, Tennessee, USA
[†]Bredesen Center for Interdisciplinary Research and Graduate Education, University of Tennessee, Knoxville, Tennessee, USA
[‡]Chemical Sciences Division, Oak Ridge National Laboratory, Oak Ridge, Tennessee, USA
[§]Department of Chemical and Biochemical Engineering, Rutgers University, Piscataway, New Jersey, USA
[¶]Department of Biochemistry and Molecular Biology, Michigan State University, East Lansing, Michigan, USA
[‖]Department of Chemistry, Michigan State University, East Lansing, Michigan, USA
[#]Biosciences Division, Oak Ridge National Laboratory, Oak Ridge, Tennessee, USA
[1]Corresponding author: e-mail address: oneillhm@ornl.gov

Contents

1. Introduction 124
   1.1 Neutrons in Biology 124
   1.2 Biological Toxicity of Deuterium Oxide 125
2. The Occurrence and Properties of Cellulose 127
3. Deuteration of Bacterial Cellulose 129
   3.1 Choice of Bacterial Strain 130
   3.2 Growth Media for Deuterated Cellulose Production 130
   3.3 Adaptation of Cell Growth in Deuterium Oxide 131
   3.4 Cellulose Purification 133
4. Characterization of Deuterated Cellulose 133
   4.1 Chemical and Physical Characterization of Bacterial Cellulose 133
   4.2 Fourier Transform Infrared Spectroscopy 134
   4.3 Mass Spectrometry 135

☆This manuscript has been authored by UT-Battelle, LLC under Contract No. DE-AC05-00OR22725 with the U.S. Department of Energy. The United States Government retains and the publisher, by accepting the article for publication, acknowledges that the United States Government retains a non-exclusive, paid-up, irrevocable, world-wide license to publish or reproduce the published form of this manuscript, or allow others to do so, for United States Government purposes. The Department of Energy will provide public access to these results of federally sponsored research in accordance with the DOE Public Access Plan (http://energy.gov/downloads/doe-public-access-plan).

4.4 SANS Analysis of Bacterial Cellulose	138
Acknowledgments	142
References	142

Abstract

Isotopic enrichment of biomacromolecules is a widely used technique that enables the investigation of the structural and dynamic properties to provide information not accessible with natural abundance isotopic composition. This study reports an approach for deuterium incorporation into bacterial cellulose. A media formulation for growth of *Acetobacter xylinus* subsp. *sucrofermentans* and *Gluconacetobacter hansenii* was formulated that supports cellulose production in deuterium (D) oxide. The level of D incorporation can be varied by altering the ratio of deuterated and protiated glycerol used during cell growth in the D_2O-based growth medium. Spectroscopic analysis and mass spectrometry show that the level of deuterium incorporation is high (>90%) for the perdeuterated form of bacterial cellulose. The small-angle neutron scattering profiles of the cellulose with different amounts of D incorporation are all similar indicating that there are no structural changes in the cellulose due to substitution of deuterium for hydrogen. In addition, by varying the amount of deuterated glycerol in the media it was possible to vary the scattering length density of the deuterated cellulose. The ability to control deuterium content of cellulose extends the range of experiments using techniques such as neutron scattering to reveal information about the structure and dynamics of cellulose, and its interactions with other biomacromolecules as well as synthetic polymers used for development of composite materials.

1. INTRODUCTION

1.1 Neutrons in Biology

Neutron scattering is a nondestructive technique that is able to probe biological materials over multiple length and time scales providing key insights into their structural properties. For low resolution studies using small-angle neutron scattering (SANS), the strength of neutron scattering is its ability to resolve the structural properties of individual components in a complex system by using the contrast variation technique. Figure 1 shows how the scattering length densities (SLDs) of classes of biomolecules are inherently different which enables structural studies of complex systems of combinations of proteins, lipids, carbohydrates, and nucleotides by varying the H_2O/D_2O ratio in the solvent. However, the only way to distinguish between components of a system in which the SLDs are all similar, such as protein–protein or protein–carbohydrate complexes, is through the use of D-labeling techniques. In these cases, studies of the structure and dynamics of biological systems using

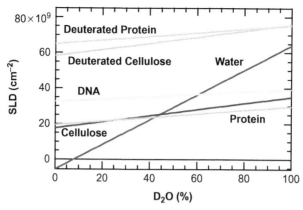

Figure 1 Scattering length densities of different classes of biomolecules in H_2O/D_2O solvents.

neutron scattering are greatly enhanced by the production of D-labeled biological macromolecules that permit selected parts of macromolecular structures to be highlighted and analyzed *in situ*. In SANS studies, D-labeling techniques provide the potential to sequentially highlight components of protein/protein, protein/carbohydrate, or protein/lipid/nucleic acid complexes and, moreover, to follow their conformational changes under near physiological conditions in solution.

1.2 Biological Toxicity of Deuterium Oxide

One of the biggest bottlenecks in realizing the full-potential of neutron scattering for studying biological systems is the ability to produce deuterium labeled biomolecules. Tolerance to D_2O is higher in prokaryotes compared to eukaryotes but in all cases cell growth is inhibited in D_2O (Crespi, 1977; Katz & Crespi, 1966). The bacterium *Escherichia coli* is well established for heterologous expression of D-labeled proteins (Graslund et al., 2008). Growth of lower order heterotrophic eukaryotes such as *Pichia pastoris* in a completely deuterated media formulation is very challenging requiring long adaption times to D_2O and has only recently been achieved for the production of deuterated lipids (de Ghellinck et al., 2014). In contrast, growth of higher-order heterotrophic eukaryotes, such as insect and mammalian cell lines used for heterologous gene expression, is limited to ~30% D_2O, which is inadequate to obtain a sufficient amount of D incorporation into biomolecules for neutron scattering structural studies (Takahashi & Shimada, 2010). Photosynthetic eukaryotes generally exhibit higher tolerance of D_2O

than heterotrophic counterparts. Several species of unicellular eukaryotic algae have successfully been adapted to growth at high levels of D_2O (Katz & Crespi, 1966). Higher plants are known to tolerate and grow in concentrations as high as 50% D_2O, though typically at slower rates than in H_2O (Katz & Crespi, 1966). Average D incorporation of 35% has been achieved for cellulosic biomass from annual ryegrass cultivated in 50% D_2O (Evans et al., 2014).

The underlying reasons for D_2O toxicity in living organisms are not completely understood. D_2O is a stable isotope of water in which both protium atoms (^1H) are replaced by deuterium (^2H). Natural water contains approximately 0.015% D_2O. Pure D_2O has a higher density, higher viscosity, and higher pD compared to water. Deuterium bonds can exhibit a higher binding energy and a shorter binding length compared to hydrogen bonds. When living organisms are exposed to D_2O, the differences in the physical properties of H_2O and D_2O can result in effects that can be divided into the solvent isotope effect of D_2O on the structure of water and macromolecules and the kinetic isotope effect (KIEs) of D replacing H in biological molecules. These give rise to complex effects on cell growth and protein expression (Kushner, Baker, & Dunstall, 1999). At the transcription level, DNA replication could be affected by inhibition of polymerases either directly by D-induced conformational changes in the biomolecule or by a decrease in the rate of DNA splitting into single bonds due to the increased strength of D-bonds compared to H-bonds (Basu, Padhy, & Mookerjee, 1990), leading to impairment of cell division and protein expression. At the translational level, D_2O has been reported to have a negative effect on formation of mitotic spindles as has been demonstrated in studies on sea-urchin eggs and grasshopper spermatocytes (Basu et al., 1990; Gross & Spindel, 1960; Itoh & Sato, 1984; Lamprecht, Schroeter, & Paweletz, 1991). The inhibitory effect of D_2O appears to be an inhibition of tubulin polymerization and microtubule-organizing centers causing cell-cycle arrest. Other effects of D_2O have been reported that include effects on energy production, such as lowered ATP/ADP ratios (Vasilescu & Katona, 1986), and on membrane receptors (Andjus, Kataev, Alexandrov, Vucelic, & Berestovsky, 1994), such as impaired Ca channels and Na-K ATPase (Andjus et al., 1994). D_2O also affects heat sensitivity (Unno & Okada, 1994; Unno, Shimba, & Okada, 1989) and the longevity of singlet oxygen (Bachor, Shea, Gillies, & Hasan, 1991; Utsumi & Elkind, 1991). Cumulatively, these studies highlight that cultivation in D_2O is stressful for cells.

2. THE OCCURRENCE AND PROPERTIES OF CELLULOSE

Cellulose, the most abundant of the carbohydrates, was originally isolated from the cell walls of higher plants. Many algae, several fungi, certain bacteria, and a few marine animals such as tunicates also synthesize cellulose (Perez & Samain, 2010). Chemically, cellulose is a biopolymer composed of linear, parallel chains of β-1,4-linked glucose units which are assembled into crystalline microfibrils during biosynthesis. Its crystalline, fibrous nature imparts special physical properties to cellulose that have made it an important material, particularly in the paper, pulp, and textile industries (Franz & Blaschek, 1990). Expansion of cellulose utilization as an alternative, sustainable feedstock for production of fuels and materials is being investigated, with an emphasis on fuel ethanol production (Ragauskas et al., 2006). In plant cell walls, cellulose microfibrils are an integral part of the laminate structure, forming strong associations with lignin, hemicellulose, pectin, and proteins. Removal of the other cell wall components from cellulose in order to utilize it requires physical and chemical treatments that vary according to the intended use and which add to the expense of the process. Strategies to overcome this cell wall recalcitrance are being investigated (Himmel et al., 2007).

Many different genera of bacteria including *Acetobacter/Gluconacetobacter*, *Agrobacterium*, *Achromobacter*, *Aerobacter*, *Alcaligenes*, *Pseudomonas*, *Rhizobium*, *Salmonella*, *Escherichia*, and *Sarcina* have certain subspecies which secrete extracellular cellulose (Lin et al., 2013). The structure and morphology of the cellulose secreted by these organisms vary widely. The best-known cellulose producers are from the acetic acid bacteria phylum. These organisms produce a type of pure cellulose as an extracellular biofilm or membrane called a pellicle (Brown, 1886; Cannon & Anderson, 1991; Iguchi, Yamanaka, & Budhiono, 2000; Klemm, Heublein, Fink, & Bohn, 2005; Schramm & Hestrin, 1954). The bacteria known to synthesize cellulose exhibit typical characteristics of the genus *Acetobacter*: aerobic, gram-negative, do not require amino acid supplementation and can utilize ammonium as a nitrogen source, but require vitamin supplementation. They have been isolated from a wide range of environments, including vinegar, wine, kombucha, nata de coco, and spoiled fruit (Bergey & Holt, 1994; Cannon & Anderson, 1991; Iguchi et al., 2000). The amount of cellulose produced varies depending on the species and strain, with three species generally cited as producing the typical thick pellicle: *Acetobacter xylinus* (synonym *xylinum*), *hansenii*, and *pasteuriani* (Bernardo, Neilan, & Couperwhite, 1998). The placement of these species in a separate genus *Gluconacetobacter* was proposed (Yamada, Hoshino, & Ishikawa, 1997). Bacterial cellulose differs in its physical properties and higher structural organization from plant cellulose. In its natural state, it is a hydrogel containing 99% water. As well as high water retention, it also possesses higher tensile strength and Young's modulus than purified plant cellulose (Iguchi et al., 2000). The physical properties of bacterial cellulose as well as the ease of purification have led to investigation of its potential as a material for a wide range of applications, including paper and pulp, electronics manufacture, and fuel cells (Cannon & Anderson, 1991; Evans, O'Neill, Malyvanh, Lee, & Woodward, 2003; Iguchi et al., 2000; Shah & Brown, 2005). Investigation of its potential use as a biomedical material for wound dressing and implants continues to expand (Czaja, Young, Kawecki, & Brown, 2007). Bacterial cellulose has been a useful model to elucidate the synthesis of cellulose. Investigation of cellulose synthesis by *A. xylinum* has established that the immediate precursor substrate used by the cellulose synthase complex is UDP-glucose, which is synthesized from glucose-1-phosphate and UTP (Cannon & Anderson, 1991; Glaser, 1958). Further investigation of the cellulose synthesis pathway established that several gene products were required to achieve high levels of production of crystalline cellulose (Kawano et al., 2002; Saxena, Kudlicka, Okuda, & Brown, 1994).

However, the production of cellulose by *Acetobacter/Gluconacetobacter* species is complicated by their multiple metabolic pathways for utilization of carbohydrates, as well as by the variations between strains and the frequently observed instability of cellulose synthesis. Static growth conditions in shallow vessels with high surface area are known to favor cellulose pellicle formation (Bernardo et al., 1998; Brown, 1886; Cannon & Anderson, 1991; Schramm & Hestrin, 1954). Agitation of the growth media often results in loss or reduction of cellulose production, with noncellulose producing (cel-) bacteria forming smooth instead of rough colonies on agar plates (Bernardo et al., 1998; Cannon & Anderson, 1991; Krystynowicz et al., 2002; Schramm & Hestrin, 1954). Similar inhibition of cellulose production was also observed with increase in medium viscosity by addition of increasing amounts of agar or gellan (Evans & O'Neill, 2005). Some strains, including *A. xylinum* subsp. *sucrofermentans*, maintain stable cellulose synthesis, producing cellulose spheres during growth under agitated conditions, while forming the typical thick pellicles under static growth conditions (Matsuoka, Tsuchida, Matsushita, Adachi, & Yoshinaga, 1996). The bacteria are able to utilize several different carbohydrate sources including glucose, fructose, mannitol, and glycerol, and extracellular products which cause acidification of the growth media have been identified (Herrmann & Neuschul, 1931). Growth on glucose as the carbon source results in extracellular production of gluconic acid which results in acidification and eventual inhibition of growth and cellulose production. The bacteria oxidize ethanol to acetic acid, which also causes acidification of the media. However, the extracellular product from glycerol in the growth media was determined to be dihydroxyacetone (Herrmann & Neuschul, 1931). Redistribution of nonexchangeable hydrogens from C6 to C1 and C2 in cellulose produced by *A. xylinum* grown on glucose deuterated at C6 was established by isotopic labeling experiments (Gagnaire & Taravel, 1975). The reported redistribution of deuterium isotopic labels is consistent with the production of triose phosphates and their eventual condensation to hexoses carried out by the pentose phosphate pathway enzymes hexose phosphate isomerase, phosphofructose kinase, triose phosphate isomerase, and aldolase inside the bacterial cell. A similar redistribution of isotopic labels was found in cellulose produced by the bacteria from ^{13}C labeled glucose (Arashida et al., 1993). This isotopic redistribution was found to be reduced when lactate or ethanol was added to growth media (Arashida et al., 1993; Minor, Greathouse, Shirk, Schwartz, & Harris, 1954). The ability to grow well on sucrose depends on the strain (Bernardo et al., 1998; Brown, 1886). Addition of ethanol, acetate, or lactate to growth media containing glucose, fructose, or sucrose increases cellulose yields (Krystynowicz et al., 2002; Matsuoka et al., 1996). Defined minimal media formulations have been developed based on investigation of nutritional needs to increase cellulose yields (Matsuoka et al., 1996; Son et al., 2003).

3. DEUTERATION OF BACTERIAL CELLULOSE

Incorporation of deuterium into bacterial cellulose from deuterated glucose and D_2O in growth medium was used previously to elucidate biosynthetic pathways. Acetylation followed by ^{1}H NMR was used to determine the incorporation of deuterium at the nonexchangeable positions in the resultant cellulose (Barnoud, Gagnaire, Odier, & Vincendo, 1971; Gagnaire & Taravel, 1975). Here, we describe methods for controlled incorporation of high levels of deuterium into bacterial cellulose. The deuterated cellulose products were characterized by Fourier transform Infrared spectroscopy (FTIR), mass spectrometry, and SANS.

3.1 Choice of Bacterial Strain

The bacterial strains *A. xylinus* subsp. *sucrofermentans* (ATCC 700178) and *Gluconacetobacter hansenii* (ATCC 10821) are available from the American Type Culture Collection (Manassas, Virginia, USA). Both strains have been successfully adapted to growth in D_2O medium. *A. xylinus* subsp. *sucrofermentans* is the preferred strain because it exhibits more robust growth in D_2O.

3.2 Growth Media for Deuterated Cellulose Production

In order to achieve controlled and reproducible incorporation of deuterium into cellulose, it is important to avoid use of undefined media components such as yeast extract or corn steep liquor in the growth medium. Commercially available deuterated carbon sources suitable for bacterial cellulose production are limited to glucose and glycerol, both of which are suitable for production of bacterial cellulose. For this work, deuterated glycerol (D_8, 99%) was obtained from Cambridge Isotope Laboratory, Andover, Massachusetts, USA. A synthetic salts formulation for cellulose production with *A. xylinum* had been described (Son et al., 2003), and supplementation with corn steep liquor, using fructose as the carbon source, had been reported to increase cellulose production (Sano, Rojas, Gatenholm, & Davalos, 2010). The stimulatory effects of corn steep liquor were further investigated for cellulose production by *A. xylinus* subsp. *sucrofermentans*, establishing that lactate was the ingredient that enabled faster growth and increased cellulose production (Matsuoka et al., 1996). The mineral and vitamin supplements for preparation of a synthetic defined medium based on these results were described (Matsuoka et al., 1996; Watanabe, Shibata, Ougiya, Hioki, & Morinaga, 2000).

Tests were carried out to determine the minimal medium additives required to obtain good cellulose formation using *A. xylinus* subsp. *sucrofermentans* (ATCC 700178) without the addition of undefined components that could not be obtained in deuterated form (He et al., 2014). For growth in H_2O, glycerol and 0.04% sodium DL-lactate are used with vitamins and minerals. For growth in D_2O media, glycerol, which is commercially available in deuterated form, is substituted for hydrogenated glycerol as the carbon source.

In detail, this defined growth medium developed for *A. xylinus* subsp. *sucrofermentans* (ATCC 700178) based on previously published formulations

(Matsuoka et al., 1996; Son et al., 2003) contained 40 g/L glycerol, 4 mL/L 10% sodium lactate in H_2O, 10 mL/L vitamin stock solution, 10 mL/L mineral salts stock solution, in a basal salt solution which consisted of 2 g/L $(NH_4)_2SO_4$, 3 g/L KH_2PO_4, 3 g/L $Na_2HPO_4 \cdot 12H_2O$, 0.8 g/L $MgSO_4$, and 0.003 g/L boric acid in deionized distilled water. The pH was adjusted to 5.5 with dilute sulfuric acid. Composition of the vitamin stock solution was 200 mg/L inositol, 40 mg/L nicotinic acid, 40 mg/L pyridoxal hydrochloride, 40 mg/L thiamine hydrochloride, 20 mg/L calcium D-panthothenate, 20 mg/L riboflavin, 20 mg/L p-amino benzoic acid, 0.2 mg/L D-biotin, and 0.2 mg/L folic acid. The mineral salts stock solution contained 360 mg/L $FeSO_4 \cdot 7H_2O$, 1470 mg/L $CaCl_2 \cdot H_2O$, 242 mg/L $Na_2MoO_4 \cdot 2H_2O$, 173 mg/L $ZnSO_4 \cdot 7H_2O$, 139 mg/L $MnSO_4 \cdot 5H_2O$, and 5 mg/L $CuSO_4 \cdot 5H_2O$. The composition of the media is provided in Table 1. The mineral salts solution and basal salts solution were sterilized by autoclaving at 121 °C and cooled to room temperature before mixing. The vitamin stock solution and sodium lactate solutions were filter-sterilized with 0.2 µm pore size Nalgene Bottle-Top Sterile Filter Units.

3.3 Adaptation of Cell Growth in Deuterium Oxide

The deuterated media was prepared by evaporating a known volume of the basal salt solution (adjusted to pD 5.9), as prepared above, using a Rotavaporator. The minimum amount of D_2O (99.7%) needed to dissolve the salts was then added and the evaporation step was repeated. The basal salts were finally dissolved in D_2O to the original starting volume. This approach ensured that the exchangeable H and hydration waters were exchanged to D_2O. The mineral salt solution in D_2O was prepared the same way. The vitamin and sodium lactate solutions were exchanged to D_2O by lyophilization. The complete growth medium was filtered using a 0.2-micron sterile filter.

Gradual adaptation of the bacteria to D_2O was carried out by sequentially increasing the D_2O concentration in the growth medium from 0% to 100% D_2O in 20% increments, with *H*-glycerol as the carbon source. Finally, the D_2O-adapted cells were grown in deuteration media with deuterated glycerol as the carbon source. For production of perdeuterated cellulose, the stock culture of D_2O-adapted cells was subcultured into 50 mL of growth medium (1:5 dilution) and grown in 32-oz polypropylene plant culture vessels with lids (PhytoTechnology Laboratories, Shawnee Mission, Kansas, USA) at 25 °C for 14 days under static conditions.

Table 1 An Optimized Minimal Salts Growth Medium Was Developed for Cultivation of *Acetobacter xylinus* subsp. *sucrofermentans*

Basal Salts	Concentration (%)
$(NH_4)_2SO_4$	0.2
KH_2PO_4	0.3
$Na_2HPO_4 \cdot 12H_2O$	0.3
$MgSO_4 \cdot 7H_2O$	0.08
H_3BO_3	0.003
Carbon Sources	**Concentration (%)**
Fructose	4
Sodium DL-lactate	0.04
Mineral Salts	**Concentration (mg/L)**
$FeSO_4 \cdot 7H_2O$	8.6
$CaCl_2 \cdot H_2O$	14.7
$Na_2MoO_4 \cdot 2H_2O$	2.4
$ZnSO_4 \cdot 7H_2O$	1.7
$MnSO_4 \cdot 5H_2O$	1.4
$CuSO_4 \cdot 5H_2O$	0.05
Vitamins	**Concentration (mg/L)**
Inositol	2
Nicotinic acid	0.4
Pyroxidine hydrochloride	0.4
Thiamin hydrochloride	0.4
Calcium pantothenate	0.2
Riboflavin	0.2
p-Amino benzoic acid	0.2
Folic acid	0.0002
D–Biotin	0.0002

3.4 Cellulose Purification

Intact cellulose pellicles were washed with 500 mL deionized H_2O at room temperature, then heated in 250 mL of water to 90 °C. The pellicles were cleaned by soaking in 1% NaOH at room temperature to remove the bacterial debris, replacing the NaOH solution several times until the A_{280} absorbance value was less than 0.01. The NaOH was removed by dialysis against 10 L deionized H_2O overnight. The final pH of the liquid surrounding the cellulose was approximately 7–8. The yield of cellulose was typically 0.72 g dry cellulose/L of culture.

In an alternative approach, the cellulose pellicles were frozen, placed in 20 mL of D_2O, and disrupted using a blender (Osterizer Mo 83-1) until a homogeneous slurry was formed (~2 min). The ground cellulose was collected by centrifugation (4500 × g for 20 min) and resuspended in 1% NaOD in D_2O to remove bacterial debris. The centrifugation/resuspension procedure was repeated, adding fresh 1% NaOD solution each time, until $A_{280}=0.01$ was obtained. The cellulose slurry was then washed with D_2O until a pH value of ~7–8 was obtained.

4. CHARACTERIZATION OF DEUTERATED CELLULOSE

4.1 Chemical and Physical Characterization of Bacterial Cellulose

Standard methods that are typically employed to characterize cellulose samples were used to investigate the chemical and physical properties of the highly deuterated bacterial cellulose obtained using the deuterated glycerol-D_2O culture medium described above. The amount and location of deuterium incorporation in the deuterated bacterial cellulose were examined by $^1H^2H$ NMR of native samples dissolved in ionic liquid and ^1H NMR of acetylated samples. NMR results were consistent with the LC–MS analysis described above, with a small amount of H detected at C-6 of the deuterated sample. Crystallinity index as determined by X-ray diffraction, proportion of Iα/Iβ allomorphs, and number-average degree of polymerization as determined by gel permeation chromatography (GPC) did not appear to be significantly changed by deuteration. The polydispersity index (PDI), a measure of the broadness of the typically multimodal distribution in molecular weights obtained from GPC, was higher for deuterated than protiated bacterial cellulose. This appeared to indicate the presence of higher molecular weight cellulose chains in the deuterated sample. Deuterated bacterial cellulose exhibited a reticulated microfibrillar structure similar to the protiated control in scanning electron microscopy (SEM) images (Bali et al., 2013).

4.2 Fourier Transform Infrared Spectroscopy

FTIR spectroscopy was used to examine D incorporation into bacterial cellulose. Purified bacterial cellulose samples were lyophilized for 3 days prior to measurement. FTIR spectra were measured on a Jasco FT/IR 6100 Series Fourier Transform Infrared Spectrometer in an attenuated total reflectance (ATR) mode. A total of 32 accumulative scans were taken, with a resolution of 1 cm^{-1}, in the range of 4000–650 cm^{-1}. The protiated and deuterated cellulose samples were equilibrated in H_2O and D_2O, respectively, before drying by lyophilization. A comparison of the FTIR spectra of deuterated cellulose and protiated cellulose between 4000 and 1800 cm^{-1} are shown in Fig. 2. The intensities of the sharp OH-stretching infrared bands located at approximately 3500–3300 cm^{-1} were decreased in the deuterated cellulose sample. This was accompanied by the appearance of an OD stretching band at 2600–2400 cm^{-1} (Hofstetter, Hinterstoisser, & Salmen, 2006). The presence of OH stretching in deuterated cellulose is attributed to the presence of hydration water at the surface of the cellulose. The band signal at 2950–2800 cm^{-1} corresponds to the stretching vibrations of alkane (sp^3 hybridized carbon) attached to hydrogen (Tashiro & Kobayashi, 1991). These bands were largely absent in the deuterated cellulose sample, while, a band appeared at 2100 cm^{-1} corresponding to the CD stretching region (Torres, Kukol, & Arkin, 2000), indicating that the level of deuterium incorporation in the cellulose was very high. The presence of a small amount of aliphatic hydrogen in the deuterated cellulose sample that can be observed in its spectrum at ~2900 cm^{-1} may be responsible for the minor differences in the shapes of the CH and CD stretching regions in their respective spectra (Fig. 1B). The presence of CH and CDH groups in the deuterated cellulose due to anomalous incorporation of hydrogen could give rise to changes in the region associated with CD/CD$_2$ vibrational stretching. The presence of a small amount of aliphatic hydrogen in the deuterated cellulose was confirmed by mass spectrometry analysis (see below), and is consistent with the results of $^1H^2H$ NMR analysis comparing protiated and deuterated bacterial cellulose (Bali et al., 2013).

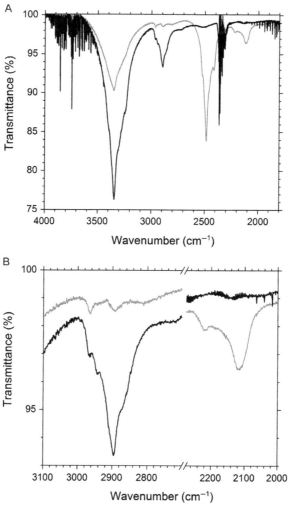

Figure 2 (A) FTIR-ATR spectra of lyophilized protiated (black) and deuterated (gray) bacterial cellulose. (B) CH and CD stretching regions of the FTIR spectra. *Reproduced from He et al. (2014) with kind permission from Springer Science and Business Media.*

4.3 Mass Spectrometry

Mass spectrometry is a useful technique to identify isotopomers of organic molecules (Biemann, 1962) and has previously has been employed to analyze deuterium incorporation in bacterial cellulose grown on deuterated precursors (Barnoud et al., 1971; Chumpitazi-Hermoza, Gagnaire, & Taravel, 1983). The purified bacterial cellulose was prepared for mass spectrometry analysis as follows: The cellulose (0.1% w/v) was hydrolyzed by commercial cellulase cocktail (Accellerase 1500, 0.1 g/g cellulose enzyme loading) in a 7 mL reaction volume (pH 4.8, 50 mM citrate buffer) for 24 h at 50 °C, and 250 rpm (Gao, Chundawat, Krishnan, Balan, & Dale, 2010). HPLC-refractive index (RI) analysis was then carried out to estimate total hydrogenated or deuterated glucose concentration in the bacterial cellulose hydrolysate as described previously (Chundawat et al., 2010). Identical retention times and RI integrated peak areas (to within 5%) were observed for both labeled and unlabeled sugar standards during the HPLC-RI analysis. The hydrolysate was further purified to remove background proteins that interfere with LC–MS analysis, as described elsewhere (McIntosh, Davis, & Matthews, 2002). LC–MS analysis was carried out using a QTRAP 3200 mass spectrometer (AB/Sciex) coupled to a binary LC-20AD pump, CTO-10Avp column oven, and Sil-HTc autosampler (Shimadzu). A 5.0 µL aliquot of each sample was injected onto a Prevail Carbohydrate ES column (Grace) that was held at 40 °C. Total mobile phase flow rates was 0.25 mL/min, using Solvent A (10 mM aqueous ammonium acetate) and Solvent B (acetonitrile) using the following gradient (A/B): initial (25/75), linear gradient to 5.0 min (50/50) followed by a hold until 6.0 min; at 6.01 min, the composition was returned to the initial condition and held there until 8 min. Mass spectrometry analyses employed electrospray ionization in negative ion mode using an ion source temperature of 450 °C. Both Gas1 and Gas2 flows were set to 45 (arbitrary units). Quantification of abundances of individual ions from m/z 179.1 to 187.1 was performed using Q1 multiple ion monitoring data acquisition with a 25 ms dwell time for each nominal mass. Confirmatory analyses were also performed in positive ion mode. Individual ion abundances were determined by manually integrating each extracted ion chromatogram using Analyst software (Ver 1.4.2, AB/SCIEX, Framingham, MA) and integrated peak areas were exported to text files for analysis. Calibration curve for protiated glucose and deuterated glucose characteristic ions (m/z 179 and 186, respectively) were linear up to 1 mM (adjusted-R^2 of 0.98) but had a poorer linear-fit at 10-fold higher concentrations (adjusted-R^2 of 0.88). The fully deuterated D_{12}-glucose is expected to rapidly exchange deuterons attached to oxygen with protons from water to give D_7H_5-glucose. Therefore, for fully labeled bacterial cellulose, we should expect to see D_7H_5-glucose after enzymatic hydrolysis in water. Enzyme blanks were included (no cellulose present) in the analysis to account for possible interference from unknown molecular species (with similar m/z between 179 and 186) present in the commercial enzyme cocktails.

Enzymatic digestion of the protiated cellulose and deuterated cellulose produced $\sim95\pm2\%$ and $\sim90\pm1\%$ of the expected glucose yield, respectively. No other sugars were detected by HPLC-RI in significant concentrations from either cellulose sample. This indicates that the glucose released by the enzymatic digestion is representative of the total labeled and unlabeled carbohydrate content of the cellulose samples analyzed. The slightly higher hydrolysis yields for protiated cellulose versus deuterated cellulose could be caused by secondary KIEs on the activity of the cellulolytic enzymes as they hydrolyze cellulose to cellobiose and finally glucose. However, as glycosidase mechanisms are difficult to discern from secondary KIE's (Vasella, Davies, & Bohm, 2002), a more detailed investigation will be needed to confirm this. The LC–MS analysis was performed using total glucose concentrations ranging from 1 to 10 mM.

A summary of the LC–MS analytical results is shown in Fig. 3. The pseudomolecular ion peaks at m/z 179 for $[M - H]^-$ of unlabeled glucose and m/z 186 of D_7-glucose are evident. Surprisingly, the relative abundance ratios A_{185}/A_{186} and A_{184}/A_{186}, which reflect percent molar ratios of D_6 to D_7 labeled glucose and D_5 to D_7 labeled glucose, differ between the deuterated cellulose hydrolysate and the deuterated glucose standard (Table 2), and are consistent with incomplete deuteration. The anomalous incorporation of varying amounts of D atoms into cellulose could be due to either slow condensation of H_2O moisture within the culture tube that exchanged slowly with the heavy water and caused protonation of intermediates involved in cellulose synthesis. Or, more interestingly, the conversion of the unlabeled lactate into pyruvate, which is finally taken up for cellulose biogenesis could also explain the anomalous levels of D atoms incorporation into cellulose. Previous work has shown lactate can increase bacterial cellulose production by providing an easily utilized energy source for cell growth (Matsuoka et al., 1996). More importantly, it has been shown that *Acetobacter* produces a unique pyruvate-phosphate dikinase that catalyzes the direct phosphorylation of pyruvate into phosphoenol-pyruvate that can be directly assimilated into the production of cellulose (Benziman & Eizen, 1971). Nevertheless, our analysis suggests a high level of deuterium incorporation into deuterated cellulose samples. The total deuterated glucose incorporated into the bacterial cellulose (calculated using the peaks likely representative of $[M - H]^-$ based D_7, D_6, and D_5 glucose and the integrated peak areas from m/z 184 to 186) indicates that 97% of glucose subunits contain at least five deuterium atoms.

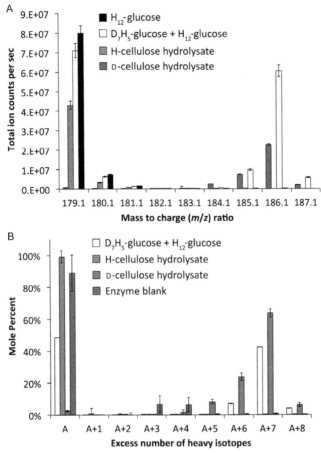

Figure 3 (A) Relative ion abundance and (B) calculated mole percent of major molecular species in hydrolysis products of deuterated and unlabeled cellulose. LC–MS was carried out under Q1 multiple ion monitoring using electrospray ionization in negative-ion mode to detect pseudomolecular ion ($[M-H]^-$) species in the range of m/z 179–187. Note that D_7H_5-glucose + H_{12}-glucose is an equimolar mixture of each. Error bars represent standard deviations for duplicate hydrolysate reaction mixtures. Percent mole of heavy isotopes relative to the unlabeled species (m/z 179) was calculated based on the procedure described by K. Biemann *in Mass spectrometry: Organic chemical applications*, McGraw-Hill, NY, 1962. *Reproduced from He et al. (2014) with kind permission from Springer Science and Business Media.* (See the color plate.)

4.4 SANS Analysis of Bacterial Cellulose

Neutron scattering techniques have recently been used for biomass characterization, in particular to carry out structural analysis and dynamic characterization of its deconstruction (Pingali et al., 2010a, 2010b). Neutron scattering, when combined with X-ray scattering and molecular dynamics

Table 2 Molar Abundance (A) Ratios of Different Mass to Charge (m/z) Species Detected in the Negative-Ion Mode LC–MS Analyses of Hydrolysates of *Protiated* and *Deuterated* Cellulose

Sample Description	A_{185}/A_{186}	A_{184}/A_{186}
D_7H_5-glucose + H_{12}-glucose	0.17	0.01
H-cellulose hydrolysate	ND	ND
D-cellulose hydrolysate	0.37	0.13

Ratios were only estimated for peaks with total ion counts greater than 10^5 counts per second as shown in Fig. 2. ND stands for not detected. Reported ratios are means from duplicate hydrolysate reaction mixtures. Reproduced from He et al. (2014) with kind permission from Springer Science and Business Media.

simulations, revealed two fundamental processes: cellulose dehydration and lignin–hemicellulose phase separation, were responsible for the morphological and structural changes in biomass, providing critical insight to understanding the basic driving forces during thermochemical deconstruction (Langan et al., 2014). This technique can also be used to study enzyme structure under physiologically relevant conditions, such as studying pH-dependent structural changes in *Trichoderma reesei* Cel7A (Pingali et al., 2011). The availability of D-labeled cellulose makes possible a detailed analysis of the structural and dynamic properties of cellulose using advanced neutron scattering methods building on previously published approaches (Byron & Gilbert, 2000; Melnichenko & Wignall, 2007; Nishiyama, Okano, Langan, & Chanzy, 1999; Pingali et al., 2010a).

SANS measurements were performed with the CG-3 Bio-SANS instrument (Heller et al., 2014) at the High Flux Isotope Reactor (HFIR) facility of Oak Ridge National Laboratory. Cylindrical Hellma cells with 1 mm path length (Model# 120-QS 1.0 mm) were used to perform SANS studies. The three different instrument configurations employed to cover the range, $0.003 < Q\ (\text{Å}^{-1}) < 0.4$, of scattering vectors with sufficient overlap were sample-to-detector distances of 2529, 6829, and 15,329 mm at 6 Å neutron wavelength. The scattering vector $Q(Q = 4\pi \sin\theta/\lambda)$ describes the relation of Q to λ, neutron wavelength, and 2θ, the scattering angle. The center of the area detector (1 m × 1 m GE-Reuter Stokes Tube Detector) was offset by 350 mm from the beam. The instrument resolution was defined using circular aperture diameters of 40 and 14 mm for source and sample, respectively, and separated by distances: 3262, 9332, and 17,430 mm. The relative wavelength spread $\Delta\lambda/\lambda$ was set to 0.15. The scattering intensity profiles $I(Q)$ versus Q, were obtained by azimuthally averaging the processed 2D images, which were normalized to incident beam monitor counts, and corrected for detector dark current, pixel sensitivity, and scattering from the quartz cell.

For the SANS experiments, the cellulose was ground into slurry using a commercial blender for 2 min. The cellulose slurry was centrifuged followed by resuspension in 0%, 25%, 50%, 75%, and 100% D_2O at a 1:10 ratio to exchange the labile hydroxyl hydrogens in the cellulose. This process was repeated three times. The final concentration of cellulose was 0.5% (w/v).

SANS was used to compare the nanoscale structural features of bacterial cellulose with different levels of deuterium incorporation. Figure 4A shows the scattering profiles recorded at a cellulose concentration of 0.5% w/v in 100% H_2O (maximum contrast between the cellulose and the solvent) for cellulose grown in different deuterated glycerol/H-glycerol ratios 100%, 50%, 40%, 30%, and 0%. The curves look similar without significant differences in structural features. A contrast variation series was performed by varying the H_2O/D_2O ratio of the solvent between 0% to 100% D_2O. At a Q value that shows enhanced contrast variation, the relation of scattering intensity versus ratio of D_2O in solvent was fit to a parabola and SLD associated with the ratio of D_2O at the vertex is the contrast match SLD. The relation between SLD, obtained by performing contrast variation study for each cellulose variant, and deuterated glycerol ratio used to produce those samples is shown in Fig. 4B. The relation demonstrates that

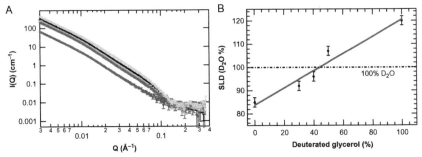

Figure 4 SANS analysis of purified bacterial cellulose with different levels of deuterium incorporation. (A) SANS profiles of celluloses grown using 100%, 50%, 40%, 30%, and 0% deuterated glycerol (orange, purple, green, blue, and red) measured in 100% H_2O. (B) Scattering length densities (SLDs) of different cellulose samples related to the fraction of deuterated glycerol present in the growth medium. *Reproduced from He et al. (2014) with kind permission from Springer Science and Business Media.* (See the color plate.)

by varying the amount of deuterated carbon source added to the growth medium, it is possible to control the level of deuterium incorporation into the cellulose molecules.

All scattering curves monotonically increase in scattering intensity for smaller wave-vectors, Q, exhibiting a power-law dependence with an exponent of -2 (Fig. 5A). An exponent of -2 implies that deuterated bacterial cellulose represents disk-like forms as seen by SEM (Fig. 5B). A large disk model was used to fit the curves of bacterial cellulose grown using 0% and 100% deuterated glycerol. The most important aspect of our fits, large disk with monodisperse or polydisperse thickness is that the thickness of the disk are comparable between the bacterial cellulose grown using 0% and 100% deuterated glycerol, suggesting minimal structural effect on deuterium incorporation into bacterial cellulose. The fits using the monodisperse function provide a lower bound to the disk radius, $R_{min} = 1132 \pm 6$ Å and 1154 ± 3 Å and the disk thickness, $T = 128 \pm 1$ Å and 83 ± 1 Å for the protiated and deuterated forms of the bacterial cellulose, respectively. Schulz's polydispersity in disk thickness also provides a lower bound for disk radius, $R = 1158 \pm 6$ Å and 1158 ± 3 Å and average disk thickness, $T_{avg} = 80 \pm 2$ Å and 62 ± 1 Å for protiated and deuterated bacterial cellulose, respectively.

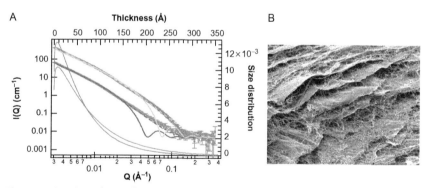

Figure 5 SANS analysis of purified bacterial cellulose with 0% and 100% deuterium incorporation. (A) SANS profiles of celluloses grown using 100% (blue open circles) and 0% (red dots) deuterated glycerol measured in 100% H_2O. Large disks (closest approximation to sheet-like forms) with monodisperse thickness (solid lines) and Schulz's polydisperse thickness (dashed lines) were used as models to fit data. The thickness distribution (100% and 0% deuterated glycerol—blue and red lines, respectively) is plotted against the thickness (top) versus size distribution (right). (B) SEM image of protiated bacterial cellulose (scale bar 50 μm). *Reproduced from He et al. (2014) with kind permission from Springer Science and Business Media.* (See the color plate.)

The PDI for both curves tend to the highest level of polydispersity (0.95 was the upper bound on that parameter). The SEM image in Fig. 4B shows features that can be related to cellulose sheets and fibers. All these features will have comparable contributions in the high-Q region ($Q > 0.05$ Å^{-1}) which explains the reason for some discrepancy in agreement between experimental data and the fits that purely models for large disks.

ACKNOWLEDGMENTS

This research is funded by the Genomic Science Program, Office of Biological and Environmental Research, U.S. Department of Energy, under FWP ERKP752. The Center for Structural Molecular Biology (Project ERKP291) and the Bio-SANS beam line is supported by the Office of Biological and Environmental Research U.S. Department of Energy. Research at the Spallation Neutron Source and High Flux Isotope Reactor at Oak Ridge National Laboratory (ORNL) is supported by the Scientific User Facilities Division, Office of Basic Energy Sciences, U.S. Department of Energy (DOE). ORNL is managed by UT-Battelle, LLC, for the U.S. DOE under Contract No. DE-AC05-00OR22725.

REFERENCES

Andjus, P. R., Kataev, A. A., Alexandrov, A. A., Vucelic, D., & Berestovsky, G. N. (1994). D$_2$O-Induced ion channel activation in Characeae at low ionic strength. *Journal of Membrane Biology, 142*(1), 43–53. http://dx.doi.org/10.1007/Bf00233382.

Arashida, T., Ishino, T., Kai, A., Hatanaka, K., Akaike, T., Matsuzaki, K., et al. (1993). Biosynthesis of cellulose from culture media containing C-13-labeled glucose as a carbon source. *Journal of Carbohydrate Chemistry, 12*(4–5), 641–649. http://dx.doi.org/10.1080/07328309308019413.

Bachor, R., Shea, C. R., Gillies, R., & Hasan, T. (1991). Photosensitized destruction of human bladder carcinoma cells treated with chlorin E$_6$-conjugated microspheres. *Proceedings of the National Academy of Sciences of the United States of America, 88*(4), 1580–1584. http://dx.doi.org/10.1073/pnas.88.4.1580.

Bali, G., Foston, M. B., O'Neill, H. M., Evans, B. R., He, J. H., & Ragauskas, A. J. (2013). The effect of deuteration on the structure of bacterial cellulose. *Carbohydrate Research, 374*, 82–88. http://dx.doi.org/10.1016/j.carres.2013.04.009.

Barnoud, F., Gagnaire, D., Odier, L., & Vincendo, M. (1971). Biosynthesis of deuterated bacterial cellulose—NMR study on levels of incorporation and localization of deuterium. *Biopolymers, 10*(11), 2269–2273. http://dx.doi.org/10.1002/bip.360101118.

Basu, J., Padhy, N., & Mookerjee, A. (1990). An insight into the structure of DNA through melting studies. *Indian Journal of Biochemistry & Biophysics, 27*(4), 202–208.

Benziman, M., & Eizen, N. (1971). Pyruvate-phosphate dikinase and control of gluconeogenesis in acetobacter-xylinum. *Journal of Biological Chemistry, 246*(1), 57–61.

Biemann, K. (1962). *Mass spectrometry: Organic chemical applications*. New York: McGraw-Hill.

Bergey, D. H., & Holt, J. G. (1994). *Bergey's manual of determinative bacteriology*. Baltimore, MD: Williams & Wilkins.

Bernardo, E. B., Neilan, B. A., & Couperwhite, I. (1998). Characterization, differentiation and identification of wild-type cellulose-synthesizing Acetobacter strains involved in Nata de Coco production. *Systematic and Applied Microbiology, 21*(4), 599–608.

Brown, A. J. (1886). XLIII.—On an acetic ferment which forms cellulose. *Journal of the Chemical Society, Transactions, 49*, 432–439. http://dx.doi.org/10.1039/ct8864900432.

Byron, O., & Gilbert, R. J. (2000). Neutron scattering: Good news for biotechnology. *Current Opinion in Biotechnology*, *11*(1), 72–80. S0958-1669(99)00057-9 [pii].

Cannon, R. E., & Anderson, S. M. (1991). Biogenesis of bacterial cellulose. *Critical Reviews in Microbiology*, *17*(6), 435–447. http://dx.doi.org/10.3109/10408419109115207.

Chumpitazi-Hermoza, B. F., Gagnaire, D., & Taravel, F. R. (1983). Bacterial biosynthesis of cellulose from d-glucose or glycerol precursors labelled with deuterium. *Carbohydrate Polymers*, *3*(1), 1–12. http://dx.doi.org/10.1016/0144-8617(83)90008-5.

Chundawat, S. P., Vismeh, R., Sharma, L. N., Humpula, J. F., da Costa Sousa, L., Chambliss, C. K., et al. (2010). Multifaceted characterization of cell wall decomposition products formed during ammonia fiber expansion (AFEX) and dilute acid based pretreatments. *Bioresource Technology*, *101*(21), 8429–8438. http://dx.doi.org/10.1016/j.biortech.2010.06.027. S0960-8524(10)01008-4 [pii].

Crespi, H. L. (1977). *Biosynthesis with deuterated micro-organisms stable isotopes in the life sciences*. (pp. 111–121). Vienna: International Atomic Energy Agency.

Czaja, W. K., Young, D. J., Kawecki, M., & Brown, R. M., Jr. (2007). The future prospects of microbial cellulose in biomedical applications. *Biomacromolecules*, *8*(1), 1–12. http://dx.doi.org/10.1021/bm060620d.

de Ghellinck, A., Schaller, H., Laux, V., Haertlein, M., Sferrazza, M., Marechal, E., et al. (2014). Production and analysis of perdeuterated lipids from *Pichia pastoris* cells. *PLoS One*, *9*(4), e92999. http://dx.doi.org/10.1371/journal.pone.0092999. PONE-D-13-53874 [pii].

Evans, B. R., Bali, G., Reeves, D. T., O'Neill, H. M., Sun, Q., Shah, R., et al. (2014). Effect of D2O on growth properties and chemical structure of annual ryegrass (Lolium multiflorum). *Journal of Agricultural and Food Chemistry*, *62*(12), 2595–2604. http://dx.doi.org/10.1021/jf4055566.

Evans, B. R., & O'Neill, H. M. (2005). Effect of surface attachment on synthesis of bacterial cellulose. *Applied Biochemistry and Biotechnology*, *121–124*, 439–450. ABAB:121:1-3:0439 [pii].

Evans, B. R., O'Neill, H. M., Malyvanh, V. P., Lee, I., & Woodward, J. (2003). Palladium-bacterial cellulose membranes for fuel cells. *Biosensors & Bioelectronics*, *18*(7), 917–923. S0956566302002129 [pii].

Franz, G., & Blaschek, W. (1990). Cellulose. In P. M. Dey (Ed.), *Methods in plant biochemistry: Vol. 2*. (pp. 291–322). New York, NY: Academic Press.

Gagnaire, D., & Taravel, F. R. (1975). Biosynthesis of bacterial cellulose from glucose selectively deuterated in position 6: NMR study. *FEBS Letters*, *60*(2), 317–321.

Gao, D., Chundawat, S. P., Krishnan, C., Balan, V., & Dale, B. E. (2010). Mixture optimization of six core glycosyl hydrolases for maximizing saccharification of ammonia fiber expansion (AFEX) pretreated corn stover. *Bioresource Technology*, *101*(8), 2770–2781. http://dx.doi.org/10.1016/j.biortech.2009.10.056. S0960-8524(09)01432-1 [pii].

Glaser, L. (1958). The synthesis of cellulose in cell-free extracts of Acetobacter xylinum. *Journal of Biological Chemistry*, *232*(2), 627–636.

Graslund, S., Nordlund, P., Weigelt, J., Bray, J., Hallberg, B. M., Gileadi, O., et al. (2008). Protein production and purification. *Nature Methods*, *5*(2), 135–146. http://dx.doi.org/10.1038/Nmeth.F.202.

Gross, P. R., & Spindel, W. (1960). Mitotic arrest by deuterium oxide. *Science*, *131*(3392), 37–39. http://dx.doi.org/10.1126/science.131.3392.37.

He, J., Pingali, S. V., Chundawat, S. P. S., Pack, A., Jones, A. D., Langan, P., et al. (2014). Controlled incorporation of deuterium into bacterial cellulose. *Cellulose*, *21*(2), 927–936.

Heller, W. T., Urban, V. S., Lynn, G. W., Weiss, K. L., O'Neill, H. M., Pingali, S. V., et al. (2014). The Bio-SANS instrument at the High Flux Isotope Reactor of Oak Ridge National Laboratory. *Journal of Applied Crystallography*, *47*, 1238–1246. http://dx.doi.org/10.1107/S1600576714011285.

Herrmann, S., & Neuschul, P. (1931). Zur Biochemie der Essigbakterien, zugleich ein Vorschlag fuer eine neue Systematik. *Biochemische Zeitschrift, 233*, 129–216.
Himmel, M. E., Ding, S. Y., Johnson, D. K., Adney, W. S., Nimlos, M. R., Brady, J. W., et al. (2007). Biomass recalcitrance: Engineering plants and enzymes for biofuels production. *Science, 315*(5813), 804–807. http://dx.doi.org/10.1126/science.1137016 315/5813/804 [pii].
Hofstetter, K., Hinterstoisser, B., & Salmen, L. (2006). Moisture uptake in native cellulose— The roles of different hydrogen bonds: A dynamic FT-IR study using Deuterium exchange. *Cellulose, 13*(2), 131–145. http://dx.doi.org/10.1007/s10570-006-9055-2.
Iguchi, M., Yamanaka, S., & Budhiono, A. (2000). Bacterial cellulose—A masterpiece of nature's arts. *Journal of Materials Science, 35*(2), 261–270. http://dx.doi.org/10.1023/A:1004775229149.
Itoh, T. J., & Sato, H. (1984). The effects of deuterium Oxide (^2H_2O) on the polymerization of tubulin in vitro. *Biochimica et Biophysica Acta, 800*(1), 21–27. http://dx.doi.org/10.1016/0304-4165(84)90089-8.
Katz, J. J., & Crespi, H. L. (1966). Deuterated organisms: Cultivation and uses. *Science, 151*(3715), 1187–1194.
Kawano, S., Tajima, K., Uemori, Y., Yamashita, H., Erata, T., Munekata, M., et al. (2002). Cloning of cellulose synthesis related genes from Acetobacter xylinum ATCC23769 and ATCC53582: Comparison of cellulose synthetic ability between strains. *DNA Research, 9*(5), 149–156. http://dx.doi.org/10.1093/dnares/9.5.149.
Klemm, D., Heublein, B., Fink, H. P., & Bohn, A. (2005). Cellulose: Fascinating biopolymer and sustainable raw material. *Angewandte Chemie-International Edition, 44*(22), 3358–3393. http://dx.doi.org/10.1002/anie.200460587.
Krystynowicz, A., Czaja, W., Wiktorowska-Jezierska, A., Goncalves-Miskiewicz, M., Turkiewicz, M., & Bielecki, S. (2002). Factors affecting the yield and properties of bacterial cellulose. *Journal of Industrial Microbiology & Biotechnology, 29*(4), 189–195. http://dx.doi.org/10.1038/sj.jim.7000303.
Kushner, D. J., Baker, A., & Dunstall, T. G. (1999). Pharmacological uses and perspectives of heavy water and deuterated compounds. *Canadian Journal of Physiology and Pharmacology, 77*(2), 79–88.
Lamprecht, J., Schroeter, D., & Paweletz, N. (1991). Derangement of microtubule arrays in interphase and mitotic PtK2 cells treated with deuterium oxide (heavy water). *Journal of Cell Science, 98*, 463–473.
Langan, P., Petridis, L., O'Neill,, H. M., Pingali, S. V., Foston, M., Nishiyama, Y., et al. (2014). Common processes drive the thermochemical pretreatment of lignocellulosic biomass. *Green Chemistry, 16*(1), 63–68. http://dx.doi.org/10.1039/C3gc41962b.
Lin, S. P., Calvar, I. L., Catchmark, J. M., Liu, J. R., Demirci, A., & Cheng, K. C. (2013). Biosynthesis, production and applications of bacterial cellulose. *Cellulose, 20*(5), 2191–2219. http://dx.doi.org/10.1007/s10570-013-9994-3.
Matsuoka, M., Tsuchida, T., Matsushita, K., Adachi, O., & Yoshinaga, F. (1996). A synthetic medium for bacterial cellulose production by *Acetobacter xylinum* subsp. *sucrofermentans*. *Bioscience, Biotechnology, and Biochemistry, 60*(4), 575–579.
McIntosh, T. S., Davis, H. M., & Matthews, D. E. (2002). A liquid chromatography-mass spectrometry method to measure stable isotopic tracer enrichments of glycerol and glucose in human serum. *Analytical Biochemistry, 300*(2), 163–169. http://dx.doi.org/10.1006/abio.2001.5455.
Melnichenko, Y. B., & Wignall, G. D. (2007). Small-angle neutron scattering in materials science: Recent practical applications. *Journal of Applied Physics, 102*(2), 021101. http://dx.doi.org/10.1063/1.2759200. Artn 021101.
Minor, F. W., Greathouse, G. A., Shirk, H. G., Schwartz, A. M., & Harris, M. (1954). Biosynthesis of C^{14}-specifically labeled cellulose by *Acetobacter xylinum*. II. From

D-Mannitol-1-C^{14} with and without Ethanol[1]. *Journal of the American Chemical Society, 76*(20), 5052–5054. http://dx.doi.org/10.1021/ja01649a012.

Nishiyama, Y., Okano, T., Langan, P., & Chanzy, H. (1999). High resolution neutron fibre diffraction data on hydrogenated and deuterated cellulose. *International Journal of Biological Macromolecules, 26*(4), 279–283. S0141-8130(99)00094-X [pii].

Perez, S., & Samain, D. (2010). Structure and engineering of celluloses. *Advances in Carbohydrate Chemistry and Biochemistry, 64,* 25–116. http://dx.doi.org/10.1016/S0065-2318(10)64003-6. S0065-2318(10)64003-6 [pii].

Pingali, S. V., O'Neill, H. M., McGaughey, J., Urban, V. S., Rempe, C. S., Petridis, L., et al. (2011). Small angle neutron scattering reveals pH-dependent conformational changes in Trichoderma reesei cellobiohydrolase I: Implications for enzymatic activity. *Journal of Biological Chemistry, 286*(37), 32801–32809. http://dx.doi.org/10.1074/jbc.M111.263004.

Pingali, S. V., Urban, V. S., Heller, W. T., McGaughey, J., O'Neill, H., Foston, M., et al. (2010a). Breakdown of cell wall nanostructure in dilute acid pretreated biomass. *Biomacromolecules, 11*(9), 2329–2335. http://dx.doi.org/10.1021/Bm100455h.

Pingali, S. V., Urban, V. S., Heller, W. T., McGaughey, J., O'Neill, H. M., Foston, M., et al. (2010b). SANS study of cellulose extracted from switchgrass. *Acta Crystallographica, Section D: Biological Crystallography, 66*(Pt. 11), 1189–1193. http://dx.doi.org/10.1107/S0907444910020408. S0907444910020408 [pii].

Ragauskas, A. J., Williams, C. K., Davison, B. H., Britovsek, G., Cairney, J., Eckert, C. A., et al. (2006). The path forward for biofuels and biomaterials. *Science, 311*(5760), 484–489. http://dx.doi.org/10.1126/science.1114736. 311/5760/484 [pii].

Sano, M., Rojas, A., Gatenholm, P., & Davalos, R. (2010). Electromagnetically controlled biological assembly of aligned bacterial cellulose nanofibers. *Annals of Biomedical Engineering, 38*(8), 2475–2484. http://dx.doi.org/10.1007/s10439-010-9999-0.

Saxena, I. M., Kudlicka, K., Okuda, K., & Brown, R. M. (1994). Characterization of genes in the cellulose-synthesizing operon (Acs Operon) of acetobacter-xylinum—Implications for cellulose crystallization. *Journal of Bacteriology, 176*(18), 5735–5752.

Schramm, M., & Hestrin, S. (1954). Factors affecting production of cellulose at the air liquid interface of a culture of acetobacter-xylinum. *Journal of General Microbiology, 11*(1), 123–129.

Shah, J., & Brown, R. M. (2005). Towards electronic paper displays made from microbial cellulose. *Applied Microbiology and Biotechnology, 66*(4), 352–355. http://dx.doi.org/10.1007/s00253-004-1756-6.

Son, H. J., Kim, H. G., Kim, K. K., Kim, H. S., Kim, Y. G., & Lee, S. J. (2003). Increased production of bacterial cellulose by Acetobacter sp V6 in synthetic media under shaking culture conditions. *Bioresource Technology, 86*(3), 215–219. http://dx.doi.org/10.1016/.S0960-8524(02)00176-1.

Takahashi, H., & Shimada, I. (2010). Production of isotopically labeled heterologous proteins in non-*E. coli* prokaryotic and eukaryotic cells. *Journal of Biomolecular NMR, 46*(1), 3–10. http://dx.doi.org/10.1007/s10858-009-9377-0.

Tashiro, K., & Kobayashi, M. (1991). Theoretical evaluation of 3-dimensional elastic-constants of native and regenerated celluloses—Role of hydrogen-bonds. *Polymer, 32*(8), 1516–1530. http://dx.doi.org/10.1016/0032-3861(91)90435-L.

Torres, J., Kukol, A., & Arkin, I. T. (2000). Use of a single glycine residue to determine the tilt and orientation of a transmembrane helix. A new structural label for infrared spectroscopy. *Biophysical Journal, 79*(6), 3139–3143.

Unno, K., & Okada, S. (1994). Deuteration causes the decreased induction of heat-shock proteins and increased sensitivity to heat denaturation of proteins in *Chlorella*. *Plant and Cell Physiology, 35*(2), 197–202.

Unno, K., Shimba, S., & Okada, S. (1989). Modification of thermal response of *Chlorella ellipsoidea* by deuteration. *Chemical & Pharmaceutical Bulletin, 37*(11), 3047–3049.

Utsumi, H., & Elkind, M. M. (1991). Caffeine and D_2O medium interact in affecting the expression of radiation-induced potentially lethal damage. *International Journal of Radiation Biology, 60*(4), 647–655. http://dx.doi.org/10.1080/09553009114552471.

Vasella, A., Davies, G. J., & Bohm, M. (2002). Glycosidase mechanisms. *Current Opinion in Chemical Biology, 6*(5), 619–629. http://dx.doi.org/10.1016/S1367-5931(02)00380-0.

Vasilescu, V., & Katona, E. (1986). Deuteration as a tool in investigating the role of water in the structure and function of excitable membranes. *Methods in Enzymology, 127*, 662–678.

Watanabe, K., Shibata, A., Ougiya, H., & Hioki, N. (2000). Method for processing bacterial cellulose. *USA Patent, 6*, 153, 413.

Yamada, Y., Hoshino, K., & Ishikawa, T. (1997). Taxonomic studies of acetic acid bacteria and allied organisms. 11. The phylogeny of acetic acid bacteria based on the partial sequences of 16S ribosomal RNA: The elevation of the subgenus Gluconoacetobacter to the generic level. *Bioscience, Biotechnology, and Biochemistry, 61*(8), 1244–1251.

CHAPTER SEVEN

Isotopic Labeling of Proteins in *Halobacterium salinarum*

Thomas E. Cleveland IV[*,‡,1], Zvi Kelman[†,‡]

*NIST Center for Neutron Research, National Institute of Standards and Technology, Gaithersburg, Maryland, USA
†Biomolecular Labeling Laboratory, Institute for Bioscience and Biotechnology Research, National Institute of Standards and Technology and the University of Maryland, Rockville, Maryland, USA
‡Biomolecular Structure and Function Group, Institute for Bioscience and Biotechnology Research, National Institute of Standards and Technology and the University of Maryland, Rockville, Maryland, USA
[1]Corresponding author: e-mail address: thomas.cleveland@nist.gov

Contents

1. Introduction — 148
2. Growth and Maintenance of *Halobacterium salinarum* — 149
   2.1 Overview — 149
   2.2 Reagents — 150
   2.3 Solutions and Buffers — 150
   2.4 Starter Culture Growth in Nonlabeled Medium — 152
   2.5 Frozen Storage of Cultures — 152
   2.6 Growth on Agar Plates — 152
   2.7 Growth of *Halobacterium salinarum* in Isotopically Labeled Medium — 153
3. Purification of Proteins from *Halobacterium salinarum* — 153
   3.1 Factors Affecting Protein Expression — 153
   3.2 Cell Growth for Protein Production — 154
   3.3 Cell Harvest — 154
   3.4 Cell Lysis — 156
   3.5 Expression and Purification of Bacteriorhodopsin — 158
   3.6 Verifying Isotopic Labeling of Bacteriorhodopsin by MALDI-MS — 162
4. Summary — 163
Acknowledgments — 164
References — 164

Abstract

It is often necessary to obtain isotopically labeled proteins containing ^{15}N, ^{13}C, or ^{2}H for nuclear magnetic resonance; and ^{2}H for small-angle neutron scattering or neutron diffraction studies. To achieve uniform isotopic labeling, protein expression is most commonly performed in *Escherichia coli* or yeast using labeled media. However, proteins from extreme halophiles sometimes require a cellular environment with high ionic strength and cannot be heterologously expressed in *E. coli* or yeast in functional form.

Methods in Enzymology, Volume 565
ISSN 0076-6879
http://dx.doi.org/10.1016/bs.mie.2015.06.002

We present here methods for the cultivation of *Halobacterium salinarum* in isotopically labeled rich media, using commercially available isotopically labeled hydrolysates. The methods described here are both technically simple and relatively inexpensive.

1. INTRODUCTION

Halobacterium salinarum (formerly known as *H. halobium*) is an extreme halophilic archaeon, commonly used as a model organism for halophilic archaea, and as a source of the membrane protein bacteriorhodopsin (bR) (Oesterhelt & Stoeckenius, 1971). This organism is easy to cultivate using standard equipment and reagents, and it can also be transformed for the overexpression of heterologous proteins (Cline, Lam, Charlebois, Schalkwyk, & Doolittle, 1989; Jaakola et al., 2005). Unfortunately, simple minimal media like those used in the culture of *Escherichia coli* are not known for *H. salinarum*, and this has limited the use of this organism in the production of isotopically labeled proteins. While a variety of complex media based on hydrolysates (acid- or enzymatically digested protein from animal, microbial, or plant sources) will support its growth, its chemically defined media require the addition of many amino acids (Robb, DasSarma, & Fleischmann, 1995), which is generally cost-prohibitive for isotopic labeling.

The organism is reported to grow in deuterated media based on deuterated algal hydrolysates (Crespi, 1982) as a nutrient source. This method has not often been exploited, however, due to the burden of having to obtain algal cultures, grow them using isotopically labeled nutrients (e.g., ^2H_2O, $^{13}CO_2$, etc.), and manually prepare hydrolysates from them (which, in the case of acid hydrolysis, requires the use of ^2HCl and NaO^2H). Furthermore, many published reports have implied that *H. salinarum* can be sensitive to the exact methods of preparation for protein hydrolysates, for reasons which are not well understood, but may involve sensitivity to toxic components present in certain hydrolysates. Some hydrolysates are not recommended for use with *H. salinarum* (Oesterhelt & Stoeckenius, 1974), while others are specifically recommended (most commonly Oxoid bacteriological peptone). The reported methods used to produce algal hydrolysates that support the growth of *H. salinarum* have varied, and at least one study has reported that additional purification steps, such as repeated rounds of organic extraction prior to protein hydrolysis, were required when preparing algal hydrolysates (Patzelt et al., 1997).

Based on this background, we could not assume that a given algal hydrolysate would necessarily be able to support the growth of *H. salinarum*, particularly in light of the additional insult of growth in ^{2}H-labeled conditions. To avoid the inconsistency and difficulty of self-preparing algal hydrolysates, we investigated the use of several commercially available, isotopically labeled algal hydrolysates for the growth of *H. salinarum*. These hydrolysates were found to be suitable for growth, and the methods for their use are presented below. Methods for the in-house preparation of hydrolysates have also been described (Crespi, 1982; Patzelt et al., 1997). We note that algal hydrolysates could also potentially be used to make complex isotopically labeled media for a variety of other organisms lacking convenient minimal media. These algal media are cost-competitive with minimal media, while allowing simpler preparation, and potentially a larger variety of organisms to be grown.

2. GROWTH AND MAINTENANCE OF *HALOBACTERIUM SALINARUM*

2.1 Overview

Most procedures for the maintenance of *H. salinarum* cultures are similar to standard methods used for *E. coli*. However, *H. salinarum* grows more slowly than *E. coli* and requires moderately higher temperatures for optimal growth. The organism will grow at any temperature from approximately 30 to 50 °C, with fastest doubling times reportedly near 50 °C (Robinson et al., 2005). We incubate plates and grow initial small-volume cultures at 45 °C. Other protocols commonly suggest 37–42 °C for these steps (Oesterhelt & Stoeckenius, 1974), but growth is slow at 37 °C, resulting in long wait times when starting suspension cultures from single colonies (up to 3 days) or when waiting for single colonies to appear on plates (a week or more). When incubated at 45 °C, growth in culture tubes inoculated from single colonies will be apparent within a day. For protein expression, cultures can be grown at 45 °C to a moderate initial density before shifting to a different temperature for expression. Expression temperature should be optimized on a case-by-case basis within a suggested range of 30–45 °C, as optimal doubling times do not imply optimal expression yields.

Typical published culture media for *H. salinarum* (Sehgal & Gibbons, 1960) contains 3 g/L sodium citrate. We found this to be unnecessary, and therefore eliminated it from our culture media in order to minimize the number of components potentially requiring isotopic labeling.

2.2 Reagents

Sources are listed when necessary. When not specified, any source may be used.

Hydrochloric acid (HCl)
10 M sodium hydroxide solution
Zinc sulfate heptahydrate ($ZnSO_4 \cdot 7H_2O$)
Manganese sulfate monohydrate ($MnSO_4 \cdot H_2O$)
Ammonium ferrous sulfate hexahydrate ((NH_4)$_2$Fe(SO_4)$_2 \cdot 6H_2O$)
Copper(II) sulfate pentahydrate ($CuSO_4 \cdot 5H_2O$)
Sodium chloride (NaCl)
Magnesium sulfate heptahydrate ($MgSO_4 \cdot 7H_2O$)
Anhydrous magnesium sulfate ($MgSO_4$)
Potassium chloride (KCl)
3-(N-morpholino)propanesulfonic acid (MOPS)
Oxoid bacteriological peptone (Thermo Scientific, #LP0037)
Celtone Plus powder, isotopically labeled as needed (Cambridge Isotope Laboratories, Inc.)
Deuterium oxide (referred to as D_2O or ^2H_2O) if labeling with deuterium.

2.3 Solutions and Buffers

Trace Metals Solution (10,000×)

$ZnSO_4 \cdot 7H_2O$	1.32 g
$MnSO_4 \cdot H_2O$	0.34 g
$(NH_4)_2Fe(SO_4)_2 \cdot 6H_2O$	0.78 g
$CuSO_4 \cdot 5H_2O$	0.14 g
Dissolve in 0.1 M HCl to a final volume of 200 mL	

Buffered Salt Solution

NaCl	250 g
$MgSO_4 \cdot 7H_2O$	20 g
KCl	2 g
Trace metals solution	0.1 mL
MOPS, no pH adjustment	2.093 g (10 mM final)
10 M NaOH	0.5 mL (5 mM final)
Dissolve in water to a final volume of 1 L	

Instead of MOPS buffer, sodium phosphate monobasic can in principle be added and carefully titrated with NaOH to pH 6.8 to 7.0. However, salts of phosphate with divalent cations will increasingly precipitate as the pH is raised above this range. We prefer to avoid this possibility. Other buffers (e.g., HEPES, Tris) could be tried, but have not been tested by us.

Deuterated Buffered Salt Solution:
Make the Following Substitutions in the above Recipe

MgSO$_4$ anhydrous	9.8 g
10 M NaOH in D$_2$O	0.5 mL

NaOD may be used, but is generally not necessary, as NaOH adds only about 0.005 atom% ^{1}H. Exchangeable hydrogens in MOPS buffer can also be replaced with deuterium by dissolution in D$_2$O and evaporation. Similar exchange procedures can be carried out if phosphate or another buffer is used. However, given typical ^{1}H levels present in any case in the D$_2$O solvent, and in the deuterated algal hydrolysates, exchange of ^{1}H for ^{2}H in minor buffer components is not likely to be justifiable.

Dissolve in D$_2$O to a final volume of 1 L

Standard growth medium

1. Dissolve 10 g/L of Oxoid bacteriological peptone in buffered salt solution.
2. Filter sterilize if necessary (using, e.g., 0.22 µm sterile syringe or bottle-top filters).

    Sterilization is not strictly required for routine cell growth and protein expression, since few contaminating organisms are able to grow in this very high-salt medium. However, sterile technique is highly advisable when storing, maintaining, or retrieving frozen stock cultures, particularly if other halophiles are being maintained in the same laboratory.

Isotopic labeling using algal growth medium

1. Add 10 g/L of appropriate isotopically labeled Celtone Plus powder (available in ^{2}H-, ^{15}N-, and/or ^{13}C-labeled as well as unlabeled forms) to buffered salt solution (deuterated if necessary).
2. Allow 30 min to dissolve with occasional mixing or gentle stirring.

    A fine precipitate sometimes remains.
3. Remove any remaining residue by filtration (using, e.g., 0.22 µm sterile syringe or bottle-top filters).

80% glycerol/medium

1. Mix 80 mL glycerol with 20 mL growth medium.
2. Filter sterilize (using, e.g., 0.22 µm syringe filters).

2.4 Starter Culture Growth in Nonlabeled Medium

1. Fill a culture tube to approximately 25% capacity with standard growth medium (e.g., 3 mL of medium in a 14 mL round-bottomed Falcon tube, such as Corning No. 352059).
2. Inoculate with one colony of *H. salinarum* from a plate.
3. Place in a shaking incubator at 45 °C and 225 rpm.
4. Growth will appear within 1 day.
5. Larger-scale (e.g., 1 L) cultures can be inoculated with starter cultures at a volume ratio of 1:100 to 1:1000 after 1–2 days of starter growth.

2.5 Frozen Storage of Cultures

1. Grow culture 1–2 days from single colony as above.
2. To three parts culture, add one part 80% glycerol/medium and mix gently.
3. Transfer 1 mL aliquots to screw-cap cryo vials.
4. Freeze in dry ice/ethanol.
5. Store at −80 °C.
6. To retrieve from storage, scratch the top of a frozen aliquot with a sterilized loop and immediately return to −80 °C without thawing. Streak plates (see preparation below) with loop to obtain single colonies.

2.6 Growth on Agar Plates

1. Add 15 g agar to 1 L of standard growth medium.
2. Autoclave for 20 min.
    Alternatively, agar may be dissolved by microwaving or by stirring on a hot plate. Small volumes of solid media (<100 mL) for two to three plates may be conveniently prepared by microwaving, but may require repeated heating/swirling and readdition of evaporated water.
3. Precipitates may appear upon heating. This will not prevent colony growth. Filtration is not necessary.
4. After dissolving agar, pour thick plates (e.g., 30 mL per 10-cm plate) to minimize salt crystal formation.
5. Allow plates to solidify.
6. Streak/spread *H. salinarum* cultures.
7. Invert plates, place in a sealed container to minimize evaporation, and incubate at 45 °C.
8. Colonies will appear in 4–6 days.
9. Plates may be sealed and stored at 4 °C for several months.

2.7 Growth of *Halobacterium salinarum* in Isotopically Labeled Medium

In our hands, gradual adaptation was absolutely required to obtain growth in ^{2}H-labeled culture media. Adaptation to growth in completely deuterated media can take an extended period of time (up to 4–6 weeks to obtain well-adapted cultures). It is recommended that this to be done in advance of anticipated experiments, and that preadapted cultures be stored as frozen glycerol stocks for future use.

1. Prepare a small-volume starter culture in standard medium as above.
2. For ^{2}H-labeled media: wean the culture into isotopically labeled algal growth medium by splitting 1:1 with labeled medium every 2–4 days. Wait for appreciable cell growth to occur, e.g., an optical density at 600 nm (OD_{600}) of 0.25–0.5, before each additional split. After 5–10 splits, additional subcultures into new media can be performed at volume ratios of 1/10, and finally 1/300 once cultures are well adapted.
3. Inoculate large volume shake flasks (e.g., 1 L) containing labeled medium 1/300 with adapted starter culture. Alternatively, start with a small-volume culture and increase the volume sequentially in fivefold increments until the desired volume is reached, waiting for appreciable growth before each increase in volume.

3. PURIFICATION OF PROTEINS FROM *HALOBACTERIUM SALINARUM*

3.1 Factors Affecting Protein Expression

While starter cultures are generally grown under fixed conditions of 45 °C and 225 rpm in culture tubes, larger cultures for protein expression may be grown under variable conditions depending on the protein. Our experience is particular to bR production, although the same expression parameters may also hold for other proteins heterologously expressed under the control of the bR gene (*bop*) promoter. However, different conditions are probably optimal for other proteins and/or promoters. For unlabeled bR production, we have found optimal conditions to be low levels of aeration (shaking at 150 rpm in 1 L nonbaffled flasks), moderate temperatures (37 °C), and high illumination (under a compact fluorescent light bulb with a brightness of 6400 lumen) carried out for 5 days after the initial appearance of growth.

When starting an expression project on a new protein, the most critical parameter to evaluate is expression temperature, which should be varied in the range of around 30–45 °C. For lower temperatures in this range, to avoid very long wait times, the culture should be grown up to density at 40–45 °C before shifting to the lower temperature. In addition to temperature, harvest time should be optimized, which can be easily done by taking small samples at fixed intervals for expression quantification, e.g., by sodium dodecyl sulfate-polyacrylamide gel electrophoresis (SDS-PAGE), Western blot, or small-scale trial purification (for instance, batch purification with a few microliters of nickel resin if using His-tagged recombinant proteins). Finally, to achieve different levels of aeration, rotation speed may be varied in the range of 150–225 rpm, or baffled flasks may be tried. We have found bR expression to be highest with less aeration, as described above; this observation is consistent with the well-known induction of bR expression by low oxygen conditions (Oesterhelt & Stoeckenius, 1973). Production of other proteins may be optimal with more aeration. Growth under high-intensity light is likewise specific to bR production under its native promoter, but may also affect expression of other proteins (and *should* be tried for other proteins expressed under the control of the *bop* promoter).

3.2 Cell Growth for Protein Production

This protocol is a suggested starting point. Growth times will likely need to be extended when ^{2}H-labeled growth medium is used.

1. Using 1–3 mL of starter culture prepared as above, inoculate 1 L of medium in a 4-L shake flask (nonbaffled for bR production, otherwise to be determined empirically).
2. Shake at 45 °C for 1–2 days until culture reaches an OD_{600} of 0.5–1.0 (longer may be needed for ^{2}H-labeled medium).
3. Change the temperature to one that is optimal for expression, as determined empirically for the protein being expressed. A suggested starting point is 37 °C, with a range of 30–45 °C.
4. Induce expression if necessary.
5. Shake for the empirically determined optimal duration to obtain maximum yield of the protein of interest. A suggested starting point is 3–5 days for growth at 37 °C.

3.3 Cell Harvest

H. salinarum is often difficult to resuspend after centrifugal harvest, possibly due to partial lysis. If the cells are to be washed and repelleted after initial harvest, we therefore add glycerol to protect against shear and allow resuspension.

Cell harvest (with cell wash)
1. After removing shake flasks from incubator, add 200 mL glycerol to each liter of culture and swirl to mix.
2. Transfer cultures to bottles and spin at $4000 \times g$ for 20 min
3. Resuspend pellet in buffered salts solution containing 20% glycerol, using about 10 mL/g packed cells.
4. Transfer to smaller vessels, e.g., 50 mL conical tubes, and spin at $4,000 \times g$ for 30 min.

    Centrifugation times are longer than is ordinary for other microorganisms such as *E. coli*, due to the high viscosity and density of the growth medium, particularly with added glycerol.
5. At this point, washed cell pellets may be frozen at $-80\,°C$ for future processing.

    Storage at $-20\,°C$ is not recommended. The high-salt concentration of the media, particularly with added glycerol, prevents complete freezing at this temperature.
6. Resuspend pellet in lysis buffer and proceed to lysis.

Cells should resuspend more easily if fresh, although it is normal for resuspension to be more difficult than for *E. coli*. If frozen, cell clumps may need to be broken up more aggressively using a homogenizer or syringe with small needle. If resuspending in low-osmotic buffer, cells will begin to lyse immediately whether fresh or frozen and will need to be broken up.

Cell harvest (without wash)
1. Transfer cultures to bottles and spin at $4000 \times g$ for 20 min.
2. Pellets may be frozen directly in bottles at $-80\,°C$ for storage.

    If frozen in this way, it is possible to remove pellets while still frozen by gently hand-warming the wall of the bottle and scraping firmly at the pellet edge with a small metal spatula. Frozen pellets may then be quickly removed from bottles, combined, and returned to $-80°C$ for more compact storage, or weighed and transferred to lysis buffer for resuspension.
3. Add lysis buffer to cell pellets in bottles. The lysis volume required will be dependent on the protein and intended method of purification, but we suggest 10 mL buffer per gram of cells as a starting point.
4. Crudely resuspend pellets by vortex, spatula, etc., and pool cell suspensions from all bottles. Pellet will be sticky and cell clumps will remain.
5. Break up cell clumps. We use a Dounce homogenizer for this purpose. Other homogenizers may be used, or cell slurry may be passed repeatedly

through a small needle. Magnetic stirring at 4 °C in the presence of DNase (see below) will also eventually result in complete resuspension, but more time is needed.
6. Proceed to lysis.

3.4 Cell Lysis
3.4.1 General Considerations
A general lysis buffer cannot be suggested since it is entirely dependent on the protein and method of purification to be used. However, the following general considerations are suggested:
- Unless there is a specific need to avoid it, add a few grains of DNase powder, or the manufacturer-specified amount of any other nonspecific nuclease product (e.g., 1 μL per 10 mL of lysis buffer of Benzonase Nuclease, Novagen) to reduce lysate viscosity. This can be omitted if sonication, which itself shears DNA, will be used for lysis. If French press or low-osmotic lysis is to be used, lysis volumes may need to be increased if DNase is to be omitted. However, DNA can interfere unpredictably with a wide spectrum of purification methods, and we strongly suggest some type of DNA removal be considered.
- We suggest adding 1 mM phenylmethylsulfonyl fluoride (PMSF) as proteinase inhibitor (from a 100 mM stock solution in methanol). Commercial inhibitor tablets can also be used as necessary, but PMSF is inexpensive and almost always adequate.
- For proteins sensitive to oxidation, we recommend adding 1 mM tris(2-carboxyethyl)phosphine (TCEP) or 1 mM dithiothreitol (DTT), as well as 1 mM ethylenediaminetetraacetic acid (EDTA), if these will not interfere with downstream purification or other activity. However, for proteins with both oxidized and reduced cysteines (such as membrane proteins with intact cytoplasmic and extracellular domains), TCEP or DTT may result in the inappropriate reduction of native disulfide bonds. In this case, we recommend the addition of EDTA alone, if possible. This protects free cysteines by chelating trace metal ions, which catalyze cysteine oxidation (Cecil & Mcphee, 1959); but since EDTA is not itself a reducing agent, existing disulfide bonds will remain.
- If purifying deuterium-labeled protein, D_2O may be substituted for ^1H_2O in lysis and purification buffers. If this is cost-prohibitive, the protein can be purified in H_2O; a final desalting step into D_2O and incubation may give satisfactory replacement of exchangeable ^1H for ^2H.

3.4.2 Low-Salt Lysis

H. salinarum will lyse spontaneously in ionic strengths of less than approximately 500 mM NaCl (Stoeckenius & Rowen, 1967). The target protein must, of course, be stable at "low" ionic strength (relative to halophiles) at the chosen buffer pH.
1. Add lysis buffer to cell pellet.
2. Resuspend cell clumps using a Dounce homogenizer or by passing through a needle.
3. If necessary, incubate cell suspension with gentle stirring until lysis is complete.
   Lysis occurs very quickly in buffers with very low ionic strength. Required incubation times will need to be empirically determined as ionic strength increases, and may also be affected by incubation temperature.
4. For soluble proteins, clear lysate by centrifugation, e.g., at 30,000 × g for 30 min. Filter lysate through 0.22 or 0.45 μm low-protein-binding syringe filters (e.g., polyethersulfone). Proceed to purification.
   For membrane proteins, general methods are beyond the scope of this article, but a protocol for the particular case of bR is given below.

3.4.3 French Press, Microfluidizer, or Similar

These methods are preferred in cases where low-osmotic lysis cannot be used (i.e., high-salt lysis buffers). Any lysis buffer may be used. If using a French press, we recommend prechilling the cell. Since hardware differs, follow manufacturer's directions for exact lysis protocol. After lysis, clear lysate in the same way as for low-osmotic lysis above.

3.4.4 Sonication Lysis

This method is not as gentle as the others (Benov & Al-Ibraheem, 2002), but can be convenient once an optimized protocol is developed. Detergents should be avoided, as they lead to foaming and protein denaturation. Sonication duty cycle and process time must be determined on a case-by-case basis, depending on the machine and characteristics of the sample and lysis buffer (sample volume, viscosity-modifying additives such as glycerol, process temperature, and ionic strength are the most important variables). We recommend trying the maximum power, using a fixed duty cycle chosen not to cause excessive heating given the particular cooling method used, and taking samples after increasing process times. To determine the minimum required process time, these samples can then be cleared by centrifugation at maximum speed in a microcentrifuge, and total released protein estimated by any method that is compatible with the lysis buffer (e.g., UV, Bradford, BCA). When the total protein concentration stops increasing, lysis is complete. Once a sonication protocol is established, the lysed sample should be cleared as above, followed by purification.

3.5 Expression and Purification of Bacteriorhodopsin

The following is a complete example expression and purification optimized for bR. Purple membrane isolation and bR purification is illustrated in Fig. 1.

3.5.1 Solutions

Lysis buffer: 10 mM MOPS, 5 mM NaOH (pH will be approximately 7.2 without additional adjustment). Add a few grains of DNase powder, or nonspecific nuclease according to the manufacturer's instructions (e.g., 1 μL of Benzonase Nuclease per 10 mL of lysis buffer). Add PMSF to a final concentration of 1 mM from stock solution of 100 mM in methanol.
Resuspension buffer: 10 mM MOPS, 5 mM NaOH.

3.5.2 Expression

1. Prepare 3 mL starter culture of *H. salinarum* as described above.
   For deuterium labeling, the starter culture must be sequentially adapted to deuterated algal medium in 100% D_2O as described.
2. Use the entire starter culture to inoculate 1 L of culture medium in a 4-L nonbaffled shake flask.
3. Grow overnight at 45 °C.
4. Move culture to 37 °C with illumination by a 6400-lumen compact fluorescent light.
5. Allow expression for 4–5 days.
6. Harvest cells by centrifugation at 4000 × g for 10 min.
7. Freeze at −80°C for later use if desired.

3.5.3 Lysis and Membrane Harvest

1. All operations are carried out at 4 °C or on ice.
2. To cell pellets, add 50 mL of lysis buffer per liter of culture.
3. Homogenize by Douncing (Fig. 1A, II).
4. Divide lysate into tubes for SW 28 ultracentrifuge rotor (Beckman Coulter) and fill to required volume with lysis buffer.

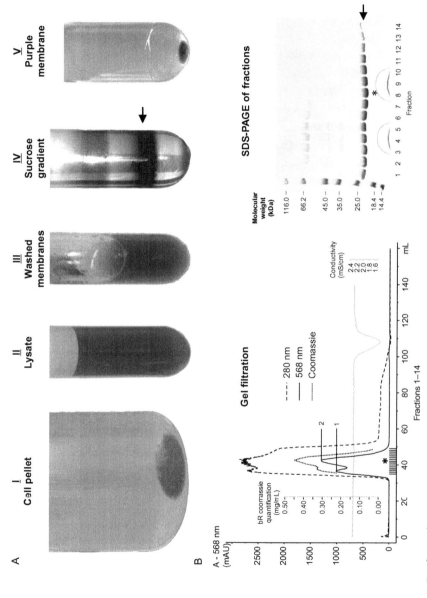

Figure 1 See legend on next page.

5. Harvest membranes by centrifugation at 22,000 rpm for 1 h.
6. Discard red supernatant.
   Membrane pellets may be frozen in their tubes for later use if desired.
7. Dislodge pellets with a jet of resuspension buffer (a syringe/needle works well).
8. Transfer pellets and buffer to Dounce homogenizer and resuspend in a minimum of 10 mL of buffer per gram of membrane pellet.
   Additional buffer will not be harmful, as it will be removed after centrifugation. Some additional buffer may be required in any case to bring the ultracentrifuge tubes to their required volume.
9. Repeat membrane wash (steps 5 through 7) for two to three cycles until supernatants are only faintly red. The remaining pellet will now have a distinctly purple color, rather than the cell pellet's initial red/pink color (Fig. 1A, III).

3.5.4 Sucrose Gradient Ultracentrifugation

Solutions: 30%, 40%, and 50% (w/v) sucrose containing 10 mM MOPS and 5 mM NaOH.

1. Using a Dounce homogenizer, resuspend washed membrane pellet in a minimal volume of resuspension buffer.

Figure 1 Purification of bR. (A) The isolation of the "purple membrane" (PM) is illustrated. The following stages of the protein purification are shown: cell harvest (I), lysis in low-osmotic buffer (II), resuspended membranes after several wash cycles with low-osmotic buffer (III), the band of PM after passage through a sucrose step gradient (IV), and the final PM pellet after washing to remove sucrose (V). The two-colored compounds visualized in the sucrose gradient step are bacterioruberin (red, upper band) and bR (purple, marked with an arrow). The preparation changes from reddish to purple as the bacterioruberin is increasingly eliminated. (B) A final gel filtration chromatography step is used to isolate soluble and monodisperse bR. The first peak (labeled "1") is aggregated bR in the void volume and is discarded, while peak "2" is retained. The peak bR fraction is labeled with an asterisk (*) on the gel filtration chromatogram (left), and on the corresponding SDS-PAGE gel of its fractions (right). UV absorbance during gel filtration was measured at 568 nm, which gives a signal specific to bR; and at 280 nm, which detects total protein as well as Triton X-100. Due to the high absorbance of Triton X-100, the detector saturates at this wavelength. In addition, the Coomassie-stained bR band (marked with arrow) on the SDS-PAGE was quantified using ImageJ (Schneider, Rasband, & Eliceiri, 2012), and the quantification overlaid on the UV trace. Finally, the conductivity trace is shown; bR is generally purified in low ionic strength buffers. (See the color plate.)

2. Layer 7 mL of each sucrose solution in any type of thick-wall SW 28 ultracentrifuge tube, starting with the 30% solution and underlaying the 40% and 50% solutions sequentially using a long needle. Prepare one sucrose gradient tube for every 4 mL of membrane suspension (a minimum of two tubes for balance).

   Solution volumes will need to be adjusted if different-sized tubes/rotors are used.
3. Layer up to 4 mL of membrane suspension on each tube. Balance tube pairs by transferring small amounts of membrane suspension between tubes.
4. Centrifuge at 22,000 rpm for about 24 h in an SW 28 rotor. Speeds of up to 28,000 rpm may be used in this rotor model, which will result in shorter centrifugation times.
5. Two bands will result from centrifugation: an upper red band, and a lower deep-purple band (which may be almost black if bR is highly concentrated). Other membranous material and debris may form a brownish pellet in the tube bottom (Fig. 1A, IV).
6. To harvest the "purple membrane" (PM), slowly and carefully collect the purple band from the underside, using suction with a syringe and blunt needle with U-shaped bend at the tip.
7. To remove sucrose, transfer the PM to a clean SW 28 tube, fill tube to maximum volume with resuspension buffer, mix, and spin a final time at 22,000 rpm for 2 h.
8. Discard the sucrose-containing supernatant and resuspend the PM pellet (Fig. 1A, V) in a minimal volume of resuspension buffer or distilled water. Snap-freeze and store at $-80\,°C$ if desired.

3.5.5 Detergent Solubilization and Gel Filtration

The following protocol illustrates solubilization and gel filtration in the presence of Triton X-100. Octyl-β-D-glucopyranoside is also commonly used for bR purification. The protocol is similar, but mass concentrations of detergent differ.

1. Measure the mass concentration of bR using absorption at 550 nm. We assume an extinction coefficient of $5.8 \times 10^4/M/cm$, which gives an absorbance of 2.16 for a mass concentration of 1 mg/mL.
2. Add a 5:1 (w/w) ratio of Triton X-100 to bR in 20 mM Tris–HCl pH 8.0. Bring to a final volume of 4 mL/10 mg of bR.
3. To solubilize, incubate overnight at room temperature in the dark.

4. Using 2 volumes of 0.03% (w/v) Triton X-100 in 20 mM Tris–HCl pH 8.0, equilibrate a suitable gel filtration column, such as Superdex 200 HiPrep 16/60 (or 26/60 for larger load volumes) (GE Healthcare).
5. Remove insoluble protein by ultracentrifugation at 100,000 × g for 1 h.
6. Apply detergent-solubilized protein to gel filtration column.

   If available, follow elution by absorption at 568 nm. Absorption at 280 nm can be followed, but will also detect peaks caused by empty micelles of Triton X-100, which elute later than bR. At 280 nm, the absorbance of the bR peak will likely exceed the limit of the detector due to the large amount of bound Triton X-100.
7. Pool fractions from the bR peak (Fig. 1B), which will elute at an apparent molecular weight of approximately 125 kDa due to the bound Triton X-100 micelle. Discard any fractions corresponding to the void volume of the column or high molecular weight oligomers.

3.6 Verifying Isotopic Labeling of Bacteriorhodopsin by MALDI-MS

We grew *H. salinarum* using triply labeled (^{15}N, ^{13}C, and ^{2}H) medium and isolated the purple membrane as outlined above. Isotopic labeling was confirmed by MALDI-MS (Fig. 2). The following protocol allows MALDI-MS

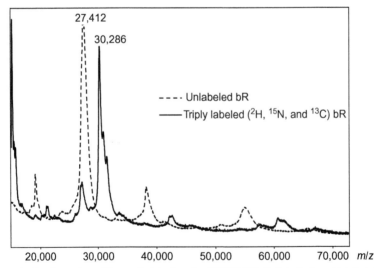

Figure 2 Confirmation of isotopic labeling by MALDI-MS. Unlabeled and triply labeled bR (^{2}H, ^{15}N, and ^{13}C) were analyzed by MALDI-MS, showing the expected mass shift.

to be performed directly on the PM to verify isotopic labeling of bR. It is not necessary to solubilize and perform gel filtration first.

Solutions

0.1% trifluoroacetic acid (TFA) in H_2O (also prepare in D_2O for analysis of deuterated proteins)

Matrix solvent: mix one part acetonitrile with two parts 0.1% TFA by volume.

Saturated matrix solution: add a small quantity of α-cyano-4-hydroxycinnamic acid (CHCA) to matrix solvent in a small tube. Vortex periodically over several minutes. A small amount of undissolved CHCA matrix should remain in the tube. Spin down and use supernatant. If deuterated proteins will be analyzed, prepare separate stocks using ^1H_2O (for calibration standards) and D_2O (for the deuterated protein).

Sinapinic acid is more typically recommended for analysis of whole proteins by MALDI-MS. In our hands, following the protocol below, better spectra were obtained using CHCA.

3.6.1 Protein Calibration Standards

Procedure

1. Prepare saturated matrix solution.
2. Dilute 1 μL of PM suspension into 10 μL of neat TFA.
3. Immediately add 1 μL of PM/TFA to 1 μL of saturated matrix solution (prepared with D_2O if analyzing deuterated protein).
4. Immediately spot 0.5 μL of PM/TFA/matrix onto a MALDI target and allow solvent to evaporate.
5. Prepare standards as recommended by the manufacturer. Mix equal parts standards solution with saturated matrix solution, spot 0.5 μL of the mixture onto MALDI target, and allow solvent evaporation.
6. Analyze spots.

4. SUMMARY

We have presented a collection of methods for the expression and purification of isotopically labeled proteins from *H. salinarum*. These methods should be useful for anyone working on the biochemistry or structural biology of proteins from halophilic organisms that needs to produce isotopically labeled proteins for techniques such as nuclear magnetic resonance, small-angle neutron scattering, or neutron diffraction. In addition to allowing the production of labeled bR, established genetic tools for *H. salinarum* (Cline & Doolittle, 1987; Cline et al., 1989; Jaakola et al., 2005; Karan, Capes, DasSarma, & DasSarma, 2013; Nomura & Harada, 1998) could allow the heterologous expression and labeling of other proteins, particularly those that cannot be expressed and labeled in functional form in *E. coli* or yeast.

ACKNOWLEDGMENTS

We would like to thank Dr. Lori Kelman for critical reading of the manuscript.

Disclaimer. Certain commercial equipment, instruments, and materials are identified in this paper in order to specify the experimental procedure. Such identification does not imply recommendation or endorsement by the National Institute of Standards and Technology, nor does it imply that the material or equipment identified is necessarily the best available for the purpose.

REFERENCES

Benov, L., & Al-Ibraheem, J. (2002). Disrupting *Escherichia coli*: A comparison of methods. *Journal of Biochemistry and Molecular Biology, 35*, 428–431.

Cecil, R., & Mcphee, J. R. (1959). The sulfur chemistry of proteins. *Advances in Protein Chemistry, 14*, 255–389.

Cline, S. W., & Doolittle, W. F. (1987). Efficient transfection of the archaebacterium *Halobacterium halobium*. *Journal of Bacteriology, 169*, 1341–1344.

Cline, S. W., Lam, W. L., Charlebois, R. L., Schalkwyk, L. C., & Doolittle, W. F. (1989). Transformation methods for halophilic archaebacteria. *Canadian Journal of Microbiology, 35*, 148–152.

Crespi, H. L. (1982). The isolation of deuterated bacteriorhodopsin from fully deuterated *Halobacterium halobium*. *Methods in Enzymology, 88*, 3–5.

Jaakola, V. P., Rehn, M., Moeller, M., Alexiev, U., Goldman, A., & Turner, G. J. (2005). G-protein-coupled receptor domain overexpression in *Halobacterium salinarum*: Long-range transmembrane interactions in heptahelical membrane proteins. *Proteins: Structure, Function, and Genetics, 60*, 412–423.

Karan, R., Capes, M. D., DasSarma, P., & DasSarma, S. (2013). Cloning, overexpression, purification, and characterization of a polyextremophilic β-galactosidase from the Antarctic haloarchaeon *Halorubrum lacusprofundi*. *BMC Biotechnology, 13*, 3.

Nomura, S., & Harada, Y. (1998). Functional expression of green fluorescent protein derivatives in *Halobacterium salinarum*. *FEMS Microbiology Letters, 167*, 287–293.

Oesterhelt, D., & Stoeckenius, W. (1971). Rhodopsin-like protein from the purple membrane of *Halobacterium halobium*. *Nature: New Biology, 233*, 149–152.

Oesterhelt, D., & Stoeckenius, W. (1973). Functions of a new photoreceptor membrane. *Proceedings of the National Academy of Sciences of the United States of America, 70*, 2853–2857.

Oesterhelt, D., & Stoeckenius, W. (1974). Isolation of the cell membrane of *Halobacterium halobium* and its fractionation into red and purple membrane. *Methods in Enzymology, 31*, 667–678.

Patzelt, H., Ulrich, A. S., Egbringhoff, H., Düx, P., Ashurst, J., Simon, B., et al. (1997). Towards structural investigations on isotope labelled native bacteriorhodopsin in detergent micelles by solution-state NMR spectroscopy. *Journal of Biomolecular NMR*, *10*, 95–106.

Robb, F. T., DasSarma, S., & Fleischmann, E. M. (1995). *Archaea: A laboratory manual*. Cold Spring Harbor Laboratory: Halophiles.

Robinson, J. L., Pyzyna, B., Atrasz, R. G., Henderson, C. A., Morrill, K. L., Burd, A. M., et al. (2005). Growth kinetics of extremely halophilic archaea (family halobacteriaceae) as revealed by arrhenius plots. *Journal of Bacteriology*, *187*, 923–929.

Schneider, C. A., Rasband, W. S., & Eliceiri, K. W. (2012). NIH Image to ImageJ: 25 years of image analysis. *Nature Methods*, *9*, 671–675.

Sehgal, S. N., & Gibbons, N. E. (1960). Effect of some metal ions on the growth of Halobacterium cutirubrum. *Canadian Journal of Microbiology*, *6*, 165–169.

Stoeckenius, W., & Rowen, R. (1967). A morphological study of *Halobacterium halobium* and its lysis in media of low salt concentration. *Journal of Cell Biology*, *34*, 365–393.

CHAPTER EIGHT

Amino Acid Selective Unlabeling in Protein NMR Spectroscopy

Chinmayi Prasanna*, Abhinav Dubey*,[†], Hanudatta S. Atreya*,[1]
*NMR Research Centre, Indian Institute of Science, Bangalore, Karnataka, India
[†]Institute Mathematics Initiative, Indian Institute of Science, Bangalore, Karnataka, India
[1]Corresponding author: e-mail address: hsatreya1@gmail.com

Contents

1. Introduction 168
2. Method Description 171
   2.1 Sample Preparation 171
   2.2 Identification of Peaks 172
   2.3 Isotope Scrambling 173
   2.4 Choice of Amino Acid Types for Selective Unlabeling 176
   2.5 A Method for Sequential Assignments Using Selectively Unlabeling 180
3. Applications of Selective Unlabeling 182
4. Conclusions 186
Acknowledgments 187
References 187

Abstract

Three-dimensional structure determination of proteins by NMR requires the acquisition of multidimensional spectra followed by assignment of chemical shifts to the respective nuclei. In order to speed up this process, resonances corresponding to individual amino acid types are often selectively identified and assigned. One of the ways of achieving this is by using the method of "selective unlabeling." In this method, resonances from one or more amino acid types are suppressed selectively in the NMR spectra, which can be achieved using both cell-based and cell-free methods. This helps not only in identifying them but also results in spectral simplification by reducing the number of peaks observed. Further, the assignments are not limited to amino acids that are specifically unlabeled. Using specially designed NMR experiments, assignments of amino acids in the neighborhood of those being selectively unlabeled can also be obtained. In this chapter, we discuss the theoretical and practical aspects of selective unlabeling focusing on how the sample is prepared, which amino acid or a combination of amino acids should be optimally chosen for unlabeling, and how this method can be used for sequential assignments of proteins.

1. INTRODUCTION

NMR-based protein structure determination typically involves sample preparation, data collection, data analysis/resonance assignments, and structure calculation/refinement. In this process, the data analysis/resonance assignments part consists of mapping the chemical shifts of all the peaks in a given NMR spectrum to their respective nuclei (Wüthrich, 1986). This in turn involves the linking of chemical shifts of the neighboring amino acids in a sequential manner to obtain what is called as sequence-specific resonance assignments. For proteins with molecular mass greater than ~10 kDa, heteronuclear multidimensional experiments become necessary, which necessitates isotope enrichment with ^{13}C, ^{15}N nuclei or with ^{13}C, ^{15}N, ^{2}H (Atreya, 2012). Sequence-specific resonance assignments of the labeled protein are then carried out using a set of two-dimensional (2D) and three-dimensional (3D) NMR experiments which establish the connectivity of chemical shifts between neighboring amino acid residues (Cavanagh, Fairbrother, Palmer, Skelton, & Rance, 2010).

While isotope labeling (^{13}C, ^{15}N, ^{2}H) and advancements in multidimensional NMR experiments have succored in achieving sequential resonance assignments and high-resolution structure determination, their applications to large proteins are constrained due to various complexities in the NMR spectrum. One such complexity is spectral overlap/crowding occurring in proteins with higher molecular weight (Frueh, 2014), those in solid state (McDermott, 2004), membrane proteins (Sanders & Sonnichsen, 2006), and intrinsically disordered proteins (IDPs) (Dyson & Wright, 1998; Konrat, 2014) hampering unambiguous resonance assignment. Increased signal overlap happens due to two reasons: (i) the increase in the number of peaks due to increase in the number of amino acids and (ii) their concomitant broadening resulting from slower overall molecular tumbling. Reducing the spectral overlap therefore becomes important for solving structures, particularly of challenging proteins.

There are various approaches in this regard: (i) identification of amino acid types using specific pulse sequences (Barnwal, Atreya, & Chary, 2008; Barnwal, Rout, Atreya, & Chary, 2008; Dotsch, Matsuo, & Wagner, 1996; Dotsch, Oswald, & Wagner, 1996; Feuerstein, Plevin, Willbold, & Brutscher, 2012; Pantoja-Uceda & Santoro, 2012; Schubert, Oschkinat, & Schmieder, 2001; Schubert, Smalla, Schmieder, & Oschkinat, 1999), (ii) selective isotope labeling of amino acids in a uniformly unlabeled background (Ohki & Kainosho, 2008), and (iii) selective unlabeling of amino acids in a uniform ^{13}C, ^{15}N-labeled background (Atreya & Chary, 2000, 2001; Jaipuria, Krishnarjuna, Mondal, Dubey, & Atreya, 2012; Krishnarjuna, Jaipuria, Thakur, D'Silva, & Atreya, 2011; Shortle, 1994; Vuister, Kim, Wu, & Bax, 1994). In the first approach, NMR experiments are specifically designed to detect signals selectively from a particular amino acid type. This works by exploiting the fact that different amino acids have different side-chain topologies and these can be selectively filtered using tailored coherence transfer pathways. Identifying amino acids using such pulse sequences often suffers loss in sensitivity with increase in protein size making it suitable only for small proteins. This is attributed to the fact that these pulse sequences comprise of long delay periods, which results in loss of sensitivity due to transverse relaxation. Another limitation of this approach is the loss of magnetization from side-chain protons if they are replaced by deuterons, which renders this method nonapplicable to deuterated proteins.

In the second approach, a given amino acid or combinations of different amino acids are selectively labeled (^{15}N, ^{13}C) over a uniformly unlabeled background (^{14}N, ^{12}C) in a cell-based or cell-free protein expression system. A major drawback of this approach is the requirement of isotopically enriched amino acids which are expensive. And the expense goes up with the drop in the protein yield which is being labeled.

Considering the above approaches, an alternative is the simple method of selective amino acid unlabeling (also referred to as reverse labeling) (Atreya & Chary, 2000, 2001; Jaipuria et al., 2012; Krishnarjuna et al., 2011; Shortle, 1994; Vuister et al., 1994). In this strategy, a single or a group of amino acids is unlabeled (^{14}N, ^{12}C) leaving the rest of the amino acids in the protein labeled with ^{13}C and/or ^{15}N. Resonances from the unlabeled residues become unobservable in the NMR spectra. Comparison of this data with a second control (uniformly labeled) sample allows the identification of chemical shifts of the unlabeled residue(s) and thereby assign the given amino acid type(s). Figure 1 depicts schematically the two labeling approaches using the 2D heteronuclear single quantum coherence (HSQC) spectrum which is typically used as a fingerprint spectrum in NMR studies of proteins.

Selective unlabeling has the advantage of being relatively inexpensive (only unlabeled amino acids are used along with ^{15}NH$_4$Cl and/or ^{13}C-D-glucose). Initial experiments involving selective unlabeling started with incorporation of naturally abundant (^{14}N) phenylalanine into uniform

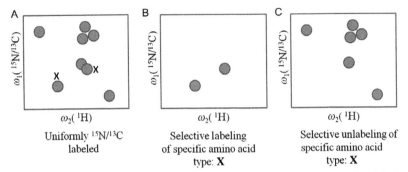

Figure 1 A schematic depiction of the method of selective labeling and unlabeling. (A) A 2D HSQC spectrum (either 2D [^{15}N, ^{1}H] or 2D [^{13}C, ^{1}H] correlation spectrum) of a uniformly labeled protein. A specific amino acid type is marked "X." (B) In the selective labeling approach, a given residue type (say X) is selectively labeled and observed, while the resonances from other amino acids are absent. (C) In the method of selective unlabeling, resonances from a given residue (say X) are unlabeled selectively (rendering it absent) while the resonances from the other amino acids are observed.

^{13}C, ^{15}N-enriched DNA-binding domain of Drosophila heat shock factor (Vuister et al., 1994). Another application involved the use of ^{14}N threonine, arginine, methionine, and histidine in staphylococcal nuclease against a ^{15}N-labeled background (Shortle, 1994). Subsequently, an approach for stereospecific NMR assignments of methyl (CH$_3$) groups of (Val, Leu) in a fractionally ^{13}C-labeled calcium-binding protein (*Eh*-CaBP) was designed by incorporation of a combination of ^{12}C-lysine, leucine, isoleucine, and threonine (Atreya & Chary, 2001) to simplify the spectrum. Since then, the method has gone on to become quite popular resulting in several applications such as spectral simplification (Atreya & Chary, 2001; Vuister et al., 1994), spin-system or amino acid-type identification (Atreya & Chary, 2000; Krishna Mohan, Barve, Chatterjee, Ghosh-Roy, & Hosur, 2008; Shortle, 1994), resonance assignments in solution (Bellstedt et al., 2013; Krishnarjuna et al., 2011) and solid state (Banigan, Gayen, & Traaseth, 2013; Heise et al., 2005; Shi et al., 2009), structure determination (Kelly et al., 1999), stereospecific assignment of the prochiral methyl groups of Val and Leu in large molecular weight proteins (Tugarinov & Kay, 2004), measurement of residual dipolar couplings (Mukherjee, Mustafi, Atreya, & Chary, 2005), and measurement of pseudo-contact shifts (Kobashigawa et al., 2012). Some of these applications are discussed more in detail in the "Application" section 3.

In addition to spectral simplification and amino acid-type identification, selective unlabeling is also useful for sequence-specific resonance

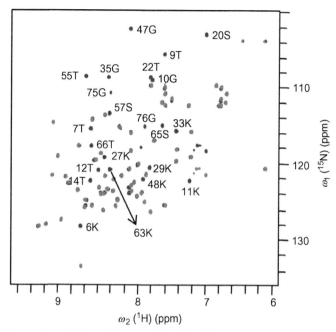

Figure 2 An overlay of selected region of 2D [^{15}N–^{1}H] HSQC spectrum of uniformly ^{15}N labeled (blue) and lysine, threonine (K, T) selectively unlabeled (red) samples of ubiquitin. Note that in addition to Lys and Thr, the resonances of Gly and Ser are also absent in the selectively unlabeled sample due to isotope scrambling as discussed in the text. (See the color plate.)

2.3 Isotope Scrambling

Isotope scrambling implies misincorporation of ^{14}N/^{13}C isotopes to undesired/nontargeted amino acids (Muchmore, McIntosh, Russell, Anderson, & Dahlquist, 1989). Scrambling leads to unlabeling of one or more amino acids apart from our choice of amino acid (Muchmore et al., 1989). This is mostly due to bacterial transaminases, which are responsible for amine group transfer between different amino acids, particularly α-amino group in the amino acid metabolism pathway (Shortle, 1994; Waugh, 1996). Isotope scrambling can lead to cross/wrong labeling. The ^{14}N isotope scrambling is more prominent than ^{13}C isotope scrambling. A detailed analysis of ^{14}N and ^{12}CO scrambling using *E. coli* BL21 (DE3) as a standard system for heterologous protein expression has been discussed (Bellstedt et al., 2013). Table 1 summaries the extent of ^{14}N isotope scrambling in selective unlabeling.

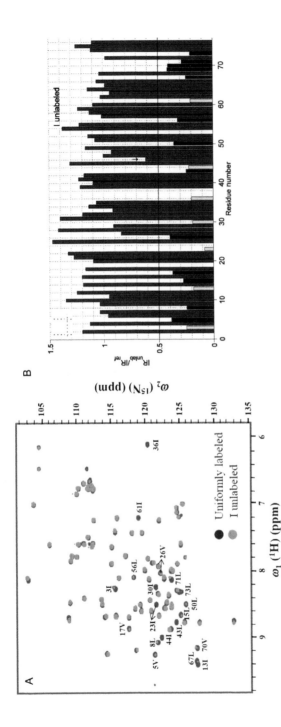

Figure 3 (A) An overlay of selected region of 2D [^{15}N–^{1}H] HSQC spectrum of uniformly ^{15}N labeled (blue) and isoleucine selectively unlabeled (red) samples of ubiquitin. Assignments for the unlabeled residues are indicated by the residue number. (B) Normalized intensity plot. Plots of $IR_{unlab,i}/IR_{ref,i}$; $IR_{unlab,i} = I_i^{unlab}/I_{control}^{unlab}$ and $IR_{ref,i} = I_i^{ref}/I_{control}^{ref}$ where i denotes the unlabeled residue number. The colored bars indicate the following: black: $IR_{unlab,i}/IR_{ref,i} < 0.5$, i.e., undergoing strong isotope scrambling (undesired residues corresponding to Leu and Val), green: desired selectively unlabeled residues (Ile), and blue: uniformly labeled residues. I denotes volume of the peak and "control" denotes reference residue which does not undergo any effect of unlabeling in both selectively unlabeled and the reference sample. Residues E24 and G53 were not assigned and hence are absent along with Pro. The control residue chosen was G47. (See the color plate.)

Table 1 Isotope Scrambling in Selective Unlabeling

Selective Unlabeled Amino Acid Residue	[14]N Isotope Scrambling[a]
Alanine	Tryptophan
Arginine	–
Asparagine	–
Aspartic acid	Uniform scrambling
Cysteine	–
Glutamine	–
Glutamic acid	Uniform scrambling
Glycine	Serine, cysteine
Histidine	–
Isoleucine	Leucine, valine
Leucine	Valine, isoleucine
Lysine	–
Methionine	–
Phenylalanine	Tyrosine
Proline	Not applicable
Serine	Glycine, cysteine
Threonine	Glycine
Tryptophan	Uniform scrambling
Tyrosine	Phenylalanine
Valine	Leucine, isoleucine

[a] [14]N scrambling based on analysis of ^{1}H, ^{15}N-HSQC peak intensities for GB1 (Bellstedt et al., 2013), APTX (Bellstedt et al., 2013), and ubiquitin (Krishnarjuna et al., 2011). Uniform scrambling implies uniform decrease in peak intensities of all other amino acids.

Isotope scrambling can be reduced to a certain extent by any of the following methods: (i) use of auxotrophic strains of bacteria (Waugh, 1996), (ii) use of inhibitors of specific amino acid synthesis pathways (Kim, Perez, Ferguson, & Campbell, 1990), (iii) use of unlabeled precursors of amino acids (Rasia, Brutscher, & Plevin, 2012), and (iv) optimal choice and selection of amino acids (Krishnarjuna et al., 2011). Using auxotrophic strains of bacteria for overexpression of proteins with selective unlabeling solves much of the problem of isotope scrambling, as these strains lack certain transaminases which bring about isotope scrambling. This comes however at the cost of undermining the overall cell growth and protein yields. To recompense, larger culture volume and longer growth periods are required to achieve the same protein yield. When a mixture of amino acids is required to be unlabeled in the same sample, constructing strains with multiple auxotrophic markers to avoid isotope scrambling seems impractical. Such strains lacking multiple transaminases may have poor viability.

Immediate metabolic precursor of an amino acid in its biosynthetic pathway can be used for unlabeling instead of the amino acid itself. This is based on the premise that the precursor is directly converted to its amino acid, that is, the final amino acid isotope composition directly relies on the isotope composition of its metabolic precursor. This is useful for site-specific selective unlabeling (Rasia et al., 2012). Since the α-amino group is added at the end of the metabolic pathway by aminotransferases for a few amino acids (e.g., Ile, Leu), adding unlabeled precursors for these residues in a uniform ^{13}C, ^{15}N-labeled medium would result in the ^{12}C-labeled amino acid with ^{15}N-labeled at the α-amino position. Table 2 gives a list of metabolic precursors as suitable candidates for unlabeling as proposed by Rasia et al.

2.4 Choice of Amino Acid Types for Selective Unlabeling

The choice of a particular or a group of amino acid type(s) for selective unlabeling has to be made judiciously and depends first on one of the two factors: (i) whether the purpose of unlabeling is to identify and assign the amino acid type(s) or (ii) if it is solely for simplifying the spectrum to reduce overlap of peaks. While majority of the applications belong to the first category (discussed under "Application" below), it is also important in some applications to simplify the spectrum for reducing overlaps

Table 2 Metabolic Precursors of Amino Acids Suitable for Selective Unlabeling (Rasia et al., 2012)

Precursors	Amino Acid
α-Ketobutyrate	Isoleucine
α-Ketoisovalerate	Leucine, valine
Phenyl pyruvate	Phenylalanine
4-Hydroxy phenylpyruvate	Tyrosine

(Atreya & Chary, 2001; Hong & Jakes, 1999; Mukherjee et al., 2005; Tugarinov & Kay, 2004).

If the purpose of selective unlabeling is resonance assignments, it is important to choose amino acids based on the following factors depending on whether (i) only one type of amino acid residue has to be identified (e.g., in ligand–protein interaction) or (ii) if the protein has to be sequentially assigned. In the case of ligand–protein interaction studies, it is sometimes useful to unlabel only a particular type of amino acid located near the ligand binding site. On the other hand, if sequential assignments are intended, it is useful to choose a combination of several amino acids for selective unlabeling in the same sample. The combination has to be chosen optimally based on several factors. This is discussed below.

2.4.1 Optimal Combination of Amino Acid Types for Selective Unlabeling

The optimal choice for the combination of amino acids to be selectively unlabeled in the same sample depends on (i) their abundance in primary sequence, (ii) the ease with which the different amino acid types can be distinguished from each other in the spectrum, and (iii) the extent of isotope scrambling.

(i) It is important to choose amino acids such that they can be distinguished from each other in the NMR spectrum. The amino acid types are usually identified based on their $^{13}C^{\alpha}$ and $^{13}C^{\beta}$ shifts (Atreya & Chary, 2000, 2001; Cavanagh et al., 2010). It is well known that the 20 amino acid types can be classified into different categories based on their $^{13}C^{\alpha}$ and $^{13}C^{\beta}$ shifts as shown in Fig. 4 (Atreya & Chary, 2000, 2001). All amino acids in a given category have similar $^{13}C^{\beta}$ shifts and cannot be distinguished from each other. Hence, the first criterion is that the amino acid types chosen for selective unlabeling should belong to different chemical shift categories. These categories are depicted in Fig. 4.

(ii) The second criterion is their abundance and distribution in the protein primary sequence. Since there exist methods that can identify tri- and tetrapeptide stretches flanking the unlabeled amino acid residue (Krishnarjuna et al., 2011), amino acid types to be chosen for selective unlabeling should be sufficiently abundant in the primary sequence. In addition, they should be dispersed throughout the sequence so that the percentage of unique assignments can be maximized.

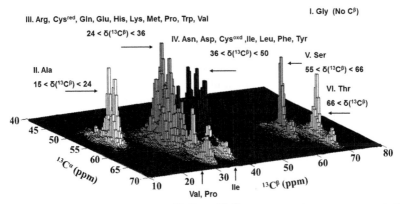

Figure 4 Amino acid types and their $^{13}C^\alpha$ and $^{13}C^\beta$ chemical shifts range (in ppm). (See the color plate.)

(iii) The third criterion takes into account the misincorporation (scrambling) of ^{14}N or ^{12}C rendering undesired amino acid types to get unlabeled. In the case of selective unlabeling, different amino acid types have varying extent of isotope scrambling (Bellstedt et al., 2013; Jaipuria et al., 2012; Krishnarjuna et al., 2011). In our approach, isotope scrambling is exploited such that only those amino acids are chosen which result in unlabeling of amino acids type that does not belong to the same category. For example, unlabeling of serine results in unlabeling of glycine and cysteine which belong to different categories based on their $^{13}C^\alpha$ and $^{13}C^\beta$ shifts (Fig. 4) and hence can be distinguished from each other.

Based on the three criteria discussed above, we can find an *optimal* set of amino acid types for selective unlabeling which can be used across all proteins in general and that will maximize the percentage of sequential assignments. We carried out a simulation to find this optimal combination. For this purpose, 1412 nonredundant polypeptide sequences (categorized into three sets: globular (structured) proteins, IDPs, and membrane proteins) varying in 33–691 residues in length were taken. The labeled amino acids were assigned unique codes which are numbers given as per the respective categories shown in Fig. 4 (marked in roman numerals for each set of amino acids). The unlabeled amino acid types were not given codes since their identity is assumed to be known exactly. The protein primary sequence was thus translated into a sequence of codes. For instance, a sequence stretch "….AVRTKNGL…" where R and N have been unlabeled will get translated to "…2-3-R-6-3-N-1-4" (see Fig. 4 for the codes). For each protein, different combination of amino acids which can easily be identified and/or do not cross metabolize (viz., Ala, Arg, Asn, Gly, His, Lys, Ser, and Thr) was considered for unlabeling. Next, for each of such combinations of unlabeled amino acids, we considered the tri- and tetrapeptide stretches flanking an unlabeled residue that can be sequentially liked (as shown below in Fig. 7; Krishnarjuna et al., 2011) and the total percentage of unambiguous assignments thus possible for the protein were computed. Given that the unlabeled amino acids should belong to different chemical shift categories (Fig. 4), no two amino acid types belonging to the same category were considered in the same set. The best percentage assignment thus possible with the different combinations of unlabeled amino acids was calculated.

As discussed above, the choice of amino acid types to be chosen for selective unlabeling is based on three criteria: their $^{13}C^{\alpha}$ and $^{13}C^{\beta}$ shifts which allow the individual amino acids in the set of chosen residues to be distinguished from each other, their abundance and distribution in the protein, and the extent of their isotope scrambling. The most suited amino acid types satisfying these criteria are Ala (A), Arg (R), Asn (N), Gly (G), His (H), Lys (K), Ser (S), and Thr (T). Each of these amino acids is usually found abundant in most proteins ($\sim$40% of the total residues), is distributed throughout the protein sequence, belongs to different chemical shift categories (Fig. 4; except Arg and Lys), and does not cross metabolize to amino acids within their group (e.g., Ser scrambles to Gly and Cys which are distinct from each other). Among these residues, a set or a combination of amino acids has to be chosen so that they can be discriminated from each other based on $^{13}C^{\beta}$ shifts (Fig. 4). This results in the following set: (Ala, Arg, Asn, Gly, Ser, Thr) [denoted as RNTGSA] or (Ala, Asn, Lys, Gly, Ser, Thr) [denoted as KNTGSA] or (Ala, Asn, His, Lys, Gly, Ser, Thr) [denoted as HNTGSA]. The four amino acids: Ala, Gly, Ser, and Thr are common to the set. These amino acids have unique $^{13}C^{\alpha}$ and $^{13}C^{\beta}$ shifts (Fig. 4). Among the remaining three amino acids Arg, Lys, and Asn, the first two belong to the same category (Fig. 4) and hence should not be chosen in the same set. The amino acid Asn does not cross metabolize to any other amino acid and has distinct $^{13}C^{\beta}$ shifts compared to others in the set.

A statistical analysis of the percentage of possible unambiguous assignments that can be obtained using the above set of amino acids reveals that the set: (Ala, Arg, Asn, Gly, Ser, Thr) or (Ala, Asn, Lys, Gly, Ser, Thr) is optimally suited for most of the proteins in general and yields $\sim$70% assignments on an average. This is depicted in Fig. 5A. The percentage of possible assignments as a function of protein size are depicted in Fig. 5B, which

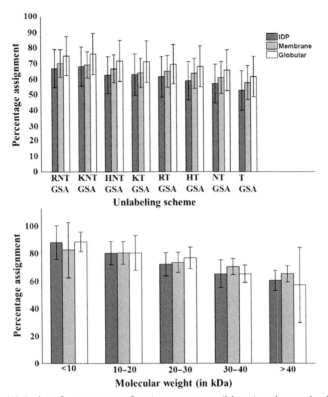

Figure 5 (A) A plot of percentage of assignments possible using the method proposed by Krishnarjuna et al. (2011), which yields assignments of tripeptide segments in proteins. Different combination of amino acids shown was simulated for selective unlabeling across 1412 proteins in the molecular weight range of 3.6–76 kDa. (B) The plot of percentage of assignments possible across different protein sizes using an optimal combination of amino acids (the optimal combination was computed for each protein separately).

indicates that even for large protein a significant percentage of assignments can be obtained using this approach. The decrease in the percentage of possible assignments with increase in the protein size arises due to the fact that as the number of residues increases, the uniqueness of the tri- and tetrapeptide stretches goes down.

2.5 A Method for Sequential Assignments Using Selectively Unlabeling

An approach for sequential assignments in selectively unlabeled proteins was proposed by Krishnarjuna et al. (2011). This involves a new NMR experiment named ($^{12}CO_i$–$^{15}N_{i+1}$)-filtered HSQC, which aids in linking the $^1H^N/^{15}N$ resonances of the selectively unlabeled residue, i, and its C-terminal neighbor, $i+1$, in H^N-detected double and triple resonance spectra. Assignment or identification of the amino acid type corresponding to $i-1$ residue can be done using the conventional 3D HNCACB or 3D CBCA(CO) NH experiments. Thus, we can assign a tripeptide segment from the knowledge of the amino acid types of residues: $i-1$, i, and $i+1$, thereby speeding up the sequential assignment process.

Note that the 2D ^{15}N–H^N plane of a 3D HNCO can also give the same information as described by Bellstedt et al. (2013). This can be seen from the 2D ^{15}N–H^N plane of a 3D HNCO spectrum acquired on a PKNA selectively unlabeled ^{13}C–^{15}N-labeled ubiquitin sample (Fig. 6). In the 2D plane of 3D HNCO, two types of ^{15}N–H^N peaks are absent: (i) the residue i which is unlabeled and (ii) the residue "$i+1$" which has ^{12}CO as its N terminal neighbor. The ^{15}N–H^N peaks corresponding to the unlabeled residue i are known based on a comparison of the HSQC of the unlabeled sample with that of the control sample. The additionally absent peaks in the 2D plane of HNCO thus belong to $i+1$ residues.

The next step is to map the resonances of residues i to that $i+1$ (i.e., identify which $i+1$ resonance corresponds to the correct residue i). This is carried out as follows. In the general sequential assignment process, a link

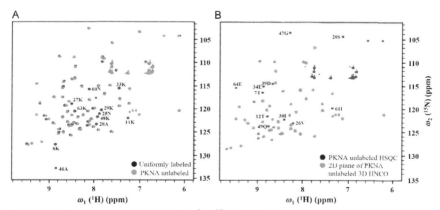

Figure 6 (A) An overlay of a 2D ^1H–^{15}N HSQC spectrum of PKNA selectively unlabeled sample (red) with that of a uniformly labeled sample of ubiquitin. (B) The overlay of a 2D ^1H–^{15}N HSQC of PKNA unlabeled sample (blue) with the 2D ^1H–^{15}N plane of a 3D HNCO. The residues which are $i+1$ to the unlabeled PKNA residues are additionally in the 2D plane of 3D HNCO when compared to the HSQC of PKNA sample. (See the color plate.)

between $^1H^N/^{15}N$ resonances of residue i and residue $i+1$ along the polypeptide chain is established by using, in concert, 3D HNCACB and 3D CBCA(CO)NH/HN(CO)CACB spectra. Starting from a spin-system i in 2D [^{15}N–^{1}H] HSQC, a search is carried out in the 3D CBCA(CO)NH/HN(CO)CACB spectrum for all spin-systems j which are identified in the ($^{12}CO_i$–$^{15}N_{i+1}$)-filtered HSQC or 2D plane of HNCO (discussed above) such that the $^{13}C^\alpha$ and $^{13}C^\beta$ chemical shifts of residue i (obtained from HNCACB) are observed on j. The spin-system j then corresponds to the C-terminal neighbor (i.e., $i+1$) of i. In the case of spectral overlap and/or larger molecular weight proteins, $^1H^\alpha$ and/or ^{13}CO shifts from 4D experiments can also be used to establish this link. This process is hindered if the search ends up with more than one choice for spin systems corresponding to $i+1$. For unambiguous assignments, it is thus preferable to reduce the "search space" such that the spin-system $i+1$ is easily tractable for a given spin-system i. Toward this end, the devised ($^{12}CO_i$–$^{15}N_{i+1}$)-filtered HSQC aids in detecting $^1H^N/^{15}N$ resonances of the C-terminus ($i+1$) residue to the amino acid unlabeled selectively (residue i) as shown in Fig. 7.

3. APPLICATIONS OF SELECTIVE UNLABELING

Amino acid unlabeling has become one of the popular methods for structural studies in protein NMR, especially for challenging systems in both solution and solid state. This is owing to the fact that it is relatively cheaper compared to amino acid selective labeling. Expenses of ^{15}N-ammonium chloride and ^{13}C glucose are much lower than ^{13}C, ^{15}N-labeled amino acids required for selective labeling. Sample preparation, heterologous overexpression is straight forward without any trade-off in cell growth and protein yields. We discuss here a few examples, which reveal the broad spectrum of applications possible with selective unlabeling.

Shortle first demonstrated the applicability of selective unlabeling of mixture of amino acids in the protein—staphylococcal nuclease in *E. coli* to identify and assign the amino acid type (Shortle, 1994). Five samples of the wild-type nuclease protein were prepared which included a uniformly labeled sample and four samples containing different mixtures of amino acid covering a total of 14 amino acids (excluding glutamic acid, glutamine, aspartic acid, asparagine, alanine) in unique combinations (Shortle, 1994). In this work, glycerol was used as a carbon source rather than glucose to achieve more efficient correlation-peak suppression. This is an example

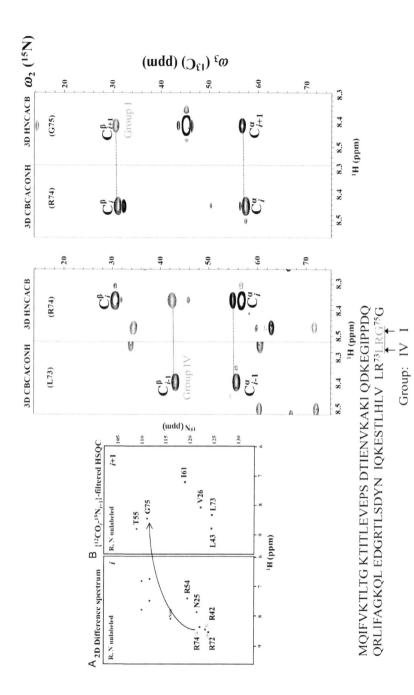

Figure 7 Illustration of sequence-specific resonance assignment of a tripeptide segment in ubiquitin. Given a peak in the 2D difference spectrum of (R, N) selectively unlabeled sample (e.g., R 74), its C-terminal neighbour ($i+1$) is identified using the procedure shown and its amino acid type is assigned using $^{13}C^{\alpha}$ and $^{13}C^{\beta}$ chemical shift in 3D HNCACB. The amino acid type of residue $i-1$ is assigned likewise from 3D CBCA(CO)NH and the tripeptide segment (containing the amino acid-type information) is then mapped onto the protein sequence (shown below) for sequence-specific assignments.

of a combinatorial approach in which one or more protein samples are prepared with different combination of amino acids unlabeled selectively in a given sample. This requires a judicious selection of amino acids such that when taken together each amino acid has a unique labeling pattern across the different samples. For example, consider two labeling modes or states for a given amino acid (e.g., ^{15}N-labeled or unlabeled). The cross peak for that amino acid will either be present or absent in one sample. By choosing the amino acids appropriately in one sample, the presence/absence pattern of an amino acid across the different samples can be rendered unique. Thus, for N samples up to 2^{N-1} amino acid types can be chosen such that each amino acid can have a unique labeling pattern (present/absent) across the N samples. Shortle distributed 14 amino acids across four samples and the amino acid types were identified based on the intensity pattern (i.e., presence/absence of the peaks) in different samples. Since the number of samples was four, a maximum of 16 amino acids (2^4) could be assigned.

In another pioneering experiment, unlabeled phenylalanine (Phe) was incorporated into an otherwise uniformly W/^{15}N-enriched DNA-binding domain of Drosophila heat shock factor, dHSF (33–155), a 123-residue domain containing 10 Phe residues. Amino acid unlabeling was particularly useful here to assign the spin system of phenylalanine, which is otherwise difficult to assign due to poor dispersion in aromatic ^{13}C chemical shifts (Vuister et al., 1994). Selective unlabeling yielded a significant 39% increase in NOEs and reduction in the spread of the NMR structures (rmsd) from 1.4 to 0.87 Å for the backbone plus ordered side-chain atoms of the protein dHSF.

A methodology for stereospecific NMR assignments of methyl (CH$_3$) groups of Val and Leu residues was demonstrated in a fractionally ^{13}C-labeled protein—the 15 kDa calcium-binding protein from *Entamoeba histolytica* (Eh-CaBP) (Atreya & Chary, 2001). This approach was based on selective "unlabeling" of specific amino acids in proteins while uniform ^{13}C-labeling the rest. The requirement for selective unlabeling stemmed from the fact that cross peaks arising from Ala(C^β–H^β), Lys(C^γ–H^γ), Ile($C^{\gamma 1}$–$H^{\gamma 1}$ and $C^{\gamma 2}$–$H^{\gamma 2}$), Met(C^ε–H^ε), and Thr(C^γ–H^γ) correlations have significant overlaps with that of Val(C^γ–H^γ) or/and Leu(C^δ–H^δ) in the methyl region (~15–25 ppm in ^{13}C; ~0–1.5 ppm in ^{1}H) of a 2D [^{13}C, ^{1}H] HSQC spectrum. This hampers stereospecific assignments of the latter. The extent of overlap of Val and Leu methyl cross peaks with those belonging to other residues is 25–40%. Further, about 20% of Val(C^γ–H^γ) cross peaks overlap with Leu(C^δ–H^δ) correlations. Such spectral overlap can be overcome with selective unlabeling. Toward this end, two samples were prepared. In one sample, Ile, Leu, Lys, and Thr were selectively unlabeled, which resulted in resonances present only for valines. This helped to simplify the spectrum significantly and thereby in obtaining unambiguous stereospecific assignment of the valine methyls. In the second sample, Ile, Lys, and Thr were selectively unlabeled together, which helped in the assignment of leucine. This is an example how selective unlabeling can be used for spectral simplification by suppressing undesired resonances from the region of interest. Such spectral simplification was also used subsequently for unambiguous stereospecific assignment of diastereotopic CH_3 groups in Val and Leu residues in a large (70 kDa) protein (Tugarinov & Kay, 2004).

A combination of selective labeling and unlabeling with experiments involving frequency-selective dephasing that dramatically improved the ability to resolve peaks in crowded spectra was demonstrated using the polytopic membrane protein EmrE—an efflux pump involved in multidrug resistance by magic angle spinning solid-state NMR spectroscopy (Banigan et al., 2013). Leu was selectively ^{15}N-labeled and Ile was selectively unlabeled with rest of the amino acids labeled uniformly with ^{13}C, ^{15}N. Resonances of residues preceding the selectively labeled amino acid were inferred using a 3D NCOCX afterglow experiment and those following were recorded using a frequency-selective dephasing experiment. The aim was to reduce the spectral congestion of this protein and provide a sensitive way to obtain chemical shift assignments for a membrane protein, where no high-resolution structure is available. They prepared isoleucine (Ile) unlabeled (reverse labeled) EmrE sample in uniform ^{13}C, ^{15}N background along with ^{15}N-Leu labeling (called as RevIL). Selective unlabeling in their case resulted in complete ^{13}C, ^{15}N incorporation at all sites except Leu (^{15}N-labeled) and Ile (selective unlabeled) residues, and created interresidue pairwise connectivity. Using this, they could successfully identify residues preceding to leucine and those following leucine and isoleucine in the primary protein sequence. With this novel method, they also point out that the methodology is widely applicable to the study of other polytopic membrane proteins in functional lipid bilayer environments.

The complete backbone assignment for apo-OBP3 was obtained using 2D and 3D methods combined with global isotopic labeling and selective unlabeling strategies to examine the effects of ligand binding on Rat OBP3 (odorant-binding protein) structure and dynamics by NMR (Portman et al., 2014). Assignments of specific residues of Rat OBP3 were greatly aided by use of "unlabeling" experiments, where various amino acids

(Lys, Arg, Thr, and Gln) were systematically supplemented during protein expression.

Resonance assignment for a particularly challenging protein—23 kDa catalytic domain of human aprataxin—was based on systematic unlabeling/reverse labeling of amino acids (Bellstedt et al., 2013). Selective unlabeling was presented here as a feasible strategy to include the amino acid-type information during the assignment process. As described above, instead of the 2D ($^{12}CO_i$–$^{15}N_{i+1}$)-filtered HSQC to identify resonances corresponding to the C-terminal ($i+1$) neighbor of selectively unlabeled residue, i, a 2D ^{15}N–$^{1}H^N$ plane of 3D HNCO was used. This in turn aided in resolving the potential ambiguities due to incomplete NMR data sets. In this study, they not only used selective unlabeling for aiding sequence-specific resonance assignment but also in classifying the extent of ^{15}N and ^{13}C isotope scrambling (metabolic interconversion) between amino acids and presented a detailed analysis. In the process, they systematically evaluated the unlabeling approach for all 20 natural amino acids for two recombinant proteins (GB1 and human aprataxin) expressed in the standard T7 polymerase-based E. coli BL21 (DE3) system.

4. CONCLUSIONS

Amino acid selective unlabeling helps in the process of unambiguous sequence-specific resonance assignments en route to solving 3D structures and studying dynamics of large and challenging proteins by NMR. In this chapter, we have discussed different approaches in implementing selective unlabeling in proteins and their diverse applications in aiding sequence-specific resonance assignments not only in solution NMR but also in solid-state NMR spectroscopy. Couple of decades since its coming into light, this technique is being widely practiced. In future, selective unlabeling can be used in concert with high-resolution NMR experiments such as those used for membrane proteins (Zhang, Atreya, Kamen, Girvin, & Szyperski, 2008) or with fast NMR techniques in solid state (Franks, Atreya, Szyperski, & Rienstra, 2010). Apart from bacterial systems, its implementation for *in-vivo/in-cell* systems, eukaryotic systems such as yeast, insect, and mammalian cells, remains to be explored. While selective amino acid labeling has been established in these systems (Atreya, 2012; Goto & Kay, 2000; Hansen et al., 1992; Wood & Komives, 1999), a step forward to selective unlabeling is expected in days to come.

ACKNOWLEDGMENTS

The facilities provided by NMR Research Centre at IISc supported by Department of Science and Technology (DST), India are gratefully acknowledged. H.S.A. acknowledges support from DAE-BRNS and DST-SERC research awards. We thank Garima Jaipuria for helping in preparing the figures.

REFERENCES

Atreya, H. S. (Ed.), (2012). *Isotope labeling in biomolecular NMR* (1 ed.). Netherlands: Springer.

Atreya, H. S., & Chary, K. V. R. (2000). Amino acid selective 'unlabelling' for residue-specific NMR assignments in proteins. *Current Science, 79*, 4–5.

Atreya, H. S., & Chary, K. V. R. (2001). Selective 'unlabeling' of amino acids in fractionally ^{13}C labeled proteins: An approach for stereospecific NMR assignments of CH_3 groups in Val and Leu residues. *Journal of Biomolecular NMR, 19*, 267–272.

Banigan, J. R., Gayen, A., & Traaseth, N. J. (2013). Combination of ^{15}N reverse labeling and afterglow spectroscopy for assigning membrane protein spectra by magic-angle-spinning solid-state NMR: Application to the multidrug resistance protein EmrE. *Journal of Biomolecular NMR, 55*, 391–399.

Barnwal, R. P., Atreya, H. S., & Chary, K. V. R. (2008b). Chemical shift based editing of CH(3) groups in fractionally (13)C-labelled proteins using GFT (3,2)D CT-HCCH-COSY: Stereospecific assignments of CH(3) groups of Val and Leu residues. *Journal of Biomolecular NMR, 42*, 149–154.

Barnwal, R., Rout, A., Atreya, H. S., & Chary, K. V. R. (2008a). Identification of C-terminal neighbours of amino acid residues without an aliphatic 13Cγ as an aid to NMR assignments in proteins. *Journal of Biomolecular NMR, 41*, 191–197.

Bellstedt, P., Seiboth, T., Hafner, S., Kutscha, H., Ramachandran, R., & Gorlach, M. (2013). Resonance assignment for a particularly challenging protein based on systematic unlabeling of amino acids to complement incomplete NMR data sets. *Journal of Biomolecular NMR, 57*, 65–72.

Cavanagh, J., Fairbrother, W. J., Palmer, A. G., Skelton, N. J., & Rance, M. (2010). *Protein NMR spectroscopy: Principles and practice*. USA: Elsevier Science.

Dotsch, V., Matsuo, H., & Wagner, G. (1996a). Amino-acid-type identification for deuterated proteins with a beta-carbon-edited HNCOCACB experiment. *Journal of Magnetic Resonance. Series B, 112*, 95–100.

Dotsch, V., Oswald, R. E., & Wagner, G. (1996b). Amino-acid-type-selective triple-resonance experiments. *Journal of Magnetic Resonance Series B, 110*, 107–111.

Dyson, H. J., & Wright, P. E. (1998). Equilibrium NMR studies of unfolded and partially folded proteins. *Natural Structural Biology, 5*(Suppl.), 499–503.

Feuerstein, S., Plevin, M. J., Willbold, D., & Brutscher, B. (2012). iHADAMAC: A complementary tool for sequential resonance assignment of globular and highly disordered proteins. *Journal of Magnetic Resonance, 214*, 329–334.

Franks, W. T., Atreya, H. S., Szyperski, T., & Rienstra, C. M. (2010). GFT projection NMR spectroscopy for proteins in the solid state. *Journal of Biomolecular NMR, 48*, 213–223.

Frueh, D. P. (2014). Practical aspects of NMR signal assignment in larger and challenging proteins. *Progress in Nuclear Magnetic Resonance Spectroscopy, 78*, 47–75.

Goto, N. K., & Kay, L. E. (2000). New developments in isotope labeling strategies for protein solution NMR spectroscopy. *Current Opinion in Structural Biology, 10*, 585–592.

Hansen, A. P., Petros, A. M., Mazar, A. P., Pederson, T. M., Rueter, A., & Fesik, S. W. (1992). A practical method for uniform isotopic labeling of recombinant proteins in mammalian cells. *Biochemistry, 31*, 12713–12718.

Heise, H., Hoyer, W., Becker, S., Andronesi, O. C., Riedel, D., & Baldus, M. (2005). Molecular-level secondary structure, polymorphism, and dynamics of full-length alpha-synuclein fibrils studied by solid-state NMR. *Proceedings of the National Academy of Sciences of the United States of America, 102*, 15871–15876.

Hong, M., & Jakes, K. (1999). Selective and extensive 13C labeling of a membrane protein for solid-state NMR investigations. *Journal of Biomolecular NMR, 14*, 71–74.

Jaipuria, G., Krishnarjuna, B., Mondal, S., Dubey, A., & Atreya, H. S. (2012). Amino acid selective labeling and unlabeling for protein resonance assignments. *Advances in Experimental Medicine and Biology, 992*, 95–118.

Kelly, M. J. S., Krieger, C., Ball, L. J., Yu, Y., Richter, G., Schmieder, P., et al. (1999). Application of amino acid type-specific ^{1}H and ^{14}N labeling in a ^{2}H-, ^{15}N-labeled background to a 47 kDa homodimer: Potential for NMR structure determination of large proteins. *Journal of Biomolecular NMR, 14*, 79–83.

Kim, H. W., Perez, J. A., Ferguson, S. J., & Campbell, I. D. (1990). The specific incorporation of labelled aromatic amino acids into proteins through growth of bacteria in the presence of glyphosate. Application to fluorotryptophan labelling to the H(+)-ATPase of Escherichia coli and NMR studies. *FEBS Letters, 272*, 34–36.

Kobashigawa, Y., Saio, T., Ushio, M., Sekiguchi, M., Yokochi, M., Ogura, K., et al. (2012). Convenient method for resolving degeneracies due to symmetry of the magnetic susceptibility tensor and its application to pseudo contact shift-based protein-protein complex structure determination. *Journal of Biomolecular NMR, 53*, 53–63.

Konrat, R. (2014). NMR contributions to structural dynamics studies of intrinsically disordered proteins. *Journal of Magnetic Resonance, 241*, 74–85.

Krishna Mohan, P. M., Barve, M., Chatterjee, A., Ghosh-Roy, A., & Hosur, R. V. (2008). NMR comparison of the native energy landscapes of DLC8 dimer and monomer. *Biophysical Chemistry, 134*, 10–19.

Krishnarjuna, B., Jaipuria, G., Thakur, A., D'Silva, P., & Atreya, H. S. (2011). Amino acid selective unlabeling for sequence specific resonance assignments in proteins. *Journal of Biomolecular NMR, 49*, 39–51.

McDermott, A. E. (2004). Structural and dynamic studies of proteins by solid-state NMR spectroscopy: Rapid movement forward. *Current Opinion in Structural Biology, 14*, 554–561.

Muchmore, D. C., McIntosh, L. P., Russell, C. B., Anderson, D. E., & Dahlquist, F. W. (1989). Expression and nitrogen-15 labeling of proteins for proton and nitrogen-15 nuclear magnetic resonance. *Methods in Enzymology, 177*, 44–73.

Mukherjee, S., Mustafi, S. M., Atreya, H. S., & Chary, K. V. R. (2005). Measurement of 1J(Ni, Cαi), 1J(Ni, C'i−1), 2J(Ni, Cαi−1), 2J(HNi, C'i−1) and 2J(HNi, Cαi) values in 13C/15N-labeled proteins. *Magnetic Resonance in Chemistry, 43*, 326–329.

Ohki, S. Y., & Kainosho, M. (2008). Stable isotope labeling methods for protein NMR spectroscopy. *Progress in Nuclear Magnetic Resonance Spectroscopy, 53*(4), 208–226.

Pantoja-Uceda, D., & Santoro, J. (2012). New amino acid residue type identification experiments valid for protonated and deuterated proteins. *Journal of Biomolecular NMR, 54*, 145–153.

Portman, K. L., Long, J., Carr, S., Briand, L., Winzor, D. J., Searle, M. S., et al. (2014). Enthalpy/entropy compensation effects from cavity desolvation underpin broad ligand binding selectivity for rat odorant binding protein 3. *Biochemistry, 53*, 2371–2379.

Rasia, R. M., Brutscher, B., & Plevin, M. J. (2012). Selective isotopic unlabeling of proteins using metabolic precursors: Application to NMR assignment of intrinsically disordered proteins. *Chembiochem, 13*, 732–739.

Sanders, C. R., & Sonnichsen, F. (2006). Solution NMR of membrane proteins: Practice and challenges. *Magnetic Resonance in Chemistry, 44*, S24–S40.

Schubert, M., Oschkinat, H., & Schmieder, P. (2001). Amino acid type-selective backbone 1H-15N-correlations for Arg and Lys. *Journal of Biomolecular NMR, 20*, 379–384.

Schubert, M., Smalla, M., Schmieder, P., & Oschkinat, H. (1999). MUSIC in triple-resonance experiments: Amino acid type-selective (1)H-(15)N correlations. *Journal of Magnetic Resonance, 141*, 34–43.

Shi, L., Ahmed, M. A. M., Zhang, W., Whited, G., Brown, L. S., & Ladizhansky, V. (2009). Three-dimensional solid-state NMR study of a seven helical integral membrane proton pump—Structural insights. *Journal of Molecular Biology, 386*, 1078–1093.

Shortle, D. (1994). Assignment of amino acid type in ^1H-^{15}N correlation spectra by labeling with 14N-amino acids. *Journal of Magnetic Resonance Series B, 105*, 88–90.

Tugarinov, V., & Kay, L. E. (2004). Stereospecific NMR assignments of prochiral methyls, rotameric states and dynamics of valine residues in malate synthase G. *Journal of the American Chemical Society, 126*, 9827–9836.

Vuister, G. W., Kim, S.-J., Wu, C., & Bax, A. (1994). 2D and 3D NMR study of phenylalanine residues in proteins by reverse isotopic labeling. *Journal of the American Chemical Society, 116*, 9206–9210.

Waugh, D. S. (1996). Genetic tools for selective labeling of proteins with alpha-15N-amino acids. *Journal of Biomolecular NMR, 8*, 184–192.

Wood, M., & Komives, E. (1999). Production of large quantities of isotopically labeled protein in Pichia pastoris by fermentation. *Journal of Biomolecular NMR, 13*(2), 149–159.

Wüthrich, K. (1986). *NMR of proteins and nucleic acids*. New York: Wiley.

Zhang, Q., Atreya, H. S., Kamen, D. E., Girvin, M. E., & Szyperski, T. (2008). GFT projection NMR based resonance assignment of membrane proteins: Application to subunit C of E. coli F(1)F(0) ATP synthase in LPPG micelles. *Journal of Biomolecular NMR, 40*, 157–163.

SECTION II

Labeling in Eukarya

CHAPTER NINE

Isotope Labeling of Eukaryotic Membrane Proteins in Yeast for Solid-State NMR

Ying Fan*,[†], Sanaz Emami*,[†], Rachel Munro*,[†], Vladimir Ladizhansky*,[†], Leonid S. Brown*,[†,1]

*Department of Physics, University of Guelph, Guelph, Ontario, Canada
[†]Biophysics Interdepartmental Group, University of Guelph, Guelph, Ontario, Canada
[1]Corresponding author: e-mail address: lebrown@uoguelph.ca

Contents

1. Introduction 194
2. Expression and Isotope Labeling in *P. pastoris*: Background 196
3. Isotope Labeling of Membrane Proteins in *P. pastoris* 197
   3.1 Protein Targets and Vectors 197
   3.2 Optimization of Small-Scale Natural Abundance Expression 198
   3.3 Large-Scale Isotope Labeling 200
   3.4 Production of Samples for ssNMR 202
4. Outlook 207
Acknowledgments 207
References 207

Abstract

Solid-state NMR (ssNMR) is a rapidly developing technique for exploring structure and dynamics of membrane proteins, but its progress is hampered by its low sensitivity. Despite the latest technological advances, routine ssNMR experiments still require several milligrams of isotopically labeled protein. While production of bacterial membrane proteins on this scale is usually feasible, obtaining such quantities of eukaryotic membrane proteins is often impossible or extremely costly. We have demonstrated that, by using isotopic labeling in yeast *Pichia pastoris*, one can inexpensively produce milligram quantities of doubly labeled functional samples, which yield multidimensional ssNMR spectra of high resolution suitable for detailed structural investigation. This was achieved by combining protocols of economical isotope labeling of soluble proteins previously used for solution NMR with protocols of expression of eukaryotic membrane proteins successfully employed for other methods. We review two cases of such isotope labeling, of fungal rhodopsin from *Leptosphaeria maculans* and human aquaporin-1.

Methods in Enzymology, Volume 565
ISSN 0076-6879
http://dx.doi.org/10.1016/bs.mie.2015.05.010

© 2015 Elsevier Inc.
All rights reserved.

193

1. INTRODUCTION

For many years, structure and dynamics of membrane proteins has been a very active, but extremely challenging, area of research. While the established structural techniques such as cryoelectron microscopy, X-ray crystallography, and solution nuclear magnetic resonance (NMR) made numerous important contributions to the field, the number of solved unique structures of membrane proteins is minuscule compared to that of soluble proteins (http://www.rcsb.org/pdb/). Besides the main technical challenges related to functional expression, stability in detergents, and crystallization, X-ray crystallography and solution NMR often face a conceptual problem of providing a native-like environment for membrane proteins, being restricted to using three-dimensional protein crystals or small detergent micelles. In this respect, an emerging powerful technique of solid-state NMR (ssNMR) provides an attractive alternative, as it allows studies of membrane proteins in the artificial native-like lipid bilayers or even in the native membranes, without having any inherent limitations on the protein molecular weight (Hong, Zhang, & Hu, 2012; Judge, Taylor, Dannatt, & Watts, 2015; Murray, Das, & Cross, 2013; Opella, 2015; Renault et al., 2012; Wang & Ladizhansky, 2014; Ward, Brown, & Ladizhansky, 2015). In the last few years, several structures of polytopic membrane proteins and their oligomers have been determined by ssNMR (Das et al., 2015; Park et al., 2012; Shahid et al., 2012; Wang et al., 2013), making it a reputable player in the team of structural biology techniques. Despite the latest technological improvements, such as dynamic nuclear polarization, fast magic-angle spinning, paramagnetic doping, and proton detection, its low sensitivity remains to be one of the main limitations for broader application and further development of ssNMR (Agarwal et al., 2014; Akbey et al., 2010; Bajaj, Mak-Jurkauskas, Belenky, Herzfeld, & Griffin, 2009; Barbet-Massin et al., 2014; Ward et al., 2014). As routine multidimensional ssNMR experiments normally require several milligrams of isotopically labeled membrane protein, which should be functional and homogeneous, the choice of expression system becomes extremely important, especially for eukaryotic targets.

Various systems for isotope labeling of membrane proteins for solution and ssNMR have been extensively discussed in the literature (Gautier, 2014; Goncalves, Ahuja, Erfani, Eilers, & Smith, 2010; Kim, Howell, Van Horn, Jeon, & Sanders, 2009; Maslennikov & Choe, 2013;

Sobhanifar et al., 2010; Takahashi & Shimada, 2010; Tapaneeyakorn, Goddard, Oates, Willis, & Watts, 2011; Verardi, Traaseth, Masterson, Vostrikov, & Veglia, 2012), so we will only provide a brief summary of the points most important for understanding our choice of the expression system. The use of bacterial expression for isotope labeling of membrane proteins (most commonly done in *Escherichia coli*) offers many advantages, such as low cost, fast cell growth, and versatility of possible labeling schemes, which allow uniform, sparse, alternate, and specific carbon labeling along with perdeuteration and selective protonation (Gautier, 2014; Kim et al., 2009; Verardi et al., 2012). While *E. coli* expression systems usually work well for bacterial membrane proteins, their use is often problematic for eukaryotic membrane targets. Among the common problems are low functional expression yields, expression of misfolded nonfunctional proteins, tendency to form inclusion bodies, and lack of proper posttranslational modifications (Gautier, 2014; Kim et al., 2009; Takahashi & Shimada, 2010). Nevertheless, *E. coli* remains to be the most popular expression host, and structures of several polytopic eukaryotic membrane proteins have been determined by NMR after isotopic labeling in *E. coli* (Berardi, Shih, Harrison, & Chou, 2011; Hiller et al., 2008; Park et al., 2012), with several more yielding promising NMR spectra (Kimura et al., 2014; Schmidt, Thomas, Muller, Scheidt, & Huster, 2014). On the opposite end of the spectrum is expression and labeling in insect and mammalian cell cultures, which can provide the most native fold and posttranslational modifications for eukaryotic membrane proteins but often suffer from high costs, low yields, slow growth, and the lack of established protocols for deuteration and uniform labeling (Gautier, 2014; Kim et al., 2009; Takahashi & Shimada, 2010; Tapaneeyakorn et al., 2011). While these systems keep improving (Egorova-Zachernyuk, Bosman, & Degrip, 2011; Kofuku et al., 2014; Opefi, Tranter, Smith, & Reeves, 2015; Shirzad-Wasei et al., 2013), they produced only a few selectively labeled eukaryotic membrane proteins for the detailed NMR studies so far, e.g., visual rhodopsin (Kimata, Reeves, & Smith, 2015). Special mention should be given to cell-free expression, which gains popularity for eukaryotic membrane protein production both for solution and ssNMR as its costs are decreasing, while yields and versatility are increasing (Klammt et al., 2012; Linser et al., 2014).

To produce samples for our multidimensional ssNMR studies, we have been looking for an isotope-labeling system, which would combine low cost, ease and speed of growth, and high expression yields of *E. coli* with the native protein fold provided by mammalian and insect cell cultures

(Kim et al., 2009; Takahashi & Shimada, 2010). In our view, these conditions are satisfied by the expression and isotopic labeling in the methylotrophic yeast *Pichia pastoris*, which will be described in this chapter using two examples, fungal rhodopsin from *Leptosphaeria maculans* and human aquaporin-1 (hAQP1) (Emami, Fan, Munro, Ladizhansky, & Brown, 2013; Fan, Shi, Ladizhansky, & Brown, 2011).

2. EXPRESSION AND ISOTOPE LABELING IN *P. pastoris*: BACKGROUND

Using *P. pastoris* for isotopic labeling of eukaryotic membrane proteins is a logical product of synthesis of the two well-developed biotechnological trends for this species of methylotrophic yeast: successful expression of many membrane proteins for X-ray crystallography and other biophysical methods and established economical protocols for isotope labeling of soluble proteins for solution NMR.

For many years, *P. pastoris* has been explored as a promising expression host for mammalian membrane proteins, especially, for G-protein-coupled receptors (GPCRs) (Andre et al., 2006; Hedfalk, 2013; Ramon & Marin, 2011; Sarramegna, Demange, Milon, & Talmont, 2002; Singh et al., 2012; Yurugi-Kobayashi et al., 2009). It was convincingly shown that functional expression of GPCRs is much more likely in *Pichia* than in *E. coli* (Lundstrom et al., 2006), and indeed, two X-ray structures of GPCRs have been already obtained using expression in the former host (Hino et al., 2012; Shimamura et al., 2011). Several X-ray structures of challenging mammalian membrane targets from other protein classes (including various ion channels, aquaporins, and ABC transporters) have been obtained using *P. pastoris* expression as well (Aller et al., 2009; Horsefield et al., 2008; Kane Dickson, Pedi, & Long, 2014; Long, Campbell, & Mackinnon, 2005; Molina et al., 2007). Finally, expression of various eukaryotic membrane proteins for studies by other biophysical techniques, such as Fourier transform infrared (FTIR) spectroscopy, Raman spectroscopy, time-resolved spectroscopy in the visible, electron paramagnetic resonance, and cryo-electron microscopy, is well established (Bieszke, Spudich, Scott, Borkovich, & Spudich, 1999; Krause, Engelhard, Heberle, Schlesinger, & Bittl, 2013; Muller, Bamann, Bamberg, & Kuhlbrandt, 2015; Waschuk, Bezerra, Shi, & Brown, 2005).

In parallel to that, carbon- and nitrogen-isotope-labeling protocols for soluble proteins for solution NMR are long-established in *Pichia*

(Laroche, Storme, De Meutter, Messens, & Lauwereys, 1994; Pickford & O'Leary, 2004; Sugiki, Ichikawa, Miyazawa-Onami, Shimada, & Takahashi, 2012; Wood & Komives, 1999) and have been optimized to become economical, remaining efficient at the same time (Rodriguez & Krishna, 2001). Possibility of deuteration has been demonstrated as well (Massou et al., 1999; Morgan, Kragt, & Feeney, 2000; Sugiki et al., 2012; Tomida et al., 2003). We further extended these economical shake-flask isotope-labeling protocols (Pickford & O'Leary, 2004; Rodriguez & Krishna, 2001) to eukaryotic membrane proteins and optimized them to produce ssNMR samples giving excellent resolution sufficient for detailed structural characterization (Emami et al., 2013; Fan, Shi, et al., 2011).

3. ISOTOPE LABELING OF MEMBRANE PROTEINS IN *P. pastoris*

3.1 Protein Targets and Vectors

Our first eukaryotic membrane protein target was fungal rhodopsin from *L. maculans*, known as *Leptosphaeria* rhodopsin (LR) or Mac in optogenetics field (Chow et al., 2010; Idnurm & Howlett, 2001; Waschuk et al., 2005). We have previously optimized expression, purification, and lipid reconstitution protocols for several fungal rhodopsins (Fan, Solomon, Oliver, & Brown, 2011; Furutani et al., 2004; Waschuk et al., 2005). Similar to other opsins, LR forms red-colored chromoprotein upon addition of external all-*trans*-retinal, making the monitoring and optimization of its expression convenient. To maximize yield of isotopically labeled LR, we recloned its gene (encoding N-terminally truncated and C-terminally 6xHis-tagged variant) into the pPICZαA vector (Invitrogen), which provides N-terminal secretion signal (α-factor) directing LR to the plasma membrane along with the ability for selection of multiple chromosomal integrations on zeocin (Cedarlane) plates.

Our second eukaryotic membrane protein target was hAQP1, constitutive selective water channel (Agre et al., 1993), whose high-yield expression in *P. pastoris* has been thoroughly optimized previously, both in shake flasks and fermenters (Norden et al., 2011; Nyblom et al., 2007; Oberg et al., 2009; Oberg & Hedfalk, 2012). We used pPICZB-hAQP1-Myc-His6 expression vector (kindly provided by Frederick Öberg and Kristina Hedfalk, Göteborg University, Sweden), which encodes full-length hAQP1 with a C-terminal Myc and 6xHis tags (Nyblom et al., 2007) and, similar to the vector used for LR, allows for selection of the transformants with multiple integration events, based on resistance to zeocin.

The vectors were linearized with the appropriate restriction enzymes (*Bst*XI or *Pme*I), desalted by the QIAquick nucleotide removal kit (Qiagen), and transformed into protease-deficient *P. pastoris* strain SMD1168H (Invitrogen) by electroporation (MicroPulser; Bio-Rad) according to the manual of the *Pichia* expression kit (Invitrogen), with small modifications described in detail earlier (Fan, Shi, et al., 2011).

3.2 Optimization of Small-Scale Natural Abundance Expression

Before proceeding with the large-scale isotope labeling, which can be quite costly, it is important to optimize the expression of protein of interest in small scale using natural abundance media. One needs not only to select the best-producing transformant colony but also to optimize the length of growth, based both on biochemical and economic considerations. First, it is important to find the optimal balance between the amount of expressed protein and the quantity of added methanol. Second, one should make sure that the cells are not overincubated with methanol, as amount of protein was shown to decrease in some cases of prolonged incubation (Emami et al., 2013; Fan, Shi, et al., 2011; Issaly et al., 2001). Finally, it is recommended to reconfirm optimal incubation time in the large-scale expression with natural abundance media, as the timing of protein synthesis in small and large scales may differ somewhat.

To find the best-producing colonies, 100–200 μL aliquots of the transformation mixtures were spread on yeast peptone dextrose (YPD, see Table 1) sorbitol plates containing zeocin at two different concentrations, 100 and 500 μg/mL. After 3–10 days of incubation at 30 °C, the transformant colonies were isolated from the plates and screened for the highest expression level. Cells were inoculated into 5 mL of buffered minimal dextrose (BMD, see Table 1) medium in a 250-mL baffled flask and were shaken at 30 °C, 300 rpm overnight. After OD_{600} reached ~ 2, another 20 mL of sterile BMD was added to the same flask, which was shaken at 300 rpm, 30 °C for 18–24 h until the OD_{600} was 10. This culture was centrifuged at $1500 \times g$ for 5 min at 4 °C, resuspended in 25 mL of buffered minimal methanol (BMM, see Table 1) to induce protein expression, and was shaken at 240 rpm, 28–30 °C for variable periods of time needed to achieve optimal expression in each case as described below.

Table 1 Media Recipes According to the Manual of the *Pichia* Expression Kit (Invitrogen Life Technologies) and Protocols Used for Solution NMR (Pickford & O'Leary, 2004; Sugiki et al., 2012)

Medium	Composition
YPD	1% yeast extract, 2% peptone, and 2% glucose
BMD	100 mM potassium phosphate, pH 6.0; 1.34% YNB with ammonium sulfate and without amino acids (or 0.34% YNB without ammonium sulfate and amino acids, and 1% ammonium sulfate); 4×10^{-5}% biotin; and 0.5% glucose
BMM	100 mM potassium phosphate, pH 6.0; 1.34% YNB with ammonium sulfate and without amino acids (or 0.34% YNB without ammonium sulfate and amino acids, and 1% ammonium sulfate); 4×10^{-5}% biotin; and 0.5% methanol

See text for modifications used for solid-state NMR sample preparation.

In the case of LR (Fan, Shi, et al., 2011), additional 175 µL of 100% methanol (final concentration 0.7%) and 6.25 µL of 10 mM all-*trans*-retinal in isopropanol (Sigma, needed for rhodopsin regeneration, final concentration 2.5 µM) were added after 24 h of growth in BMM. At different time points (24, 40, 48, and 52 h), 1 mL of the expression culture was withdrawn and centrifuged at $1500 \times g$ for 5 min at 4 °C. The expression level of LR and its optimal timing were evaluated by the intensity of the red color of the yeast pellet, and the colonies showing the most intense red color were selected for large-scale expression. It was also found that 40 h may be the optimal postinduction time, as longer incubation led to decrease in the color intensity, similar to what was observed earlier for *Neurospora* rhodopsin (Furutani et al., 2004).

For hAQP1 (Emami et al., 2013), which is not colored, evaluation of the expression levels required a different approach. The cells were harvested and resuspended in cell resuspension buffer (CRB) (20 mM Tris–HCl, pH 7.6, 100 mM NaCl, 0.5 mM EDTA, 5% (w/v) glycerol) (Nyblom et al., 2007). To digest the cell walls, 0.5–2 mg of lyticase (from *Arthrobacter luteus*; Sigma) was added to the resuspended pellets and slowly shaken for 3 h at 4 °C. The cells were centrifuged at $1500 \times g$ for 5 min at 4 °C and immediately resuspended in one pellet volume of CRB, with the addition of the same volume of ice-cold acid-washed glass beads (Fisher, 420–600 µm diameter), and then the cells were disrupted using vigorous repeated vortexing. The cell debris were removed by centrifugation at $700 \times g$ for 5 min at 4 °C, and the cell lysate was collected (vortexing and centrifugation were repeated up to 8× to achieve complete breakage of the cells). All cell supernatants containing the membrane fractions were combined for each colony and centrifuged at 40,000 ×g for 30 min at 4 °C. The relative hAQP1 content of these membranes was monitored by immunoblotting (using anti-His$_6$ primary antibody; Clontech, and antimouse IgG HRP conjugate secondary antibody; Bio-Rad) or InVision™ His-tag In-gel Stain (Invitrogen), and the colony giving the most intense band corresponding to the His-tagged protein (molecular weight of ∼30 kDa) was chosen for the large-scale growth and isotope labeling. For optimization of the postinduction growth length (the time between transfer to BMM and harvesting), using a colony with the highest level of hAQP1 expression, the cells were harvested after 6, 12, 24, 36, and 44 h after the induction (additional 0.5% methanol was added after 24 h), and the level of expression was monitored by immunoblotting or InVision™ gel staining. We found that the best time for harvesting the cells is 24 h as longer incubation did not produce appreciable increase in the hAQP1 yield but required the addition of extra methanol. On the other hand, the amount of hAQP1 substantially decreased upon longer induction and became barely detectable at 44 h. Note that this expression timing is quite different from that of LR, stressing the need for individual optimization for each protein.

3.3 Large-Scale Isotope Labeling

For both LR and hAQP1, the large-scale protein expression followed the established shake-flask protocols for secreted soluble proteins (Pickford & O'Leary, 2004; Rodriguez & Krishna, 2001) with small modifications and protein-to-protein variations.

For LR (Fan, Shi, et al., 2011), material from a cell colony with the highest protein expression level found in small-scale culture as described above was inoculated into 50 mL of ^{13}C,^{15}N-BMD, with 0.5% ^{13}C-glucose and 0.8% ^{15}NH$_4$Cl (Cambridge Isotope Laboratories), in a sterile 250-mL baffled flask. Note that for isotope labeling, yeast nitrogen base used in preparation of BMD and BMM should not contain natural abundance ammonium sulfate. This culture was shaken at 30 °C (300 rpm) for 18–24 h, until the OD$_{600}$ exceeded 2, and inoculated into a sterile 2-L baffled flask containing 200 mL of ^{13}C,^{15}N-BMD, which was shaken at 29–30 °C (270 rpm) for 18–24 h, until the OD$_{600}$ reached 10. To induce LR expression, the cells were pelleted in sterile containers at 1500 ×g for 5 min at 4 °C and gently resuspended in 0.8 L of $^{13}C,^{15}N$-BMM (with 0.5% ^{13}C-methanol and 0.8% $^{15}NH_4Cl$), which was placed into 2.8 L Fernbach flask and shaken at 29–30 °C (240 rpm). Note that in our experience, the expression yields are quite sensitive to oxygenation level, so it is important not to put more than 1 L of the medium into each flask (with 0.8 L being the optimal volume for LR). The temperature was kept at the standard value, as there was no evidence of any significant protein misfolding or heterogeneity warranting expression at lower temperatures. It should be noted that incubation temperature may need to be lowered in cases of other, more temperature-sensitive proteins. A volume of 10 mM isopropanol stock of all-*trans*-retinal (final concentration 5 µM) and 100% filtered ^{13}C-methanol (final concentration 0.5%) were added to the growth medium after 24 h of induction. The 0.5% concentration of labeled methanol (and glucose), lower than that in several labeling protocols, was used following the recommendation of the protocol for economical labeling (Rodriguez & Krishna, 2001), where it showed additional benefits of more complete incorporation of isotopes and lack of cell lysis by high concentration of methanol. We evaluated growth at 1% ^{13}C-methanol, but did not find any improvement of the protein yield. The red-colored cells were collected by centrifugation at $1500 \times g$ for 5 min at 4 °C after 40 h of induction, as the protein yield was found to be lower upon longer (48–52 h) and shorter (24 h) incubation times, similar to what was observed in small scale (see above). The protein yields at 24 h were less than 20% of those at 40 h, showing much lower yield per unit of ^{13}C-methanol, making the collection at 40 h the most economical. The final postpurification yield of doubly isotope-labeled LR produced in shake flasks exceeded 5 mg/L of culture (estimated spectrophotometrically, Cary 50; Varian). With such yields, and since only 0.5% ^{13}C-methanol was used, replenished just once (after 24 h of incubation), the cost of LR sample is close to that for bacterial proteins of the same class (microbial rhodopsins) produced by us in *E. coli* earlier (Shi et al., 2009; Shi, Kawamura, Jung, Brown, & Ladizhansky, 2011).

In a similar fashion, for the large-scale expression and isotope labeling of hAQP1 (Emami et al., 2013), the best producing colony was used to inoculate 50 mL of $^{13}C,^{15}N$–BMD containing 0.8% ($^{15}NH_4)_2SO_4$ and 0.5% ^{13}C glucose in a 250-mL baffled flask. The cells were grown at 300 rpm and 30 °C for 18–24 h and transferred to a 2-L baffled flask with an additional 200 mL of isotope-labeled BMD. The cells were then shaken at 275 rpm and 29 °C for 18–24 h before being centrifuged at $1500 \times g$, 4 °C for 10 min. The cells were resuspended in 1 L of $^{13}C,^{15}N$-BMM (0.5% ^{13}C-

methanol and 0.8% ($^{15}NH_4)_2SO_4$) in a sterile 2.8-L flask and grown for 24 h at 240 rpm and 28 °C before being collected by centrifugation at 1500 × g, 4 °C for 10 min. The final postpurification yield of ~6 mg of isotope-labeled hAQP1 per liter of culture (determined as described below) is on par with that found for LR (Fan, Shi, et al., 2011) and is very economical, as no second addition of ^{13}C-methanol was needed.

3.4 Production of Samples for ssNMR

Even though many of the procedures described in this section are not isotope labeling specific, they are crucial for making high-quality ssNMR samples. Further steps for the production of eukaryotic membrane protein samples for ssNMR include protein purification, quantification, lipid reconstitution, functional assays, and spectroscopic verification. While protein purification followed different protein-specific protocols in each case, described in detail earlier (Emami et al., 2013; Fan, Shi, et al., 2011), it is important to mention a few common points. As the isotopically labeled material is much more valuable than natural abundance one, it is advisable to take several extra steps to minimize protein losses upon purification and reconstitution. First, one should make sure that cell breakage is complete and not much protein remained in the debris, which can be verified spectroscopically (for chromoproteins) or by Western blotting. Second, it is important to optimize the solubilization conditions, by choosing the best detergent and then screening for the most appropriate time, pH, ionic strength, and temperature of solubilization. One should carefully optimize not only the detergent concentration but also the ratio of volumes of the detergent and membrane pellets. Third, one should make sure that His-tag affinity protein purification (in our case, using Ni^{2+}-NTA resin from Qiagen) is not affected by protein proteolysis (by adding effective protease inhibitors which at the same time do not interfere with the resin binding) and protein denaturation (by choosing appropriate detergent, pH, and temperature). It is imperative to verify efficiencies of solubilization, resin binding, washing, and protein elution either by spectroscopically (for chromoproteins) or by SDS–PAGE (combined with Western blotting or InVision™ gel staining as needed). Finally, the purity of the final product needs to be assessed spectroscopically (making sure that there is no observable absorption corresponding to membrane-bound cytochromes at 410 nm), as well as by SDS–PAGE and mass spectrometry (MALDI-TOF in our case, but other types can be used depending on the protein). The latter procedure can also verify the extent of isotope labeling and reveal whether posttranslational processing, such as glycosylation and N-terminal cleavage, is present. It should be noted that we have not observed any glycosylation of our protein targets, which was later confirmed by ^{13}C ssNMR spectroscopy (see below).

Once the purification is complete, it is important to precisely quantify the purified isotope-labeled protein before attempting lipid reconstitution, as protein-to-lipid ratio may critically affect the quality of ssNMR spectra (Shi et al., 2009; Yang, Aslimovska, & Glaubitz, 2011). While quantification of purified LR and other microbial rhodopsins is trivial, as it can be done spectrophotometrically, we used several concurrent methods to determine an exact amount of hAQP1. The final yield of purified solubilized hAQP1 was estimated by SDS–PAGE, Bradford assay with bovine serum albumin standard (Bio-Rad DC Protein assay), as well as by UV–Vis spectroscopy after removing imidazole, using absorbance at 280 nm of 0.1% protein of 0.944 (estimated from PROTPARAM). The ultimate postreconstitution verification of protein quantity was done by measuring the amplitudes of amide I bands of FTIR spectra (Fig. 1) and relative intensities of 1D ^{15}N ssNMR signals of hAQP1 proteoliposomes.

Lipid reconstitution optimization is another key step that needs to be performed to obtain stable homogeneous functional membrane protein

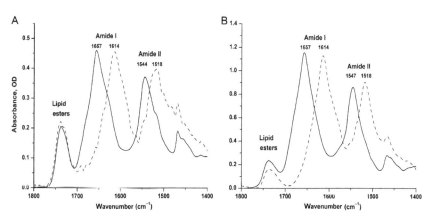

Figure 1 Static FTIR spectra (Bruker IFS66vs) of lipid-reconstituted LR (A) and hAQP-1 (B) expressed in *Pichia pastoris*. Large isotope shifts of the amide I and amide II bands can be observed by comparing spectra of natural abundance (solid lines) and ^{15}N-,^{13}C-labeled proteins (dashed lines), positions of the maxima are indicated. Additionally, by comparing the intensities of lipid ester vibrations and protein amide I bands, one can estimate the actual postreconstitution protein-to-lipid ratios (daCosta & Baenziger, 2003). *This figure was constructed from data reported by us earlier (Emami et al., 2013; Fan, Shi, et al., 2011).*

samples yielding good spectral resolution in ssNMR. Among the parameters to optimize are lipid compositions, protein-to-lipid ratios, and conditions of reconstitution (removal of detergent by dialysis or Bio-Beads, rate of detergent removal, pH, ionic strength, and temperature). Lipid compositions should be chosen in such a way that the reconstituted protein is stable and retains its native structure and function at high protein-to-lipid ratios needed to achieve good signal-to-noise ratios of ssNMR signals, and yields high spectral resolution. In principle, the higher protein-to-lipid ratios can lead to both better spectral resolution and higher signal amplitudes of ssNMR, which in some cases can be related to formation of 2D crystals, e.g., in the case of *Anabaena* sensory rhodopsin (Ward, Wang, et al., 2015). On the other hand, for some proteins, such as green proteorhodopsin, having somewhat lower protein-to-lipid ratios not leading to 2D crystallization produces more homogeneous samples with better spectral resolution (Shi et al., 2009; Yang et al., 2011). It should be also taken into the account that the final actual protein-to-lipid ratio can be somewhat different from the intended one, depending on how much of endogenous lipids is retained by the protein. The outcome of optimization of lipid reconstitution is certainly protein specific, so that no universal recommendations can be given, but we will mention that in the case of LR (Fan, Shi, et al., 2011), we used DMPC/DMPA (Avanti Polar Lipids) lipid mixture (9:1 w/w), 1:1 protein-to-lipid ratio, reconstitution was mediated by Triton X-100, and detergent was withdrawn by Bio-Beads SM2 (Bio-Rad). For hAQP1 (Emami et al., 2013), egg PC/brain PS (Avanti Polar Lipids) lipid mixture (9:1 w/w) was used, at 2:1 protein-to-lipid ratio, and the detergent (β-octylglucoside; ThermoFisher) was slowly withdrawn by dialysis.

Many of the parameters mentioned above can be conveniently screened by taking static FTIR spectra of a small fraction of a sample (50–150 μg) avoiding costly optimization by ssNMR, which needs milligram quantities of protein. Figure 1 shows typical FTIR absorption spectra of lipid-reconstituted natural abundance and doubly isotope-labeled LR and hAQP1 in mid-infrared range, obtained from dried liposomes identical to those used for ssNMR. Much useful information can be extracted from these spectra, including the amount of protein, its secondary structure, extent of isotope labeling, actual protein-to-lipid ratio, and, sometimes, functionality. Specifically, the amount of protein can be determined from the amplitude of amide I band, provided that the extinction coefficient and film geometry (such as thickness) are known. The position of amide I band, which mainly reflects C=O vibrations of the backbone atoms, is very sensitive to its secondary structure (e.g., the 1657 cm^{-1} maxima observed for the natural abundance LR and hAQP1 are typical for α-helical proteins) (Barth & Zscherp, 2002; Tamm & Tatulian, 1997). If one observes broadening or appearance of new shoulders of this spectral band in some of the lipid mixtures, most likely, the protein lost its native conformation and these lipids should not be used (e.g., DMPC/DMPA mixture was not optimal for hAQP1 in this respect). Additionally, the amide I vibrational bands are sensitive to ^{13}C labeling, while amide II bands, which include C–N vibrations are sensitive to both ^{13}C and ^{15}N labeling, and their large and nearly complete downshifts (Fig. 1) indicate high extent of the isotope labeling. The actual protein-to-lipid ratio of the reconstituted sample can be determined from comparing the amplitudes (or more precisely, areas) of the amide I bands and vibrations of lipid esters centered around 1738 cm^{-1} (daCosta & Baenziger, 2003). Often, functionality of the sample can be tested by FTIR as well, but this may require time-resolved reaction-induced measurements, such as those performed by us for LR (Fan, Shi, et al., 2011). For hAQP1, we conducted additional standard functionality tests using proteoliposomes with lower protein-to-lipid ratios (liposome shrinkage stopped-flow assays under hypertonic shock) (Emami et al., 2013).

Even though prescreening of samples by FTIR can be very useful in terms of time and cost savings, the final sample optimization has to be done by ssNMR. The simplest and cheapest test one can employ in such optimization is to look at the resolution of 1D ^{15}N spectra, which eliminates the need for more expensive ^{13}C labeling at this stage. Both lipid compositions and protein-to-lipid ratios can be optimized this way, and we were able to achieve ^{15}N linewidths of 0.5–0.7 ppm for our protein targets. Ultimately, 2D ssNMR experiments on doubly (^{13}C and ^{15}N) labeled samples have to be conducted to ensure that the spectra have not only good resolution (about 0.5 ppm for ^{13}C for LR and hAQP1) and dispersion but also good completeness of spectral coverage. This can be done by integrating peak intensities of easily identifiable amino acid types, such as glycines (from N/Cα correlations in the NCA spectrum) and alanines (from Cα/Cβ correlations in the ^{13}C–^{13}C correlation spectrum), and comparing the results to the counts expected from the primary structure. Figure 2 shows examples of well-resolved and well-dispersed ^{13}C–^{13}C correlations for LR and hAQP1, from which completeness of spectral coverage for alanines could be determined. These ssNMR spectra also report on other characteristics of the sample, such as homogeneity, native secondary structure, and lack of glycosylation (judging from the lack of characteristic carbohydrate signals at 70–80 ppm).

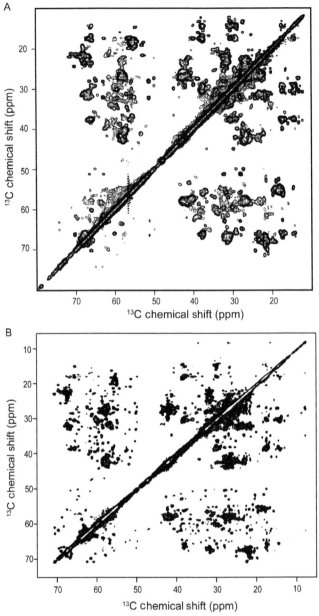

Figure 2 Resolution of MAS ssNMR spectra of eukaryotic membrane proteins expressed in *Pichia pastoris*. Two-dimensional ^{13}C-^{13}C DARR (dipolar-assisted rotational resonance) (Morcombe, Gaponenko, Byrd, & Zilm, 2004; Takegoshi, Nakamura, & Terao, 2001) chemical shift correlation spectra of lipid-reconstituted ^{15}N-,^{13}C-labeled LR (A) and hAQP-1 (B). (A) Spectra of ^{13}C,^{15}N-LR proteoliposomes were recorded at a field of 600 MHz (proton frequency), at a spinning frequency of 12 kHz, and at 5 °C (Fan, Shi, et al., 2011). (B) Spectra of ^{13}C,^{15}N-hAQP-1 proteoliposomes recorded at a field of 800 MHz, at a spinning frequency of 14.3 kHz, and at 5 °C (Emami et al., 2013). *Panels (A) and (B) are reproduced with kind permission from Springer Science and Business Media.*

Repeated recording of ssNMR spectra over the course of several weeks is advisable to verify protein stability under employed experimental conditions. The obtained spectral quality (Fig. 2) allows for the extensive resonance assignments, proving that eukaryotic membrane proteins expressed in *P. pastoris* can be used for detailed studies of structure and dynamics by ssNMR.

4. OUTLOOK

In summary, we demonstrated that *P. pastoris* can serve as an economical host for uniform isotope labeling of eukaryotic membrane proteins and can be used to produce excellent ^{15}N,^{13}C-labeled samples for multi-dimensional ssNMR. For the future, extending the existing deuteration protocols to membrane proteins looks very realistic. Additionally, possibility of amino acid selective labeling has been explored and looks promising (Austin, Kuestner, Chang, Madden, & Martin, 2011; Chen et al., 2006; Whittaker, 2007).

One limitation of isotope labeling in *P. pastoris* is inability for sparse labeling, often required for better spectral resolution in ssNMR, as the main carbon source is ^{13}C-methanol. In the future, this shortcoming may be rectified by using other species of yeast, which can metabolize more complex carbon sources amenable to alternate labeling (Miyazawa-Onami et al., 2013; Nars et al., 2014; Sugiki et al., 2012).

ACKNOWLEDGMENTS

The research was supported by grants from the Canada Foundation for Innovation/Ontario Innovation Trust/Ontario Ministry of Research and Innovation, the Natural Sciences and Engineering Research Council of Canada (NSERC), and the University of Guelph.

REFERENCES

Agarwal, V., Penzel, S., Szekely, K., Cadalbert, R., Testori, E., Oss, A., et al. (2014). De novo 3D structure determination from sub-milligram protein samples by solid-state 100 kHz MAS NMR spectroscopy. *Angewandte Chemie (International Ed. in English)*, 53(45), 12253–12256.

Agre, P., Preston, G. M., Smith, B. L., Jung, J. S., Raina, S., Moon, C., et al. (1993). Aquaporin CHIP: The archetypal molecular water channel. *The American Journal of Physiology*, 265(4 Pt. 2), F463–F476.

Akbey, U., Lange, S., Trent Franks, W., Linser, R., Rehbein, K., Diehl, A., et al. (2010). Optimum levels of exchangeable protons in perdeuterated proteins for proton detection in MAS solid-state NMR spectroscopy. *Journal of Biomolecular NMR*, 46(1), 67–73.

Aller, S. G., Yu, J., Ward, A., Weng, Y., Chittaboina, S., Zhuo, R., et al. (2009). Structure of P-glycoprotein reveals a molecular basis for poly-specific drug binding. *Science*, *323*(5922), 1718–1722.

Andre, N., Cherouati, N., Prual, C., Steffan, T., Zeder-Lutz, G., Magnin, T., et al. (2006). Enhancing functional production of G protein-coupled receptors in Pichia pastoris to levels required for structural studies via a single expression screen. *Protein Science*, *15*(5), 1115–1126.

Austin, R. J., Kuestner, R. E., Chang, D. K., Madden, K. R., & Martin, D. B. (2011). SILAC compatible strain of Pichia pastoris for expression of isotopically labeled protein standards and quantitative proteomics. *Journal of Proteome Research*, *10*(11), 5251–5259.

Bajaj, V. S., Mak-Jurkauskas, M. L., Belenky, M., Herzfeld, J., & Griffin, R. G. (2009). Functional and shunt states of bacteriorhodopsin resolved by 250 GHz dynamic nuclear polarization-enhanced solid-state NMR. *Proceedings of the National Academy of Sciences of the United States of America*, *106*(23), 9244–9249.

Barbet-Massin, E., Pell, A. J., Retel, J. S., Andreas, L. B., Jaudzems, K., Franks, W. T., et al. (2014). Rapid proton-detected NMR assignment for proteins with fast magic angle spinning. *Journal of the American Chemical Society*, *136*(35), 12489–12497.

Barth, A., & Zscherp, C. (2002). What vibrations tell us about proteins. *Quarterly Reviews of Biophysics*, *35*(4), 369–430.

Berardi, M. J., Shih, W. M., Harrison, S. C., & Chou, J. J. (2011). Mitochondrial uncoupling protein 2 structure determined by NMR molecular fragment searching. *Nature*, *476*(7358), 109–113.

Bieszke, J. A., Spudich, E. N., Scott, K. L., Borkovich, K. A., & Spudich, J. L. (1999). A eukaryotic protein, NOP-1, binds retinal to form an archaeal rhodopsin-like photochemically reactive pigment. *Biochemistry*, *38*(43), 14138–14145.

Chen, C. Y., Cheng, C. H., Chen, Y. C., Lee, J. C., Chou, S. H., Huang, W., et al. (2006). Preparation of amino-acid-type selective isotope labeling of protein expressed in Pichia pastoris. *Proteins*, *62*(1), 279–287.

Chow, B. Y., Han, X., Dobry, A. S., Qian, X., Chuong, A. S., Li, M., et al. (2010). High-performance genetically targetable optical neural silencing by light-driven proton pumps. *Nature*, *463*(7277), 98–102.

daCosta, C. J., & Baenziger, J. E. (2003). A rapid method for assessing lipid:protein and detergent:protein ratios in membrane-protein crystallization. *Acta Crystallographica. Section D, Biological Crystallography*, *59*(Pt. 1), 77–83.

Das, N., Dai, J., Hung, I., Rajagopalan, M. R., Zhou, H. X., & Cross, T. A. (2015). Structure of CrgA, a cell division structural and regulatory protein from Mycobacterium tuberculosis, in lipid bilayers. *Proceedings of the National Academy of Sciences of the United States of America*, *112*(2), E119–E126.

Egorova-Zachernyuk, T. A., Bosman, G. J., & Degrip, W. J. (2011). Uniform stable-isotope labeling in mammalian cells: Formulation of a cost-effective culture medium. *Applied Microbiology and Biotechnology*, *89*(2), 397–406.

Emami, S., Fan, Y., Munro, R., Ladizhansky, V., & Brown, L. S. (2013). Yeast-expressed human membrane protein aquaporin-1 yields excellent resolution of solid-state MAS NMR spectra. *Journal of Biomolecular NMR*, *55*(2), 147–155.

Fan, Y., Shi, L., Ladizhansky, V., & Brown, L. S. (2011). Uniform isotope labeling of a eukaryotic seven-transmembrane helical protein in yeast enables high-resolution solid-state NMR studies in the lipid environment. *Journal of Biomolecular NMR*, *49*(2), 151–161.

Fan, Y., Solomon, P., Oliver, R. P., & Brown, L. S. (2011). Photochemical characterization of a novel fungal rhodopsin from Phaeosphaeria nodorum. *Biochimica et Biophysica Acta*, *1807*(11), 1457–1466.

Furutani, Y., Bezerra, A. G., Waschuk, S. A., Sumii, M., Brown, L. S., & Kandori, H. (2004). FTIR spectroscopy of the K photointermediate of Neurospora rhodopsin: Structural changes of the retinal, protein, and water molecules after photoisomerization. *Biochemistry, 43*(30), 9636–9646.

Gautier, A. (2014). Structure determination of alpha-helical membrane proteins by solution-state NMR: Emphasis on retinal proteins. *Biochimica et Biophysica Acta, 1837*(5), 578–588.

Goncalves, J. A., Ahuja, S., Erfani, S., Eilers, M., & Smith, S. O. (2010). Structure and function of G protein-coupled receptors using NMR spectroscopy. *Progress in Nuclear Magnetic Resonance Spectroscopy, 57*(2), 159–180.

Hedfalk, K. (2013). Further advances in the production of membrane proteins in Pichia pastoris. *Bioengineered, 4*(6), 363–367.

Hiller, S., Garces, R. G., Malia, T. J., Orekhov, V. Y., Colombini, M., & Wagner, G. (2008). Solution structure of the integral human membrane protein VDAC-1 in detergent micelles. *Science, 321*(5893), 1206–1210.

Hino, T., Arakawa, T., Iwanari, H., Yurugi-Kobayashi, T., Ikeda-Suno, C., Nakada-Nakura, Y., et al. (2012). G-protein-coupled receptor inactivation by an allosteric inverse-agonist antibody. *Nature, 482*(7384), 237–240.

Hong, M., Zhang, Y., & Hu, F. (2012). Membrane protein structure and dynamics from NMR spectroscopy. *Annual Review of Physical Chemistry, 63*, 1–24.

Horsefield, R., Norden, K., Fellert, M., Backmark, A., Tornroth-Horsefield, S., Terwisscha van Scheltinga, A. C., et al. (2008). High-resolution x-ray structure of human aquaporin 5. *Proceedings of the National Academy of Sciences of the United States of America, 105*(36), 13327–13332.

Idnurm, A., & Howlett, B. J. (2001). Characterization of an opsin gene from the ascomycete Leptosphaeria maculans. *Genome, 44*(2), 167–171.

Issaly, N., Solsona, O., Joudrier, P., Gautier, M. F., Moulin, G., & Boze, H. (2001). Optimization of the wheat puroindoline-a production in Pichia pastoris. *Journal of Applied Microbiology, 90*(3), 397–406.

Judge, P. J., Taylor, G. F., Dannatt, H. R., & Watts, A. (2015). Solid-state nuclear magnetic resonance spectroscopy for membrane protein structure determination. *Methods in Molecular Biology, 1261*, 331–347.

Kane Dickson, V., Pedi, L., & Long, S. B. (2014). Structure and insights into the function of a Ca(2+)-activated Cl(-) channel. *Nature, 516*(7530), 213–218.

Kim, H. J., Howell, S. C., Van Horn, W. D., Jeon, Y. H., & Sanders, C. R. (2009). Recent advances in the application of solution NMR spectroscopy to multi-span integral membrane proteins. *Progress in Nuclear Magnetic Resonance Spectroscopy, 55*(4), 335–360.

Kimata, N., Reeves, P. J., & Smith, S. O. (2015). Uncovering the triggers for GPCR activation using solid-state NMR spectroscopy. *Journal of Magnetic Resonance, 253*, 111–118.

Kimura, T., Vukoti, K., Lynch, D. L., Hurst, D. P., Grossfield, A., Pitman, M. C., et al. (2014). Global fold of human cannabinoid type 2 receptor probed by solid-state (13) C-, (15) N-MAS NMR and molecular dynamics simulations. *Proteins, 82*(3), 452–465.

Klammt, C., Maslennikov, I., Bayrhuber, M., Eichmann, C., Vajpai, N., Chiu, E. J., et al. (2012). Facile backbone structure determination of human membrane proteins by NMR spectroscopy. *Nature Methods, 9*(8), 834–839.

Kofuku, Y., Ueda, T., Okude, J., Shiraishi, Y., Kondo, K., Mizumura, T., et al. (2014). Functional dynamics of deuterated beta(2)-adrenergic receptor in lipid bilayers revealed by NMR spectroscopy. *Angewandte Chemie (International Ed. in English), 53*(49), 13376–13379.

Krause, N., Engelhard, C., Heberle, J., Schlesinger, R., & Bittl, R. (2013). Structural differences between the closed and open states of channelrhodopsin-2 as observed by EPR spectroscopy. *FEBS Letters, 587*(20), 3309–3313.

Laroche, Y., Storme, V., De Meutter, J., Messens, J., & Lauwereys, M. (1994). High-level secretion and very efficient isotopic labeling of tick anticoagulant peptide (TAP) expressed in the methylotrophic yeast, Pichia pastoris. *Biotechnology (N Y), 12*(11), 1119–1124.

Linser, R., Gelev, V., Hagn, F., Arthanari, H., Hyberts, S. G., & Wagner, G. (2014). Selective methyl labeling of eukaryotic membrane proteins using cell-free expression. *Journal of the American Chemical Society, 136*(32), 11308–11310.

Long, S. B., Campbell, E. B., & Mackinnon, R. (2005). Crystal structure of a mammalian voltage-dependent Shaker family K+ channel. *Science, 309*(5736), 897–903.

Lundstrom, K., Wagner, R., Reinhart, C., Desmyter, A., Cherouati, N., Magnin, T., et al. (2006). Structural genomics on membrane proteins: Comparison of more than 100 GPCRs in 3 expression systems. *Journal of Structural and Functional Genomics, 7*(2), 77–91.

Maslennikov, I., & Choe, S. (2013). Advances in NMR structures of integral membrane proteins. *Current Opinion in Structural Biology, 23*(4), 555–562.

Massou, S., Puech, V., Talmont, F., Demange, P., Lindley, N. D., Tropis, M., et al. (1999). Heterologous expression of a deuterated membrane-integrated receptor and partial deuteration in methylotrophic yeasts. *Journal of Biomolecular NMR, 14*(3), 231–239.

Miyazawa-Onami, M., Takeuchi, K., Takano, T., Sugiki, T., Shimada, I., & Takahashi, H. (2013). Perdeuteration and methyl-selective (1)H, (13)C-labeling by using a Kluyveromyces lactis expression system. *Journal of Biomolecular NMR, 57*(3), 297–304.

Molina, D. M., Wetterholm, A., Kohl, A., McCarthy, A. A., Niegowski, D., Ohlson, E., et al. (2007). Structural basis for synthesis of inflammatory mediators by human leukotriene C-4 synthase. *Nature, 448*(7153), 613–616.

Morcombe, C. R., Gaponenko, V., Byrd, R. A., & Zilm, K. W. (2004). Diluting abundant spins by isotope edited radio frequency field assisted diffusion. *Journal of the American Chemical Society, 126*(23), 7196–7197.

Morgan, W. D., Kragt, A., & Feeney, J. (2000). Expression of deuterium-isotope-labelled protein in the yeast pichia pastoris for NMR studies. *Journal of Biomolecular NMR, 17*(4), 337–347.

Muller, M., Bamann, C., Bamberg, E., & Kuhlbrandt, W. (2015). Light-induced helix movements in channelrhodopsin-2. *Journal of Molecular Biology, 427*(2), 341–349.

Murray, D. T., Das, N., & Cross, T. A. (2013). Solid state NMR strategy for characterizing native membrane protein structures. *Accounts of Chemical Research, 46*(9), 2172–2181.

Nars, G., Saurel, O., Bordes, F., Saves, I., Remaud-Simeon, M., Andre, I., et al. (2014). Production of stable isotope labelled lipase Lip2 from Yarrowia lipolytica for NMR: Investigation of several expression systems. *Protein Expression and Purification, 101*, 14–20.

Norden, K., Agemark, M., Danielson, J. A., Alexandersson, E., Kjellbom, P., & Johanson, U. (2011). Increasing gene dosage greatly enhances recombinant expression of aquaporins in Pichia pastoris. *BMC Biotechnology, 11*(1), 47.

Nyblom, M., Oberg, F., Lindkvist-Petersson, K., Hallgren, K., Findlay, H., Wikstrom, J., et al. (2007). Exceptional overproduction of a functional human membrane protein. *Protein Expression and Purification, 56*(1), 110–120.

Oberg, F., Ekvall, M., Nyblom, M., Backmark, A., Neutze, R., & Hedfalk, K. (2009). Insight into factors directing high production of eukaryotic membrane proteins; production of 13 human AQPs in Pichia pastoris. *Molecular Membrane Biology, 26*(4), 215–227.

Oberg, F., & Hedfalk, K. (2012). Recombinant production of the human aquaporins in the yeast Pichia pastoris (Invited Review). *Molecular Membrane Biology, 30*, 15–31.

Opefi, C. A., Tranter, D., Smith, S. O., & Reeves, P. J. (2015). Construction of stable Mammalian cell lines for inducible expression of g protein-coupled receptors. *Methods in Enzymology, 556*, 283–305.

Opella, S. J. (2015). Solid-state NMR and membrane proteins. *Journal of Magnetic Resonance*, *253*, 129–137.
Park, S. H., Das, B. B., Casagrande, F., Tian, Y., Nothnagel, H. J., Chu, M., et al. (2012). Structure of the chemokine receptor CXCR1 in phospholipid bilayers. *Nature*, *491*(7426), 779–783.
Pickford, A. R., & O'Leary, J. M. (2004). Isotopic labeling of recombinant proteins from the methylotrophic yeast Pichia pastoris. *Methods in Molecular Biology*, *278*, 17–33.
Ramon, A., & Marin, M. (2011). Advances in the production of membrane proteins in Pichia pastoris. *Biotechnology Journal*, *6*(6), 700–706.
Renault, M., Tommassen-van Boxtel, R., Bos, M. P., Post, J. A., Tommassen, J., & Baldus, M. (2012). Cellular solid-state nuclear magnetic resonance spectroscopy. *Proceedings of the National Academy of Sciences of the United States of America*, *109*(13), 4863–4868.
Rodriguez, E., & Krishna, N. R. (2001). An economical method for (15)N/(13)C isotopic labeling of proteins expressed in Pichia pastoris. *Journal of Biochemistry*, *130*(1), 19–22.
Sarramegna, V., Demange, P., Milon, A., & Talmont, F. (2002). Optimizing functional versus total expression of the human mu-opioid receptor in Pichia pastoris. *Protein Expression and Purification*, *24*(2), 212–220.
Schmidt, P., Thomas, L., Muller, P., Scheidt, H. A., & Huster, D. (2014). The G-protein-coupled neuropeptide Y receptor type 2 is highly dynamic in lipid membranes as revealed by solid-state NMR spectroscopy. *Chemistry*, *20*(17), 4986–4992.
Shahid, S. A., Bardiaux, B., Franks, W. T., Krabben, L., Habeck, M., van Rossum, B. J., et al. (2012). Membrane-protein structure determination by solid-state NMR spectroscopy of microcrystals. *Nature Methods*, *9*(12), 1212–1217.
Shi, L., Ahmed, M. A., Zhang, W., Whited, G., Brown, L. S., & Ladizhansky, V. (2009). Three-dimensional solid-state NMR study of a seven-helical integral membrane proton pump—Structural insights. *Journal of Molecular Biology*, *386*(4), 1078–1093.
Shi, L., Kawamura, I., Jung, K. H., Brown, L. S., & Ladizhansky, V. (2011). Conformation of a seven-helical transmembrane photosensor in the lipid environment. *Angewandte Chemie (International Ed. in English)*, *50*(6), 1302–1305.
Shimamura, T., Shiroishi, M., Weyand, S., Tsujimoto, H., Winter, G., Katritch, V., et al. (2011). Structure of the human histamine H1 receptor complex with doxepin. *Nature*, *475*(7354), 65–70.
Shirzad-Wasei, N., van Oostrum, J., Bovee-Geurts, P. H., Wasserman, M., Bosman, G. J., & Degrip, W. J. (2013). Large scale expression and purification of mouse melanopsin-L in the baculovirus expression system. *Protein Expression and Purification*, *91*(2), 134–146.
Singh, S., Gras, A., Fiez-Vandal, C., Martinez, M., Wagner, R., & Byrne, B. (2012). Large-scale production of membrane proteins in Pichia pastoris: The production of G protein-coupled receptors as a case study. *Methods in Molecular Biology*, *866*, 197–207.
Sobhanifar, S., Reckel, S., Junge, F., Schwarz, D., Kai, L., Karbyshev, M., et al. (2010). Cell-free expression and stable isotope labelling strategies for membrane proteins. *Journal of Biomolecular NMR*, *46*(1), 33–43.
Sugiki, T., Ichikawa, O., Miyazawa-Onami, M., Shimada, I., & Takahashi, H. (2012). Isotopic labeling of heterologous proteins in the yeast Pichia pastoris and Kluyveromyces lactis. *Methods in Molecular Biology*, *831*, 19–36.
Takahashi, H., & Shimada, I. (2010). Production of isotopically labeled heterologous proteins in non-E. coli prokaryotic and eukaryotic cells. *Journal of Biomolecular NMR*, *46*(1), 3–10.
Takegoshi, K., Nakamura, S., & Terao, T. (2001). C-13-H-1 dipolar-assisted rotational resonance in magic-angle spinning NMR. *Chemical Physics Letters*, *344*(5–6), 631–637.
Tamm, L. K., & Tatulian, S. A. (1997). Infrared spectroscopy of proteins and peptides in lipid bilayers. *Quarterly Reviews of Biophysics*, *30*(4), 365–429.

Tapaneeyakorn, S., Goddard, A. D., Oates, J., Willis, C. L., & Watts, A. (2011). Solution- and solid-state NMR studies of GPCRs and their ligands. *Biochimica et Biophysica Acta, 1808*(6), 1462–1475.

Tomida, M., Kimura, M., Kuwata, K., Hayashi, T., Okano, Y., & Era, S. (2003). Development of a high-level expression system for deuterium-labeled human serum albumin. *The Japanese Journal of Physiology, 53*(1), 65–69.

Verardi, R., Traaseth, N. J., Masterson, L. R., Vostrikov, V. V., & Veglia, G. (2012). Isotope labeling for solution and solid-state NMR spectroscopy of membrane proteins. *Advances in Experimental Medicine and Biology, 992*, 35–62.

Wang, S., & Ladizhansky, V. (2014). Recent advances in magic angle spinning solid state NMR of membrane proteins. *Progress in Nuclear Magnetic Resonance Spectroscopy, 82*, 1–26.

Wang, S., Munro, R. A., Shi, L., Kawamura, I., Okitsu, T., Wada, A., et al. (2013). Solid-state NMR spectroscopy structure determination of a lipid-embedded heptahelical membrane protein. *Nature Methods, 10*(10), 1007–1012.

Ward, M. E., Brown, L. S., & Ladizhansky, V. (2015). Advanced solid-state NMR techniques for characterization of membrane protein structure and dynamics: Application to Anabaena Sensory Rhodopsin. *Journal of Magnetic Resonance, 253*, 119–128.

Ward, M. E., Wang, S., Krishnamurthy, S., Hutchins, H., Fey, M., Brown, L. S., et al. (2014). High-resolution paramagnetically enhanced solid-state NMR spectroscopy of membrane proteins at fast magic angle spinning. *Journal of Biomolecular NMR, 58*(1), 37–47.

Ward, M. E., Wang, S., Munro, R., Ritz, E., Hung, I., Gor'kov, P. L., et al. (2015). In situ structural studies of anabaena sensory rhodopsin in the E. coli membrane. *Biophysical Journal, 108*(7), 1683–1696.

Waschuk, S. A., Bezerra, A. G., Shi, L., & Brown, L. S. (2005). Leptosphaeria rhodopsin: Bacteriorhodopsin-like proton pump from a eukaryote. *Proceedings of the National Academy of Sciences of the United States of America, 102*(19), 6879–6883.

Whittaker, J. W. (2007). Selective isotopic labeling of recombinant proteins using amino acid auxotroph strains. *Methods in Molecular Biology, 389*, 175–188.

Wood, M. J., & Komives, E. A. (1999). Production of large quantities of isotopically labeled protein in Pichia pastoris by fermentation. *Journal of Biomolecular NMR, 13*(2), 149–159.

Yang, J., Aslimovska, L., & Glaubitz, C. (2011). Molecular dynamics of proteorhodopsin in lipid bilayers by solid-state NMR. *Journal of the American Chemical Society, 133*(13), 4874–4881.

Yurugi-Kobayashi, T., Asada, H., Shiroishi, M., Shimamura, T., Funamoto, S., Katsuta, N., et al. (2009). Comparison of functional non-glycosylated GPCRs expression in Pichia pastoris. *Biochemical and Biophysical Research Communications, 380*(2), 271–276.

CHAPTER TEN

Development of Approaches for Deuterium Incorporation in Plants

Barbara R. Evans[*,1], Riddhi Shah[†,‡]

[*]Chemical Sciences Division, Oak Ridge National Laboratory, Oak Ridge, Tennessee, USA
[†]Bredesen Center for Interdisciplinary Research and Graduate Education, University of Tennessee, Knoxville, Tennessee, USA
[‡]Biology and Soft Matter Division, Oak Ridge National Laboratory, Oak Ridge, Tennessee, USA
[1]Corresponding author: e-mail address: evansb@ornl.gov

Contents

1. Introduction 214
   1.1 Applications of Deuterium Labeling in Plants 214
2. Challenges of Plant Cultivation in D_2O 215
   2.1 Inhibitory Effects of D_2O on Plants 215
   2.2 Chemical and Physical Properties of D_2O 216
   2.3 Differences Between Plant Growth in D_2O and H_2O 217
   2.4 Natural Variation of D_2O Tolerance in Plant Species 219
3. Analysis of Deuterium-Labeled Plant Biomass 220
   3.1 Analysis of Deuterium Substitution 220
   3.2 Chemical and Physical Characterization of Lignocellulosic Biomass 221
4. Deuterium Labeling of Plants for Metabolic Studies 223
   4.1 Deuteration Studies of Duckweed 223
   4.2 Isotopic Labeling of Plants for Nutritional Tracer Studies 224
   4.3 Development of Arabidopsis as a Platform for Deuterium Tracing 225
5. Production of Deuterated Plants for Structural Studies 225
   5.1 Methods for Cultivation of Plants in High Concentrations of D_2O 225
   5.2 Multiple-Chamber Perfusion System for Long-Term Cultivation 229
   5.3 Production of Deuterated Annual Ryegrass 230
   5.4 Production of Deuterated Switchgrass 232
   5.5 Deuterium Labeling Winter Grain Rye with D_2O and Deuterated Phenylalanine 234
Acknowledgments 239
References 239

Abstract

Soon after the discovery of deuterium, efforts to utilize this stable isotope of hydrogen for labeling of plants began and have proven successful for natural abundance to 20% enrichment. However, isotopic labeling with deuterium (^2H) in higher plants at the level of 40% and higher is complicated by both physiological responses, particularly water exchange through transpiration, and inhibitory effects of D$_2$O on germination, rooting, and growth. The highest incorporation of 40–50% had been reported for photoheterotrophic cultivation of the duckweed *Lemna*. Higher substitution is desirable for certain applications using neutron scattering and nuclear magnetic resonance (NMR) techniques. ^{1}H^{2}H NMR and mass spectroscopy are standard methods frequently used for determination of location and amount of deuterium substitution. The changes in infrared (IR) absorption observed for H to D substitution in hydroxyl and alkyl groups provide rapid initial evaluation of incorporation. Short-term experiments with cold-tolerant annual grasses can be carried out in enclosed growth containers to evaluate incorporation. Growth in individual chambers under continuous air perfusion with dried sterile-filtered air enables long-term cultivation of multiple plants at different D$_2$O concentrations. Vegetative propagation from cuttings extends capabilities to species with low germination rates. Cultivation in 50% D$_2$O of annual ryegrass and switchgrass following establishment of roots by growth in H$_2$O produces samples with normal morphology and 30–40% deuterium incorporation in the biomass. Winter grain rye (*Secale cereale*) was found to efficiently incorporate deuterium by photosynthetic fixation from 50% D$_2$O but did not incorporate deuterated phenylalanine-*d8* from the growth medium.

1. INTRODUCTION

Investigation of the effects of deuterium (^{2}H) and deuterium oxide (heavy water, ^{2}H$_2$O, D$_2$O) on plants started soon after the discovery of this isotope. Plants are the source of many chemicals and materials for which production of deuterated forms would be advantageous. Deuterium tracing methods at natural abundance or slightly enriched levels provide vital information on water usage and biosynthesis by plants in the natural environment and in greenhouse and laboratory studies. Here, we summarize earlier studies of deuterium incorporation in higher plants and our efforts to attain higher levels of deuterium substitution in lignocellulosic biomass from grasses.

1.1 Applications of Deuterium Labeling in Plants

Isotopic substitution with deuterium, ^{2}H or D, the heavier stable isotope of hydrogen, has proven extremely useful for investigation of biochemical processes. Deuterium substitution enables elucidation of enzyme mechanisms and investigation of nutrient metabolism. Deuterated solvents and deuterium substitution are extensively used for nuclear magnetic resonance (NMR) studies of chemical bonding and structure (Katz & Crespi, 1966; Thomas, 1971). Deuterium oxide (D$_2$O) and other deuterated solvents are used for neutron imaging techniques, particularly small-angle neutron scattering (SANS), which investigate mesoscale hierarchical structures and molecular associations at the scale of 1–100 nm. In particular, SANS offers the capability to examine the cell walls of higher plants which exhibit those characteristics, while the intricate chemical bonding and physical associations render its component polymers difficult to separate without modification of their native structures (Langan et al., 2012, 2014; Martínez-Sanz, Gidley, & Gilbert, 2015; Pingali et al., 2010; Stuhrmann & Miller, 1978). Use of deuterated solvents, particularly D_2O, provides reduction of background, assisting component visualization. The scattering properties of component materials in a mixture differ based on their chemical composition and physical properties. Deuterium and hydrogen interact with neutrons very differently. Substituting deuterium for hydrogen atoms in the chemical composition of a component alters its scattering length density (SLD), enabling separation of scattering patterns from components of similar elemental composition (Langan et al., 2012; Perkins, 1981). The D_2O/H_2O ratios in the solvent can be adjusted to match the scattering lengths of individual components, allowing them to be differentiated.

2. CHALLENGES OF PLANT CULTIVATION IN D_2O
2.1 Inhibitory Effects of D_2O on Plants

The first attempts to cultivate plants in D_2O were carried out soon after the discovery of deuterium by Harold Urey (Lewis, 1933). Soon thereafter, photosynthetic incorporation of deuterium in algal biomass was demonstrated for algae grown in 12% D_2O (Reitz & Bonhoeffer, 1934). It was found that unicellular algae are able to carry out photosynthesis in D_2O, with several species able to adapt to and grow in concentrations up to 99.8% D_2O. Experiments with algae transferred to 99.8% D_2O found that the light reactions of photosynthesis were not inhibited, while the dark reaction, i.e., carbon fixation, was 40% slower than in H_2O (Pratt & Trelease, 1938). Protocols for photoautotrophic cultivation of both eukaryotic green algae and cyanobacteria in 99.6% D_2O were developed and successfully used for mass production of algal biomass to provide deuterated biochemicals and biopolymers (Crespi, Daboll, & Katz, 1970; Crespi & Katz, 1972; DaBoll, Crespi, & Katz, 1962).

It therefore appeared both logical and useful to cultivate higher plants in similar concentrations of D_2O. However, from the first experiments carried out with tobacco seeds (Lewis, 1933) and wheat (Pratt & Curry, 1937), it has been found that D_2O at concentrations greater than 50% severely impacts vascular plants, inhibiting seed germination and root growth, while growth rates were slowed even at 50% D_2O (Katz, 1960; Katz & Crespi, 1966; Yang et al., 2010). Of course, plants are normally exposed to low levels of D_2O in the environment, with natural abundance in surface waters ranging from 0.0145 to 0.0149 mole percent (Katz, 1960). Even at the natural abundance levels, incorporation of D into plant tissues, particularly cellulose, has been used to examine differences in carbon fixation and transpiration between plants (Sternberg & DeNiro, 1983). Tracer studies with higher amounts of D_2O have been used to establish partitioning between different plant tissues and to differentiate between cellulose derived from a fixed carbon source such as sucrose and from photosynthesis (Yakir & DeNiro, 1990).

2.2 Chemical and Physical Properties of D_2O

As the concentration of D_2O is increased, the physical and chemical differences between D_2O and H_2O begin to affect the metabolism and growth of living organisms (Table 1). Although the boiling point of D_2O is only 1.41 °C higher than that of water, its melting point is 3.79 °C higher, while temperature of maximum density increases from 3.98 for H_2O to 11.23 °C for D_2O. The viscosity of D_2O is 25% higher than that of H_2O, but the ionization constant is one-fifth that of H_2O. Salts and gases have lower solubility in D_2O, while amino acids show a 0.5-pK unit increase in their apparent ionization constants. This directly impacts availability of CO_2 for photosynthesis, as it reduces both dissolution of CO_2 and formation of carbonic acid, as well as

Table 1 Physical and Chemical Properties of H_2O and D_2O (Kirshenbaum, 1951)

Property	H_2O	D_2O
Molecular weight (g/mol)	18.0	20.03
Melting point (°C)	0.00	3.8
Temperature of maximum density (°C)	3.98	11.23
Boiling point (°C)	100.0	101.4
Viscosity at 25 °C (millipoise)	8.93	11.00
Dissociation constant	1.00×10^{-14}	1.95×10^{-15}
IR vibrational bands (cm^{-1})	3825.03	2757.85
	1653.78	1212.17
	3935.32	2883.48

ionization of carbonic acid to bicarbonate. The kinetic isotope effects are due to the difference in chemical bond reactivity for bonds to deuterium compared to bonds to hydrogen, resulting in slower reaction rates. Reactions are generally 7- to 10-fold slower for molecules with deuterium substitution due to the change in bond strength and activation energy (Katz, 1960; Wibert, 1955). In living organisms, this results in a lag time in growth, which can be overcome for algae and bacteria by adaptation to increasing amounts of D_2O (Katz, 1960; Katz & Crespi, 1966). The greater strength of the bonds of deuterium to other atoms results in band shifts in the absorption spectra. The differences in the infrared spectra of H_2O and D_2O provided another analytical method for quantification (Table 1). With the increased sensitivity and availability of instruments using Fourier Transform Infrared spectroscopy (FTIR), this technique provides a convenient method to determine chemistry and approximate ratios of deuterium in biological materials (Table 2). However, analyses are complicated by the impact of the differences in bond strengths on sample ionization for mass spectrometry, chemical coupling constants and peak height in NMR, and peak heights in FTIR (Evans et al., 2014; Evans et al., 2015; Thomas, 1971).

2.3 Differences Between Plant Growth in D_2O and H_2O

Several species of higher plants are reported to grow in 30–50% D_2O, with adverse effects increasing with D_2O concentration. Intracellular water content is generally found to be 10–15% below the concentration of the watering solution or liquid medium. Typically, plants incorporate deuterium into biomass components at weight percent about 10% lower than that of the D_2O weight percent in the feed solution. Partially deuterated carrots, kale, and spinach are grown hydroponically in solutions of 15–30% D_2O to obtain isotopically labeled vegetables for nutritional studies. These plants are germinated in vermiculite watered with H_2O then grown for 4 days before the seedlings are transferred to the hydroponic deuteration chambers (Grusak, 2000; Putzbach et al., 2005). For nutritional studies of human vitamin A metabolism, stable isotopic labeling of β-carotene with deuterium was carried out by cultivation of spinach and carrots in growth solutions containing 25% D_2O. The analysis of the β-carotene in the partially deuterated plants by mass spectroscopy found a range of partially deuterated isotopomers, with the highest incorporation and mass peak being that corresponding to replacement by deuterium of 10 of the 56 hydrogen atoms of the β-carotene molecule, corresponding to approximately 18% deuteration (Tang, Qin,

Table 2 FTIR Wavenumber Assignments Reported for Alkyl and Hydroxyl Absorption Bands Used for Detection of Deuterium Substitution

Wavenumber (cm^{-1})	Assignment	Reference
2900	C–H stretching (β-carotene and chlorophyll)	Katz (1960)
2150–2250	C–D stretching (β-carotene and chlorophyll)	Katz (1960)
2800–3000	C–H stretching (cellulose)	He et al. (2014)
2100	C–D stretching (cellulose)	He et al. (2014)
3300–3500	O–H stretching (cellulose)	He et al. (2014)
2400	O–D stretching (cellulose)	He et al. (2014)
2924–2954	C–H stretching (cellulose)	Bali et al. (2013)
2139–2251	C–D stretching (cellulose)	Bali et al. (2013)
3430–3435	O–H stretching (cellulose)	Bali et al. (2013)
2495	O–D stretching (cellulose)	Bali et al. (2013)
2098	C–D symmetric stretching (glycine)	Torres, Kukol, and Arking (2000)
2245	C–D asymmetric stretching (glycine)	Torres et al. (2000)
2900	C–H stretching (cellulose)	Hofstetter, Hinterstoisser, and Salmén (2006)
2490	O–D stretching (cellulose)	Hofstetter et al. (2006)
3340	O–H stretching (cellulose)	Hofstetter et al. (2006)
2486	O–D stretching (cellulose and hemicellulose)	Evans et al. (2014)
3328, 3331	O–H stretching (cellulose and hemicellulose)	Evans et al. (2014)
2168	C–D stretching (cellulose and hemicelluloses)	Evans et al. (2014)
2912–2946	C–H stretching (cellulose and hemicelluloses)	Evans et al. (2014)

Dolnikowski, Russell, & Grusak, 2005). Abnormal morphology and delayed development are noted at concentrations of D_2O above 50% for many species (Bhatia & Smith, 1968; Katz & Crespi, 1966). Seed germination and root elongation are adversely affected as D_2O concentrations are increased with wheat (Burgess & Northcote, 1969), winter rye (Siegel, Halpern, & Giumaro, 1964; Waber & Sakai, 1974), and *Arabidopsis* (Bhatia & Smith, 1968; Yang et al., 2010). The decrease in root elongation caused by D_2O may be due to the inhibition of proton exchange by the plasma membrane ATPase (Sacchi & Cocucci, 1992). Flowering and seed set are inhibited at concentrations lower than germination and growth (Bhatia & Smith, 1968; Katz & Crespi, 1966). High-throughput genomics have been used to examine the effects of growth in 30% D_2O on gene expression in Arabidopsis (Yang et al., 2010).

2.4 Natural Variation of D_2O Tolerance in Plant Species

Seed germination experiments indicated that two properties, (1) large seed size and (2) cold tolerance, may be indicative of higher D_2O tolerance (Blake, Crane, Uphaus, & Katz, 1968; Siegel et al., 1964). Of species studied that exhibited higher D_2O tolerance, winter grain rye has very high cold tolerance and large seeds that sprout rapidly following imbibing, even at temperatures as low as 4 °C, while *Arabidopsis* has small seeds and good cold tolerance (Bhatia & Smith, 1968). The large-seeded species winter grain rye, maize (corn), peas, and fescue grass were reported to germinate in 90–100% D_2O following treatment with fungicide in H_2O, while the large-seeded species morning glory and squash did not (Blake et al., 1968).

To avoid the problems due to inhibition of germination, vegetative propagation was first attempted using buds from the succulent *Kalaenchoe*, but root elongation was inhibited by 86.7% D_2O, initially at 10% of H_2O controls, and completely stopping after 54 h (Pratt & Curry, 1937). Later studies examined the small aquatic plant duckweed (*Lemna* spp.), which reproduces by budding from fronds. Duckweed plants (*Lemna peripusilla*) were reported to grow in 50–63% D_2O if supplied with 0.5% glucose as a fixed carbon source (Cope, Bose, Crespi, & Katz, 1965). However, abnormal morphology was noted at concentrations higher than 50%, with smaller air spaces, larger cells, and decrease in root length during the 10-day growth period. Addition of kinetin, a cytokine which stimulates cell division and delays senescence, resulted in improved morphology and prevented bleaching of fronds. Later studies found that the tonoplast membranes of the fronds of *L. minor* undergo disruptive rearrangements during initial exposure to 50% D_2O followed by adaptation and return to more normal morphology (Cooke & Davies, 1980; Cooke, Grego, Olivier, & Davies, 1979; Cooke, Grego, Roberts, & Davies, 1980).

Species that require warm temperatures for growth and seedling establishment such as rice (*Oryza sativa*) tend to have low tolerance of D_2O, but by careful experimental design, can be used for deuterium labeling (Tang, Qin, Dolnikowski, Russell, & Grusak, 2009). In addition to inhibition of germination and root elongation, stunting and slower growth are typically observed for plants grown in higher concentrations of D_2O. Comparative studies found lower transport of D_2O and greater inhibition of cation transport in leaf shoots of deuterium-sensitive rice seedlings than in deuterium-resistant winter rye seedlings (Shibabe & Yoda, 1984a,1984b, 1985).

3. ANALYSIS OF DEUTERIUM-LABELED PLANT BIOMASS

3.1 Analysis of Deuterium Substitution

Several methods have been used for analysis of deuterium incorporation in plant biomass. Isotopic analysis of combustion water has been used to determine deuterium content of dried plant samples (Cope et al., 1965; Crespi & Katz,1961; Yakir & DeNiro, 1990). In the case of small molecules such as carotene, mass spectroscopy (MS) is typically the method of choice (Grusak, 2000; Putzbach et al., 2005; Yang et al., 2010), as it allows determination of the isotopomer distribution. FTIR provides information on the relative amount and chemistry of deuterium substitution, with the large band shifts observed for alkyl and hydroxyl groups being particularly valuable. A compilation of waveband values reported in the literature (Table 2) shows the consistency of the alkyl and hydroxyl deuterium substitution shifts for cellulose (He et al., 2014; Hofstetter et al., 2006), hemicelluloses (Evans et al., 2014), and the amino acid glycine (Torres et al., 2000). However, in the case of the cell wall biopolymers cellulose and hemicelluloses, component polymers need to be separated and hydrolyzed, as MS is not suitable for analysis of the intact polymers of this composition and size range. Highly sensitive determination of environmental variations in deuterium content can be determined by isolation of cellulose with the sodium chlorite-acetic acid method, followed by nitration. The resultant cellulose nitrate is then analyzed by mass spectroscopy or solution phase NMR. Alternatively, isolated cellulose can be hydrolyzed to glucose and acetylated before analysis.

Substitution of deuterium for hydrogen changes spectral properties of chemical compounds as a result of changes in bond strength. The type of spectroscopy and site of substitution determine the ease and reliability of detection. Shifts of UV–visible absorption maxima for biochemicals such as chlorophyll are generally on the order of a few nanometers even for completely deuterated molecules (Katz, 1960).

Infrared (IR) spectroscopy readily detects the large band shifts observed for the stretching bands of O–H compared to O–D, and in C–H compared to C–D (Table 2). IR has been used to estimate deuterium in water of combustion as well as in chlorophyll (Katz & Crespi, 1966; Katz, 1960). The wave band shifts observed in FTIR for O–H versus O–D (3400 vs. 2500 cm^{-1}) and C–H to C–D (2800–3000 vs. 2000–2300 cm^{-1}) enable rapid detection and semiquantitative comparison of deuterium incorporation between intact biomass samples (Bali et al., 2013; Evans et al., 2014, 2015). NMR techniques enable quantitative determination of deuterium incorporation and localization to carbohydrates, lignin, and lipids. Although ^1H NMR has been used alone to estimate deuterium substitution from loss or reduction of proton peaks compared to protiated controls (Perkins, 1981; Putzbach et al., 2005), combination of ^1H and ^2H NMR enables more accurate determination of incorporation and location of deuterium (^2H). For direct NMR analysis, an ionic liquid solvent mixture consisting of 1:4 pyridinium chloride: dimethyl sulfoxide for ^2H and 1:4 deuterated pyridinium chloride: deuterated dimethyl sulfoxide for ^1H detection is used to dissolve samples of cellulose and whole biomass at 60 °C under a nitrogen atmosphere (Bali et al., 2013; Evans et al., 2014; Foston, McGaughey, O'Neill, Evans, & Ragauskas, 2012). An external standard composed of trifluoroacetic acid and deuterated trifluoroacetic acid (TFA/d-TFA) enables quantification of total proton and deuteron nuclei. A solid-state NMR method has been described that utilizes ^1H single-pulse and ^2H solid echo to estimate deuterium content of freeze-dried biomass by comparison to standard mixtures of glucose and deuterated glucose-1,2,3,4,5,6,6-*d7* (Foston et al., 2012).

3.2 Chemical and Physical Characterization of Lignocellulosic Biomass

The chemical and physical properties of biomass and component biomolecules of interest from plants cultivated in D_2O mixtures are best compared to those from control plants that are grown in H_2O solutions under the same conditions to validate experimental utilization of the deuterated versions.

Analytical techniques such as FTIR, $^1H^2H$ NMR, and mass spectroscopy that are used to determine the amount and location of deuterium incorporation also provide information on the chemical properties and composition of the samples. The intended use of the deuterated plant material will determine the specific properties that will need to be characterized and if the biomolecules of interest need to be purified for the intended application. The properties of lignocellulosic biomass that are relevant to biofuel production are characterized by an array of established methods (Foston & Ragauskas, 2012). Protocols for characterization of individual cell wall components lignin, cellulose, and hemicelluloses can require sequential extractions to separate them. Lignin content is often estimated as Klaason lignin by a spectrophotometric method. Carbohydrate composition is typically determined by acid hydrolysis followed by identification of component sugars by HPLC retention times. Due to its crystalline nature, cellulose is often characterized by X-ray diffraction, while ^{13}C cross-polarization magic angle spinning (CP/MAS) NMR provides more detailed information on both degree of crystallinity and proportions of crystalline isoforms.

Molecular weight and degree of polymerization are important physical parameters for characterization of the component biopolymers of the plant cell wall. Gel permeation chromatography (GPC) provides estimates of molecular weight distribution from which degree of polymerization (DP) and polydispersity index (PDI) can be calculated. Heterogeneity in chain length and hydrodynamic properties as well as poor solubility and hydrolysis or other reactions with concentrated acid and alkali complicate GPC analysis of the major cell wall components cellulose, hemicellulose, and lignin. Plant cell walls are composed of multiple composite layers formed during the course of cell growth and lignification. Component polymer properties will also vary between cell types. Prior functionalization and composition of mobile phase solvent vary dependent on the polymeric component being analyzed (cellulose, hemicelluloses, or lignin) and the type of separation column being used. Cellulose is often converted from its insoluble crystalline native form to the trianthranilate derivative to enable GPC analysis. The multimodal, broad elution peaks observed in GPC profiles for these biopolymers are characterized by the first statistical moment that elutes, termed the weight-average molecular weight (M_w), while the second statistical moment during the elution is the number-average molecular weight (M_n). Molecular weight estimation for M_w and M_n is based on polystyrene calibration standards. The DP is calculated by dividing the M_w by the monomeric unit molecular weight, taking into account changes caused by chemical modification (i.e., increase due to trianthranilate functionalization of hydroxyl groups of glucose units of cellulose) if done. The PDI is calculated by dividing M_w by the M_n, thus providing a measure of the spread in molecular weights observed for a particular sample of a biopolymer such as cellulose. Thus, these two parameters, DP and PDI, can be used to compare cellulose, hemicellulose, and lignin properties between different plants (Foston & Ragauskas, 2010), between deuterated and control bacterial cellulose (Bali et al., 2013), and between deuterated and control plants (Evans et al., 2014, 2015).

4. DEUTERIUM LABELING OF PLANTS FOR METABOLIC STUDIES

4.1 Deuteration Studies of Duckweed

The duckweed *Lemna* comprises a genus of small, rapidly growing aquatic plants which are widely used in science education (Environmental Inquiry, 2015; National Academy of Science & Committee on Biology Teacher Inservice Program, 1996; Robinson, 1988) and in freshwater toxicology tests (ISO, 2005; Moody & Miller, 2005; Sinkkonen, Myyr, Penttinen, & Rantalainen, 2011). Duckweed is increasingly under development for a range of biotechnological, bioenergy, and agricultural applications (Appenroth, Crawford, & Les, 2015; Appenroth et al., 2015; Stomp, 2005). *Lemna* species have been used for several studies of deuteration and the effects of D_2O. These plants, although renowned as the smallest flowering plants in North America, typically reproduce vegetatively through production of buds that separate from the mother plant, with each plant eventually possessing two to three leaves. They produce simple roots with length varying dependent on growth conditions. It was therefore thought that they would be able to tolerate higher D_2O concentrations, since seed germination and root elongation were not essential for growth. However, photoheterotrophic growth with supplementation of a fixed carbon source, glucose, was needed for growth in 50–60% D_2O (Cope et al., 1965). Addition of kinetin improved survival and prevented bleaching in 50–63% D_2O. Deuterium incorporation in whole biomass was achieved of 35% for growth with glucose in 50–63% D_2O and 59% for growth with 0.5% deuterated glucose in 50–63% D_2O. Later investigations by Cooke and coworkers utilized 50% D_2O for comparison of stress responses, particularly protein degradation, in *Lemna minor*. Isotopic stress induced by transfer to 50% D_2O exhibited features similar to nitrate starvation and other stressors, such as tonoplast membrane disruption and increased protein degradation. After exposure for 5 h, chloroplast and vacuole membranes showed signs of disruption, with maximum loss of proteins and amino acids, but after 25 h, plasma, tonoplast, and chloroplast membranes had started to recovered, and after 1 week were normal in appearance (Cooke & Davies, 1980; Cooke et al., 1979, 1980).

4.2 Isotopic Labeling of Plants for Nutritional Tracer Studies

Stable isotope labeling is an important technique for elucidation of metabolic pathways in whole organisms. The relative safety of low levels of stable isotopes has enabled studies in human patients. Hydroponic cultivation is a useful method for production of isotopically enriched plants for nutritional studies, as the amount of the isotope of interest can be more easily controlled and the amount needed minimized (Weaver, 1984). In the case of deuterium labeling, D_2O or deuterated compounds will be diluted by the large pool of ^1H hydrogen in the environmental atmosphere, the growth matrix, and the watering solutions, requiring a higher concentration to be used than is needed for isotopes of inorganic salt components such as iron and magnesium (Grusak, 2000). Vegetables of interest for nutritional studies, including kale, carrots, and spinach, were started from seed with H_2O in Vermiculite, a commercial soil substitute, to avoid the inhibitory effects of D_2O on seed germination and root elongation. After 4 days, seedlings were transferred to the hydroponic system and grown in solutions containing D_2O. Since D_2O concentrations of 20–30% give sufficient levels of isotopic labeling (15–30%) for nutritional tracing, deleterious effects of D_2O on plant growth were minimized. To limit water vapor exchange, hydroponic systems with aeration and circulation were enclosed with plastic curtains and perfused with compressed air. This approach was successfully used for the challenging task of production of isotopically labeled golden rice grain (Tang et al., 2009). Rice (*Oryza sativa*) is known to have low tolerance of D_2O compared to winter rye and other plants (Siegel et al., 1964), and flowering and seed set are known to be more sensitive to D_2O than vegetative growth (Bhatia & Smith, 1968). To avoid these problems, the rice was grown in H_2O under hydroponic conditions until after flowering and initiation of seed formation before transfer to 23% D_2O. Cultivation was continued until the rice grain was ready for harvest. An alternative to the large hydroponic tanks is the pot culture technique which used 2-L pots without cycling for improved efficiency in labeling of edible plants with stable isotopes (Weaver, 1984).

4.3 Development of Arabidopsis as a Platform for Deuterium Tracing

Widely used as a model plant for studies of genetics and systematic genomics, it is natural that the germination and growth of *Arabidopsis thaliana* in D_2O has been investigated. The growth habit of *Arabidopsis*, with seeds requiring cold stratification, would appear indicative of ability to adapt to higher ratios of D_2O/H_2O. Multigenerational adaptation was reported to be successful in adapting Arabidopsis to growth at 70% D_2O, although viable seed production was only achieved up to 50% (Bhatia & Smith, 1968). For metabolic tracer studies, *Arabidopsis* has been grown in 30% D_2O (Astot, Dolezal, Moritz, & Sandberg, 2000). The effects of D_2O on Arabidopsis grown in hydroponic culture were later examined and defined by microarray assay of total RNA and mass spectroscopy of 15 amino acids as part of development of a system for metabolic analysis. Initial screening of germination and seedling growth found that root elongation decreased as D_2O concentration was increased from 0% to 40%, with a sharp downward deflection between 30% and 40%. A concentration of 30% D_2O was therefore used for examination of gene expression under long-term and short-term (4 h) conditions. Proteolysis and mass spectroscopy were used to analyze incorporation of deuterium in proteins and amino acids. Adaptation to growth in 30% D_2O was evident, as seedlings continuously grown in 30% D_2O showed upregulation of 122 and downregulation of 99 proteins, while short-term exposure induced upregulation of 509 genes and downregulation of 258 genes (Yang et al., 2010). The largest group of upregulated genes corresponded to those known to be induced in response to environmental stressors, including heat, salt, oxidation, and wounding. For comparison, another study using a variation of the SILAC (stable isotope labeling by amino acids in cell culture) technique for proteomics in which Arabidopsis seedlings were exposed to mild salt stress (80 mM NaCl) for 7 days detected 92 upregulated proteins and 123 downregulated proteins (Lewandowska et al., 2013).

5. PRODUCTION OF DEUTERATED PLANTS FOR STRUCTURAL STUDIES

5.1 Methods for Cultivation of Plants in High Concentrations of D_2O

The cultivation conditions need to balance healthy plant growth with achievement of the desired deuterium incorporation levels. For most experimental investigations, it will be desirable to grow the plants in conditions that approximate greenhouse or field conditions as closely as feasible. This means photoautotrophic growth on solid matrices or in hydroponic solutions. The basic requirements are appropriate illumination and adequate supply of CO_2 and O_2 for plant growth, prevention of dilution of D_2O by transpiration exchange with ambient water vapor, and avoidance of microbial contamination. Algae, cyanobacteria, bacteria, and fungi tolerate high concentrations of D_2O much better than do plants (Katz, 1960; Katz & Crespi, 1966), and contaminate plant deuteration experiments (Evans et al., 2015; Siegel et al., 1964). Addition of any carbon source such as glucose renders maintenance of sterile conditions even more important, and use of nutrient agar or gellan media with vitamins requires stringent tissue culture methods. Soil and soil pellets need to be preswollen in water or fertilizer solution before steam sterilization to avoid hornification that inhibits subsequent swelling. To avoid back-exchange of D_2O during steam sterilization, it is preferable to swell preweighed soil pellets in known, minimal amounts of H_2O solution prior to sterilization. After sterilization and cooling, filter-sterilized D_2O solution is added to the H_2O-swollen soil to obtain the desired final D_2O concentration. A sterile cabinet or biological hood should be used for preparation of seeds and plants and transfer of plants and solutions. Solutions containing D_2O should be filter sterilized and sealed to avoid exchange with water vapor. Purified reagent grade or distilled water with resistivity of 17–18 $M\Omega\ cm^{-1}$ should be used for preparation of solutions. Several different growth media have been used in deuteration experiments, including modified Hoagland (Gorham, 1945), Murashige and Skoog (1962), and Schenk and Hildebrandt (1972). Premixed basal salts formulated for addition to distilled water can be obtained commercially, offering a convenient means to prepare growth solutions with varying D_2O/H_2O ratios. Photosynthesis by higher plants requires photoexcitation of chlorophyll, which absorbs red and blue light but reflects green (Bickford & Dunn, 1972; Bowyer & Leegood, 1997). Illumination should be with a light source that includes both blue (420–460 nm) and red light (650–700 nm) at intensities and diurnal period suitable for the species of interest and the purpose of the experiments. Many higher plants require a daily dark period for normal growth. Some grasses and other plants require illumination with yellow as well as red and blue light (Bickford & Dunn, 1972). Light fixtures with lamps emitting optimal wavelength ranges for plant growth can be obtained through greenhouse and scientific supply companies. Parameters used for our studies of grasses and other plants are given in Table 3.

Table 3 Parameters Used for Cultivation of Grasses in Phytotron Perfusion Chambers

Parameter	Cultivation of Grasses
Chamber volume	1.2 L, 4.5 L
Chamber interior height	42 cm, 50 cm
Chamber inner diameter	6 cm, 10 cm
Maximum D_2O volume	<100 mL
Growth solution volumes	20–200 mL
Growth solution composition	Schenk and Hildebrandt's basal salts[a]
Growth solution starting pH	4.2
Hydroponic support	Glass fiber filters in 2-in. plastic hydroponic baskets
Soil medium	Jiffy soil pellets in 2-in. hydroponic baskets
Light sources	Sunsystem 2 metal halide lamp, Sylvania GRO-LUX lamps
Light intensity[b]	Sunsystem 2:10,660–33,400 lux (45–15 cm from lamp) Sylvania: 1120–2620 lux (74–15 cm from lamp)
Diurnal cycle	12 h dark/12 h light
Temperature in growth area	22–29 °C

[a]Purchased as a dry mixture from Phytotechnology Laboratories, Shawnee Mission, Kansas and dissolved at 3.2 g L^{-1}.
[b]Measured with a Traceable Light Meter (International LLC, Radnor, PA).

To avoid dilution of D_2O by transpiration and water vapor exchange, the plants need to be grown enclosed in some manner to limit water vapor exchange. Closed culture dishes, resealable plastic bags, plant jars, and commercial plant growth containers can be used for germination and short-term growth. However, for most experimental investigations, it will be desirable to grow the plants for longer periods of time under photoautotrophic conditions, which requires a continual supply of CO_2. Although certain plants, for example, winter grain rye (*Secale cereale*), germinate under anoxic conditions (Siegel, Rosen, & Giumarro, 1962), the seeds of many species require O_2 for germination as well. Additionally, high humidity in sealed chambers promotes microbial growth and can affect leaf and stem morphology. It is preferable to use dry air rather than to humidify the air stream with D_2O/H_2O mixtures due to easier maintenance and avoidance of microbial contamination. Water vapor can be removed from the air stream used for perfusion by in-line drying tubes. A desiccant should be chosen that does not absorb CO_2. Silica gel desiccants with color indicator (Natrosorb, Eagle Silica Gel) do not absorb CO_2 and have good physical stability, enabling repetitive drying for reuse. Caution should be used when employing compressed air supply lines installed in research buildings, as CO_2 scrubbers may have been installed for other experimental activities. An economical and clean solution when a specific air feed composition is not needed is the use of aquarium pumps to perfuse the plant chambers with ambient air dried with silica gel desiccant in in-line drying tubes. Small in-line valves designed for use with aquarium pumps can be used to step down the air flow rate and T-connectors can be used to connect multiple chambers to the pump. Perfusion chambers can be assembled by fitting glass flasks, jars, or graduated cylinders with closures that have two inserted or attached tubing lengths or ports suitable for attachment of inflow and outflow air lines. Short lengths of silicone tubing can be used to attach sterile in-line filters, such as 0.2 μ pore size syringe filters, to the glass tubing lengths to filter out microbial contamination and particulates. The lids need to be tightly sealed to the chambers due to the air pressure even at relatively low perfusion rates of 100 mL min^{-1}, which can be accomplished by sealing the junction first with Teflon plumbers' tape, then securing with a layer of duct tape.

All procedures should first be tried with H_2O solutions to verify that the growth conditions are correct for the species of interest and that seeds, cuttings, or young plants that will be used are sufficiently axenic for enclosed cultivation. D_2O is expensive and can be expected to adversely affect growth of higher plants to a greater extent than that of many microbes. Surface sterilization of seeds with bleach may be necessary. Seeds prepared commercially for food use or planting may have low enough bacterial and fungal counts to use after surface decontamination with 70% ethanol followed by washing with sterile water. Treatment with fungicides has been reported in the literature (Blake et al., 1968). Testing seeds for germination in a range of D_2O/H_2O ratios can be used as an initial screening method to estimate D_2O tolerance of a species of interest. However, due to the higher impact of D_2O on root elongation, it is preferable to start seeds in H_2O and transfer to 50% or higher D_2O after roots have been established (3–10 days, depending on the species). This has the additional advantage that the seedlings can be screened for fungal or bacterial infection before starting deuteration. Choice of support matrix for germination depends on the preferences of the species being studied. Glass fiber filters are preferable to filter paper. Filter paper is composed of cellulose, and therefore contains exchangeable hydrogen atoms that will reduce the effective D_2O concentration, complicates analysis of roots, and provides a carbon source for any microbial contamination. If soil is necessary for germination, commercial soil pellets are convenient and can be autoclaved after swelling. Dry weight of soil pellets can be used to estimate the possible impact of the exchangeable hydrogen from the soil on the final D_2O concentration.

5.2 Multiple-Chamber Perfusion System for Long-Term Cultivation

For production of plant biomass with incorporation of deuterium approaching, the levels required to change the match point for SANS experiments and in quantities required for chemical and physical characterization, we devised a phytotron system, in which separate, modular growth chambers are illuminated under the same lamp fixtures and perfused from a common air source (Fig. 1). Parameters of the system are described in Table 3. In-line sterilization filters and drying tubes keep environmental microbes and ambient humidity from contaminating the system. Any pathogens that may infest an individual chamber are prevented from spreading to other experiments. Using this system, multiple experimental cultures can be maintained simultaneously with different concentrations of D_2O or other deuterated compounds and different growth matrices. Due to the low volume of hydroponic solution per chamber and to minimize loss due to evaporation, the solutions were not aerated, but were perfused at the top of the chambers. To recover evaporated D_2O, air outflow from an individual chamber can be routed to a water trap kept at 6 °C with a refrigerated water bath (Fig. 1). Tubing should be placed so as to avoid contact with the lamp fixture, as the fixture surfaces may become hot during illumination to melt or ignite plastic tubing. Electrical equipment should be plugged into outlets above the level of the water bath and the growth chambers to avoid electrical shock hazard from spills or leakage. Air pumps need to be placed above the level of the drying tubes and chambers to avoid siphoning of water or particulates. To accommodate the growth habit of grasses and provide lignocellulosic biomass approximating field conditions, 1-L and 3-L glass graduated cylinders were used to fabricate growth chambers as described previously (Evans et al., 2014). The width of standard 1-L graduated cylinders accommodates standard 2-in. plastic plant baskets used for hydroponic cultivation. By limiting volume of growth solution in the cylinders to 20–80 mL, aeration was not required for growth, and amount of D_2O required was minimized. Commercial soil plugs or pellets are available in

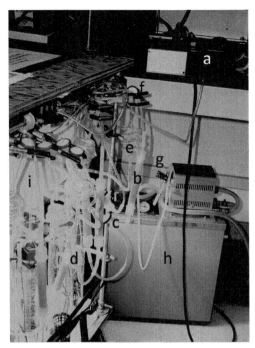

Figure 1 Multiple plants can be grown in separate chambers under perfusion with dried sterile-filtered air. Ambient air from the pump (a) is routed through individual drying tubes (b) to a sterile inflow filter (c) on the inflow tube to an individual chamber (d). To reduce loss from evaporation, air leaving the chamber passes through a condensation coil (e) before exiting through the sterile outflow filter (f). To capture water vapor for D_2O recycle or other purposes, a glass water trap (g) cooled in a refrigerated water bath (h) can be attached to the outflow filter (f) by a suitable tube. Chambers are illuminated by a metal halide lamp (i) plugged into a timer in an AC wall receptacle (not shown). All electrical equipment is plugged into outlets located above the water bath and growth chambers to avoid possible electrical shock hazards from spills and leakage. The air pump is placed above the level of the growth chambers and drying tubes to avoid siphoning water or particulates if power is lost.

sizes that fit the 2-in. baskets, simplifying use of a soil matrix in a perfusion chamber for germination and growth of plants that require soil. Using this system, we have succeeded in producing deuterated biomass from annual ryegrass and switchgrass with deuterium incorporation levels of ~35% that will potentially be sufficient for improvement of contrast in SANS experiments (Evans et al., 2014, 2015).

5.3 Production of Deuterated Annual Ryegrass

We investigated annual ryegrass (*Lolium multiflorum*), a cold-tolerant species widely grown as a forage and amenity crop, for growth and deuterium incorporation (Evans et al., 2014). Initial screening found inhibition of germination and growth above 50% D_2O, with little increase in deuterium incorporation. Coleoptile (leaf shoot) growth was not inhibited as highly as root growth by D_2O at 50% and higher concentrations, consistent with results reported for wheat (Pratt & Curry, 1937). Germination and growth of seeds in H_2O for 11 days to establish roots before transfer to 50% D_2O improved survival without appreciable decrease in deuterium incorporation. Following transfer to the perfusion chambers, the growth rate of annual ryegrass in 50% D_2O was 45% of the growth rate of the controls growing in H_2O, and the deuterated plants remained stunted, never attained the height of controls even when grown for several months (Fig. 2).

Vernalization, or cold-hardening, by chilling seed at 4 °C for 1 day, followed by 10 days at 10 °C, improved germination but not subsequent seedling survival in 70% D_2O. Methods of NMR analysis of whole biomass by dissolution in ionic liquids were adapted to analyze amount and site of deuterium incorporation. An internal standard of TFA/d-TFA was used to quantify the total proton and deuteron nuclei detected with $^1H^2H$ NMR for plants grown in 50% D_2O and in H_2O. The annual ryegrass grown in 50% D_2O had an average deuterium incorporation of 36.9%. Analysis of both commercially obtained 31% deuterated kale (Foston et al., 2012) and annual rye grass grown in 50% D_2O found higher levels of deuterium incorporation in the carbohydrates than in the lignin (Evans et al., 2014). These differences likely result from the combination of

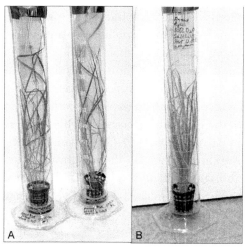

Figure 2 Annual ryegrass was grown in perfusion chambers for 33 days in H_2O (A) and for 85 days in 50% D_2O (B). Leaf shoot elongation rate in 50% D_2O was 45% that of H_2O-grown controls.

cumulative kinetic isotope effects dependent on the number of reactions in the biosynthetic pathways as well as isotopic differences in intracellular transport. Chemical composition and physical properties of the partially deuterated annual ryegrass leaf shoots were characterized by standard methods that are used for bioenergy crops (Evans et al., 2014). Cellulose and hemicelluloses were isolated and their molecular weights analyzed by gel permeation chromatography (GPC). The crystalline index (CrI) determined by ^{13}C NMR was similar for cellulose isolated from annual ryegrass grown in H_2O and 50% D_2O, although the molecular weight M_w determined by GPC was somewhat higher for the partially deuterated cellulose. Carbohydrate composition and lignin content were comparable.

5.4 Production of Deuterated Switchgrass

Following the successful production of deuterated and control annual ryegrass, the perfusion chamber system has been used for production of deuterated switchgrass (*Panicum virgatum*) by long-term cultivation in 50% D_2O (Evans et al., 2015). Switchgrass seed requires stratification and a soil matrix for germination and establishment of plants, and even then exhibits variable and inconsistent germination, typically about 5–10% (Zegada-Lizarazu, Wullschleger, & Nair, 2012). Young switchgrass seedlings are sensitive to chilling and do not tolerate D_2O. Switchgrass grown from seed in soil with H_2O for 2–4 months could be watered with D_2O solutions, but water absorbed by the soil matrix complicated determination of final concentration and plants survived only a few months. To provide better control of D_2O concentration, hydroponic switchgrass plants were established from tiller cuttings. Young tillers were cut at the tiller crown and incubated in growth solution until rooting and shoot elongation occurred. After establishment of roots in H_2O, a hydroponic plant was transferred to 50% D_2O and grown with periodic harvest of leaf biomass for 2 years. In contrast to the results with the C3 annual grasses, switchgrass tillers of mature plants are not stunted by growth in 50% D_2O, growing at rates similar to H_2O-grown controls and showing similar tillering growth habit (Fig. 3A and B). The D incorporation in the stems and leaves of the hydroponic switchgrass grown in 50% D_2O was determined by $^1H^2H$ NMR analysis to be 32–41%, staying at approximately the same level during long-term cultivation. The average incorporation level of ~35% D/H in switchgrass tiller biomass is similar to that obtained with the annual ryegrass. Plant gross morphology (Fig. 3A and B) and leaf morphology appeared unchanged by

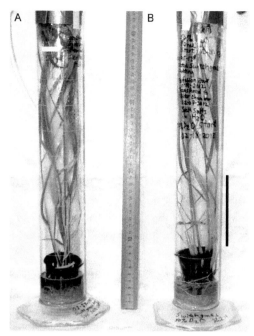

Figure 3 Long-term hydroponic cultures of the perennial switchgrass can be established from tiller cuttings. The plants can be maintained for 2 years and show normal tillering behavior when grown in H_2O (A) and in 50% D_2O (B). No stunting was observed for this chamber height (42 cm). Size bar = 10 cm.

growth in 50% D_2O (Evans et al., 2015). The total hemicellulose and glucan contents in the switchgrass grown in 50% D_2O were not statistically different from those in the control samples. Cellulose isolated from deuterated plants had weight-average molecular weight (M_w) and number-average molecular weight (M_n) comparable to its control counterpart, indicating that the molecular weight distribution profile was not changed. Cellulose crystallinity determined with ^{13}C NMR was similar for control and deuterated switchgrass. In contrast, hemicellulose isolated from deuterated switchgrass had a 6% higher arabinose content than hemicellulose from control plants, with a wider molecular weight distribution range with fractions 5% and 35% higher than in the controls, as analyzed by gel permeation chromatography (GPC). These results show that effective ^2H incorporation can be accomplished via long-term hydroponic cultivation of switchgrass and other perennials, which will ultimately expedite SANS experimentation.

5.5 Deuterium Labeling Winter Grain Rye with D_2O and Deuterated Phenylalanine

As described previously (Evans et al., 2014, 2015; Foston et al., 2012), deuterium appeared to be preferentially incorporated into the carbohydrates of kale, annual ryegrass, and switchgrass grown in 50% D_2O, while little substitution could be detected in the lignin. An alternative route to nonspecific deuterium labeling by growth in D_2O is the addition of a chemical precursor of the product of interest to the growth media. Deuterium-labeled phenylalanine had been used to label decursin intermediates in hairy roots of *Angelica gigas* (Ji, Huh, & Kim, 2008), while deuterated coniferin was used to label lignin in poplar saplings (Rolando, Daubresse, Pollet, Jouanin, & Lapierre, 2004). Deuterium-labeled coniferyl alcohol was reported to be incorporated more directly, resulting in higher deuterium substitution, into lignin than its glycoside coniferin in eucalyptus and magnolia (Tsuji, Chen, Yasuda, & Fukushima, 2004). Since coniferyl alcohol and the other monolignol precursors for lignin biosynthesis are derived from phenylalanine (Fig. 4) produced by the shikimate pathway (Strack, 1997), we hypothesized that deuterium labeling could be targeted to lignin by supplementing growth solutions with deuterated phenylalanine-*d8*. The assumption is that, if deuterated phenylalanine is absorbed by the roots and translocated into the leaf shoots, it will be utilized for both protein and monolignol synthesis, then subsequently incorporated into the lignin.

We carried out a study to evaluate this route for deuterium labeling of lignin by growing winter grain rye (*Secale cereale*) in H_2O and in 50% D_2O with and without supplementation with deuterated phenylalanine-*d8*. The plants were grown from seed in 32-oz glass plant jars on glass fiber filter supports. Deuterium oxide (D_2O), 99.8%, and deuterated phenylalanine-*d8* were purchased from Cambridge Isotope Laboratories (Cambridge, MA). All plant growth solutions were made with Schenk and Hildebrandt's basal salts (Phytotechnology Laboratories, Shawnee Mission, KS). Composition of Schenk and Hildebrandt's basal salts (Schenk & Hildebrandt, 1972) prepared at 3.2 g L^{-1} is given by the manufacturer as: 300 mg L^{-1} ammonium phosphate monobasic, 5 mg L^{-1} boric acid, 151 mg L^{-1} anhydrous calcium chloride, 0.1 mg L^{-1} cobalt chloride hexahydrate, 0.2 mg L^{-1} cupric sulfate pentahydrate, 20 mg L^{-1} Na_2EDTA dihydrate, 15 mg L^{-1} ferrous sulfate heptahydrate, 195.4 mg L^{-1} anhydrous magnesium sulfate, 10 mg L^{-1} manganese sulfate monohydrate, 0.1 mg L^{-1} sodium molybdate dehydrate, 1 mg L^{-1} potassium iodide, 2500 mg L^{-1}

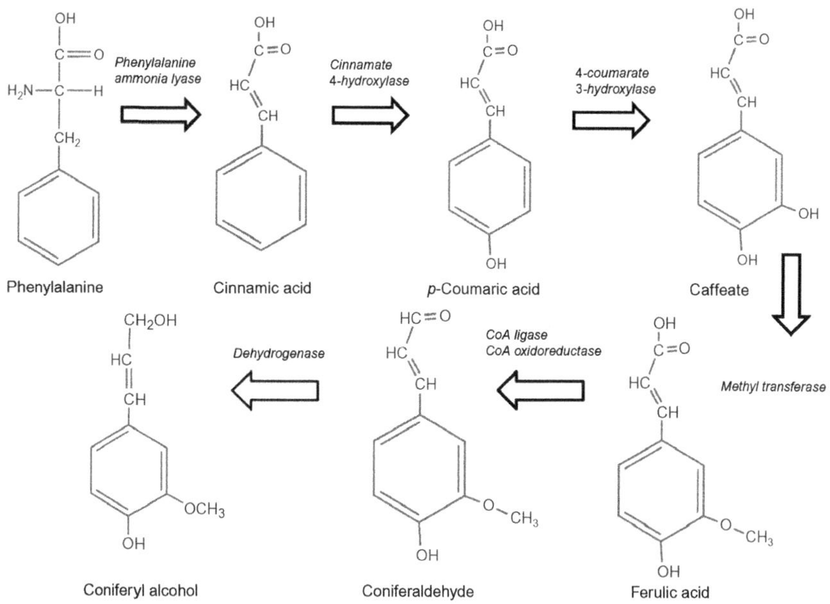

Figure 4 The shikimate biosynthetic pathway produces the aromatic amino acids phenylalanine and tyrosine, which are precursors of the monolignols that are polymerized to form lignin. A possible pathway from phenylalanine to coniferyl alcohol based on published information on monolignol synthesis (Strack, 1997) is shown. Introduction of deuterium-labeled phenylalanine is thus a possible route to specifically label lignin as well as proteins.

potassium nitrate, 1 mg L^{-1} zinc sulfate heptahydrate, pH 4.2. No vitamins or hormones were added to the hydroponic solutions. Plant jars and both unvented and 0.2 μ filter vented polypropylene lids were also purchased from Phytotechnology Laboratories. Distilled water was further purified with an E-Pure water purification system (Barnstead Thermolyne, Dubuque, IA) before use. All solutions and equipment were steam sterilized in an autoclave before use, except for solutions containing D_2O which were filter sterilized.

Growth data comparing shoot elongation found no difference in growth rate when phenylalanine-*d8* was added to the growth solution. The typical slower growth rate in 50% D_2O is evident (Figs. 5 and 6; Table 1). Winter rye grows rapidly and the plants were harvested after reaching the top of the growth containers at 15 days for H_2O and at 18 days for 50% D_2O. The seedlings grown in 50% D_2O show the typical stunting of roots and leaf shoots (Fig. 5C and D). The addition of 2 m*M* deuterated phenylalanine

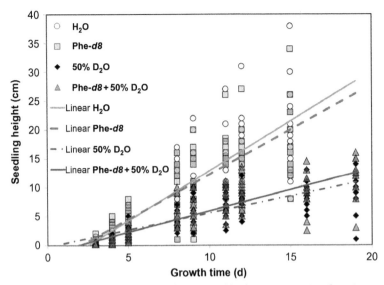

Figure 5 Comparison of growth rates determined by linear regression for winter grain rye found that deuterated phenylalanine (Phe-*d8*) had no effect on growth rate, while the rate in 50% D_2O was 40% of the rate in H_2O.

to the growth solution did not appear to affect growth or morphology as compared to the solution without it (Figs. 5 and 6; Table 4).

Analysis by FTIR was used to rapidly detect deuterium uptake and incorporation in the winter grain rye seedlings. In addition to the deuterium shifts, the spectra show features typical of lignocellulosic samples (Fig. 7). These include bands corresponding to absorbed water associated with cellulose at 1438 cm^{-1}, carbonyls of hemicelluloses at 1737 and 1733 cm^{-1}, and the large peak from the glycosidic C–O–C from 900 to 1130 cm^{-1} indicative of cellulose (Liu, Yu, & Huang, 2005; Stewart, Wilson, Hendra, & Morrison, 1995). A peak appearing at 1438 cm^{-1} in the samples grown in 50% D_2O most probably corresponds to D_2O associated with cellulose (Kirshenbaum, 1951). The region from 680 to 860 cm^{-1} corresponding to the aromatic C–H bending does not show an obvious shift in the samples grown in 50% D_2O and 2 mM deuterated phenylalanine-*d8*. Analysis of the incorporation by FTIR shows deuterium incorporation in hydroxyl and alkyl groups due to growth in D_2O, but no evidence for deuterium incorporation from deuterated phenylalanine is apparent (Fig. 7). A Jasco FT/IR-6300 instrument was used to examine leaf sections from fresh-frozen leaves. The large shifts in wave number resulting for D compared to H enable detection of deuteration of hydroxyl (O–D)

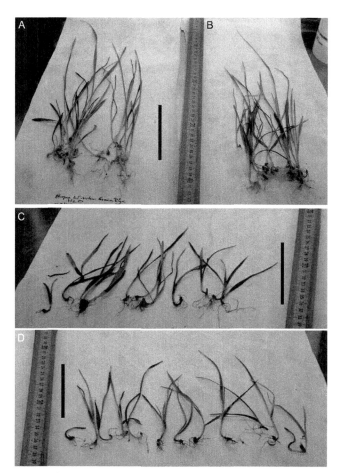

Figure 6 Compared to winter grain rye seedlings grown for 15 days in H_2O (A) and in 2 m*M* deuterated phenylalanine-*d8* (B), seedlings grown for 18 days in 50% D_2O (C), and in 2 m*M* deuterated phenylalanine-*d8* in 50% D_2O (D) are shorter and have stunted roots. Size bars = 10 cm. (See the color plate.)

and alkyl (C–D) groups in fresh-frozen leaf samples (Fig. 7). Wavebands from C–H at 2851 and 2920 cm^{-1} visible in H_2O-grown rye are reduced in magnitude in plants grown in 50% D_2O, while a band appears at 2128 cm^{-1} corresponding to C–D (Table 2; Fig. 7). Similarly, the appearance of the O–D band at 2483 cm^{-1} corresponds to a reduction in the O–H band at 3310 cm^{-1}. Examination of the spectra shows that D_2O as well as H_2O was taken up by the roots and transported to the leaves, and that D as well as H was incorporated into alkyl and hydroxyl groups in the lignocellulosic biomass through photosynthesis (Fig. 7B and D). Since there is no

Table 4 Growth and Yield from Deuterium Labeling Experiment with Winter Grain Rye

Parameter	H_2O	H_2O+Phe-d8	50% D_2O	50% D_2O+Phe-d8
Number of seeds	20	20	20	20
Growth time (d)	15	15	19	19
Germination (%)	85	75	65	70
Growth rate (cm/d)	1.7	1.5	0.58	0.74
Total leaf weight (g)	2.3618	2.2483	1.3961	1.2556
Total root weight (g)	0.5144	1.1291	0.3181	0.2761
Leaf weight/seedling (g)	0.1389	0.1499	0.1074	0.09712
Leaf root/seedling (g)	0.03026	0.07527	0.02447	0.02124

Phe-d8, 2 mM phenylalanine-d8.

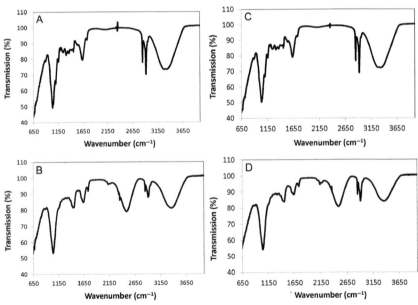

Figure 7 Compared to winter rye seedlings grown in H_2O (A), FTIR spectra of leaves from seedlings grown in 50% D_2O (B) show both C–D and O–D bands indicative of uptake and fixation of D_2O. In contrast, no D bands are apparent in seedlings grown in 2 mM deuterated phenylalanine-d8 in H_2O (C), and, compared to seedlings grown in 50% D_2O (B), there is no additional increase in D bands in leaves from seedlings grown in 2 mM deuterated phenylalanine-d8 in 50% D_2O (D).

apparent difference between the FTIR spectra from leaves of seedlings grown in 2 mM phenylalanine-$d8$ in H_2O (Fig. 7C) and those grown in H_2O (Fig. 7A), the deuterated amino acid was neither transported into the leaves nor incorporated into lignin and proteins. Spectra of seedlings grown in 2 mM deuterated phenylalanine-$d8$ in 50% D_2O (Fig. 7D) did not show the presence of any additional deuterium compared to those grown in 50% D_2O (Fig. 7B). These results indicate that the seedlings are not utilizing the phenylalanine-$d8$ for biosynthesis, most likely because they are not absorbing it and transporting it into the leaf shoots. The presence of 50% D_2O and the resultant partial deuteration of cellulose and other components did not appear to have a synergistic effect on uptake and incorporation of deuterated phenylalanine.

ACKNOWLEDGMENTS

This research was supported by the Genomic Science Program, Office of Biological and Environmental Research, U.S. Department of Energy, under Contract FWP ERKP752. The research at Oak Ridge National Laboratory's Center for Structural Molecular Biology (CSMB) was supported by the Office of Biological and Environmental Research under Contract FWP ERKP291, using facilities supported by the Office of Basic Energy Sciences, U.S. Department of Energy. R.S. was partly supported by the graduate fellowship program of the Bredesen Center for Interdisciplinary Research and Graduate Education, University of Tennessee, Knoxville. Oak Ridge National Laboratory is managed by UT-Battelle, LLC, for the U.S. Department of Energy under Contract DE-AC05-00OR22725.

This manuscript has been authored by UT-Battelle, LLC under Contract No. DE-AC05-00OR22725 with the U.S. Department of Energy. The United States Government retains and the publisher, by accepting the article for publication, acknowledges that the United States Government retains a non-exclusive, paid-up, irrevocable, world-wide license to publish or reproduce the published form of this manuscript, or allow others to do so, for United States Government purposes. The Department of Energy will provide public access to these results of federally sponsored research in accordance with the DOE Public Access Plan (http://energy.gov/downloads/doe-public-access-plan).

REFERENCES

Appenroth, K. -J., Cheng, J. J., Fakhoorian, T., Mercovich, E., Morikawa, M., & Hai, Z. (2015). *International Steering Committee on Duckweed Research and Applications Newsletter, 3*, 34–80. (http://lemnapedia.org/wiki/ISCDRA)

Appenroth, K.-J., Crawford, D. J., & Les, D. H. (2015). After the genome sequencing of duckweed—How to proceed with research on the fastest growing angiosperm? *Plant Biology, 17*, 1–4. Special Issue: Duckweed Research and Application.

Astot, C., Dolezal, K., Moritz, T., & Sandberg, G. J. (2000). Deuterium in vivo labelling of cytokinins in *Arabidopsis thaliana* analysed by capillary liquid chromatography/frit-fast atom bombardment mass spectrometry. *Mass Spectrometry, 35*, 13–22.

Bali, G., Foston, M. B., O'Neill, H. M., Evans, B. R., He, J., & Ragauskas, A. J. (2013). The effect of deuteration on the structure of bacterial cellulose. *Carbohydrate Research, 374*, 82–88.

Bhatia, C. R., & Smith, H. H. (1968). Adaptation and growth response of *Arabidopsis thaliana* to deuterium. *Planta, 80*, 176–184.

Bickford, E. D., & Dunn, S. (1972). *Lighting for plant growth*. Kent, OH: The Kent State University Press.

Blake, M. I., Crane, F. A., Uphaus, R. A., & Katz, J. J. (1968). Effect of heavy water on the germination of a number of species of seeds. *Planta, 78*, 35–38.

Bowyer, J. R., & Leegood, R. C. (1997). Chapter 2. Photosynthesis. In P. M. Dey & J. B. Harborne (Eds.), *Plant biochemistry* (pp. 49–110). San Diego, CA: Academic Press.

Burgess, J., & Northcote, D. H. (1969). Action of colichine and heavy water on the polymerization of microtubules in wheat root meristem. *Journal of Cell Science, 5*, 433–451.

Cooke, R. J., & Davies, D. D. (1980). General characteristics of normal and stress-enhanced protein degradataion in *Lemna minor* (duckweed). *Biochemical Journal, 192*, 499–506.

Cooke, R. J., Grego, S., Olivier, J., & Davies, D. D. (1979). The effect of deuterium oxide on protein turnover in *Lemna minor*. *Planta, 146*, 229–236.

Cooke, R. J., Grego, S., Roberts, K., & Davies, D. D. (1980). The mechanism of deuterium oxide-induced protein degradation in *Lemna minor*. *Planta, 148*, 374–380.

Cope, B. T., Bose, S., Crespi, H. L., & Katz, J. J. (1965). Growth of *Lemna* in H_2O-D_2O mixtures: Enhancement by kinetin. *Botanical Gazette, 126*, 214–221.

Crespi, H. L., Daboll, H. F., & Katz, J. J. (1970). Production of deuterated algae. *Biochimica et Biophysica Acta, 200*, 26–33.

Crespi, H. L., & Katz, J. J. (1961). The Determination of Deuterium in Biological Fluids. *Analytical Biochemistry, 2*, 247–279.

Crespi, H. L., & Katz, J. J. (1972). Chapter 27. Preparation of deuterated proteins and enzymes. In C. H. W. Hirs & S. N. Timasheff (Eds.), *Enzyme structure part C: Vol. 26. Methods in enzymology* (pp. 627–637). New York, NY: Academic Press.

DaBoll, H. F., Crespi, H. L., & Katz, J. J. (1962). Mass cultivation of algae in pure heavy water. *Biotechnology and Bioengineering, 4*, 281–297.

Environmental Inquiry. (2015). *Environmental inquiry: Authentic scientific research for high school students (Cornell University). Toxicology: Bioassays*. http://ei.cornell.edu/. Accessed on 14 May 2015.

Evans, B. R., Bali, G., Foston, M., Ragauskas, A. J., O'Neill, H. M., Shah, R., et al. (2015). Production of deuterated switchgrass by hydroponic cultivation. *Planta, 242*, 215–222.

Evans, B. R., Bali, G., Reeves, D., O'Neill, H., Sun, Q., Shah, R., & Ragauskas, A. J. (2014). Effect of D_2O on growth properties and chemical structure of annual ryegrass (*Lolium multiflorum*). *Journal of Agricultural and Food Chemistry, 62*, 2592–2604.

Foston, M. B., McGaughey, J., O'Neill, H., Evans, B. R., & Ragauskas, A. J. (2012). Deuterium incorporation in biomass cell wall components by NMR analysis. *Analyst, 137*, 1090–1093.

Foston, M., & Ragauskas, A. J. (2010). Changes in lignocellulosic supramolecular and ultrastructure during dilute acid pretreatment of *Populus* and switchgrass. *Biomass and Bioenergy, 34*, 1885–1895.

Foston, M., & Ragauskas, A. J. (2012). Biomass characterization: Recent progress in understanding biomass recalcitrance. *Industrial Biotechnology, 8*, 191–208.

Gorham, R. P. (1945). Growth factor studies with *Sprirodela polyrrhiza* (L) Schleid. *American Journal of Botany, 32*, 496–505.

Grusak, M. A. (2000). Intrinsic stable isotope labeling of plants for nutritional investigations in humans. *Journal of Nutritional Biochemistry, 71*, 1555–1562.

He, J., Pingali, S. V., Chundawat, S. P. S., Pack, A., Jones, A. D., Langan, P., et al. (2014). Controlled incorporation of deuterium into bacterial cellulose. *Cellulose, 21,* 927–936.

Hofstetter, K., Hinterstoisser, B., & Salmén, L. (2006). Moisture uptake in native cellulose—The roles of different hydrogen bonds: A dynamic FT-IR study using Deuterium exchange. *Cellulose, 13,* 131–145.

ISO. (2005). *Water quality—Determination of the toxic effect of water constituents and waste water on duckweed (Lemna minor)—Duckweed growth inhibition test. ISO guideline 20079.* Geneva, Switzerland: International Organization for Standardization.

Ji, X., Huh, B., & Kim, S. U. (2008). Determination of biosynthetic pathway of decursin in hairy root culture of *Angelica gigas. Journal of Korean Society for Applied Biological Chemistry, 51,* 1738–2203.

Katz, J. J. (1960). Chemical and biological studies with deuterium. *American Scientist, 48,* 544–580.

Katz, J. J., & Crespi, H. (1966). Deuterated organisms: Cultivation and uses. *Science, 151,* 1187–1194.

Kirshenbaum, I. (1951). Physical properties and analysis of heavy water. In H. C. Urey & G. M. Murphy (Eds.), *National nuclear energy series: Vol. 4A* (1st ed.). New York, NY: McGraw-Hill Book Company, Inc.

Langan, P., Evans, B. R., Foston, M., Heller, W. T., O'Neill, H. M., Petridis, L., et al. (2012). Neutron technologies for bioenergy research. *Industrial Biotechnology, 8,* 209–216.

Langan, P., Petridis, L., O'Neill, H. M., Pingali, S. V., Foston, M., Nishiyama, Y., et al. (2014). Common processes drive the thermochemical pretreatment of lignocellulosic biomass. *Green Chemistry, 16,* 63–69.

Lewandowska, D., ten Have, S., Hodge, K., Tillemans, V., Lamond, A. I., & Brown, J. W. (2013). Plant SILAC: Stable-isotope labeling with amino acids of Arabidopsis seedlings for quantitative genomics. *PLoS One, 8,* e72207.

Lewis, G. N. (1933). The biology of heavy water. *Science, 79,* 151–153.

Liu, R., Yu, H., & Huang, Y. (2005). Structure and morphology of cellulose in wheat straw. *Cellulose, 12,* 25–34.

Martínez-Sanz, M., Gidley, M. J., & Gilbert, E. P. (2015). Application of X-ray and neutron small angle scattering techniques to study the hierarchical structure of plant cell walls: A review. *Carbohydrate Polymers, 125,* 120–134.

Moody, M., & Miller, J. (2005). Lemna minor growth inhibition test. In C. Blaise & J.-F. Férard (Eds.), *Small scale freshwater toxicity investigations* (pp. 271–298). Amsterdam, The Netherlands: Springer Netherlands.

Murashige, T., & Skoog, F. (1962). A revised medium for rapid growth and bioassays with tobacco tissue cultures. *Physiology of the Plant, 15,* 473–497.

National Academy of Science, & Committee on Biology Teacher Inservice Program. (1996). *The role of scientists in the professional development of science teachers. Appendix H1 ecology of duckweed* (pp. 221–224). Washington, DC: National Academy Press. Appendix H2 effect of pH on the growth of duckweed, pp. 226–227.

Perkins, S. J. (1981). Estimation of deuteration levels in whole cells and cellular proteins by ^{1}H n.m.r. spectroscopy and neutron scattering. *Biochemical Journal, 199,* 163–170.

Pingali, S. V., Urban, V. S., Heller, W. T., McGaughey, J., O'Neill, H. M., Foston, M., Myles, D. A., Ragauskas, A., & Evans, B. R. (2010). Breakdown of cell wall nanostructure in dilute acid pretreated biomass. *Biomacromolecules, 11,* 2329–2335.

Pratt, R., & Curry, J. (1937). Growth of roots in deuterium oxide. *American Journal of Botany, 24,* 412–416.

Pratt, R., & Trelease, S. F. (1938). Influence of deuterium oxide on photosynthesis in flashing and continuous light. *American Journal of Botany, 25,* 133–139.

Putzbach, K., Krucker, M., Albert, K., Grusak, M. A., Tang, F., & Dolnkowski, G. G. (2005). Structure determination of partially deuterated carotenoids from intrinsically labeled vegetables by HPLC-MS and ^{1}H NMR. *Journal of Agriculture and Food Chemistry*, *53*, 671–677.

Reitz, O., & Bonhoeffer, K. F. (1934). Über dem Einbau von schwerem Wasserstoff in wachsenden Organismen. *Naturwissenschaften*, *22*, 744.

Robinson, G. (1988). Experimental method and biological concepts demonstrated using duckweed. *School Science Review*, *69*, 505–508.

Rolando, C., Daubresse, N., Pollet, B., Jouanin, L., & Lapierre, C. (2004). Lignification in poplar plantlets fed with deuterium-labelled lignin precursors. *Comptes Rendus Biologies*, *327*, 799–807.

Sacchi, G. A., & Cocucci, M. (1992). Effects of deuterium oxide on growth proton extrusion, potassium influx, and in vitro plasma membrane activities in maize root segments. *Plant Physiology*, *100*, 1962–1967.

Schenk, R. M., & Hildebrandt, A. C. (1972). Medium and techniques for induction and growth of monocotyledonous and dicotyledonous plant cell cultures. *Canadian Journal of Botany*, *50*, 199–204.

Shibabe, S., & Yoda, K. (1984a). Transport of calcium, germanium, and rubidium ions in rice seedlings in deuterium oxide. *Radioisotopes*, *33*, 606–610.

Shibabe, S., & Yoda, K. (1984b). Hydrogen isotope effect on transport of potassium ion in rice seedlings equilibrated with deuterium oxide. *Radioisotopes*, *33*, 675–678.

Shibabe, S., & Yoda, K. (1985). Water and potassium ion absorption by deuterium resistant winter rye seedlings. *Radioisotopes*, *34*, 266–269.

Siegel, S. M., Halpern, L. A., & Giumaro, C. (1964). Germination and seedling growth of winter rye in deuterium oxide. *Nature*, *201*, 1244–1245.

Siegel, S. M., Rosen, L. A., & Giumarro, C. (1962). Effects of reduced oxygen tension on vascular plants. IV. Winter rye germination under near-Martian conditions and in other nonterrestrial environments. *Proceedings of the National Academy of Sciences of the United States of America*, *48*, 725–728.

Sinkkonen, A., Myyr, M., Penttinen, O.-P., & Rantalainen, A.-L. (2011). Selective toxicity at low doses: Experiments with three plant species and toxicants. *Dose Response*, *9*, 130–143.

Sternberg, L., & DeNiro, M. J. (1983). Isotopic composition of cellulose from C3, C4, and CAM plants growing near one another. *Science*, *220*, 947–949.

Stewart, D., Wilson, H. M., Hendra, P. J., & Morrison, I. M. (1995). Fourier-transform infrared and Raman-spectroscopic study of biochemical and chemical treatments of oak wood (*Quercus rubra*) and Barley (*Hordeum vulgare*) Straw. *Journal of Agricultural and Food Chemistry*, *43*, 2219–2225.

Stomp, A. M. (2005). Duckweeds a valuable plant for biomanufacturing. *Biotechnology Annual Review*, *11*, 69–99.

Strack, D. (1997). Chapter 10. Phenolic metabolism. In P. M. Dey & J. B. Harborne (Eds.), *Plant biochemistry* (pp. 387–416). San Diego, CA: Academic Press.

Stuhrmann, H. B., & Miller, A. (1978). Small-angle scattering of biological structures. *Journal of Applied Crystallography*, *11*, 325–345.

Tang, G., Qin, J., Dolnikowski, G. G., Russell, R. M., & Grusak, M. A. (2005). Spinach or carrots can supply significant amounts of vitamin A as assessed by feeding with intrinsically deuterated vegetables. *American Journal of Clinical Nutrition*, *82*, 821–828.

Tang, G., Qin, J., Dolnikowski, G. G., Russell, R. M., & Grusak, M. A. (2009). Golden rice is an effective source of vitamin A. *American Journal of Clinical Nutrition*, *89*, 1776–1783.

Thomas, A. F. (1971). *Deuterium labeling in organic chemistry*. New York, NY: Meredith Corporation.

Torres, J., Kukol, A., & Arking, I. T. (2000). Use of a single glycine residue to determine the tilt and orientation of a transmembrane helix. A new structural label for infrared spectroscopy. *Biophysical Journal, 79,* 3139–3143.

Tsuji, Y., Chen, F., Yasuda, S., & Fukushima, K. (2004). The behavior of deuterium-labeled monolignol and monolignol glucosides in lignin biosynthesis in angiosperms. *Journal of Agricultural and Food Chemistry, 52,* 131–134.

Waber, J., & Sakai, W. S. (1974). Effect of growth in 99.8% deuterium oxide on ultrastructure of winter rye. *Plant Physiology, 53,* 128–130.

Weaver, C. M. (1984). Intrinsic labeling of edible plants with stable isotopes. In J. R. Turnland & P. E. Johnson (Eds.), *ACS symposium series: Vol. 258. Stable isotopes in nutrition* (pp. 61–75). Washington, DC: American Chemical Society.

Wibert, K. B. (1955). The Deuterium Isotope Effect. *Chemical Reviews, 55,* 713–743.

Yakir, D., & DeNiro, M. J. (1990). Oxygen and hydrogen isotope fractionation during cellulose metabolism in *Lemna gibba* L. *Plant Physiology, 93,* 325–332.

Yang, X.-Y., Chen, W.-P., Rendahl, A. K., Hegeman, A. D., Gray, W. M., & Cohen, J. D. (2010). Measuring the turnover rates of Arabidopsis proteins using deuterium oxide: An auxin signaling case study. *The Plant Journal, 63,* 680–695.

Zegada-Lizarazu, W., Wullschleger, S. D., & Nair, S. S. (2012). Chapter 3. Crop physiology. In A. Monti (Ed.), *Switchgrass: A valuable biomass crop for energy* (pp. 55–86). London: Springer London.

CHAPTER ELEVEN

Isotope Labeling of Proteins in Insect Cells

Lukasz Skora, Binesh Shrestha, Alvar D. Gossert[1]

Novartis Institutes of BioMedical Research, Novartis Campus, Basel, Switzerland
[1]Corresponding author: e-mail address: alvar.gossert@novartis.com

Contents

1. Insect Cells as Expression System	246
1.1 Comparison to Other Expression Systems	246
1.2 Insect Cell Cultures and Their Use as Expression System	247
1.3 The Baculovirus as a Vehicle for Efficient Transfection of Insect Cells	248
1.4 Optimization of Yields	249
2. General Considerations for Isotope Labeling in Insect Cells	250
2.1 Sources of Amino Acids in Insect Cell Media	250
2.2 Factors Influencing Isotope Incorporation Levels	251
3. Amino Acid Type-Specific Isotope Labeling in Insect Cells	252
3.1 Scrambling Versus Label Dilution	252
3.2 Experimental Approaches	252
4. Uniform Isotope Labeling in Insect Cells	256
4.1 Experimental Approaches	256
4.2 Uniform ^{15}N Labeling	259
4.3 Uniform ^{13}C Labeling	262
4.4 Uniform ^{2}H Labeling	265
5. Applications	268
5.1 Uniform Labeling for Structural Studies	269
5.2 Membrane Proteins	270
5.3 In-Cell NMR	271
5.4 Applications in Drug Discovery	273
6. Protocols	274
6.1 Materials	274
6.2 Culturing of Insect Cells	275
6.3 Transfection and Infection of Insect Cells with Baculovirus	277
6.4 General Protocol for Amino Acid-Type Selective Isotope Labeling in Insect Cells	279
6.5 General Protocol for Uniform Isotope Labeling in Insect Cells	281
References	284

Abstract

Protein targets of contemporary research are often membrane proteins, multiprotein complexes, secreted proteins, or other proteins of human origin. These are difficult to express in the standard expression host used for most nuclear magnetic resonance (NMR) studies, *Escherichia coli*. Insect cells represent an attractive alternative, since they have become a well-established expression system and simple solutions have been developed for generation of viruses to efficiently introduce the target protein DNA into cells. Insect cells enable production of a larger fraction of the human proteome in a properly folded way than bacteria, as insect cells have a very similar set of cytosolic chaperones and a closely related secretory pathway. Here, the limited and defined glycosylation pattern that insect cells produce is an advantage for structural biology studies. For these reasons, insect cells have been established as the most widely used eukaryotic expression host for crystallographic studies. In the past decade, significant advancements have enabled amino acid type-specific as well as uniform isotope labeling of proteins in insect cells, turning them into an attractive expression host for NMR studies.

This chapter is divided into six relatively independent sections. Detailed step-by-step protocols are found in Section 6. Section 1 gives a general introduction to insect cells as an expression system, focusing on baculovirus-mediated expression of heterologous proteins. Sections 2–4 deal with aspects of selective and uniform isotope labeling in insect cells, with in-depth description of relevant metabolic pathways. Readers with prior knowledge and experience in insect cell expression system may go directly to these. Section 5, which precedes the protocols, presents an overview on several applications of isotope labeling using insect cells as expression system, and can be read independently of the rest of the chapter.

1. INSECT CELLS AS EXPRESSION SYSTEM

1.1 Comparison to Other Expression Systems

The most widely used system for expressing labeled proteins for nuclear magnetic resonance (NMR) is in fact not insect cells, but *Escherichia coli* (*E. coli*). *E. coli* grows fast, its simple genetics allow easy manipulation in order to introduce foreign target genes and it has a powerful metabolism permitting growth in defined minimal media that enable a wealth of isotope-labeling patterns. However, the center of interest of contemporary research are human proteins, often membrane proteins or large protein complexes, and very frequently, these cannot be expressed in *E. coli* because of improper folding or the lacking secretion machinery. In drug discovery research, for example, protein targets from humans are required and therefore preference is given to eukaryotic expression systems. The most popular eukaryotic expression system used in structural biology are insect cells, as revealed by analysis of entries to the RCSB's protein databank (Research collaboratory for structural bioinformatics; http://www.rcsb.org/pdb/). For a review, see Assenberg, Wan, Geisse, & Mayr (2013).

The major advantage of insect cells is their ability to produce large amounts of properly folded cytosolic proteins and protein complexes, because they possess similar chaperones and folding cofactors as human cells do. For secreted proteins, insect cells offer homogenous and limited glycosylation patterns, facilitating structural studies. Therefore, a very large fraction of the human proteome is accessible with insect cells as expression system.

On the flip side, culturing, insertion of target genes and isotope labeling are time consuming and labor intensive. Growth of insect cells is relatively slow, with doubling times in the order of a day. Consequently, all processes are lengthy: for example, a typical expression experiment takes 2–3 days. Direct transfection with DNA plasmids is inefficient and will only allow expression of minute quantities of protein. Therefore, either a virus is used as vehicle to introduce target DNA or stable cell lines are created. Fortunately, virus generation is well established nowadays and easily achieved using commercial kits in a matter of days to weeks.

Finally, isotope labeling presents its own challenges in these organisms. Amino acid type-specific labeling is well established and often yields better results than in *E. coli* in terms of label incorporation and scrambling. Uniform labeling in turn is more expensive in insect cells and isotope incorporation is lower with 75–80% depending on the labeling pattern.

1.2 Insect Cell Cultures and Their Use as Expression System

The insect cell lines commonly used as expression systems are derived from the fall armyworm *Spodoptera frugiperda* (Sf9, Sf21), the cabbage looper *Trichoplusia ni* (High Five), and the fruit fly *Drosophila melanogaster* (S2). S2 cells are primarily used as stable cell lines, which will not be covered in this text. Here, we focus on virus-mediated protein overexpression. Sf9 and Sf21 cells can be used for all insect cell-based work from transfection to protein production. We recommend using High Five cells for protein production only. For protein production, a suspension culture is preferred over an adherent one; hence the protocols explained here are designed for cell culture in suspension. Details of insect cell culture methods have been described in two laboratory manuals (Invitrogen, 2002; O'Reilly, Miller, & Luckow, 1992).

1.3 The Baculovirus as a Vehicle for Efficient Transfection of Insect Cells

Baculovirus-mediated expression is widely used for recombinant protein production in a large variety of eukaryotic cell cultures. Well-established protocols allow safe and easy propagation of baculoviruses (Vialard, Arif, & Richardson, 1995). Most expression vectors currently used in the baculovirus expression vector system (BEVS) are based on *Autographa californica* multicapsid nucleopolyhedrovirus (*Ac*MNPV) infection of *S. frugiperda*.

1.3.1 The Baculovirus Life Cycle

Understanding the viral life cycle is a prerequisite to use it as a tool for heterologous protein overexpression. The natural baculovirus life cycle starts with ingestion into the gut of a suitable host, for example, an insect larva, and primary infection of cells of the gut lumen. It takes about 30 min for the virus to enter the cytosol through endosomes. In the first 10–20 h post-infection, the viral DNA is replicated and new virions are produced, which are subsequently released and infect further cells of the host in a secondary wave of infection. Finally, virus particles are released by dying cells into the outside world for eventual infection of a new host. However, virus particles are not stable outside of the host and need to be protected in so-called occlusion bodies. These consist of a large protein matrix, where virions are embedded. Therefore, during the very late stage of infection, 24–48 h after the primary infection, large amounts of these matrix proteins are produced, namely polyhedrin and p10.

1.3.2 Virus Generation Protocols

For heterologous protein expression, polyhedrin is typically replaced with the target protein at DNA level. Insect cells infected with such a modified baculovirus will therefore produce the target protein as the most abundant protein. Expression is typically detected 24 h after infection, and after 48–72 h the culture is stopped, as cells start dying.

Baculoviruses have many attractive features for production of proteins and protein complexes of eukaryotic origin in larger quantities. The baculovirus capsid can freely extend to accommodate larger double-stranded DNA (Fraser, 1986; O'Reilly et al., 1992; Summers & Anderson, 1972), which makes it one of the preferred systems for production of large proteins and protein complexes (Bieniossek, Imasaki, Takagi, & Berger, 2012; Fitzgerald et al., 2006). Multiple high-titer viruses can be generated and titered in parallel (Kärkkäinen et al., 2009; O'Reilly et al., 1992) and recombinant viruses may be stored for long periods at 4 °C or as TIPS (titerless infected-cells preservation and scale-up; Jarvis & Garcia, 1994; Wasilko & Lee, 2006). AcMNPV can be manipulated in biosafety level 1 facilities due to its host specificity and no known relation to disease in vertebrates (Burges, Croizier, & Huger, 1980; Kost & Condreay, 2002). The advantages of BEVS along with its practical aspects have been reviewed by several authors (Airenne et al., 2013; Cremer, Bechtold, Mahnke, & Assenberg, 2014; Shrestha, Smee, & Gileadi, 2008).

The increasing demand for BEVS in past decades led to the commercialization of the system, making it easier to use, especially for those with little experience in eukaryotic cell culture. One of the most commonly used systems in BEVS is the Bac-to-Bac system (commercialized by Invitrogen), which uses site-specific transposition of an expression cassette into a baculovirus genome maintained in *E. coli* (Luckow, Lee, Barry, & Olins, 1993). The advancement of the system is extensively reviewed by Kost, Condreay, & Jarvis (2005) and van Oers, Pijlman, & Vlak (2015). Additionally, the *in vivo* recombination method is gaining popularity, which offers an alternative method for efficient generation of recombinant virus (Possee et al., 2008).

1.4 Optimization of Yields

Expression of target protein depends on various factors. Some of the common parameters to be tested are: addition of fetal bovine serum (FBS), different virus concentrations, various time points of harvest ranging from 36 to 96 h, different cell lines, or even different temperatures. For some target proteins, coexpression with specific binding partners or small-molecule inhibitors has proven to be most effective. Working with insect cells is labor intensive; media and reagents are not as cheap as for *E. coli*. Hence, it is advised to carry out the initial expression optimization tests in small volumes (3–4 mL) using the 24-well block format (Shrestha et al., 2008). The optimum condition can then be used to scale-up expression. We have found that conditions optimized in small scale can be used more effectively when cells are grown in reagent bottles rather than conical flasks (Rieffel et al., 2014).

It is noteworthy here that if the target protein can already be expressed in "mg per liter" quantity, optimization may not increase the yield dramatically. However, we have found it beneficial in case of targets that are expressed only in "microgram per liter" quantity, where improvement in yields can be several fold.

2. GENERAL CONSIDERATIONS FOR ISOTOPE LABELING IN INSECT CELLS

2.1 Sources of Amino Acids in Insect Cell Media

Insect cells require very different strategies for isotope labeling than *E. coli*. Bacteria can synthesize all molecules of life such as amino acids, sugars, nucleotides, vitamins, lipids, etc. from the simplest precursors: mineral salts, ammonia, and a carbon source like glucose or glycerol. In contrast, insect cells, similarly to many higher eukaryotic cells, have lost the ability of synthesizing many of these molecules and need to take them up with their diet. Growth media for insect cells are therefore very complex because they contain many essential substances supporting cell viability. In other words, a "minimal medium" with single defined carbon and nitrogen sources cannot sustain growth of insect cells. For isotope labeling, in principle, all medium components including lipids and vitamins would need to be replaced with labeled ones, rendering the media extremely expensive. However, many substances are only present in trace amounts and can be safely ignored for isotope-labeling purposes (Gossert & Jahnke, 2012). However, amino acids, sugars, fetal bovine serum (FBS), and pluronic are present in high amounts. For these ingredients, it must be determined whether they are integrated into proteins, and if this is the case they must either be left away or supplied in labeled form. Pluronic, a mixture of detergents and lipids important for maintaining cell integrity in suspension cultures, is not being metabolized and can therefore be left unchanged in the medium. FBS contains amino acids, which would be integrated into proteins. Fortunately, cells can be adapted to growth in serum-free media. Therefore, for isotope-labeling purposes, such adapted cell stocks should be used and FBS left away in the labeling medium.

Sugars, mainly glucose, are present in amounts of 4–10 g/L, depending on the medium. Sugars are incorporated into amino acids, that are directly related to glycolysis and citric acid cycle, namely Ala, Glu, Asp, Gln, and Asn (Strauss et al., 2005). Therefore, for most applications, unlabeled glucose can be used for the medium, but if carbon labeling of these amino acids is intended, glucose needs to be supplied in labeled form.

Amino acids are present in the media from two main sources: amino acids which are added in pure form and yeast extract which contains free amino acids as well as oligopeptides. If yeast extract is left away, cell growth and protein expression become much less reproducible. This is probably due to the fact that yeast extract contains a large variety of nucleotides, vitamins, and other factors that promote robust cell growth and facilitate viral infection. Yeast extract should therefore not be completely removed from the media, but its amount can be reduced up to 10-fold without adverse effects for an expression culture (Gossert et al., 2011). In summary, for isotope labeling in insect cells, free amino acids need to be replaced by labeled ones, yeast extract needs to be reduced, and FBS needs to be left away.

2.2 Factors Influencing Isotope Incorporation Levels

As examined above, unlabeled components in the actual expression medium are not the primary cause for reduced isotope incorporation levels. The three main factors are discussed in the following.

2.2.1 Carry Over from Preculture

When transferring insect cells from a preculture to labeling medium, cells cannot be diluted by large factors like 100–1000 due to their slow growth. Therefore, in order to minimize carryover of unlabeled components from the preculture, cells need to be centrifuged gently and resuspended in labeling medium. Usually, cells are washed one time with amino acid-free medium or phosphate-buffered saline in order to remove unlabeled medium from the preculture. Still, cells contain a large pool of unlabeled amino acids. Cells can in principle be starved, but this will strongly compromise their performance in protein expression (Meola et al., 2014). For mass spectrometry (MS) studies, cells are often grown for several passages in labeling medium before actual expression, leading to very high isotope incorporation levels (>95%) (Ong & Mann, 2007). However, for most labeling patterns at the scale needed for NMR studies, this procedure is prohibitively expensive.

2.2.2 Unlabeled Medium from Baculovirus Suspension

As described above (Section 1.3), baculoviruses are used as vehicle for inserting the target DNA into insect cells. Baculoviruses (BV) are generated in unlabeled media and are added as a suspension, which therefore contains unlabeled amino acids. It is thus important to produce virus stocks with the highest possible titers in order to minimize the volume of BV suspension added to the expression culture to a few milliliters.

2.2.3 Adverse Effects from Insect Cell Metabolism

Depending on the amino acid type and the labeling pattern, insect cell metabolism may introduce unlabeled atoms into a protein. For example, alanine is derived from glucose and the amino group of glutamate. If these two precursors are not labeled, most of the alanines in a protein will not be either. The implications of insect cell metabolism on isotope incorporation of individual amino acids are discussed in the following sections.

3. AMINO ACID TYPE-SPECIFIC ISOTOPE LABELING IN INSECT CELLS

3.1 Scrambling Versus Label Dilution

Insect cells are in fact better suited for amino acid type-specific labeling than *E. coli*. This is due to the reduced metabolic capabilities of insect cells. In amino acid type-specific labeling, metabolism can lead to introduction of the isotope label into other, unwanted amino acids, which is termed isotope scrambling. On the flip side, isotope incorporation can be lowered by metabolism producing unlabeled versions of the labeled amino acid, termed isotope dilution. For insect cells, eight amino acids are essential—namely Ile, Leu, Lys, Met, Phe, Thr, Trp, and Val (Mitsuhashi, 1982)—for these no isotope dilution should happen. While, in principle, all other amino acids can be synthesized by insect cells (Lehninger, 1975), some of the nonessential amino acids are synthesized at very low rates, enabling scrambling-free labeling (Fig. 1).

3.2 Experimental Approaches

3.2.1 Replacing Amino Acids

Amino acid-type selective labeling is achieved by formulating a medium containing all amino acids in unlabeled form, but one, which will be added with the desired isotope label (Creemers et al., 1999; DeLange et al., 1998; Gossert et al., 2011; Kofuku et al., 2012; Nygaard et al., 2013; Strauss et al., 2003). Media can either be prepared from individual substances (Brüggert, Rehm, Shanker, Georgescu, & Holak, 2003) or nowadays can be ordered from vendors as so-called amino acid dropout media. Incorporation of up to 94% has been reported for selected amino acids; in general, 90% is usually achievable for nonmetabolized amino acids (Fig. 2; Gossert et al., 2011; Strauss et al., 2003). These high incorporation levels also allow dual ^{15}N, ^{13}C$_1$ amino-acid-type selective labeling which enables resonance assignments by HNCO experiments (Fig. 2C; Kainosho & Tsuji, 1982; Vajpai et al., 2008a; Yabuki et al., 1998).

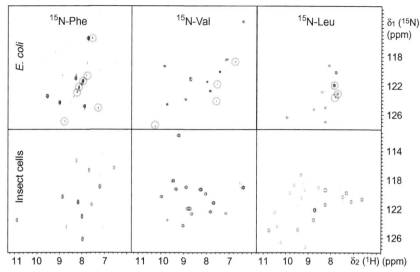

Figure 1 *Isotope scrambling in insect cells is much reduced compared to E. coli.* In the top row, (^{1}H,^{15}N)-HSQCs of selectively labeled protein samples produced in *E. coli* are shown, where secondary signals arising from isotope scrambling are clearly visible (gray circles). Phe scrambles primarily to Tyr, Val to Ala, and Leu to Val and Ile. In the spectra of other proteins with the same labeling pattern, but produced in insect cells, no other signals than those of the desired amino acid are visible, i.e., the count of signals corresponds to the expected number (11 out of 11 expected signals are observed for Phe, 16 out of 18 for Val, and 25 out of 26 for Leu). The respective pathways are either absent (Val to Ala, Leu to Val and Ile) or inactive (Phe to Tyr). Note that the *E. coli* expression lasted for only 4 h, while insect cell expression ran for 72 h, clearly demonstrating strongly reduced metabolism.

3.2.2 Excess Labeling

A very simple alternative approach for amino acid-type selective labeling is based on adding a sufficiently large excess of a labeled amino acid to full growth media. This approach requires knowledge on the approximate amount of the unlabeled target amino acid in the full medium. The diagram in Fig. 3 can serve as a rough guidance, but individual media may have differing quantities of amino acids. The labeled amino acid needs to be added in 5- to 10-fold excess in order to obtain 80–90% labeled protein. Typically, amounts larger than 1 g/L will be needed. Only few labeled amino acids are therefore affordable enough, in order to be used for this approach. Another limitation is that at such high concentrations, toxicity levels of certain amino acids may be reached. Nevertheless, we obtained 84% incorporation of ^{15}N-Leu (Fig. 3) by excess labeling using 2 g/L of the labeled amino acid (Gossert et al., 2011).

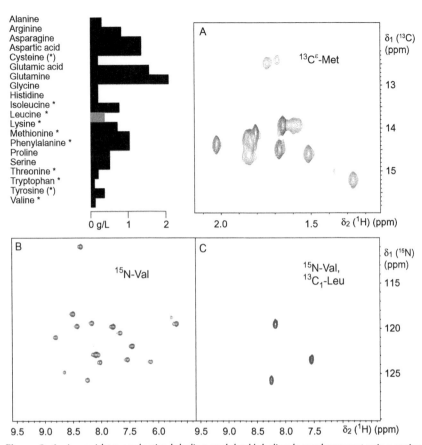

Figure 2 *Amino acid-type selective labeling and dual labeling by replacement using amino acid dropout media.* In the left diagram, the amino acid composition of a customized medium for Leu labeling is shown. Essential and semi-essential amino acids are indicated by asterisks and asterisks in parentheses, respectively. Amino acid amounts derived from pure sources are shown with green (black in the print version) bars, labeled amino acids are indicated by magenta (gray in the print version) bars. Note that yeast extract and unlabeled Leu are left away in the customized labeling medium. (A–C) (^{15}N,^{1}H)-HSQC spectra of protein kinases are shown, labeled with the amino acids indicated in the figure. (C) Dual labeling, signals from an 2D HN(CO) are superimposed on the spectrum of ^{15}N-Valine-labeled Abl kinase. For all the shown amino acids, incorporation levels of >90% were achieved using labeling by replacement of the target amino acid in dropout media. *Figure adapted from Gossert et al. (2011) with kind permission from Springer Science and Business Media.*

3.2.3 Chemical Labeling of Expressed Proteins

In this chapter, we focus on isotope labeling *in vivo*. However, for difficult to express proteins, posttranslational chemical labeling of proteins produced in unlabeled form in insect cells is a valid approach, leading to similarly suitable

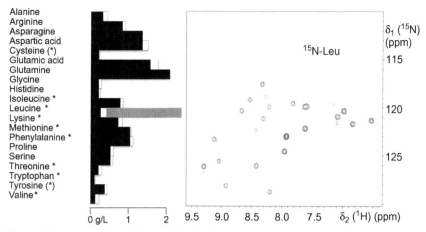

Figure 3 *Amino acid-type selective labeling by excess.* In the left diagram, the amino acid composition of a customized medium for Leu labeling is shown. Amino acid amounts derived from pure sources and yeast extract are shown with green and brown bars, respectively. Labeled amino acids are indicated by magenta bars. Abl kinase was expressed in full SF-4 medium where 2 g/L of ^{15}N-Leucine was added. The final incorporation level of ^{15}N-Leu was determined to be 84% by mass spectrometry. *Figure reproduced from Gossert et al. (2011) with kind permission from Springer Science and Business Media.* (See the color plate.)

samples for many types of functional studies. The most accessible amino acid residues for chemical modification under mild conditions are lysine and cysteine. Typically, $^{13}CH_3$ or CF_3 groups are introduced, as they are excellent NMR probes in terms of signal intensity and relaxation properties. CF_3 groups additionally yield extremely simplified spectra. The amine groups of lysine can be derivatized with ^{13}C formaldehyde by reductive methylation to yield dimethylated lysine (Abraham, Hoheisel, & Gaponenko, 2008; Dick, Sherry, Newkirk, & Gray, 1988; Lee, Abraham, & Gaponenko, 2012; Moore et al., 1998; Rayment, 1997) or with CF_3 derivatives to introduce CF_3 groups (Adriaensens et al., 1988; Mehta, Kulkarni, Mason, Constantinescu, & Antich, 1994). Free cysteine residues have been derivatized with different trifluoromethyl groups (Horst, Liu, Stevens, & Wüthrich, 2013; Liu, Horst, Katritch, Stevens, & Wüthrich, 2012; Nelson, 1978). In order to obtain a close analog amino acid to methionine, which is ideally suited for studying very large proteins, cysteine residues have been derivatized with ^{13}C methyl methanethiosulfonide (Religa, Ruschak, Rosenzweig, & Kay, 2011). Finally, it has also been reported that tyrosine can be monofluorinated ortho to the hydroxyl group (Hebel, Kirk, Cohen, & Labroo, 1990).

4. UNIFORM ISOTOPE LABELING IN INSECT CELLS
4.1 Experimental Approaches
4.1.1 Commercial Media
For uniform isotope labeling in insect cells, all sources of amino acids in the medium need to be replaced by labeled ones. Several approaches have been devised to achieve this. All amino acids could be supplied in pure form, which however is the most expensive strategy. The next best alternative are commercial media based on this approach, such as Bioexpress 2000 from Cambridge Isotope Laboratories (CIL). These media are well established and can be used to produce u-^{15}N and ^{13}C,^{15}N-labeled proteins at incorporation levels close to 90% with expression yields comparable to nonlabeled media (Strauss et al., 2005). However, these media have not found widespread applications, because they are very expensive and to date limited to the above two labeling patterns. In order to extend the isotope labeling patterns and to reduce costs, other approaches for supplying amino acids have been developed in the past several years.

4.1.2 Economic Media Based on Amino Acid Extracts
To economize, instead of using pure amino acids, cell extracts from yeast, bacteria, or algae can be used as a less expensive source of labeled amino acids (Fig. 4; Hansen et al., 1992). These simpler organisms can be grown on inexpensive labeled substrates, and then amino acid extracts can be prepared in various ways. Two main approaches have been mostly used: autolysis, where cells are incubated at 50 °C and internal proteases degrade cellular proteins to amino acids (Egorova-Zachernyuk, Bosman, & DeGrip, 2011; Egorova-Zachernyuk, Bosman, Pistorius, & DeGrip, 2009) or acidic hydrolysis, where purified protein extracts are subjected to high temperatures (120 °C) at low pH (LeMaster & Richards, 1982). Autolysis is the milder approach conserving most amino acid types, but yields lower amounts of free amino acids and it is more difficult to obtain high batch-to-batch reproducibility. Acid hydrolysis will lead to destruction of the amino acids Gln and Asn, as well as Cys and Trp (Fig. 5). Additionally, if deuterated extracts are prepared, several amino acids are protonated if the hydrolysis is not carried out in D_2O (Markley, Putter, & Jardetzky, 1968). A less commonly used third approach is enzymatic hydrolysis, where a mixture of endo- and exopeptidases like pronase are used to produce amino acid extracts under mild conditions (Hansen et al., 1992; Sweeney & Walker, 1993a, 1993b). All these

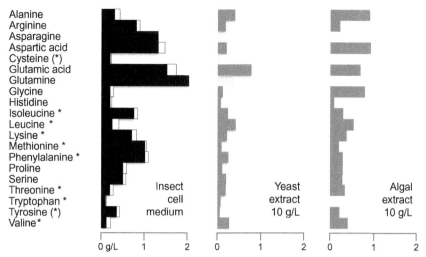

Figure 4 *Amino acid composition of different insect cell media: commercial nonlabeled medium compared to yeast and algal extract-based ones.* On the left diagram, the amino acid composition of a typical insect cell medium is shown, with amounts added in pure form in green (black in the print version) and amino acid amounts from yeast extract in brown (white in the print version). The asterisks on the amino acid name mark essential amino acids. Cys and Tyr are semiessential, as they can be synthesized by insect cells if Met/Ser and Phe are present, respectively. On the center and the right diagram, the amino acid amounts from using 10 g/L of yeast and algal extracts, respectively, are shown. The magenta (gray in the print version) color indicates labeled amino acids. Note that in algal hydrolysates, the overall amino acid content is higher, but Asn, Gln, Cys, and Trp are missing due to acidic hydrolysis. If yeast extracts are prepared by autolysis, these amino acids are partially conserved, as indicated by transparent bars.

approaches result in amino acid extracts containing free amino acids and residual oligopeptides. The amino acid composition depends on the source organism and on the susceptibility of peptide bonds between different amino acid types to the method of hydrolysis.

Yeast, bacteria, and algae have been used as source organisms for amino acid extracts. Yeast-based extracts enable stable growth and are also part of standard insect cell media (Meola et al., 2014). While bacterial extracts have not been employed successfully yet, algal extracts seem to be a viable alternative (Hansen et al., 1992; Kofuku et al., 2014; Sitarska et al., 2015). Isotope-labeled algal extracts are fundamentally less expensive to produce than yeast extract, because algae can use the simplest molecules as sources for building up amino acids, i.e., $^{15}NH_3$, $^{13}CO_2$, and D_2O. Yeast require more complex carbohydrates like glucose, which renders ^{13}C and ^{2}H labeling expensive, considering that only about 10% of the employed glucose is

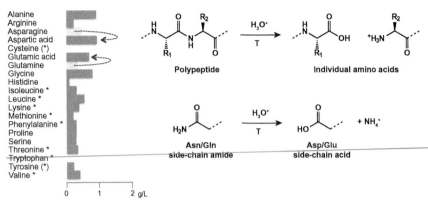

Figure 5 *Products of acidic protein hydrolysis.* On the left diagram, the amino acid composition of 10 g/L of algal hydrolysate is shown, which results from acidic hydrolysis. Polypeptides are broken down into individual amino acids, but some amino acid side chains are altered. Asn and Gln are converted into Asp and Glu, respectively, as indicated by the dashed arrows. The thiol of Cys is derivatized into an inhomogeneous range of products and tryptophan only remains in trace amounts due to instability under acidic hydrolysis.

converted into amino acids, i.e., 30–40% is converted to biomass, of which about 30% consists of amino acids (Egorova-Zachernyuk et al., 2009). Yeast autolysates in ^{15}N and ^{15}N,^{13}C-labeled forms are commercially available (Protein labeling innovation and CortecNet; Meola et al., 2014). Algal hydrolysates are less expensive and are commercially available in all labeling patterns, i.e., ^{15}N; ^{13}C,^{15}N; ^{2}H,^{15}N; and ^{2}H,^{13}C,^{15}N from different vendors (Sigma–Aldrich and Cambridge Isotope Laboratories). Additionally, algal extracts have a higher amino acid content than yeast extracts of 65% versus 35%, respectively. Usually, about half of the amino acids are present in their free form and the rest is present as small oligopeptides.

Standard media contain roughly 16 g of amino acids per liter, 14 g in pure form, and about 2 g from yeast extract (Fig. 4). These need to be replaced with labeled ones. However, the osmolarity of the medium sets an upper limit to the amount of amino acid extract that can be added. We found that addition of 10–12 g/L of algal extract leads to media with maximum osmolarity tolerated by insect cells (320 mOsm/kg; Sitarska et al., 2015). The total amino acid amount of approximately 6 g/L supplied in this way is enough—by a fair margin—to promote robust growth and protein expression. Highest possible amounts of labeled amino acids are chosen for the following reasons: (i) maximization of label incorporation, (ii) reduction of unwanted amino acid metabolism by product inhibition, and (iii) high supply of substrate amino acids for desired metabolism producing nonessential amino acids missing in the amino acid extract.

4.2 Uniform ^{15}N Labeling

Uniform ^{15}N-labeling can be achieved using media based on algal or yeast extract (Meola et al., 2014; Sitarska et al., 2015), if amino acids missing in the extract are supplied. Gln and Asn are present in standard insect cell media in amounts of 2 and 1.5 g/L, respectively, and Cys and Trp are present at 200 mg/L (Fig. 4). Adding these four amino acids in labeled form would lead to prices in the order of the commercial medium. Luckily, Gln, Asn, and Cys are nonessential amino acids that can be synthesized by insect cells. Trp, however, needs to be supplied, as it is an essential amino acid. But its amount can be reduced 10-fold to 20 mg/L without affecting isotope incorporation or cell growth (Hansen et al., 1992; Sitarska et al., 2015). Cys can be synthesized by insect cells from Met and Ser (Fig. 6C; Doverskog, Han, & Häggström, 1998) and does not need to be supplied, because Ser and Met are present in amino acid hydrolysates in high enough amounts. Gln can be synthesized by insect cells from Glu and ammonia (Fig. 6A; Doverskog, Jacobsson, Chapman, Kuchel, & Häggström, 2000; Drews et al., 2000; Öhman et al., 1996). The side-chain nitrogen of such synthesized Gln is in turn used to produce Asn from Asp (Fig. 6B; Hansen et al., 1992). Since Asp and Glu are abundant in extracts—they are the products of Asn and Gln hydrolysis—Asn and Gln will be produced by insect cells in labeled form if ^{15}N ammonia is supplied (Hansen et al., 1992; Meola et al., 2014; Sitarska et al., 2015). Ammonia is toxic to insect cells at elevated concentration, but it has been determined in several studies that 5 mM is a nontoxic concentration at which enough amino acids are produced to promote robust growth and protein expression (Hansen et al., 1992; Meola et al., 2014; Sitarska et al., 2015).

Due to the coincidence that the amino acids missing in cell extracts are produced actively by insect cells, affordable media for ^{15}N labeling can be formulated based on yeast and algal extracts, provided that tryptophan and ammonia are added in labeled form. There is one complication arising from this medium composition: ammonia inhibits the entry of baculovirus from endosomes into the cytoplasm leading to lower and less reproducible target protein expression (Dabydeen & Meneses, 2009; Dong et al., 2010; Hefferon, Oomens, Monsma, Finnerty, & Blissard, 1999). Fortunately, target gene expression in BV systems only starts 24–48 h post-infection.

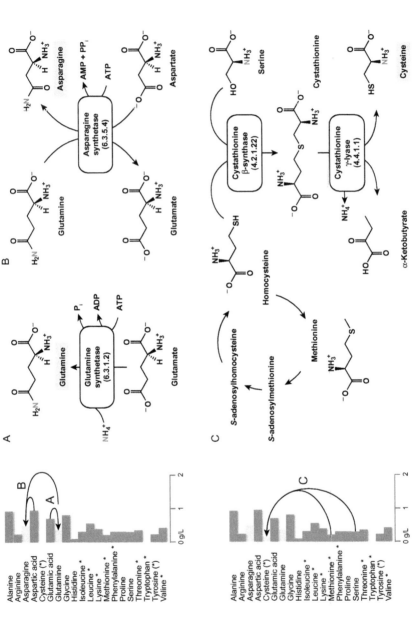

Figure 6 *Biochemical pathways restoring missing amino acids in hydrolyzed extracts.* On the left diagram, the amino acid composition of 10 g/L of algal hydrolysate is shown. Arrows indicate the substrate and product amino acids from the relevant biochemical reactions. The respective reaction schemes identified with A, B, and C are shown on the right with emphasis on nitrogen incorporation. Key enzymes are indicated by rounded squares with their respective names and enzyme commission number indicated.

Target genes are inserted into the BV genome under control of the polyhedrin promoter, which is only activated in the very late stage of infection, because its natural function is to produce a protein that is used to embed mature virus particles in a protein matrix. Therefore, viral infection does not necessarily need to be carried out in the labeling medium. The protocol for labeling in ammonia-containing media is therefore adapted in the following way: cells are infected in nonlabeled medium and transferred to labeling medium only 16 h post-infection (Fig. 7). With this protocol expression yields as high as in nonlabeled medium can be achieved and isotope incorporation levels are only slightly lowered, typically about 3%. This is well acceptable in light of at least fivefold increased yields and high reproducibility. Figure 8 shows a spectrum of Abelson kinase (Abl), produced according to a protocol based on algal extracts, confirming uniform incorporation of ^{15}N into all amino acid types, including the ones missing in the algal extract. The overall ^{15}N incorporation is typically 80% as determined by MS.

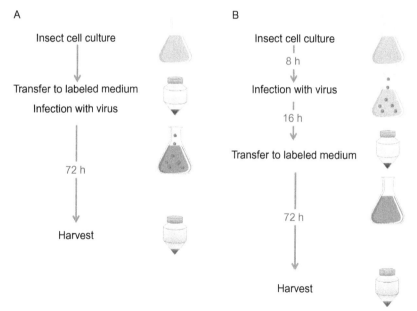

Figure 7 *Comparison of cell culturing protocols for amino acid-specific and uniform labeling in insect cells.* (A) For amino acid-type specific labeling, cells are infected with baculovirus in labeling medium. The same applies for labeling in commercial Bioexpress 2000 media. (B) For uniform labeling in media containing free ammonia, virus entry is inhibited. Therefore, cells are infected with baculovirus in nonlabeled medium and incubated for about 16 h. Although virus entry only takes about half an hour, highest protein yields are obtained if cells are incubated in nonlabeled medium for up to 16 h.

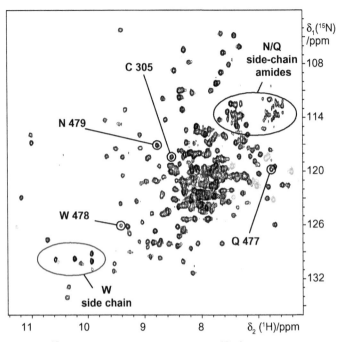

Figure 8 *Uniform ^{15}N labeling of all amino acid types.* ($^{15}N,^1H$)-TROSY spectrum of u-^{15}N-labeled Abl (80% incorporation) produced in Sf9 insect cells, in complex with the inhibitor imatinib (Vajpai et al., 2008a). All expected correlation signals are seen, including backbone and side chain signals of the amino acids not present in the algal hydrolysate: asparagine, cysteine, glutamine, and tryptophan. In order to also detect NH_2 moieties in the TROSY experiment, the INEPT delay was shortened to 2.1 ms.

4.3 Uniform ^{13}C Labeling

Uniform ^{13}C or $^{15}N,^{13}C$ labeling can be performed using commercial Bioexpress 2000 (CIL; Hamatsu et al., 2013; Strauss et al., 2005) or homemade media (Sitarska et al., 2015; Walton, Kasprzak, Hare, & Logan, 2006). For carbon labeling, algal extracts are the most economic source of amino acids, as they can be produced from $^{13}CO_2$ instead of more complex and more expensive carbohydrates. When using the same approach as described above, 75% ^{13}C incorporation can be achieved. The lowered isotope incorporation is due to isotope dilution by insect cell metabolism, which produces a number of amino acids from unlabeled glucose present in the medium, namely, Ala, Glu, and Asp, are produced from intermediates of glucose metabolism in simple one-step reaction by transamination or reductive amidation (Fig. 9; Drews et al., 2000; Öhman et al., 1995). Because Asn and Gln are synthesized from Glu and Asp, respectively, the amino acids obtained

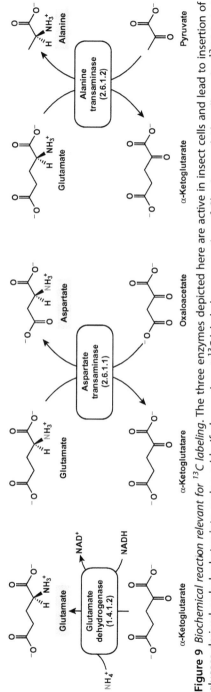

Figure 9 *Biochemical reaction relevant for ^{13}C labeling.* The three enzymes depicted here are active in insect cells and lead to insertion of glucose-derived carbohydrates into amino acids. If glucose is not ^{13}C-labeled, some portion of Glu, Asp, and Ala will not be ^{13}C-labeled as well—this also applies for the amino acids synthesized from these like Gln and Asn.

from these reactions will be fully ^{15}N labeled, but the carbon chain will inherit the label of glucose. ^{13}C labeling efficiency can be brought to 80% by replacing unlabeled glucose with ^{13}C-labeled one. However, considering the amounts of glucose of 5–10 g/L, the costs of media increase in this way. Analysis of carbon incorporation from glucose into amino acids reveals that biochemical pathways for most nonessential amino acids are not active, as there are no significant amounts of Ser, Gly, Pro, Met, or Arg produced by insect cells from glucose. Asx and Glx only suffer from 10–20% label dilution, while nearly 80% of Ala is synthesized from glucose. This single metabolic pathway can be selectively inhibited by L-cycloserine (Wong, Fuller, & Molloy, 1973). Therefore, it is possible to economize and use unlabeled glucose if alanine transaminase is being inhibited (Fig. 10) and

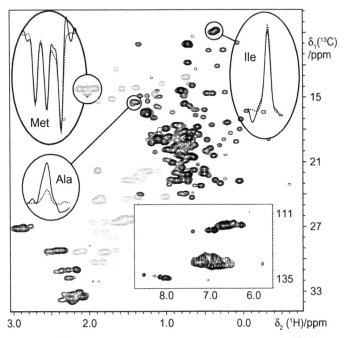

Figure 10 *Uniform ^{13}C labeling of all methyl groups including alanine.* (^{13}C,^{1}H)-ct-HSQC of u-^{13}C,^{15}N-labeled Abl (76% incorporation) produced in Sf9 insect cells. 1D traces show signals of selected amino acid types (scaled according to Met and Ile signals) after expression using unlabeled glucose with (blue, solid line in the print version) or without (red, dotted line in the print version) 50 μM L-cycloserine. Alanine signal intensity can be fourfold increased by this method and a complete methyl spectrum can be obtained in an affordable way. In the excerpt on the lower right, the aromatic region of a conventional (^{13}C,^{1}H)-HSQC is shown, demonstrating labeling of aromatic side chains.

still obtain ^{13}C-HSQCs with complete methyl signals and only marginally reduced intensities for Asx and Glx.

For practical reasons, it should be noted that although the overall incorporation is 75%, each individual labeled amino acid, except for Asx and Glx, has a ^{13}C incorporation level of 98%. This is because fully labeled amino acids from the 98%-labeled algal extract are integrated into proteins as a whole, without being metabolized. The lower overall incorporation is due to integration of fully unlabeled amino acids present in the cells, either from earlier growth media or from degradation of unlabeled proteins. As a consequence, in NMR experiments, intraresidual magnetization transfer steps (as in HSQC and TOCSY) are very efficient (98%) and interresidual ones (in triple resonance experiments and NOESY) have a lower efficiency (75%). In an HNCA, for example, the intraresidual HNCA pathway will yield an overall transfer efficiency of 94% (0.98^3) for the approximately 75% labeled amino acids in a protein. The interresidual HNCA$_{-1}$ pathway, however, will be 73% efficient ($0.98 \times 0.98 \times 0.75$) because only 75% of neighboring residues will statistically be labeled. Therefore, it is advisable to use labeled glucose for experiments involving interresidual magnetization transfer, to increase transfer efficiency to close to 80% also for Asx and Glx.

4.4 Uniform ^{2}H Labeling

4.4.1 Approaches

Deuteration in eukaryotic cells remains one of the most challenging tasks. Although it was employed in the very early days of NMR in order to simplify spectra (Markley et al., 1968), no other approaches were published in the next 40 years. One principal challenge is that while for most purposes, HSQC spectra of 80% or 99% ^{15}N- and ^{13}C-labeled proteins are of the same quality, deuteration should be as complete as possible in order to enable studies of very large proteins. Additionally, deuteration in insect cells is less uniform across different amino acids than ^{13}C labeling, not to mention ^{15}N labeling. This is due to insect cell metabolism and because deuterated glucose cannot be used in insect cell media as it impairs cell growth and protein expression.

Two protocols have been published based on deuterated algal extracts. Kofuku et al. have developed a protocol optimized for membrane proteins in order to study methyl signals of Met residues in a highly deuterated background (Kofuku et al., 2014). The protocol did not aim at uniform deuteration, but the majority of hydrophobic amino acids were highly labeled. Relatively low amounts of algal extract were used and the medium was additionally supplied with specifically labeled and unlabeled amino acids at different time points of the culture according to a refined schedule. Details on the resulting labeling pattern are described in the supplementary material of the aforementioned publication. An alternative, and probably more general approach is based on adding the maximal possible amount of deuterated algal extract, following the same principles as described above for uniform ^{15}N and ^{13}C labeling. In this way, overall incorporation levels of 76% and 73% for $^{2}H,^{15}N$ and $^{2}H,^{13}C,^{15}N$ labeling can be obtained, respectively (Sitarska et al., 2015). This results in significantly improved spectra and enables triple resonance experiments of large proteins (Fig. 11). However, also in this case, deuteration is not uniform and incorporation patterns are different for individual amino acids.

4.4.2 Relevant Metabolism for ^{2}H Labeling

The relevant biochemical reactions involved in reducing deuteration levels are in part the ones already discussed for ^{15}N and ^{13}C labeling. The scheme of central amino acid metabolism in Fig. 12 summarizes these reactions. Since glucose cannot be added in deuterated form to insect cell media, Glx, Asx, and Ala will be synthesized with protonated carbon chains at similar levels as determined for the case of ^{13}C labeling. Also here, Ala synthesis can be controlled with the inhibitor L-cycloserine. However, in addition to these initial reactions, the alpha positions of the amino acids involved in transaminase reactions will be protonated. Pro, Arg, Gly, Ser, Thr, and Met do not seem to be synthesized at relevant levels from glucose. However, Gly is heavily protonated at the alpha position in an important housekeeping reaction, which produces N^5,N^{10}-methylene tetrahydrofolate that is used in numerous methylation reactions (Fig. 13; Kofuku et al., 2014). All these reactions contribute to lower and less uniform incorporation of deuterium than ^{13}C or ^{15}N. Additionally, deuterated media reduce protein yields to 30–60% of what is obtained in nonlabeled media. In contrast, for ^{13}C and ^{15}N labeling, similar protein amounts are obtained as in nonlabeled media. Reduced yields are probably also due to reactions like 3, 5, and 4 (reversed) in Fig. 12. These lead to deuterated carbohydrates, which are introduced into the citric acid cycle. These deuterated intermediates slow down reactions involving the breakage of covalent hydrogen bonds sevenfold, therefore acting as weak inhibitors of the citric acid cycle. This probably leads to decreased performance of the cells, which ultimately results in lowered protein amounts. Inhibiting enzymes 3–5 (Fig. 12) could potentially result in increased protein yields and at the same time, higher and more uniform

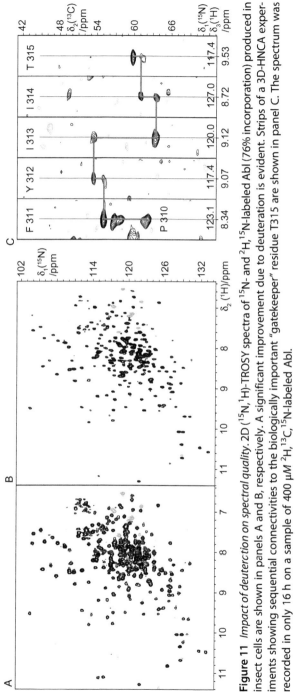

Figure 11 *Impact of deuteration on spectral quality.* 2D (^{15}N,^{1}H)-TROSY spectra of ^{15}N- and ^{2}H,^{15}N-labeled Abl (76% incorporation) produced in insect cells are shown in panels A and B, respectively. A significant improvement due to deuteration is evident. Strips of a 3D-HNCA experiments showing sequential connectivities to the biologically important "gatekeeper" residue T315 are shown in panel C. The spectrum was recorded in only 16 h on a sample of 400 μM ^{2}H,^{13}C,^{15}N-labeled Abl.

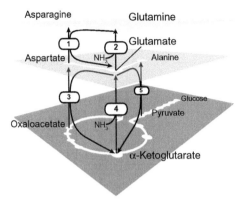

Figure 12 *Overview of most relevant metabolic pathways for labeling in insect cells.* The bottom plane (black) depicts glycolysis and the citric acid cycle; selected metabolites are indicated with their names. The middle level (blue, gray in the print version) shows amino acids that are synthesized in a one-step reaction from the respective carbohydrates by addition of one nitrogen. The top level (light blue, light gray in the print version) shows amino acids with yet an additional nitrogen. Enzymatic reactions are marked with arrows and the enzymes are enumerated as follows: 1: asparagine synthetase (Fig. 6B), 2: glutamine synthetase (Fig. 6A), 3: aspartate transaminase (Fig. 9, middle), 4: glutamate dehydrogenase (Fig. 9, left), 5: alanine transaminase (Fig. 9, right).

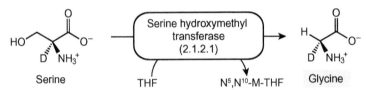

Figure 13 *Alpha-protonation of glycine.* The reaction producing N^5,N^{10}-methylene tetrahydrofolate, an important vitamin for methyl transfer reactions, leads to protonation of Gly at the alpha position.

isotope incorporation. Initial experiments with the inhibitors L-cycloserine (inhibits enzyme 5 in Fig. 12; Wong et al., 1973), aminooxyacetic acid (3) (Morita et al., 2004; Rej, 1977), and bithionol (4) (Li, Smith, Walker, & Smith, 2009), however, showed no improved effect over using L-cycloserine alone (Sitarska et al., 2015).

5. APPLICATIONS

Isotope labeling in insect cells enabled studies on several novel and challenging systems, which were otherwise not amenable to NMR spectroscopy. In the following, we highlight merely a few possible applications that were reported in the literature. These span a wide range: from uniform labeling for structural studies, through selective labeling approaches for functional studies and drug discovery, to observation of isotope-labeled proteins in living insect cells.

5.1 Uniform Labeling for Structural Studies

Preparation of selectively labeled proteins in insect cells is relatively straightforward, and several studies describing such an approach to studying protein function can be found in the literature (Sections 5.2–5.4). On the other hand, uniform labeling efforts were hindered by the lack of appropriate labeling media. The introduction of Bioexpress 2000 (CIL), the first commercially available labeling medium for insect cells, enabled uniform labeling in this expression system. Abelson tyrosine kinase (Abl) was the first labeled protein expressed in insect cells (Strauss et al., 2005), for which nearly complete resonance assignments were reported (Vajpai et al., 2008a). Nevertheless, although uniformly ^{13}C,^{15}N-labeled protein could be obtained using Bioexpress 2000, the assignment procedure was not trivial. The relatively large size of the catalytic domain of Abl (~33 kDa) and unavailability of media allowing for deuteration resulted in poor quality of the 3D HNCACB experiment. To aid the HNCA- and HN(CO)CA-based assignments, a set of 15 selectively labeled samples with various combinatorial patterns, including dual labeling (Kainosho & Tsuji, 1982) was prepared, resulting in nearly complete (96%) assignment of backbone ^{1}H^{N}, ^{15}N, ^{13}C$^{\alpha}$, and ^{13}CO resonances (Vajpai et al., 2008a). Importantly, despite the conformational flexibility of the activation loop of Abl, many of the residues within this region were observable. This allowed Vajpai and coworkers to investigate the active and inactive states of Abl complexed with different inhibitors, and prove that the conformations of the activation loop observed in crystal structures are also present in solution and do not result from crystal-packing artifacts (Vajpai et al., 2008b). Despite the success of the Abl study, reports of uniform labeling in insect cells remain scarce. Although it may seem surprising at first, it is understandable considering the very high price of the commercial labeling medium, which severely limited its use. However, this situation may change with protocols describing more affordable media formulations for uniform labeling in insect cells (Meola et al., 2014; Sitarska et al., 2015). Notably, these media also allow for deuteration, expanding their applicability to larger systems which cannot otherwise be produced in bacteria (Sitarska et al., 2015).

5.2 Membrane Proteins

For many membrane proteins, insect cells are the host of choice, offering both good yields and functionally active proteins. Furthermore, strategic selection of amino acid types for labeling often allows investigating functionally relevant residues without the need for sequential assignment. Instead, key residues may be assigned by mutagenesis.

The first membrane protein selectively labeled in insect cells was the visual G-protein-coupled receptor (GPCR) rhodopsin. The bovine protein was selectively labeled with ring-^{2}H$_4$-tyrosine in order to study the involvement of tyrosines in the photoactivation process by Fourier transform infrared spectroscopy (DeLange et al., 1998). Later, rhodopsin was labeled with α,ε-^{15}N$_2$-lysine, and subsequent solid-state NMR experiments combined with computation revealed a Schiff-base between a nitrogen atom of Lys296 and retinal that exists in a protonated state stabilized by a complex counterion (Creemers et al., 1999).

A popular strategy for selective labeling of recombinant proteins relies on incorporation of isotopically labeled methionine. Practical use of methionine labeling for NMR applications dates back to early 1980s (Kainosho & Tsuji, 1982). While this study utilized only 1D NMR, methionine labeling became commonly used largely due to its benefits for 2D experiments. The characteristic chemical shift and favorable relaxation properties of methionine methyl groups lead to sharp and easily identifiable resonances in a 2D (^{13}C,^{1}H)-HSQC spectrum. This method was applied by Kofuku and coworkers to characterize conformational changes of the transmembrane regions of the β$_2$-adrenergic receptor (β$_2$AR). Monitoring the signal of Met82 in complexes with neutral and partial agonists, the authors could show that these states exist as equilibria between the inverse-agonist- and full-agonist-bound states, depending on the efficacy of the ligand (Kofuku et al., 2012). These observations provided insights into the mechanism of signal transduction mediated by β$_2$AR and other GPCRs. In a subsequent study, the same research group combined α,β,γ-^{2}H-, methyl-^{13}C-methionine labeling with targeted partial deuteration. By selecting nine amino acid types in the proximity of the ^{13}C-methyl groups, 80–90% of ^{1}H–^{1}H dipole–dipole interactions could be removed, resulting in approximately fivefold increase in the sensitivity of the NMR experiment (Fig. 14; Kofuku et al., 2014). Interestingly, it was shown that the exchange rates between the different conformational states of β$_2$AR in lipid bilayers were significantly different from those measured previously

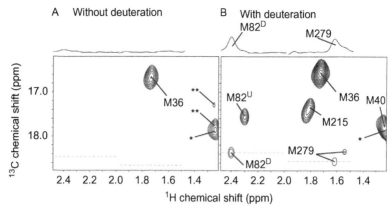

Figure 14 Sensitivity enhancements of methionine methyl resonances of β2-adrenergic receptor upon fractional deuteration in ($^{13}C,^{1}H$)-HMQC spectra. Selective deuteration of nine amino acid types leads to fivefold sensitivity increase in signals of methionine methyl groups in the right spectrum compared to the left spectrum of protonated β2-adrenergic receptor (see text). Single and double asterisks identify natural-abundance signals from the ligand and from lipids, respectively. *Figure reproduced with permission from Kofuku et al. (2014).*

in nonphysiological detergent micelles. Moreover, based on simulations of Met82 resonances using a three-state exchange model, it was found that GPCR signaling occurs on timescales faster than for receptor tyrosine kinases, presumably enabling rapid neurotransmission and sensory perception.

5.3 In-Cell NMR

Observation of proteins inside living cells was first reported for the bacterially overexpressed 7 kDa domain of mercuric ion reductase (Serber et al., 2001). To minimize the background signal of the cellular environment, bacteria were first grown in unlabeled medium and then transferred to ^{15}N-labeled minimal medium for protein overexpression. An alternative approach was developed for in-cell NMR applications using oocytes of the African clawed frog *Xenopus laevis*. The germ cells are large enough that the protein can be introduced by microinjection. Therefore, the labeled protein of interest can be conventionally purified from a different expression system (e.g., *E. coli*), resulting in no background signal in subsequent NMR experiments (Sakai et al., 2006; Selenko, Serber, Gadea, Ruderman, & Wagner, 2006). In-cell NMR studies were also reported for proteins expressed in the yeast *Pichia pastoris* (Bertrand, Reverdatto, Burz,

Zitomer, & Shekhtman, 2012). Remarkably, depending on the carbon source in the growth medium, the protein of interest could be either localized in the cytosol or sequestered into storage vesicles.

In-cell NMR was successfully demonstrated using insect cells as the expression host (Hamatsu et al., 2013). Four proteins, ranging in length from 57 to 148 amino acids, were expressed uniformly $^{13}C,^{15}N$-labeled in Sf9 cells using the Bioexpress 2000 medium (CIL). To improve the quality of the NMR spectra, insect cells were infected with the baculovirus in an unlabeled medium and after 24 h transferred to the labeling medium where expression carried on for another 24 h. Furthermore, background signals were reduced by subtraction of a spectrum recorded on cells infected with a control baculovirus not harboring the respective gene. Notably, although the authors found the cell preparation to be stable for only 3.5 h (as indicated by 83–90% viability), three-dimensional HNCA, HNCO, and HN(CO)CA could be obtained within this time by taking advantage of nonuniform sampling schemes in the indirect dimension (Fig. 15).

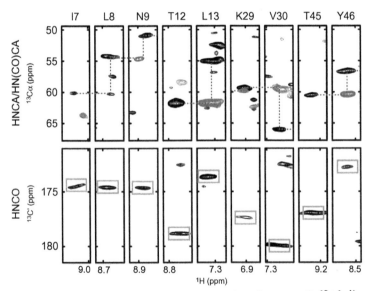

Figure 15 *In-cell triple resonance experiments obtained in Sf9 insect cells.* $^{13}C,^{1}H^{N}$ strips of a 3D HNCA (black) and HN(CO)CA (red) spectra (overlaid, upper panel) and the 3D HNCO spectrum (lower panel) of Sf9 cells expressing the protein GB1. Cells were grown in $^{13}C,^{15}N$-labeled Bioexpress 2000 medium (Cambridge isotope labs) for protein expression. *Reprinted with permission from Hamatsu et al. (2013). Copyright 2013 American Chemical Society.* (See the color plate.)

5.4 Applications in Drug Discovery

NMR is routinely used in drug discovery to validate binding of small molecules to a protein target, which can be monitored by changes in the spectra of either the small molecule (ligand observation) or the protein. As ligand observation does not require labeled protein, we will focus on the latter approach. Often, the goal of the NMR experiment is to confirm the interaction rather than provide detailed structural information on its nature. In such cases, selective labeling is used (Gossert et al., 2011; Strauss et al., 2003), and the type of amino acid to be isotopically labeled depends on the studied system. If the protein of interest contains methionine residues in the vicinity of the expected binding site, ^{13}C labeling of their methyl groups is particularly attractive. This approach has been applied, among others, to several kinases, which were produced in insect cells by supplementing a SF-4 based medium with $^{13}C^{\varepsilon}$-Met (Gossert et al., 2011). In protein kinases, a conserved Asp-Phe-Gly (DFG) motif determines the active or inactive conformation of the enzyme. Because of the proximity of the DFG motif to the ATP binding site targeted by kinase inhibitors, protein kinases were also labeled with ^{15}N-phenylalanine. Subsequent NMR spectra recorded in the presence of various ligands revealed significant changes in the peak pattern between the complexes of proteins with active- and inactive-state inhibitors (Gossert et al., 2011). In a different study, selective labeling in insect cells was applied to develop an NMR-based conformational assay. Abelson tyrosine kinase was labeled with ^{15}N-valine and using two constructs with different lengths, the C-terminal Val525 could be assigned. The signal intensity of this residue was shown to report on the bending of helix I, resulting from binding of allosteric inhibitors (Jahnke et al., 2010).

NMR is also a powerful tool for fragment-based drug discovery (FBDD). For a review of the technique and a discussion of its advantages over high-throughput screening, see Harner, Frank, & Fesik (2013). Since at early stages of the discovery process, FBDD is principally a hit-finding activity, it is beneficial that the protein is uniformly labeled. If protein supply is not limiting, fragment-based approaches can be applied on an unconventionally large scale, as illustrated by a study describing the discovery of small-molecule inhibitors of the bromodomain of ATAD2 (Harner, Chauder, Phan, & Fesik, 2014). The authors screened a library of approximately 13,800 fragments (in mixtures of 12) by protein observation using uniformly ^{15}N-labeled bromodomain of ATAD2. The approach required in total more than 2 g of protein and while the bromodomain was bacterially expressed, it is immediately clear that expression on such a scale would not be economically viable if the protein was to be produced in insect cells using commercially available labeling media. However, given the reports of affordable media formulations for uniform isotope labeling (Meola et al., 2014; Sitarska et al., 2015) and the typically high expression yields, preparation of gram amounts of labeled protein from insect cell cultures seems no longer inconceivable.

6. PROTOCOLS

6.1 Materials

Materials needed for all the following protocols are listed in this section.

6.1.1 Insect Cell Lines
- Sf9 (Invitrogen, Life Technologies, Cat No 11496-015)
- Sf21 (Invitrogen, Life Technologies, Cat No 11497-013)

6.1.2 Media
Various commercial media are available. All media work well for growth and maintenance of cells. Some media such as Sf-900III and ExCell420 support higher cell densities, i.e., $8-9 \times 10^6$ cells/mL and beyond, whereas the SF-4 medium supports cell densities of $4-5 \times 10^6$ cells/mL. For the following media, we have tested custom amino acid dropout versions or amino acid-free and glucose-free media.
- Sf-900II and Sf-900III SFM (Invitrogen, Life Technologies). Custom Sf-900 III media can be ordered through a web-based media configurator.
    http://www.lifetechnologies.com/ch/en/home/life-science/bioproduction/custom-cell-culture-media-and-services/gibco-custom-media-configurator.html (last accessed March 31, 2015)
- SF-4 Baculo Express ICM (BioConcept). Custom SF-4 media can be ordered by email: info@bioconcept.ch
- ExCell420 (Sigma–Aldrich)

6.1.3 Reagents
- Disinfectants: 70% ethanol solution and Virkon or Decon (10% v/v)
- Fetal bovine serum (FBS; Sigma–Aldrich)

- Antibiotics: penicillin/streptomycin
- Yeast extract filtrate 6% (Yeast extract powder Sigma-Aldrich, Cat. No. 70161-500G)
- ISOGRO® with desired isotope-labeling pattern (Sigma–Aldrich)
- $^{15}NH_4Cl$ (Cambridge Isotope Laboratories or Sigma–Aldrich)
- D-Glucose or u-^{13}C D-glucose (Cambridge Isotope Laboratories or Sigma–Aldrich)
- Tryptophan with desired isotope-labeling pattern (Cambridge Isotope Laboratories or Sigma–Aldrich)
- L-cycloserine (Sigma–Aldrich, Cat. No. C1159)

6.1.4 Consumables
- Sterile culturing bottles or Erlenmeyer flasks with vented caps for 250 mL and 1 L culture volumes
- Sterile vials and pipettes
- 24-well tissue culture plates
- Container for disposal of used media
- Sterile container and magnetic bar for mixing the media
- Sterile filtering devices with 0.22 μm pores (Millipore, Cat. No. SCGPU11RE)
- Sterile conical 250 and 500 mL centrifuge bottles (Corning, Cat No. 430776 and 431123, or similarly shaped bottles, that allow decanting)

6.1.5 Equipment
- Sterile laminar flow hood
- Shaker incubator
- Cell viability analyzer, e.g., Vi-CELL (Beckmann Coulter) with appropriate reagents or inverted microscope and Neubauer's counting chamber
- Centrifuge with capacity for 500 mL bottles

6.2 Culturing of Insect Cells
6.2.1 Starting an Insect Cell Culture
Strict aseptic technique is required to handle and manipulate all cell lines for which a sterile laminar flow hood is required. Anything taken inside the hood should be surface treated with 70% ethanol to decontaminate it, unless stated otherwise. All vessels and pipettes used with live cells or viruses must be decontaminated properly using autoclaving, Virkon (10% w/v), or

diluted Decon (10% v/v). Wipe work surfaces with Virkon/bleach followed by 70% ethanol to immediately treat any spills.

1. Starting of cell culture, Day 1
   1.1 Before starting the manipulation, bring the medium, FBS, and antibiotic (penicillin/streptomycin) solution to 27 °C. Transfer them into a laminar flow hood. Supplement the medium with a final concentration of 1% FBS (v/v), 50 U of penicillin, and 50 µg of streptomycin per mL of medium. Mix by gently inverting the medium bottle three times. Take a 250-mL Erlenmeyer flask with a vented cap and transfer 29 mL of supplemented medium into it.
   1.2 Take a cryotube containing frozen stock of Sf9 cells from a liquid nitrogen tank and thaw rapidly in a 37 °C water bath. When completely thawed, spray the outer surface of the tube with 70% ethanol, and take it into the hood. Transfer the contents of the cryotube into a sterile Eppendorf tube. Spin the tube in a tabletop centrifuge at $500 \times g$ for 5 min. Discard the supernatant (the storage medium contains dimethyl sulfoxide, DMSO, that may slow down cell growth).
   1.3 Resuspend the cell pellet in 1 mL of supplemented medium. Use gentle pipetting in all cell-handling steps. Transfer resuspended cells to the flask containing the medium.
   1.4 Incubate the culture at 27 °C with 90 rpm in a shaker incubator.
2. Monitoring of cells, Days 2–5
   2.1 After 24 h of incubation, take out an appropriate volume (100 µL) of culture and place in an Eppendorf tube. Analyze the samples either in a Vi-CELL cell analyzer or follow the manual procedure: Add 10 µL of 0.4% Trypan blue stain, mix well, and transfer a small volume into a Neubauer's chamber. Count the cells using an inverted microscope. Count the number of live cells (appear white) and dead cells (appear blue). Calculate the total number of viable cells per mL of culture and, hence, the percentage viability.
   2.2 As soon as cells are growing in logarithmic phase, passage them twice a week by diluting the culture 10-fold using fresh medium. Keep starting cell density between 0.8×10^6 and 1.2×10^6 cells/mL. When cells are in earlier passages, it is worth preparing a stock cell line for future use.
   2.3 If the viability remains less than 95% for three consecutive passages, the culture is not suitable for further work.

6.2.2 Generation of Stock Cell Lines

Cell lines can be stored for long periods of time if grown and stored properly. We recommend to back up stocks every year in case of problems that may arise in the running stock. Preparation of 10 vials of stock, each with 1 mL of culture, is described here. It can be expanded according to need.

1. Preparation of cell stock aliquots for freezing, Day 1
   1.1 Examine the Sf9 culture as described previously (see Section 6.2.1). The viability of the culture should be higher than 95% and in an early exponential growth, that is, $1-1.5 \times 10^6$ cells/mL, for the purpose of storage. If cells are overgrown, dilute back to 0.8×10^6 cells/mL and incubate to achieve desired density.
   1.2 Prepare 10 mL of freezing medium by mixing 6 mL (60% v/v) Grace's insect (Invitrogen, Life Technologies) or other medium, 3 mL (30% v/v) FBS, and 1 mL (10% v/v) DMSO. Bring the medium to room temperature.
   1.3 Transfer 100 mL of cell culture into two 50-mL Falcon tubes. Centrifuge at $500 \times g$ for 5 min. During this time, take 10 cryovials and mark each with the name of the cell line, the date, expected cell density (e.g., 1×10^7 cells), and original batch number from the supplier. Keep the vials on ice.
   1.4 Discard the medium aseptically after centrifugation. Resuspend the pellets from both tubes in 10 mL of freezing medium; count should be around 1×10^7 cells/mL. Dispense 1 mL of the suspension into each cryovial.
   1.5 Place the vials in a CryoCane and freeze gradually by placing at $-20\ °C$ for 1–2 h, transfer to $-80\ °C$ for 2 days, and finally transfer to liquid nitrogen for long-term storage. Slow freezing is a critical step for the long-term viability of insect cells.
2. Quality control of stock cell lines, 2 weeks later
   2.1 After 2 weeks, thaw one vial according to Section 6.2.1 to check viability and absence of contamination.

6.3 Transfection and Infection of Insect Cells with Baculovirus

1. Transfection of cells, Day 1
   1.1 Grow Sf9 cells to mid-log phase ($3-4 \times 10^6$ cells/mL) in a medium supplemented with 1% (v/v) FBS.

1.2 Dilute to 2×10^5 cells/mL. Dispense 1 mL (2×10^5 cells) into each well of four 24-well tissue culture (TC) plates. Include one well for cellfectin and another for untreated cell controls. Incubate the plate at 27 °C for 0.5–1 h to allow cell attachment.

1.3 Bring the Grace's insect medium or other medium (serum free) and cellfectin to room temperature.

1.4 Take a sterile flat-bottomed 96-well microtiter plate (that can hold ∼200 µL of sample per well). Dispense 50 µL of Grace's insect medium into each well. Transfer 10 µL of recombinant bacmid DNA (containing in total 0.5–1 µg of DNA) for each transfection and mix by shaking the plate gently or pipetting.

1.5 Mix cellfectin thoroughly by tapping the tube gently. In a tube, mix 50 µL of medium and 5 µL of cellfectin per well. Mix thoroughly and add to the mix prepared in step 1.4. For control, do not add cellfectin mixture but add an equivalent amount of medium. Mix briefly by tapping the plates. Cover the plate and incubate for 45 min at room temperature.

1.6 Dilute the solution by adding 100 µL of medium without serum.

1.7 Take out the 24-well tissue culture plate containing cells from step 1.2, aspirate the medium using a pipette. Wash once with 1 mL of serum-free medium. Remove the medium and overlay the cells with diluted lipid–DNA complex. Add a further 100 µL of medium and incubate cells for 5 h at 27 °C.

1.8 Remove the transfection mixture and add 0.7 mL of Sf-900 III insect medium containing 1% (v/v) FBS and antibiotics (50 U penicillin and 50 µg streptomycin per mL medium) to each well. Incubate cells at 27 °C for 72–96 h.

2. Monitoring of cells, Days 2–4

   2.1 Look for signs of infection in the transfected cells 72 h post-transfection by comparing with the control cells under an inverted microscope. Confluent growth of cells will be seen in control wells, whereas areas of clearing will be prominent in wells with infected cells. Infected cells usually are larger and deformed or elongated compared to uninfected cells. Collect viruses when the cells are well infected; this may take up to 96 h or more.

   2.2 Transfer the liquid contents from the 24-well tissue culture plate into a sterile container and centrifuge at $1500 \times g$ for 20 min at room temperature. Collect the clear supernatant in another sterile 96-deep-well block or suitable tubes. This is the P0 viral stock. Store at 4 °C, protected from light.

6.3.1 Virus Amplification

The P0 virus stock obtained after successful transfection usually has a low viral titer. One or two rounds of virus amplification are recommended in order to obtain a large enough volume of high-titer virus for expression and protein production.

1. Generation of P1 viruses, Day 1
   1.1. Take 3 mL of Sf9 cells grown to $1.5–1.8 \times 10^6$ cells/mL for each amplification. We routinely use 120 µL of P0 virus to infect 3 mL culture.
   1.2. Incubate infected cells at 27 °C at 220 rpm for 48 h.
2. Monitoring of infection process, Days 2–3
   2.1 Monitor the infected cells under inverted microscope or Vi-CELL device for increase in cell diameter and decrease in viability. Typically, cell diameter increases by 2–3 µm and viability decreases below 90% with no or slight change in cell density.
   2.2 Centrifuge at $1500 \times g$ for 20 min at room temperature. Collect the clear supernatant in sterile container. This is the P1 viral stock. Store at 4 °C, protected from light.
3. Generation of P2 virus, Days 4–5
   3.1 Take 100 mL of Sf9 cells grown to $1.5–1.8 \times 10^6$ cells/mL in an Erlenmeyer flask with vented cap for each amplification.
   3.2 Add 100 µL of P1 virus and incubate at 27 °C at 90 rpm for 48 h.
   3.3 Observe the infected cells under an inverted microscope or Vi-CELL device for signs of infection.
   3.4 Centrifuge at $1500 \times g$ for 20 min at room temperature. Collect the clear supernatant in a sterile container. This is the P2 viral stock. Store at 4 °C, protected from light. This stock serves as the working stock for all necessary expression tests.

6.4 General Protocol for Amino Acid-Type Selective Isotope Labeling in Insect Cells

The following protocol is based on a previously published one (Gossert et al., 2011). It has however been simplified through the use of so-called dropout media which are now available from different vendors (BioConcept and Gibco/Life Technologies). Dropout media are custom media, where one or several amino acids are removed from the formulation. For the following protocol, a medium should be ordered, where the amino acids to be labeled, as well as yeast extract, are missing. For the more general protocol, where all amino acids are added to an amino acid-free medium the reader is referred to the supplementary material of Gossert et al. (2011).

This method is applicable for amino acid-specific labeling of proteins and proved to be usable also for dual (^{13}CO-AAy/^{15}N-AAx–) or multiple (^{15}N-AAn–) labeling of proteins.

6.4.1 Recipe for 1 L of Medium for Amino Acid-Type Selective Labeling

990	mL	Amino acid dropout medium without yeast extract[a,b]
10	mL	Yeast extract filtrate 6%[c]
x	mg	Labeled amino acid[d]

Add a clean magnetic stirrer to the medium and stir for approximately 10 min until amino acids are dissolved. If needed, heat to 30 °C.

Check pH 6.2 ± 0.2 by measuring a 5 mL aliquot offline. If needed adjust with NaOH or HCl.

Filter-sterilize the medium through 0.22 μm pores into a sterile bottle and store closed at 4 °C.

[a]See materials above for ordering information.
[b]Experiment-dependent additives, e.g., compounds (if additives containing amino acids, e.g., FBS are necessary, the influence on the incorporation rate of the labeled amino acid must be determined before in small-scale experiments.)
[c]Stock solution prepared from 3 g of yeast extract powder in 50 mL of amino acid dropout medium and sterile filtered.
[d]Use amounts given in Fig. 4 as a guide. Depending on the cost of the labeled amino acid, the amount can be reduced taking into consideration the content of the unlabeled amino acid in yeast extract and amino acid metabolism.

6.4.2 Protocol

0. Preparation of starter culture, Day −2 (Friday)
    0.1 Prepare a starter culture (Sf9 or Sf21 cells in SF-4, ExCell420, or Sf-900 III) by passaging to a density of $0.7–1.2 \times 10^6$ cells/mL in 250 mL of working volume.
    0.2 Incubate at 27 °C over the weekend.
1. Start of culture, Day 1 (Monday)
    1.1 Prepare the medium according to the recipe given above.
    1.2 Prewarm the medium to 27 °C for at least 15 min.
    1.3 In a sterile hood, prepare the culture flasks with the amount of medium you would like to use.
    1.4 Measure the Vi-CELL parameters of your precultures. Ideally, the cell density is between 2 and 6×10^6 cells/mL and viability is 92–96%.

1.5 Transfer 1.5×10^6 cells into sterile 250 ml Corning centrifuge bottles.

1.6 Spin down for 3 min at $300\text{–}400 \times g$. Do not spin for too long, as cell viability suffers.

1.7 Carefully decant the medium immediately after centrifugation and discard it. (The pellet is not compact, but enough so, to be able to decant the supernatant.)

1.8 Resuspend the cell pellet in 50 mL per centrifuge bottle of warm (27 °C) amino acid dropout medium or phosphate buffered saline (PBS) in order to wash away remaining unlabeled medium. Centrifuge again for 5 min at $300 \times g$ and 27 °C.

This step improves the labeling ratio by about 5%, depending on how efficiently the culture medium can be removed from the pellet in the previous step. Cell viability and likewise the productivity of the culture may however be compromised by this additional handling. This step may therefore be omitted.

1.9 Carefully decant supernatant and discard it.

1.10 Immediately resuspend cells in 50–100 mL of prewarmed (27 °C) labeling medium from the culture flask.

1.11 Transfer cells into culture flask. Rinse centrifuge bottle with labeling medium if needed.

1.12 Add virus to multiplicity of infection (MOI) needed.

The more concentrated the virus stock the better, as virus stock contains unlabeled amino acids. Therefore, virus titer should be as high as possible, e.g., $0.5\text{–}3.0 \times 10^8$ pfu/ml. The optimum amount of virus or MOI should be identified in a small-scale experiment.

2. Expression, Days 2–4

    2.1 Let cells grow and express for your preferred time, typically 72 h.
    Expression time course should be monitored previously in a small-scale experiment.

3. Harvest, Day 4 (Thursday)

    3.1 Harvest the cells by centrifugation for 15 min at $1500 \times g$ and 4 °C.

    3.2 Freeze the cell pellet at -80 °C or process supernatant if the target protein is secreted.

6.5 General Protocol for Uniform Isotope Labeling in Insect Cells

The following protocol is based on our in-house protocol, closely following the one published by Sitarska et al. (2015).

6.5.1 Recipe for 1 L of Medium for Uniform Isotope Labeling in Insect Cells

1	L	Amino acid-free basal medium[a]
10	g	ISOGRO® algal amino acid extract with desired labeling pattern
10	g	D-Glucose (or 5 g of ^{13}C-glucose)[b]
2	mL	Tryptophan stock solution[c] (10 mg/mL) with desired labeling pattern
833	µL	^{15}NH$_4$Cl (300 mg/mL)
1	mL	L-cycloserine (5 mg/mL)[d]

Dissolve ISOGRO powder and glucose in basal medium by stirring with a magnetic bar for 15 min at room temperature. If needed, sonicate medium to fully dissolve any solids.

Add remaining stock solutions and stir for another minute.

Filter-sterilize the medium through 0.22 µm pores into a sterile bottle and store closed at 4 °C.

[a]See materials above for ordering information.
[b]In order to improve ^{13}C labeling ratio from 75% to 80%, i.e., for backbone assignment experiments, ^{13}C-labeled glucose can be used. It is incorporated primarily not only into Ala but also into Glu, Gln, Asp, and Asn by insect cell metabolism. Only 5 g of glucose may be used in this case. For SF-4-based media, generally only 5 g of glucose is used.
[c]The tryptophan stock solution should always be stored protected from light at 4 °C. The stock solution is at the limit of solubility and should be prepared at room temperature.
[d]For ^{13}C labeling and ^{2}H labeling, the use of L-cycloserine is encouraged when nonlabeled glucose is used. It will ensure full ^{13}C and ^{2}H labeling of alanine by inhibiting the enzyme (alanine transaminase) producing alanine from unlabeled pyruvate derived from glucose. For ^{2}H labeling, it will additionally reduce the amount of deuterated α-ketoglutarate produced in this reaction, leading to less inhibition of the citric acid cycle and therefore improve cell viability.

6.5.2 Protocol

0. Preparation of starter culture, Day −2 (Friday)
  0.1 Prepare a starter culture (Sf9 or Sf21 cells in SF-4 medium or in Sf-900 III) by passaging to a density of 0.7–1.2×10^6 cells/mL in 250 mL of working volume.
  0.2 Incubate at 27 °C over the weekend.

1. Infection, Day 1 (Monday)
  1.1 9:00 AM: Dilute starter cell culture (density around 8.0×10^6 cells/mL) with fresh full medium to a final volume of 1 L and approximately 1.5×10^6 cells/mL in a culturing bottle with vented cap.
  1.2 Incubate for 8 h under the same conditions as for the starter culture.
  1.3 Meanwhile, prepare the expression medium for isotope labeling according to the recipe above.

1.4 5:00 PM: Take the preincubated 1 L cell culture and infect it with the respective baculovirus. Infection ratio (multiplicity of infection, MOI) depends on the potency of a particular viral construct, which should be evaluated experimentally beforehand.

Infection has to be performed in the full culturing medium. Ammonium chloride, which is a component of the labeling medium, inhibits the viral entry and renders viral infection inefficient.

1.5 Incubate the infected culture overnight (16 h) under the same parameters.

Protein overexpression induced by the polyhedrin promoter starts 24–48 h post-infection. Therefore, 16 h of incubation post-infection in full medium does not significantly compromise the incorporation of isotopic label. However, it greatly increases protein yields and ensures reproducibility by allowing efficient viral entry.

2. Medium change, Day 2 (Tuesday)

2.1 9:00 AM: Prewarm the labeling medium to 27 °C in a water bath for at least 15 min.

If washing of the cells for increased isotope incorporation is planned, also prewarm basal medium or PBS to 27 °C. The amount needed is about one fourth of the volume of the expression culture.

2.2 Check cell culture parameters (cell count, cell diameter, and viability) of the infected culture before centrifuging.

2.3 Transfer the cell suspension into four 500 mL centrifuge bottles and spin them for 5 min at $300 \times g$ and 27 °C.

When 500 mL centrifuge bottles are only half filled, the cell suspension spin lasts shorter and requires lower centrifugal force. The cell pellet, while still not truly compact, is yet more stable while handling.

2.4 Carefully decant supernatant and discard it.

2.5 Resuspend the cell pellet in 50 mL per centrifuge bottle of warm (27 °C) basal medium or PBS in order to wash away remaining unlabeled medium. Centrifuge again for 5 min at $300 \times g$ and 27 °C.

This step improves the labeling ratio by about 5%, depending on how efficiently the culture medium can be removed from the pellet in the previous step. Cell viability and likewise the productivity of the culture may however be compromised by this additional handling. This step may therefore be omitted.

2.6 Carefully decant supernatant and discard it.
2.7 Resuspend the cells immediately in prewarmed labeling medium (27 °C).
2.8 Check cell count, cell diameter, and viability of the infected culture after these manipulations.
2.9 Incubate for additional 72 h or until viability drops below 60%.
3. Monitoring of culture, Days 3–5 (Wednesday–Friday)
3.1 Regularly monitor cell culture parameters (cell count, cell diameter, and viability).

Viability drop and increased cell diameter over time indicate an efficient viral infection. For cytosolic proteins, viability should not fall below 60%, because subcellular compartmentalization in dying cells will be weakened, leading to elevated proteolysis in the cytosol. The cell count should also not drop significantly, because lysed cells will release protein into the supernatant. Therefore, a decrease in culture viability compromises protein yields. For optimal results, expression should be continued for 56–80 h from the medium change point, which is 72–96 h post-infection. The infection ratio with the particular baculovirus should be adjusted accordingly.

4. Harvest, Day 5 or earlier (Friday)
4.1 5:00 PM (or when viability drops below 60%): Harvest the cells by centrifugation for 15 min at $1500 \times g$ and 4 °C.
4.2 Freeze the cell pellet at -80 °C or process supernatant if the target protein is secreted.

REFERENCES

Abraham, S. J., Hoheisel, S., & Gaponenko, V. (2008). Detection of protein–ligand interactions by NMR using reductive methylation of lysine residues. *Journal of Biomolecular NMR, 42,* 143–148.

Adriaensens, P., Box, M. E., Martens, H. I., Onkelinx, E., Put, J., & Gelan, J. (1988). Investigation of protein structure by means of ^{19}F-NMR. *European Journal of Biochemistry, 177,* 383–394.

Airenne, K. J., Hu, Y.-C., Kost, T. A., Smith, R. H., Kotin, R. M., Ono, C., et al. (2013). Baculovirus: An insect-derived vector for diverse gene transfer applications. *Molecular Therapy, 21,* 739–749.

Assenberg, R., Wan, P. T., Geisse, S., & Mayr, L. M. (2013). Advances in recombinant protein expression for use in pharmaceutical research. *Current Opinion in Structural Biology, 23,* 393–402.

Bertrand, K., Reverdatto, S., Burz, D. S., Zitomer, R., & Shekhtman, A. (2012). Structure of proteins in eukaryotic compartments. *Journal of the American Chemical Society, 134,* 12798–12806.

Bieniossek, C., Imasaki, T., Takagi, Y., & Berger, I. (2012). MultiBac: Expanding the research toolbox for multiprotein complexes. *Trends in Biochemical Sciences, 37,* 49–57.

Brüggert, M., Rehm, T., Shanker, S., Georgescu, J., & Holak, T. A. (2003). A novel medium for expression of proteins selectively labeled with ^{15}N-amino acids in Spodoptera frugiperda (Sf9) insect cells. *Journal of Biomolecular NMR, 25*, 335–348.

Burges, H., Croizier, G., & Huger, J. (1980). A review of safety tests on baculoviruses. *Entomaphaga, 25*, 329–340.

Creemers, A. F. L., Klaassen, C. H. W., Bovee-Geurts, P. H. M., Kelle, R., Kragl, U., Raap, J., et al. (1999). Solid state ^{15}N NMR evidence for a complex Schiff base counterion in the visual G-protein-coupled receptor rhodopsin. *Biochemistry, 38*, 7195–7199.

Cremer, H., Bechtold, I., Mahnke, M., & Assenberg, R. (2014). Efficient processes for protein expression using recombinant baculovirus particles. In R. Pörtner (Ed.), *Animal cell biotechnology: Vol. 1104* (pp. 395–417). Totowa, NJ: Humana Press.

Dabydeen, S. A., & Meneses, P. I. (2009). The role of NH$_4$Cl and cysteine proteases in human papillomavirus type 16 infection. *Virology Journal, 6*, 109.

DeLange, F., Klaassen, C. H., Wallace-Williams, S. E., Bovee-Geurts, P. H., Liu, X.-M., DeGrip, W. J., et al. (1998). Tyrosine structural changes detected during the photoactivation of rhodopsin. *Journal of Biological Chemistry, 273*, 23735–23739.

Dick, L. R., Sherry, A. D., Newkirk, M. M., & Gray, D. M. (1988). Reductive methylation and ^{13}C NMR studies of the lysyl residues of fd gene 5 protein. Lysines 24, 46, and 69 may be involved in nucleic acid binding. *Journal of Biological Chemistry, 263*, 18864–18872.

Dong, S., Wang, M., Qiu, Z., Deng, F., Vlak, J. M., Hu, Z., et al. (2010). Autographa californica multicapsid nucleopolyhedrovirus efficiently infects Sf9 cells and transduces mammalian cells via direct fusion with the plasma membrane at low pH. *Journal of Virology, 84*, 5351–5359.

Doverskog, M., Han, L., & Häggström, L. (1998). Cystine/cysteine metabolism in cultured Sf9 cells: Influence of cell physiology on biosynthesis, amino acid uptake and growth. *Cytotechnology, 26*, 91–102.

Doverskog, M., Jacobsson, U., Chapman, B. E., Kuchel, P. W., & Häggström, L. (2000). Determination of NADH-dependent glutamate synthase (GOGAT) in Spodoptera frugiperda (Sf9) insect cells by a selective ^{1}H/^{15}N NMR in vitro assay. *Journal of Biotechnology, 79*, 87–97.

Drews, M., Doverskog, M., Öhman, L., Chapman, B. E., Jacobsson, U., Kuchel, P. W., et al. (2000). Pathways of glutamine metabolism in Spodoptera frugiperda (Sf9) insect cells: Evidence for the presence of the nitrogen assimilation system, and a metabolic switch by ^{1}H/^{15}N NMR. *Journal of Biotechnology, 78*, 23–37.

Egorova-Zachernyuk, T. A., Bosman, G. J., & DeGrip, W. J. (2011). Uniform stable-isotope labeling in mammalian cells: Formulation of a cost-effective culture medium. *Applied Microbiology and Biotechnology, 89*, 397–406.

Egorova-Zachernyuk, T. A., Bosman, G. J., Pistorius, A. M., & DeGrip, W. J. (2009). Production of yeastolates for uniform stable isotope labelling in eukaryotic cell culture. *Applied Microbiology and Biotechnology, 84*, 575–581.

Fitzgerald, D. J., Berger, P., Schaffitzel, C., Yamada, K., Richmond, T. J., & Berger, I. (2006). Protein complex expression by using multigene baculoviral vectors. *Nature Methods, 3*, 1021–1032.

Fraser, M. (1986). Ultrastructural observations of virion maturation in Autographa californica nuclear polyhderosis virus infected Spodoptera frugiperda cell cultures. *Journal of Ultrastructure and Molecular Structure Research, 95*, 189–195.

Gossert, A. D., Hinniger, A., Gutmann, S., Jahnke, W., Strauss, A., & Fernández, C. (2011). A simple protocol for amino acid type selective isotope labeling in insect cells with improved yields and high reproducibility. *Journal of Biomolecular NMR, 51*, 449–456.

Gossert, A. D., & Jahnke, W. (2012). Isotope labeling in insect cells. In H. S. Atreya (Ed.), *Isotope labeling in biomolecular NMR: Vol. 992* (pp. 179–196). Dordrecht: Springer.

Hamatsu, J., O'Donovan, D., Tanaka, T., Shirai, T., Hourai, Y., Mikawa, T., et al. (2013). High-resolution heteronuclear multidimensional NMR of proteins in living insect cells using a baculovirus protein expression system. *Journal of the American Chemical Society, 135*, 1688–1691.

Hansen, A. P., Petros, A. M., Mazar, A. P., Pederson, T. M., Rueter, A., & Fesik, S. W. (1992). A practical method for uniform isotopic labeling of recombinant proteins in mammalian cells. *Biochemistry, 31*, 12713–12718.

Harner, M. J., Chauder, B. A., Phan, J., & Fesik, S. W. (2014). Fragment-based screening of the bromodomain of ATAD2. *Journal of Medicinal Chemistry, 57*, 9687–9692.

Harner, M. J., Frank, A. O., & Fesik, S. W. (2013). Fragment-based drug discovery using NMR spectroscopy. *Journal of Biomolecular NMR, 56*, 65–75.

Hebel, D., Kirk, K. L., Cohen, L. A., & Labroo, V. M. (1990). First direct fluorination of tyrosine-containing biologically active peptides. *Tetrahedron Letters, 31*, 619–622.

Hefferon, K. L., Oomens, A. G. P., Monsma, S. A., Finnerty, C. M., & Blissard, G. W. (1999). Host cell receptor binding by baculovirus GP64 and kinetics of virion entry. *Virology, 258*, 455–468.

Horst, R., Liu, J. J., Stevens, R. C., & Wüthrich, K. (2013). β2-adrenergic receptor activation by agonists studied with ^{19}F NMR spectroscopy. *Angewandte Chemie, International Edition, 52*, 10762–10765.

Invitrogen. (2002). Growth and maintenance of insect cell lines. In *Invitrogen life technologies*, Paisley, UK.

Jahnke, W., Grotzfeld, R. M., Pellé, X., Strauss, A., Fendrich, G., Cowan-Jacob, S. W., et al. (2010). Binding or bending: Distinction of allosteric Abl kinase agonists from antagonists by an NMR-based conformational assay. *Journal of the American Chemical Society, 132*, 7043–7048.

Jarvis, D. L., & Garcia, A. J. (1994). Long-term stability of baculoviruses stored under various conditions. *BioTechniques, 16*, 508–513.

Kainosho, M., & Tsuji, T. (1982). Assignment of the three methionyl carbonyl carbon resonances in streptomyces subtilisin inhibitor by a carbon-13 and nitrogen-15 double-labeling technique. A new strategy for structural studies of proteins in solution. *Biochemistry, 21*, 6273–6279.

Kärkkäinen, H.-R., Lesch, H. P., Määttä, A. I., Toivanen, P. I., Mähönen, A. J., Roschier, M. M., et al. (2009). A 96-well format for a high-throughput baculovirus generation, fast titering and recombinant protein production in insect and mammalian cells. *BMC Research Notes, 2*, 63.

Kofuku, Y., Ueda, T., Okude, J., Shiraishi, Y., Kondo, K., Maeda, M., et al. (2012). Efficacy of the β2-adrenergic receptor is determined by conformational equilibrium in the transmembrane region. *Nature Communications, 3*, 1045.

Kofuku, Y., Ueda, T., Okude, J., Shiraishi, Y., Kondo, K., Mizumura, T., et al. (2014). Functional dynamics of deuterated β2-adrenergic receptor in lipid bilayers revealed by NMR spectroscopy. *Angewandte Chemie, International Edition, 53*, 13376–13379.

Kost, T. A., & Condreay, J. P. (2002). Innovations-biotechnology: Baculovirus vectors as gene transfer vectors for mammlian cells: Biosafety considerations. *Journal of the American Biological Safety Association, 7*, 167–169.

Kost, T. A., Condreay, J. P., & Jarvis, D. L. (2005). Baculovirus as versatile vectors for protein expression in insect and mammalian cells. *Nature Biotechnology, 23*, 567–575.

Lee, Y., Abraham, S. J., & Gaponenko, V. (2012). The use of reductive methylation of lysine residues to study protein-protein interactions in high molecular weight complexes by solution NMR. In J. Cai (Ed.), *Protein Interactions* (pp. 45–52). Shanghai: InTech.

Lehninger, A. L. (1975). Chapter 22: Biosynthesis of amino acids, nucleotides, and related molecules. In *Principles of Biochemistry* (pp. 833–880). New York, NY: Worth Publishers Inc.

LeMaster, D. M., & Richards, F. M. (1982). Preparative-scale isolation of isotopically labeled amino acids. *Analytical Biochemistry, 122*, 238–247.

Li, M., Smith, C. J., Walker, M. T., & Smith, T. J. (2009). Novel inhibitors complexed with glutamate dehydrogenase: Allosteric regulation by control of protein dynamics. *Journal of Biological Chemistry, 284*, 22988–23000.

Liu, J. J., Horst, R., Katritch, V., Stevens, R. C., & Wüthrich, K. (2012). Biased signaling pathways in β2-adrenergic receptor characterized by ^{19}F-NMR. *Science, 335*, 1106–1110.

Luckow, V. A., Lee, S. C., Barry, G. F., & Olins, P. O. (1993). Efficient generation of infectious recombinant baculoviruses by site-specific transposon-mediated insertion of foreign genes into a baculovirus genome propagated in Escherichia coli. *Journal of Virology, 67*, 4566–4579.

Markley, J. L., Putter, I., & Jardetzky, O. (1968). High-resolution nuclear magnetic resonance spectra of selectively deuterated staphylococcal nuclease. *Science, 161*, 1249–1251.

Mehta, V. D., Kulkarni, P. V., Mason, R. P., Constantinescu, A., & Antich, P. P. (1994). Fluorinated proteins as potential 19F magnetic resonance imaging and spectroscopy agents. *Bioconjugate Chemistry, 5*, 257–261.

Meola, A., Deville, C., Jeffers, S. A., Guardado-Calvo, P., Vasiliauskaite, I., Sizun, C., et al. (2014). Robust and low cost uniform ^{15}N-labeling of proteins expressed in Drosophila S2 cells and Spodoptera frugiperda Sf9 cells for NMR applications. *Journal of Structural Biology, 188*, 71–78.

Mitsuhashi, J. (1982). Determination of essential amino acids for insect cell lines. In K. Maramorosch & J. Mitsuhashi (Eds.), *Invertebrate cell culture applications* (pp. 9–52). New York, NY: Academic Press Inc.

Moore, G., Cox, M., Crowe, D., Osborne, M., Rosell, F., Bujons, J., et al. (1998). Nepsilon, nepsilon-dimethyl-lysine cytochrome c as an NMR probe for lysine involvement in protein–protein complex formation. *The Biochemical Journal, 332*, 439–449.

Morita, E. H., Shimizu, M., Ogasawara, T., Endo, Y., Tanaka, R., & Kohno, T. (2004). A novel way of amino acid-specific assignment in ^{1}H-^{15}N HSQC spectra with a wheat germ cell-free protein synthesis system. *Journal of Biomolecular NMR, 30*, 37–45.

Nelson, D. J. (1978). Fluorine-19 magnetic resonance of muscle calcium binding parvalbumin: pH dependency of resonance position and spin-lattice relaxation time. *Inorganica Chimica Acta, 27*, L71–L74.

Nygaard, R., Zou, Y., Dror, R. O., Mildorf, T. J., Arlow, D. H., Manglik, A., et al. (2013). The dynamic process of β2-adrenergic receptor activation. *Cell, 152*, 532–542.

O'Reilly, D. R., Miller, L. K., & Luckow, V. A. (1992). *Baculovirus expression vectors: A laboratory manual*. New York: W.H. Freeman and Co.

Öhman, L., Alarcon, M., Ljunggren, J., Ramqvist, A.-K., & Häggström, L. (1996). Glutamine is not an essential amino acid for Sf-9 insect cells. *Biotechnology Letters, 18*, 765–770.

Öhman, L., Ljunggren, J., & Häggström, L. (1995). Induction of a metabolic switch in insect cells by substrate-limited fed batch cultures. *Applied Microbiology and Biotechnology, 43*, 1006–1013.

Ong, S.-E., & Mann, M. (2007). A practical recipe for stable isotope labeling by amino acids in cell culture (SILAC). *Nature Protocols, 1*, 2650–2660.

Possee, R. D., Hitchman, R. B., Richards, K. S., Mann, S. G., Siaterli, E., Nixon, C. P., et al. (2008). Generation of baculovirus vectors for the high-throughput production of proteins in insect cells. *Biotechnology and Bioengineering, 101*, 1115–1122.

Rayment, I. (1997). Reductive alkylation of lysine residues to alter crystallization properties of proteins. *Methods in Enzymology, 276*, 171–179.

Rej, R. (1977). Aminooxyacetate is not an adequate differential inhibitor of aspartate aminotransferase isoenzymes. *Clinical Chemistry, 23*, 1508–1509.

Religa, T. L., Ruschak, A. M., Rosenzweig, R., & Kay, L. E. (2011). Site-directed methyl group labeling as an NMR probe of structure and dynamics in supramolecular protein systems: Applications to the proteasome and to the ClpP protease. *Journal of the American Chemical Society, 133*, 9063–9068.

Rieffel, S., Roest, S., Klopp, J., Carnal, S., Marti, S., Gerhartz, B., et al. (2014). Insect cell culture in reagent bottles. *MethodsX*, *1*, 155–161.
Sakai, T., Tochio, H., Tenno, T., Ito, Y., Kokubo, T., Hiroaki, H., et al. (2006). In-cell NMR spectroscopy of proteins inside Xenopus laevis oocytes. *Journal of Biomolecular NMR*, *36*, 179–188.
Selenko, P., Serber, Z., Gadea, B., Ruderman, J., & Wagner, G. (2006). Quantitative NMR analysis of the protein G B1 domain in Xenopus laevis egg extracts and intact oocytes. *Proceedings of the National Academy of Sciences of the United States of America*, *103*, 11904–11909.
Serber, Z., Keatinge-Clay, A. T., Ledwidge, R., Kelly, A. E., Miller, S. M., & Dötsch, V. (2001). High-resolution macromolecular NMR spectroscopy inside living cells. *Journal of the American Chemical Society*, *123*, 2446–2447.
Shrestha, B., Smee, C., & Gileadi, O. (2008). Baculovirus expression vector system: An emerging host for high-throughput eukaryotic protein expression. In Mike Starkey & Ramnath Elaswarapu (Eds.), *Genomics Protocols* (pp. 269–289). Totowa, NJ: Humana Press Inc.
Sitarska, A., Skora, L., Klopp, J., Roest, S., Fernández, C., Shrestha, B., et al. (2015). Affordable uniform isotope labeling with ^{2}H, ^{13}C and ^{15}N in insect cells. *Journal of Biomolecular NMR*, *62*, 191–197.
Strauss, A., Bitsch, F., Cutting, B., Fendrich, G., Graff, P., Liebetanz, J., et al. (2003). Amino-acid-type selective isotope labeling of proteins expressed in Baculovirus-infected insect cells useful for NMR studies. *Journal of Biomolecular NMR*, *26*, 367–372.
Strauss, A., Bitsch, F., Fendrich, G., Graff, P., Knecht, R., Meyhack, B., et al. (2005). Efficient uniform isotope labeling of Abl kinase expressed in Baculovirus-infected insect cells. *Journal of Biomolecular NMR*, *31*, 343–349.
Summers, M., & Anderson, D. (1972). Characterization of deoxyribonucleic acid isolated from the granulosis viruses of the cabbage looper, Trichoplusia ni and the fall armyworm, Spodoptera frugiperda. *Virology*, *50*, 459–471.
Sweeney, P. J., & Walker, J. M. (1993a). Aminopeptidases. In M. M. Burrell (Ed.), *Enzymes of molecular biology* (pp. 319–329). Totowa, NJ: Humana Press Inc.
Sweeney, P. J., & Walker, J. M. (1993b). Pronase (EC 3.4. 24.4). In M. M. Burrell (Ed.), *Enzymes of molecular biology* (pp. 271–276). Totowa, NJ: Humana Press Inc.
Vajpai, N., Strauss, A., Fendrich, G., Cowan-Jacob, S. W., Manley, P. W., Grzesiek, S., et al. (2008a). Solution conformations and dynamics of ABL kinase-inhibitor complexes determined by NMR substantiate the different binding modes of imatinib/nilotinib and dasatinib. *Journal of Biological Chemistry*, *283*, 18292–18302.
Vajpai, N., Strauss, A., Fendrich, G., Cowan-Jacob, S. W., Manley, P. W., Jahnke, W., et al. (2008b). Backbone NMR resonance assignment of the Abelson kinase domain in complex with imatinib. *Biomolecular NMR Assignments*, *2*, 41–42.
van Oers, M. M., Pijlman, G. P., & Vlak, J. M. (2015). Thirty years of baculovirus-insect cell protein expression: from dark horse to mainstream technology. *The Journal of General Virolology*, *96*, 6–23.
Vialard, J., Arif, B., & Richardson, C. D. (1995). Introduction to the molecular biology of baculoviruses. *Methods in Molecular Biology*, *39*, 1–24.
Walton, W. J., Kasprzak, A. J., Hare, J. T., & Logan, T. M. (2006). An economic approach to isotopic enrichment of glycoproteins expressed from Sf9 insect cells. *Journal of Biomolecular NMR*, *36*, 225–233.
Wasilko, D., & Lee, S. (2006). TIPS: Titerless infected-cells preservation and scale up. *BioProcessing Journal*, *5*, 29–32.
Wong, D. T., Fuller, R. W., & Molloy, B. B. (1973). Inhibition of amino acid transaminases by L-cycloserine. *Advances in Enzyme Regulation*, *11*, 139–154.
Yabuki, T., Kigawa, T., Dohmae, N., Takio, K., Terada, T., Ito, Y., et al. (1998). Dual amino acid-selective and site-directed stable-isotope labeling of the human c-Ha-Ras protein by cell-free synthesis. *Journal of Biomolecular NMR*, *11*, 295–306.

CHAPTER TWELVE

Effective Isotope Labeling of Proteins in a Mammalian Expression System

Mallika Sastry[*,1], Carole A. Bewley[†,1], Peter D. Kwong[*,1]

[*]Vaccine Research Center, National Institute of Allergy and Infectious Diseases, National Institutes of Health, Bethesda, Maryland, USA
[†]Laboratory of Bioorganic Chemistry, National Institute of Diabetes and Digestive and Kidney Diseases, National Institutes of Health, Bethesda, Maryland, USA
[1]Corresponding authors: e-mail address: sastrym@niaid.nih.gov; CaroleB@mail.nih.gov; pdkwong@nih.gov

Contents

1. Introduction	290
2. Overview of Mammalian Expression	291
2.1 Transient Protein Expression in Mammalian Cells	292
2.2 Transient Protein Expression Using Mammalian Viruses	295
3. Protein Expression	299
3.1 Selective Labeling of Specific Amino Acids	300
4. NMR Characterization of Expressed Protein	300
5. Conclusions	302
6. Materials	303
Acknowledgments	303
References	303

Abstract

Isotope labeling of biologically interesting proteins is a prerequisite for structural and dynamics studies by NMR spectroscopy. Many of these proteins require mammalian cofactors, chaperons, or posttranslational modifications such as myristoylation, glypiation, disulfide bond formation, or N- or O-linked glycosylation; and mammalian cells have the necessary machinery to produce them in their functional forms. Here, we describe recent advances in mammalian expression, including an efficient adenoviral vector-based system, for the production of isotopically labeled proteins. This system enables expression of mammalian proteins and their complexes, including proteins that require posttranslational modifications. We describe a roadmap to produce isotopically labeled ^{15}N and ^{13}C posttranslationally modified proteins, such as the outer domain of HIV-1 gp120, which has four disulfide bonds and 15 potential sites of N-linked glycosylation. These methods should allow NMR spectroscopic analysis of the structure and function of posttranslationally modified and secreted, cytoplasmic, or membrane-bound proteins.

ABBREVIATIONS

ATCC American Type Culture Collection
BGHpA bovine growth hormone polyadenylation signal
BHK21 baby-hamster-kidney cell line
CAR coxsackie-adenovirus receptor
CHO Chinese hamster ovary cells
CMV cytomegalovirus
DMEM Dulbecco's minimal eagle media
FBS dialyzed fetal bovine serum
HEK human embryonic kidney cells
HIV-1 human immunodeficiency virus type 1
PBS phosphate-buffered saline
PEI polyethyleneimine
SPR surface plasmon resonance

1. INTRODUCTION

NMR spectroscopic characterization of proteins and protein domains generally requires the incorporation of the stable isotopes ^{15}N, ^{13}C, and/or ^{2}H. Incorporating these isotopes is most easily done in prokaryotic systems; therefore, in most cases, full-length proteins or their individual domains are expressed and purified from prokaryotic expression systems, and a number of these methods are discussed in other chapters in this volume. Owing to structural genomics initiatives, large-scale expression trials have shown that only 10% of eukaryotic proteins can be expressed in their functional forms in *Escherichia coli* (Braun & LaBaer, 2003). This observation likely reflects the necessity of eukaryotic cofactors, chaperons, or posttranslational modifications such as myristoylation, disulfide bond formation, and glycosylation for proper folding and activity. Traditional eukaryotic systems such as insect, lower eukaryotes, and mammalian cells often produce low yields of the desired protein; in addition, these systems can be time consuming, prohibitively expensive, or both. Lower eukaryotes such as *Dictyostelium discoideum* (*D. discoideum*) have been used to produce isotopically enriched proteins in milligram quantities to allow resonance assignments from triple-resonance experiments (Cubeddu et al., 2000). *D. discoideum* has the potential to be an attractive expression system because isotopically enriched bacteria can be used as a nutrient source, thereby reducing the overall cost of isotopic labeling. However, heterologous expression of proteins in *D. discoideum* has its own challenges (Arya, Bhattacharya, & Saini, 2008); to date, there is only one published report describing successful production of an isotopically labeled protein using this system (Swarbrick et al., 2009). Thus, expression of isotopically enriched eukaryotic proteins in their stable, correctly folded form remains a bottleneck for spectroscopic characterization of many proteins. The development of large-scale mammalian transient expression systems along with advances in cell culture technologies have allowed milligram-to-gram-scale production of proteins suitable for structural analysis from Chinese hamster ovary (CHO) and human embryonic kidney (HEK) cells. Despite these advances, the absence of a suitable expression system than can incorporate labeling with high yield at a reasonable cost has stymied the production of isotopically enriched proteins suitable for NMR spectroscopy. We reported the adaptation of an efficient adenoviral vector-based mammalian expression system for the production of isotopically enriched cytoplasmic proteins and secreted glycoproteins (Sastry et al., 2011). This chapter gives a brief overview of mammalian expression systems and provides details for an adenoviral vector-based expression system through a case study of the human immunodeficiency virus type 1 (HIV-1) gp120 outer domain (OD) comprising 230 amino acids, 15 potential sites of N-linked glycosylation, and four disulfide bonds.

2. OVERVIEW OF MAMMALIAN EXPRESSION

Unlike prokaryotic systems, mammalian cells have the necessary cofactors, chaperons, and cellular machinery to produce correctly folded, posttranslationally modified functional proteins. Historically, heterologous expression of proteins using mammalian cells has proven difficult. However, the development of several eukaryotic expression systems— including yeast, *D. discoideum*, insect cells including Schneider cells, SF9, and silkworm-based baculovirus expression systems (Kato, Kajikawa, Maenaka, & Park, 2010; Unger & Peleg, 2012), and mammalian cells with their associated vectors which allow for the transient, inducible, or constitutive expression of a target gene (Kriz et al., 2010; Nettleship, Assenberg, Diprose, Rahman-Huq, & Owens, 2010; Trowitzsch, Klumpp, Thoma, Carralot, & Berger, 2011)—has allowed researchers to investigate the possibility of overexpressing proteins in eukaryotic systems. Large-scale protein production for therapeutic purposes generally uses constitutive or inducible expression from CHO, HEK, baby hamster kidney (BHK21), or human lymphoma (Namalwa) cells. Therefore, a number of research groups have succeeded in producing isotopically labeled proteins from mouse hybridoma or CHO stable cell lines using either a mixture of algal/bacterial hydrolysates (Hansen et al., 1992, 1994; Yamaguchi et al., 2006) or a mixture of labeled amino acids (Hinck et al., 1996; Lustbader et al., 1996; Shindo, Masuda, Takahashi, Arata, & Shimada, 2000; Wyss et al., 1995, 1993; Wyss, Dayie, & Wagner, 1997). However, these methods can be time consuming and prohibitively expensive leaving room for improvement in methodology (Table 1).

2.1 Transient Protein Expression in Mammalian Cells

Protein production using transient transfection (Kingston, Chen, & Okayama, 2001; Potter & Heller, 2003) is an attractive alternative method and has been used routinely in the production of posttranslationally modified proteins including antibodies, glycoproteins, and membrane proteins. Transient transfection of HEK293 cells with polyethyleneimine (PEI)-mediated transfections has been used to obtain milligram-to-gram quantities of secreted proteins (Backliwal et al., 2008; Coleman et al., 1997). Transient transfection can be achieved in adherent as well as suspension cells (Geisse & Fux, 2009; Hopkins, Wall, & Esposito, 2012) to produce biologically important proteins that are functionally active and has been used extensively in biophysical and crystallographic studies (Kwon et al., 2015; Pancera et al., 2014; Zhou et al., 2015). A number of articles describing transient transfections and stable cell line development have been published (Longo, Kavran, Kim, & Leahy, 2013a, 2013b). Factors governing

Table 1 Mammalian Expression Systems Used to Obtain Isotopically Enriched Proteins

Media	Cell Line	Protein Expression	^{15}N Media (mg/L)	^{15}N/^{13}C Media (mg/L)
Algal and bacterial mixture of amino acids[a]	Sp2/0	Stable	30	30
Algal mixture of amino acids[b]	CHO	Stable	10	10
CIL Bioexpress 6000 (^{15}N/^{15}N,^{13}C GKLQSTVW)[c]	HEK293	Stable	2	2.12
Commercial Media (CIL) CGM6000[d]	A549	Transient Adenoviral Expression	50	43

[a]Hansen et al. (1992).
[b]Lustbader et al. (1996).
[c]Werner, Richter, Klein-Seetharaman, and Schwalbe (2008).
[d]Sastry et al. (2011). *Adapted from Sastry et al. (2011) with permission from Springer.*

gene expression in eukaryotic cells are choice of expression vector, the presence of an optimized Kozak sequence upstream of the start codon, codon-optimization of your gene of interest and inclusion of targeting sequences such as signal peptide or endoplasmic reticulum localization sequences similar to requirements for prokaryotic expression systems discussed in preceding chapters. Protein expression is driven by a strong promoter such as the SV40 early promoter, the Rous sarcoma virus (RSV) promoter, or the cytomegalovirus (CMV) very early promoter (Xia et al., 2006). We provide a roadmap for overexpression and isotopic labeling of proteins from mammalian cells (Fig. 1) and discuss individual steps in the following sections.

2.1.1 Procedure for Transient Expression of Proteins and Protein Domains in Mammalian Cells

We outline a procedure for transient expression in this section and include relevant references and discuss details for production of isotopically enriched proteins using an adenoviral vector-based mammalian expression system with the HIV-1 gp120 OD as an example.

Design and clone the gene for the protein of interest (POI) in a mammalian expression vector of choice (Li et al., 2007; Van Craenenbroeck, Vanhoenacker, & Haegeman, 2000). A designed construct should include an upstream Kozak sequence such as gccgccA/GccATGG (Kozak, 1984, 1987), localization sequences particular to the protein (for example, a signal peptide for a secreted protein), a codon-optimized gene, and a poly A sequence 3' to the gene of interest (GOI). For our case study, the HIV-1 gp120 OD was cloned into a CMV/R vector (pVRC8400) (Wu et al., 2006), protein expression was under the control of the CMV promoter. For easier tracking of an expressed protein, individual researchers may utilize C-terminal fusion of POI with green fluorescent protein (GFP) or yellow fluorescent protein. As with bacterial expression optimizing, the N- and C-termini may influence expression levels and protein stability. It is also highly recommended to generate varying lengths of a construct and test each for optimal levels of expression and ideally, function.

2.1.2 Protein Expression and Construct Selection

It is recommended to use high-quality endotoxin-free plasmid DNA preferably at $1\,\mu g/\mu l$ concentration to transfect HEK293T or HEK293FS cells. Small-scale expression to screen constructs can be performed using Lipofectamine, PEI, True Fect max, or 293 Fectin using established methods

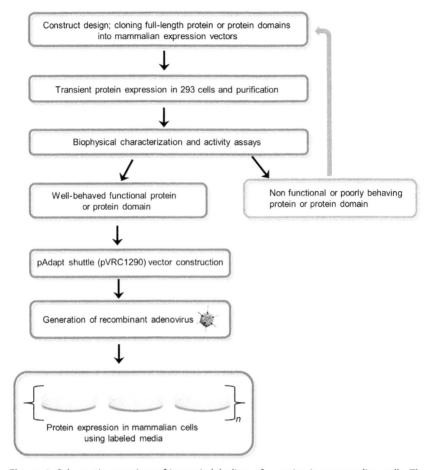

Figure 1 Schematic overview of isotopic labeling of proteins in mammalian cells. The gene of interest (GOI) is initially cloned into a suitable mammalian expression vector. The protein of interest is expressed using transient transfection and its functional and biophysical properties are analyzed. A construct either enters the adenoviral-based mammalian expression pathway to generate recombinant adenovirus followed by protein production or enters an iterative loop for further modifications to obtain an optimal construct.

(Longo et al., 2013b; McLellan et al., 2013). The amount of DNA and the ratio of DNA to transfection reagent may need to be screened for optimal expression of the POI. Protein expression of the target gene can be monitored and confirmed by western blots or BioLayer Interferometry during the duration of expression. Once protein expression is confirmed, the crude supernatant or cell lysate can be further evaluated for the presence of a correctly folded POI by surface plasmon resonance (SPR) analysis (Raghavan & Bjorkman, 1995; Rich & Myszka, 2010), BioLayer Interferometry, ELISAs, or an equivalent technique using, for example, a structure-dependent antibody.

An individual researcher may also choose to characterize the expressed protein further using $^{1}H-^{15}N$ and $^{1}H-^{13}C$ HSQC spectra at natural abundance (Chen, Freedberg, & Keire, 2015) as protein expression from adherent cells using ^{15}N, $^{15}N/^{13}C$-labeled media is feasible although potentially prohibitively expensive. Yet, additional essential steps in obtaining isotopically labeled proteins from mammalian cells include choosing optimal growth media for your cell line and tracking cell viability during the duration of protein expression (Sastry et al., 2011). Furthermore, a researcher may choose to investigate suitable media and cell types in the searchable database (http://www.goodcellculture.com). Despite the advances in cell culture technology and the high yields obtained with transient transfection, lack of suitable labeling media for suspension cells complicates production and screening of labeled proteins for NMR spectroscopic measurements. As a result, we chose to adapt and optimize transient protein production using a mammalian expression system that utilizes mammalian viruses to deliver the GOI.

2.2 Transient Protein Expression Using Mammalian Viruses

Mammalian viruses such as adenovirus, herpesviruses, poxviruses, papillomaviruses, and SV40 have been used to express heterologous proteins for almost four decades (Gluzman, Reichl, & Solnick, 1982; Howley, Sarver, & Law, 1983; Mulligan, Howard, & Berg, 1979; Southern & Berg, 1982; Warnock, Daigre, & Al-Rubeai, 2011). Adenoviruses used in our expression system are double-stranded DNA viruses (Berk, 1986) with a deletion of the E1 gene that renders it replication defective (Aoki, Barker, Danthinne, Imperiale, & Nabel, 1999; Sastry, Bewley, & Kwong, 2012). Recombinant adenovirus 5 (rAd5) initially binds cell surface coxsackie-adenovirus receptor (CAR), followed by an interaction with cellular integrins and subsequent internalization via receptor-mediated endocytosis. Adenovirus infection is epichromosomal in all known cells, and the replication-defective virus results in a very effective transfection system for introducing a functional gene into cells. Transfection with recombinant adenovirus is quick and simple, with a postinfection viability ~100%. Thus, a replication-defective recombinant adenovirus has the ability to transfer genetic material into cells with negligible apparent toxic effects. Viral promoters drive protein expression at the expense of production of cellular proteins and can increase the overall yield of the heterologous protein (Babich, Feldman, Nevins, Darnell, & Weinberger, 1983; Huang & Schneider, 1990). The following section focuses on the adenoviral expression system that we have adapted to produce isotopically enriched glycoproteins.

2.2.1 Construction of the Adenoviral Shuttle Vector

A schematic outline of the composition of an adenoviral shuttle vector is shown in Fig. 2. Once an optimal construct for the GOI is identified from the transient transfection study, the target gene is subcloned into pVRC1290 shuttle vector using suitable restriction sites such that the Kozak sequence, localization sequences, and purification tags are retained (Aoki et al., 1999; Sastry et al., 2012). Protein expression using this shuttle vector can once again be tested using small-scale transient transfection (McLellan et al., 2013) in HEK293 adherent cells or any other cell line of choice.

2.2.2 Generation of Recombinant Adenoviral Genome Containing the GOI

A schematic outline of the generation of a recombinant adenoviral genome is shown in Fig. 2. The adenoviral cosmid (pVRC1194) consists of 9.2–100 mu (map units) of the adenoviral genome, with a deletion in the E1 region and a loxP site at 9.2 mu. The cosmid DNA ($\sim$26 kb) is digested with the restriction enzyme *Cla*I (New England Biolabs), and reaction is monitored by agarose gel electrophoresis. *Cla*I-cleaved cosmid DNA is purified using standard phenol/chloroform purification followed by ethanol precipitation. The shuttle plasmid (pVRC1290) containing the GOI–GFP or GOI is linearized with restriction enzyme *Pac*I (New England Biolabs), reaction is monitored by agarose gel electrophoresis, and the linearized product is purified by phenol/chloroform purification (Moore & Dowhan, 2002; Voytas, 2001). Equimolar amounts of the shuttle plasmid and the adenoviral cosmid are recombined *in vitro* using Cre recombinase (New England Biolabs), recombination reaction is monitored using gel electrophoresis and upon completion, the recombinant adenoviral genome is purified by standard phenol/chloroform extraction and ethanol precipitation. The recombinant adenoviral DNA obtained from the Cre–Lox recombination reaction contains the GOI flanked by the adenoviral 5′

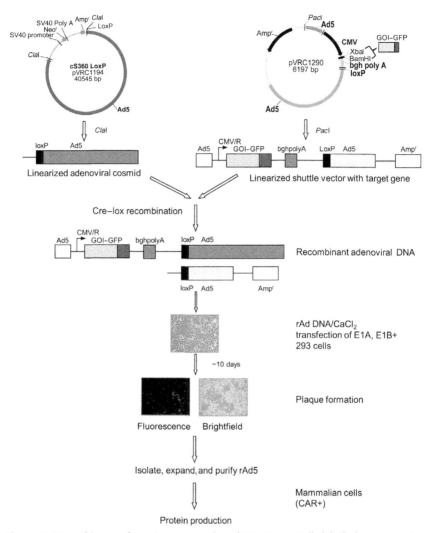

Figure 2 Recombinant adenovirus as a tool to obtain isotopically labeled proteins. The target gene is cloned into a shuttle vector (pVRC1290) using the restriction sites (e.g., Xba1, BamH1) in the multiple cloning site; adenoviral cosmid DNA (pVRC1194) and shuttle vector are each linearized with ClaI and PacI, respectively, then recombined in vitro with Cre—Lox recombinase to obtain recombinant adenoviral genome (Aoki et al., 1999). The recombined adenoviral type 5 DNA is transfected into 293 adherent helper mammalian cells and recombinant adenovirus is isolated and purified using well-established methods. Target protein production is achieved by infecting CAR+ mammalian cells, such as A549 or CHO(CAR+). (See the color plate.)

inverted terminal repeat (ITR), a 0- to 1-mu packaging signal, the bovine growth hormone polyadenylation signal (bghpolyA), and 9.2–100 mu of the adenoviral genome. These flanking sequences consist of DNA packaging sequences as well as sites for recombination with the rest of the viral DNA to reconstitute replication-defective adenoviruses. Generation of recombinant adenovirus can generally be a lengthy process. Therefore, prior to embarking on recombinant adenoviral production, traditional transient transfection into either HEK293 adherent or suspension cells is highly recommended in order to optimize the construct and perform functional analysis of the expressed protein.

2.2.3 Generation of Adenoviruses Containing GOI

Recombinant adenoviral DNA obtained from the Cre–Lox reaction is transfected into helper 293 cells (Fig. 2) using calcium phosphate transfection methodology. HEK293 cells are transfected with varying amounts of rAd5 DNA and monitored over a time course of nearly 10 days. HEK293 cell line with its indigenous E1 protein complements the absence of E1 gene in the recombinant adenoviral DNA (Graham, Smiley, Russell, & Nairn, 1977) and allows for the production of recombinant adenovirus. If the recombination reaction and calcium phosphate-mediated transfection are successful, rAd5 viruses are released via lysis of HEK293 cells resulting in a near circular clearing in the monolayer of cells. This region of visible lysis seen clearly in the case of a domain of HIV-1 gp120-GFP construct in Fig. 2 is known as a plaque and is indicative of virus formation. Upon plaque formation cells are harvested, the recombinant crude virus is isolated and protein expression of the desired gene tested by infecting A549 (CAR+) lung carcinoma cells. Once protein expression is confirmed, the crude virus preparation is used to infect low-passage HEK293 cells to produce recombinant adenovirus as described previously (Sastry et al., 2012). Briefly, HEK293 cells are seeded at 2.0×10^7 cells per 15-cm plate 24 h prior to infection. Cells are infected with the crude virus. Nearly 30 h postinfection, cells and culture media are collected and spun down at $\sim 300 \times g$ for 10 min, the cell pellet is washed twice with phosphate-buffered saline (PBS) resuspended in 10 mM Tris–HCl, pH 8.0. Recombinant virus is released by a series of freeze–thaw cycles, and the virus is purified using well-established CsCl gradient centrifugation methodology (Chillon & Alemany, 2011; Duffy, O'Doherty, O'Brien, & Strappe, 2005; Moore & Dowhan, 2002; Tan, Li, Jiang, & Ma, 2006). The purified adenovirus is quantitated, aliquoted aseptically, and stored at $-20\ °C$ until needed for protein expression.

3. PROTEIN EXPRESSION

Once the pure recombinant adenovirus has been obtained, small-scale protein expression can be initiated in six-well plates. This is necessary prior to embarking on large-scale protein production to assess the expression conditions as well as proper folding and activity of the desired protein. In our case study of HIV-1 gp120 OD, the glycoprotein is secreted into the culture medium. A researcher may choose to direct expression of a POI into the secretory pathway; however, attention must be paid to monitor inadvertent glycosylation of a cytoplasmic protein when it is directed into the Golgi secretory pathway. Additionally, expression of glycoproteins from mammalian cells in contrast to insect or yeast hosts (Jenkins, Parekh, & James, 1996) can result in heterogeneous glycosylation due to variation in the occupancy as well as the nature of glycan. The glycosylation pattern obtained from mammalian cells can be high mannose, hybrid, or complex in nature, and this may have an impact on stability and folding as well as *in vivo* functional activity (Weigel & Yik, 2002). Thus, a researcher may choose to design and direct POI production based on the source of a recombinant gene. In our case study, we used a combination of kifunensine, a potent inhibitor of α-mannosidase I (Elbein, Tropea, Mitchell, & Kaushal, 1990) as well as swainsonine a potent inhibitor of α-mannosidase II (Elbein, 1991) to obtain endoglycosidase H-sensitive high-mannose glycans (Kong et al., 2010; Magnelli, Bielik, & Guthrie, 2012).

Small-scale protein expression using the adenoviral expression system can be performed in six-well plates using A549 adherent cells as described previously (Sastry et al., 2012). Typically, cells are seeded at 0.8×10^6 cell/well in a six-well plate on Day 1 in fresh Dulbecco's minimal eagle medium (DMEM) containing 10% heat-inactivated-dialyzed fetal bovine serum (FBS), 1% penicillin/streptomycin. Cells are allowed to grow overnight at 37 °C and 5% CO_2. The following day (Day 2), media is replaced with labeled ^{15}N, $^{15}N/^{13}C$ CGM6000, or fresh DMEM containing 10% heat-inactivated-dialyzed FBS and 1% penicillin/streptomycin. A549 cells are then infected with rAd5 containing the GOI to a final concentration of 2500 particles/cell. Protein expression of a cytoplasmic or secreted protein is monitored 72–96 h postinfection by either harvesting cells or the culture media, respectively, and testing for the presence of the protein using ELISAs, SPR, BioLayer Interferometry or an equivalent technique.

Large-scale production of isotopically enriched glycoprotein is performed using the protocol described previously (Sastry et al., 2011, 2012) and purified using multistep affinity chromatography followed by size-exclusion chromatography. For cytoplasmic proteins, culture supernatant is removed, and adherent A549 cells are harvested following treatment with Trypsin-EDTA. Cells are pelleted at $300 \times g$, washed twice with PBS, and subsequently lysed with cell lysis buffer (cell signaling) and the POI purified by the method of choice.

3.1 Selective Labeling of Specific Amino Acids

CHO cell lines are most commonly used for production of therapeutic reagents. As a result, protein production for structural studies was focused on obtaining labeled material from stable CHO and HEK293 cell lines. Due to the exorbitant cost of labeled media and low yields, researches interested in studying biologically interesting proteins by NMR spectroscopy turned to partial (Werner et al., 2008) or amino acid type-specific labeling (Anglister, Frey, & McConnell, 1984; Arata, Kato, Takahashi, & Shimada, 1994; Klein-Seetharaman et al., 2002, 2004) of proteins from mammalian expression systems. Insect cells have proven to be a good source for obtaining specifically labeled proteins (Gossert et al., 2011). The adenoviral expression system is particularly suitable to obtaining proteins with specific amino acids labeled. We demonstrated this for ^{15}N glycine and ^{15}N valine (Sastry et al., 2012). Herein, we extend the methodology to incorporating ^{15}N/^{13}C proline. The methodology to produce OD selectively labeled with ^{15}N/^{13}C proline is similar to that described previously and in Section 3. CGM-6750-CUSTOM media containing ^{15}N/^{13}C-labeled proline and ^{15}N labels in all other amino acids and media components can be used to produce protein specifically labeled with ^{15}N/^{13}C proline. We estimate yields of 33 mg/L of pure glycosylated his-tagged ^{15}N/^{13}C proline-enriched OD protein, thus an NMR sample can be obtained from only 200–300 mL culture. These yields balance out the high cost of the labeled media and allow for the production of selectively labeled proteins.

4. NMR CHARACTERIZATION OF EXPRESSED PROTEIN

Initially, an unlabeled sample of a full-length protein or a protein domain of interest expressed either transiently or by the adenoviral-induced expression can be characterized using 1D or 2D NMR spectroscopy. Thus, in our example, the HIV-1 gp120 OD obtained from the adenovirus expression system exhibits a one-dimensional ^{1}H NMR spectrum (Fig. 3A)

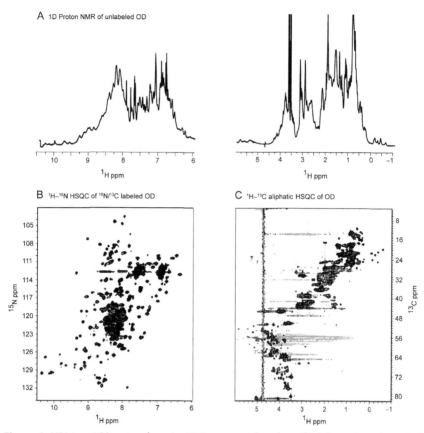

Figure 3 HIV-1 gp120 outer domain (OD) expressed and purified from the adenoviral expression system is functionally active and well-folded protein. One-dimensional proton spectra of unlabeled deglycosylated HIV-1 gp120 outer domain acquired at 600 MHz and 25 °C are shown in (A). ^{1}H–^{15}N and ^{1}H–^{13}C HSQC of ^{15}N/^{13}C-labeled outer domain acquired at 900 MHz and 25 °C are shown in (B) and (C), respectively. *Reproduced from Sastry et al. (2011) with permission from Springer.*

that is well dispersed with well resolved, upfield-shifted methyl protons as well as a nicely dispersed amide region, indicative of a folded protein. The ^{1}H–^{15}N HSQC spectra of the OD are also of very high quality and exhibits resolved backbone and side-chain amides (Fig. 3B). The ^{1}H–^{13}C HSQC (Fig. 3C) exhibits very good chemical shift dispersion along with upfield-shifted methyl resonances, indicative of a structured protein.

To assess selective ^{15}N/^{13}C proline incorporation necessary for assignment of backbone and side-chain atoms, two data sets were recorded.

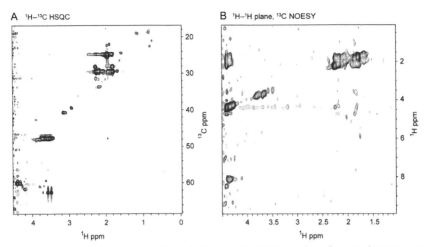

Figure 4 Single amino acid labeling of HIV-1 gp120 OD from the adenoviral expression system is feasible. (A) ^{1}H–^{13}C HSQC spectra of HIV-1 gp120 OD selectively labeled with ^{15}N/^{13}C proline acquired at 900 MHz and 25 °C. (B) ^{1}H–^{13}C NOESY HSQC spectra of HIV-1 gp120 OD selectively labeled with ^{15}N/^{13}C proline in a ^{15}N background acquired at 900 MHz and 25 °C. An ^{1}H–^{1}H NOESY plane at C_α 60.0 ppm is shown. The ^{1}H–^{13}C aliphatic HSQC and NOESY of selectively labeled outer domain spectrum demonstrate that the adenoviral system can be used to obtain selectively labeled proteins.

In Fig. 4A, a high-quality ^{1}H–^{13}C aliphatic HSQC spectrum acquired at 900 MHz and 25 °C shows selective incorporation of isotopically labeled proline residues. The ^{1}H–^{1}H plane of a ^{1}H–^{13}C 3D NOESY spectrum at a C_α chemical shift of 60 ppm also shows nOes to amide residues in the 8–9 ppm range, demonstrating conclusively that the adenoviral expression system provides sufficient isotope enrichment to allow acquisition of triple-resonance and NOESY-type experiments to obtain full backbone and side-chain resonance assignments. Selective labeling of specific amino acids allows the study of specific interactions and enables confirmation of assignments in extremely crowded regions of 3D spectra. Selectively labeled OD spectra enriched with ^{15}N/^{13}C proline (Fig. 4A and B) along with our previously published data provide further proof that the adenoviral expression system is suitable for obtaining isotopically enriched proteins for structural characterization of proteins and protein complexes by NMR spectroscopy.

5. CONCLUSIONS

Isotope labeling of biologically interesting proteins is a prerequisite for structural and dynamics studies by NMR spectroscopy. Despite the technological advances in NMR spectroscopy, difficulties in obtaining uniformly and specifically labeled protein samples of posttranslationally modified proteins have limited the breadth and scope of solution NMR measurements. In this chapter, we described a mammalian expression system that is based on the delivery of transgenes using mammalian viruses with nearly 100% efficiency and low toxicity. The high level of protein expression offsets the cost of the labeled media, and the versatility of the expression system should allow solution NMR studies of difficult to express proteins and their complexes.

6. MATERIALS

We report materials used for selective $^{15}N/^{13}C$ labeling of Proline residues. Materials for summarized experiments can be found in the respective primary publications (Aoki et al., 1999; Sastry et al., 2011). pVRC1194 and pVRC1290 vectors are available upon request.

Specifically labeled $^{15}N/^{13}C$ proline-CGM6750 CUSTOM media, in which all other amino acids are uniformly ^{15}N labeled while other components are unlabeled, were obtained from Cambridge Isotope Laboratories, Inc. (Andover, MA) (CIL). High-glucose-containing DMEM with Hepes and $NaHCO_3$ was obtained from Life Technologies. Kifunensine and swainsonine were obtained from Enzo Life Sciences. Nickel-NTA was obtained from Qiagen Inc. HEK293 and A549 cells were obtained from american tissue culture collection (ATCC).

ACKNOWLEDGMENTS

We thank the NMR staff at the New York Structural Biology Consortium for assistance with instrumentation. We also thank the members of the Structural Biology Section and Structural Bioinformatics Section at the Vaccine Research Center for insightful comments and discussions. Support for this work was provided by the Intramural Program of the NIH (NIAID and NIDDK). 900 MHz spectrometers were purchased with funds from NIH, USA, the Keck Foundation, New York State, and the NYC Economic Development Corporation.

REFERENCES

Anglister, J., Frey, T., & McConnell, H. M. (1984). Magnetic resonance of a monoclonal anti-spin-label antibody. *Biochemistry, 23*(6), 1138–1142.

Aoki, K., Barker, C., Danthinne, X., Imperiale, M. J., & Nabel, G. J. (1999). Efficient generation of recombinant adenoviral vectors by Cre-lox recombination in vitro. *Molecular Medicine, 5*(4), 224–231.

Arata, Y., Kato, K., Takahashi, H., & Shimada, I. (1994). Nuclear magnetic resonance study of antibodies: A multinuclear approach. *Methods in Enzymology, 239*, 440–464.

Arya, R., Bhattacharya, A., & Saini, K. S. (2008). Dictyostelium discoideum—A promising expression system for the production of eukaryotic proteins. *The FASEB Journal, 22*(12), 4055–4066.

Babich, A., Feldman, L. T., Nevins, J. R., Darnell, J. E., Jr., & Weinberger, C. (1983). Effect of adenovirus on metabolism of specific host mRNAs: Transport control and specific translational discrimination. *Molecular and Cellular Biology, 3*(7), 1212–1221.

Backliwal, G., Hildinger, M., Chenuet, S., Wulhfard, S., De Jesus, M., & Wurm, F. M. (2008). Rational vector design and multi-pathway modulation of HEK 293E cells yield recombinant antibody titers exceeding 1 g/l by transient transfection under serum-free conditions. *Nucleic Acids Research, 36*(15), e96.

Berk, A. J. (1986). Adenovirus promoters and E1A transactivation. *Annual Review of Genetics, 20*, 45–79 (Review).

Braun, P., & LaBaer, J. (2003). High throughput protein production for functional proteomics. *Trends in Biotechnology, 21*(9), 383–388.

Chen, K., Freedberg, D. I., & Keire, D. A. (2015). NMR profiling of biomolecules at natural abundance using 2D ^{1}H-^{15}N and ^{1}H-^{13}C multiplicity-separated (MS) HSQC spectra. *Journal of Magnetic Resonance, 251*, 65–70.

Chillon, M., & Alemany, R. (2011). Methods to construct recombinant adenovirus vectors. *Methods in Molecular Biology, 737*, 117–138.

Coleman, T. A., Parmelee, D., Thotakura, N. R., Nguyen, N., Bürgin, M., Gentz, S., et al. (1997). Production and purification of novel secreted human proteins. *Gene, 190*(1), 163–171.

Cubeddu, L., Moss, C. X., Swarbrick, J. D., Gooley, A. A., Williams, K. L., Curmi, P. M., et al. (2000). Dictyostelium discoideum as expression host: isotopic labeling of a recombinant glycoprotein for NMR studies. *Protein Expression and Purification, 19*(3), 335–342.

Duffy, A. M., O'Doherty, A. M., O'Brien, T., & Strappe, P. M. (2005). Purification of adenovirus and adeno-associated virus: Comparison of novel membrane-based technology to conventional techniques. *Gene Therapy, 12*(S1), S62–S72.

Elbein, A. D. (1991). Glycosidase inhibitors: Inhibitors of N-linked oligosaccharide processing. *The FASEB Journal, 5*(15), 3055–3063.

Elbein, A. D., Tropea, J. E., Mitchell, M., & Kaushal, G. P. (1990). Kifunensine, a potent inhibitor of the glycoprotein processing mannosidase I. *Journal of Biological Chemistry, 265*(26), 15599–15605.

Geisse, S., & Fux, C. (2009). Chapter 15 recombinant protein production by transient gene transfer into mammalian cells. In R. B. Richard & P. D. Murray (Eds.), *Methods in enzymology: Vol. 463* (pp. 223–238). San Diego: Academic Press.

Gluzman, Y., Reichl, H., & Solnick, D. (1982). Helper-free adenovirus type-5 vectors. In Y. Gluzman (Ed.), *Eukaryotic viral vectors* (pp. 187–192). Cold Spring Harbor, NY: Cold Spring Harbor Laboratory.

Gossert, A., Hinniger, A., Gutmann, S., Jahnke, W., Strauss, A., & Fernández, C. (2011). A simple protocol for amino acid type selective isotope labeling in insect cells with improved yields and high reproducibility. *Journal of Biomolecular NMR, 51*(4), 449–456.

Graham, F. L., Smiley, J., Russell, W. C., & Nairn, R. (1977). Characteristics of a human cell line transformed by DNA from human adenovirus type 5. *Journal of General Virology, 36*(1), 59–72.

Hansen, A. P., Petros, A. M., Mazar, A. P., Pederson, T. M., Rueter, A., & Fesik, S. W. (1992). A practical method for uniform isotopic labeling of recombinant proteins in mammalian cells. *Biochemistry, 31*(51), 12713–12718.

Hansen, A. P., Petros, A. M., Meadows, R. P., Nettesheim, D. G., Mazar, A. P., Olejniczak, E. T., et al. (1994). Solution structure of the amino-terminal fragment of urokinase-type plasminogen activator. *Biochemistry, 33*(16), 4847–4864.

Hinck, A. P., Archer, S. J., Qian, S. W., Roberts, A. B., Sporn, M. B., Weatherbee, J. A., et al. (1996). Transforming growth factor β1: Three-dimensional structure in solution and comparison with the X-ray structure of transforming growth factor β2†,‡. *Biochemistry, 35*(26), 8517–8534.

Hopkins, R. F., Wall, V. E., & Esposito, D. (2012). Optimizing transient recombinant protein expression in mammalian cells. In J. L. Hartley (Ed.), *Protein expression in mammalian cells: Vol. 801* (pp. 251–268). New York: Humana Press.

Howley, P. M., Sarver, N., & Law, M.-F. (1983). Eukaryotic cloning vectors derived from bovine papillomavirus DNA. In L. G. K. M. Ray Wu (Ed.), *Methods in enzymology: Vol. 101* (pp. 387–402): Academic Press.

Huang, J. T., & Schneider, R. J. (1990). Adenovirus inhibition of cellular protein synthesis is prevented by the drug 2-aminopurine. *Proceedings of the National Academy of Sciences of the United States of America, 87*(18), 7115–7119.

Jenkins, N., Parekh, R. B., & James, D. C. (1996). Getting the glycosylation right: Implications for the biotechnology industry. *Nature Biotechnology, 14*(8), 975–981.

Kato, T., Kajikawa, M., Maenaka, K., & Park, E. (2010). Silkworm expression system as a platform technology in life science. *Applied Microbiology and Biotechnology, 85*(3), 459–470.

Kingston, R. E., Chen, C. A., & Okayama, H. (2001). Calcium phosphate transfection. In *Current protocols in immunology: Vol 31* (pp. 10.13.1–10.13.9). New York, NY: John Wiley & Sons, Inc.

Klein-Seetharaman, J., Reeves, P. J., Loewen, M. C., Getmanova, E. V., Chung, J., Schwalbe, H., et al. (2002). Solution NMR spectroscopy of [α-^{15}N]lysine-labeled rhodopsin: The single peak observed in both conventional and TROSY-type HSQC spectra is ascribed to Lys-339 in the carboxyl-terminal peptide sequence. *Proceedings of the National Academy of Sciences of the United States of America, 99*(6), 3452–3457.

Klein-Seetharaman, J., Yanamala, N. V. K., Javeed, F., Reeves, P. J., Getmanova, E. V., Loewen, M. C., et al. (2004). Differential dynamics in the G protein-coupled receptor rhodopsin revealed by solution NMR. *Proceedings of the National Academy of Sciences of the United States of America, 101*(10), 3409–3413.

Kong, L., Sheppard, N. C., Stewart-Jones, G. B. E., Robson, C. L., Chen, H., Xu, X., et al. (2010). Expression-system-dependent modulation of HIV-1 envelope glycoprotein antigenicity and immunogenicity. *Journal of Molecular Biology, 403*(1), 131–147.

Kozak, M. (1984). Point mutations close to the AUG initiator codon affect the efficiency of translation of rat preproinsulin in vivo. *Nature, 308*(5956), 241–246. http://dx.doi.org/10.1038/308241a0.

Kozak, M. (1987). An analysis of 5′-noncoding sequences from 699 vertebrate messenger RNAs. *Nucleic Acids Research, 15*(20), 8125–8148.

Kriz, A., Schmid, K., Baumgartner, N., Ziegler, U., Berger, I., Ballmer-Hofer, K., et al. (2010). A plasmid-based multigene expression system for mammalian cells. *Nature Communications, 1*, 120.

Kwon, Y. D., Pancera, M., Acharya, P., Georgiev, I. S., Crooks, E. T., Gorman, J., et al. (2015). Crystal structure, conformational fixation and entry-related interactions of mature ligand-free HIV-1 Env. *Nature Structural & Molecular Biology, 22*(7), 522–531.

Li, J., Menzel, C., Meier, D., Zhang, C., Dübel, S., & Jostock, T. (2007). A comparative study of different vector designs for the mammalian expression of recombinant IgG antibodies. *Journal of Immunological Methods, 318*(1–2), 113–124.

Longo, P. A., Kavran, J. M., Kim, M.-S., & Leahy, D. J. (2013a). Generating mammalian stable cell lines by electroporation. In J. Lorsch (Ed.), *Methods in Enzymology: Vol. 529*. (pp. 209–226). Waltham, MA: Academic Press.

Longo, P. A., Kavran, J. M., Kim, M.-S., & Leahy, D. J. (2013b). Transient mammalian cell transfection with Polyethylenimine (PEI). In J. Lorsch (Ed.), *Methods in Enzymology: Vol. 529*. (pp. 227–240). Waltham, MA: Academic Press.

Lustbader, J. W., Birken, S., Pollak, S., Pound, A., Chait, B. T., Mirza, U. A., et al. (1996). Expression of human chorionic gonadotropin uniformly labeled with NMR istopes in

Chinese hamster ovary cells: An advance toward rapid determination of glycoprotein structures. *Journal of Biomolecular NMR, 7*(4), 295–304.

Magnelli, P., Bielik, A., & Guthrie, E. (2012). Identification and characterization of protein glycosylation using specific endo- and exoglycosidases. In J. L. Hartley (Ed.), *Protein expression in mammalian cells: Vol. 801* (pp. 189–211). New York: Humana Press.

McLellan, J. S., Chen, M., Leung, S., Graepel, K. W., Du, X., Yang, Y., et al. (2013). Structure of RSV fusion glycoprotein trimer bound to a prefusion-specific neutralizing antibody. *Science (New York, N.Y.), 340*(6136), 1113–1117.

Moore, D., & Dowhan, D. (2002). Purification and concentration of DNA from aqueous solutions. *Current protocols in molecular biology: Vol. 59* (pp. 2.1.1–2.1.10). New York, NY: John Wiley & Sons, Inc.

Mulligan, R. C., Howard, B. H., & Berg, P. (1979). Synthesis of rabbit [beta]-globin in cultured monkey kidney cells following infection with a SV40 [beta]-globin recombinant genome. *Nature, 277*(5692), 108–114.

Nettleship, J. E., Assenberg, R., Diprose, J. M., Rahman-Huq, N., & Owens, R. J. (2010). Recent advances in the production of proteins in insect and mammalian cells for structural biology. *Journal of Structural Biology, 172*(1), 55–65.

Pancera, M., Zhou, T., Druz, A., Georgiev, I. S., Soto, C., Gorman, J., et al. (2014). Structure and immune recognition of trimeric prefusion HIV-1 Env. *Nature, 514*(7523), 455–461.

Potter, H., & Heller, R. (2003). Transfection by electroporation. In *Current protocols in molecular biology: Vol. 62* (pp. 9.3.1–9.3.6). New York, NY: John Wiley & Sons, Inc.

Raghavan, M., & Bjorkman, P. J. (1995). BIAcore: A microchip-based system for analyzing the formation of macromolecular complexes. *Structure, 3*(4), 331–333.

Rich, R. L., & Myszka, D. G. (2010). Grading the commercial optical biosensor literature—Class of 2008: 'The Mighty Binders'. *Journal of Molecular Recognition, 23*(1), 1–64.

Sastry, M., Bewley, C., & Kwong, P. (2012). Mammalian expression of isotopically labeled proteins for NMR spectroscopy. In H. S. Atreya (Ed.), *Isotope labeling in biomolecular NMR: Vol. 992* (pp. 197–211). The Netherlands: Springer.

Sastry, M., Xu, L., Georgiev, I., Bewley, C., Nabel, G., & Kwong, P. (2011). Mammalian production of an isotopically enriched outer domain of the HIV-1 gp120 glycoprotein for NMR spectroscopy. *Journal of Biomolecular NMR, 50*(3), 197–207.

Shindo, K., Masuda, K., Takahashi, H., Arata, Y., & Shimada, I. (2000). Letter to the Editor: Backbone 1H, 13C, and 15 N resonance assignments of the anti-dansyl antibody Fv fragment. *Journal of Biomolecular NMR, 17*(4), 357–358.

Southern, P., & Berg, P. (1982). Transformation of mammalian cell to antibiotic resistance with a bacterial gene under control of the SV40 early region promoter. *Journal of Molecular and Applied Genetics, 1,* 327–341.

Swarbrick, J., Cubeddu, L., Ball, G., Curmi, P. G., Gooley, A., Williams, K., et al. (2009). NMR assignment of prespore specific antigen—A cell surface adhesion glycoprotein from Dictyostelium discoideum. *Biomolecular NMR Assignments, 3*(1), 1–3.

Tan, R., Li, C., Jiang, S., & Ma, L. (2006). A novel and simple method for construction of recombinant adenoviruses. *Nucleic Acids Research, 34*(12), e89.

Trowitzsch, S., Klumpp, M., Thoma, R., Carralot, J.-P., & Berger, I. (2011). Light it up: Highly efficient multigene delivery in mammalian cells. *BioEssays, 33*(12), 946–955.

Unger, T., & Peleg, Y. (2012). Recombinant protein expression in the baculovirus-infected insect cell system. In E. D. Zanders (Ed.), *Chemical genomics and proteomics: Vol. 800* (pp. 187–199). New York: Humana Press.

Van Craenenbroeck, K., Vanhoenacker, P., & Haegeman, G. (2000). Episomal vectors for gene expression in mammalian cells. *European Journal of Biochemistry, 267*(18), 5665–5678.

Voytas, D. (2001). Agarose gel electrophoresis. *Current protocols in molecular biology: Vol. 51* (pp. 2.5A.1–2.5A.9). New York, NY: John Wiley & Sons, Inc.

Warnock, J. N., Daigre, C., & Al-Rubeai, M. (2011). Introduction to viral vectors. *Methods in Molecular Biology, 737*, 1–25.

Weigel, P. H., & Yik, J. H. N. (2002). Glycans as endocytosis signals: The cases of the asialoglycoprotein and hyaluronan/chondroitin sulfate receptors. *Biochimica et Biophysica Acta (BBA)—General Subjects, 1572*(2–3), 341–363.

Werner, K., Richter, C., Klein-Seetharaman, J., & Schwalbe, H. (2008). Isotope labeling of mammalian GPCRs in HEK293 cells and characterization of the C-terminus of bovine rhodopsin by high resolution liquid NMR spectroscopy. *Journal of Biomolecular NMR, 40*(1), 49–53.

Wu, L., Yang, Z.-Y., Xu, L., Welcher, B., Winfrey, S., Shao, Y., et al. (2006). Cross-clade recognition and neutralization by the V3 region from clade C human immunodeficiency virus-1 envelope. *Vaccine, 24*(23), 4995–5002.

Wyss, D. F., Choi, J., Li, J., Knoppers, M., Willis, K., Arulanandam, A., et al. (1995). Conformation and function of the N-linked glycan in the adhesion domain of human CD2. *Science, 269*(5228), 1273–1278.

Wyss, D. F., Dayie, K. T., & Wagner, G. (1997). The counterreceptor binding site of human CD2 exhibits an extended surface patch with multiple conformations fluctuating with millisecond to microsecond motions. *Protein Science, 6*(3), 534–542.

Wyss, D. F., Withka, J. M., Knoppers, M. H., Sterne, K. A., Recny, M. A., & Wagner, G. (1993). Proton resonance assignments and secondary structure of the 13.6 kDa glycosylated adhesion domain of human CD2. *Biochemistry, 32*(41), 10995–11006.

Xia, W., Bringmann, P., McClary, J., Jones, P. P., Manzana, W., Zhu, Y., et al. (2006). High levels of protein expression using different mammalian CMV promoters in several cell lines. *Protein Expression and Purification, 45*(1), 115–124.

Yamaguchi, Y., Nishimura, M., Nagano, M., Yagi, H., Sasakawa, H., Uchida, K., et al. (2006). Glycoform-dependent conformational alteration of the Fc region of human immunoglobulin G1 as revealed by NMR spectroscopy. *Biochimica et Biophysica Acta (BBA)—General Subjects, 1760*(4), 693–700.

Zhou, T., Lynch, R. M., Chen, L., Acharya, P., Wu, X., Doria-Rose, N. A., et al. (2015). Structural repertoire of HIV-1-neutralizing antibodies targeting the CD4 supersite in 14 donors. *Cell, 161*(6), 1280–1292.

SECTION III

In Vitro Labeling

CHAPTER THIRTEEN

Escherichia coli Cell-Free Protein Synthesis and Isotope Labeling of Mammalian Proteins

Takaho Terada, Shigeyuki Yokoyama[1]
RIKEN Structural Biology Laboratory, Yokohama, Japan
[1]Corresponding author: e-mail address: yokoyama@riken.jp

Contents

1. Introduction	312
2. The *E. coli* Cell-Free Protein Synthesis Method	316
2.1 Coupled Transcription–Translation	316
2.2 Reaction Modes	317
2.3 Coding Region	318
2.4 Tags	319
2.5 Template DNA	319
2.6 T7 RNA Polymerase	320
2.7 S30 Extract and tRNAs	320
2.8 Low-Molecular Mass Components	321
2.9 Disulfide Bonds	321
2.10 Molecular Chaperones	322
2.11 Ligand Complexes	322
3. Stable Isotope Labeling of Proteins	323
3.1 Stable Isotope-Labeled Amino Acids	323
3.2 Inhibitors of Amino Acid Metabolism	324
3.3 Site-Directed Labeling	326
3.4 Constructs and Cell-Free Reaction Conditions	327
3.5 Medium-Scale Production of ^{15}N-Labeled Proteins	328
3.6 Large-Scale Production of ^{13}C/^{15}N-Labeled Proteins	331
3.7 Purification of Synthesized Proteins	332
3.8 WWE Domains	333
3.9 Zinc-Binding Proteins	334
3.10 Peptide Complexes	336
Acknowledgments	339
References	340

Abstract

This chapter describes the cell-free protein synthesis method, using an *Escherichia coli* cell extract. This is a cost-effective method for milligram-scale protein production and is particularly useful for the production of mammalian proteins, protein complexes, and membrane proteins that are difficult to synthesize by recombinant expression methods, using *E. coli* and eukaryotic cells. By adjusting the conditions of the cell-free method, zinc-binding proteins, disulfide-bonded proteins, ligand-bound proteins, etc., may also be produced. Stable isotope labeling of proteins can be accomplished by the cell-free method, simply by using stable isotope-labeled amino acid(s) in the cell-free reaction. Moreover, the cell-free protein synthesis method facilitates the avoidance of stable isotope scrambling and dilution over the recombinant expression methods and is therefore advantageous for amino acid-selective stable isotope labeling. Site-specific stable isotope labeling is also possible with a tRNA molecule specific to the UAG codon. By the cell-free protein synthesis method, coupled transcription–translation is performed from a plasmid vector or a PCR-amplified DNA fragment encoding the protein. A milligram quantity of protein can be produced with a milliliter-scale reaction solution in the dialysis mode. More than a thousand solution structures have been determined by NMR spectroscopy for uniformly labeled samples of human and mouse functional domain proteins, produced by the cell-free method. Here, we describe the practical aspects of mammalian protein production by the cell-free method for NMR spectroscopy.

1. INTRODUCTION

In protein biosynthesis, the base sequences are transcribed from the genes encoding proteins to the messenger RNAs (mRNAs), and then the coding regions are translated into the amino acid sequences of the proteins. In the cell-free, or *in vitro*, protein synthesis methods, unlike the recombinant DNA methods using host cells, the proteins of interest are synthesized in cell extracts obtained from *Escherichia coli* (Kigawa, 2010a; Kigawa, Muto, & Yokoyama, 1995; Kigawa et al., 2004, 1999; Kigawa & Yokoyama, 1991; Kim & Choi, 1996; Kim et al., 2006; Kim, Kigawa, Choi, & Yokoyama, 1996; Kim & Swartz, 1999, 2000; Makino, Goren, Fox, & Markley, 2010; Pratt, 1984; Spirin, Baranov, Ryabova, Ovodov, & Alakhov, 1988; Yabuki et al., 1998; Zubay, 1973), wheat germ (Madin, Sawasaki, Ogasawara, & Endo, 2000; Takai & Endo, 2010; Takai, Sawasaki, & Endo, 2010), insect cells (Suzuki, Ezure, Ito, Shikata, & Ando, 2009; Tarui, Imanishi, & Hara, 2000; Wakiyama, Kaitsu, & Yokoyama, 2006), human cells (Mikami, Kobayashi, Masutani, Yokoyama, & Imataka, 2008; Mikami, Masutani, Sonenberg, Yokoyama, & Imataka,

2006), and other sources. The cell-free protein synthesis methods have well been established and can therefore be applied to large-scale protein synthesis for stable isotope labeling. The cell-free protein synthesis system based on the *E. coli* cell extract has been used in the largest number of cases of stable isotope labeling (Kigawa, 2010b; Kigawa, Inoue, et al., 2008; Kigawa, Matsuda, Yabuki, & Yokoyama, 2008; Kigawa et al., 1995, 2004, 1999; Matsuda et al., 2007; Ozawa, Wu, Dixon, & Otting, 2006; Takeda & Kainosho, 2012; Yabuki et al., 1998). The cell-free protein synthesis system contains ribosomes, transfer RNAs (tRNAs), translation factors, and molecular chaperones. The mRNA used for translation is usually generated by transcription from the template DNA, by RNA polymerase added to the cell-free system (designated as "coupled transcription–translation") (Pratt, 1984; Zubay, 1973). Thus, the reaction solution contains the cell-free protein synthesis system, the template DNA for coupled transcription–translation, the low-molecular mass substrates for translation and transcription, such as amino acids and ribonucleotide triphosphates, and the ATP regeneration system (Fig. 1). The cell-free protein synthesis reaction continues for about 1 h in a container (test tube, etc.; the batch mode, Fig. 1) but can be prolonged for several hours by dialyzing the reaction

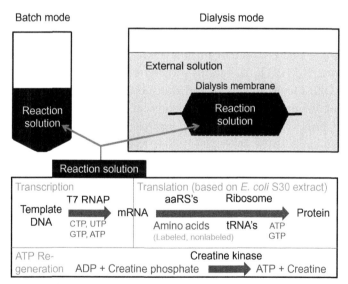

Figure 1 The batch and dialysis modes of coupled transcription–translation for the *E. coli* cell-free protein synthesis method. aaRS, aminoacyl-tRNA synthetase; RNAP, RNA polymerase.

solution against an external solution containing the low-molecular mass substrates (the dialysis mode, Fig. 1). The cell-free method is convenient and cost-effective, as milligram quantities of protein can be produced (about 1 mg/mL cell-free reaction solution, corresponding to about 50 mL of *E. coli* cell culture) in a few hours, not only from a plasmid but also from a PCR-amplified linear DNA template (Aoki et al., 2009; Ahn et al., 2005; Carlson, Gan, Hodgman, & Jewett, 2012; Hahn & Kim, 2006; Jackson, Boutell, Cooley, & He, 2004; Kigawa, Inoue, et al., 2008; Kigawa, Matsuda, et al., 2008; Yabuki et al., 2007). Thus, the cell-free protein synthesis method has become the standard for protein preparation.

The cell-free protein synthesis method has many advantages over the conventional recombinant expression methods with host cells. First, proteins toxic to cells can be synthesized by the cell-free method. Second, components of the cell-free protein synthesis system can be easily added or eliminated. In fact, *E. coli* cells can be genetically engineered to tag a nuclease or protease, for example, for removal from the cell extract (Seki, Matsuda, Yokoyama, & Kigawa, 2008), even if the nuclease/protease is essential for cell growth. *E. coli* and/or eukaryotic molecular chaperones can be added to the cell-free system, in order to achieve proper folding of the protein products. Proteins can also be synthesized in complexes with ligand(s), such as metal ions (Kuchenreuther, George, Grady-Smith, Cramer, & Swartz, 2011; Matsuda et al., 2006), low-molecular mass cofactors, substrates, inhibitors, peptide binders, and/or nucleic acids (Terada, Murata, Shirouzu, & Yokoyama, 2014). Third, various parameters, including the reaction temperature, the incubation time, and the substrate and template concentrations, can be optimized, independently of the cell growth conditions. Disulfide bonds can be formed in the reaction solution, by controlling the redox status (Kim & Swartz, 2004; Matsuda, Watanabe, & Kigawa, 2013; Oh, Kim, Kim, Park, & Choi, 2006). Fourth, the cell-free protein synthesis method is suitable for the production of protein complexes consisting of two or more different components or subunits (Terada et al., 2014). The stoichiometry of the components can be more strictly controlled in cell-free coexpression than in cell-based recombinant coexpression, simply by adjusting the amounts of their template DNAs in the reaction solution. Otherwise, protein complexes may be prepared in a stepwise manner, and one of the components can be synthesized in the presence of the others (Terada et al., 2014). Fifth, the cell-free synthesis method is more advantageous than the recombinant cell-based methods for the production of membrane proteins, in terms of both quantity and quality (Henrich, Hein, Dötsch, & Bernhard, 2015; Ishihara et al., 2005; Junge et al., 2011; Kimura-Someya, Shirouzu, & Yokoyama, 2014; Shimono et al., 2009; Wada et al., 2011).

Many mammalian proteins are too difficult to express by the conventional recombinant method, using living *E. coli* cells as the host (Saul et al., 2014; Terpe, 2006). Such difficult-to-express mammalian proteins are usually tested for recombinant expression in eukaryotic cells, such as budding and methylotrophic yeasts and insect and mammalian cell cultures. However, because of the aforementioned advantages, the cell-free protein synthesis method using the *E. coli* cell extract is highly suitable for mammalian protein production, with respect to both quality and quantity (Terada et al., 2014) and is also naturally useful for bacterial protein production. In other words, most of the problems associated with the production of mammalian proteins by the recombinant expression in *E. coli* cells can be solved, by using the *E. coli* cell-free method. Cell-free synthesis methods with wheat germ, insect, and human cell extracts are also used for mammalian protein expression (Rosenblum & Cooperman, 2014), but the protein synthesis rates are much slower, and the yields are concomitantly lower, than those of the *E. coli* cell-free method. Thus, it should be emphasized here that in many cases, the *E. coli* cell-free method is even more suitable for mammalian protein production than the eukaryotic recombinant and cell-free expression methods.

Considering all of these advantages, the *E. coli* cell-free method is ideal for the stable isotope labeling of mammalian proteins. In particular, stable isotope labeling of proteins with ^{15}N, ^{13}C, and/or ^{2}H for NMR measurements can easily be performed by cell-free protein synthesis (Aoki et al., 2009; Kigawa, 2010a, 2010b; Kigawa, Inoue, et al., 2008; Kigawa, Matsuda, et al., 2008; Kigawa et al., 2004, 1999; Kigawa & Yokoyama, 1991; Matsuda et al., 2007; Yabuki et al., 1998). Uniform stable isotope labeling of proteins may be accomplished by using a mixture of uniformly labeled amino acids (Kigawa, 2010a, 2010b; Kigawa, Inoue, et al., 2008; Kigawa, Matsuda, et al., 2008). In fact, our groups have determined 1333 solution structures of human/mouse proteins (Table 1, mostly functional domains), by using the *E. coli* cell-free method for the preparation of uniformly labeled protein samples (Kigawa, Inoue, et al., 2008; Yokoyama, Terwilliger, Kuramitsu, Moras, & Sussman, 2007). Moreover, a variety of selective labeling techniques have been developed, by utilizing the

Table 1 The Numbers of PDB-Deposited Structures of Proteins Produced by the *E. coli* Cell-Free Protein Synthesis Method in Our Group (Deposited from April 2001 to June 2014) and by the Wheat Germ Cell-Free Synthesis Method (Available from http://www.uwstructuralgenomics.org/structures.htm, as of June 22, 2015)

Source Organism	E. coli	Wheat Germ
Vertebrate		
Homo sapiens	1029	5
Mus musculus	261	1
Rattus rattus	1	
Danio rerio		2
Invertebrate	4	
Yeast	2	
Plant	33	9
Bacteria	3	
Virus		
Total	1333	17

advantages of the cell-free synthesis method (Kigawa et al., 1995; Yabuki et al., 1998). The stepwise synthesis of multicomponent complexes should be useful for the selective stable isotope labeling of one of the components. In this chapter, we describe the practical aspects of the cell-free production of stable isotope-labeled protein samples.

2. THE *E. COLI* CELL-FREE PROTEIN SYNTHESIS METHOD

2.1 Coupled Transcription–Translation

The basics of protein production by the *E. coli* cell-free protein synthesis method, described in this section, are applicable for stable isotope labeling as well as other structural biology purposes. The *E. coli* cell-free protein synthesis system is fundamentally based on the Zubay system (Zubay, 1973), which uses the *E. coli* cell extract. Transcription from DNA to mRNA and translation from mRNA to protein are performed in a coupled manner. We use an exogenous RNA polymerase from T7 phage for transcription

(Davanloo, Rosenberg, Dunn, & Studier, 1984). Preliminary experiments may easily be performed on a small scale, to optimize various conditions. Using these optimized conditions, the reaction scale can be increased to the large-scale protein production. The reaction solution for *E. coli* cell-free coupled transcription–translation contains the *E. coli* S30 extract, the DNA template, T7 RNA polymerase, and the substrates for transcription and translation.

2.2 Reaction Modes

The components of the standard reaction solution are listed in Table 2. There are two reaction modes, the batch and dialysis modes, for cell-free coupled transcription–translation (Fig. 1). The *E. coli* cell-free synthesis reaction is performed at 15–30 °C in both modes. In the batch mode, the cell-free protein synthesis reaction is performed by incubating the reaction solution in a container, such as a test tube or a well on a multi-well plate. The batch-mode reaction reaches a plateau in a few hours, because the low-molecular mass substrates for coupled transcription–translation and ATP regeneration become exhausted, and by-products accumulate. In the dialysis mode, the cell-free synthesis reaction is performed by placing the reaction solution in a container equipped with a dialysis membrane (Fig. 1), and

Table 2 Standard Composition of the *E. coli* Cell-Free Protein Synthesis Reaction Solution

Reagent	Stock Conc.	Final Conc.
LMCP(Y)		37.33% (v/v)
A.A.(-Y)	20 mM	1.5 mM each
NaN$_3$	5% (w/v)	0.05% (w/v)
Mg(OAc)$_2$	1.6 M	9.28 mM
tRNA	17.5 mg/mL	0.175 mg/mL
Creatine kinase	3.75 mg/mL	0.25 mg/mL
S30 extract		30% (v/v)
T7 RNA polymerase	10 mg/mL	66.7 µg/mL
Other factors		
Milli-Q water		To the desired volume

simultaneously incubating it and dialyzing against the external solution, which contains the same set of low-molecular mass components as that in the reaction solution. In our standard conditions, we use a 10-fold larger volume of the external solution relative to that of the reaction solution. In the dialysis mode, the substrates and the by-products are continuously exchanged between the reaction and external solutions, through the dialysis membrane with a molecular weight cutoff of usually 10–15 kDa. Therefore, the cell-free synthesis reaction in the dialysis mode continues much longer than that in the batch mode, although in practice the reaction is stopped at 3–4 h, to avoid denaturation of the products. Due to the longer duration of the reaction, higher yields are achieved in the dialysis mode than in the batch mode. In the dialysis mode, the yield at 25–30 °C may reach 1–5 mg/mL reaction in 3–4 h. For a reaction at a lower temperature, such as 15 °C, the reaction may be continued up to overnight.

2.3 Coding Region

The coding region for the targeted mammalian protein is either obtained from its cDNA or chemically synthesized. In the latter case, optimization of the codon usage and minimization of hairpin structures may be performed. Mammalian proteins frequently contain one or more functional domains. When a cDNA fragment encoding the domain(s) of interest is obtained from the full-length cDNA, the 5′- and 3′-termini should be adjusted to properly correspond to the N- and C-terminal ends of the domain fold. The cell-free protein synthesis is particularly useful for screening numerous protein fragments for optimization of the ends, and identifying the constructs exhibiting excellent properties, such as solubility and stability. Screening based on NMR spectroscopy is described in Section 3, where stable isotope labeling is also discussed. Many functional domains bind peptides including the specific sequences of their binding partner proteins. By fusing the peptide through the linker to the N- or C-terminus of the binding domain fragment, their complex sample may be prepared without chemical synthesis of the peptide. The binding of the domain part to the peptide part may facilitate the proper folding of the protein. For segmental stable isotope labeling (Liu, Xu, & Cowburn, 2009; Ludwig et al., 2009; Michel, Skrisovska, Wuthrich, & Allain, 2013; Yamazaki et al., 1998), the cell-free method may be applied for the preparation of fragments for intein-based mechanisms.

2.4 Tags

Tags, such as 6 × histidine (6His), streptavidin-binding peptide (SBP), glutathione-S-transferase (GST), maltose-binding protein (MBP), and small ubiquitin-related modifier (SUMO), are used for the purification and/or improvement of folding/solubility. The N-terminal tags are useful to standardize the translation initiation rates among different constructs for the target protein, to facilitate the selection of the best constructs in the optimization of the domain termini, as the N-terminal part of the protein is involved in translation initiation. The natural polyhistidine tag (NHis-tag) is suitable for the efficient preparation of stable isotope-labeled samples (Aoki et al., 2009). The SUMO tag is chosen not only to improve folding/solubility but also to generate the natural N-terminus of the protein, with no additional sequence remaining after the tag cleavage. The N11–SUMO tag (Numata, Motoda, Watanabe, Osanai, & Kigawa, 2015), in which a modified version of the NHis-tag (N11-tag) and the SUMO tag are fused in tandem, may also be used.

2.5 Template DNA

For cell-free coupled transcription–translation, the template DNA must contain the flanking sequences for transcription, translation, and purification, in addition to the coding region. A typical template DNA is shown in Fig. 2. The designed template DNA may be constructed by PCR, with primers containing the flanking regions. The two-step PCR method (Yabuki et al., 2007) is useful for the preparation of a number of different constructs for screening. The template DNA prepared by PCR may be subcloned into a plasmid vector for large-scale preparation. The plasmid DNA must be purified well, for example, with a commercially available DNA purification kit (Qiagen, Promega, etc.). The PCR-amplified linear DNA can be used directly in the coupled transcription–translation, with the cell-free system with low nuclease activity. This is one of the

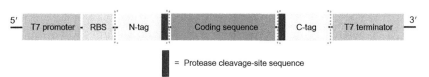

Figure 2 Typical organization of the template DNA for the E. coli cell-free protein synthesis method.

advantages of the cell-free method over the recombinant expression methods. The optimal DNA template concentration for coupled transcription–translation should be determined by preliminary small-scale, cell-free experiments.

2.6 T7 RNA Polymerase

We use an exogenous RNA polymerase from T7 phage for transcription, as it is highly productive (Kigawa et al., 1995). The template DNA contains the promoter sequence for T7 RNA polymerase (Fig. 2). T7 RNA polymerase, prepared as reported in Davanloo et al. (1984), is added to the cell-free reaction solution for coupled transcription–translation. When the *E. coli* cell-free protein synthesis system is supplemented by adding a tRNA with a special function, its *in vitro* transcript may be prepared with T7 RNA polymerase. Other productive RNA polymerases, such as SP6 phage polymerase, may also be used for coupled transcription–translation.

2.7 S30 Extract and tRNAs

The cell extract for cell-free protein synthesis is prepared as the supernatant fraction, after the disruption of *E. coli* cells and centrifugation of the lysate at $30,000 \times g$ (the S30 extract). The S30 extract contains the endogenous tRNAs from *E. coli* cells but is supplemented with *E. coli* MRE600-derived tRNA (Roche Applied Sciences, 109550). Furthermore, minor tRNA species are also needed for the efficient translation of mRNAs bearing consecutive minor codons. We use the S30 fraction of *E. coli* strain BL21 CodonPlus-RIL (Agilent Technologies), containing extra copies of the genes encoding minor tRNAs (*argU, ileY, leuW*) (http://www.genomics.agilent.com/article.jsp?pageId=484). The detailed protocols for *E. coli* cell extract preparation have been published (Kigawa, 2010a; Kigawa et al., 2004; Seki et al., 2008), and many laboratories prepare their own cell extracts. However, kits for cell-free protein synthesis with *E. coli* cell extracts are commercially available. The kits suitable for structural biology purposes are the Cell-free Protein Expression Kit "Musaibo-Kun" (Taiyo Nippon Sanso, Japan), the "*i*PE Kit" (Sigma-Aldrich, USA), and the RYTS (Remarkable Yield Translation System) Kit (Protein Express, Japan), which are based on the method by Kigawa et al. (2004), as well as the RTS 100 *E. coli* HY Kit (Biotechrabbit GmbH, Germany), the S30 T7 High-Yield

System (Promega, USA), etc. Some products are optimized for special purposes, such as the use of a linear template DNA or disulfide bond formation.

2.8 Low-Molecular Mass Components

Table 2 lists the components of the standard reaction solution, in the order roughly corresponding to that used to set up the reaction solution. Most of the low-molecular mass components are prepared in two forms, a mixture of 19 amino acids (A.A.(-Y)) and the low-molecular mass component mixture solution (Low-Molecular-weight Creatine Phosphate tYrosine, LMCPY). The A.A.(-Y) contains 10 mM dithiothreitol (DTT) and 20 mM each of the 19 amino acids other than L-tyrosine, while L-tyrosine is included in the LMCPY. The use of amino acids for cell-free stable isotope labeling is described below, in Sections 3.5 and 3.6. The LMCPY solution contains 160 mM HEPES-KOH buffer (pH 7.5), 4.13 mM L-tyrosine, 534 mM potassium L-glutamate, 5 mM DTT, 3.47 mM ATP, 2.40 mM GTP, 2.40 mM CTP, 2.40 mM UTP, 0.217 mM folic acid, 1.78 mM cAMP, 74 mM ammonium acetate, and 214 mM creatine phosphate. The optimal magnesium concentration depends to some extent on the target proteins, and it should therefore be optimized for each target protein, in the range of 5–20 mM. For ATP regeneration, creatine kinase and its substrate, creatine phosphate, are used.

2.9 Disulfide Bonds

In the standard conditions, the cell-free synthesis reaction is performed in the presence of DTT, and thus under reducing conditions. Therefore, for the production of disulfide-bonded proteins, such as secreted proteins and membrane proteins, cell-free synthesis is performed under more oxidative redox conditions than the standard conditions (Kim & Swartz, 2004; Matsuda et al., 2013; Oh et al., 2006). In place of the standard DTT-containing LMCPY, the DTT-free LMCPY is used, and the redox conditions are controlled with reduced glutathione (GSH) and oxidized glutathione (GSSG). In our standard protocol, the ratio between GSH and GSSG is optimized, by testing ratios from 1:9 to 9:1. The *E. coli* periplasmic protein DsbC exhibits disulfide isomerase and chaperone activities (Kadokura, Katzen, & Beckwith, 2003) and is used during the cell-free synthesis of disulfide-bonded proteins, to facilitate their proper oxidative folding (Kim & Swartz, 2004; Matsuda et al., 2013). In addition to DsbC, the *E. coli* periplasmic chaperone Skp (Muller, Koch, Beck, & Schafer, 2001; Weski & Ehrmann, 2012) may be used in the cell-free reaction.

2.10 Molecular Chaperones

Molecular chaperones facilitate protein folding and are particularly required for large proteins and protein complexes. Therefore, the *E. coli* cell-free protein synthesis system is supplemented with *E. coli* or eukaryotic molecular chaperones, to achieve the correct folding of the product protein. The molecular chaperones can be added in two ways. One is the use of a mixture of appropriate molecular chaperones (Thomas, Ayling, & Baneyx, 1997), prepared separately. The other is the use of the S30 extract prepared from *E. coli* cells overexpressing the molecular chaperones. The molecular chaperone sets DnaK/DnaJ/GrpE and GroEL/GroES are considered to function in the early and late stages of chaperone-assisted protein folding in *E. coli*, respectively. Therefore, one or both of the *E. coli* chaperone sets is used in the cell-free protein synthesis. Molecular chaperones are usually included to solubilize precipitating or aggregating proteins. Moreover, the quality of the produced protein may be improved by the molecular chaperones, with respect to uniform folding.

2.11 Ligand Complexes

Many proteins tightly bind a specific ligand for the formation and maintenance of their correct tertiary structures. A typical example is the Zn ion coordinated by a Zn-binding protein. Zn-binding proteins tend to misfold and precipitate upon cell-free synthesis in the absence of Zn ion. Therefore, to achieve the correct folding of Zn-binding proteins during cell-free synthesis, an appropriate concentration (usually around 50 μM) of $ZnCl_2$ or $ZnSO_4$ should be added (Matsuda et al., 2007; Terada et al., 2014). When proteins have other types of ligands, such as cofactors and inhibitors, the addition of such ligands to the cell-free reaction solution is expected to exert similar effects to that of the Zn ion for Zn-binding proteins. Proteins that form a heterocomplex often must be coexpressed for their soluble synthesis. For such protein complex formation, two or more DNA templates are simultaneously used, in the ratio corresponding to that of the proteins composing the heterocomplex, in the cell-free synthesis reaction (Terada et al., 2014). Examples of the cell-free production of Zn-binding proteins and protein heterocomplexes in the soluble forms are described in Sections 3.9 and

3.10. Furthermore, membrane proteins can be produced, simply by adding detergents and/or lipids to the cell-free synthesis reaction solution (Ishihara et al., 2005; Kimura-Someya et al., 2014; Shimono et al., 2009; Wada et al., 2011).

3. STABLE ISOTOPE LABELING OF PROTEINS
3.1 Stable Isotope-Labeled Amino Acids

Stable isotope labeling of proteins can be performed by the *E. coli* cell-free protein synthesis method, by using stable isotope-labeled amino acid(s) in place of the nonlabeled one(s). For uniform stable isotope labeling of proteins by the conventional recombinant method using *E. coli* cells, stable isotopes ^{15}N, ^{13}C, and/or ^{2}H are provided in the culture medium as ammonium chloride (or sulfate), glucose, and/or deuterium oxide, respectively, and labeled amino acids are biosynthesized within the cells. In contrast, uniform stable isotope labeling of proteins is accomplished with a mixture of the 20 acids uniformly labeled with ^{15}N, ^{13}C, and/or ^{2}H. Mixtures of the uniformly ^{15}N and/or ^{13}C labeled amino acids are commercially available (Taiyo Nippon Sanso, Japan; Sigma-Aldrich, USA; Cambridge Isotope Laboratories, Inc., USA). The use of ^{2}H$_2$O in the cell-free method affects protein synthesis to a much smaller extent than the presence of ^{2}H$_2$O in the *E. coli* cell growth medium.

Amino acid-selective stable isotope labeling with respect to one or several kinds of amino acids can be performed more easily by the cell-free method than by the conventional recombinant method, because the stable isotope scrambling between amino acids is minimized in the cell-free reaction. Inhibitors of amino acid metabolism may be used to prevent residual isotope scrambling and dilution. In contrast, the recombinant expression method requires the use of the corresponding amino acid auxotroph for amino acid-selective labeling. This is analogous to the selenomethionine substitution of proteins for multiwavelength anomalous diffraction phasing in X-ray crystallography, which can be performed by the cell-free method simply through the replacement of L-methionine with L-selenomethionine (Kigawa, Inoue, et al., 2008; Kigawa et al., 2002; Terada et al., 2014; Wada et al., 2003), but requires a methionine auxotrophic strain of *E. coli* for protein expression by the recombinant method.

When Tyr is included in the labeling, the low-molecular mass component mixture solution without L-tyrosine (*Low-Molecular-weight Creatine*

Phosphate, LMCP) is used, in place of the LMCPY in the cell-free reaction. Potassium, which is required for cell-free translation, is provided as potassium L-glutamate at a much higher concentration than those of the other 19 amino acids, without interfering with translation (Kigawa et al., 1995). However, the high concentration of potassium L-glutamate is not suitable for uniform or glutamate-selective stable isotope labeling. Therefore, the cell-free method in which potassium D-glutamate is used, in place of potassium L-glutamate (the D-Glu method), was developed to achieve productive cell-free production suitable for stable isotope labeling (Matsuda et al., 2007).

Isotope labeling of selected atom(s) within an amino acid is a powerful strategy to simplify NMR spectra. An elegant example is the stereo-array isotope labeling (SAIL) method (Kainosho et al., 2006; http://www.sail-technologies.com/prod/technology.html), developed by Kainosho, which is useful for reducing the problems encountered in isotope-aided NMR spectroscopy; e.g., the spin–spin or transverse relaxation of the NMR signals and the complexity of the NMR spectra due to spin–spin coupling and signal overlapping. The SAIL amino acids are designed to minimize such problems and to provide all essential information for protein NMR, by endowing each of the 20 amino acids with a particular pattern of isotopes: ^1H or ^2H, ^{12}C or ^{13}C, and ^{14}N or ^{15}N. The SAIL method is particularly powerful for large proteins, for which the aforementioned problems are often serious. The cell-free protein synthesis method is suitable for the SAIL method, as smaller amounts of the isotope-labeled amino acids are consumed, as compared with the conventional recombinant method. Other types of atom-directed labeling by the recombinant method, such as methyl-specific labeling, are economically performed by metabolic conversion of an inexpensive precursor in the cell (Gardner & Kay, 1997; Goto, Gardner, Mueller, Willis, & Kay, 1999; Velyvis, Ruschak, & Kay, 2012). Similar labeling profiles by the cell-free synthesis method require the chemical or biochemical preparation of labeled amino acid(s) for use in the cell-free reaction.

3.2 Inhibitors of Amino Acid Metabolism

In the cell-free protein synthesis method, most of the stable isotope scrambling is negligible. However, non-labile ^2H dilution with ^1H still occurs in the *E. coli* cell-free system. Furthermore, ^{15}N scrambling and dilution also remain among aspartate, asparagine, glutamate, and glutamine in the *E. coli* cell-free system. For such problems, Yokoyama, Kigawa, and their colleagues reported the successful use of chemical inhibitors of the metabolic enzymes, which prevented the scrambling and dilution of the stable isotope (Su, Loh, Qi, & Otting, 2011; Yokoyama, Matsuda, Koshiba, Tochio, & Kigawa, 2011). The non-labile ^{2}H dilutions at the α- and β-positions in certain amino acids are catalyzed by pyridoxal 5′-phosphate (PLP)-requiring enzymes, in the ordinary cell-free system based on ^{1}H$_2$O. Therefore, the cell-free synthesis of perdeuterated proteins from perdeuterated amino acids was achieved by the cell-free protein synthesis reaction in which all of the reagents, including the *E. coli* cell extract, were prepared with ^{2}H$_2$O (Etezady-Esfarjani, Hiller, Villalba, & Wüthrich, 2007). However, for all such amino acids, except methionine and cysteine, the non-labile ^{2}H dilution can be suppressed by using two PLP-requiring enzyme inhibitors (John, Charteris, & Fowler, 1978; Koizumi, Hiratake, Nakatsu, Kato, & Oda, 1999; Shi, Pelton, Cho, & Wemmer, 2004; Yokoyama, Matsuda, Koshiba, & Kigawa, 2010), aminooxyacetate (20 m*M*) and D-malate (20 m*M*), in the cell-free reaction. Therefore, this technique enables the cell-free synthesis of perdeuterated proteins from perdeuterated amino acids with the ^{1}H$_2$O-based reaction (Yokoyama et al., 2011).

The exchange of the main-chain NH$_3$ group of aspartate with the NH$_3$ from the ammonium ion [L-aspartate ↔ fumarate + NH$_3$] by aspartate ammonia-lyase (Falzone, Karsten, Conley, & Viola, 1988) is efficiently suppressed by adding 20 m*M* D-malate. The exchange of the main-chain NH$_3$ group between aspartate and glutamate, through the formation of glutamate from aspartate and vice versa (glutamate to aspartate) [L-aspartate + α-ketoglutarate ↔ oxaloacetate + L-glutamate] by aspartate aminotransferase, is dramatically suppressed by adding 20 m*M* aminooxyacetate to the cell-free reaction. As for the side-chain NH$_3$ groups, the aspartate-to-asparagine conversion [L-aspartate + NH$_3$ → L-asparagine] by asparagine synthase and the glutamate-to-glutamine conversion [L-glutamate + NH$_3$ → glutamine] by glutamine synthase are efficiently suppressed by adding 20 m*M* S-methyl-L-cysteine sulfoximine (Koizumi et al., 1999) and 0.5 m*M* L-methionine sulfoximine (Manning, Moore, Rowe, & Meister, 1969), respectively, to the cell-free reaction. The converse conversions, namely, the asparagine-to-aspartate conversion [L-asparagine → L-aspartate + NH$_3$] by asparaginase and the glutamine-to-glutamate conversion [L-glutamine → L-glutamate + NH$_3$] by glutaminase, should be suppressed by using enzyme inhibitors.

Actually, 5-diazo-4-oxo-L-norvaline (Handschumacher, Bates, Chang, Andrews, & Fischer, 1968; Willis & Woolfolk, 1974) and 6-diazo-5-oxo-L-norleucine (Brown et al., 2008; Hartman, 1968) irreversibly inhibit asparaginase and glutaminase, respectively, through covalent attachment to the active site, and the asparagine-to-aspartate and glutamine-to-glutamate conversions can be nearly completely suppressed by the use of an S30 extract pretreated with 20 mM 5-diazo-4-oxo-L-norvaline and 10 mM 6-diazo-5-oxo-L-norleucine, and further prevented by adding them to the cell-free reaction (Yokoyama et al., 2011).

3.3 Site-Directed Labeling

One of the advantages of cell-free protein synthesis is that site-directed stable isotope labeling is possible. Usually, one of the stop codons, UAG, is used to direct the targeted site in the mRNA. An engineered tRNA bearing the anticodon CUA (UAG or amber "suppressor tRNA") should be esterified with the stable isotope-labeled amino acid and then added to the cell-free reaction. There are two different methods to prepare the stable isotope-labeled aminoacyl-tRNA. One is the "chemical acylation" method, in which a tRNA lacking the terminal nucleotide (C–A or dC–A in most cases) is enzymatically ligated with the chemically synthesized aminoacyl nucleotide. The other is the aminoacylation of the tRNA with an aminoacyl-tRNA synthetase. For example, Tyr32 of the human H-Ras protein was site-specifically ^{15}N-labeled, which resulted in a single cross peak in the ^{1}H–^{15}N heteronuclear single quantum coherence (HSQC) spectrum (Yabuki et al., 1998).

For the chemical and enzymatic methods of tRNA aminoacylation, after the tRNA is used on the ribosome, it should not be reused through aminoacylation by the endogenous aminoacyl-tRNA synthetases in the *E. coli* cell-free protein synthesis system, to avoid isotope dilution with the nonlabeled amino acid in the reaction solution. This property of the UAG suppressor tRNA and the corresponding aminoacyl-tRNA synthetase is called "orthogonality" to the *E. coli* protein synthesis system. The eukaryotic and archaeal tyrosine-specific tRNA (tRNA^{Tyr}) and tyrosyl-tRNA synthetase (TyrRS) are "orthogonal" to those from the bacterial systems (Kobayashi, Yanagisawa, Sakamoto, & Yokoyama, 2009) and are useful for cell-free, site-directed, stable isotope labeling of a Tyr residue in a protein. Several orthogonal tRNA and aminoacyl-tRNA synthetase pairs have been developed for the site-directed incorporation of non-natural amino acids into proteins (Hirao et al., 2002; Kiga et al., 2002; Mukai et al., 2010, 2011; Sakamoto et al., 2002; Wang, Brock, Herberich, & Schultz, 2001; Yanagisawa et al., 2014).

3.4 Constructs and Cell-Free Reaction Conditions

The designs of the coding region and the tags are based on those described in Section 2. As for mammalian proteins that contain one or more functional domains with terminal tail(s) and/or linker(s), cDNA fragments encoding the domain of interest are prepared from the full-length cDNA, with a series of PCR primers corresponding to the N- and C-terminal ends of the domain fragments. First, the expression levels and properties of the constructs are examined, by small-scale cell-free protein synthesis reactions with nonlabeled amino acids, rather than the stable isotope-labeled ones. As the medium- and large-scale cell-free productions of stable isotope-labeled proteins are performed in the dialysis mode, ideally the small-scale cell-free synthesis reactions should also be performed in the dialysis mode. For this purpose, a 30-µL reaction solution in a Slide-A-Lyzer MINI Dialysis Unit with a 10 kDa MWCO dialysis membrane (Thermo Scientific Pierce) is inserted into a cavity of 24-well V-bottom plate, which holds the 1000-µL external solution as a reservoir (Fig. 3). The 24-well plate is wrapped with

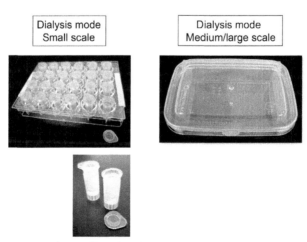

Figure 3 Apparatuses for small- and medium/large-scale dialysis-mode cell-free protein synthesis. The reaction solution, in a dialysis tube (Spectra/Por 7, MWCO: 15 kDa, Spectrum, USA), is dialyzed against the external solution with gentle shaking.

plastic wrap and incubated at 15–30 °C for 3–24 h. Thus, the reaction conditions, such as the template DNA concentration, the incubation temperature, the incubation time, and the necessity of additives (molecular chaperones, ligands, etc.), are optimized, and promising constructs are selected, with respect to the synthesis level and the solubility, by SDS-PAGE analysis of the products.

Cell-free protein synthesis is particularly useful for screening a large number of fragments for optimization of the ends, and isotope-aided one- or two-dimensional NMR spectroscopy is a powerful method for selecting the best ones (the cell-free production protocol is described in Sections 3.5 and 3.6). A few examples of such applications are described in Sections 3.8–3.10. In the case of a complex of a functional domain with a peptide from its binding partner protein, a fully labeled sample can be obtained by isotope labeling of the domain fragment fused at its N- or C-terminus with the peptide through a linker, while the chemical synthesis of stable isotope-labeled peptides is usually expensive. Examples of the cost-effective preparation of fully labeled peptide–protein complexes are described in Section 3.10.

For site-directed stable isotope labeling, the codon corresponding to the specified site in the coding region is replaced by TAG for the UAG suppressor tRNA. The preparation of the stable isotope-labeled fragment for the segmental labeling of a large protein, by the mechanisms derived from intein-mediated *trans* splicing (Liu et al., 2009; Ludwig et al., 2009; Yamazaki et al., 1998), can be efficiently performed by the cell-free synthesis method and may be combined with the cell-free amino acid-type selective labeling.

3.5 Medium-Scale Production of ^{15}N-Labeled Proteins

Milligram quantities of proteins for NMR analyses are prepared by the medium- and large-scale dialysis-mode cell-free synthesis methods. To evaluate the suitability of the protein sample for NMR structure determination, ^{1}H–^{15}N HSQC spectra of ^{15}N-labeled proteins provide direct and reliable information. A series of ^{15}N-labeled protein samples are efficiently produced from the DNA constructs described above (Section 3.4), by the medium-scale dialysis-mode cell-free synthesis method with ^{15}N-labeled amino acid(s), instead of nonlabeled amino acid(s). The typical protocol uses a

Table 3 Standard Composition of the Reaction Solution for Stable Isotope Labeling by the Medium- and Large-Scale *E. coli* Cell-Free Protein Synthesis Methods

	Reaction Solution (mL)	External Solution (mL)
LMCP(D-Glu)	1.12 (3.36)	11.2 (33.6)
20 mM labeled amino acid mixture	0.225 (0.675)	2.25 (6.75)
5% NaN$_3$	0.03 (0.09)	0.3 (0.9)
1.6 M Mg(OAc)$_2$	0.017 (0.052)	0.17 (0.52)
17.5 mg/mL tRNA	0.03 (0.09)	
3.75 mg/mL creatine kinase	0.2 (0.6)	
S30 buffer		9 (27)
S30 extract	0.9 (2.7)	
10 mg/mL T7 RNA polymerase	0.02 (0.06)	
Other factors		
Template DNA		
Milli-Q water	To 3 (9)	To 30 (90)
Total	3 (9)	30 (90)

The volumes for the large-scale synthesis are shown in parentheses.

3-mL reaction solution and a 30-mL external solution (Fig. 3 and Table 3), as follows:

1. *Preparation of the dialysis tube for the 3-mL reaction solution*: Wash an 11 cm-long dialysis tube (Spectra/Por 7, MWCO: 15,000, Spectrum; Cat. 132124) with Milli-Q water and keep it submerged in Milli-Q water in a beaker until use.
2. *Thawing of stock reagents*: Thaw the cell-free synthesis reagents in the tubes on ice. Gently shake each tube on ice, until the solution becomes homogeneous. Keep all of the reagents on ice until use.
3. *Preparation of the 30-mL external solution*: Combine the 20 mM ^{15}N-labeled amino acid solution (2.25 mL) and the LMCP(D-Glu) solution (11.2 mL) in a 50-mL centrifuge tube or 50-mL polypropylene container and mix it thoroughly. To this solution, add the sodium azide solution (0.3 mL), the magnesium acetate solution (0.17 mL), the S30 buffer (10 mM Tris–acetate buffer (pH 8.2) containing 60 mM

potassium acetate, 16 mM magnesium acetate, and 1 mM DTT) (9 mL), and other factor(s) as needed, such as $ZnSO_4/ZnCl_2$, the appropriate ratio of GSH and GSSG, and/or detergents. Bring the volume of the mixture solution to 30 mL with Milli-Q water. After mixing the solution well by turning the 50-mL tube or container upside down several times, transfer the solution to a 150-mL square-shaped polystyrene case (Fig. 3).

4. *Preparation of the 3-mL reaction solution*: Transfer the 20 mM ^{15}N-labeled amino acid solution (0.225 mL) into the LMCP(D-Glu) solution (1.12 mL) in a 15-mL centrifuge tube and mix it thoroughly. To this solution, add the sodium azide solution (30 µL), the magnesium acetate solution (17 µL), and other low-molecular mass factor(s) as needed, such as $ZnSO_4/ZnCl_2$ or the appropriate ratio of GSH and GSSG. Subsequently, add the tRNA solution (90 µL), the creatine kinase solution (200 µL), the S30 extract (900 µL), the T7 RNA polymerase solution (20 µL), and other factor(s) as needed, such as chaperones (*E. coli* DnaK/DnaJ/GrpE, GroEL/GroES, DsbC, Skp, etc.), ligands, lipids, and/or detergents. Finally, add the template DNA and bring the volume of the mixture to 3 mL with Milli-Q water. Mix the solution well, by gently turning the tube or container upside down several times.

5. *Setting up the reaction solution in a dialysis tube*: Close one end of the dialysis tube with a closure (Spectrum). Transfer the 3-mL reaction solution into the dialysis tube, remove as much air from the tube as possible, and seal the open end of the tube with another closure.

6. *The dialysis-mode reaction*: Immerse the dialysis tube in the external solution in the polystyrene case. Wrap the case with plastic wrap and start shaking it reciprocally with an incubator shaker (50 mm amplitude, 50 rpm), using the optimized temperature and reaction time.

7. *Recovery of the reaction solution*: After the reaction, remove the reaction solution from the dialysis tube and place it into a 35-mL conical Oak Ridge tube (Nalgene). Transfer a 20-µL aliquot into an Eppendorf tube for SDS-PAGE analysis. Centrifuge the 35-mL conical Oak Ridge tube at 16,000 rpm, 4 °C, for 20 min, transfer the supernatant to a 50-mL centrifuge tube, and keep it on ice until purification.

8. *Estimation of the expression level*: Centrifuge the 20-µL aliquot of the reaction mixture in the Eppendorf tube, at 16,000 rpm for 10 min. Transfer the supernatant to another Eppendorf tube. Suspend the precipitate in an

appropriate buffer, using the same volume as that of the supernatant. Analyze 1 μL each of the total reaction mixture, the supernatant, and the suspended precipitate by SDS-PAGE.

3.6 Large-Scale Production of ^{13}C/^{15}N-Labeled Proteins

For NMR structure determination, uniformly ^{13}C/^{15}N-labeled proteins are produced by the large-scale dialysis-mode cell-free synthesis method, with ^{13}C/^{15}N-labeled amino acid(s) instead of the nonlabeled one(s). The typical protocol uses a 9-mL reaction solution and a 90-mL external solution.

1. *Preparation of the dialysis tube for the 9-mL reaction solution*: Wash a 14 cm-long dialysis tube (Spectra/Por 7, MWCO: 15,000, Spectrum; Cat. 132124) with Milli-Q water, and submerge it in Milli-Q water in a beaker until use.
2. *Thawing of stock reagents*: Thaw the cell-free synthesis reagents in the tubes on ice. Gently shake each tube on ice, until the solution becomes homogeneous. Keep all of the reagents on ice until use.
3. *Preparation of the 90-mL external solution*: Combine the 20 mM ^{13}C/^{15}N-labeled amino acid solution (6.75 mL) and the LMCP(D-Glu) solution (33.6 mL) in a 100-mL polypropylene bottle and mix it thoroughly. To this solution, add the sodium azide solution (0.9 mL), the magnesium acetate solution (0.52 mL), the S30 buffer (27 mL), and other factor(s) as needed, such as ZnSO$_4$/ZnCl$_2$, the appropriate ratio of GSH and GSSG, and/or detergents. Bring the volume of the mixture solution to 90 mL with Milli-Q water. After mixing the solution well, by turning the 100-mL bottle upside down several times, transfer the solution from the 100-mL bottle to a 400-mL square-shaped polystyrene case (Fig. 3).
4. *Preparation of the 9-mL reaction solution*: Transfer the 20 mM ^{13}C/^{15}N-labeled amino acid solution (0.675 mL) into the LMCP(D-Glu) solution (3.36 mL) in a 50-mL centrifuge tube or 50-mL polypropylene container and mix it thoroughly. To this solution, add the sodium azide solution (90 μL), the magnesium acetate solution (52 μL), and other low-molecular mass factor(s) as needed, such as ZnSO$_4$/ZnCl$_2$ and/or the appropriate ratio of GSH and GSSG. Subsequently, add the tRNA solution (90 μL), the creatine kinase solution (600 μL), the S30 extract (2.7 mL), the T7 RNA polymerase solution (60 μL), and other factor(s) as needed, such as chaperones (*E. coli* DnaK/DnaJ/GrpE, GroEL/GroES, DsbC, Skp, etc.), ligands, lipids, and/or detergents.

Finally, add the template DNA and bring the volume of the mixture to 9 mL with Milli-Q water. Mix the solution well, by gently turning the tube or container upside down several times.

5. *Seting up the reaction solution in a dialysis tube*: Close one end of the dialysis tube with a closure (Spectrum). Transfer the 9-mL reaction solution into the dialysis tube, remove as much air from the tube as possible, and seal the open end of the tube with another closure.
6. *The dialysis-mode reaction*: Immerse the dialysis tube in the external solution in the polystyrene case. Wrap the case with plastic wrap and start shaking it reciprocally with an incubator shaker (50-mm amplitude, 50 rpm), using the optimized temperature and reaction time.
7. *Recovery of the reaction solution*: After the reaction, remove the reaction solution from the dialysis tube and place it into a 35-mL conical Oak Ridge tube (Nalgene). Transfer a 20-μL aliquot into an Eppendorf tube for SDS-PAGE analysis. Centrifuge the 35-mL conical Oak Ridge tube at 16,000 rpm, 4 °C, for 20 min, transfer the supernatant to a 50-mL centrifuge tube, and keep it on ice until purification.
8. *Estimation of the expression level*: Centrifuge the 20-μL aliquot of the reaction mixture in the Eppendorf tube at 16,000 rpm for 10 min. Transfer the supernatant to another Eppendorf tube. Suspend the precipitate in an appropriate buffer, using the same volume as that of the supernatant. Analyze 1 μL each of the total reaction mixture, the supernatant, and the suspended precipitate by SDS-PAGE.

3.7 Purification of Synthesized Proteins

After the medium- or large-scale protein synthesis reaction, the produced protein is purified by affinity column chromatography in the first step. The supernatant of the reaction solution is filtered through a 25-mm diameter syringe filter with a 0.45-μm pore size PVDF membrane, diluted threefold with the first affinity column buffer (20 mM Tris–HCl buffer (pH 8.0) containing 1 M NaCl and 20 mM imidazole for His-tag affinity purification), and applied to the affinity purification column equilibrated with the same buffer. The protein is eluted with a linear gradient up to 500 mM imidazole, in 20 mM Tris–HCl buffer (pH 8.0) containing 500 mM NaCl. The protein-containing fraction is incubated with the specific protease (e.g., tobacco etch virus or SUMO protease) to cleave the affinity tag, and the released affinity tag peptide is removed with the same affinity column. When the protease is permanently fused with the same affinity tag, the affinity-tagged protease can be removed together with the affinity tag peptides in one step. After buffer exchange by desalting chromatography or dialysis, the fraction containing the produced protein is subjected to ion-exchange chromatography. If necessary, the produced protein is further purified by gel-filtration chromatography.

3.8 WWE Domains

As an example of the cell-free production of stable isotope-labeled proteins, we describe the case of the WWE domains of poly-ADP-ribose polymerases (PARPs) (He et al., 2012). The DNA templates for cell-free reactions were constructed for the WWE domains of human PARP11 and mouse PARP14, as well as that of mouse RNF146 for comparison (corresponding to residues 15–105, 1542–1618, and 83–179, respectively). The uniformly ^{13}C- and ^{15}N-labeled WWE domain fragments were synthesized by the cell-free dialysis-mode protein synthesis method with a ^{13}C/^{15}N-labeled amino acid mixture, as described above (Section 3.6). Thus, the solution structures of these three WWE domain fragments were determined, as shown in Fig. 4A–C. The ADP-ribose titration ^{1}H–^{15}N HSQC experiment was performed for the PARP11 (Fig. 5A) and RNF146 (Fig. 5B) WWE domain fragments, and the perturbed residues were mapped on the model structures (Fig. 5C and D). The structures of the PARP11 and RNF146 WWE domains revealed their different specificities for the ADP-ribose moieties within the poly-ADP-ribose chain.

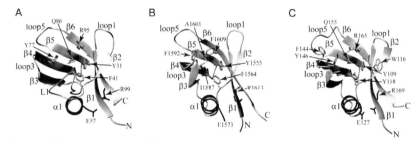

Figure 4 Solution structures (ribbon representations) of the WWE domains from human PARP11 (residues 25–103) (A), mouse PARP14 (residues 1549–1617) (B), and mouse RNF146 (residues 103–173) (C). The β-strands and α-helical structures are colored cyan (gray in the print version) and red (dark gray in the print version), respectively. The side chains of the highly conserved residues among the WWE domains are colored green (gray in the print version). *Adopted with modifications from He et al. (2012).*

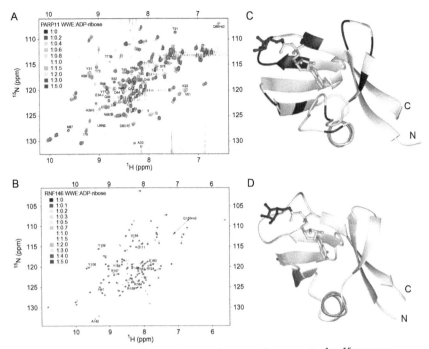

Figure 5 NMR chemical shift perturbations by ADP-ribose on the ^1H–^{15}N HSQC spectra of the WWE domains of PARP11 (A) and RNF146 (B), and color-coded mapping of the degrees of chemical shift changes of the perturbed residues on the model structures ((C) and (D), respectively). *Adopted with modifications from He et al. (2012).* (See the color plate.)

3.9 Zinc-Binding Proteins

Zinc-binding proteins require zinc for their structures and functions. The solubility and/or stability of the proteins significantly increase in the presence of an appropriate concentration of zinc (50–100 μM $ZnSO_4$ or $ZnCl_2$), during the cell-free protein synthesis. The correctly folded proteins are synthesized by the cell-free method, as confirmed by NMR analyses. Therefore, the cell-free method can be used to synthesize correctly folded and functional zinc-binding proteins (Matsuda et al., 2006).

One of the zinc-binding proteins produced by the cell-free synthesis method is the zinc-finger gene in the autoimmune thyroid disease (AITD) susceptibility region (ZFAT) (Tochio et al., 2015). ZFAT is a transcriptional regulator that contains 18 C_2H_2-type zinc-fingers and 1

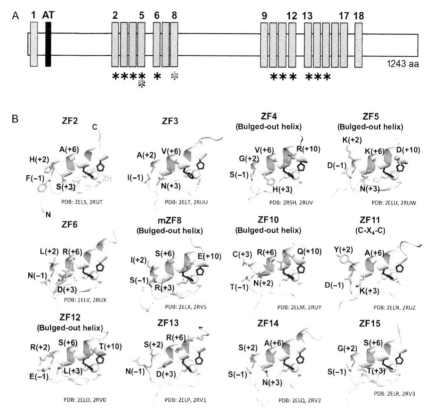

Figure 6 Zinc-finger motifs of ZFAT. (A) The positions of the 18 C_2H_2-type zinc-finger motifs (ZF1–ZF18) and the AT-hook motif (AT) are indicated by the green and blue boxes, respectively, along the primary structure of ZFAT. The black and violet asterisks indicate the zinc-finger motifs with solution structures determined for the human and mouse ZFAT proteins, respectively. (B) Ribbon diagrams of the solution structures of the isolated zinc-fingers, ZF2–ZF6, ZF8, ZF10–ZF15, of ZFAT. The zinc ion and the zinc-coordinating His and Cys residues of each zinc-finger motif are shown as a yellow ball and magenta and cyan stick models, respectively. The residues potentially involved in DNA recognition are shown as green stick models. *Adopted with modifications from Tochio et al. (2015).* (See the color plate.)

AT-hook (Fig. 6A) and is involved in AITD, apoptosis, and immune-related cell survival. The individual zinc-fingers, ZF2–ZF5, ZF6, ZF8, ZF10, ZF11, and ZF13, of ZFAT (Fig. 6A), were produced by the cell-free method, and their solution structures were determined by NMR spectroscopy (Fig. 6B; Tochio et al., 2015).

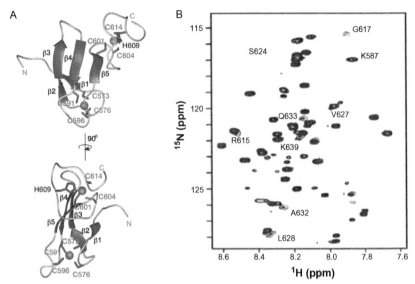

Figure 7 The zinc-binding region (ZBR) of the Emi2 protein. (A) Solution structure of the Emi2 ZBR domain (two different views are shown in ribbon diagram representations). The β-strands are colored blue (dark gray in the print version). The two zinc ions are shown as magenta (gray in the print version) balls, and the zinc-coordinating Cys and His residues are labeled in orange (gray in the print version) and violet (dark gray in the print version), respectively. (B) Superimposition of the ^{1}H–^{15}N HSQC spectra of the ZBR-RL fragment (mouse Emi2; residues 547–641), in the presence (magenta; black in the print version) and absence (yellow; light gray in the print version) of the ANAPC2$_{CW}$ fragment (mouse ANAPC2 cullin-Winged-helix subdomain; residues 512–837). *Adopted with modifications from Shoji et al. (2014).*

Anaphase-promoting complex or cyclosome (APC/C) is a multisubunit ubiquitin ligase E3 that targets cell cycle regulators and is activated by Cdc20 in M phase. The Emi2 protein inhibits APC/C-Cdc20 and causes long-term M phase arrest of mature oocytes. The zinc-binding region (ZBR) domain of Emi2 is important for the inhibition. A fragment corresponding to the ZBR domain was produced by the cell-free method, and its solution structure was determined by NMR spectroscopy (Fig. 7A). The interaction of the Emi2 ZBR domain with APC/C (Fig. 7B) was analyzed by chemical shift perturbation experiments (Shoji et al., 2014).

3.10 Peptide Complexes

The phosphotyrosine interaction domain 2 (PID2) of Fe65L1 (Fig. 8A) interacts with a 32-mer peptide derived from the cytoplasmic domain of

Alzheimer amyloid precursor protein (APP) (Fig. 8B). Three different constructs, in which the N- or C-terminus of the PID2 domain of Fe65L1 was fused to the 32-mer peptide of APP through a 14- or 23-residue linker (Fig. 8C), were $^{13}C/^{15}N$-labeled by the cell-free protein synthesis method, and their solution structures were determined by NMR spectroscopy (Fig. 9A–C). This protein fusion stabilized the PID2–peptide complex, while enabling the $^{13}C/^{15}N$-labeling of both the PID2 domain and the peptide. On the basis of these NMR analyses, the solution structure was determined for the complex between the $^{13}C/^{15}N$-labeled PID2 domain and the chemically synthesized, nonlabeled 32-mer peptide (Fig. 9D and E). These complex structures revealed the novel and extensive mode of peptide binding by the PID domain (Li et al., 2008).

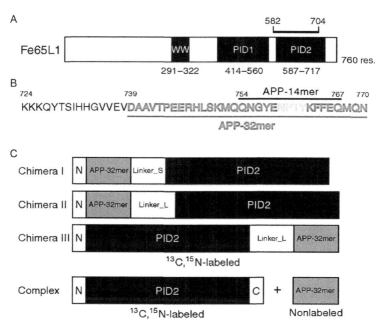

Figure 8 The complex of the second PID domain of the Fe65L1 protein with the 32-mer peptide from APP. (A) The domains of Fe65L1. (B) The amino acid sequence of the 32-mer peptide that binds to the PID2 domain of Fe65L1. (C) Chimeras and the complex of the Fe65L1 PID2 domain and the APP 32-mer peptide. Chimeras I–III, in which the uniformly $^{13}C,^{15}N$-labeled PID2 domain of Fe65L1 was connected through the N- or C-terminal linker S/L to the APP 32-mer peptide. The complex is formed between the Fe65L1 PID2 domain and the APP 32-mer peptide, which were $^{13}C,^{15}N$-labeled and nonlabeled, respectively. *Adopted with modifications from Li et al. (2008)*.

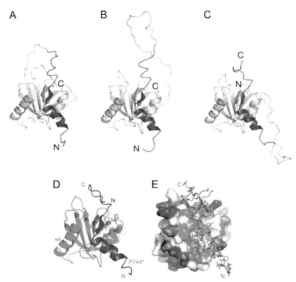

Figure 9 Solution structures of chimeras I–III and the complex between the Fe65L1 PID2 domain and the APP 32-mer peptide. (A–C) Ribbon representations of chimeras I–III, respectively. The Fe65L1 PID2 domain, the APP 32-mer peptide, and the linker are shown in gray, magenta, and yellow, respectively. (D) Ribbon representation of the complex between the Fe65L1 PID2 domain and the APP 32-mer peptide. The PID2 domain and the APP 32-mer peptide are shown in green and magenta, respectively. (E) Surface presentation of the Fe65L1 PID2 domain in complex with the APP 32-mer peptide, shown as a stick model. The PID2 surface is colored with the hydrophobic residues (Ala, Val, Leu, Ile, Met, Phe, and Trp) in green, the positively charged residues (Arg and Lys) in blue, the negatively charged residues (Asp and Glu) in red, and the other residues in gray. *Adopted with modifications from Li et al. (2008).* (See the color plate.)

DOCK2 (dedicator of cytokinesis 2) activates the small GTP-binding protein Rac and thereby plays a critical role in cellular signaling events. The N-terminal SH3 domain of DOCK2 binds to the C-terminal Pro-rich region of ELMO1 (engulfment and cell motility 1) (Fig. 10A), which is important for Rac activation. First, the precise regions of DOCK2 and ELMO1 required for their association were elucidated, by expressing 87 different fused complexes by the cell-free protein synthesis method (Fig. 11A and B). Thus, the solution structure of one of the fusion proteins [an ELMO1 peptide (residues 697–722) fused to the N-terminus of the DOCK2 SH3 domain (residues 8–70)] (Fig. 11A) was determined by

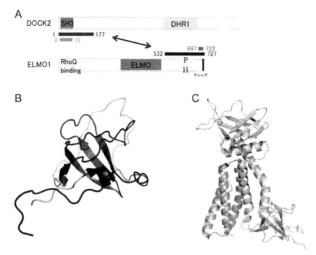

Figure 10 The complex of the SH3 domain of DOCK2 and the Pro-rich region of ELMO1. (A) Domain organizations of the N-terminal parts of DOCK2 and ELMO1. The regions involved in the association between DOCK2 and ELMO1 are residues 8–70 (red) and residues 697–722 (blue), respectively, for NMR spectroscopy and residues 1–177 (orange) and residues 532–727 (green) for crystallography. (B) The solution structure of the ELMO1 (697–722)–DOCK2 (8–70) fusion protein. The vector-derived peptide, the ELMO1 peptide, the linker, and the DOCK2 SH3 domain are colored light green, blue, gray, and red, respectively. (C) The crystal structure of the DOCK2 (1–177)·ELMO1 (532–727) complex. DOCK2 and ELMO1 are shown in orange and green, respectively. *Adopted with modifications from Hanawa-Suetsugu et al. (2012).* (See the color plate.)

cell-free $^{13}C/^{15}N$ labeling (Fig. 10B). On the basis of the structural information, a larger construct of the DOCK2–ELMO complex was designed, and its crystal structure was solved (Fig. 10C). Thus, these structures revealed an exceptionally intensive interaction between the SH3 domain of DOCK2 and the C-terminal region of ELMO (Hanawa-Suetsugu et al., 2012).

ACKNOWLEDGMENTS

We thank Ms. T. Imada and Ms. T. Nakayama for their assistance in manuscript preparation. This work was supported by the RIKEN Structural Genomics/Proteomics Initiative (RSGI), the National Project on Protein Structural and Functional Analyses, the Targeted Proteins Research Program (TPRP), and the Platform for Drug Discovery, Informatics, and Structural Life Science, of the Ministry of Education, Culture, Sports, Science and Technology (MEXT) of Japan (to S.Y.).

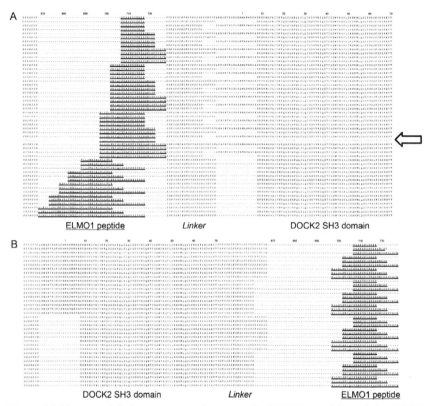

Figure 11 The tested fusion constructs between the ELMO1 peptide and the DOCK2 SH3 domain. (A) ELMO1 peptide—linker—DOCK2 SH3 domain constructs. (B) DOCK2 SH3 domain—linker—ELMO1 peptide constructs. All of the constructs have seven extra N-terminal residues derived from the expression vector. The residue numbers are shown at the top. The construct used for the solution structure determination is colored red (gray in the print version). *Adopted with modifications from Hanawa-Suetsugu et al. (2012).*

REFERENCES

Ahn, J. H., Chu, H. S., Kim, T. W., Oh, I. S., Choi, C. Y., Hahn, G. H., et al. (2005). Cell-free synthesis of recombinant proteins from PCR-amplified genes at a comparable productivity to that of plasmid-based reactions. *Biochemical and Biophysical Research Communications, 338*(3), 1346–1352.

Aoki, M., Matsuda, T., Tomo, Y., Miyata, Y., Inoue, M., Kigawa, T., et al. (2009). Automated system for high-throughput protein production using the dialysis cell-free method. *Protein Expression and Purification, 68*, 128–136.

Brown, G., Singer, A., Proudfoot, M., Skarina, T., Kim, Y., Chang, C., et al. (2008). Functional and structural characterization of four glutaminases from *Escherichia coli* and *Bacillus subtilis*. *Biochemistry, 47*, 5724–5735.

Carlson, E. D., Gan, R., Hodgman, C. E., & Jewett, M. C. (2012). Cell-free protein synthesis: Applications come of age. *Biotechnology Advances, 30*, 1185–1194.

Davanloo, P., Rosenberg, A. H., Dunn, J. J., & Studier, F. W. (1984). Cloning and expression of the gene for bacteriophage T7 RNA polymerase. *Proceedings of the National Academy of Sciences of the United States of America, 81*, 2035–2039.

Etezady-Esfarjani, T., Hiller, S., Villalba, C., & Wüthrich, K. (2007). Cell-free protein synthesis of perdeuterated proteins for NMR studies. *Journal of Biomolecular NMR, 39*, 229–238.

Falzone, C. J., Karsten, W. E., Conley, J. D., & Viola, R. E. (1988). L-Aspartase from *Escherichia coli*: Substrate specificity and role of divalent metal ions. *Biochemistry, 27*, 9089–9093.

Gardner, K. H., & Kay, L. E. (1997). Production and incorporation of ^{15}N, ^{13}C, ^{2}H (^{1}H-δ1 methyl) isoleucine into proteins for multidimensional NMR studies. *Journal of the American Chemical Society, 119*, 7599–7600.

Goto, N. K., Gardner, K. H., Mueller, G. A., Willis, R. C., & Kay, L. E. (1999). A robust and cost-effective method for the production of Val, Leu, Ile (d1) methyl-protonated ^{15}N-, ^{13}C-, ^{2}H-labeled proteins. *Journal of Biomolecular NMR, 13*, 369–374.

Hahn, G. H., & Kim, D. M. (2006). Production of milligram quantities of recombinant proteins from PCR-amplified DNAs in a continuous-exchange cell-free protein synthesis system. *Analytical Biochemistry, 355*(1), 151–153.

Hanawa-Suetsugu, K., Kukimoto-Niino, M., Mishima-Tsumagari, C., Akasaka, R., Ohsawa, N., Sekine, S., et al. (2012). Structural basis for mutual relief of the Rac guanine nucleotide exchange factor DOCK2 and its partner ELMO1 from their auto inhibited forms. *Proceedings of the National Academy of Sciences of the United States of America, 109*, 3305–3310.

Handschumacher, R. E., Bates, C. J., Chang, P. K., Andrews, A. T., & Fischer, G. A. (1968). 5-Diazo-4-oxo-L-norvaline: Reactive asparagine analog with biological specificity. *Science, 161*, 62–63.

Hartman, S. C. (1968). Glutaminase of *Escherichia coli*: III. Studies on there action mechanism. *The Journal of Biological Chemistry, 243*, 870–878.

He, F., Tsuda, K., Takahashi, M., Kuwasako, K., Terada, T., Shirouzu, M., et al. (2012). Structural insight into the interaction of ADP-ribose with the PARP WWE domains. *FEBS Letters, 586*, 3858–3864.

Henrich, E., Hein, C., Dötsch, V., & Bernhard, F. (2015). Membrane protein production in *Escherichia coli* cell-free lysates. *FEBS Letters, 589*(15), 1713–1722.

Hirao, I., Ohtsuki, T., Fujiwara, T., Mitsui, T., Yokogawa, T., Okuni, T., et al. (2002). An unnatural base pair for incorporating amino acid analogs into proteins. *Nature Biotechnology, 20*, 177–182.

Ishihara, G., Goto, M., Saeki, M., Ito, K., Hori, T., Kigawa, T., et al. (2005). Expression of G protein coupled receptors in a cell-free translational system using detergents and thioredoxin-fusion vectors. *Protein Expression and Purification, 41*, 27–37.

Jackson, A. M., Boutell, J., Cooley, N., & He, M. (2004). Cell-free protein synthesis for proteomics. *Briefings in Functional Genomics & Proteomics, 2*, 308–319.

John, R. A., Charteris, A., & Fowler, L. J. (1978). The reaction of amino-oxyacetate with pyridoxal phosphate-dependent enzymes. *The Biochemical Journal, 171*, 771–779.

Junge, F., Haberstock, S., Roos, C., Stefer, S., Proverbio, D., Dötsch, V., et al. (2011). Advances in cell-free protein synthesis for the functional and structural analysis of membrane proteins. *Nature Biotechnology, 28*, 262–271.

Kadokura, H., Katzen, F., & Beckwith, J. (2003). Protein disulfide bond formation in prokaryotes. *Annual Review of Biochemistry, 72*, 111–135.

Kainosho, M., Torizawa, T., Iwashita, Y., Terauchi, T., Ono, M. A., & Güntert, P. (2006). Optimal isotope labeling for NMR protein structure determinations. *Nature, 440*, 52–57.

Kiga, D., Sakamoto, K., Kodama, K., Kigawa, T., Matsuda, T., Yabuki, T., et al. (2002). An engineered *Escherichia coli* tyrosyl-tRNA synthetase for site-specific incorporation of an unnatural amino acid into proteins in eukaryotic translation and its application in a wheat germ cell-free system. *Proceedings of the National Academy of Sciences of the United States of America, 99*, 9715–9720.

Kigawa, T. (2010a). Cell-free protein preparation through prokaryotic transcription–translation methods. *Methods in Molecular Biology, 607*, 1–10.

Kigawa, T. (2010b). Cell-free protein production system with the *E. coli* crude extract for determination of protein folds. *Methods in Molecular Biology, 607*, 101–111.

Kigawa, T., Inoue, M., Aoki, M., Matsuda, T., Yabuki, T., Seki, E., et al. (2008). The use of the *Escherichia coli* cell-free protein synthesis for structural biology and structural proteomics. In A. S. Spirin & J. R. Swartz (Eds.), *Methods and protocols: Cell-free protein synthesis* (pp. 99–109). Weinheim: Wiley-VCH Verlag GmbH & Co. KGaA.

Kigawa, T., Matsuda, T., Yabuki, T., & Yokoyama, S. (2008). Bacterial cell-free system for highly efficient protein synthesis. In A. S. Spirin & J. R. Swartz (Eds.), *Methods and protocols: Cell-free protein synthesis* (pp. 83–97). Weinheim: Wiley-VCH Verlag GmbH & Co. KGaA.

Kigawa, T., Muto, Y., & Yokoyama, S. (1995). Cell-free synthesis and amino acid-selective stable isotope labeling of proteins for NMR analysis. *Journal of Biomolecular NMR, 6*, 129–134.

Kigawa, T., Yabuki, T., Matsuda, N., Matsuda, T., Nakajima, R., Tanaka, A., et al. (2004). Preparation of *Escherichia coli* cell extract for highly productive cell-free protein expression. *Journal of Structural and Functional Genomics, 5*, 63–68.

Kigawa, T., Yabuki, T., Yoshida, Y., Tsutsui, M., Ito, Y., Shibata, T., et al. (1999). Cell-free production and stable-isotope labeling of milligram quantities of proteins. *FEBS Letters, 442*, 15–19.

Kigawa, T., Yamaguchi-Nunokawa, E., Kodama, K., Matsuda, T., Yabuki, T., Matsuda, N., et al. (2002). Selenomethionine incorporation into a protein by cell-free synthesis. *Journal of Structural and Functional Genomics, 2*, 29–35.

Kigawa, T., & Yokoyama, S. (1991). A continuous cell-free protein synthesis system for coupled transcription–translation. *Journal of Biochemistry, 110*, 166–168.

Kim, D. M., & Choi, C. H. (1996). A semicontinuous prokaryotic coupled transcription/translation system using a dialysis membrane. *Biotechnology Progress, 12*, 645–649.

Kim, T. W., Keum, J. W., Oh, I. S., Choi, C. Y., Park, C. G., & Kim, D. M. (2006). Simple procedures for the construction of a robust and cost-effective cell-free protein synthesis system. *Journal of Biotechnology, 126*(4), 554–561.

Kim, D. M., Kigawa, T., Choi, C.-Y., & Yokoyama, S. (1996). A highly efficient cell-free protein synthesis system from *Escherichia coli*. *European Journal of Biochemistry, 239*, 881–886.

Kim, D. M., & Swartz, J. R. (1999). Prolonging cell-free protein synthesis with a novel ATP regeneration system. *Biotechnology and Bioengineering, 66*(3), 180–188.

Kim, D. M., & Swartz, J. R. (2000). Prolonging cell-free protein synthesis by selective reagent additions. *Biotechnology Progress, 16*(3), 385–390.

Kim, D. M., & Swartz, J. R. (2004). Efficient production of a bioactive, multiple disulfide-bonded protein using modified extracts of *Escherichia coli*. *Biotechnology and Bioengineering, 85*(2), 122–129.

Kimura-Someya, T., Shirouzu, M., & Yokoyama, S. (2014). Cell-free membrane protein expression. *Methods in Molecular Biology, 1118*, 267–273.

Kobayashi, T., Yanagisawa, T., Sakamoto, K., & Yokoyama, S. (2009). Recognition of non-α-amino substrates by pyrrolysyl-tRNA synthetase. *Journal of Molecular Biology, 385*(5), 1352–1360.

Koizumi, M., Hiratake, J., Nakatsu, T., Kato, H., & Oda, J. I. (1999). A potent transition-state analogue inhibitor of *Escherichia coli* asparagine synthetase A. *Journal of the American Chemical Society*, *121*, 5799–5800.

Kuchenreuther, J. M., George, S. J., Grady-Smith, C. S., Cramer, S. P., & Swartz, J. R. (2011). Cell-free H-cluster synthesis and [FeFe] hydrogenase activation: All five CO and CN^- ligands derive from tyrosine. *PLoS One*, *6*(5), e20346.

Li, H., Koshiba, S., Hayashi, F., Tochio, N., Tomizawa, T., Kasai, T., et al. (2008). Structure of the C-terminal phosphotyrosine interaction domain of Fe65L1 complexed with the cytoplasmic tail of amyloid precursor protein reveals a novel peptide binding mode. *The Journal of Biological Chemistry*, *183*(40), 27165–27178.

Liu, D., Xu, R., & Cowburn, D. (2009). Segmental isotopic labeling of proteins for nuclear magnetic resonance. *Methods in Enzymology*, *462*, 151–175.

Ludwig, C., Schwarzer, D., Zettler, J., Garbe, D., Janning, P., Czeslik, C., et al. (2009). Semisynthesis of proteins using split inteins. *Methods in Enzymology*, *462*, 77–96.

Madin, K., Sawasaki, T., Ogasawara, T., & Endo, Y. (2000). A highly efficient and robust cell-free protein synthesis system prepared from wheat embryos: Plants apparently contain a suicide system directed at ribosomes. *Proceedings of the National Academy of Sciences of the United States of America*, *97*, 559–564.

Makino, S., Goren, M. A., Fox, B. G., & Markley, J. L. (2010). Cell-free protein synthesis technology in NMR high-throughput structure determination. *Methods in Molecular Biology*, *607*, 127–147.

Manning, J. M., Moore, S., Rowe, W. B., & Meister, A. (1969). Identification of L-methionine S-sulfoximine as the diastereoisomer of L-methionine S, R-sulfoximine that inhibits glutamine synthetase. *Biochemistry*, *8*, 2681–2685.

Matsuda, T., Kigawa, T., Koshiba, S., Inoue, M., Aoki, M., Yamasaki, K., et al. (2006). Cell-free synthesis of zinc-binding proteins. *Journal of Structural and Functional Genomics*, *7*, 93–100.

Matsuda, T., Koshiba, S., Tochio, N., Seki, E., Iwasaki, N., Yabuki, T., et al. (2007). Improving cell-free protein synthesis for stable-isotope labeling. *Journal of Biomolecular NMR*, *37*, 225–229.

Matsuda, T., Watanabe, S., & Kigawa, T. (2013). Cell-free synthesis system suitable for disulfide-containing proteins. *Biochemical and Biophysical Research Communications*, *431*, 296–301.

Michel, E., Skrisovska, L., Wuthrich, K., & Allain, F. H. (2013). Amino acid-selective segmental isotope labeling of multidomain proteins for structural biology. *Chembiochem*, *14*, 457–466.

Mikami, S., Kobayashi, T., Masutani, M., Yokoyama, S., & Imataka, H. (2008). A human cell-derived *in vitro* coupled transcription/translation system optimized for production of recombinant proteins. *Protein Expression and Purification*, *62*, 190–198.

Mikami, S., Masutani, M., Sonenberg, N., Yokoyama, S., & Imataka, H. (2006). An efficient mammalian cell-free translation system supplemented with translation factors. *Protein Expression and Purification*, *46*, 348–357.

Mukai, T., Hayashi, A., Iraha, F., Sato, A., Ohtake, K., Yokoyama, S., et al. (2010). Codon reassignment in the *Escherichia coli* genetic code. *Nucleic Acids Research*, *38*, 8188–8195.

Mukai, T., Yanagisawa, T., Ohtake, K., Wakamori, M., Adachi, J., Hino, N., et al. (2011). Genetic-code evolution for protein synthesis with non-natural amino acids. *Biochemical and Biophysical Research Communications*, *411*, 757–761.

Muller, M., Koch, H. G., Beck, K., & Schafer, U. (2001). Protein traffic in bacteria: Multiple routes from the ribosome to and across the membrane. *Progress in Nucleic Acid Research and Molecular Biology*, *66*, 107–157.

Numata, K., Motoda, Y., Watanabe, S., Osanai, T., & Kigawa, T. (2015). Co-expression of two polyhydroxyalkanoate synthase subunits from *Synechocystis* sp. PCC 6803 by cell-free synthesis and their specific activity for polymerization of 3-hydoxy 3-hydroxybutyryl-coenzyme A. *Biochemistry, 54*, 1401–1407.

Oh, I. S., Kim, D. M., Kim, T. W., Park, C. G., & Choi, C. Y. (2006). Providing an oxidizing environment for the cell-free expression of disulfide-containing proteins by exhausting the reducing activity of *Escherichia coli* S30 extract. *Biotechnology Progress, 22*(4), 1225–1228.

Ozawa, K., Wu, P. S., Dixon, N. E., & Otting, G. (2006). ^{15}N-labeled proteins by cell-free protein synthesis. Strategies for high-throughput NMR studies of proteins and protein-ligand complexes. *The FEBS Journal, 273*, 4154–4159.

Pratt, J. M. (1984). Coupled transcription–translation in prokaryotic cell-free system. In B. D. Hames & S. J. Higgins (Eds.), *Transcription and translation* (pp. 179–209). Oxford, Washington, DC: IRL Press.

Rosenblum, G., & Cooperman, B. S. (2014). Engine out of the chassis: Cell-free protein synthesis and its uses. *FEBS Letters, 588*, 261–268.

Sakamoto, K., Hayashi, A., Sakamoto, A., Kiga, D., Nakayama, H., Soma, A., et al. (2002). Site-specific incorporation of an unnatural amino acid into proteins in mammalian cells. *Nucleic Acids Research, 30*(21), 4692–4699.

Saul, J., Petritis, B., Sau, S., Rauf, F., Gaskin, M., Ober-Reynolds, B., et al. (2014). Development of a full-length human protein production pipeline. *Protein Science, 23*, 1123–1135.

Seki, E., Matsuda, N., Yokoyama, S., & Kigawa, T. (2008). Cell-free protein synthesis system from *Escherichia coli* cells cultured at decreased temperatures improves productivity by decreasing DNA template degradation. *Analytical Biochemistry, 377*, 156–161.

Shi, J., Pelton, J. G., Cho, H. S., & Wemmer, D. E. (2004). Protein signal assignments using specific labeling and cell-free synthesis. *Journal of Biomolecular NMR, 28*, 235–247.

Shimono, K., Goto, M., Kikukawa, T., Miyauchi, S., Shirouzu, M., Kamo, N., et al. (2009). Production of functional bacteriorhodopsin by an *Escherichia coli* cell-free protein synthesis system supplemented with steroid detergent and lipid. *Protein Science, 18*, 2160–2171.

Shoji, S., Muto, Y., Ikeda, M., He, F., Tsuda, K., Ohsawa, N., et al. (2014). The zinc-binding region (ZBR) fragment of Emi2 can inhibit APC/C by targeting its association with the coactivator Cdc20 and UBE2C-mediated ubiquitylation. *FEBS Open Bio, 4*, 689–703.

Spirin, A. S., Baranov, V. I., Ryabova, L. A., Ovodov, S. Y., & Alakhov, Y. B. (1988). A continuous cell-free translation system capable of producing polypeptides in high yield. *Science, 242*, 1162–1164.

Su, X. C., Loh, C. T., Qi, R., & Otting, G. (2011). Suppression of isotope scrambling in cell-free protein synthesis by broadband inhibition of PLP enzymes for selective ^{15}N-labelling and production of perdeuterated proteins in H$_2$O. *Journal of Biomolecular NMR, 50*, 35–42.

Suzuki, T., Ezure, T., Ito, M., Shikata, M., & Ando, E. (2009). An insect cell-free system for recombinant protein expression using cDNA resources. *Methods in Molecular Biology, 577*, 97–108.

Takai, K., & Endo, Y. (2010). The cell-free protein synthesis system from wheat germ. *Methods in Molecular Biology, 607*, 23–30.

Takai, K., Sawasaki, T., & Endo, Y. (2010). Practical cell-free protein synthesis system using purified wheat embryos. *Nature Protocols, 5*(2), 227–238.

Takeda, M., & Kainosho, M. (2012). Cell-free protein synthesis using *E. coli* cell extract for NMR studies. *Advances in Experimental Medicine and Biology, 992*, 167–177.

Tarui, H., Imanishi, S., & Hara, T. (2000). A novel cell-free translation/glycosylation system prepared from insect cells. *Journal of Bioscience and Bioengineering, 90*, 508–514.

Terada, T., Murata, T., Shirouzu, M., & Yokoyama, S. (2014). Cell-free expression of protein complexes for structural biology. *Methods in Molecular Biology, 1091*, 151–159.
Terpe, K. (2006). Overview of bacterial expression systems for heterologous protein production: From molecular and biochemical fundamentals to commercial systems. *Applied Microbiology and Biotechnology, 72*, 211–222.
Thomas, J. G., Ayling, A., & Baneyx, F. (1997). Molecular chaperones, folding catalysts, and the recovery of active recombinant proteins from *E. coli*. *Applied Biochemistry and Biotechnology, 66*, 197–238.
Tochio, N., Umehara, T., Nakabayashi, K., Yoneyama, M., Tsuda, K., Shirouzu, M., et al. (2015). Solution structures of the DNA-binding domains of immune-related zinc-finger protein ZFAT. *Journal of Structural and Functional Genomics, 16*, 55–65.
Velyvis, A., Ruschak, A. M., & Kay, L. E. (2012). An economical method for production of ^{2}H, ^{13}CH$_{3}$-threonine for solution NMR studies of large protein complexes: Application to the 670 kDa proteasome. *PLoS One, 7*(9), e43725.
Wada, T., Shimono, K., Kikukawa, T., Hato, M., Shinya, N., Kim, S. Y., et al. (2011). Crystal structure of the eukaryotic light-driven proton-pumping rhodopsin, *Acetabularia* rhodopsin II, from marine alga. *Journal of Molecular Biology, 411*(5), 986–998.
Wada, T., Shirouzu, M., Terada, T., Ishizuka, Y., Matsuda, T., Kigawa, T., et al. (2003). Structure of a conserved CoA-binding protein synthesized by a cell-free system. *Acta Crystallographica Section D: Biological Crystallography, 59*, 1213–1218.
Wakiyama, M., Kaitsu, Y., & Yokoyama, S. (2006). Cell-free translation system from *Drosophila* S2 cells that recapitulates RNAi. *Biochemical and Biophysical Research Communications, 343*, 1067–1071.
Wang, L., Brock, A., Herberich, B., & Schultz, P. G. (2001). Expanding the genetic code of *Escherichia coli*. *Science, 292*(5516), 498–500.
Weski, J., & Ehrmann, M. (2012). Genetic analysis of 15 protein folding factors and proteases of the *Escherichia coli* cell envelope. *Journal of Bacteriology, 194*, 3225–3233.
Willis, R. C., & Woolfolk, C. A. (1974). Asparagine utilization in *Escherichia coli*. *Journal of Bacteriology, 118*, 231–241.
Yabuki, T., Kigawa, T., Dohmae, N., Takio, K., Terada, T., Ito, Y., et al. (1998). Dual amino acid-selective and site-directed stable-isotope labeling of the human c-Ha-Ras protein by cell-free synthesis. *Journal of Biomolecular NMR, 11*, 295–306.
Yabuki, T., Motoda, Y., Hanada, K., Nunokawa, E., Saito, M., Seki, E., et al. (2007). A robust two-step PCR method of template DNA production for high-throughput cell-free protein synthesis. *Journal of Structural and Functional Genomics, 8*, 173–191.
Yamazaki, T., Otomo, T., Oda, N., Kyogoku, Y., Uegaki, K., Ito, N., et al. (1998). Segmental isotope labeling for protein NMR using peptide splicing. *Journal of the American Chemical Society, 120*, 5591–5592.
Yanagisawa, T., Takahashi, M., Mukai, T., Sato, S., Wakamori, M., Shirouzu, M., et al. (2014). Multiple site-specific installations of N^{ε}-monomethyl-L-lysine into histone proteins by cell-based and cell-free protein synthesis. *Chembiochem, 15*(12), 1830–1838.
Yokoyama, J., Matsuda, T., Koshiba, S., & Kigawa, T. (2010). An economical method for producing stable-isotope labeled proteins by the *E. coli* cell-free system. *Journal of Biomolecular NMR, 48*(4), 193–201.
Yokoyama, J., Matsuda, T., Koshiba, S., Tochio, N., & Kigawa, T. (2011). A practical method for cell-free protein synthesis to avoid stable isotope scrambling and dilution. *Analytical Biochemistry, 411*(2), 223–229.
Yokoyama, S., Terwilliger, T. C., Kuramitsu, S., Moras, D., & Sussman, J. L. (2007). RIKEN aids international structural genomics efforts. *Nature, 445*, 21.
Zubay, G. (1973). In vitro synthesis of protein in microbial systems. *Annual Review of Genetics, 7*, 267–287.

CHAPTER FOURTEEN

Rapid Biosynthesis of Stable Isotope-Labeled Peptides from a Reconstituted *In Vitro* Translation System for Targeted Proteomics

Feng Xian[*,†,‡], Shuwei Li[§,1], Siqi Liu[*,†,‡,1]

[*]CAS Key Laboratory of Genome Sciences and Information, Beijing Institute of Genomics, Chinese Academy of Sciences, Beijing, PR China
[†]BGI-Shenzhen, Shenzhen, PR China
[‡]Sino-Danish Center/Sino-Danish College, University of Chinese Academy of Sciences, Beijing, PR China
[§]Institute for Bioscience and Biotechnology Research, University of Maryland, Rockville, Maryland, USA
[1]Corresponding authors: e-mail address: sli@umd.edu; siqiliu@genomics.cn

Contents

1. Introduction	348
2. Equipment, Materials, and Buffers	350
3. Section 1: DNA Template Preparation	351
3.1 Overview	351
3.2 PCR Amplification of Double-Strand DNA Template for Peptide Expression	353
3.3 Tips	355
4. Section 2: Peptide Synthesis with PURE System	355
4.1 Overview	355
4.2 Translation Using PURE System	356
4.3 Tips	356
5. Section 3: Enrichment and Digestion of Synthesized Peptide	358
5.1 Overview	358
5.2 Purification and Digestion of PURE-Synthesized Peptide	359
5.3 Tips	360
6. Section 4: Quantification of PURE-Synthesized Peptide	360
6.1 Overview	360
6.2 Quantification of PURE-Expressed Peptides for Absolute Quantification	361
6.3 Tips	362
7. An Example	363
8. Summary and Discussion	365
Acknowledgment	365
References	365

Abstract

Stable isotope-labeled peptides are routinely used as internal standards (a.k.a. reference peptides) for absolute quantitation of proteins in targeted proteomics. These peptides can either be synthesized chemically on solid supports or expressed biologically by concatenating multiple peptides together to a large protein. Neither method, however, has required versatility, convenience, and economy for making a large number of reference peptides. Here, we describe the biosynthesis of stable isotope-labeled peptides from a reconstituted *Escherichia coli in vitro* translation system. We provide a detailed protocol on how to express these peptides with high purity and how to determine their concentrations with easiness. Our strategy offers a general, fast, and scalable approach for the easy preparation of labeled reference peptides, which will have broad application in both basic research and translational medicine.

1. INTRODUCTION

Multiple reaction monitoring (MRM) coupled with new generation triple quadrupole mass spectrometers has been developed to detect specific peptides in complex biological mixtures such as human plasma and serum (Anderson & Hunter, 2006). In such targeted assays, peptides of interest are selected based on their parent ion mass in first quadrupole (Q1) and fragmented in the second quadrupole (Q2 acted as a collision cell) to generate product ions. Preselected pairs of parent and product ions (a.k.a. transitions) are then monitored by the third quadrupole (Q3). After this two-step filtering mechanism, biological background is mostly removed, leading to the detection of preselected peptides with high specificity and sensitivity. A modern triple quadrupole mass spectrometer is capable of scanning hundreds of transitions in a single experiment to detect multiple peptides simultaneously. In addition, if stable isotope-labeled peptides with known quantity are spiked into a sample as internal standards, the absolute quantity of these preselected peptides can be determined with high accuracy.

Apparently, the success of MRM experiments relies on the availability of isotope-labeled peptides, which share the same ionization efficiency and give the same MS signal as their unlabeled native forms, but differ in their mass by a few daltons. With the advance of instrumentation, stable isotope-labeled peptides have become a gold standard for MS-based targeted proteomics. Currently, there are several approaches to incorporate stable isotopes into peptides or proteins for quantitative analysis, which basically fall into four categories: (1) adding labeled forms directly; (2) incorporating through an enzyme during digestion; (3) introducing a labeled chemical tag through chemical derivatization; and (4) integrating metabolically during cell growth (Ong & Mann, 2005).

For MRM-based target proteomics, the most popular way to introduce stable isotope labels is to spike synthetic labeled peptides into samples with known concentrations. For instance, Stemmann et al. employed this approach to quantify the separase phosphorylation status during *Xenopus* cell cycle, in which the reference peptides were obtained from solid-phase peptide synthesis (SPPS) (Stemmann, Zou, Gerber, Gygi, & Kirschner, 2001). Generally, SPPS peptides need to be purified to high homogeneity and quantified by multiple-step amino acid analysis, which can be tedious and time consuming. Whether a peptide could be synthesized successfully also depends on its sequence and length. In addition, synthesizing multiple stable isotope-labeled peptides is prohibitively expensive. To solve these problems, Partt et al. introduced a new technology called quantitative concatemers (QconCATs). They first assembled a set of oligonucleotides encoding multiple peptides into a concatenated gene and then expressed the gene in *Escherichia coli* cells that grew in a medium containing $^{15}NH_4Cl$ as the sole nitrogen source (Pratt et al., 2006). During cell growth, ^{15}N isotope is completely incorporated into the overexpressed concatenated protein, which generates several stable isotope-labeled peptides at equimolar concentrations after trypsin digestion. Despite many advantages of the QconCAT technology, this method also suffers from some limitations. For example, not all QconCAT proteins can be expressed successfully (Mirzaei, McBee, Watts, & Aebersold, 2008). Furthermore, since one QconCAT gene yields equal amount of reference peptides for multiple proteins, they are not suitable for simultaneous detection of multiple proteins with large concentration difference. Although QconCAT proteins can also be expressed in cell-free translational systems based on cell extracts from *E. coli*, wheat germ, and rabbit reticulocytes (Hino et al., 2008), their drawbacks in quantitative proteomics are still present.

With a large amount of data accumulating in shotgun proteomics, which is a robust tool for biomarker discovery, verification of such data has become increasingly important. As a result, there are urgent needs to generate isotope-labeled forms of peptides that can be used for the quantification of all proteins, especially for low-abundance proteins (Stergachis, MacLean, Lee, Stamatoyannopoulos, & MacCoss, 2011). Neither SPPS nor QconCAT is suitable for such large-scale attempts (Mirzaei et al., 2008). Herein, we introduce a novel approach based on PURE (Protein synthesis Using Recombinant Elements) system for the preparation of stable isotope-labeled reference peptides (Shimizu et al., 2001).

PURE system is an *E. coli*-based reconstituted cell-free protein translation system that offers several benefits not shared by conventional *in vitro* protein expression systems based on cell extracts. PURE system is composed of purified enzymes and chemically defined factors necessary for protein expression, such as 20 aminoacyl-tRNA synthetases, ribosome, tRNA molecules, and amino acids. Because PURE lacks nuclease and protease activity, the stability of both RNA templates and expressed proteins is significantly improved (Shimizu et al., 2001). This system is commonly employed for recombinant protein expression, but is rarely used for peptide generation. We demonstrated for the first time that PURE system was able to synthesize stable isotope-labeled peptides. Our data showed that the yield and purity of synthesized peptide were satisfactory for both relative and absolute quantitative proteomics.

2. EQUIPMENT, MATERIALS, AND BUFFERS

Equipment, Materials, and Buffers	Section Required
PCR thermocycler	Section 1
Water bath	Section 2
Vortex	Sections 1–4
Thermomixer Comfort (Eppendorf)	Section 3
NanoDrop (Thermo Scientific)	Section 1
Centrifuge	Sections 1–4
SpeedVac	Sections 1 and 3
Microfuge	Section 1
Electrophoresis instrument	Section 1
EasyPfu polymerase (TransGen Biotech)	Section 1
Agarose	Section 1
Ethidium bromide	Section 1
Nuclease-free double-distilled H_2O	Sections 1 and 2
DNA extraction kit (GalaxyBio)	Section 1

−20 °C freezer	Section 1
−80 °C freezer	Section 2
18 Native amino acids except arginine and lysine (Sigma-Aldrich)	Section 2
U-$^{13}C_6$, U-$^{15}N_4$ arginine and U-$^{13}C_6$, U-$^{15}N_2$ lysine (Cambridge Isotope Inc.)	Section 2
0.2 M NaOH	Section 2
0.1 M HCl	Section 2
PURExpress® Δ (aa, tRNA) Kit (E6840S, New England Biolabs)	Section 2
RNase Inhibitor (Life Technology)	Section 2
Strep-Tactin Magnetic Beads (IBA)	Section 3
0.2- and 1.5-mL tubes	Sections 1–4
Magnetic separator	Section 3
Washing buffer (50 mM NH$_4$HCO$_3$, pH 8.0)	Section 3
Elution buffer (50 mM NH$_4$HCO$_3$, 10 mM D-dethiobiotin, pH 8.0)	Section 3
Dithiothreitol (DTT, 1 M)	Section 3
Iodoacetamide (IAM)	Section 3
Trypsin (sequencing grade, Promega)	Section 3
0.1% Formic acid solution (2% acetonitrile, 98% H$_2$O)	Section 4
Mobile phase A: 98% H$_2$O + 2% ACN + 0.1% formic acid	Section 4
Mobile phase B: 98% ACN + 2% H$_2$O + 0.1% formic acid	Section 4
Nano-HPLC (Eksigent NanoLC Ultra)	Section 4
Triple-quadrupole mass spectrometry (QTRAP 5500)	Section 4

3. SECTION 1: DNA TEMPLATE PREPARATION

3.1 Overview

A DNA template for peptide expression is designed with several important elements (Fig. 1), including a T7 promoter, a ribosome-binding site (RBS), and a coding sequence, which encodes a peptide containing a N-terminus

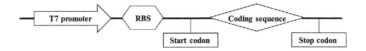

T7 promoter	Ribosome binding site	Start codon

5'-GAAAT <u>TAATACGACTCACTATA</u> GGGTAACTTTAAG <u>AAGGAG</u> ATATACCA <u>ATG</u> GGT

ᶠMet Gly

Target peptide
GCG GGTCGT <u>GAAGTGGTGACCCCGGGCATTCCGGCGGAAGAAATTCCGAAA</u>
Ala Gly Arg Glu Val Val Thr Pro Gly Ile Pro Ala Glu Glu Ile Pro Lys

Sterp-tag	Stop codon

<u>TGGAGCCATCCGCAGTTTGAAAAAGGTGGCGAT</u> <u>TAATGA</u> ATA-3'
Trp Ser His Pro Gln Phe Glu Lys Gly Gly Asp

Figure 1 The design of PURE-expressed isotope-labeled peptides. The upper panel shows critical elements on a DNA template for the biosynthesis of peptides in PURE system. The lower panel gives the whole sequence coding an example peptide of EVVTPGIPAEEIPK.

constant sequence ᶠMGAGR (ᶠM representing formylmethionine), the variable target peptide, and a C-terminus constant sequence WSHPQFEKGGD. WSHPQFEK is the *Strep*-tag, a short peptide binding with streptavidin tightly that has dual functions for both the purification and quantification of the expressed peptide. GGD is added to prevent premature truncation of the *Strep*-tag. As PURE includes T7 RNA polymerase and four ribonucleotides required for RNA transcription, the double-strand DNA template can be added directly into the expression system without being transcribed into RNA first.

A double-strand DNA template for peptide expression is generated by PCR amplification from synthetic oligonucleotides, which can be reversely translated from the peptide sequence according to the codon usage of *E. coli*. The size of synthetic oligonucleotides is limited to less than 60 residues, which are cheaper to prepare and easier to be of high purity than longer ones. As a result, for the target peptides less than nine amino acids, one synthetic oligonucleotide is enough; for those between nine to 25 amino acids, two synthetic oligonucleotides with a short overlapping sequence (10–14 nucleotides) are used to prepare the double-strand DNA template. For PCR amplification, the forward primer contains the T7 promoter and RBS, while the reverse primer has sequence encoding C-terminus constant peptide. When two synthetic oligonucleotides are used for PCR, they are mixed at equal ratio and amplified first for five cycles without the forward and reverse primers, and then further amplified for 35 cycles with both primers. An example of two synthetic oligonucleotides for the preparation of a target peptide is illustrated in Fig. 2A and the two-step PCR procedure is shown in Fig. 2B. The forward and reverse primers are listed in Table 1.

3.2 PCR Amplification of Double-Strand DNA Template for Peptide Expression

1. Assemble the reaction on ice in a clear tube as detailed in Table 2. We generally divide the mixture into five tubes for PCR (48.7 μL/tube).
2. After brief vortex and centrifugation, place the tubes in a PCR thermocycler and run a PCR program as following (Table 3) for five cycles.

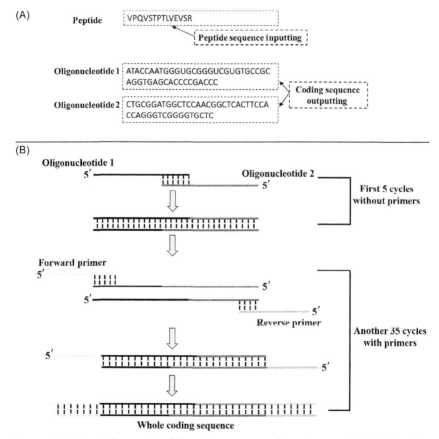

Figure 2 (A) A typical example of the coding sequence for a given peptide. (B) Two-step PCR procedure when two synthetic oligonucleotides are used.

Table 1 Design of PCR Primers

Forward primer (5′–3′)
GAAATTAATACGACTCACTATAGGGTAACT
TTAAGAAGGAGATATACCAATGGGTGCGGGTCGT

Reverse primer (5′–3′)
TATTCATTAATCGCCACCTTTTTCAAACTGCGGATGGCTCCA

The underline sequences are overlapping with those in synthetic oligonucleotides.

Table 2 Reaction Mixture for PCR

Reagent	Volume (µL)
H_2O	186
10 × Pfu buffer	25
dNTP (10 mM)	6.5
Oligonucleotide 1 (10 nM)	6.5
Oligonucleotide 2 (10 nM)	6.5
DMSO	6.5
EasyPfu polymerase (5 U/µL)	6.5
Total volume	243.5

Table 3 The First PCR Program Without Primers

Initial denaturation/activation	94 °C	5 min
Denaturation	94 °C	30 s
Annealing	58 °C	30 s
Elongation	72 °C	10 s
Final elongation	72 °C	5 min

3. Take out the tubes and add 1.3 µL/tube primers mixture (10 µM). Run the PCR for another 35 cycles as following (Table 4).
4. Pool PCR products together and load onto 1–1.5% agarose gel and run for 20 min at 130 V. Stain the gel with ethidium bromide and cut the product band in a UV light box.
5. Follow the GalaxyBio MinElute Gel Extraction kit for small DNA fragments manual. Elute DNA with 150 µL nuclease-free H_2O twice.

Table 4
The Second PCR Program with Primers

Initial denaturation/activation	94 °C	5 min
Denaturation	94 °C	30 s
Annealing	58 °C	30 s
Elongation	72 °C	20 s
Final elongation	72 °C	10 min

6. Concentrate the purified DNA by SpeedVac and add 50 μL nuclease-free H_2O to redissolve the DNA.
7. Use NanoDrop 2000 Spectrophotometer to quantify the purified DNA template.
8. Store the purified DNA at −20 °C for further use.

3.3 Tips

i. For most reference peptides used in proteomics studies, the length is within the range of 6–25 amino acids.
ii. For short peptides only need one oligonucleotide, use additional 1.25 μL H_2O to replace the other oligonucleotide in the reaction mixture.
iii. The high-fidelity *EasyPfu* polymerase is used for PCR amplification. DMSO is added to improve the activity of enzyme for maximum yields.
iv. The concentration of agarose gel (1–1.5%) used to separate PCR products was recommended by the purification kit for the maximum recovery of DNA.
v. Purification kit obtained from GalaxyBio is more effective for small DNA fragment, and the running condition for electrophoresis is variable depending on different instruments.
vi. Use nuclease-free H_2O rather than TE buffer (10 mM Tris, 1 mM EDTA, pH 8.0) to elute and redissolve DNA for further experiment consideration.

4. SECTION 2: PEPTIDE SYNTHESIS WITH PURE SYSTEM
4.1 Overview

With the DNA template added into the PURE system, the resulting parent peptide starts from a constant ^{f}MGAGR and ends with a constant

WSHPQFEKGGD. Here, the N-terminus ^{f}M is a formylmethionine which is required for the initiation of protein translation in E. coli, while the extra three residues (GGD) are added to prevent potential premature truncation at C-terminus. It is worth noting that formylmethionine is not included in the amino acid mixture because it is generated by methionyl-tRNA formyltransferase present in the PURE system. Between these two constant sequences is a variable reference peptide. All arginine and lysine residues in the parent peptide are isotopically labeled by supplementing with U-^{13}C$_6$, U-^{15}N$_4$ arginine (R*) and U-^{13}C$_6$, U-^{15}N$_2$ lysine (K*), together with other unlabeled 18 native amino acids.

4.2 Translation Using PURE System

1. Prepare amino acids stock solution separately. Use suitable solutions (Table 5) to dissolve 20 amino acids to final concentration 100 mM, and store at -20 °C for future use.
2. Mix 20 amino acids into a working solution (220 μL nuclease-free H$_2$O + 30 μL/each native amino acid + 120 μL U-^{13}C$_6$, U-^{15}N$_4$ arginine + 120 μL U-^{13}C$_6$, U-^{15}N$_2$ lysine). The final concentration of 18 native amino acid is 3 mM and labeled lysine and arginine are at 12 mM.
3. Thaw the necessary solutions from a commercial PURExpress® Δ (aa, tRNA) kit on ice, which provides tRNA and amino acid mixture separately. Pulse-spin in microfuge to collect solutions at the bottom of tubes.
4. Assemble the reaction on ice in a new tube in the following order (Table 6).
5. Mix gently and pulse-spin in microfuge to collect mixture at the bottom of the tube.
6. Incubate at 37 °C for 6 h to achieve maximum yield.
7. Stop the reaction by placing the tube on ice.

4.3 Tips

i. The PURE kits should be stored at -80 °C.
ii. Certain components in solution A may precipitate during storage. Be sure to mix it well prior to assembling reactions. Do not vortex solution B, mix gently.
iii. Formulations in Table 6 allow an increase in the "user-added" volume (for template, supplements, etc.). We strongly recommend adding 20 U

Table 5 The Recommended Solutions for 20 Amino Acids

Amino Acid	Dissolving Solution
Ala	H_2O
Asn	H_2O
Asp	0.2 M NaOH
Cys	H_2O
Gln	H_2O (50 °C water bath)
Glu	0.2 M NaOH
Gly	H_2O
His	H_2O
Ile	0.1 M HCl
Leu	0.1 M HCl
Met	H_2O
Phe	0.1 M HCl
Pro	H_2O
Ser	H_2O
Thr	H_2O
Trp	0.1 M HCl
Tyr	0.2 M NaOH
Val	0.1 M HCl
Lys (U-^{15}N2, U-^{13}C6)	H_2O
Arg (U-^{15}N4, U-^{13}C6)	H_2O

RNase Inhibitor to avoid RNase contamination and improve RNA stability. Add solution B to solution A, not the reverse order.

iv. 400 ng DNA template and incubation for 6 h are optimal conditions for maximum peptide expression. Since the PURE system may contain trace amount of lysine and arginine, large excess of labeled arginine and lysine (12 mM) are used to make sure >98% incorporation.

v. We recommend using an incubator rather than a water bath to prevent evaporation. Incubating reactions below 37 °C will likely reduce yield.

vi. It is possible to use other isotope-labeled amino acids to replace labeled arginine and lysine if necessary.

Table 6 Components for PURE Expression

Solution A (minus aa, tRNA)	5 μL
aa Mixture	2.5 μL
tRNA	2.5 μL
Solution B	7.5 μL
RNase Inhibitor	0.5 μL (20 U)
Nuclease-free H$_2$O	12−x μL
Template DNA	x μL to make 400 ng
Maximum total volume	30 μL

5. SECTION 3: ENRICHMENT AND DIGESTION OF SYNTHESIZED PEPTIDE

5.1 Overview

Using affinity tags to achieve the purification and detection of recombinant proteins has become indispensable. *Strep*-tag was originally selected from a random peptide library as an eight-amino acid peptide (WRHPQFGG) that specifically binds to streptavidin (Pahler, Hendrickson, Kolks, Argarana, & Cantor, 1987; Schmidt & Skerra, 1993). Because of its extraordinary affinity for biotin and its high intrinsic stability, streptavidin is widely used for the detection and purification of macromolecules (Laitinen, Hytonen, Nordlund, & Kulomaa, 2006).

With a good specificity to cleave C-terminal to Arg and Lys residues (except Arg-Pro and Lys-Pro bonds) and minimal autoproteolysis, trypsin is one of the most popular proteases used in proteomics studies. For many proteins, cleavage with trypsin generates a good distribution of peptides within the mass range 600–4000 Da, which are suitable for MS-based peptide analysis. Moreover, tryptic peptides easily generate doubly charged ions that extend the usefulness from MS to MS/MS analysis for peptide identification and increase the scope of quantification.

As demonstrated in Section 1, PURE-expressed peptides have the form of ^{f}MGAGR-XXXXXX-WSHPQFEKGGD, in which the underlined segment represents a reference peptide between 6 and 25 residues. In this design, the C-terminal *Strep*-tag has dual functions. First, it allows the full-length parent peptide to be purified with *Strep*-Tactin beads, which is a streptavidin mutant with higher affinity toward eight-residue *Strep*-tag than the wild type. In addition, since each purified peptide generates equimolar mixture of the labeled *Strep*-tag (WSHPQFEK*) and a labeled reference peptide after trypsin digestion, the reference peptide can be quantified indirectly based on the amount of the labeled *Strep*-tag, which is determined by spiking a known concentration of unlabeled *Strep*-tag (note: unlabeled *Strep*-tag itself can be quantified by traditional amino acid analysis).

5.2 Purification and Digestion of PURE-Synthesized Peptide

1. Vortex magnetic *Strep*-Tactin beads briefly before use because the beads may precipitate during storage.
2. Take out 50 μL suspension solution into a clear 0.2 mL tube and place on a magnetic separator for 10 s and carefully remove the supernatant.
3. Add 100 μL washing buffer (50 mM NH$_4$HCO$_3$, pH 8.0) into the tube to equilibrate the beads. Briefly vortex and place on a magnetic separator for 10 s and carefully remove the supernatant.
4. Repeat step 3 once.
5. Add 70 μL washing buffer into 30 μL PURE expression mixture. Vortex the beads and place the tube on a shaker at 4 °C to incubate for 1 h.
6. Place the tube on the magnetic separator and remove the supernatant (flow-through) carefully.
7. Vortex the beads after adding 100 μL washing buffer. Place the tube on magnetic separator and remove the supernatant.
8. Repeat step 7 twice to remove background proteins.
9. Add 50 μL elution buffer (50 mM NH$_4$HCO$_3$, 10 mM D-dethiobiotin, pH 8.0) into the tube, gently vortex to suspend the beads, and incubate for 5 min at 4 °C on a shaker.
10. Place the tube on magnetic separator and collect the supernatant (eluate) into a new tube.
11. Repeat step 10 two more times.
12. Three eluates (150 μL in total) are pooled together and stored at −20 °C for next use.

* Steps 13–15 are only performed for reference peptides containing cysteine.

13. Add 1.5 μL DTT (1 M) into 150 μL elute product, briefly vortex, and incubate at 56 °C for 45 min.
14. Collect the reaction mixture at the bottom of the tube by short centrifuging.
15. Add IAM into the reaction mixture to 55 mM final concentration and keep the tube in the dark at room temperature for 30 min.
16. Briefly centrifuge and add 2 μL trypsin (1 μg/μL) into the reaction mixture. Vortex, centrifuge, and place the tube on 37 °C in water bath for 4 h.
17. Add another 2 μL trypsin (1 μg/μL) and incubate at 37 °C for extra 6–10 h to achieve complete digestion.
18. Centrifuge at 12,000 rpm for 1 min to collect tryptic peptides.
19. Remove the solvent by SpeedVac. The tryptic peptides should be stored at −20 °C for further use.

5.3 Tips

i. Two factors are considered when determining the volume of *Strep*-Tactin beads used to enrich synthesized peptide: (1) the binding capacity of the *Sterp*-Tactin and (2) the recommended minimum volume of 20 μL. The yield of enriched peptide may increase if larger volume of beads was used.

ii. During equilibration and washing step, we suggest to use Bradford working solution to detect proteins in supernatant to determine equilibration and washing times.

iii. There may still have some background proteins from PURE in enriched peptide even after affinity beads enrichment, but it is not critical if the enriched peptide is contaminated with minor proteins because only the expressed peptide is isotopically labeled.

iv. We suggest that the pH of solutions used in this section should be kept at 8.0 because the Strep-Tactin beads perform optimally at this pH and tryptic digestion can be conducted without exchanging to a different buffer.

6. SECTION 4: QUANTIFICATION OF PURE-SYNTHESIZED PEPTIDE

6.1 Overview

When a target peptide in a complex biological sample needs to be quantified, the respective reference peptide with a known concentration is added as an internal standard. Therefore, it is critical to get the accurate quantity of reference peptides. For chemically synthesized peptides, amino acid analysis is usually used, but it is a tedious multistep procedure in which a peptide needs to be hydrolyzed completely into free amino acids, labeled by a UV absorbance or fluorescence tag, and analyzed by liquid chromatography. In addition, acid hydrolysis may introduce some side reactions that make the quantification less accurate. For example, peptide bonds with valine or isoleucine are less hydrolyzable, while asparagine and glutamine can be oxidized during hydrolysis (Pratt et al., 2006). This approach is also inconvenient with large-scale experiments in which multiple reference peptides are needed. For QconCAT technology, the concentration of purified proteins is usually determined by protein assay like Bradford assay. However, even highly purified proteins still contain contamination that can affect accuracy of quantification.

In our design, every PURE-expressed parent peptides contain a fixed *Strep*-tag sequence, which is at 1:1 ratio to the reference peptide upon complete trypsin digestion. Therefore, this *Strep*-tag is quantified and used to deduce the concentration of the reference peptide, rather than to determine the concentration of the reference peptide itself. As a result, any PURE-synthesized reference peptide can be quantified by spiking a universal unlabeled *Strep*-tag with defined concentration.

6.2 Quantification of PURE-Expressed Peptides for Absolute Quantification

1. Redissolve the tryptic peptides in 30 µL of 0.1% formic acid solution (2% acetonitrile, 98% H_2O) and vortex for 1 min.
2. Centrifuge at 12,000 rpm for 10 min.
3. Take out several 1 µL aliquots mix with different quantities of *Strep*-tag (e.g., 2 fmol, 5 fmol, 10 fmol, 20 fmol, and 30 fmol). Vortex and centrifuge these mixtures.
4. Deliver each mixture into a reverse-phase nano-HPLC (Eksigent NanoLC ultra, AB SCIEX). The peptides are first enriched on a trap column by running mobile phase A at 5 µL/min for 8 min and then eluted into analytical column with gradient as listed in Table 7 at 300 nL/min for 35 min. The eluent from nano-HPLC is directly delivered into a QTRAP 5500 MS (AB SCIEX) for MRM detection. The parameters for MS acquisition are listed in Table 8.
5. Use MultiQuant software to calculate the MRM peak areas of *Strep*-tags (labeled and unlabeled form) in each analyte.

Table 7 The Gradient of Liquid Chromatography (Eksigent NanoLC Ultra)

Time (min)	Mobile Phase B%
0	2
20	30
23	80
27	80
27.1	2
35	2

Table 8 Parameters for MRM Detection in AB SCIEX QTRAP 5500 MS

Parameter	Value/Type
Ion source	Nanospray
Curtain gas (CUR)	35
Collision gas (CAD)	High
Ionspray voltage (IS)	2300
Ion source gas 1	23
Ion source gas 2	0
Interface heater temperature (IHT)	150
Resolution Q1 and Q3	Unit
Pause between mass range	5.007 ms

6. Generate a calibration curve by the MRM peak area ratio of (light/heavy) against the quantity of unlabeled *Strep*-tag.
7. Ensure the linear regression curve is qualified to estimate the concentration of synthesized reference peptide ($R^2 > 0.97$).
8. On the basis of the equation of the curve, calculate the concentration of target peptide by finding the reading of x-axis when the signal ratio (y-axis) is equal to 1 (Fig. 3).

6.3 Tips

i. In order to redissolve the peptides completely, we strongly suggest vortex for enough time to achieve more accurate quantification.

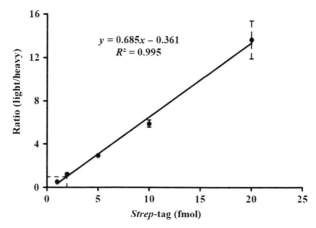

Figure 3 The quantitative calibration curve of *Strep*-tag used to estimate the concentration of a PURE-synthesized example peptide (AILNYIASK*). The error bar was generated from three independent experiments.

ii. In step 3, several 1-μL aliquots of a purified labeled peptide (total 150 μL) were mixed with different concentrations of unlabeled *Strep*-tag. This 1 μL volume can be either increased or decreased for different peptides because of two reasons: (1) the LOQ (limit of quantitation) of peptides may be different in different LC-MS system and (2) the yields of different peptides are usually different.

iii. Before running MRM detection, it may be necessary to take out 1 μL synthesized peptide mixture to verify the transition list for MRM experiment. There are multiple ways to obtain the theoretical transition list: (1) searching online databases like MRMaltas and PeptideAtlas; (2) based on the preliminary MS/MS data of target peptides; and (3) using software for the prediction of MRM transitions based on empirical algorithms like Skyline (https://skyline.gs.washington.edu/labkey/project/home/software/Skyline/begin.view).

7. AN EXAMPLE

To assess the accuracy of PURE-synthesized and -quantified peptides, we performed a proof-of-concept experiment to quantify recombinant immunoglobulin J chain protein from reference peptides generated from PURE expression. The detail protocol of this experiment is described below.

1. The immunoglobulin J chain protein was expressed in *E. coli*, purified, and evaluated by SDS-PAGE.
2. Three peptides were selected from the trypsin-digested immunoglobulin J chain protein (Pep1: IIVPLNNR, Pep2: SSEDPNEDIVER, Pep3: ENISDPTSPLR).
3. Three isotope-labeled and quantified peptides (Peps 1–3) were obtained by following the protocol described above.
4. Unknown amount of the recombinant immunoglobulin J chain protein was digested with trypsin.
5. Serially diluted the tryptic peptides of the recombinant immunoglobulin J chain protein (50-, 100-, 200-, 400-, 500-, and 1000-fold dilutions) and spiked with known amount of isotope-labeled peptides, respectively.
6. Used MRM-MS to detect both the heavy and light forms of Peps 1–3. The ratios of the MRM peak areas of target peptides (L/H) were calculated and used to generate calibration curves of three peptides against their dilution factors.
7. Estimated the quantification results according to the three calibration curves.

From these curves (Fig. 4A), the concentrations of three peptides were determined as 305.8, 268.6, and 253.2 fmol/μL, respectively (Fig. 4B), and their average was 275.8 fmol/μL with a CV (coefficient of variation) of 10% for three individual peptides, which was comparable to results obtained from SPPS or QconCAT peptides.

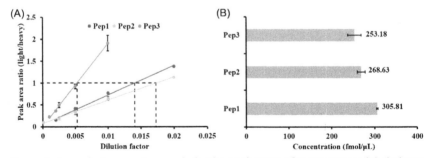

Figure 4 A proof-of-concept example for the application of PURE-express-labeled peptides. (A) The calibration curve of Peps 1–3. (B) The quantification of the immunoglobulin J chain protein obtained from three PURE-synthesized isotope-labeled peptides.

8. SUMMARY AND DISCUSSION

In this chapter, we describe a straightforward procedure for the synthesis and quantification of stable isotope-labeled peptide with PURE system, which is a rapid, scalable, and cost-effective approach for targeted proteomics. Compared to QconCAT, our approach can generate highly purified peptides within 1–2 days from the target gene design to final peptide quantification, whereas QconCAT approach takes relatively longer period (Pratt et al., 2006). This system is especially suitable for the production of MRM reference peptides because of several reasons. First, quantification of reference peptides is based on accompanying *Strep*-tag at equimolar ratio, which avoids tedious amino acid analysis. Second, the PURE system is nuclease- and protease-free, so short unstructured peptides can be expressed without degradation. Third, any heavy amino acid can be incorporated at >98% isotopic purity to fit different applications. Finally, multiple peptides with diverse length and sequences can be expressed, purified, and quantified rapidly. As MRM-based targeted proteomics allows protein quantification from complex samples with high sensitivity and specificity, this approach will have broad applications for biomarker validation and clinical diagnostics.

ACKNOWLEDGMENT

This work was supported by the National High Technology Research and Development Program of China (2012AA020206) and the National Basic Research Program of China (2011CB910704).

REFERENCES

Anderson, L., & Hunter, C. L. (2006). Quantitative mass spectrometric multiple reaction monitoring assays for major plasma proteins. *Molecular & Cellular Proteomics*, *5*(4), 573–588.

Hino, M., Kataoka, M., Kajimoto, K., Yamamoto, T., Kido, J.-I., Shinohara, Y., et al. (2008). Efficiency of cell-free protein synthesis based on a crude cell extract from Escherichia coli, wheat germ, and rabbit reticulocytes. *Journal of Biotechnology*, *133*(2), 183–189.

Laitinen, O. H., Hytonen, V. P., Nordlund, H. R., & Kulomaa, M. S. (2006). Genetically engineered avidins and streptavidins. *Cellular and Molecular Life Sciences*, *63*(24), 2992–3017.

Mirzaei, H., McBee, J. K., Watts, J., & Aebersold, R. (2008). Comparative evaluation of current peptide production platforms used in absolute quantification in proteomics. *Molecular & Cellular Proteomics*, 7(4), 813–823.

Ong, S. E., & Mann, M. (2005). Mass spectrometry-based proteomics turns quantitative. *Nature Chemical Biology*, *1*(5), 252–262.

Pahler, A., Hendrickson, W. A., Kolks, M. A., Argarana, C. E., & Cantor, C. R. (1987). Characterization and crystallization of core streptavidin. *Journal of Biological Chemistry*, *262*(29), 13933–13937.

Pratt, J. M., Simpson, D. M., Doherty, M. K., Rivers, J., Gaskell, S. J., & Beynon, R. J. (2006). Multiplexed absolute quantification for proteomics using concatenated signature peptides encoded by QconCAT genes. *Nature Protocols*, *1*(2), 1029–1043.

Schmidt, T. G. M., & Skerra, A. (1993). The random peptide library-assisted engineering of a c-terminal affinity peptide, useful for the detection and purification of a functional Ig Fv fragment. *Protein Engineering*, *6*(1), 109–122.

Shimizu, Y., Inoue, A., Tomari, Y., Suzuki, T., Yokogawa, T., Nishikawa, K., et al. (2001). Cell-free translation reconstituted with purified components. *Nature Biotechnology*, *19*(8), 751–755.

Stemmann, O., Zou, H., Gerber, S. A., Gygi, S. P., & Kirschner, M. W. (2001). Dual inhibition of sister chromatid separation at metaphase. *Cell*, *107*(6), 715–726.

Stergachis, A. B., MacLean, B., Lee, K., Stamatoyannopoulos, J. A., & MacCoss, M. J. (2011). Rapid empirical discovery of optimal peptides for targeted proteomics. *Nature Methods*, *8*(12), 1041–1043.

CHAPTER FIFTEEN

Labeling of Membrane Proteins by Cell-Free Expression

Aisha Laguerre[1], Frank Löhr, Frank Bernhard, Volker Dötsch

Institute of Biophysical Chemistry, Centre for Biomolecular Magnetic Resonance, J.W. Goethe-University, Frankfurt-am-Main, Germany
[1]Corresponding author: e-mail address: laguerre@bpc.uni-frankfurt.de

Contents

1. Introduction	368
2. Core Considerations for the Cell-Free Generation of MP Samples	369
3. Specific Challenges of NMR Studies with MPs	372
4. An Emerging Perspective: NMR with NDs	374
5. Labeling of Cell-Free Synthesized MPs with Stable Isotopes	375
6. Reducing Scrambling Problems	376
7. Perdeuteration of Cell-Free Synthesized MPs	380
8. Conclusion	382
Acknowledgments	382
References	382

Abstract

The particular advantage of the cell-free reaction is that it allows a plethora of supplementation during protein expression and offers complete control over the available amino acid pool in view of concentration and composition. In combination with the fast and reliable production efficiencies of cell-free systems, the labeling and subsequent structural evaluation of very challenging targets, such as membrane proteins, comes into focus. We describe current methods for the isotopic labeling of cell-free synthesized membrane proteins and we review techniques available to the practitioner pursuing structural studies by nuclear magnetic resonance spectroscopy. Though isotopic labeling of individual amino acid types appears to be relatively straightforward, an ongoing critical issue in most labeling schemes for structural approaches is the selective substitution of deuterons for protons. While few options are available, the continuous refinement of labeling schemes in combination with improved pulse sequences and optimized instrumentation gives promising perspectives for extended applications in the structural evaluation of cell-free synthesized membrane.

1. INTRODUCTION

Production of membrane proteins (MPs) in conventional recombinant protein expression systems, such as *Escherichia coli*, insect, or yeast cells, often overcrowds the membrane or causes toxic effects, resulting in growth problems or death of the expression host. Further, common methods of extraction from the membrane or refolding from inclusion bodies with harsh detergents or denaturants destroy native folds and render final protein samples insufficient in terms of yield, quality, and functionality.

Low yields of functional MPs have posed significant barriers for many structural projects in the past. Approaches by nuclear magnetic resonance (NMR) spectroscopy require high protein yields, efficient isotopic labeling, and homogenous samples (Opella & Marassi, 2004). In addition, NMR studies of MPs are hampered by a variety of further challenges that are currently being addressed by advances in the generation of isotopic labels, labeling schemes, and membrane mimetics, in addition to substantial developments in NMR instrumentation and pulse sequence design (Tamm & Liang, 2006).

The open accessibility of the CF reaction offers numerous advantages for labeling and straightforward manipulation of MP sample quality that make it a promising choice when pursuing structural investigations by NMR. The available CF systems based on lysates originating from various pro- or eukaryotic cells differ in the options they provide for studies with MPs, namely (i) expression yield, (ii) posttranslational modifications, and (iii) tolerances for supplemented additives. Currently, CF systems based on *E. coli* lysates are most commonly employed for structural studies of MPs (Bernhard & Tozawa, 2013). *E. coli* lysates provide reliable expression yields sufficient for NMR experiments and have a high tolerance for additives, e.g., ligands that stabilize MP conformations or membrane mimetics that allow the cotranslational solubilization of synthesized MPs (Hein, Henrich, Orbán, Dötsch, & Bernhard, 2014). Furthermore, *E. coli* lysates show a significantly reduced level of metabolism that enable the isotopic labeling of selected amino acid types with low isotopic scrambling. These main characteristics make CF systems an excellent option for pursuing NMR studies of MPs and in the paragraphs later we discuss a number of practical considerations to be taken into account.

2. CORE CONSIDERATIONS FOR THE CELL-FREE GENERATION OF MP SAMPLES

The foremost consideration when approaching structural studies is expression yield. The continuous exchange cell-free configuration is the most efficient and detailed protocol for the production of MPs in *E. coli* lysates has been described (Kigawa et al., 1999; Reckel et al., 2010). Protocols for the preparation of *E. coli* lysates are detailed in numerous publications elsewhere (Henrich, Dötsch, & Bernhard, 2015; Schwarz et al., 2007). Fast and standardized strategies that address insufficient expression efficiencies relying on the reduced complexity of *E. coli* lysates have been developed (Haberstock et al., 2012). Success rates for obtaining a MP target in milligram amounts are high and there is no current evidence that certain types of MPs are preferably expressed in CF systems with regard to topology, function, or size (Junge et al., 2011; Sachse, Dondapati, Fenz, Schmidt, & Kubick, 2014).

A second and more challenging consideration is protein functionality and subsequent sample quality. In cellular expression systems, relatively harsh detergents are typically used either to extract the synthesized MPs from endogenous membranes or to refold from inclusion bodies. These procedures can have detrimental effects on MP functionality and can be avoided by using CF systems. A new and unique property of CF expression is that numerous modes can be employed in order to apply either co- or posttranslational folding strategies and to solubilize the MPs in a large variety of different environments (Hein et al., 2014). Systematic screening of additives, stabilizers, or redox conditions can significantly boost expression and improve sample quality (Kai, Dötsch, Kaldenhoff, & Bernhard, 2013; Michel & Wüthrich, 2012; Schwarz et al., 2007). The low volume of CF reactions allows screening of expensive and/or rare compounds. In particular, supplementation of the CF reaction with binding partners can have a significant impact on MP folding and dynamics, thus resulting in improved spectral dispersion and subsequent resonance assignment (Fig. 1). Line broadening can be dramatically addressed by reducing internal conformational exchange dynamics with a stabilizing binding partner for the MP (Jaremko, Jaremko, Giller, Becker, & Zweckstetter, 2014; Yabuki et al., 1998). Small compounds and inhibitors that do not significantly contribute to the overall size of the complex are most attractive.

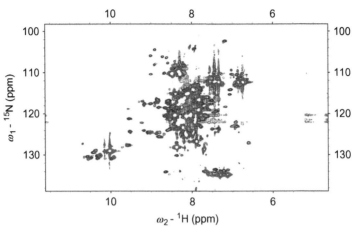

Figure 1 ^{15}N–^{1}H BEST-TROSY of a uniformly ^{2}H,^{15}N-labeled four TMS integral membrane protein with (blue) and without (red) binding partner. TROSY spectra were measured on MP samples of 300 μM in 20 mM sodium acetate buffer pH 4 at 45° on an 800 MHz spectrometer. (See the color plate.)

MPs can be synthesized in the precipitate (P-CF) forming mode in the absence of any supplied hydrophobic compounds. Although expression efficiencies are target and protocol dependent, MP yields of up to several mg/mL can routinely be achieved. CF protocol development usually starts with the P-CF mode and optimization of ion concentrations or variations in the DNA template design are standardized steps in order to obtain efficient MP production (Haberstock et al., 2012). Further, P-CF-generated MP precipitates may retain folded structures that support their efficient posttranslational solubilization in certain detergents (Maslennikov et al., 2010). High quality samples of P-CF-generated MPs can be obtained. The membrane-integrated MraY translocase of *Bacillus subtilis* had the highest reported specific activity after P-CF production and subsequent solubilization in the detergent, n-dodecyl β-D-maltoside (DDM) (Ma et al., 2011). The crystal structure of P-CF-synthesized diacylglycerolkinase DgkA was solved at 2.28 Å diffraction (Boland et al., 2014). In addition, several NMR structures of P-CF-synthesized MPs have been solved (Klammt et al., 2012; Maslennikov et al., 2010; Sobhanifar et al., 2010).

In the detergent-based (D-CF) mode, the MPs are expressed in the presence of a detergent or other hydrophobic compound in order to facilitate their cotranslational solubilization and functional folding. *E. coli* lysates are quite robust and tolerate a variety of detergents significantly above their critical micellar concentrations (CMCs). However, tolerated detergents are relatively mild with low CMCs. Long-chain Brij derivatives, the steroid detergent digitonin, triton-X-100, or DDM are some frequently employed detergents for D-CF expressions (Hein et al., 2014). Other options include fluorinated surfactants, peptide surfactants, or noncharged amphipols (Bazzacco et al., 2012; Koutsopoulos, Kaiser, Eriksson, & Zhang, 2012; Park et al., 2007; Wang et al., 2011; Zheng et al., 2014). Mixed detergent micelles or detergent/lipid mixtures can further improve the stability of D-CF-expressed MPs. The solution NMR structure of D-CF-expressed proteorhodopsin was solved with samples synthesized in the presence of a mixture of digitonin and the short chain lipid di-heptanoyl-phosphocholine (DHPC) (Reckel et al., 2011). Likewise, the crystal structure of *Acetabularia* rhodopsin II was obtained after D-CF expression in presence of a similar mixture (Wada et al., 2011).

In the lipid-based (L-CF) expression mode, lipid bilayers are supplied for the cotranslational solubilization of MPs. The obvious appeal of L-CF expression is the higher potential of MPs to adopt functional folds in more native like environments. Specific lipid compositions for a particular MP can be selected and modified to influence functionality (Roos et al., 2012). Lipid bilayers are provided either as preformed liposomes, as bicelles in appropriate mixtures with detergents, or as preassembled nanodiscs (NDs). However, L-CF expression is prone to lowered expression yields and often does not meet the basic criteria pursuant to structural studies.

Methods to assess MP functionality are critical in order to develop expression and purification protocols and validate final sample conditions. Synthesized MPs are usually purified by affinity chromatography and by taking advantage of small terminal purification tags. Primary detergents used for the initial solubilization of the MPs in the P-CF or D-CF modes can be exchanged to more NMR suitable detergents upon immobilization of the MPs on affinity matrices. The purified MP samples are concentrated to final concentrations between 100 and 500 μM for NMR measurements. Low salt and low pH buffers are desirable for high sensitivity NMR experiments. Variable detergent concentrations can have effects on viscosity and sample quality and must therefore be kept consistent between samples. Hydrophobic environments useful for solution NMR studies of MPs are quite restricted and confined to detergents, isotropic bicelles, and NDs. Identifying the appropriate environment for a MP is subject to intense screening and acceptable compromises in view of spectral dispersion, solubilization efficiency, and MP stability have to be found. However, irrespective of the media, it must maintain MP functionality and remain stable for

Table 1 Membrane Protein Structures Determined by NMR Spectroscopy

Protein	Environment	Size (kDa)	References
CrgA from *Mycobacterium tuberculosis*	POPC/POPG lipid bilayers	40	Das et al. (2015)
Rocker—designed Zn^{2+}/Co^{2+} transporter	DPC micelles/DMPC liposomes	29.7	Joh et al. (2014)
Transmembrane domain of insulin receptor from *Homo sapiens*	DPC micelles	20	Li, Wong, and Kang (2014)
Vascular endothelial growth factor receptors—CF expressed from *H. sapiens*	DCP micelles	8.2	Manni et al. (2014)
α7 Nicotinic acetylcholine receptor (nAChR) from *H. sapiens*	LDAO micelles	14.5	Bondarenko et al. (2014)
YgaP—α-helical integral membrane protein from *E. coli*	DHPC/1-myristoyl-2-hydroxy-*sn*-glycero-3-phospho-(1′-*rac*-glycerol)	14.3	Eichmann et al. (2014)
Mitochondrial translocator protein TSPO from *Mus musculus*	DPC micelles	19.2	Jaremko et al. (2014)
β-Barrel OmpX from *E. coli*	DMPC nanodisc	16.4	Hagn, Etzkorn, Raschle, and Wagner (2013)
Sensor rhodopsin from *Anabeana nostoc*	DMPC/DMPA liposomes	82.5	Wang et al. (2013)

long NMR experiments. Among the currently most popular detergents for NMR studies are 1-myristoyl-2-hydroxy-*sn*-glycero-3-phosphate (LMPG) and dodecylphosphocholine (DPC). Detergents have remained popular due to their ease of use and fast tumbling in complex with MP; however, the functionality of several classes of MPs such as transporters or channels is difficult to assess (Table 1).

3. SPECIFIC CHALLENGES OF NMR STUDIES WITH MPs

NMR spectra of MPs, and in particular of predominantly α-helical MPs, share some characteristics with large, unstructured proteins. Unlike unfolded proteins which exhibit high internal flexibility, slow relaxation, and are subject to pulse sequences with long transfer delays, MPs are characterized by slow tumbling resulting in fast relaxation and broadened resonances. This is due to the often large size of MPs, their frequent tendency to form oligomers and to the surrounding hydrophobic environment that is necessary to keep the protein in solution. Complex size, rotational correlation time, and NMR line widths are directly proportional. The relevant particle size of the total complex and its resulting rotational correlation time are influenced by both the type of selected membrane mimetic and the individual MP size. Selecting the "smallest" molecular weight membrane mimetic, such as detergent micelles with low aggregation numbers or organic solvents, can be an effective strategy in order to address this issue, but often compromises MP functionality. Some improvement in line broadening can be seen by increasing the tumbling rate of the complex by measuring at elevated temperatures; however, this is often insufficient and can lead to protein instability.

Furthermore, homo-oligmerization of MPs often requires higher concentrations of detergent or lipids for solubilization which again increases the complex sizes. Even small MPs consisting of 100–300 amino acid residues can thus behave as large particles in complex with hydrophobic environments. In addition, approximately 60% of the amino acids typically present in the hydrophobic transmembrane helices (TMS) of MPs consist mainly of the six residues alanine, phenylalanine, glycine, isoleucine, leucine, and valine (Reckel et al., 2008). This redundancy significantly contributes to signal overlap and resonance degeneracy. TMS specific and amino acid combinatorial labeling schemes utilizing CF expression have been designed in order to address this issue (Löhr et al., 2012; Maslennikov et al., 2010; Reckel et al., 2008).

The combination of perdeuteration and transverse relaxation optimized spectroscopy (TROSY) experiments which ease transverse relaxation have addressed line broadening of large complexes (Pervushin, Riek, Wider, & Wüthrich, 1997), but introduce additional obstacles when obtaining nuclear Overhauser effects (NOEs) necessary to measure distances and ultimately structure determination. In addition, the use of deuterated amino acids in CF expression has been shown to lead to variable levels of protonation at the Hα and Hβ positions, due to endogenous metabolic enzymes in the E. coli lysate (Etezady-Esfarjani, Hiller, Villalba, & Wüthrich, 2007). However, the open nature of the CF reaction provides tools to address these obstacles as discussed further later.

These general features result in serious problems for resonance assignment and spectra analysis as well as difficulties with the detection of long-range interactions that are important for structural calculations. Much trial and error is required in the search for conditions that maintain native structured and stable MPs while still yielding high-quality NMR spectra. Improving NMR spectra mainly requires screening of different membrane mimetics and other MP solubilization compounds. However, liquid-state NMR studies limit the practical use of most classical membrane mimetics due to size and viscosity restrictions. Therefore, the search and development of MP-solubilizing compounds appropriate for liquid NMR studies such as amphipols or NDs is ongoing and essential (Etzkorn et al., 2013; Raschle et al., 2009; Warschawski et al., 2011; Zoonens, Catoire, Giusti, & Popot, 2005).

4. AN EMERGING PERSPECTIVE: NMR WITH NDs

NDs are monodisperse complexes consisting of a lipid bilayer encapsulated by two membrane scaffold proteins (MSPs) (Denisov, Grinkova, Lazarides, & Sligar, 2004). NDs are highly soluble and can be added to the CF reaction in final concentrations up to 100 μM without inhibiting expression (Katzen et al., 2008; Lyukmanova et al., 2012; Roos et al., 2014). The membranes within NDs can be designed out of a large variety of lipids or lipid mixtures and currently popular are 1,2-dimyristoyl-*sn*-glycero-3-phosphocholine (DMPC) and 1,2-dimyristoyl-*sn*-glycero-3-phosphoglycerol (Roos et al., 2014).

The use of NDs as a MP hydrophobic environment is gaining attraction in the liquid-state NMR field. The channels V-DAC and OmpX have been subject of NMR investigations in NDs with DMPC (Etzkorn et al., 2013; Hagn et al., 2013). MPs can either be cotranslationally expressed into preformed NDs by L-CF expression or be posttranslationally assembled in a mixture of the desired lipids and corresponding MSPs. Homogeneity of the resulting MP/ND complexes can be assessed by size exclusion chromatography profiling and the complexes can be concentrated by ultrafiltration to high concentrations. Larger MPs inserted into NDs can be visualized by cryo-electron microscopy (Hagn et al., 2013). The mechanism of MP insertion into NDs remains unclear and a significant concern is complex heterogeneity and methods to probe this.

NMR studies of MP/ND complexes are immediately accompanied by issues of large complex size and slow tumbling in liquid-state NMR.

Engineered MSP derivatives have been constructed that result in NDs with different diameters ranging from 6.3 nm by using the ΔH4-H6 MSP to up to 12 nm with the MSP1E3D1 derivative (Denisov et al., 2004; Hagn et al., 2013). Small NDs formed by the ΔH5-MSP derivative of approximately 8 nm are currently the most popular in NMR approaches with MPs due to its stability (Hagn et al., 2013). The assembly of ΔH5-MSP-NDs is performed with selected sodium cholate-solubilized lipids such as DMPC in a ratio of 1:50:100 (ΔH5-MSP:DMPC:sodium cholate). The detergent is removed by slow dialysis and proper ND formation is subsequently analyzed by size exclusion profiling.

5. LABELING OF CELL-FREE SYNTHESIZED MPs WITH STABLE ISOTOPES

CF systems enable the labeling of almost any amino acid type either individually or in desired combinations. Particular labeling schemes can be used in order to map protein interaction interfaces as shown for soluble proteins expressed *in vivo* (Reese & Dötsch, 2003), to facilitate backbone assignments (Löhr et al., 2012), and to determine three-dimensional NMR structures (Reckel et al., 2011).

The first goal of an NMR project is obtaining sequential assignments of backbone resonances. With well dispersed proteins, triple-resonance experiments such as the HNCACB, CBCACONH, or CBCANH of ^{15}N/^{13}C-labeled proteins are sufficient. However, due to the large effective size of MPs, only the most robust NMR pulse sequences with short coherence transfer delays can be used. In addition, the crowded nature of MP spectra demands the use of particular labeling strategies that is largely facilitated by CF expression. Comprehensive labeling strategies that reduce spectral overlap have been developed in the past (Abdine et al., 2010; Parker, Aulton-Jones, Hounslow, & Craven, 2004; Reckel et al., 2008; Wu et al., 2006; Yabuki et al., 1998) The number of individually labeled MP samples has been reduced by the development of combinatorial labeling strategies (Löhr et al., 2012). A number of programs are available in order to assist the design of labeling strategies using ^{15}N- and ^{13}C-labeled amino acids that cover the highest proportion of the backbone sequence (Hefke et al., 2011). However, a manual assessment based on the number of unique amino acid pairs is often sufficient. Combinatorial labeling schemes which generate three to five samples are used to obtain assignment information (Löhr et al., 2014). Two-dimensional NH-detected experiments (i.e., TROSY,

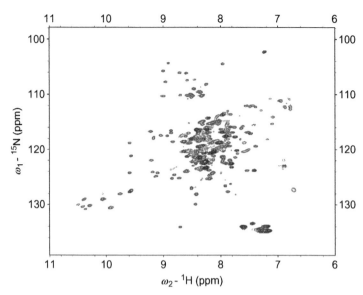

Figure 2 Overlay of ^{15}N–^{1}H BEST-TROSY spectra of a uniformly and selectively labeled four TMS integral MP. The protein was uniformly labeled by CF expression in presence of either a complete set of ^{2}H,^{15}N-labeled amino acids (red) or a combination of the ^{2}H,^{15}N-labeled amino acids arginine, lysine, threonine, and tyrosine (blue). Amino acids were supplied at 4 mM final concentrations. TROSY spectra were measured on MP samples of 200 μM in 20 mM sodium acetate buffer pH 4 at 45° on an 800 MHz spectrometer. (See the color plate.)

HNCA, HNCO) in combination with standard 3D experiments such as HNCA, HNCO, and HNCACB of uniformly ^{13}C/^{15}N-labeled proteins are used to identify amide and carbon shifts. ^{15}N-TROSY spectra of amino acid selectively labeled MPs display far less overlap when compared with those of fully labeled samples (Fig. 2).

6. REDUCING SCRAMBLING PROBLEMS

Selective labeling strategies are notoriously hindered by isotopic scrambling caused by metabolic conversion of supplied labeled amino acids. This problem is less significant in CF systems due to the reduced complexity of the processed cell lysates but remains a hindrance to combinatorial labeling schemes. Scrambling frequently originates from residual pyridoxal 5′-phosphate (PLP)-requiring metabolic enzymes present in the *E. coli* lysates that catabolize and rearrange isotopically labeled amino acids resulting in heterogeneous label incorporation in the synthesized MP (Table 2). Mixed signals and loss of signal intensity are the main outcomes of

Table 2 Amino Acid Scrambling Enzymes and Inhibition Strategies

Enzyme	Amino Acid Scrambler	Inhibitor	References
Aspartate aminotransferase, alanine transaminase	Glu to Asp; Glu to Ala	Aminooxyacetate (AOA), β-chloro-L-alanine	Mahon, Graber, Christen, and Malthouse (1999) and Morita et al. (2004)
Asparagine synthetase; glutamine synthetase	Transamination of Asp to Glu, Gln, and Asn; transamination of Glu to Asp and Gln	S-methyl-L-cysteine sulfoximine; L-methionine sulfoximine	Koizumi, Hiratake, Nakatsu, Kato, and Oda (1999); Manning, Moore, Rowe, and Meister (1969), and Tonelli et al. (2011)
Aspartate ammonia-lyase	Exchange b/w main-chain and side-chain amide groups of Asp and Asn	D-Malic acid	Falzone, Karsten, Conley, and Viola (1988)
Glutaminase	Amide of Gln off to give Glu	6-Diazo-5-oxo-L-norleucine (DON)	Hartman (1968)
Asparaginase	Amide of Asn off to give Asp	5-Diazo-4-oxo-L-norvaline (DONV)	Handschumacher, Bates, Chang, Andrews, and Fischer (1968)
General pyridoxal 5′-phosphate (PLP)-requiring enzymes	Back protonation—at α-position of Ala, Asp, Glu, Phe, Tyr, Trp, and β position of Asp and Asn	AOA and D-malic acid; NaBH$_4$ to reduce Schiff base formed between PLP and amino groups	Yokoyama, Matsuda, Koshiba, Tochio, and Kigawa (2011) and Su, Loh, Qi, and Otting (2011)
Methionine γ-lyase	Exchange of α- and β-hydrogens of L-methionine and S-methyl-L-cysteine with deuterium from solvent	L-Cycloserine	Esaki, Nakayama, Sawada, Tanaka, and Soda (1985) and Kuznetsov et al. (2015)

Continued

Table 2 Amino Acid Scrambling Enzymes and Inhibition Strategies—cont'd

Enzyme	Amino Acid Scrambler	Inhibitor	References
Methionyl-tRNA synthetase and cysteinyl-tRNA synthetase	Back protonation at the α-position of Met and Cys in ^2H_2O	REP8839—not tested in CF	Eriani, Dirheimer, and Gangloff (1991), Green et al. (2009), and Williams and Rosevear (1991)
Tryptophanase	Back protonation at α-position of many amino acids	L-Bishomotryptophan—not CF tested	Do, Nguyen, Celis, and Phillips (2014) and Faleev et al. (1990)
Glutamine synthetase	Glutamate and ammonia to glutamine	L-2-Amino-4-(hydroxymethylphosphinyl)butanoic acid (PPT)—not CF tested	Logusch et al. (1990)
Cystathionine gamma-synthase	Hydrogen exchange by PLP-requiring enzymes	β-Cyanoalanine, propargylglycine, L-aminoethoxyvinylglycine	Asimakopoulou et al. (2013), Clausen, Huber, Prade, Wahl, and Messerschmidt (1998), and Homer, Kim, and LeMaster (1993)
Serine hydroxymethyltransferase	Hydrogen exchange by PLP-requiring enzymes	5-Formyltetrafolate—not CF tested	Fitzpatrick and Malthouse (1998) and Stover and Schirch (1991)
Tryptophan synthase	Hydrogen exchange by PLP-requiring enzymes	None known	Milne and Malthouse (1996)
Tyrosine phenol-lyase	Hydrogen exchange by PLP-requiring enzymes	None for *E. coli* known	Faleev, Demidkina, Tsvetikova, Phillips, and Yamskov (2004)

scrambling which defeat the initial purpose of using selective labels. The first objective in pursuing selective labeling schemes of CF produced MP for backbone assignment is to address scrambling inhibition within amide positions, with the worst offenders being asparagine, aspartate, glutamine, and glutamate. The addition of specific inhibitors to CF reactions or chemical pretreatment of the CF lysates can significantly improve the labeling efficiency and specificity of expressed MPs (Table 2; Ozawa et al., 2004; Wu et al., 2006; Yokoyama, Matsuda, Koshiba, & Kigawa, 2010). Many scrambling enzymes utilize similar reaction mechanisms, e.g., transaminases require PLP as an electron acceptor. This makes them susceptible to broad-spectrum inhibition and pretreatment of the CF lysate, i.e., $NaBH_4$ treatment has been shown to inactivate PLP-dependent enzymes (Su et al., 2011). Similar approaches in *E. coli* cells are limited as broad-spectrum inhibition of amino acid metabolic enzymes would severely compromise the overall protein expression and cell viability. A further potent inhibitor of PLP-requiring enzymes is aminooxyacetate (AOA) that suppresses the transamination of aspartic acid, asparagine, glutamic acid, and glutamine (Yokoyama et al., 2011). D-Malic acid is an inhibitor of the *E. coli* aspartate ammonia-lyase (Falzone et al., 1988). A combination of 20 mM AOA, 20 mM D-cycloserine, and 3 mM D-malic acid was selected as a cocktail in order to repress amide and side-chain scrambling. The solid powders were added directly to the feeding mixture and did not reduce the expression by more than 10–20%. Figure 3 shows the ^{15}N-TROSY spectra of a selectively labeled MP synthesized in the presence or absence of the scrambling inhibitor cocktail. In particular, the scrambling of ^{15}N aspartic acid residues was largely reduced by the inhibitor cocktail. However, while ^{15}N and ^{13}C scrambling can successfully be reduced by taking advantage of inhibitors, the back protonation at Hα positions as well as strategies for selective and specific deuteration have not been fully addressed yet.

On the other hand, the absence of intact metabolic pathways in CF lysates has drawbacks. A significant advantage of labeling in *E. coli* is the manipulation of enzymatic pathways in order to yield selective and mixed deuteration (Hilty, Wider, Fernández, & Wüthrich, 2003; Kerfah et al., 2015). For large complexes, the strategic introduction of protons is the most viable route in order to obtain distance restraints for structure calculations. In CF systems, expensive selectively/stereo-isotopically labeled amino acids remain the main route for introducing protons at selected sites. Unfortunately, exploiting the metabolic enzymes present in *E. coli* lysates that are able to manipulate labels has received little attention as of yet (Loscha & Otting, 2013).

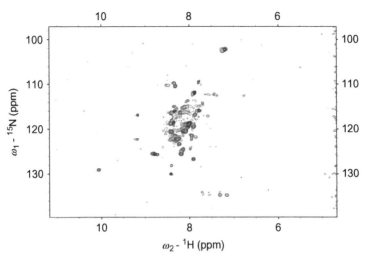

Figure 3 Overlay of ^{15}N–^{1}H BEST-TROSY spectra of a four TMS integral membrane protein selectively labeled with ^{13}C, ^{15}N, ^{2}H aspartate and threonine. Blue: expression with scrambling inhibitors, aminooxyacetate, D-malic acid, and D-cycloserine; red: expression without scrambling inhibitors. TROSY spectra were measured on MP samples of 200 μM in 20 mM sodium acetate buffer pH 4 at 45° on a 600 MHz spectrometer. (See the color plate.)

7. PERDEUTERATION OF CELL-FREE SYNTHESIZED MPs

Structural characterization of MPs by liquid-state NMR often demands the use of perdeuteration. Labeled and perdeuterated proteins can be synthesized with nearly no reduction in expression yields by CF expression. In addition, deuterated or otherwise labeled detergents or lipids can be implemented for the solubilization of CF-synthesized MPs without risking contaminations of endogenous lipids. A valuable advantage upon perdeuteration in CF systems is that the protein amide positions remain protonated resulting in ideal samples for many NMR experiments. Such labeling schemes are often hard to achieve by conventional expression in *E. coli* cells due to reduced growth and expression efficiencies in deuterated media. Moreover, proton exchange at amide positions is often incomplete and requires partial denaturation in order to make them solvent accessible (Su et al., 2011).

Deuteration increases signal intensities and resolution due to reduction in relaxation caused by surrounding protons. However, experiments that rely on magnetization transfer between side-chain protons are not possible. The NOESY experiment is the most important for obtaining distance restraints for structure calculation and is useless for fully deuterated MPs, with the exception of measuring NOEs between amide protons. Hence, selective deuteration and mixed labels are an absolute requirement when pursuing large MP structures by NMR. Full deuteration in the CF system can be achieved by using deuterated labels or by CF expression using deuterated *E. coli* extract (Etezady-Esfarjani et al., 2007). Selecting one method is target specific but both approaches remain subject to back protonation scrambling. Back protonation or back exchange occurs when CF-synthesized proteins in the presence of deuterated amino acids undergo exchange with ^{1}H from the solvent introducing ^{1}H at deuterated Cα and Cβ positions (Etezady-Esfarjani et al., 2007). Several compounds have been shown to inhibit back protonation with variable success (Table 2). NaBH$_4$ treatment of the cell lysates has also been shown to reduce back exchange at the Hα position (Su et al., 2011). NaBH$_4$ dissolved in dimethylforamide is added to 1 mL of thawed *E. coli* lysate and gently shaken for 10 min to aid hydrogen gas release. Dialyzing the cell lysate against the CF buffer before the reaction is further recommended but in our experience may reduce yield and cause precipitation (Su et al., 2011).

AOA and L-methionine sulfoximine have been found to block proton exchange at Cα sites, with the exception of glycine and Cβ sites of alanines (Tonelli et al., 2011). We have found that L-methionine sulfoximine has a negative effect on CF expression yield although this may be target specific. The discussed inhibitors are used to nonspecifically reduce back protonation of CF-expressed proteins. However, selectively labeled or stereospecific-labeled proton positions in combination with resonance assignments are necessary in order to obtain useful distance restraints. Strategies that avoid the difficulty in attaining full assignment of large deuterated systems and selective deuteration have been developed and include methyl-selective protonation of isoleucine, leucine, and valine residues (Rosen et al., 1996) and the stereo-array isotope labeling method (Kainosho et al., 2006). Another promising method with great potential uses extracts from MPs produced in *E. coli* with selective methyl labeling in CF (Linser et al., 2014). These methods have led to resonance assignments in proteins that are mostly deuterated. However, methyl-selective protonation of isoleucine, leucine, and valine residues relies on the full component of *E. coli* metabolic enzymes during expression in deuterated media and are thus currently not useful for CF expression. An interesting publication (Loscha & Otting, 2013) describes a CF method for the generation of stereoselectively deuterated glycine residues by exploiting the scrambling enzyme serine hydroxymethyltransferase present in *E. coli* extracts. However, these approaches are only currently being explored and require detailed proteomics analysis of the *E. coli* lysates. Supplying chemically synthesized amino acids with selective protonated sites is currently the only straightforward method for producing selectively protonated MPs by CF expression. Hence, scrambling remains an issue to be addressed in CF systems.

8. CONCLUSION

CF expression offers numerous advantages in the production of isotopically labeled MPs for solution NMR studies. We have not discussed the numerous labeling techniques appropriate for applications beyond NMR spectroscopy, such as site-specific labeling and the incorporation of non-natural amino acids. The reader is referred to other excellent publications detailing these innovative techniques (Chen et al., 2015; Zimmerman et al., 2014). These methods in addition to those in the development of pulse sequences which demand specific labeling schemes are actively being developed to expand the use of CF to produce labeled MPs. The accessible nature of the CF system and the manipulation of the CF *E. coli* extract remains the key to the power of this system and the future for labeling of MPs.

ACKNOWLEDGMENTS

This work was funded by the Collaborative Research Center (SFB) 807 of the German Research Foundation (DFG) and by a research grant of the DFG to A.L. Support was further provided by Instruct, part of the European Strategy Forum on Research Infrastructures (ESFRI).

REFERENCES

Abdine, A., Verhoeven, M. A., Park, K.-H., Ghazi, A., Guittet, E., Berrier, C., et al. (2010). Structural study of the membrane protein MscL using cell-free expression and solid-state NMR. *Journal of Magnetic Resonance (San Diego, Calif.: 1997), 204*(1), 155–159.

Asimakopoulou, A., Panopoulos, P., Chasapis, C. T., Coletta, C., Zhou, Z., Cirino, G., et al. (2013). Selectivity of commonly used pharmacological inhibitors for cystathionine β synthase (CBS) and cystathionine γ lyase (CSE). *British Journal of Pharmacology, 169*(4), 922–932.

Bazzacco, P., Billon-Denis, E., Sharma, K. S., Catoire, L. J., Mary, S., Le Bon, C., et al. (2012). Nonionic homopolymeric amphipols: Application to membrane protein folding, cell-free synthesis, and solution nuclear magnetic resonance. *Biochemistry, 51*(7), 1416–1430.

Bernhard, F., & Tozawa, Y. (2013). Cell-free expression—Making a mark. *Current Opinion in Structural Biology, 23*(3), 374–380.

Boland, C., Li, D., Shah, S. T. A., Haberstock, S., Dötsch, V., Bernhard, F., et al. (2014). Cell-free expression and in meso crystallisation of an integral membrane kinase for structure determination. *Cellular and Molecular Life Sciences: CMLS*, *71*(24), 4895–4910.

Bondarenko, V., Mowrey, D. D., Tillman, T. S., Seyoum, E., Xu, Y., & Tang, P. (2014). NMR structures of the human α7 nAChR transmembrane domain and associated anesthetic binding sites. *Biochimica et Biophysica Acta*, *1838*(5), 1389–1395.

Chen, W.-N., Kuppan, K. V., Lee, M. D., Jaudzems, K., Huber, T., & Otting, G. (2015). O-tert-Butyltyrosine, an NMR tag for high-molecular weight systems and measurements of submicromolar ligand binding affinities. *Journal of the American Chemical Society*, *137*(13), 4581–4586.

Clausen, T., Huber, R., Prade, L., Wahl, M. C., & Messerschmidt, A. (1998). Crystal structure of *Escherichia coli* cystathionine gamma-synthase at 1.5 A resolution. *The EMBO Journal*, *17*(23), 6827–6838.

Das, N., Dai, J., Hung, I., Rajagopalan, M. R., Zhou, H.-X., & Cross, T. A. (2015). Structure of CrgA, a cell division structural and regulatory protein from Mycobacterium tuberculosis, in lipid bilayers. *Proceedings of the National Academy of Sciences of the United States of America*, *112*(2), E119–E126.

Denisov, I. G., Grinkova, Y. V., Lazarides, A. A., & Sligar, S. G. (2004). Directed self-assembly of monodisperse phospholipid bilayer nanodiscs with controlled size. *Journal of the American Chemical Society*, *126*(11), 3477–3487.

Do, Q. T., Nguyen, G. T., Celis, V., & Phillips, R. S. (2014). Inhibition of *Escherichia coli* tryptophan indole-lyase by tryptophan homologues. *Archives of Biochemistry and Biophysics*, *560*, 20–26.

Eichmann, C., Tzitzilonis, C., Bordignon, E., Maslennikov, I., Choe, S., & Riek, R. (2014). Solution NMR structure and functional analysis of the integral membrane protein YgaP from *Escherichia coli*. *The Journal of Biological Chemistry*, *289*(34), 23482–23503.

Eriani, G., Dirheimer, G., & Gangloff, J. (1991). Cysteinyl-tRNA synthetase: Determination of the last E. coli aminoacyl-tRNA synthetase primary structure. *Nucleic Acids Research*, *19*(2), 265–269.

Esaki, N., Nakayama, T., Sawada, S., Tanaka, H., & Soda, K. (1985). 1H NMR studies of substrate hydrogen exchange reactions catalyzed by L-methionine gamma-lyase. *Biochemistry*, *24*(15), 3857–3862.

Etezady-Esfarjani, T., Hiller, S., Villalba, C., & Wüthrich, K. (2007). Cell-free protein synthesis of perdeuterated proteins for NMR studies. *Journal of Biomolecular NMR*, *39*(3), 229–238.

Etzkorn, M., Raschle, T., Hagn, F., Gelev, V., Rice, A. J., Walz, T., et al. (2013). Cell-free expressed bacteriorhodopsin in different soluble membrane mimetics: Biophysical properties and NMR accessibility. *Structure (London, England: 1993)*, *21*(3), 394–401.

Faleev, N. G., Demidkina, T. V., Tsvetikova, M. A., Phillips, R. S., & Yamskov, I. A. (2004). The mechanism of alpha-proton isotope exchange in amino acids catalysed by tyrosine phenol-lyase. What is the role of quinonoid intermediates? *European Journal of Biochemistry/FEBS*, *271*(22), 4565–4571.

Faleev, N. G., Ruvinov, S. B., Saporovskaya, M. B., Belikov, V. M., Zakomyrdina, L. N., Sakharova, I. S., et al. (1990). Preparation of α-deuterated L-amino acids using E. coli B/It7-A cells containing tryptophanase. *Tetrahedron Letters*, *31*(48), 7051–7054.

Falzone, C. J., Karsten, W. E., Conley, J. D., & Viola, R. E. (1988). L-aspartase from *Escherichia coli*: Substrate specificity and role of divalent metal ions. *Biochemistry*, *27*(26), 9089–9093.

Fitzpatrick, T. B., & Malthouse, J. P. (1998). A substrate-induced change in the stereospecificity of the serine-hydroxymethyltransferase-catalysed exchange of the alpha-protons of amino acids—Evidence for a second catalytic site. *European Journal of Biochemistry/FEBS*, *252*(1), 113–117.

Green, L. S., Bullard, J. M., Ribble, W., Dean, F., Ayers, D. F., Ochsner, U. A., et al. (2009). Inhibition of methionyl-tRNA synthetase by REP8839 and effects of resistance mutations on enzyme activity. *Antimicrobial Agents and Chemotherapy, 53*(1), 86–94.

Haberstock, S., Roos, C., Hoevels, Y., Dötsch, V., Schnapp, G., Pautsch, A., et al. (2012). A systematic approach to increase the efficiency of membrane protein production in cell-free expression systems. *Protein Expression and Purification, 82*(2), 308–316.

Hagn, F., Etzkorn, M., Raschle, T., & Wagner, G. (2013). Optimized phospholipid bilayer nanodiscs facilitate high-resolution structure determination of membrane proteins. *Journal of the American Chemical Society, 135*(5), 1919–1925.

Handschumacher, R. E., Bates, C. J., Chang, P. K., Andrews, A. T., & Fischer, G. A. (1968). 5-Diazo-4-oxo-L-norvaline: Reactive asparagine analog with biological specificity. *Science (New York, N.Y.), 161*(3836), 62–63.

Hartman, S. C. (1968). Glutaminase of *Escherichia coli*. 3. Studies on the reaction mechanism. *The Journal of Biological Chemistry, 243*(5), 870–878.

Hefke, F., Bagaria, A., Reckel, S., Ullrich, S. J., Dötsch, V., Glaubitz, C., et al. (2011). Optimization of amino acid type-specific 13C and 15N labeling for the backbone assignment of membrane proteins by solution- and solid-state NMR with the UPLABEL algorithm. *Journal of Biomolecular NMR, 49*(2), 75–84.

Hein, C., Henrich, E., Orbán, E., Dötsch, V., & Bernhard, F. (2014). Hydrophobic supplements in cell-free systems: Designing artificial environments for membrane proteins. *Engineering in Life Sciences, 14*(4), 365–379.

Henrich, E., Dötsch, V., & Bernhard, F. (2015). Screening for lipid requirements of membrane proteins by combining cell-free expression with nanodiscs. *Methods in Enzymology, 556*, 351–369.

Hilty, C., Wider, G., Fernández, C., & Wüthrich, K. (2003). Stereospecific assignments of the isopropyl methyl groups of the membrane protein OmpX in DHPC micelles. *Journal of Biomolecular NMR, 27*(4), 377–382.

Homer, R. J., Kim, M. S., & LeMaster, D. M. (1993). The use of cystathionine gamma-synthase in the production of alpha and chiral beta deuterated amino acids. *Analytical Biochemistry, 215*(2), 211–215.

Jaremko, Ł., Jaremko, M., Giller, K., Becker, S., & Zweckstetter, M. (2014). Structure of the mitochondrial translocator protein in complex with a diagnostic ligand. *Science, 343*(6177), 1363–1366.

Joh, N. H., Wang, T., Bhate, M. P., Acharya, R., Wu, Y., Grabe, M., et al. (2014). De novo design of a transmembrane Zn^{2+}-transporting four-helix bundle. *Science (New York, N.Y.), 346*(6216), 1520–1524.

Junge, F., Haberstock, S., Roos, C., Stefer, S., Proverbio, D., Dötsch, V., et al. (2011). Advances in cell-free protein synthesis for the functional and structural analysis of membrane proteins. *New Biotechnology, 28*(3), 262–271.

Kai, L., Dötsch, V., Kaldenhoff, R., & Bernhard, F. (2013). Artificial environments for the co-translational stabilization of cell-free expressed proteins. *PloS One, 8*(2), e56637.

Kainosho, M., Torizawa, T., Iwashita, Y., Terauchi, T., Mei Ono, A., & Güntert, P. (2006). Optimal isotope labelling for NMR protein structure determinations. *Nature, 440*(7080), 52–57.

Katzen, F., Fletcher, J. E., Yang, J.-P., Kang, D., Peterson, T. C., Cappuccio, J. A., et al. (2008). Insertion of membrane proteins into discoidal membranes using a cell-free protein expression approach. *Journal of Proteome Research, 7*(8), 3535–3542.

Kerfah, R., Plevin, M. J., Pessey, O., Hamelin, O., Gans, P., & Boisbouvier, J. (2015). Scrambling free combinatorial labeling of alanine-β, isoleucine-δ1, leucine-proS and valine-proS methyl groups for the detection of long range NOEs. *Journal of Biomolecular NMR, 61*(1), 73–82.

Kigawa, T., Yabuki, T., Yoshida, Y., Tsutsui, M., Ito, Y., Shibata, T., et al. (1999). Cell-free production and stable-isotope labeling of milligram quantities of proteins. *FEBS Letters*, *442*(1), 15–19.

Klammt, C., Maslennikov, I., Bayrhuber, M., Eichmann, C., Vajpai, N., Chiu, E. J. C., et al. (2012). Facile backbone structure determination of human membrane proteins by NMR spectroscopy. *Nature Methods*, *9*(8), 834–839.

Koizumi, M., Hiratake, J., Nakatsu, T., Kato, H., & Oda, J. (1999). A potent transition-state analogue inhibitor of *Escherichia coli* asparagine synthetase A. *Journal of The American Chemical Society*, *121*(24), 5799–5800.

Koutsopoulos, S., Kaiser, L., Eriksson, H. M., & Zhang, S. (2012). Designer peptide surfactants stabilize diverse functional membrane proteins. *Chemical Society Reviews*, *41*(5), 1721–1728.

Kuznetsov, N. A., Faleev, N. G., Kuznetsova, A. A., Morozova, E. A., Revtovich, S. V., Anufrieva, N. V., et al. (2015). Pre-steady-state kinetic and structural analysis of interaction of methionine γ-lyase from Citrobacter freundii with inhibitors. *The Journal of Biological Chemistry*, *290*(1), 671–681.

Li, Q., Wong, Y. L., & Kang, C. (2014). Solution structure of the transmembrane domain of the insulin receptor in detergent micelles. *Biochimica et Biophysica Acta*, *1838*(5), 1313–1321.

Linser, R., Gelev, V., Hagn, F., Arthanari, H., Hyberts, S. G., & Wagner, G. (2014). Selective methyl labeling of eukaryotic membrane proteins using cell-free expression. *Journal of the American Chemical Society*, *136*(32), 11308–11310.

Logusch, E. W., Walker, D. M., McDonald, J. F., Franz, J. E., Villafranca, J. J., DiIanni, C. L., et al. (1990). Inhibition of *Escherichia coli* glutamine synthetase by alpha- and gamma-substituted phosphinothricins. *Biochemistry*, *29*(2), 366–372.

Löhr, F., Laguerre, A., Bock, C., Reckel, S., Connolly, P. J., Abdul-Manan, N., et al. (2014). Time-shared experiments for efficient assignment of triple-selectively labeled proteins. *Journal of Magnetic Resonance (San Diego, Calif.: 1997)*, *248*, 81–95.

Löhr, F., Reckel, S., Karbyshev, M., Connolly, P. J., Abdul-Manan, N., Bernhard, F., et al. (2012). Combinatorial triple-selective labeling as a tool to assist membrane protein backbone resonance assignment. *Journal of Biomolecular NMR*, *52*(3), 197–210.

Loscha, K. V., & Otting, G. (2013). Biosynthetically directed ^{2}H labelling for stereospecific resonance assignments of glycine methylene groups. *Journal of Biomolecular NMR*, *55*(1), 97–104.

Lyukmanova, E. N., Shenkarev, Z. O., Khabibullina, N. F., Kopeina, G. S., Shulepko, M. A., Paramonov, A. S., et al. (2012). Lipid-protein nanodiscs for cell-free production of integral membrane proteins in a soluble and folded state: Comparison with detergent micelles, bicelles and liposomes. *Biochimica et Biophysica Acta*, *1818*(3), 349–358.

Ma, Y., Münch, D., Schneider, T., Sahl, H.-G., Bouhss, A., Ghoshdastider, U., et al. (2011). Preparative scale cell-free production and quality optimization of MraY homologues in different expression modes. *The Journal of Biological Chemistry*, *286*(45), 38844–38853.

Mahon, M. M., Graber, R., Christen, P., & Malthouse, J. P. (1999). The aspartate aminotransferase-catalysed exchange of the alpha-protons of aspartate and glutamate: The effects of the R386A and R292V mutations on this exchange reaction. *Biochimica et Biophysica Acta*, *1434*(1), 191–201.

Manni, S., Mineev, K. S., Usmanova, D., Lyukmanova, E. N., Shulepko, M. A., Kirpichnikov, M. P., et al. (2014). Structural and functional characterization of alternative transmembrane domain conformations in VEGF receptor 2 activation. *Structure (London, England: 1993)*, *22*(8), 1077–1089.

Manning, J. M., Moore, S., Rowe, W. B., & Meister, A. (1969). Identification of L-methionine S-sulfoximine as the diastereoisomer of L-methionine SR-sulfoximine that inhibits glutamine synthetase. *Biochemistry*, *8*(6), 2681–2685.

Maslennikov, I., Klammt, C., Hwang, E., Kefala, G., Okamura, M., Esquivies, L., et al. (2010). Membrane domain structures of three classes of histidine kinase receptors by cell-free expression and rapid NMR analysis. *Proceedings of the National Academy of Sciences of the United States of America, 107*(24), 10902–10907.
Michel, E., & Wüthrich, K. (2012). Cell-free expression of disulfide-containing eukaryotic proteins for structural biology. *The FEBS Journal, 279*(17), 3176–3184.
Milne, J. J., & Malthouse, J. P. (1996). The effect of different amino acid side chains on the stereospecificity and catalytic efficiency of the tryptophan synthase-catalysed exchange of the alpha-protons of amino acids. *The Biochemical Journal, 314*(Pt. 3), 787–791.
Morita, E. H., Shimizu, M., Ogasawara, T., Endo, Y., Tanaka, R., & Kohno, T. (2004). A novel way of amino acid-specific assignment in (1)H-(15)N HSQC spectra with a wheat germ cell-free protein synthesis system. *Journal of Biomolecular NMR, 30*(1), 37–45.
Opella, S. J., & Marassi, F. M. (2004). Structure determination of membrane proteins by NMR spectroscopy. *Chemical Reviews, 104*(8), 3587–3606.
Ozawa, K., Headlam, M. J., Schaeffer, P. M., Henderson, B. R., Dixon, N. E., & Otting, G. (2004). Optimization of an *Escherichia coli* system for cell-free synthesis of selectively N-labelled proteins for rapid analysis by NMR spectroscopy. *European Journal of Biochemistry/FEBS, 271*(20), 4084–4093.
Park, K.-H., Berrier, C., Lebaupain, F., Pucci, B., Popot, J.-L., Ghazi, A., et al. (2007). Fluorinated and hemifluorinated surfactants as alternatives to detergents for membrane protein cell-free synthesis. *The Biochemical Journal, 403*(1), 183–187.
Parker, M. J., Aulton-Jones, M., Hounslow, A. M., & Craven, C. J. (2004). A combinatorial selective labeling method for the assignment of backbone amide NMR resonances. *Journal of the American Chemical Society, 126*(16), 5020–5021.
Pervushin, K., Riek, R., Wider, G., & Wüthrich, K. (1997). Attenuated T2 relaxation by mutual cancellation of dipole-dipole coupling and chemical shift anisotropy indicates an avenue to NMR structures of very large biological macromolecules in solution. *Proceedings of the National Academy of Sciences of the United States of America, 94*(23), 12366–12371.
Raschle, T., Hiller, S., Yu, T.-Y., Rice, A. J., Walz, T., & Wagner, G. (2009). Structural and functional characterization of the integral membrane protein VDAC-1 in lipid bilayer nanodiscs. *Journal of the American Chemical Society, 131*(49), 17777–17779.
Reckel, S., Gottstein, D., Stehle, J., Löhr, F., Verhoefen, M.-K., Takeda, M., et al. (2011). Solution NMR structure of proteorhodopsin. *Angewandte Chemie (International Ed. in English), 50*(50), 11942–11946.
Reckel, S., Sobhanifar, S., Durst, F., Löhr, F., Shirokov, V. A., Dötsch, V., et al. (2010). Strategies for the cell-free expression of membrane proteins. *Methods in Molecular Biology (Clifton, N.J.), 607*, 187–212.
Reckel, S., Sobhanifar, S., Schneider, B., Junge, F., Schwarz, D., Durst, F., et al. (2008). Transmembrane segment enhanced labeling as a tool for the backbone assignment of alpha-helical membrane proteins. *Proceedings of the National Academy of Sciences of the United States of America, 105*(24), 8262–8267.
Reese, M. L., & Dötsch, V. (2003). Fast mapping of protein-protein interfaces by NMR spectroscopy. *Journal of the American Chemical Society, 125*(47), 14250–14251.
Roos, C., Kai, L., Haberstock, S., Proverbio, D., Ghoshdastider, U., Ma, Y., et al. (2014). High-level cell-free production of membrane proteins with nanodiscs. In K. Alexandrov & W. A. Johnston (Eds.), *Methods in Molecular Biology: Vol. 1118* (pp. 109–130). Clifton, N.J.: Humana Press.
Roos, C., Zocher, M., Müller, D., Münch, D., Schneider, T., Sahl, H.-G., et al. (2012). Characterization of co-translationally formed nanodisc complexes with small multidrug transporters, proteorhodopsin and with the E. coli MraY translocase. *Biochimica et Biophysica Acta, 1818*(12), 3098–3106.

Rosen, M. K., Gardner, K. H., Willis, R. C., Parris, W. E., Pawson, T., & Kay, L. E. (1996). Selective methyl group protonation of perdeuterated proteins. *Journal of Molecular Biology*, *263*(5), 627–636.

Sachse, R., Dondapati, S. K., Fenz, S. F., Schmidt, T., & Kubick, S. (2014). Membrane protein synthesis in cell-free systems: From bio-mimetic systems to bio-membranes. *FEBS Letters*, *588*(17), 2774–2781.

Schwarz, D., Junge, F., Durst, F., Frölich, N., Schneider, B., Reckel, S., et al. (2007). Preparative scale expression of membrane proteins in *Escherichia coli*-based continuous exchange cell-free systems. *Nature Protocols*, *2*(11), 2945–2957.

Sobhanifar, S., Reckel, S., Junge, F., Schwarz, D., Kai, L., Karbyshev, M., et al. (2010). Cell-free expression and stable isotope labelling strategies for membrane proteins. *Journal of Biomolecular NMR*, *46*(1), 33–43.

Stover, P., & Schirch, V. (1991). 5-Formyltetrahydrofolate polyglutamates are slow tight binding inhibitors of serine hydroxymethyltransferase. *The Journal of Biological Chemistry*, *266*(3), 1543–1550.

Su, X.-C., Loh, C.-T., Qi, R., & Otting, G. (2011). Suppression of isotope scrambling in cell-free protein synthesis by broadband inhibition of PLP enzymes for selective 15N-labelling and production of perdeuterated proteins in H2O. *Journal of Biomolecular NMR*, *50*(1), 35–42.

Tamm, L. K., & Liang, B. (2006). NMR of membrane proteins in solution. *Progress in Nuclear Magnetic Resonance Spectroscopy*, *48*(4), 201–210.

Tonelli, M., Singarapu, K. K., Makino, S., Sahu, S. C., Matsubara, Y., Endo, Y., et al. (2011). Hydrogen exchange during cell-free incorporation of deuterated amino acids and an approach to its inhibition. *Journal of Biomolecular NMR*, *51*(4), 467–476.

Wada, T., Shimono, K., Kikukawa, T., Hato, M., Shinya, N., Kim, S. Y., et al. (2011). Crystal structure of the eukaryotic light-driven proton-pumping rhodopsin, Acetabularia rhodopsin II, from marine alga. *Journal of Molecular Biology*, *411*(5), 986–998.

Wang, X., Corin, K., Baaske, P., Wienken, C. J., Jerabek-Willemsen, M., Duhr, S., et al. (2011). Peptide surfactants for cell-free production of functional G protein-coupled receptors. *Proceedings of the National Academy of Sciences of the United States of America*, *108*(22), 9049–9054.

Wang, S., Munro, R. A., Shi, L., Kawamura, I., Okitsu, T., Wada, A., et al. (2013). Solid-state NMR spectroscopy structure determination of a lipid-embedded heptahelical membrane protein. *Nature Methods*, *10*(10), 1007–1012.

Warschawski, D. E., Arnold, A. A., Beaugrand, M., Gravel, A., Chartrand, É., & Marcotte, I. (2011). Choosing membrane mimetics for NMR structural studies of transmembrane proteins. *Biochimica et Biophysica Acta*, *1808*(8), 1957–1974.

Williams, J. S., & Rosevear, P. R. (1991). A novel alpha-proton exchange reaction catalyzed by *Escherichia coli* methionyl-tRNA synthetase. *Biochemistry*, *30*(26), 6412–6416.

Wu, P. S. C., Ozawa, K., Jergic, S., Su, X.-C., Dixon, N. E., & Otting, G. (2006). Amino-acid type identification in 15N-HSQC spectra by combinatorial selective 15N-labelling. *Journal of Biomolecular NMR*, *34*(1), 13–21.

Yabuki, T., Kigawa, T., Dohmae, N., Takio, K., Terada, T., Ito, Y., et al. (1998). Dual amino acid-selective and site-directed stable-isotope labeling of the human c-Ha-Ras protein by cell-free synthesis. *Journal of Biomolecular NMR*, *11*(3), 295–306.

Yokoyama, J., Matsuda, T., Koshiba, S., & Kigawa, T. (2010). An economical method for producing stable-isotope labeled proteins by the E. coli cell-free system. *Journal of Biomolecular NMR*, *48*(4), 193–201.

Yokoyama, J., Matsuda, T., Koshiba, S., Tochio, N., & Kigawa, T. (2011). A practical method for cell-free protein synthesis to avoid stable isotope scrambling and dilution. *Analytical Biochemistry*, *411*(2), 223–229.

Zheng, X., Dong, S., Zheng, J., Li, D., Li, F., & Luo, Z. (2014). Expression, stabilization and purification of membrane proteins via diverse protein synthesis systems and detergents involving cell-free associated with self-assembly peptide surfactants. *Biotechnology Advances, 32*(3), 564–574.

Zimmerman, E. S., Heibeck, T. H., Gill, A., Li, X., Murray, C. J., Madlansacay, M. R., et al. (2014). Production of site-specific antibody-drug conjugates using optimized non-natural amino acids in a cell-free expression system. *Bioconjugate Chemistry, 25*(2), 351–361.

Zoonens, M., Catoire, L. J., Giusti, F., & Popot, J.-L. (2005). NMR study of a membrane protein in detergent-free aqueous solution. *Proceedings of the National Academy of Sciences of the United States of America, 102*(25), 8893–8898.

CHAPTER SIXTEEN

Selective Amino Acid Segmental Labeling of Multi-Domain Proteins

Erich Michel[1,2], Frédéric H.-T. Allain[1]

Institute of Molecular Biology and Biophysics, ETH Zürich, Zürich, Switzerland
[1]Corresponding authors: e-mail address: erich.michel@chem.uzh.ch; allain@mol.biol.ethz.ch

Contents

1. Introduction — 390
    1.1 Overview — 390
    1.2 Segmental Isotope Labeling — 391
    1.3 Amino Acid-Specific Labeling — 395
    1.4 Segmental Amino Acid-Type Isotope Labeling — 396
2. Methods — 396
    2.1 Methodological Approach — 396
    2.2 Important Requirements for Expressed Protein Ligation — 397
    2.3 Choosing an Optimal Ligation Site — 398
    2.4 Production of Precursor Constructs — 400
    2.5 Ligation of Precursor Fragments — 409
    2.6 Practical Example: Amino Acid-Selective Segmental Labeling of a Multi-Domain RNA-Binding Protein — 415
3. Conclusion — 417
Acknowledgments — 419
References — 419

Abstract

The steady technical advances of nuclear magnetic resonance (NMR) over the past decades enabled a significant increase in the molecular size of protein particles that can be subjected to a structural and functional characterization in solution. The larger molecular weight of such proteins is accompanied with an increase in NMR signals that complicate spectral interpretation due to signal overlap. The application of segmental isotope labeling to selected domains in multi-domain proteins can significantly facilitate spectral interpretation by reducing the number of observable signals. However, severe signal overlap may persist within individual domains that show low signal dispersion. To further reduce the number of signals and spectral complexity in such systems, we developed a procedure for selective amino acid-type labeling in individual domains of multi-domain proteins. This strategy combines efficient amino acid-type labeling amenable

---

[2] Current address: Department of Chemistry, University of Zürich, Zürich, Switzerland.

by cell-free protein expression with near-seamless domain ligation achievable by expressed protein ligation. By application of simple dual labeling schemes, this approach further allows residue-specific isotope labeling to position NMR-observable probes at desired sites within segments of multi-domain proteins. This chapter describes a detailed protocol for selective amino acid-type segmental labeling of multi-domain proteins and illustrates its application to a multi-domain RNA-binding protein. The applied ligation approach is further suitable for efficient ligation of unlabeled and/or uniformly labeled domains produced solely by recombinant *in vivo* expression.

1. INTRODUCTION

1.1 Overview

Nuclear magnetic resonance (NMR) is an important and powerful tool to study structure, dynamics, and interactions of biomolecules in solution. It complements structural analysis by X-ray crystallography and allows investigation of protein dynamics, unstructured polypeptide stretches, and proteins that do not readily crystallize. However, NMR faces a molecular weight limitation that aggravates the study of biomolecules larger than ca. 25 kDa due to a slower overall molecular tumbling rate which increases transverse relaxation-induced signal decay (Wider & Wüthrich, 1999). Fortunately, the technical advancement of the spectrometer hardware increased available magnetic fields and provided the cryogenic probe technology which significantly increased the sensitivity of NMR. The introduction of the TROSY (transverse relaxation-optimized spectroscopy) (Pervushin, Riek, Wider, & Wüthrich, 1997, 1998; Tugarinov, Hwang, Ollerenshaw, & Kay, 2003) and CRINEPT (cross-correlated relaxation-enhanced polarization transfer) (Riek, Wider, Pervushin, & Wüthrich, 1999) NMR techniques together with preparation of partially or fully deuterated proteins further pushed the size boundaries of NMR several-fold (Kay & Gardner, 1997) to allow the investigation of proteins in particles with molecular sizes of several hundred kDa (Horst, Fenton, Englander, Wüthrich, & Horwich, 2007; Kay, 2011; Pervushin et al., 1997; Religa, Sprangers, & Kay, 2010; Riek et al., 1999). However, the rising molecular size of proteins that become amenable to investigation by NMR is accompanied by progressively larger numbers of resonances that result in higher spectral complexity, increased signal overlap, and the consequential emergence of new limitations.

1.2 Segmental Isotope Labeling

Spectral interpretation of challenging large systems can be significantly facilitated by application of advanced isotope-labeling methods. An elegant solution to resolve signal overlap in larger proteins was proposed by segmental isotope labeling of selected domains in multi-domain proteins (Xu, Ayers, Cowburn, & Muir, 1999; Yamazaki et al., 1998). This approach involves ligation of a labeled fragment harboring the desired NMR-observable domain to an unlabeled fragment resembling the NMR-invisible remainder of the target protein. The required fragments are produced as ligation-competent precursors by either recombinant protein expression (Muir, Sondhi, & Cole, 1998), solid-phase peptide synthesis (SPPS) (Dawson, Muir, Clark-Lewis, & Kent, 1994), or cell-free protein synthesis (Michel, Skrisovska, Wüthrich, & Allain, 2013) and are subsequently ligated together using either native chemical ligation (NCL) (Dawson et al., 1994), expressed protein ligation (EPL) (Muir et al., 1998), protein *trans*-splicing (PTS) (Wu, Hu, & Liu, 1998; Yamazaki et al., 1998), or sortase A-mediated domain ligation (Mao, Hart, Schink, & Pollok, 2004; Fig. 1).

1.2.1 Native Chemical Ligation

Protein ligation by NCL involves the chemo-selective reaction of an N-terminal cysteine (N-cysteine) located on an unprotected synthetic C-terminal polypeptide with a C-terminal α-thioester (α-thioester) of an unprotected synthetic N-terminal polypeptide in aqueous solution (Fig. 1A; Dawson & Kent, 2000). The ligation reaction proceeds via a reversible thiol/thioester exchange reaction of the α-thioester with the N-cysteine and the resulting peptide-thioester intermediate spontaneously undergoes a highly favored intramolecular S → N-acyl rearrangement which leads to the irreversible formation of a native peptide bond (Fig. 1A). The remarkable chemo-selectivity of NCL originates from the reversible nature of the transthioesterification reaction and the exclusive and irreversible formation of an amide bond with an N-cysteine. NCL follows bimolecular reaction kinetics and can provide near-quantitative product yields at high reactant concentrations (Dawson & Kent, 2000). The compulsory N-cysteine and α-thioester functionalities are introduced during synthesis of the polypeptides by SPPS, which provides additional flexibility for the incorporation of unnatural and/or modified amino acids (Kent, 2009). However, the synthetic production of milligram amounts of precursor becomes uneconomical for polypeptides larger than ca. 50 residues and

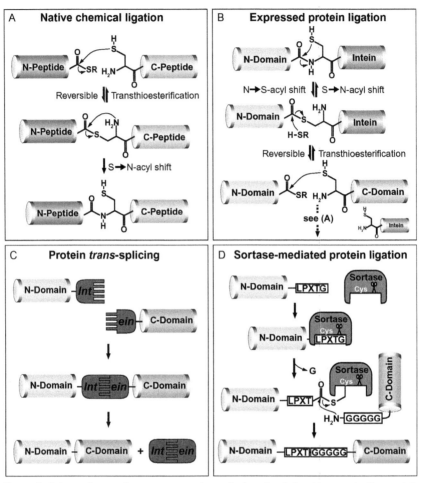

Figure 1 Schematic overview on commonly applied methods for chemo-selective protein ligation: (A) native chemical ligation (NCL), (B) expressed protein ligation (EPL), (C) protein *trans*-splicing (PTS), and (D) sortase-mediated protein ligation. (See the color plate.)

limits the application of NCL to smaller proteins of up to ca. 15 kDa (Dawson et al., 1994; Muir et al., 1998). Furthermore, the use of denatured synthetic precursors often necessitates native refolding of the ligated product and imposes a serious limitation for numerous proteins.

1.2.2 Expressed Protein Ligation
Ligation of proteins by EPL is mechanistically identical to NCL (Fig. 1A) and requires two reactive fragments containing an N-cysteine and an α-thioester, respectively. However, as opposed to NCL, EPL relies on

recombinant production of ligation-competent precursors that can be converted into reactive fragments containing the α-thioester and N-cysteine. The mechanistic basis that enables α-thioester formation is derived from a naturally occurring protein-splicing process that combines the auto-catalytic excision of an "intein" polypeptide from its multi-domain precursor with the simultaneous covalent joining of the flanking N- and C-terminal "extein" domains (Xu & Perler, 1996). These intein domains possess the remarkable ability to promote the otherwise unfavorable $N \rightarrow S$- (or $N \rightarrow O$-) acyl rearrangement at their N-terminal cysteine (or serine) residue (Evans, Benner, & Xu, 1999a; Xu & Perler, 1996), which leads to an α-thioester intermediate between the cysteine thiol and the C-terminal residue of the adjacent N-terminal extein fragment (Fig. 1B). This splice junction intermediate can be retained in intein mutants which are incapable to advance their self-excision beyond this step (Evans et al., 1999a; Xu & Perler, 1996). EPL uses recombinant production of fusions between the desired N-terminal fragment and a mutant intein (Fig. 1B) such as the *Mycobacterium xenopi* DNA gyrase subunit A (*Mxe* GyrA intein) (Evans et al., 1999a; Telenti et al., 1997) or the *Saccharomyces cerevisiae* V-type ATPase subunit A (*Sce* VMA intein) (Muir et al., 1998). Incubation of the resulting precursor molecules with an excess of thiol reagents such as 2-mercaptoethanesulfonate (MESNA) initiates a reversible thiol/thioester exchange reaction that converts the protein-thioester intermediate into the reactive N-terminal fragment containing the α-thioester (Fig. 1B; Evans et al., 1999a; Muir et al., 1998).

The reactive C-terminal fragments for EPL are prepared by recombinant expression of precursors from which the N-terminal cysteine is liberated by either self-excision of modified inteins (Evans et al., 1999a), methionine aminopeptidase-treatment (Iwai & Plückthun, 1999), or specific proteolytic cleavage (Tolbert & Wong, 2002; Xu et al., 1999; Fig. 2). Ligation of the reactive α-thioester and N-cysteine fragments by EPL follows the same bimolecular mechanism as NCL which favors mM reactant concentrations for high reaction yields. EPL is typically conducted under nondenaturing conditions which usually avoids the need for native refolding of the ligated product; however, it may also naturally prohibit the achievement of mM reactant concentrations for weakly soluble precursors and can thus result in reduced ligation yields. EPL further permits the sequential ligation of three or more fragments and thereby enables the selective isotope labeling of central domains in target proteins (Blaschke, Cotton, & Muir, 2000). A great advantage of EPL compared to NCL is the unrestricted molecular

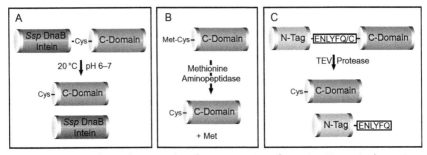

Figure 2 Commonly used approaches for preparation of reactive N-terminal cysteines. (A) N-terminal fusion of the desired C-terminal fragment to modified inteins allows triggered self-excision under certain conditions, i.e., changes in pH and temperature, which results in the liberation of an N-terminal cysteine. (B) The precursor fragment is produced starting with an N-terminal Met-Cys dipeptide, which can be processed by the endogenous methionine aminopeptidase to remove the N-terminal Met. (C) The C-terminal precursor construct is produced with an N-terminal fusion tag that contains a protease recognition site. Addition of the corresponding protease results in proteolytic liberation of the N-terminal cysteine.

size of reactive fragments that are amenable by recombinant expression of their precursors; on the other hand, SPPS of precursors for NCL offers an unrivaled flexibility for the incorporation of modified, labeled, or unnatural amino acids into the ligated protein.

1.2.3 Protein trans-Splicing

PTS is based on either naturally or artificially split inteins that are capable of self-assembly into an active form that catalyzes its self-excision and the concomitant ligation of its flanking segments (Fig. 1C; Ludwig et al., 2009; Wu et al., 1998). The individual N- and C-terminal target fragments are separately expressed as a fusion to the respective complementary part of a split intein. Mixing of the two precursor fusion constructs initiates self-assembly of the split intein followed by its self-excision and ligation of the flanking regions (Fig. 1C). The efficient self-complementation of split inteins supports good ligation yields even in reactions containing lower μM concentrations of precursor fragments (Volkmann & Iwai, 2010). In addition to the common *in vitro* ligation of fragments, PTS can also be conducted *in vivo* by shifted expression of the two precursor fragments in one cell culture grown sequentially on two different media (Züger & Iwai, 2005). The simultaneous use of two sets of split inteins with different complementarity further allows regio-selective ligation of three fragments and enables isotope labeling of central segments (Busche et al., 2009). However, PTS imposes extensive sequence constraints on the residues flanking the site of ligation and commonly introduces several non-native residues into the ligated target protein (Volkmann & Iwai, 2010).

1.2.4 Sortase-Mediated Protein Ligation

Sortase A-mediated ligation is an enzymatically catalyzed reaction and thus distinguishes itself from the other protein ligation methods. Sortases are bacterial transpeptidases that recognize the sequence motif LPXTG (where X is any amino acid) in C-terminal regions of bacterial cell wall proteins (Fig. 1D) to subsequently cleave it after the threonine. This results in a covalent intermediate between the threonine and the active site cysteine (Mao et al., 2004), which is resolved in a transpeptidation reaction involving the N-terminal amino group of pentaglycine motifs found in peptidoglycans on the bacterial cell wall (Fig. 1D). Sortases can be applied to catalyze the *in vitro* ligation of fragments containing a C-terminal LPETG motif with fragments containing several N-terminal glycines (Kobashigawa, Kumeta, Ogura, & Inagaki, 2009). However, the stringent requirement for the LPXTG motif and the N-terminal glycines leads to the introduction of several non-native residues which can significantly alter the properties of the resulting target protein.

1.3 Amino Acid-Specific Labeling

Segmental isotope labeling is a powerful approach to reduce the spectral complexity of larger proteins. However, the uniform labeling of selected domains can maintain the problem of signal overlap in domains with small signal dispersion (Dyson & Wright, 1998; Wüthrich, 1986) and in particularly large domains. An alternative approach that facilitates spectral interpretation is the selective labeling of full-length proteins with one or several amino acid types (Fiaux, Bertelsen, Horwich, & Wüthrich, 2004; Kainosho & Tsuji, 1982; Trbovic et al., 2005; Tugarinov, Kanelis, & Kay, 2006), which reduces the number of observed NMR signals and thereby decreases signal overlap. However, the efficient implementation of such labeling schemes by recombinant expression in *Escherichia coli* (*E. coli*) is often complicated by an inefficient incorporation and metabolic conversion of desired labeled amino acids. Cell-free expression is an excellent method to realize amino acid-type labeling schemes in proteins and provides significant advantages compared to *in vivo* protein expression (Kainosho et al., 2006; Kainosho & Tsuji, 1982; Trbovic et al.,

2005). It benefits from an economical incorporation of labeled amino acids into the target protein (Michel & Wüthrich, 2012b) and the low metabolic activity in cell extracts reduces isotope scrambling (Staunton, Schlinkert, Zanetti, Colebrook, & Campbell, 2006). The residual metabolic activity can be further minimized by supplementing the cell-free reaction mixture with inhibitors of these pathways (Su, Loh, Qi, & Otting, 2011).

1.4 Segmental Amino Acid-Type Isotope Labeling

However, limitations of amino acid-type labeling emerge with larger proteins as the peak positions of particular amino acid types tend to cluster within defined spectral regions (Wüthrich, 1986) which generates an increased likeliness of signal overlap. Segmental isotope labeling on the other hand enables uniform labeling of individual domains in larger proteins but significant signal overlap can prevail in troublesome domains with low signal dispersion. The combination of amino acid-type with segmental isotope labeling offers an advanced solution to minimize the signal overlap and spectral complexity of selected domains in challenging multi-domain proteins (Michel et al., 2013). This chapter describes a practical approach for amino acid-type labeling of selected domains in multi-domain proteins which is accomplished by amino acid-type labeling of the desired domain with cell-free expression and by EPL to ligate the unlabeled and labeled parts of the target protein. The presented protocol further enables ligation of unlabeled and/or uniformly labeled domains obtained from conventional recombinant expression in *E. coli* and/or cell-free expression and allows the application of dual $^{13}C/^{15}N$-combinatorial labeling schemes (Kainosho & Tsuji, 1982; Yabuki et al., 1998) to unambiguously assign desired reporter residues in selected domains. We illustrate the described protocol with applications to two homologous helix–turn–helix domains of the human glutamyl-prolyl-tRNA synthetase.

2. METHODS

2.1 Methodological Approach

Selective amino acid-type labeling of individual domain in multi-domain proteins requires efficient implementation of both the amino acid-type labeling of the precursor fragment and its chemo-selective ligation to the unlabeled fragment. We have chosen to employ an *E. coli*-based cell-free expression system for selective amino acid-type labeling of the precursors because it provides significantly reduced isotope scrambling and very economic amino acid incorporation yields compared to recombinant expression in *E. coli* (Michel & Wüthrich, 2012b; Staunton et al., 2006). The direct access to the cell-free reaction mixture further allows facile supplementation with inhibitors of transaminases and other metabolic enzymes to suppress the residual scrambling activity (Yokoyama, Matsuda, Koshiba, Tochio, & Kigawa, 2011) and enables the addition of factors that support the immediate stabilization and/or modification of the synthesized target protein (Kang, Kim, Kim, Jun, & Lee, 2005; Michel & Wüthrich, 2012a). However, due to economic reasons, we produce well-expressing unlabeled or uniformly labeled precursors by standard recombinant expression in *E. coli*. The subsequent chemo-selective ligation of the corresponding precursor fragments is achieved by EPL because its relaxed sequence requirements minimize the introduction of non-native residues into the final ligated protein (Muir et al., 1998). PTS and sortase-mediated protein ligation have more extensive sequence requirements and typically introduce several non-native residues to the ligation site (Volkmann & Iwai, 2010) while NCL is limited to ligation of rather short fragments which are amenable to preparation by SPPS (Muir et al., 1998). However, the size of the investigated protein system and its functional tolerance to non-native residues may justify the alternative use of PTS, sortase, or NCL over EPL. We would like to emphasize that the presented approach is equally suitable for ligation of fragments produced solely by recombinant *in vivo* expression and thus allows efficient preparation of multi-domain proteins containing one or multiple uniformly labeled domains.

2.2 Important Requirements for Expressed Protein Ligation

The successful implementation of the described ligation procedure imposes certain requirements:

(i) The precursor fragments should be available in at least low milligram amounts from the employed expression system. Starting with only limited amounts of precursors increases the likeliness of unsatisfactory low product yields.

(ii) The precursors should be readily convertible into the reactive species containing the α-thioester and N-cysteine, respectively. Production of the precursor fragments as an insoluble inclusion body or a soluble protein aggregate requires prior refolding into a conversion-competent state.

(iii) The required reactive functionalities of the precursor fragments should be well accessible. Soluble aggregation of the reactive precursor fragments commonly results in occlusion of the reactive ends and prevents ligation.
(iv) The ligation should proceed with highest efficiency to ensure the best possible product yield. Ligation yields depend on multiple factors including the amino acid type forming the C-terminal α-thioester and a sufficiently high concentration of the ligation-competent precursors in the ligation reaction.

The following sections in this chapter provide a practical guideline that should help to meet these requirements to successfully apply segmental isotope labeling to a target protein.

2.3 Choosing an Optimal Ligation Site
2.3.1 Importance of the Ligation Site
The first step in segmental isotope labeling of proteins involves the determination of the desired boundaries of the respective labeled and unlabeled segments of the target protein. This decision is very important and can have major consequences on subsequent steps and the overall success of the protocol due to the following reasons:

(i) The stability and solubility of the precursor can be significantly influenced by the chosen fragment boundaries. For example, choosing truncation sites within compactly folded domains increases the likeliness of precursor fragments to aggregate or precipitate. This frequently leads to expression into insoluble inclusion bodies in *E. coli* or precipitates by cell-free expression and requires subsequent refolding. However, a soluble and monomeric state of these refolded constructs can often only be maintained in presence of higher concentrations of chaotropic agents such as urea (Otomo, Teruya, Uegaki, Yamazaki, & Kyogoku, 1999), which in turn may adversely affect the intein-mediated α-thioester generation and the proteolytic liberation of the N-terminal cysteine.

(ii) The choice of fragment boundaries can alter the precursor susceptibility to proteolytic degradation due to variations of the construct stability and the exposure of protease recognition motifs.

(iii) The intein-mediated generation of the reactive α-thioester on the N-terminal fragment is largely influenced by the amino acid type at its C-terminus (Table 1). An unfavorable C-terminal residue at the

Table 1 Influence of the C-Terminal Amino Acid of the N-Terminal Fragment on the Intein-Mediated α-Thioester Generation and the Reactivity of the Resulting α-Thioester

System	Recommended Residues	Tolerated Residues	Unfavored Residues	References
Mxe GyrA (generation of C-terminal α-thioester)	Met, Tyr, Phe, Gly, Leu, His, Trp, Ala, Val, Asn, Gln, Cys	Thr, Ile (reduced thiol-induced intein cleavage) Arg, Lys (partial premature intein cleavage)	Pro, Glu, Ser (inefficient thiol-induced intein cleavage) Asp (premature intein cleavage)	Evans, Benner, and Xu (1998), Michel et al. (2013), Southworth, Amaya, Evans, Xu, and Perler (1999), Xu, Paulus, and Chong (2000)
Sce VMA (generation of C-terminal α-thioester)	Gly, Ala, Leu, Ser, Phe, Tyr, Trp, Met, Lys, Gln	Val, Ile (mediocre thiol-induced intein cleavage) Thr (increased premature intein cleavage)	Asp, His, Arg, Glu (premature intein cleavage) Pro, Asn, Cys (inefficient thiol-induced intein cleavage)	Chong, Williams, Wotkowicz, and Xu (1998), Muir et al. (1998), Xu et al. (2000)
NCL (ligation reactivity)	Gly, Cys, His, Phe, Met, Tyr, Ala, Trp	Asn, Ser, Asp, Gln, Glu, Lys, Arg	Pro, Ile, Val (inefficient ligation) Leu, Thr (slow ligation)	Hackeng, Griffin, and Dawson (1999)

ligation site can inhibit formation of the α-thioester or induce excessive premature intein cleavage during protein expression (Southworth et al., 1999). Amino acid types that promote the efficient intein-mediated conversion into the reactive precursor are indicated in Table 1.

(iv) The C-terminal residue of the N-terminal fragment affects the reactivity of the resulting α-thioester and thereby influences the achievable ligation yields. Amino acid types that form highly reactive α-thioesters in EPL correlate with those in NCL and are given in Table 1.

2.3.2 Guidelines to an Optimal Ligation Site

This leads to the following recommendations for choosing the site of ligation:

1. The ligation site is ideally contained within a flexible linker region connecting two domains of the target protein. The use of multiple sequence alignment and secondary structure prediction algorithms can help to identify flexible regions in uncharacterized proteins (Edgar & Batzoglou, 2006; Jones, 1999).
2. Protein ligation requires an N-terminal cysteine at the C-terminal fragment which is either naturally occurring in the target protein sequence or is introduced by mutagenesis. Typical residues for a mutation that minimally alters the functional properties of the target protein are endogenous serine or alanine residues that frequently occur throughout the sequence of most proteins. A multiple sequence alignment can reveal highly conserved residues with a potential functional significance which should not be mutated (Edgar & Batzoglou, 2006). Note: Desulfurization of cysteine to alanine in the final ligated protein enables a traceless ligation (Yan & Dawson, 2001); however, this reaction converts all cysteine residues into alanine and should ideally only be applied to proteins devoid of endogenous cysteine residues.
3. The C-terminal residue of the N-terminal fragment crucially influences both the intein-mediated generation of the α-thioester and its subsequent reactivity in the ligation reaction (Table 1) and ideally combines high suitability for both processes. Commonly used amino acid types that meet these criteria include Met, Tyr, Phe, and Gly for the *Mxe* GyrA intein-and the *Sce* VMA intein-based reactions (Table 1). Absence of recommended amino acid types at the desired ligation site may necessitate their introduction by mutagenesis; however, it is important to keep mutagenesis as conservative as possible to avoid functional alterations of the target protein.

2.4 Production of Precursor Constructs

2.4.1 Overview

Both uniform and selective amino acid-type segmental isotope labeling requires milligram amounts of ligation-competent precursor fragments containing the desired isotope-labeling pattern. Ligation-competence implies that the precursors are readily convertible into the corresponding reactive fragments containing the α-thioester and N-cysteine, respectively. The reactive N-terminal fragment can be obtained by its fusion to modified inteins which enable the facile intein-mediated generation of the required α-thioester functionality. Commonly used systems for that purpose include C-terminal fusions to the *Mxe* GyrA intein (Southworth et al., 1999), the *Sce* VMA intein (Muir et al., 1998), and the *Mth* RIR1 intein (Evans, Benner, & Xu, 1999b). The present protocol relies on preparation of α-thioesters from fusions to the *Mxe* GyrA mini-intein because it constitutes a soluble and stable intein which operates under a wide range of conditions, refolds efficiently and usually allows good expression of the desired fusion construct (Michel et al., 2013; Skrisovska & Allain, 2008; Vitali et al., 2006). However, additionally available intein systems can provide suitable alternatives for the generation of α-thioesters. The reactive N-cysteine at the C-terminal fragment can be obtained using either N-terminal fusions to modified inteins, such as the *Ssp* DnaB intein, that self-excise under certain conditions (Fig. 2A; Evans et al., 1999a), by methionine aminopeptidase-treatment of fragments starting with Met-Cys (Fig. 2B; Iwai & Plückthun, 1999) or by proteolytic cleavage of precursor fragments containing a protease recognition site with a cysteine in the P1′ position (Fig. 2C; Michel et al., 2013). The here-described approach uses Tobacco Etch Virus (TEV) protease for specific and efficient liberation of the N-cysteine from precursor fragments containing the recognition motif ENLYFQ/C. This avoids premature *in vivo* cleavage observed with intein-based approaches (Evans et al., 1999a) and allows the unrestricted use of N-terminal fusion tags to enhance translation and solubility of the C-terminal precursor fragment (Michel et al., 2013). Other proteases such as factor Xa provide alternatives to proteolytically liberate the N-cysteine (Blaschke et al., 2000); however, TEV protease is particularly attractive due to its high specificity and activity over a wide range of conditions, the traceless generation of N-cysteines and because milligram amounts of active protease are easily obtained from recombinant expression in *E. coli* (Tropea, Cherry, & Waugh, 2009).

Sufficient amounts of unlabeled and uniformly labeled ligation-competent precursors can usually be obtained from recombinant expression in *E. coli* growing in a complex medium or in a defined minimal medium (MM) with correspondingly labeled nutrients, respectively. However, certain instances may require cell-free expression to obtain sufficient amounts of unlabeled or uniformly labeled precursors: (i) Toxic precursors that inhibit bacterial growth may express better in a cell-free system (Spirin, 2004). (ii) Degradation-prone precursors may benefit from cell-free expression in presence of protease inhibitors. (iii) Precursors that tend to aggregate or precipitate and which cannot be efficiently refolded may profit from cell-free expression in presence of stabilizing agents (Kang et al., 2005). Selective amino acid-type labeling of the individual precursors on the other hand is best achieved by cell-free expression because of its favorable amino acid incorporation efficiency and the low metabolic activity in cell extracts which avoids excessive isotope scrambling.

2.4.2 Cloning of Target Protein Fragments

The next step after carefully deciding for an appropriate site of ligation (see Section 2.3) involves preparation of expression vectors that encode for the corresponding N- and C-terminal precursor fragments. The designated vectors pEM9B/pEM5B and pCFX7D/pCFX11B provide expression cassettes that are optimized for the recombinant *in vivo* (pEM) and cell-free expression (pCFX) of precursor molecules, respectively (Fig. 3; Michel et al., 2013). Expression of the desired N-terminal fragment in either pEM9B or pCFX7D results in N-terminal fusion to a (His)$_6$-tagged GB1 domain to enhance translational efficiency and construct solubility, and in C-terminal fusion to the *Mxe* GyrA (N198A) intein succeeded by a chitin-binding domain (CBD) (Fig. 3A). Cloning of the desired C-terminal fragment into pEM5B or pCFX11B results in N-terminal fusion to a (His)$_6$-tagged CBD (Fig. 3B). The TEV protease cleavage site ENLYFQ/C preceding the C-terminal fragment is introduced with

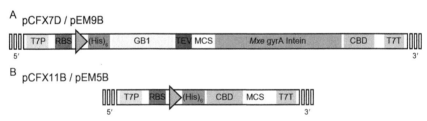

Figure 3 Schematic representation of plasmids for cell-free and recombinant *E. coli* expression of ligation-competent precursor fragments. (A) pEM9B and pCFX7D vectors for *in vivo* and cell-free expression of the N-terminal precursor, respectively. The resulting fusion construct contains a C-terminal *Mxe* GyrA intein for facile generation of the α-thioester and a CBD for simple purification on chitin beads. The N-terminal (His)$_6$-GB1 domain enhances precursor translation yields and is optionally removed by subsequent TEV protease treatment. (B) pEM5B and pCFX11B vectors for *in vivo* and cell-free expression of the C-terminal precursor, respectively. The required TEV-cleavage site at the N-terminus of the C-terminal fragment is traceless introduced via the PCR primers during cloning of the construct. T7P, T7 promoter; RBS, ribosome-binding site; TEV, TEV protease cleavage site; MCS, multiple cloning site; CBD, chitin-binding domain; T7T, T7 terminator.

oligonucleotide primers during PCR amplification of the target DNA. The use of CBD-containing precursor molecules permits their simple purification on chitin beads and allows the application of an on-column protein ligation protocol (Section 2.5.3). Alternatives to the described pEM-based vectors are offered by commercially available vector systems which provide suitable solutions for production of target protein fusions in *E. coli*; however, the efficient cell-free expression of the N-terminal precursor molecule relies on the use of the pCFX7D vector (Michel et al., 2013).

2.4.3 Precursor Production in Escherichia coli

Preparation of unlabeled and uniformly labeled precursors can be readily achieved by recombinant expression in *E. coli* which usually provides milligram amounts of protein in a cost-efficient manner. Unlabeled proteins are produced in *E. coli* growing on a rich and complex nutrient source such as LB medium while uniformly labeled proteins require growth on a defined MM supplied with ^{15}N-labeled and/or ^{13}C-labeled nutrients (Table 2).

Table 2 Minimal Medium for Recombinant Production of Uniformly Labeled Proteins in *E. coli*

Final Medium[a] (per L)		Solution Q (1000×)[b]		Vitamin Mixture (100×)[c]	
NaHPO$_4$·2H$_2$O	8.5 g	Fuming HCl	8 mL	Thiamine	500 mg
KH$_2$PO$_4$	3 g	FeCl$_2$·4H$_2$O	5 g	D-Biotin	100 mg
NaCl	0.5 g	CaCl$_2$·2H$_2$O	184 mg	Choline chloride	100 mg
MgSO$_4$ (1 M)	2 mL	H$_3$BO$_3$	64 mg	Folic acid	100 mg
Solution Q (1000×)	1 mL	CoCl$_2$·6H$_2$O	18 mg	Niacin amide	100 mg
Vitamin mixture (100×)	10 mL	CuCl$_2$·2H$_2$O	4 mg	D-Pantothenic acid	100 mg
NH$_4$Cl	1 g	ZnCl$_2$	340 mg	Pyridoxal	100 mg
D-Glucose	3 g	Na$_2$MoO$_4$·2H$_2$O	605 mg	Riboflavin	10 mg
H$_2$O or D$_2$O[d]	To 1 L	MnCl$_2$·4H$_2$O	40 mg	H$_2$O or D$_2$O[d]	To 1 L
Ampicillin[e]	100 mg	H$_2$O or D$_2$O[d]	To 1 L		

[a]Sterilization by filtration through a sterile membrane with 0.2-μm pore size.
[b]Stored at room temperature.
[c]Stored at −20 or −80 °C.
[d]Preparation of deuterated proteins requires growth in D$_2$O.
[e]The required antibiotic(s) depends on the employed strain and plasmid.

Deuterated precursor fragments can be obtained by growing the cell culture on a D_2O-containing MM while perdeuteration requires the additional use of a deuterated carbon source and necessitates preparation of D_2O-based stock solutions (Table 2).

The following protocol can be applied for recombinant precursor expression in *E. coli*:

(1) The vector encoding the desired precursor molecule is transformed into an appropriate *E. coli* host such as the commonly used BL21 (DE3) strain. More stringent control of protein expression is offered with the BL21 (DE3) pLysS or Tuner strains (Novagen) which can be beneficial for expression of precursor molecules that adversely affect bacterial cell growth due to cytotoxicity. After transformation, the cells are streaked out on a LB agar plate containing an antibiotic that selects for cells containing the vector.

(2) Single colonies from this LB agar plate are picked to inoculate a 10 mL LB medium pre-culture containing the appropriate antibiotic, which is subsequently grown with shaking at 37 °C to an OD_{600} of ca. 1–2.

(3) (A) Expression of unlabeled precursors: the pre-culture from step 2 is directly used as inoculum for the 1 L LB main culture containing the appropriate antibiotic. The main culture volume should be adjusted according to the expression levels and the required amounts of the desired precursor molecule.

(B) Expression of uniformly labeled precursors: the pre-culture from step 2 is briefly harvested by centrifugation (1 min at $4000 \times g$ and 37 °C) and the obtained cell pellet is resuspended in 25 mL of a MM that contains the desired labeled nutrients (i.e., ^{15}N-labeled NH_4Cl and ^{13}C-labeled D-glucose). This culture is grown with shaking at 37 °C to an OD_{600} of ca. 1–2 and is then used to inoculate 1 L MM main culture (the culture volume should be adjusted according to the expression levels and required amounts of precursor).

(4) The main cultures are grown with shaking at 37 °C to an OD_{600} of ca. 0.6–0.8. Addition of 0.5–1 mM IPTG induces expression of the precursor molecules and cell growth is continued for 4–6 h at 37 °C. Expression of temperature-sensitive precursors can alternatively be conducted for 8–12 h at 30 °C or for 16–24 h at 20 °C. However, certain amino acid types at the C-terminal α-thioester position of the N-terminal fragment lead to premature *in vivo* intein cleavage at lower temperatures and should be avoided for expression at reduced temperatures (Southworth et al., 1999).

(5) The cells are harvested by centrifugation (10 min at $5000 \times g$ and 4 °C) and are resuspended in lysis buffer (50 mM HEPES–NaOH at pH 8.0, 500 mM NaCl, 1 mM EDTA, and 0.5 mM TCEP) on ice. Lysis of the resuspended cells is performed at 4 °C or on ice using either cell cracking, a French Press or sonication. Commonly used thiol-based reducing agents such as 2-mercaptoethanol or DTT can cause significant premature intein cleavage and should be replaced in the lysis buffer with the phosphine-based reducing agent TCEP. Furthermore, EDTA should be omitted from the lysis buffer if the purification of the precursor by Ni-NTA is envisaged (Section 2.5.2). Supplementation of the lysis buffer with protease inhibitor cocktails can help to reduce proteolytic degradation of sensitive precursors.

The desired precursor molecules in the cell lysates are further processed as described in Sections 2.5.2 and 2.5.3.

2.4.4 Precursor Production with Cell-Free Expression

Cell-free expression benefits from efficient amino acid incorporation and low metabolic activity in cell extracts which greatly reduces isotope scrambling of labeled amino acids. In addition, the residual enzymatic activity that causes isotope scrambling can be further suppressed by supplementing the cell-free reaction with specific inhibitors (Table 3; Yokoyama et al., 2011). These features render cell-free expression highly suitable for selective amino acid-type labeling of precursor molecules. The cell extract provides the translation machinery and therefore strongly influences the performance of cell-free expression reactions. To ensure a high activity, we rely on the use of home-made *E. coli* cell extracts (Michel & Wüthrich, 2012b); however, commercially available kits provide a convenient but more expensive alternative for occasional users. The initiation of a cell-free expression reaction requires mixing of the cell extract with a template plasmid encoding the protein of interest, a DNA-dependent RNA-polymerase for transcription of the template mRNA, nucleotides for transcription and translation, amino acids, tRNA, an energy-regeneration system that ensures availability of the required nucleotides, and a buffer system that maintains constant conditions throughout the reaction (Table 4). This reaction mixture is then incubated at 30 °C either in batch mode (BM) or in continuous exchange cell-free (CECF) mode where the reaction mixture dialyzes against a feeding solution (Table 4) that prolongs the supply of required nucleotides and amino acids. CECF results in prolonged reaction times of up to 20 h and provides up to fivefold increased protein yields compared to BM; however,

Table 3 Amino Acid Isotope Scrambling in an *E. coli*-Based Cell-Free Expression System

Extend of Isotope Scrambling[a]	Amino Acid Type	Scrambling Inhibitors
<2%	Ala, Arg, His, Ile, Leu, Lys, Met, Val, Thr	Not required
2–10%	Cys, Ser, Gly	NaBH$_4$ (Ser) (Su et al., 2011) Amino-oxyacetate (John, Charteris, & Fowler, 1978)
10–20%	Phe, Tyr, Trp	Amino-oxyacetate (John et al., 1978) NaBH$_4$ (Su et al., 2011)
>20%	Asp, Asn, Glu, Gln	6-Diazo-5-oxo-L-norleucine (Gln → Glu) (Hartman, 1968) 5-Diazo-4-oxo-L-norvaline (Asn → Asp) (Handschumacher, Bates, Chang, Andrews, & Fischer, 1968) Methionine sulfoximine (Glu → Gln) (Manning, Moore, Rowe, & Meister, 1969) S-Methyl-L-cysteine sulfoximine (Asp → Asn) (Koizumi, Hiratake, Nakatsu, Kato, & Oda, 1999) D-Maleate (Asp) (Falzone, Karsten, Conley, & Viola, 1988) Amino-oxyacetate (John et al., 1978)

[a]The indicated values combine scrambling of only the ^{15}N-label of a particular amino acid and the complete metabolic conversion into other amino acid types. The values are derived by comparison of assigned [^{15}N,^{1}H]-HSQC cross-peak volumes corresponding to the selectively [u-^{15}N]-labeled amino acid type with cross-peak volumes originating from isotope scrambling.

the requirements of amino acids and NTPs in CECF are more than 10-fold higher than in BM and typically make CECF economically less favorable, especially for labeling with expensive amino acids (Michel & Wüthrich, 2012b). Typical yields range from 1 to 10 mg of purified protein from a 10 mL BM reaction mixture which requires a total of 60 mg amino acids. Selective amino acid-type labeling of target proteins requires manual preparation of amino acid mixtures that contain the desired labeled and unlabeled amino acids. The final concentration of each amino acid type in this mixture can be adjusted according to its occurrence in the target protein (Michel et al., 2013). A few amino acids such as Glu, Gln, Asn, and Asp can cause

Table 4 Composition of (A) Cell-Free Reaction Mixture, (B) Feeding Solution, and (C) Energy Regeneration-, Cofactor-, and Nucleotide-Solution (ERCN)

Component	Volume
(A) Cell-Free Reaction Mixture (10 mL)	
ERCN solution	3.73 mL
DEPC-treated water	1.41 mL
E. coli tRNA (17.5 mg mL^{-1})	100 µL
NaN$_3$ (0.77 M)	100 µL
Mg(OAc)$_2$ (1.6 M)	42 µL
Amino acid mix (20 mM)	750 µL
Creatine kinase (87 µM)	667 µL
T7 RNA-polymerase (65 µM)	100 µL
S30 extract	3 mL
Plasmid DNA (1 mg mL^{-1})	100 µL
(B) Feeding Solution (100 mL)	
ERCN solution	37.3 mL
DEPC-treated water	23.8 mL
NaN$_3$ (0.77 M)	1 mL
S30 buffer	30 mL
Amino acid mix (20 mM)	7.5 mL
Mg(OAc)$_2$ (1.6 M)	0.43 mL
(C) ERCN Solution (50 mL)	
HEPES–KOH, pH 7.5 (1 M)	1.55 mL
PEG-8000 (40% (w/v))	2.71 mL
KOAc (6 M)	890 µL
DTT (0.55 M)	87 µL
ATP, pH 7 (400 mM)	81 µL
GTP, pH 7 (400 mM)	58 µL
CTP, pH 7 (400 mM)	58 µL
UTP, pH 7 (400 mM)	58 µL
Folinic acid·Ca^{2+} (5.27 mM)	345 µL
cAMP·Na$^+$ (100 mM)	173 µL
Creatine phosphate (1 M)	2.16 mL
DEPC-treated water	1.83 mL

significant isotope scrambling during cell-free expression and require supplementation of the reaction mixture with corresponding inhibitors to minimize these effects (Table 3). However, the majority of amino acids shows little or no isotope scrambling during cell-free expression (Table 3).

The preparation of a cell-free expression reaction requires a range of stock solutions (Table 4) that need to be carefully prepared with DEPC-treated water to eliminate water-borne RNase activity. Ideal reaction settings that ensure highest attainable cell-free expression yields require the use of highly active cell extracts and the elimination of RNase contamination in the employed cell-free reaction components (Michel & Wüthrich, 2012b). However, limited protein yields due to inefficient cell-free translation of desired target genes cannot be excluded. The use of N-terminal fusion tags that enhance translational efficiency can significantly improve the yields of initially weakly expressing targets and provides the reasoning for the N-terminal (His)$_6$-tagged GB1 domain used in the pCFX7D vector (Fig. 3A; Michel et al., 2013).

The production of amino acid-type labeled precursor molecules by cell-free expression is achieved as follows:

(1) The required solutions for the cell-free reaction are individually prepared according to Table 4 and are stored at $-20\,°C$ until needed. S30 cell extract is prepared from *E. coli* BL21 (DE3) Star cells grown in phosphate-yeast extract-glucose (PYG) medium ($5.6\,g\,L^{-1}$ KH_2PO_4, $28.9\,g\,L^{-1}$ K_2HPO_4, $10\,g\,L^{-1}$ yeast extract, 1% (w/v) D-glucose) according to a previously described protocol (Michel & Wüthrich, 2012b) and is stored at $-80\,°C$ until needed. Extreme care should be taken to avoid RNase contamination during preparation of the stock solutions. The cell extract and the required components for the reaction mixture can alternatively be purchased in commercially available kits.

(2) The cell-free expression vector (pCFX7D or pCFX11B) encoding the desired precursor is obtained from a plasmid midi or maxi prep. The plasmid concentration of the stock solution should be ca. $1–2\,mg\,mL^{-1}$ and can be stored at $-20\,°C$ until needed.

(3) The stock mixture containing the desired labeled and unlabeled amino acid types is prepared by adding individual constituents to a final of $20\,mM$ for each amino acid type constituting up to 3% of the proteins' total amino acids, $40\,mM$ for those between 3% and 6% and $60\,mM$ for those above 6%.

(4) Small-scale cell-free expressions of the precursor fragments are conducted in 50 μL BM reactions which are prepared by mixing the required components according to Table 4A. The reaction mixtures are then incubated in 1.5 mL tubes with gentle agitation for 3 h at 30 °C and are followed by centrifugation (5 min at 12,000 × g). The yield and solubility of the desired target protein can be estimated by SDS–PAGE analysis of the supernatant and pellet fractions (Michel & Wüthrich, 2012b).

(5) Large-scale expression of the precursor fragments is conducted in either BM or CECF. A typical preparative cell-free expression contains 10 mL of reaction mixture and can be easily scaled up or down (Table 4A) according to the expression level of the individual target protein. In the CECF mode, the reaction mixture is transferred into 50 kDa molecular weight cut-off (MWCO) SpectraPor Float-A-Lyzer dialysis device (Spectrum Labs) which is completely immersed in a 10-fold excess of the feeding solution (Table 4B). Reactions in BM are conducted for 3 h and those in CECF for 20 h at 30 °C with gentle agitation.

(6) The reaction mixture is then centrifuged for 10 min at 4000 × g and the supernatant is immediately collected for subsequent purification as described in Sections 2.5.2 and 2.5.3.

Typical cell-free reaction temperatures range from 25 to 37 °C; however, the system usually provides best expression yields at ca. 30 °C and progressively decreases performance at lower temperatures (Shirokov, Kommer, Kolb, & Spirin, 2007). Optimization of reaction conditions to increase expression yields can include variations of the expression temperature (e.g., 25, 30, and 37 °C), the final magnesium concentration in the reaction mixture (e.g., 2 mM increments from 8 to 20 mM) and the template plasmid concentration (e.g., 5 μg increments from 5 to 40 μg DNA per mL of reaction mixture). We occasionally observe enhanced expression yields by optimization of the template plasmid concentration and rarely from variation of temperature and magnesium concentration.

2.5 Ligation of Precursor Fragments

2.5.1 Overview

Purification and ligation of the fragments should immediately be initiated once the required N- and C-terminal precursor molecules are available in either the crude cell lysate and/or the cell-free reaction supernatant.

Prolonged incubation times in the crude state can cause increased premature intein cleavage and/or precursor degradation and should be avoided. It is very important that both prepared precursor molecules are in a ligation-competent state that enables their conversion into the reactive N- and C-terminal fragments. Soluble aggregates of precursors are generally not considered to be ligation-competent and will need native refolding into the corresponding convertible state. Insoluble precursor molecules that are obtained as inclusion bodies or cell-free precipitates require native refolding as well. This is usually achieved by complete unfolding of the target protein in a buffer containing chaotropic agents such as urea or guanidine hydrochloride and is followed by slow dialysis or rapid dilution into a buffer that supports native folding into a ligation-competent state (Burgess, 2009; Middelberg, 2002; Skrisovska & Allain, 2008).

Ligation of the respective competent precursor molecules is initiated either after separate purification and conversion of each individual precursor into the reactive fragment or by simultaneous co-purification and conversion of both precursors on an affinity-column (on-column approach).

2.5.2 Ligation of Purified Precursors

The individual purification and conversion of each precursor molecule into the respective reactive fragment is more time-intensive but allows excellent control of the ligation reaction. This enables very accurate analysis of each individual step and is ideally suited to optimize the conditions for an increased ligation efficiency.

I. Purification of N-terminal precursors and conversion into the reactive α-thioester fragment
   (1) The N-terminal precursor in the cell lysate or the cell-free reaction supernatant is applied on a 5 mL HisPrep HP column (GE Healthcare) equilibrated with buffer A (50 mM Na$_2$HPO$_4$–HCl at pH 7.5, 30 mM imidazole, 500 mM NaCl, 0.5 mM TCEP, 10 μM NaN$_3$), is washed with 10 column volumes buffer A and is eluted in a linear gradient from 30 to 500 mM imidazole in buffer A.
   (2) Optional removal of the N-terminal (His)$_6$-GB1 domain: The fractions containing the desired precursor molecule are pooled and the protein concentration is determined based on the UV absorption at 280 nm. The protein solution is supplemented with 0.1 mg of TEV protease per mg of precursor and is dialyzed for 16 h in a 3.5 kDa MWCO dialysis membrane at 4 °C against

2 L of TEV-cleavage buffer (50 mM Na$_2$HPO$_4$–HCl at pH 7.5, 200 mM NaCl and 0.5 mM TCEP). Increased incubation temperatures (e.g., 16–37 °C), prolonged proteolysis and higher concentrations of the protease can enhance the reaction in cases where low cleavage rates are observed. The protein solution is then passed over a 5 mL HisPrep HP column equilibrated with buffer A and the flow-through containing the processed precursor devoid of the N-terminal (His)$_6$-GB1 domain is collected. Absence of cleaved target protein in the flow-through can indicate soluble aggregates of the precursor molecule or an inactive TEV protease batch.

(3) The collected fractions containing the desired precursor molecule are pooled and are dialyzed for 12 h in a 3.5 kDa MWCO dialysis membrane at 4 °C against 2 L of EPL buffer (50 mM HEPES–NaOH at pH 8.0, 500 mM NaCl, 1 mM EDTA, and 0.5 mM TCEP).

(4) The purified precursor protein is removed from the dialysis bag and is converted into the reactive α-thioester fragment by addition of 50–100 mM MESNA and incubation at either 4 or 20 °C for 12–16 h. The MESNA-induced generation of the α-thioester is usually most efficient at 20–25 °C and is largely influenced by the C-terminal residue of the N-terminal fragment (Table 1; Southworth et al., 1999); however, the conversion at 4 °C is a suitable alternative for sensitive proteins. SDS–PAGE analysis of the conversion reaction at various time points provides valuable information on the reaction efficiency and enables the identification of optimal conversion conditions. The loading buffer for SDS–PAGE should be devoid of reducing agents to prevent artificial intein cleavage.

II. Purification of C-terminal precursors and conversion into reactive N-cysteine fragments

(1) The C-terminal precursor in the cell lysate or the cell-free reaction supernatant is purified on a 5 mL HisPrep HP column as described in step 1 for purification of the N-terminal precursor.

(2) The fractions containing the desired C-terminal precursor molecule are pooled, the N-terminal (His)$_6$-CBD is removed by TEV protease cleavage and the reactive N-cysteine fragment is collected from the flow-through as described in step 2 for purification of the N-terminal precursor.

(3) The reactive N-cysteine fragment is dialyzed for 12 h in a 3.5 kDa MWCO dialysis membrane at 4 °C against 2 L of EPL buffer (50 mM HEPES–NaOH at pH 8.0, 500 mM NaCl, 1 mM EDTA, and 0.5 mM TCEP).

III. Ligation of purified reactive fragments

(1) Mixing of the two correspondent reactive fragments initiates their ligation. This reaction follows bimolecular kinetics and should be conducted with highest possible reactant concentrations that prevent their aggregation; recommended concentrations range from 0.1 to 2 mM for each fragment.

(2) Using an excess of one fragment will increase the consumption of the other fragment which provides an opportunity to economize the ligation yields of a valuable labeled fragment. This is typically achieved by using a 5–10-fold excess of an unlabeled fragment over the labeled fragment; however, ligations of two valuable labeled fragments are preferably conducted at stoichiometric ratios to ensure an equal consumption of both reactants.

(3) The ligation reaction is carried out for at least 16 h at 25 °C. Higher temperatures can increase ligation rates but may also destabilize the reactants and induce aggregation. We recommend to investigate the ligation efficiency at various temperatures (16–37 °C) and to follow the progress of ligation by SDS–PAGE analysis of samples taken at various reaction times.

(4) The use of chaotropic agents (e.g., 8 M urea or 6 M guanidine hydrochloride) in the reaction solution suspends aggregates, releases occluded reactive ends, allows higher reactant concentrations and can thereby significantly increase ligation yields. However, the use of such agents commonly requires subsequent native refolding of the ligated target protein.

(5) The ligated product is separated from the *Mxe* GyrA intein by a slow passage over chitin beads and is isolated from unreacted precursors by size-exclusion chromatography (e.g., on a Superdex 75 HiLoad 26/60 Prep Grade column).

The sequential ligation of three (or more) fragments is also feasible and enables isotopic labeling of central domains in target proteins (Blaschke et al., 2000). This approach starts with ligation of the C-terminal fragment to the central fragment containing a reactive α-thioester and an embedded unliberated cysteine. The ligated intermediate construct is then purified and the embedded cysteine in its central segment is converted into an N-cysteine by either proteolysis or intein self-excision (Fig. 2). Addition of the N-terminal fragment containing a reactive α-thioester initiates the second ligation reaction which results in formation of the final target protein.

2.5.3 On-Column Ligation

An on-column protocol that combines the simultaneous purification and direct conversion of both precursor molecules into reactive fragments (Fig. 4) provides an alternative to the individual preparation of each reactive fragment. This approach requires fusion of each precursor construct to a CBD, which is provided when using the pEM9B/pEM5B and pCFX7D/pCFX11B vectors described in Section 2.4.2, and enables facile purification of the precursor molecules and rapid initiation of the ligation reaction directly on the column.

Procedure for the on-column purification, activation, and ligation of fragments (Fig. 4)

(1) The cell lysates and/or cell-free reaction supernatants containing both required precursors (Fig. 4A) are pooled at an appropriate ratio and the resulting mixture is slowly passed over chitin beads equilibrated in EPL buffer (50 mM HEPES–NaOH at pH 8.0, 500 mM NaCl, 1 mM EDTA, and 0.5 mM TCEP) (Fig. 4B). Recommended ratios of the two precursors include a 5–10-fold excess of unlabeled fragments over labeled fragments and an equimolar ratio for ligation of two labeled domains (see Section 2.5.2.III.2). The quantities of the precursor molecules can be estimated by SDS–PAGE analysis of the cell lysate and/or the cell-free reaction supernatant. We recommend to provide 1 mL of a 50% (v/v) chitin bead suspension per mg of estimated precursor protein.

(2) The chitin beads are washed with five column volumes of EPL buffer and the intein-mediated α-thioester generation is initiated by addition of one column volume of reaction buffer (50 mM HEPES–NaOH at pH 8.0, 500 mM NaCl, 1 mM EDTA, 100 mM MESNA). The column is immediately emptied by gravity flow, closed and supplemented with 0.1 mg of TEV protease (2 mg mL^{-1} in reaction buffer) per mg of estimated precursor to initiate the generation of the N-cysteine (Fig. 4C).

(3) The column is properly sealed and is incubated for 16–24 h at 25 °C with gentle agitation. The intein-mediated α-thioester generation is usually optimal at 20–25 °C but the ligation reaction may proceed more efficiently at 37 °C; increasing the temperature from 25 to 37 °C after 8–16 h and continuing ligation for another 16 h may improve ligation yields.

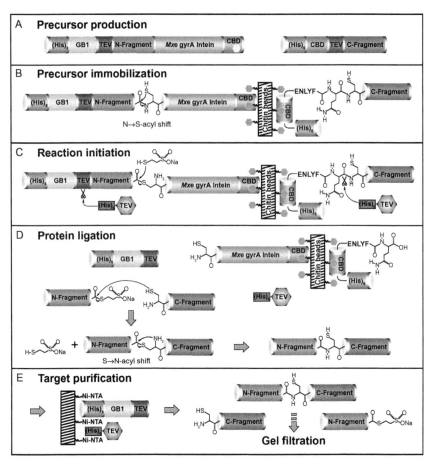

Figure 4 Schematic representation of the on-column protein ligation protocol. (A) Crude solutions containing ligation-competent N- and C-terminal precursor molecules prepared by cell-free and/or recombinant expression in *E. coli* are mixed and passed over chitin beads. (B) The encoded chitin-binding domain retains the precursor molecules on the column. (C) The ligation reaction is initiated by addition of MESNA and TEV protease, which induces intein-mediated α-thioester generation and liberation of the N-terminal cysteine, respectively. (D) After ligation for at least 16 h at 25 °C, the reaction mixture is eluted (E) and is passed over Ni-NTA beads to remove the (His)$_6$-tagged protein constructs. The ligated protein is finally separated from unreacted fragments by gel filtration. (See the color plate.)

(4) The protein mixture containing the ligated product, the unreacted fragments and the (His)$_6$-GB1 domain is eluted in two column volumes of Ni-NTA buffer (50 mM Na$_2$HPO$_4$–HCl at pH 7.5, 30 mM imidazole, 500 mM NaCl, 0.5 mM TCEP, 10 μM NaN$_3$) and is passed three times over a column packed with 10 mL of a 50% (v/v) Ni-NTA beads suspension in Ni–NTA buffer. Residual ligated product on the column is collected in two column volumes of Ni–NTA buffer.

(5) The combined flow-through of step 4 is subsequently concentrated in ultrafiltration devices to ca. 1–5 mg protein per mL and the ligated product is separated from unreacted precursors by size-exclusion chromatography (e.g., using a Superdex 75 HiLoad 26/60 Prep Grade column). A MWCO of the ultrafiltration device corresponding to ca. 65–80% of the size of the ligated product (e.g., 10 kDa MWCO for a 14 kDa protein) can be permissive for unreacted fragments and enriches the target protein prior to size-exclusion chromatography.

2.6 Practical Example: Amino Acid-Selective Segmental Labeling of a Multi-Domain RNA-Binding Protein

Human glutamyl-prolyl tRNA synthetase (EPRS) is an important subunit of the IFN-γ-activated inhibitor of translation complex (GAIT) that regulates translation of vascular endothelial growth factor-A (VEGFA) (Mukhopadhyay, Jia, Arif, Ray, & Fox, 2009; Ray et al., 2009). This regulation is mediated by the interaction of two helix–turn–helix-type domains R1 and R2 located in the linker region of EPRS with the 3′-untranslated region of the VEGFA mRNA, which induces and stabilizes a translation-silent mRNA conformation (Ray et al., 2009). To understand the structural details of this interaction, we started the resonance assignment of EPRS R1–R2 using conventional three-dimensional triple resonance NMR experiments. However, the high sequence identity between domains R1 and R2 (Fig. 5A) resulted in extensive signal overlap in the [^{15}N-^{1}H]-HSQC spectrum and complicated the unambiguous assignment of many lysine, alanine, and glutamate amide resonances. We therefore decided to apply selective amino acid-type labeling to the individual domains of EPRS R1–R2 to resolve these ambiguities.

The domain boundaries of EPRS R1–R2 were derived from available structural data (Jeong et al., 2000), a sequence alignment and secondary structure prediction (Jones, 1999) and indicated an unstructured linker region between domains R1 and R2 (Fig. 5A). The alanine of a Ser-Ala dipeptide in this region was mutated to cysteine as site of ligation, and vectors encoding the corresponding precursor constructs in pEM9B/pEM5B and pCFX7D/pCFX11B were prepared. However, the *Mxe* GyrA-mediated α-thioester formation reached only ca. 65% with the C-terminal Ser74, which is why we further analyzed the α-thioester generation with Met, Phe, Tyr, Lys, and Arg variants at this position.

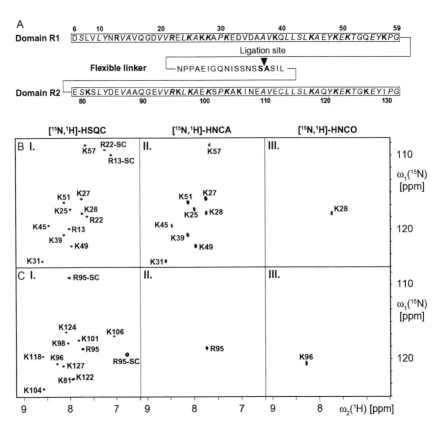

Figure 5 Amino acid-selective segmental isotope labeling of the helix–turn–helix domains R1 and R2 of human bifunctional aminoacyl-tRNA synthetase (EPRS). (A) Sequence alignment of the EPRS domains R1–R2. The chosen site of ligation (Ser-Ala) is contained in the flexible linker region between domains R1 and R2. Sequence identities between both domains are indicated by italic letters; bold letters correspond to labeled residues in the prepared NMR samples. (B, C) NMR analysis of the [^{15}N-Arg, ^{13}C,^{15}N-Lys]R1–R2 (B) and R1–[^{15}N-Lys, ^{13}C,^{15}N-Arg]R2 (C) samples containing selective amino acid-type isotope labeling in individual domains. Shown are 2D [^{15}N,^{1}H]-HSQC (I.), 2D [^{15}N,^{1}H]-HNCA (II.), and 2D [^{15}N,^{1}H]-HNCO (III.) spectra which reveal [^{15}N]- and [^{13}C,^{15}N]-labeled, [^{13}C,^{15}N]-labeled, and [^{15}N]/[^{13}C,^{15}N]-labeled residues with a [^{13}C,^{15}N]-labeled predecessor, respectively. SC indicates arginine guanidinium resonances.

Replacement with Lys and Arg resulted in ca. 50% premature intein cleavage during cell-free expression but the Met, Phe, and Tyr variants showed only ca. 10% premature cleavage and near-quantitative conversion into the reactive α-thioester upon addition of MESNA. We therefore chose to use the S74M variant for segmental isotope labeling.

Expression of (His)$_6$-GB1-R1-(S74M)-*Mxe* GyrA-CBD in 10 mL BM cell-free reactions with [^{15}N-Arg]- and [^{13}C,^{15}N-Lys]-labeling provided ca. 3.5 mg of [^{15}N-Arg, ^{13}C,^{15}N-Lys]-(His)$_6$-GB1-R1-(S74M)-*Mxe* GyrA-CBD, which was ligated with a fivefold excess of unlabeled (His)$_6$-CBD-(A75C)-R2 using the on-column approach (Section 2.5.3). Subsequent purification of the ligated product by Ni-NTA and SEC yielded 0.6 mg of ligated [^{15}N-Arg, ^{13}C,^{15}N-Lys]R1–R2 (sample A; Fig. 5B) which corresponds to ca. 55% recovery of the starting material of labeled R1. Analogous cell-free expression of (His)$_6$-CBD-(A75C)-R2 with [^{15}N-Lys]- and [^{13}C,^{15}N-Arg]-labeling provided 5 mg of ligation-competent [^{15}N-Lys, ^{13}C,^{15}N-Arg]-(His)$_6$-CBD-(A75C)-R2 from a 10 mL BM reaction. Ligation with a fivefold excess of unlabeled (His)$_6$-GB1-R1-(S74M)-*Mxe* GyrA-CBD resulted in 2.6 mg of purified R1–[^{15}N-Lys, ^{13}C,^{15}N-Arg]R2 (sample B; Fig. 5C) which also corresponds to ca. 55% recovery of the starting material of labeled R1.

Standard and selectively filtered 2D [^{15}N,^{1}H]-correlation spectra of the ligated samples showed well-resolved amide resonances that allowed the unambiguous and domain-specific identification of Arg and Lys resonances (Fig. 5B and C). NMR signals in the 2D [^{15}N,^{1}H]-HNCA spectrum originate exclusively from [^{13}C,^{15}N]-labeled residues which allows their differentiation from [^{15}N]-labeled residues of the same domain. This enabled the unambiguous discrimination of [^{13}C,^{15}N]Lys- from [^{15}N]Arg-residues within domain R1 of sample A (Fig. 5B, I. and II.) and of [^{13}C,^{15}N]Arg- from [^{15}N]Lys-residues within domain R2 of sample B (Fig. 5C, I. and II.). Furthermore, NMR signals in the 2D [^{15}N,^{1}H]-HNCO spectrum originate only from sequential pairs of [^{15}N]- or [^{13}C,^{15}N]-labeled residues preceded by either [^{13}C,^{15}N]- or [^{13}C]-labeled residues. This dual labeling enabled the unambiguous identification of Lys28 in domain R1 of sample A (Fig. 5B, III.) and of Lys96 in domain R2 of sample B (Fig. 5C, III.). The employed selective amino acid-type segmental isotope labeling thereby greatly assisted in the unambiguous assignment of all backbone amide resonances of R1–R2, especially in crowded spectral regions.

3. CONCLUSION

The steadily advancing progress in NMR enables the investigation of increasingly larger biomolecules which creates new limitations due to the larger numbers of resonances in these systems that consequently complicate spectral interpretation due to signal overlap. This development requires the allocation of advanced isotope-labeling techniques that facilitate the spectral interpretation of these increasingly complex systems. This chapter describes an approach that enables selective amino acid-type segmental isotope labeling of individual domains in multi-domain proteins which helps to reduce signal overlap in crowded spectral regions. It further allows application of domain-selective dual ^{13}C/^{15}N-labeling schemes (Yabuki et al., 1998) for the sequence-specific positioning of NMR-observable probes to monitor molecular interactions in solution without the requirement of complete NMR assignments. In the described protocol, amino acid-type labeling of ligation-competent precursor molecules is achieved in an efficient and economical manner by using cell-free expression while unlabeled and uniformly labeled precursors are economically produced by recombinant expression in *E. coli*. The resulting precursor molecules are then chemoselectively ligated using EPL to minimize the introduction of non-native residues at the site of ligation.

Selective amino acid-type labeling by *in vivo* expression in *E. coli* is feasible but typically requires auxotrophic strains and the supplementation of tens to hundreds of milligrams of the desired labeled amino acid per liter of cell culture medium (Tong, Yamamoto, & Tanaka, 2008). The simultaneous labeling of two or more amino acid types by recombinant expression is complicated by endogenous metabolic enzymes which usually cause significant isotope scrambling. Cell-free expression of the other hand benefits from low metabolic activity in the employed cell extracts which greatly reduces isotope scrambling; residual scrambling can be minimized by direct addition of inhibitors of the corresponding pathways into the cell-free reaction mixture (Yokoyama et al., 2011). Furthermore, the incorporation of amino acids into cell-free produced proteins is very efficient which significantly lowers the requirements of valuable amino acids and makes labeling very cost-efficient (Michel & Wüthrich, 2012b). The ligation of precursor fragments by EPL in turn provides a near-seamless method that minimizes introduction of non-native residues at the ligation site (Muir et al., 1998). Alternative methods such as PTS and sortase-mediated ligation have more extensive requirements at the site of ligation and result in the introduction of several non-native residues (Ludwig et al., 2009) which can alter the functional properties of the target protein. The use of truncated intein domains in PTS often decreases sample stability and increases aggregation and degradation propensities of the resulting fusion constructs (Zettler, Schutz, & Mootz, 2009). However, PTS provides very favorable ligation kinetics and typically results in higher yields of ligated product compared to EPL. The relative merits of each method should therefore be evaluated for different applications.

It is important to note that the presented approach is equally suitable for ligation of precursor fragments which are exclusively prepared by conventional recombinant expression in *E. coli* and that it further enables the selective labeling of cytotoxic or degradation-prone fragments which cannot be efficiently prepared *in vivo*.

ACKNOWLEDGMENTS

We thank the ETH Zürich and the Swiss National Science Foundation for financial support through the National Center of Competence in Research (NCCR) Structural Biology.

REFERENCES

Blaschke, U. K., Cotton, G. J., & Muir, T. W. (2000). Synthesis of multi-domain proteins using expressed protein ligation: Strategies for segmental isotopic labeling of internal regions. *Tetrahedron, 56*, 9461–9470.

Burgess, R. R. (2009). Refolding solubilized inclusion body proteins. *Methods in Enzymology, 463*, 259–282.

Busche, A. E., Aranko, A. S., Talebzadeh-Farooji, M., Bernhard, F., Dötsch, V., & Iwai, H. (2009). Segmental isotopic labeling of a central domain in a multidomain protein by protein *trans*-splicing using only one robust DnaE intein. *Angewandte Chemie International Edition, 48*(33), 6128–6131.

Chong, S., Williams, K. S., Wotkowicz, C., & Xu, M. Q. (1998). Modulation of protein splicing of the *Saccharomyces cerevisiae* vacuolar membrane ATPase intein. *Journal of Biological Chemistry, 273*(17), 10567–10577.

Dawson, P. E., & Kent, S. B. (2000). Synthesis of native proteins by chemical ligation. *Annual Review of Biochemistry, 69*, 923–960.

Dawson, P. E., Muir, T. W., Clark-Lewis, I., & Kent, S. B. (1994). Synthesis of proteins by native chemical ligation. *Science, 266*(5186), 776–779.

Dyson, H. J., & Wright, P. E. (1998). Equilibrium NMR studies of unfolded and partially folded proteins. *Nature Structural Biology, 5*, 499–503.

Edgar, R. C., & Batzoglou, S. (2006). Multiple sequence alignment. *Current Opinion in Structural Biology, 16*, 368–373.

Evans, T. C., Benner, J., & Xu, M. Q. (1998). Semisynthesis of cytotoxic proteins using a modified protein splicing element. *Protein Science, 7*, 2256–2264.

Evans, T. C., Benner, J., & Xu, M. Q. (1999a). The cyclization and polymerization of bacterially expressed proteins using modified self-splicing inteins. *Journal of Biological Chemistry, 274*(26), 18359–18363.

Evans, T. C., Jr., Benner, J., & Xu, M. Q. (1999b). The *in vitro* ligation of bacterially expressed proteins using an intein from *Methanobacterium thermoautotrophicum*. *Journal of Biological Chemistry, 274*(7), 3923–3926.

Falzone, C. J., Karsten, W. E., Conley, J. D., & Viola, R. E. (1988). L-Aspartase from *Escherichia coli*: Substrate specificity and role of divalent metal ions. *Biochemistry, 27*, 9089–9093.

Fiaux, J., Bertelsen, E. B., Horwich, A. L., & Wüthrich, K. (2004). Uniform and residue-specific ^{15}N-labeling of proteins on a highly deuterated background. *Journal of Biomolecular NMR, 29*(3), 289–297.

Hackeng, T. M., Griffin, J. H., & Dawson, P. E. (1999). Protein synthesis by native chemical ligation: Expanded scope by using straightforward methodology. *Proceedings of the National Academy of Sciences of the United States of America, 96*, 10068–10073.

Handschumacher, R. E., Bates, C. J., Chang, P. K., Andrews, A. T., & Fischer, G. A. (1968). 5-Diazo-4-oxo-L-norvaline: Reactive asparagine analog with biological specificity. *Science, 161*, 62–63.

Hartman, S. C. (1968). Glutaminase of *Escherichia coli*: I. Purification and general catalytic properties. *Journal of Biological Chemistry, 243*(5), 853–863.

Horst, R., Fenton, W. A., Englander, S. W., Wüthrich, K., & Horwich, A. L. (2007). Folding trajectories of human dihydrofolate reductase inside the GroEL–GroES chaperonin cavity and free in solution. *Proceedings of the National Academy of Sciences of the United States of America, 104*(52), 20788–20792.

Iwai, H., & Plückthun, A. (1999). Circular beta-lactamase: Stability enhancement by cyclizing the backbone. *FEBS Letters, 459*, 166–172.

Jeong, E. J., Hwang, G. S., Kim, K. H., Kim, M. J., Kim, S., & Kim, K. S. (2000). Structural analysis of multifunctional peptide motifs in human bifunctional tRNA synthetase: Identification of RNA-binding residues and functional implications for tandem repeats. *Biochemistry, 39*(51), 15775–15782.

John, R. A., Charteris, A., & Fowler, L. J. (1978). The reaction of amino-oxyacetate with pyridoxal phosphate-dependent enzymes. *Biochemical Journal, 171*, 771–779.

Jones, D. T. (1999). Protein secondary structure prediction based on position-specific scoring matrices. *Journal of Molecular Biology, 292*, 195–202.

Kainosho, M., Torizawa, T., Iwashita, Y., Terauchi, T., Mei Ono, A., & Güntert, P. (2006). Optimal isotope labelling for NMR protein structure determinations. *Nature, 440*(7080), 52–57.

Kainosho, M., & Tsuji, T. (1982). Assignment of the three methionyl carbonyl carbon resonances in *Streptomyces* subtilisin inhibitor by a carbon-13 and nitrogen-15 double-labeling technique. A new strategy for structural studies of proteins in solution. *Biochemistry, 21*(24), 6273–6279.

Kang, S. H., Kim, D. M., Kim, H. J., Jun, S. Y., & Lee, K. Y. (2005). Cell-free production of aggregation-prone proteins in soluble and active forms. *Biotechnology Progress, 21*(5), 1412–1419.

Kay, L. E. (2011). Solution NMR spectroscopy of supra-molecular systems, why bother? A methyl-TROSY view. *Journal of Magnetic Resonance, 210*(2), 159–170.

Kay, L. E., & Gardner, K. H. (1997). Solution NMR spectroscopy beyond 25 kDa. *Current Opinion in Structural Biology, 7*(5), 722–731.

Kent, S. B. (2009). Total chemical synthesis of proteins. *Chemical Society Reviews, 38*(2), 338–351.

Kobashigawa, Y., Kumeta, H., Ogura, K., & Inagaki, F. (2009). Attachment of an NMR-invisible solubility enhancement tag using a sortase-mediated protein ligation method. *Journal of Biomolecular NMR, 43*, 145–150.

Koizumi, M., Hiratake, J., Nakatsu, T., Kato, H., & Oda, J. I. (1999). A potent transition-state analogue inhibitor of *Escherichia coli* asparaginase synthetase A. *Journal of the American Chemical Society, 121*, 5799–5800.

Ludwig, C., Schwarzer, D., Zettler, J., Garbe, D., Janning, P., Czeslik, C., et al. (2009). Semisynthesis of proteins using split inteins. *Methods in Enzymology, 462*, 77–96.

Manning, J. M., Moore, S., Rowe, W. B., & Meister, A. (1969). Identification of L-methionine S-sulfoximine as the diastereoisomer of L-methionine S, R-sulfoximine that inhibits glutamine synthetase. *Biochemistry, 8*, 2681–2685.

Mao, H., Hart, S. A., Schink, A., & Pollok, B. A. (2004). Sortase-mediated protein ligation: A new method for protein engineering. *Journal of the American Chemical Society, 126*(9), 2670–2671.

Michel, E., Skrisovska, L., Wüthrich, K., & Allain, F. H. (2013). Amino acid-selective segmental isotope labeling of multidomain proteins for structural biology. *ChemBioChem, 14*(4), 457–466.

Michel, E., & Wüthrich, K. (2012a). Cell-free expression of disulfide-containing eukaryotic proteins for structural biology. *FEBS Journal, 279*(17), 3176–3184.

Michel, E., & Wüthrich, K. (2012b). High-yield *Escherichia coli*-based cell-free expression of human proteins. *Journal of Biomolecular NMR, 53*(1), 43–51.

Middelberg, A. P. (2002). Preparative protein refolding. *Trends in Biotechnology*, 20(10), 437–443.
Muir, T. W., Sondhi, D., & Cole, P. A. (1998). Expressed protein ligation: A general method for protein engineering. *Proceedings of the National Academy of Sciences of the United States of America*, 95(12), 6705–6710.
Mukhopadhyay, R., Jia, J., Arif, A., Ray, P. S., & Fox, P. L. (2009). The GAIT system: A gatekeeper of inflammatory gene expression. *Trends in Biochemical Sciences*, 34(7), 324–331.
Otomo, T., Teruya, K., Uegaki, K., Yamazaki, T., & Kyogoku, Y. (1999). Improved segmental isotope labeling of proteins and application to a larger protein. *Journal of Biomolecular NMR*, 14(2), 105–114.
Pervushin, K., Riek, R., Wider, G., & Wüthrich, K. (1997). Attenuated T2 relaxation by mutual cancellation of dipole-dipole coupling and chemical shift anisotropy indicates an avenue to NMR structures of very large biological macromolecules in solution. *Proceedings of the National Academy of Sciences of the United States of America*, 94(23), 12366–12371.
Pervushin, K., Riek, R., Wider, G., & Wüthrich, K. (1998). Transverse relaxation optimized spectroscopy (TROSY) for NMR studies of aromatic spin systems in ^{13}C-labeled proteins. *Journal of the American Chemical Society*, 120, 6394–6400.
Ray, P. S., Jia, J., Yao, P., Majumder, M., Hatzoglou, M., & Fox, P. L. (2009). A stress-responsive RNA switch regulates VEGFA expression. *Nature*, 457(7231), 915–919.
Religa, T. L., Sprangers, R., & Kay, L. E. (2010). Dynamic regulation of archaeal proteasome gate opening as studied by TROSY NMR. *Science*, 328(5974), 98–102.
Riek, R., Wider, G., Pervushin, K., & Wüthrich, K. (1999). Polarization transfer by cross-correlated relaxation in solution NMR with very large molecules. *Proceedings of the National Academy of Sciences of the United States of America*, 96(9), 4918–4923.
Shirokov, V., Kommer, A., Kolb, V., & Spirin, A. (2007). Continuous-exchange protein-synthesizing systems. In G. Grandi (Ed.), *In vitro transcription and translation protocols: Vol. 375* (pp. 19–55). Totowa, NJ: Humana Press.
Skrisovska, L., & Allain, F. H. (2008). Improved segmental isotope labeling methods for the NMR study of multidomain or large proteins: Application to the RRMs of Npl3p and hnRNP L. *Journal of Molecular Biology*, 375(1), 151–164.
Southworth, M. W., Amaya, K., Evans, T. C., Xu, M. Q., & Perler, F. B. (1999). Purification of proteins fused to either the amino or carboxy terminus of the *Mycobacterium xenopi* gyrase A intein. *BioTechniques*, 27(1), 110–120.
Spirin, A. S. (2004). High-throughput cell-free systems for synthesis of functionally active proteins. *Trends in Biotechnology*, 22(10), 538–545.
Staunton, D., Schlinkert, R., Zanetti, G., Colebrook, S. A., & Campbell, I. D. (2006). Cell-free expression and selective isotope labelling in protein NMR. *Magnetic Resonance in Chemistry*, 44(S1), S2–S9.
Su, X. C., Loh, C. T., Qi, R., & Otting, G. (2011). Suppression of isotope scrambling in cell-free protein synthesis by broadband inhibition of PLP enzymes for selective ^{15}N-labelling and production of perdeuterated proteins in H$_2$O. *Journal of Biomolecular NMR*, 50(1), 35–42.
Telenti, A., Southworth, M., Alcaide, F., Daugelat, S., Jacobs, W. R., Jr., & Perler, F. B. (1997). The *Mycobacterium xenopi* GyrA protein splicing element: Characterization of a minimal intein. *Journal of Bacteriology*, 179(20), 6378–6382.
Tolbert, T. J., & Wong, C.-H. (2002). New methods for proteomic research: Preparation of proteins with N-terminal cysteines for labeling and conjugation. *Angewandte Chemie International Edition*, 41, 2171–2174.
Tong, K. I., Yamamoto, M., & Tanaka, T. (2008). A simple method for amino acid selective isotope labeling of recombinant proteins in *E. coli*. *Journal of Biomolecular NMR*, 42(1), 59–67.

Trbovic, N., Klammt, C., Koglin, A., Löhr, F., Bernhard, F., & Dötsch, V. (2005). Efficient strategy for the rapid backbone assignment of membrane proteins. *Journal of the American Chemical Society, 127*(39), 13504–13505.

Tropea, J., Cherry, S., & Waugh, D. (2009). Expression and purification of soluble His$_6$-tagged TEV protease. In S. Doyle (Ed.), *High throughput protein expression and purification: Vol. 498* (pp. 297–307). Totowa, NJ: Humana Press.

Tugarinov, V., Hwang, P. M., Ollerenshaw, J. E., & Kay, L. E. (2003). Cross-correlated relaxation enhanced ^{1}H–^{13}C NMR spectroscopy of methyl groups in very high molecular weight proteins and protein complexes. *Journal of the American Chemical Society, 125*(34), 10420–10428.

Tugarinov, V., Kanelis, V., & Kay, L. E. (2006). Isotope labeling strategies for the study of high-molecular-weight proteins by solution NMR spectroscopy. *Nature Protocols, 1*(2), 749–754.

Vitali, F., Henning, A., Oberstrass, F. C., Hargous, Y., Auweter, S. D., Erat, M., et al. (2006). Structure of the two most C-terminal RNA recognition motifs of PTB using segmental isotope labeling. *EMBO Journal, 25*(1), 150–162.

Volkmann, G., & Iwai, H. (2010). Protein *trans*-splicing and its use in structural biology: Opportunities and limitations. *Molecular BioSystems, 6*(11), 2110–2121.

Wider, G., & Wüthrich, K. (1999). NMR spectroscopy of large molecules and multimolecular assemblies in solution. *Current Opinion in Structural Biology, 9*(5), 594–601.

Wu, H., Hu, Z., & Liu, X. Q. (1998). Protein *trans*-splicing by a split intein encoded in a split DnaE gene of *Synechocystis* sp. PCC6803. *Proceedings of the National Academy of Sciences of the United States of America, 95*(16), 9226–9231.

Wüthrich, K. (1986). *NMR of proteins and nucleic acids*. New York: Wiley.

Xu, R., Ayers, B., Cowburn, D., & Muir, T. W. (1999). Chemical ligation of folded recombinant proteins: Segmental isotopic labeling of domains for NMR studies. *Proceedings of the National Academy of Sciences of the United States of America, 96*(2), 388–393.

Xu, M. Q., Paulus, H., & Chong, S. (2000). Fusions to self-splicing inteins for protein purification. *Methods in Enzymology, 326*, 376–418.

Xu, M. Q., & Perler, F. B. (1996). The mechanism of protein splicing and its modulation by mutation. *EMBO Journal, 15*(19), 5146–5153.

Yabuki, T., Kigawa, T., Dohmae, N., Takio, K., Terada, T., Ito, Y., et al. (1998). Dual amino acid-selective and site-directed stable-isotope labeling of the human c-Ha-Ras protein by cell-free synthesis. *Journal of Biomolecular NMR, 11*(3), 295–306.

Yamazaki, T., Otomo, T., Oda, N., Kyogoku, Y., Uegaki, K., Nobutoshi, I., et al. (1998). Segmental isotope labeling for protein NMR using peptide splicing. *Journal of the American Chemical Society, 120*, 5591–5592.

Yan, L. Z., & Dawson, P. E. (2001). Synthesis of peptides and proteins without cysteine residues by native chemical ligation combined with desulfurization. *Journal of the American Chemical Society, 123*(4), 526–533.

Yokoyama, J., Matsuda, T., Koshiba, S., Tochio, N., & Kigawa, T. (2011). A practical method for cell-free protein synthesis to avoid stable isotope scrambling and dilution. *Analytical Biochemistry, 411*(2), 223–229.

Zettler, J., Schutz, V., & Mootz, H. D. (2009). The naturally split *Npu* DnaE intein exhibits an extraordinarily high rate in the protein *trans*-splicing reaction. *FEBS Letters, 583*(5), 909–914.

Züger, S., & Iwai, H. (2005). Intein-based biosynthetic incorporation of unlabeled protein tags into isotopically labeled proteins for NMR studies. *Nature Biotechnology, 23*(6), 736–740.

CHAPTER SEVENTEEN

Labeling Monosaccharides With Stable Isotopes

Wenhui Zhang*, Shikai Zhao[†], Anthony S. Serianni*,[1]

*Department of Chemistry and Biochemistry, University of Notre Dame, Notre Dame, Indiana, USA
[†]Omicron Biochemicals, Inc., South Bend, Indiana, USA
[1]Corresponding author: e-mail address: aserianni@nd.edu

Contents

1. Introduction	424
2. Terminology to Describe Different Monosaccharide Isotopomers	425
3. Introducing ^{13}C into Monosaccharides	427
3.1 Cyanohydrin Reduction	428
3.2 Permutations of Cyanohydrin Reduction	433
3.3 Molybdate-Catalyzed Epimerization of Aldoses Accompanied by C1–C2 Transposition	435
4. Multiple Labeling of Aldoses Via Chain Inversion	438
5. Labeling at the Internal Carbons of Aldoses	439
6. Extension to Biologically Important Aldoses	442
7. Relative Carbonyl Reactivities in Osones—Synthesis of Labeled 2-Ketoses	444
8. Manipulation of Three-Carbon Building Blocks in Enzyme-Mediated Aldol Condensation	445
9. Manipulation of Isotopically Labeled D-Fructose 17 and L-Sorbose 25	452
10. Concluding Remarks	452
References	455

Abstract

Chemical and chemi-enzymic methods are discussed for the preparation of monosaccharide isotopomers that are singly and multiply labeled with ^{13}C, ^{2}H, $^{17/18}$O, and ^{15}N isotopes. The discussion focuses primarily on chemical methods to incorporate stable isotopes into monosaccharides and not on methods to assemble labeled monosaccharides into more complex biomolecules such as oligosaccharides or oligonucleotides. Two primary isotope insertion reactions are considered: cyanohydrin reduction (CR) and molybdate-catalyzed epimerization (MCE). Both methods are described in detail, including discussions of their mechanistic features, and their advantages and limitations. The integration of CR, MCE, and other chemical synthetic processes with enzyme-mediated synthesis is also discussed to illustrate how a wide range of singly and multiply labeled monosaccharides can be prepared for subsequent use in the assembly of more complex isotopically labeled biomolecules.

1. INTRODUCTION

Carbohydrates are found throughout nature, serving as major energy sources in biological catabolism; molecular scaffolds for cell rigidity; barriers against bacterial and viral invasion in respiratory, intestinal and urinary tracts of vertebrates; defense molecules against tissue damage from freezing; mediators of cell–cell recognition during cell development; and modulators of protein folding, stability, and turnover in eukaryotic cells to name only a few of their biological functions (Taylor & Drickamer, 2011). These functions are made possible by their diverse structures and structural properties. The latter diversity poses significant challenges for the labeling of saccharides with isotopes; unlike amino acids and nucleotides, where a relatively small number of building block molecules are involved (20 and 8, respectively, to assemble native polynucleotides and oligonucleotides), many different types of monosaccharides are observed in nature, and they are assembled in multiple ways to give an incredibly large number of different types of oligo- and polysaccharides. Not only do labeling methods for saccharides need to have maximal flexibility to accommodate a large chemical space, but decisions on where to label monosaccharide building blocks and with which isotopes are further complicated by the type of scientific question being posed. There are no magic formulas to achieve labeling simplicity, except possibly to label all sites (e.g., uniformly label with ^{13}C), and hope this strategy yields the desired information. Often it does not. Uniform labeling is attractive because it is comparatively fast and inexpensive. In contrast, deliberate labeling at one or more specific sites in a saccharide is more demanding of time, money, and synthetic expertise, but often leads to very precise information. The high cost of selective labeling is thus often more than offset by the simplicity of data analysis and interpretation.

Given the above constraints, it is not possible to discuss in one paper the many challenges faced by researchers who have a need for isotopically labeled saccharides. This chapter will thus focus narrowly on chemical methods to *introduce* stable isotopes into monosaccharides, with the ^{13}C, ^{2}H, $^{17/18}$O, and ^{15}N isotopes treated exclusively. To make the treatment manageable, only monosaccharides commonly observed *in vivo* will be considered: D-glucose **1**, D-mannose **2**, D-galactose **3**, D-arabinose **4**, D-ribose **5**, D-xylose **6**, N-acetyl-D-glucosamine (GlcNAc) **7**, N-acetyl-D-galactosamine (GalNAc) **8**, N-acetyl-neuraminic acid (Neu5NAc) **9**, and 6-deoxy-L-galactose (L-fucose) **10** (Scheme 1). This approach has advantages

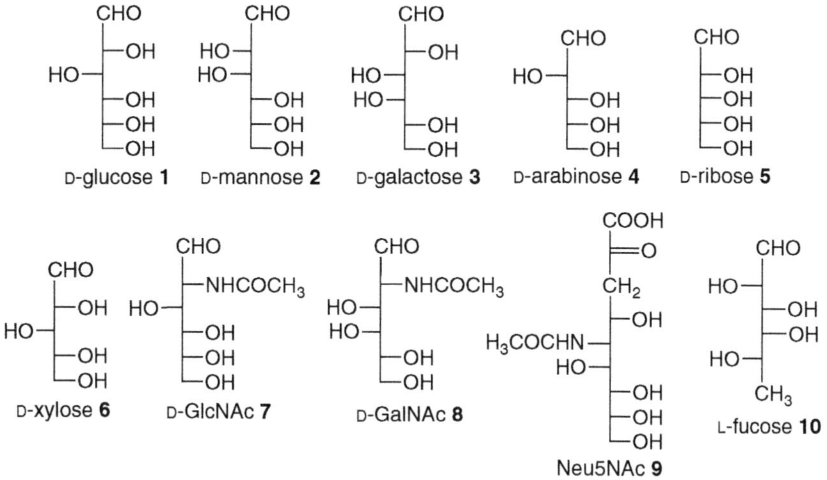

Labeling Monosaccharides With Stable Isotopes

[Structures shown: D-glucose 1, D-mannose 2, D-galactose 3, D-arabinose 4, D-ribose 5, D-xylose 6, D-GlcNAc 7, D-GalNAc 8, Neu5NAc 9, L-fucose 10]

Scheme 1 Chemical structures of monosaccharides that are commonly found *in vivo*.

because, as will be shown, chemical reactions to introduce stable isotopes into monosaccharides are relatively few in number, whereas reactions (both chemical and enzyme-catalyzed) to *transform* labeled monosaccharides into other labeled biomolecules, including other labeled monosaccharides, are divergent and too numerous to treat comprehensively here, although discussion of representative transformations is provided to illustrate some points. Transformative reactions, especially those that are enzyme-catalyzed, grow more numerous and practical over time, as new chemical processes are discovered, and new enzymes are isolated and made commercially available at reasonable cost.

2. TERMINOLOGY TO DESCRIBE DIFFERENT MONOSACCHARIDE ISOTOPOMERS

Simple monosaccharides contain multiple carbon, hydrogen, and oxygen atoms, and sometimes nitrogen atoms (e.g., 7–9), making possible a wide range of different isotopomers of any given monosaccharide that include ^{13}C, ^{2}H, ^{17}O, ^{18}O, and ^{15}N stable isotopes. A monosaccharide can be singly (site-selective), multiply, or uniformly labeled in a given isotope. More than one type of isotope can be present in a labeled monosaccharide. Given this diversity, a general method to describe these isotopomers is helpful and such a method is described in Scheme 2. The upper case "O" and "E" symbols denote hOmonuclear and hEteronuclear labeling,

> O = homonuclear labeling = labeling with only one type of stable isotope
> E = heteronuclear labeling = labeling with two or more different stable isotopes; subscripts list the isotopes in order of increasing mass
>
> s = singly labeled isotopomer
> m = multiply labeled isotopomer
> u = uniformly labeled isotopomer

Scheme 2 Nomenclature to describe different isotopomers of monosaccharides.

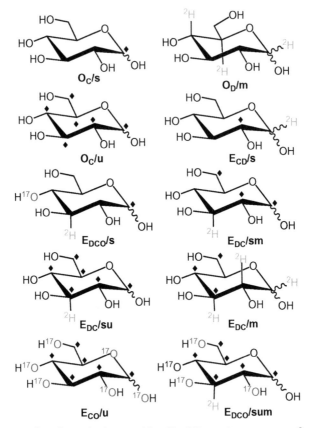

Scheme 3 Examples of terminology to identify different isotopomers of D-glucose **1**.

respectively. These designations are modified with subscripts to indicate the types of isotopes present in the molecule, listed in order of increasing mass. The lower case "s", "m," and "u" symbols designate single, multiple, or uniform labeling, respectively, of each different isotope in the molecule. This nomenclature is illustrated for different isotopomers of D-glucose **1** in Scheme 3. For example, D-[1-^{13}C]glucose **1**[1] (where ◆ denotes a ^{13}C-labeled carbon in the molecule in Scheme 3 and throughout the chapter) is denoted O_C/s, since it is labeled with one type of isotope (^{13}C) and contains only one labeled site. D-[3-^{2}H; 1-^{13}C; 4-^{17}O]glucose is denoted as E_{DCO}/s, since it contains three different isotopes (^{2}H, ^{13}C, ^{17}O) and each is present at a single site. D-[3-^{2}H; 1,2,3,4,5,6-$^{13}C_6$; 2,3,4,6-$^{17}O_4$]glucose is denoted as E_{DCO}/sum since it contains three different isotopes (^{2}H, ^{13}C, ^{17}O), with single labeling by ^{2}H, uniform labeling by ^{13}C, and multiple labeling by ^{17}O.

Synthetic strategies for site-specific labeling (single and multiple) differ from those for uniform labeling, with the former generally achieved via chemical and/or chemo-enzymic processes, and the latter achieved by *in vivo* biological processes. Thus, the synthesis of D-[1-^{13}C]glucose is likely to involve a specific chemical process to introduce the ^{13}C atom specifically at C1, whereas growing photosynthetic algal cells in a $^{13}CO_2$ atmosphere leads to full incorporation of ^{13}C atoms at all carbon sites (Behrens, Sicotte, & Delente, 1994). For example, in the latter case, the algal cells produce uniformly ^{13}C-labeled intracellular starch, which is isolated and subsequently hydrolyzed chemically and/or enzymically to give D-[1,2,3,4,5,6-$^{13}C_6$]glucose or uniformly ^{13}C-labeled DNA, which is hydrolyzed to give uniformly ^{13}C-labeled 2′-deoxyribonucleosides (Chandrasegaran, Kan, Sillerud, Skoglund, & Bothner-By, 1985). This chapter focuses mainly on purposeful (selective) methods of introducing stable isotopes into monosaccharides.

3. INTRODUCING ^{13}C INTO MONOSACCHARIDES

Several factors must be weighed when selecting chemical reactions to introduce ^{13}C site-selectively into monosaccharides. These factors include (a) the simplicity of the reaction; (b) the reliability of the reaction; (c) the reaction yield; (d) the scalability of the reaction; (e) the availability and cost of the ^{13}C-labeled reagent; and (f) options for recycling labeled by-products. Factors (a)–(c) address the ability of the reaction to give consistently acceptable yields of the desired product with minimal investment of labor and reagents. Factor (d) is relevant if large quantities of a labeled product are required. Some reactions work well on smaller scales (milligram to gram amounts) but are not easily adapted to prepare labeled compounds in kilogram or greater amounts. Factor (e) is often overlooked but is crucial in minimizing the cost of labeling. For ^{13}C, the most cost-effective reagent is ^{13}CO, since it is the primary form in which ^{13}C is enriched (Scheme 4); all other ^{13}C-labeled one-carbon (secondary) reagents derive from ^{13}CO. The farther downstream the secondary labeled reagent is from ^{13}CO, the more

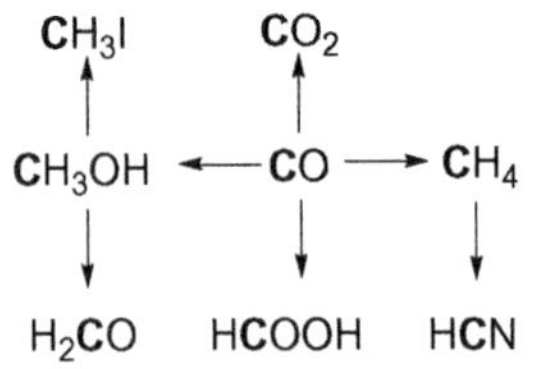

Scheme 4 Central role of ^{13}CO in the production of secondary ^{13}C-labeled one-carbon precursors.

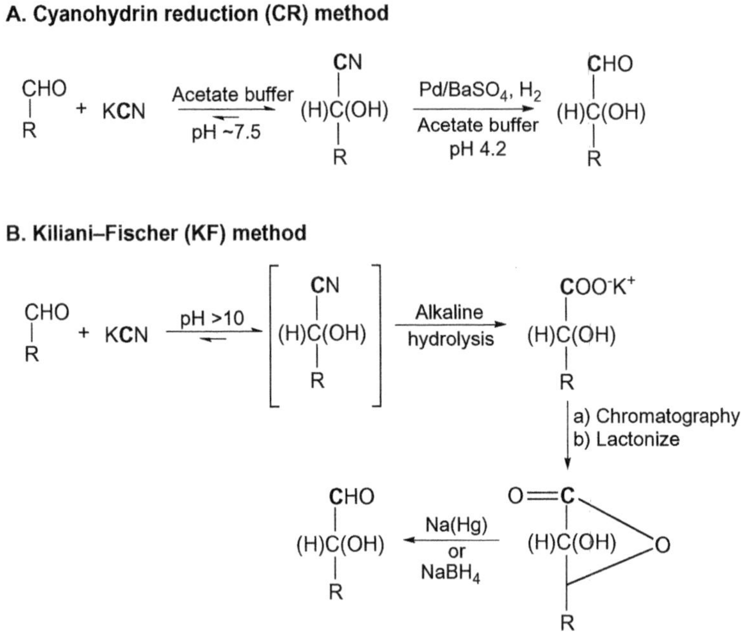

A. Cyanohydrin reduction (CR) method

B. Kiliani–Fischer (KF) method

Scheme 5 The cyanohydrin reduction (CR) and Kiliani–Fischer (KF) methods to introduce carbon isotopes into aldoses.

costly it tends to be on a molar basis. Factor (f) is also often overlooked, but due to the relatively high cost of labeling, labeled by-products ought to be convertible into the desired compound, or recycled into a reagent for subsequent use in other labeled syntheses.

3.1 Cyanohydrin Reduction

One of the most convenient reactions to introduce ^{13}C into aldoses is cyanohydrin reduction (CR) (Scheme 5) (Serianni, Clark, & Barker, 1979; Serianni, Nunez, & Barker, 1979). This method utilizes $K^{13}CN$ (or $Na^{13}CN$) as the labeled reagent, which is relatively inexpensive and commerically available in large quantities to support large-scale-labeled synthesis. The CR method is often confused with the well-known Kiliani–Fischer (KF) synthesis of aldoses (Fischer, 1889; Kiliani, 1885), which is also shown in Scheme 5. In the KF method, a starting aldose reacts with KCN under alkaline conditions to produce a C2-epimeric pair of cyanohydrins (α-hydroxynitriles or aldononitriles) that subsequently undergo *in situ* hydrolysis to the corresponding C2-epimeric aldonate salts. The epimeric aldonates are separated by anion-exchange chromatography (Angelotti, Krisko, O'Connor, & Serianni, 1987), the aldonic acids are lactonized, and the lactones are reduced to aldoses with sodium amalgam (Na(Hg)) (Fischer, 1889) or sodium borohydride ($NaBH_4$) at pH 3–4 (Wolfrom & Wood, 1951). Early on, the KF method was used to synthesize [1-^{13}C]aldoses (Walker, London, Whaley, Barker, & Matwiyoff, 1976), but it was soon realized that the method suffers from several limitations: (a) aldonate yields after chromatography vary widely; (b) some aldonates cannot be lactonized; (c) lactone reduction with Na(Hg) is cumbersome and often gives low yields; and (d) over-reduction to alditols often occurs with both Na(Hg) and $NaBH_4$.

The CR method eliminates the limitations of the KF method by conducting the cyanide condensation at an optimal solution pH that allows >90% conversion to the aldononitriles while keeping aldononitrile hydrolysis to a minimum (Serianni, Nunez, & Barker, 1980). The pH-stabilized aldononitriles are then reduced catalytically with a Pd/$BaSO_4$ catalyst under H_2 to give the C2-epimeric aldoses directly (Serianni, Nunez, et al., 1979). The pH of the heterogeneous catalytic reduction (hydrogenolysis) reaction is critical; the reaction is conducted at low pH (~pH 4.2; adjusted with acetic acid) to promote the hydrolysis of the putative imine intermediate before it has an opportunity to be reduced to the amine. In most syntheses, the cyanohydrins are not isolated, but are reduced *in situ* to the C2-epimeric aldoses, which are separated by chromatography on a cation-exchange resin with Ca^{2+} as the counter-ion and distilled water as solvent (Angyal, Bethell, & Beveridge, 1979). In some cases, the C2-epimeric aldononitriles can be separated by chromatography prior to reduction (Serianni, Pierce, & Barker, 1979).

Since lactonization is not required by the CR method, short-chain aldononitriles can be reduced to short-chain aldoses. For example, formaldehyde and glycolaldehyde can serve as electrophiles, producing [1-^{13}C] glycolaldehyde **11**[1] and DL-[1-^{13}C]glyceraldehyde **12**[1] (superscripts denote the position of the ^{13}C label), respectively, if K^{13}CN is used as the labeled reactant (Serianni, Clark, et al., 1979). In these cases, it was found that conducting the Pd/$BaSO_4$ reduction at a pH below 4.3 is necessary to achieve a high-yield conversion to the aldose. Usually a pH of ~1.7 is used for these reductions. Presumably, the 2- and 3-carbon acyclic imines are more susceptible to over-reduction to amines compared to longer chain imines, the latter being partially protected from over-reduction by cyclization to glycosylamines (Scheme 6). The lower pH ensures full protonation of the acyclic imines, thus increasing their electrophilic character and making them more susceptible to attack by solvent water.

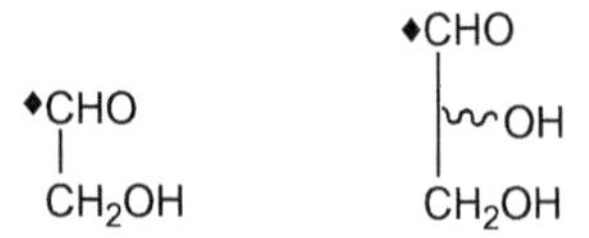

[1-^{13}C] Glycolaldehyde **11**[1]   DL-[1-^{13}C] Glyceraldehyde **12**[1]

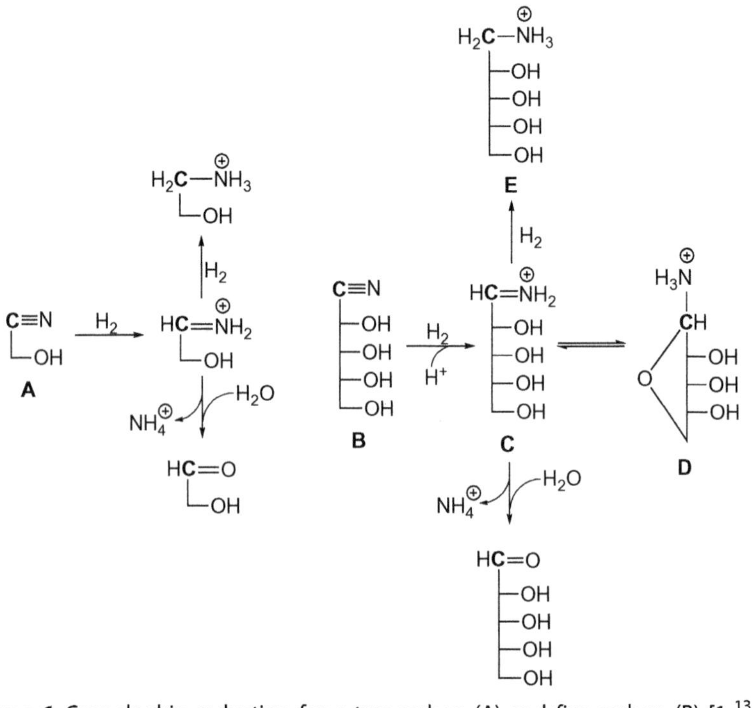

Scheme 6 Cyanohydrin reduction for a two-carbon (A) and five-carbon (B) [1-^{13}C]α-hydroxynitrile (cyanohydrin). For C5 nitrile (B), cyclization of the acyclic imine intermediate (C) to give a cyclic glycosylamine (D) appears to protect the imine from over-reduction to the 1-amino-1-deoxy-alditol (E).

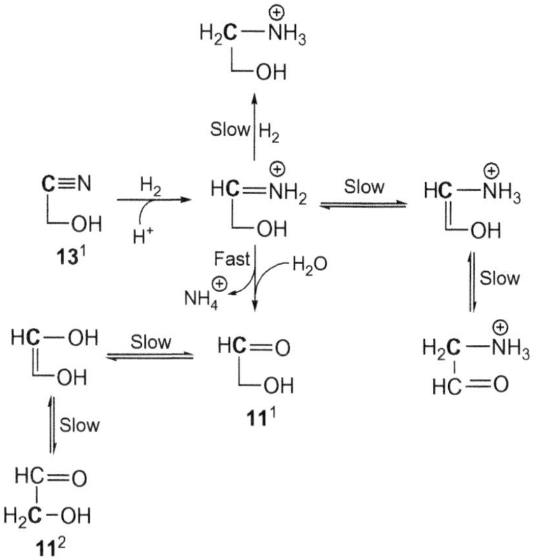

Scheme 7 Potential skeletal rearrangements during reduction of the two-carbon nitrile, **13^1**.

Complications can arise when unlabeled formaldehyde is used as the electrophile in CR with K^{13}CN (Scheme 7). Two-electron reduction of the two-carbon nitrile **13^1** gives an intermediate protonated imine, which can partition into three reaction pathways. One leads to the undesired over-reduction product (primary amine). The second leads to the desired hydrolysis product, the 2-carbon aldehyde **11^1**. However, **11^1** appears to undergo subsequent slow isomerization under the reaction conditions via a putative symmetric *ene*-diol intermediate to give [2-^{13}C]glycolaldehyde **11^2**. This isomerization leads to ^{13}C-label scrambling in the desired product. In a similar manner, the protonated imine may also rearrange to give [2-^{13}C] 2-aminoacetaldehyde. Reactions of this type often show ^{13}C-label scrambling in the product glycolaldehyde if not conducted properly, and by-products containing ^{13}C-labeled amino carbons are often observed. Label scrambling is avoided when [^{13}C]formaldehyde and K^{13}CN are used in CR.

The CR method is general in that almost any aldose can be used as the electrophile. For example, 5-deoxy-L-lyxose (Snyder & Serianni, 1987a) can be used to prepare 6-deoxy-L-[1-^{13}C]galactose (**10^1**) (L-[1-^{13}C] fucose) and 6-deoxy-L-[1-^{13}C]talose using the same reaction protocol as that used for the simple aldopentose, D-lyxose. 2-Deoxy-D-ribose

(2-deoxy-D-*erythro*-pentose) can be used to prepare 3-deoxy-D-[1-^{13}C]glucose and 3-deoxy-D-[1-^{13}C]mannose (Scensny, Hirschhorn, & Rasmussen, 1983). While not extensively investigated, aldoses having >6 carbons can serve as reactants, but excess KCN is often needed to drive cyanohydrin formation to completion. In the same vein, the CR reaction can be applied to 2-ketoses, but excess KCN is needed to drive cyanohydrin formation. For example, 1,3-dihydroxypropanone (dihydroxyacetone) 14 can serve as an electrophile to produce the branched-chain aldose, 15[1], which is subsequently used to prepare DL-[2-^{13}C]apiose 16[2] (Snyder & Serianni, 1987b) (Scheme 8). In general, since the cyanide condensation reaction is bimolecular, conducting the reaction at high aldose and cyanide concentrations helps drive cyanohydrin formation to completion, even when a 1:1 aldose:cyanide ratio is used.

Since excess K^{13}CN is sometimes used to drive cyanohydrin formation, recovery of unreacted K^{13}CN becomes a priority, especially when reactions are run on large scales. This recovery is achieved by purging reaction mixtures after cyanide addition is complete and the solution pH has been lowered from ~7.3 to ~4.2 with acetic acid. The reaction solution is aerated with N_2 through a glass frit and the gas stream containing H^{13}CN is flowed through methanolic KOH traps to recover the excess cyanide. The alcoholic K^{13}CN solution can then be concentrated and the K^{13}CN recovered for reuse.

The CR method has been automated through the design and use of a special reaction vessel and ancillary software to control it (Stafford, Serianni, & Varma, 1990). This work led to the recognition that CR reactions conducted in formic acid (rather than acetic acid) do not require the addition of H_2 gas during the reduction step. The Pd/BaSO$_4$ catalyst is able to cleave HCOOH *in situ* to give CO_2 and H_2, and the latter is subsequently consumed in CR.

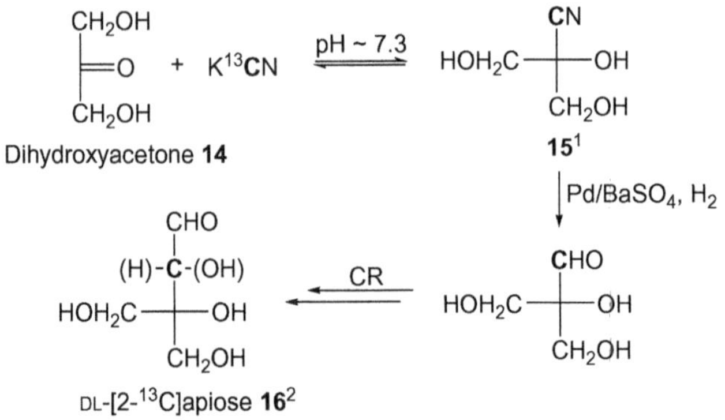

Scheme 8 The use of ketosugars as electrophiles in cyanohydrin reduction (CR).

3.2 Permutations of Cyanohydrin Reduction

The CR method can be adapted to introduce hydrogen isotopes at C1 of aldoses using isotopically labeled hydrogen gas in the hydrogenolysis step (Serianni & Barker, 1979). For example, the use of $^2H_2$ gas gives [1-^{2}H] aldoses (Scheme 9A). The reaction must be conducted in ^2H_2O solvent because ^1H–^2H exchange with solvent water occurs during reduction (Serianni & Barker, 1979). This approach to incorporate ^2H at C1 complements that involving the reduction of aldonolactones to corresponding aldoses with NaB^2H_4 at pH 3–4 (Wolfrom & Wood, 1951). The latter method requires a lactone and reaction yields depend heavily on lactone ring size (1,4- vs. 1,5-lactones) and configuration. For example, the reduction of D-mannono-1,4-lactone with NaB^2H_4 gives a high yield of D-[1-^{2}H]mannose (>80%), whereas the same reaction applied to D-ribono-1,4-lactone gives D-[1-^{2}H]ribose in <30% yield. The lower yields are caused by poor reactivity of the lactone and/or by over-reduction to the alditol.

Aldoses undergo oxygen atom exchange at C1 in solvent water via involvement of the acyclic 1,1-*gem*-diol (hydrate) tautomer (Scheme 9B) (Mega & Vann Etten, 1990). If the exchange reaction is conducted in oxygen-labeled water, then the oxygen isotope can be captured by subsequent cyanide addition and aldononitrile reduction to give [2-^{17,18}O]aldoses (Scheme 9B) (Clark & Barker, 1986). To ensure a high level of enrichment, the starting aldose is exchanged more than once with the labeled water; the number of exchanges depends on the reaction scale and the desired level of enrichment

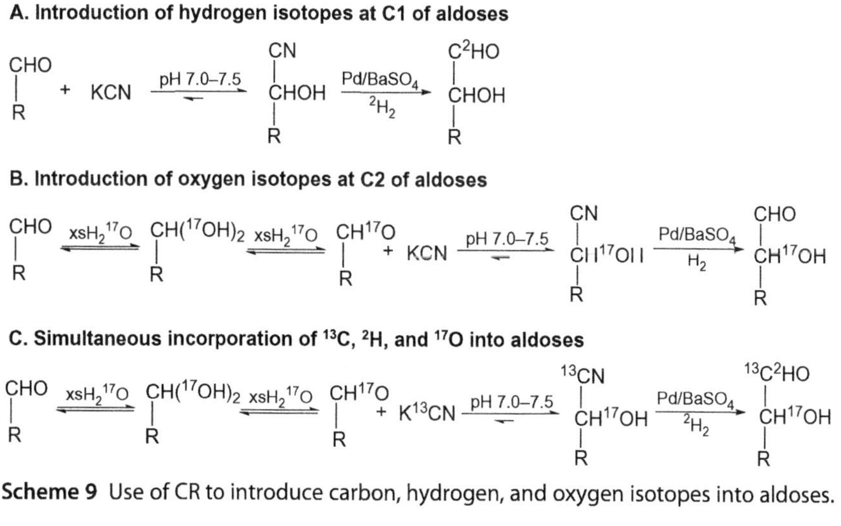

Scheme 9 Use of CR to introduce carbon, hydrogen, and oxygen isotopes into aldoses.

in the product. The extent of the oxygen exchange can be conveniently monitored by ^{13}C NMR to ensure proper enrichment prior to entrapment of the labeled oxygen via CR (Mega & Vann Etten, 1990).

A single cycle of CR applied to a starting aldose can introduce carbon isotopes at C1, hydrogen isotopes at C1, and/or oxygen isotopes at C2 of the product aldoses (Scheme 9C) (Zhu, Zajicek, & Serianni, 2001).

While the CR permutations shown in Scheme 9 for the incorporation of hydrogen and oxygen isotopes into saccharides have merit, the CR method is most powerful for the insertion of carbon isotopes. Hydrogen and oxygen labeling via CR are most practical when they are coupled to carbon insertion. In the absence of carbon insertion, hydrogen and oxygen isotopes can be introduced via other routes since both atom types are terminal. There are many examples in the literature of this approach. Typically an alcohol is oxidized to a ketone, which is then subjected to exchange with oxygen-labeled water and/or reduced with an unlabeled or deuterated reducing agent (e.g., NaB^{2}H$_4$).

3.2.1 Generalized Laboratory Protocol—Cyanohydrin Reduction Reaction

To a 300-mL three-neck round-bottom flask in a well-ventilated fume hood is added 7.93 g (0.12 mol) of K^{13}CN, followed by distilled water (30 mL). A thermometer is secured via an adaptor into one of the necks and a magnetic stir bar is added to the reaction vessel. Another neck is fitted with a pH electrode. The solution is stirred at room temperature until all of the K^{13}CN has dissolved. The solution is then cooled to ~4 °C in an ice bath and glacial acetic acid is added dropwise to the cooled reaction solution until a pH of ~8.5 is reached. The solution temperature is maintained near 4 °C during the addition of the acid [*Caution*: The pK_a of HCN is 9.2, so the cyanide solution at pH ~8.5 contains a significant amount of HCN, which is a highly poisonous gas. Care should be taken to avoid exposure to HCN by conducting the reaction in a fume hood, and by wearing an HCN monitor while conducting this part of the CR process].

In a separate reaction flask, the starting aldose (0.1 mol) is dissolved in ~40 mL of distilled water and the solution is cooled to 4 °C in an ice bath.

The cooled aldose solution is added slowly to the cooled [^{13}C]cyanide solution via the third neck of the reaction vessel, making sure the reaction temperature remains below 10–15 °C during the addition. After addition of the aldose solution is complete, the reaction solution is adjusted to pH ~7.5 (HOAc or KOH) and the reaction mixture is warmed to room temperature over ~5 min using a warm water bath. After ~15 min at pH ~7.5 and room temperature, the solution pH is lowered to ~4.2 with the addition of glacial acetic acid.

Excess labeled cyanide can be recovered by purging the reaction vessel with N_2 gas and passing the $H^{13}CN$-laden gas stream through two or three methanolic KOH traps connected in series by Tygon™ tubing [*Note*: Excess cyanide must be removed from the reaction mixture prior to catalytic reduction because cyanide will poison the palladium catalyst]. After excess cyanide has been removed, the solution pH is adjusted to 4.2 with either HOAc or KOH.

Into a 250 or 500-mL Parr glass reduction vessel is added 6.2 g of 5% palladium on barium sulfate. Distilled water (~20 mL) is added, the vessel is evacuated and filled with H_2 gas, and the catalyst is pre-reduced at ~30 psi until it turns from a brown to a gray color. The reduction vessel is opened, the reaction solution containing the [1-^{13}C]aldononitriles is added, the vessel is evacuated and filled with H_2 gas, and the reduction is conducted at ~30 psi until H_2 gas consumption ceases (3–4 h at room temperature with good shaking by the Parr apparatus).

After reduction is complete, the suspension is filtered to remove the catalyst, and the filtrate is treated batchwise and successively with excess Dowex 50-X8 (H^+) cation-exchange resin and with excess Dowex 1-X8 (OAc^-) anion-exchange resin. The deionized solution is concentrated at 30 °C *in vacuo* on a rotary evaporator to a volume of ~20 mL. This solution, which contains the product C2-epimeric [1-^{13}C]-labeled aldoses, is applied to a chromatography column containing Dowex 50-X8 (200–400 mesh) cation-exchange resin in the Ca^{2+} form (Angyal et al., 1979). The column is eluted with distilled water and the eluting aldoses are detected with a reducing sugar assay (Hodge & Hofreiter, 1962).

3.3 Molybdate-Catalyzed Epimerization of Aldoses Accompanied by C1–C2 Transposition

A disadvantage of CR is that, since cyanide addition to the carbonyl is not stereospecific, two C2-epimeric aldoses form. The ratio of C2-epimeric aldononitriles depends on the solution pH of the cyanide addition reaction (Serianni et al., 1980). For example, in the KF synthesis under alkaline conditions using D-arabinose as the electrophile, the final *gluco/manno* ratio of *aldonates* favors the *gluco* epimer (~65%). In contrast, if cyanide addition is conducted at pH ~7.3, the *gluco/manno* ratio of aldononitriles favors the *manno* epimer (~70%) (Serianni, Nunez, et al., 1979). Experimental data indicate that the KF ratio of aldonates is determined by the relative rates of hydrolysis of cyclic C2-epimeric imidolactone intermediates that form during the conversion of aldononitrile to aldonate (since CN addition is reversible and rapid under alkaline conditions), whereas the CR ratio of aldononitriles is determined by CN addition only (Serianni et al., 1980). Thus, from a biological viewpoint, CR applied to D-arabinose gives the arguably more desirable *gluco* epimer as the *minor* product.

Efforts to convert less-desirable C2-epimers obtained from CR into more-useful labeled products include chemical- and enzyme-catalyzed isomerization. For example, base-catalyzed aldose-2-ketose isomerization in the presence of boronic acids provides a means to chemically equilibrate an aldose with its C2 epimer and 2-ketose without significant alkaline degradation (Barker, Chopra, Hatt, & Somers, 1973; Barker, Hatt, & Somers, 1973). Aldose–ketose isomerases catalyze aldose-2-ketose exchange. For example, D-[1-^{13}C]mannose $\mathbf{2^1}$ is converted into an equilibrium mixture of D-[1-^{13}C]mannose $\mathbf{2^1}$, D-[1-^{13}C]glucose $\mathbf{1^1}$, and D-[1-^{13}C]fructose $\mathbf{17^1}$ by the action of xylose (glucose) isomerase (GI) (E.C. 5.3.1.5). In these cases, labeling integrity remains intact; the position of label in the starting aldose is faithfully transferred to the analogous position in the product(s).

The discovery of a remarkable carbon skeletal rearrangement in aldoses radically changed the strategies to prepare isotopically labeled monosaccharides and enhanced the applicability and overall efficiency of the CR method. Molybdate-catalyzed C2-epimerization (MCE) interconverts aldoses with their C2-epimers in high yield, as first shown by Bilik and coworkers (Bilik & Stankovic, 1973). It was assumed that backbone C–C bonds remain intact during the epimerization, and that breaking and forming C–H bonds underpin the mechanism (Bilik, Petrus, & Farkas, 1975). However, subsequent mechanism studies using ^{13}C-labeled substrates revealed unexpectedly that MCE proceeds via C1–C2 transposition that occurs in a stereospecific manner (Hayes, Pennings, Serianni, & Barker, 1982) (Scheme 10). The reaction can be viewed as an internal redox process where, in the forward direction, C1 is reduced and C2 is oxidized. Since C–H and C–O bonds in the C1–C2 fragment remain intact during the transposition, isotopic labels present at C1 are transferred intact to C2 and vice versa. Thus, as shown in Scheme 10, D-[1-^{13}C;1-^{2}H;1-^{18}O]mannose equilibrates with D-[2-^{13}C;2-^{2}H;2-^{18}O]glucose. Only catalytic amounts of sodium molybdate are required, the reaction is conducted in water, reaction times are normally <5 h, and the ratio of C2-epimeric aldoses is determined by their relative thermodynamic stabilities. Close inspection of

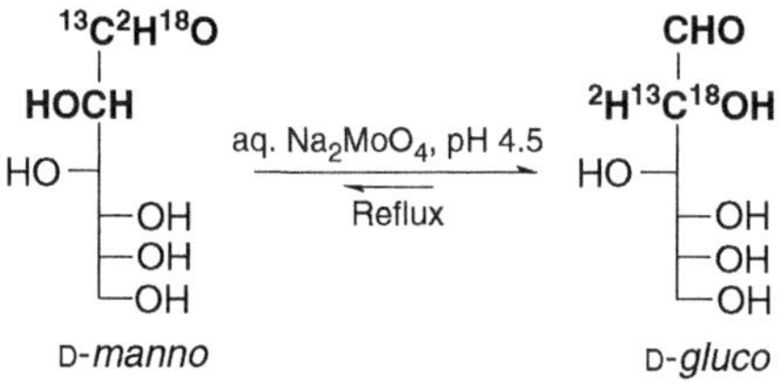

$^{13}C^2H^{18}O$ → CHO (with $^2H^{13}C^{18}OH$)

D-*manno* ⇌ (aq. Na₂MoO₄, pH 4.5, Reflux) ⇌ D-*gluco*

Scheme 10 The molybdate-catalyzed C2-epimerization (MCE) reaction of aldoses showing C1–C2 transposition.

reaction mixtures reveals the presence of minor by-products that appear to arise from conventional isomerization/epimerization. Conducting MCE with resin-bound molybdate reduces the amounts of these nonspecific by-products, especially when tetroses and pentoses are used as reactants (Clark, Hayes, & Barker, 1986).

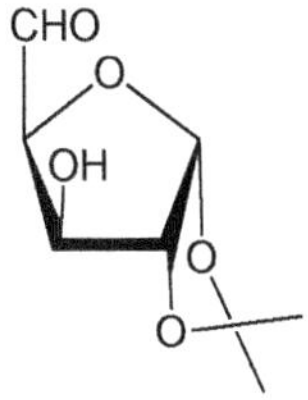

1,2-O-isopropylidene-α-D-*xylo*-pentodialdo-1,4-furanose **18**

Studies show that hydroxyl groups at C1, C2, and C3 of the aldose are absolutely required for the reaction; that the acyclic hydrate form of the aldose is the reactive form; and that molybdate binds to the acyclic hydrate in a dimolybdate complex (Hayes et al., 1982). Inspection of these putative Mo–Mo complexes shows that H1 in the starting aldose and H2 in the product aldose are exposed and peripheral in the complex. This structural feature can be exploited in MCE by replacing H1 of the aldose with an *R*-group, thus allowing interconversion of a 2-ketose and a C2-branched-chain aldose (Hricoviniova-Bilikova, Hricovini, Petrusova, Serianni, & Petrus, 1999; Wu, Pan, Zhao, Imker, & Serianni, 2007; Zhao, Petrus, & Serianni, 2001). Two examples of this type of MCE reaction are shown in Scheme 11.

3.3.1 Generalized Laboratory Protocol—Molybdate-Catalyzed Epimerization of Aldoses

An aqueous solution containing an aldose (0.1 M) and molybdic acid (85%) (5 mM) at pH 4.5 (adjusted with acetic acid) is incubated at 90 °C for 2–13 h

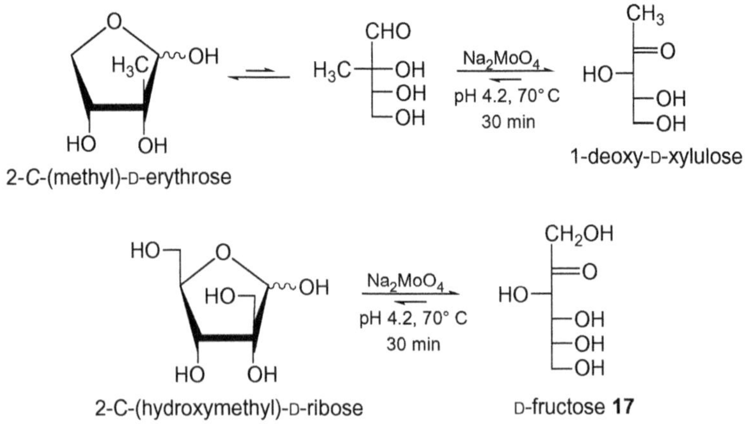

Scheme 11 The application of molybdate-catalyzed epimerization (MCE) to branched-chain aldoses.

depending on the aldose. The solution is cooled and treated batchwise and separately with excess Dowex 50-X8 (H$^+$) cation and Dowex 1-X2 (OAc$^-$) anion ion-exchange resins, and the resulting deionized solution is concentrated to ~10 mL at 30 °C *in vacuo* in a rotary evaporator. This solution is applied to a chromatography column containing Dowex 50-X8 (200–400 mesh) cation-exchange resin in the Ca^{2+} form (Angyal et al., 1979). The column is eluted with distilled water and the eluting aldoses are detected with a reducing sugar assay (Hodge & Hofreiter, 1962).

4. MULTIPLE LABELING OF ALDOSES VIA CHAIN INVERSION

CR and MCE are attractive and economical processes when there is need to introduce isotopic labels at C1 and/or C2 of an aldose. Labeling at terminal hydroxymethyl (CH$_2$OH) carbons of aldoses can also be achieved using aldose derivatives such as **18** (King-Morris, Bondo, Mrowca, & Serianni, 1988) in conjunction with CR. It is more time-consuming and costly to introduce isotopic labels at *internal* positions of aldoses (i.e., secondary alcoholic carbons). This problem can be overcome in some cases by chain inversion (Wu, Serianni, & Bondo, 1992). An example of this approach is shown in Scheme 12. D-[1,2-^{13}C$_2$]mannose **2**^{1,2} is prepared from two rounds of CR starting from D-erythrose, with MCE used to improve overall yield by converting the D-[1,2-^{13}C$_2$]glucose **1**^{1,2} by-product to D-[1,2-^{13}C$_2$]mannose **2**^{1,2}. Methyl glycosidation followed by C6 oxidation and glycoside hydrolysis gives mannuronic acid, which is reduced to give D-[5,6-^{13}C$_2$]mannonic acid. The labeled mannonic acid is reduced via the 1,4-lactone to give D-[5,6-^{13}C$_2$]mannose **2**^{5,6}, or cleaved to give D-[4,5-^{13}C$_2$]arabinose **4**^{4,5}, which can be used in CR and MCE to give D-[1,5,6-^{13}C$_3$]glucose **1**^{1,5,6} and D-[2,5,6-^{13}C$_3$]glucose **1**^{2,5,6}.

5. LABELING AT THE INTERNAL CARBONS OF ALDOSES

Integration of CR, MCE, and enzymic reactions permits the introduction of stable isotopes at internal positions of aldoses. These approaches are illustrated in Schemes 13 and 14. In Scheme 13, D-[1-^{13}C]ribose **5**^{1} and D-[1-^{13}C]arabinose **4**^{1} are produced by the application of CR with K^{13}CN to D-erythrose. MCE applied to D-[1-^{13}C]ribose **5**^{1} gives D-[2-^{13}C]arabinose **4**^{2}, which when subjected to CR with K^{13}CN gives D-[1,3-^{13}C$_2$]glucose **1**^{1,3} and D-[1,3-^{13}C$_2$]mannose **2**^{1,3}. MCE applied to the latter gives D-[2,3-^{13}C$_2$]glucose **1**^{2,3}. The by-product D-[1-^{13}C]arabinose **4**^{1} can be subjected to CR with K^{13}CN to give D-[1,2-^{13}C$_2$]glucose **1**^{1,2} and D-[1,2-^{13}C$_2$]mannose **2**^{1,2}, with the yield of the former improved by the application of MCE to the latter. Alternatively, D-[1-^{13}C]arabinose **4**^{1} can be subjected to MCE to give D-[2-^{13}C]ribose **5**^{2}, which can be used in the synthesis of labeled ribo- and 2′-deoxyribonucleosides. The initial D-[1-^{13}C]ribose **5**^{1} can also be funneled in ribo- and 2′-deoxyribonucleosides labeled at C1′. The routes shown in Scheme 13 illustrate the value of coupling CR with MCE; even though CR produces two C2-epimeric products, MCE can be used to funnel *both* epimers into useful and valuable labeled compounds, thereby eliminating waste and maximizing benefit.

Aldolase-catalyzed reactions offer the opportunity to insert stable isotopes into internal carbons of aldoses (Serianni, Cadman, Pierce, Hayes, & Barker, 1982). For example, in Scheme 14, [1-^{2}H]glycolaldehyde is subjected to CR with K^{13}CN to give DL-[1-^{13}C;2-^{2}H]glyceraldehyde. The labeled aldotriose is reacted with DHAP (dihydroxyacetone phosphate; 1,3-dihydroxypropanone phosphate) in the presence of rabbit muscle aldolase (RMA) (fructose bisphosphate aldolase) (E.C. 4.1.2.13) to give D-[4-^{13}C;5-^{2}H]fructose 1P (F1P) and L-[4-^{13}C;5-^{2}H]sorbose 1P (S1P). RMA is used to resolve an enantiomeric mixture of the aldotriose into two diastereomers. After treatment with acid phosphatase (E.C. 3.1.3.2), the resulting labeled D-fructose can be converted into D-[4-^{13}C;5-^{2}H]glucose by the action of glucose (xylose) isomerase (GI) (E.C. 5.3.1.5). The by-product L-[4-^{13}C;5-^{2}H]sorbose 1P can either be re-equilibrated with

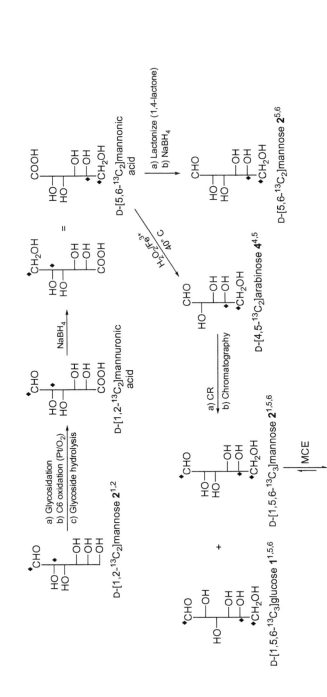

Scheme 12 A synthetic route to prepare D-[2,5,6-$^{13}C_3$]glucose $1^{2,5,6}$ from D-[1,2-$^{13}C_2$]mannose $2^{1,2}$ by chain inversion.

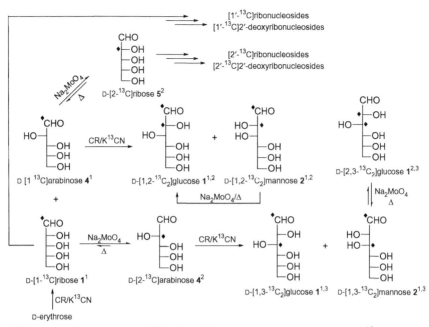

Scheme 13 The integration of the CR and MCE reactions to prepare D-[1,3-$^{13}C_2$]glucose $\mathbf{1}^{1,3}$ and D-[2,3-$^{13}C_2$]glucose $\mathbf{1}^{2,3}$.

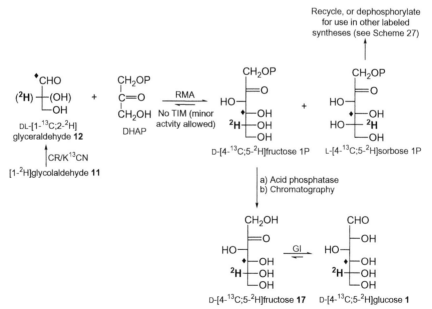

Scheme 14 The use of rabbit muscle aldolase (RMA) to prepare D-[4-^{13}C;5-^{2}H]glucose **1** from triose precursors.

RMA to improve the yield of the desired *fructo* or *gluco* products, or dephosphorylated to give L-[4-^{13}C;5-^{2}H]sorbose for use in other labeled syntheses (e.g., fructose **17**; see Scheme 27). The RMA preparation should be free of contaminating triose phosphate isomerase (TIM) (E.C. 5.3.1.1) activity in order to avoid the formation of unlabeled D-fructose 1,6-bisphosphate (FBP), which would lower the reaction yield. However, if TIM is present, the FBP can be separated from the desired monophosphates by anion-exchange chromatography.

6. EXTENSION TO BIOLOGICALLY IMPORTANT ALDOSES

The preceding discussion demonstrates that CR and MCE can be used to prepare a wide range of isotopically labeled biologically important aldoses and 2-ketoses. Correlations between the starting aldose and CR product aldoses (underlined) in Scheme 1 are as follows:
- (a) D-arabinose **4** → D-glucose **1** and D-mannose **2**
- (b) D-lyxose → D-galactose **3** and D-talose
- (c) D-erythrose → D-arabinose **4** and D-ribose **5**
- (d) D-threose → D-lyxose and D-xylose **6**
- (e) 5-deoxy-L-lyxose (Snyder & Serianni, 1987a) → 6-deoxy-L-galactose (L-fucose) **10** and 6-deoxy-L-talose

The *N*-acetylated 2-amino-2-deoxyaldoses, GlcNAc **7** and GalNAc **8**, can be prepared using CR, with glycosylamines instead of aldoses serving as electrophiles (Scheme 15) (Zhu et al., 2006). D-Aldopentoses are converted in high yield with NH$_3$ gas into glycosylamines, which are relatively stable in alkaline solution. Glycosylamines serve as good substrates in CR (as acyclic

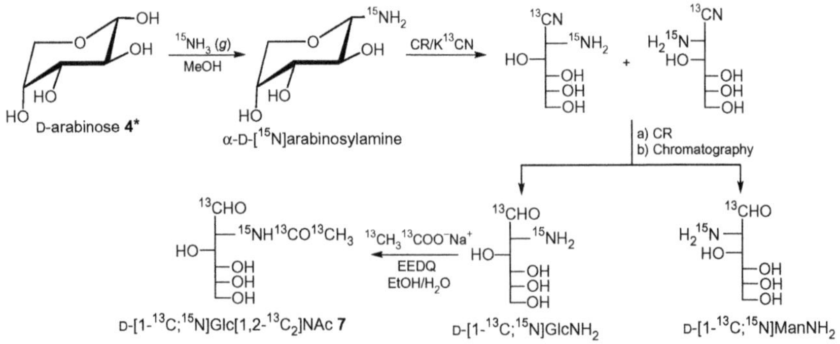

*D-lyxose, D-ribose **5**, or D-xylose **6** can be substituted for D-arabinose **4**.

Scheme 15 The application of CR to the synthesis of isotopically labeled GlcNAc **7**.

protonated imines) to give C2-epimeric aminonitriles, and the latter reduce in high yield to give a mixture of C2-epimeric 2-amino-2-deoxyaldoses. The latter are separated by cation-exchange chromatography and converted to the N-acetylated derivatives by EEDQ-mediated acetylation or by other methods. This approach allows the incorporation of ^{13}C, ^{2}H, and/or ^{15}N isotopes in a single CR cycle as shown in Scheme 15. Use of appropriately labeled aldopentoses broadens the options for labeling GlcNAc **7** and Gal-NAc **8**. If arabinose **4** is used at the starting pentose, then GlcNAc **7** (and ManNAc **19**) are produced (Scheme 15). If D-lyxose is used as starting aldopentose, then GalNAc **8** (and TalNAc) are produced.

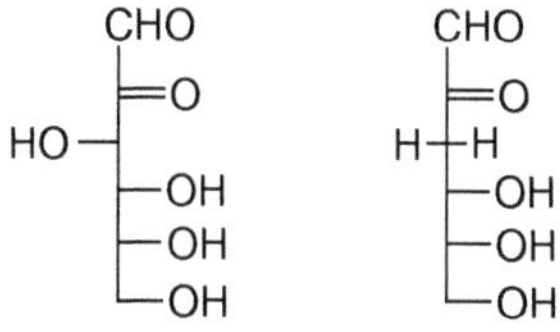

D-*arabino*-hexos-2-ulose **21**

3-deoxy-D-*erythro*-hexos-2-ulose **22**

The CR route shown in Scheme 15 leads to the production of ManNAc **19**, which along with pyruvate, can be used in the enzymic synthesis of Neu5Ac **9** (Scheme 1) (Klepach, Carmichael, & Serianni, 2008; Klepach, Zhang, Carmichael, & Serianni, 2008; Lin, Lin, & Wong, 1997). Since pyruvate is available commercially in different isotopomeric forms, and sialic acid (Neu5NAc) aldolase (E.C. 4.1.3.3) is available commercially, stable isotopically labeled **9** can be prepared as shown in Scheme 16. Purification of **9** is achieved by chromatography on an anion-exchange column (Klepach, Carmichael, et al., 2008).

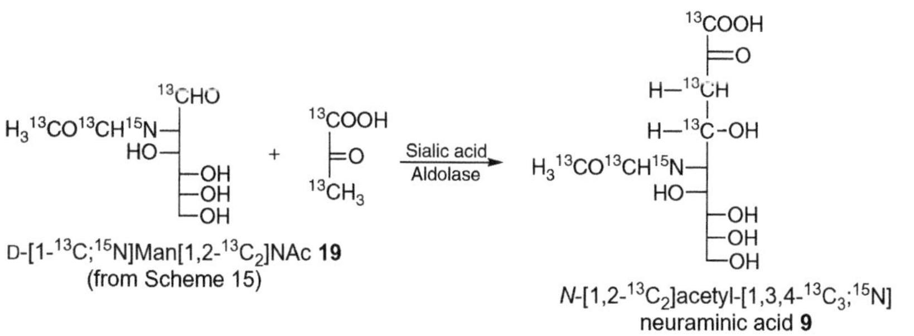

D-[1-^{13}C;^{15}N]Man[1,2-^{13}C$_2$]NAc **19**
(from Scheme 15)

N-[1,2-^{13}C$_2$]acetyl-[1,3,4-^{13}C$_3$;^{15}N] neuraminic acid **9**

Scheme 16 Enzymic synthesis of isotopically labeled Neu5Ac **9** using sialic acid aldolase.

7. RELATIVE CARBONYL REACTIVITIES IN OSONES—SYNTHESIS OF LABELED 2-KETOSES

The reactivity of ketoses as electrophiles in CR is generally lower than that of aldoses, presumably because of the greater steric hindrance at the internal carbonyl carbon in the former. This reduced reactivity can be exploited in the reduction of 1,2-dicarbonyl sugars, commonly denoted as osones. Osones can be generated chemically (Vuorinen & Serianni, 1990a) or enzymically (Freimund, Huwig, Giffhorn, & Köpper, 1998). For example, the 1,2-dicarbonyl sugar, L-*threo*-pentos-2-ulose **20**, is produced from L-xylose via oxidation with $Cu(OAc)_2$ in aqueous methanol (Scheme 17). The osone, D-glucosone **21**, is prepared by the action of pyranose 2-oxidase (E.C. 1.1.3.10) on D-glucose **1** (Zhang & Serianni, 2012). The oxidase displays broad specificity for the aldose substrate (Freimund et al., 1998), thus lending itself to the synthesis of a wide range of osones [e.g., 3-deoxy-D-glucosone **22** (Zhang, Carmichael, & Serianni, 2011)].

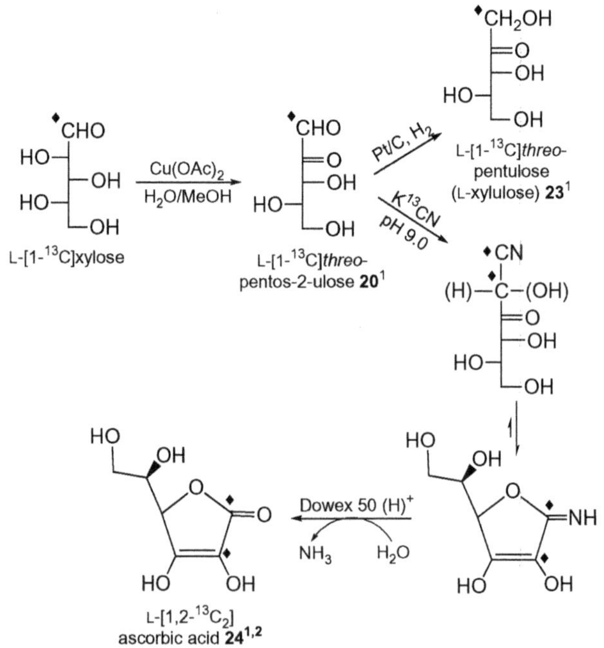

Scheme 17 Selective reactivity of the carbonyl groups in osone **20** to reduction (to give ketose **23**) and to cyanide addition/hydrolysis (to give vitamin C **24**).

Osones can be converted in high yield to 2-ketoses by reduction with Pt/C and H_2 (Vuorinen & Serianni, 1990b) (Scheme 17). The latter reaction exploits the relative susceptibilities of C1 and C2 to nucleophilic attack, either by hydride and cyanide anions. The reduction proceeds almost exclusively at C1, giving 2-ketose products. For example, as shown in Scheme 17, **20**[1] can be reduced in high yield to L-[1-^{13}C]*threo*-2-pentulose **23**[1] (L-[1-^{13}C]xylulose). If only slightly more than one equivalent of hydride donor is used, only very minor amounts of aldose and/or alditol by-products are produced. This approach provides access to isotopically labeled 2-ketosugars via labeled aldoses that may not be available via enzymic methods (see below).

The selectivity of C1 in osones to attack by cyanide anion is nicely demonstrated in the synthesis of ^{13}C-labeled vitamin C **24** (Scheme 17) (Drew, Church, Basu, Vuorinen, & Serianni, 1996). This process is not CR *per se*, but rather KF in that the initially formed aldononitriles are permitted to cyclize to imidolactone intermediates (Serianni et al., 1980), which subsequently undergo hydrolysis under acidic conditions to give [1,2-^{13}C$_2$] ascorbic acid **24**[1,2]. Imidolactones have been implicated as key intermediates in the KF reaction (Serianni et al., 1980), and in this case form rapidly because of the conformational (flexibility) constraints in the acyclic nitrile backbone imposed by an sp^2-hybridized C3 carbon. This process yields the reduced form of **24**, an α,β-unsaturated γ-lactone.

8. MANIPULATION OF THREE-CARBON BUILDING BLOCKS IN ENZYME-MEDIATED ALDOL CONDENSATION

Aldose phosphates serve as viable electrophiles in CR. For example, glycolaldehyde phosphate can be converted into DL-[1-^{13}C]glyceraldehyde 3P (G3P) by CR using K^{13}CN as a reactant (Serianni, Pierce, et al., 1979). DL-[1-^{13}C]G3P can be treated with TIM and RMA to give a single 2-ketohexose product after the reaction reaches equilibrium: L-[3,4-^{13}C$_2$] sorbose 1,6-bisphosphate (Scheme 18). Only the L-*sorbo* product is obtained because only D-G3P is a substrate for TIM, thus creating the DHAP that becomes available for RMA-mediated condensation with L-G3P.

By applying the RMA reaction to an enantiomeric mixture of DL-glyceraldehyde and unlabeled DHAP in the absence of TIM, stable isotopes can be introduced into the *bottom-half* (C4–C6 carbons) of F1P and S1P (Scheme 19). DL-[1,2-^{13}C$_2$]glyceraldehdye **12**[1,2] is generated from

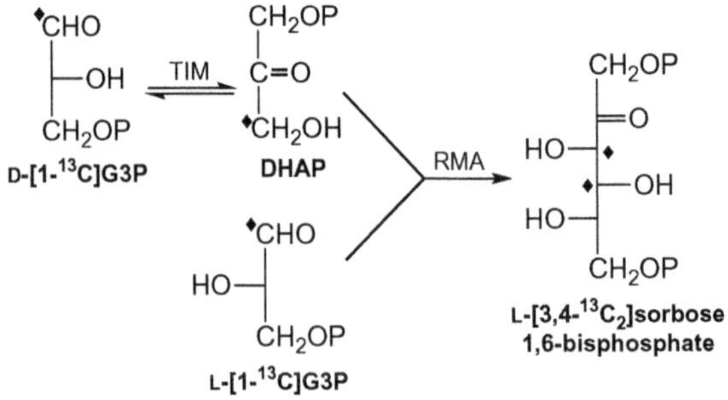

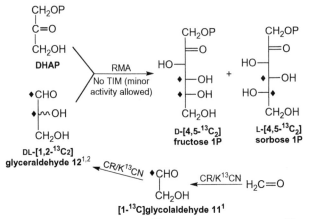

Scheme 18 Chemo-enzymic synthesis of L-[3,4-$^{13}C_2$]sorbose 1,6-bisphosphate using TIM.

Scheme 19 Use of RMA to prepare D-[4,5-$^{13}C_2$]fructose 1P and L-[4,5-$^{13}C_2$]sorbose 1P from DL-[1,2-$^{13}C_2$]glyceraldehyde **12**^{1,2}.

two rounds of CR. The dilabeled aldotriose is condensed with DHAP in the presence of RMA to give D-[4,5-$^{13}C_2$]fructose 1P and L-[4,5-$^{13}C_2$]sorbose 1P. Including a strong (transition state) inhibitor of TIM (e.g., 2-phosphoglycolate) in the reaction mixture minimizes the formation of unlabeled D-G3P, which would lead to the formation of unlabeled FBP as a by-product and lower the reaction yield of the desired labeled 2-ketose 1-phosphates. This problem can occur because some commercial preparations of RMA often contain contaminating TIM activity. However, if this side reaction does occur, it is possible to separate the labeled monophosphates from the unlabeled bisphosphate by anion-exchange chromatography (Serianni et al., 1982). Separation of the monophosphates is more easily achieved as the free 2-ketoses by acid phosphatase treatment (Serianni et al., 1982) and chromatography on Dowex 50 cation-exchange resin with Ca^{2+} as the counter-ion (Angyal et al., 1979).

The labeled D-fructose **17** and L-sorbose **25** are then converted into other useful labeled products (see Scheme 27).

While the routes shown in Schemes 18 and 19 are useful, better control over RMA-mediated isotope labeling of saccharides can be gained by exploiting ways to resolve the DL-mixture of glyceraldehyde generated by the application of CR to glycolaldehyde. The following discussion describes approaches to achieve this control.

Lin and coworkers (Hayashi & Lin, 1967) have shown that the enzyme, glycerol kinase (E.C. 2.7.1.30), accepts L-glyceraldehyde hydrate as a substrate, producing L-glycerol 3P. The kinase also accepts D-glyceraldehyde hydrate, but appears to bind it in an inverted orientation that positions one of the hemiacetal hydroxyl groups for phosphorylation (Scheme 20). The resulting phosphomonoester is labile upon release from the enzyme, and thus the kinase appears to behave as an "ATPase" when D-glyceraldehyde is the substrate. This behavior has been confirmed by NMR studies of reaction mixtures using ^{13}C-labeled substrates (Serianni, Clark, et al., 1979). Glycerol kinase is thus able to resolve CR-generated enantiomers of DL-glyceraldehyde **12**, a property that can be exploited as shown in the following reaction schemes.

It should be noted that an alternative approach to resolving DL-glyceraldehyde **12** enzymically might involve the use of triokinase (Frandsen & Grunnet, 1971; Rodrigues et al., 2014) (Scheme 21), which purportedly accepts only D-glyceraldehyde as substrate to give D-G3P, but this enzyme has not been fully characterized (it has been implicated in the metabolism of fructose in human liver) and is not commercially available.

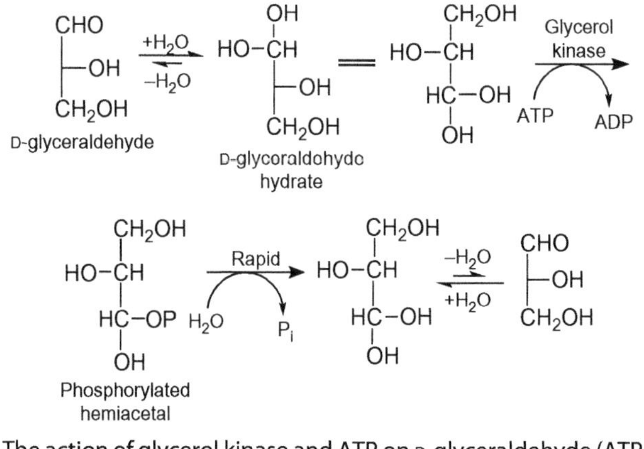

Scheme 20 The action of glycerol kinase and ATP on D-glyceraldehyde (ATPase activity).

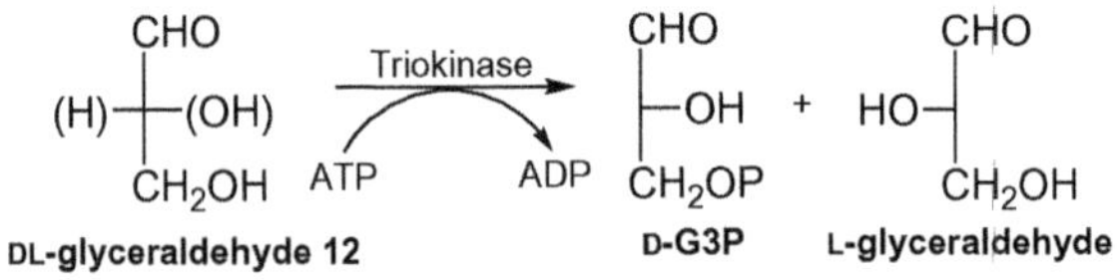

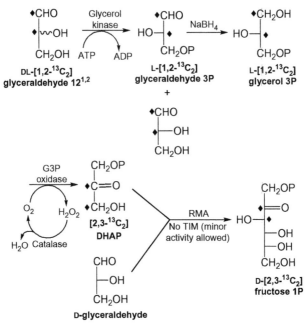

Scheme 21 The action of triokinase and ATP on D-glyceraldehyde to give D-G3P.

Scheme 22 Integration of glycerol kinase and glycerol 3P oxidase reactions to prepare D-[2,3-^{13}C$_2$]fructose 1P (top-half labeling).

Top-half labeling (C1–C3) of 2-ketohexoses can be accomplished as shown in Scheme 22. DL-[1,2-^{13}C$_2$]glyceraldehyde **12**^{1,2} is prepared chemically by CR (Scheme 19), and glycerol kinase converts the L-enantiomer into L-[1,2-^{13}C$_2$]G3P, which is separated by anion-exchange chromatography from the unphosphorylated D-enantiomer. L-[1,2-^{13}C$_2$]G3P is chemically reduced to L-[1,2-^{13}C$_2$]glycerol 3P which serves as a substrate for glycerol 3P oxidase (E.C. 1.1.3.21) to give [2,3-^{13}C$_2$]DHAP. The labeled ketotriose is then condensed with unlabeled D-glyceraldehyde (produced chemically from D-fructose; Perlin, 1962) in the presence of RMA to give one product, D-[2,3-^{13}C$_2$]fructose 1P. As shown in Scheme 18, TIM activity in the last step should be kept to a minimum. To drive the oxidase reaction to completion, catalase (E.C. 1.11.1.6) is added to the reaction mixture to convert H$_2$O$_2$ into O$_2$ and H$_2$O. While glycerol 3P dehydrogenase (E.C.

1.1.1.8) could be substituted for glycerol 3P oxidase, the former enzyme requires the coenzyme, NAD^+, which requires recycling. Use of the oxidase (commercially available) eliminates this complication.

Top- and bottom-half labeling of 2-ketohexoses mediated by RMA can be achieved in multiple ways depending on the desired labeling pattern in the product. The simplest route is shown in Scheme 23 and involves a *symmetrically labeled glycerol intermediate*. In the example given, DL-[1,3-$^{13}C_2$] glyceraldehyde **12**^{1,3} is chemically reduced to [1,3-$^{13}C_2$]glycerol, which is acted on by glycerol kinase to give L-[1,3-$^{13}C_2$]glycerol 3P. The latter is treated with glycerol 3P oxidase, RMA, and TIM to give D-[1,3,4,6-$^{13}C_4$]FBP. Unlike the route shown in Scheme 22, TIM is required to generate the labeled G3P required by RMA. Key to this sequence is "symmetric labeling" in the DL-glyceraldehyde **12** starting material, whose reduction gives *a symmetrically labeled glycerol*. Since the latter triol is prochiral, lack of label symmetry would lead to label scrambling in the product ketose phosphate, as shown in Scheme 24. The need for labeling symmetry applies not only to ^{13}C (as shown), but also to other isotopes (e.g., ^2H and/or $^{17,18}O$).

A permutation of the route shown in Scheme 23 is shown in Scheme 25. In this pathway, label scrambling complications arising from the use of "asymmetrically labeled" DL-glyceraldehyde (e.g., Scheme 24) are eliminated by employing glycerol kinase early in the pathway to resolve the glyceraldehyde enantiomers. In the illustrated route, DL-[1,2-$^{13}C_2$]glyceraldehyde is converted to L-[1,2-$^{13}C_2$]G3P, which, after separation from the unphosphorylated D-glyceraldehyde enantiomer, is chemically reduced

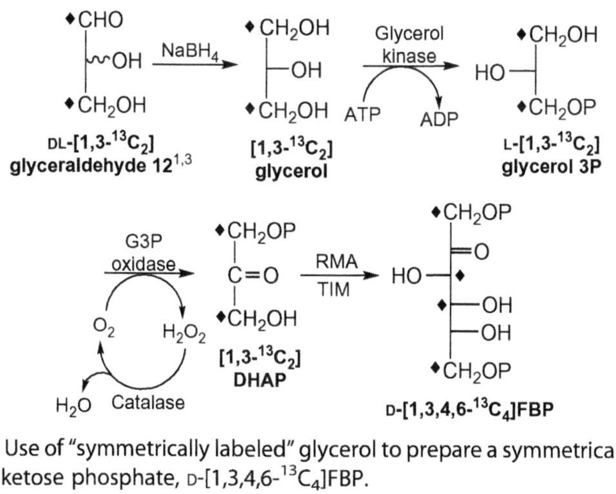

Scheme 23 Use of "symmetrically labeled" glycerol to prepare a symmetrically, multiply ^{13}C-labeled ketose phosphate, D-[1,3,4,6-$^{13}C_4$]FBP.

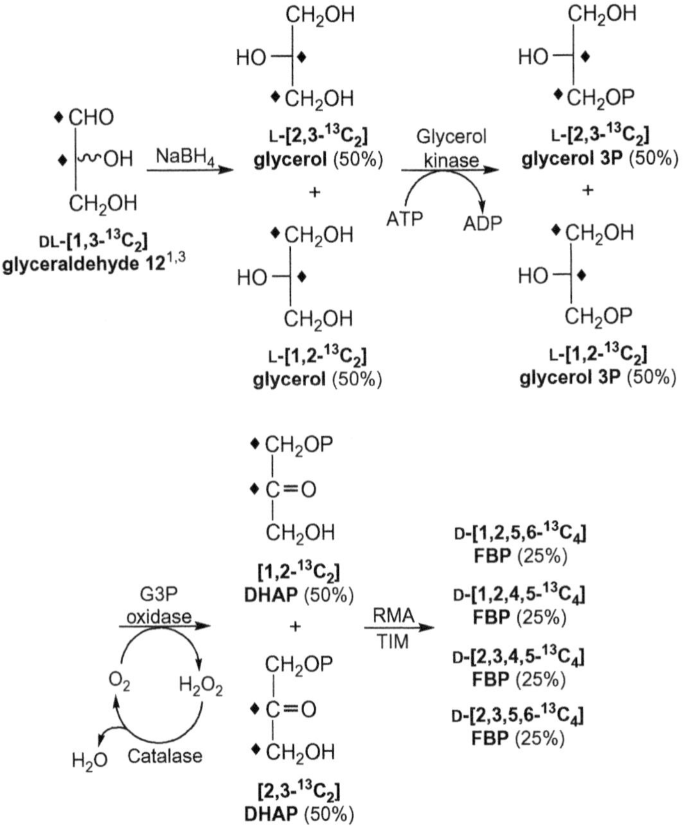

Scheme 24 Illustration of ^{13}C-label scrambling in the synthesis of tetra-^{13}C-labeled isotopomers of FBP when disymmetrically labeled glycerols are employed.

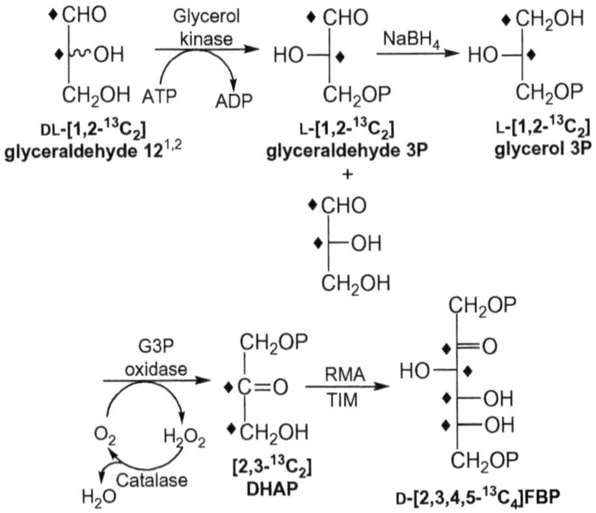

Scheme 25 The use of glycerol kinase, NaBH$_4$ reduction, and glycerol 3P oxidase to prepare a single symmetrically, multiply ^{13}C-labeled ketose phosphate, D-[2,3,4,5-^{13}C$_4$]FBP.

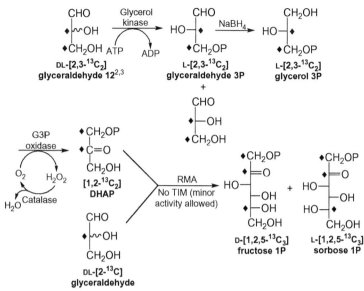

Scheme 26 Chemi-enzymic route to prepare the asymmetrically ^{13}C-labeled ketose phosphates, D-[1,2,5-^{13}C$_3$]fructose 1P and L-[1,2,5-^{13}C$_3$]sorbose 1P.

to L-[1,2-^{13}C$_2$]glycerol 3P. The latter is converted to D-[2,3,4,5-^{13}C$_4$]FBP by the concerted action of glycerol 3P oxidase, catalase, RMA, and TIM. This route, like that shown in Scheme 23, produces "symmetrically labeled" 2-ketohexoses.

Further elaboration of the general labeling strategies shown in Schemes 22, 23, and 25 is shown in Scheme 26, where "asymmetric labeling" of the product 2-ketohexose is achieved. In this route, DL-[2,3-^{13}C$_2$]glyceraldehyde **12**^{2,3} is treated with glycerol kinase to give L-[2,3-^{13}C$_2$]G3P and unphosphorylated D-[2,3-^{13}C$_2$]glyceraldehyde. After purification of the phosphoester, chemical reduction with NaBH$_4$ gives L-[2,3-^{13}C$_2$]glycerol 3P, which is converted to [1,2-^{13}C$_2$]DHAP with glycerol 3P oxidase and condensed with DL-[2-^{13}C] glyceraldehyde in the presence of RMA to give D-[1,2,5-^{13}C$_3$]fructose 1P and L-[1,2,5-^{13}C$_3$]sorbose 1P. To achieve optimal yields, TIM activity should be eliminated, but like before (Scheme 19), this activity can be tolerated. Further processing of the D-fructose 1P and L-sorbose 1P may involve treatment with acid phosphatase (E.C. 3.1.3.2), separation of the 2-ketohexoses by chromatography, and/or the application of chemical or enzymic reactions to give the desired products.

It should be noted that, since RMA exhibits broad specificity with regard to the aldose substrate, the reaction routes shown in Schemes 22 and 26 can be modified to produce, for example, 2-ketopentoses if glycolaldehyde **11** is used instead of D- or L-glyceraldehyde in the RMA reaction. Thus, for example, using [1-^{13}C]glycolaldehyde **11**^{1} instead of DL-[2-^{13}C]glyceraldehyde in the RMA-catalyzed step in Scheme 26 would give only one product, D-[1,2,5-^{13}C$_3$]xylulose 1P.

9. MANIPULATION OF ISOTOPICALLY LABELED D-FRUCTOSE 17 AND L-SORBOSE 25

The enzymic routes shown in Schemes 18–19 and 22–26 use RMA to catalyze stereospecific aldol condensation between DHAP and G3P or glyceraldehyde. Absolute configurations at C3 and C4 in the 2-ketohexose products are specified by the enzyme, such that only D-*fructo* and/or L-*sorbo* configurations (C3–C4 D-*threo*) form. Analogous DHAP-aldolases have been isolated and characterized that produce the other three C3–C4 absolute configurations (D-*erythro*; L-*threo*; L-*erythro*) (Falcicchio, Wolterink-Van Loo, Franssen, & van der Oost, 2014), and thus the other six 2-ketohexoses (L-*fructo*; D-*sorbo*; D-*psico*; L-*psico*; D-*tagato*; L-*tagato*) are accessible enzymically via the same strategies discussed for RMA. Because RMA is the most common Class I aldolase and is commercially available at reasonable cost, there is a strong incentive to use this enzyme in the three-carbon precursor assembly of 2-ketohexoses. In addition, the D-*fructo* and L-*sorbo* products can be converted into other saccharides, as shown in Scheme 27. After dephosphorylation, D-fructose **17** is converted enzymically into D-glucose **1** by glucose (xylose) isomerase (E.C. 5.3.1.5). L-Sorbose **25** is reduced to L-sorbitol (D-glucitol) chemically, and the latter oxidized to D-fructose by D-sorbitol dehydrogenase (E.C. 1.1.1.14), with the required NAD$^+$ coenzyme recycled using the lactate dehydrogenase system. Note that in the conversion of **25** to **17**, the carbon chain is inverted; for example, ^{13}C labeling at C1 and C2 of **25** gives **17** with ^{13}C-labeling at C5 and C6.

10. CONCLUDING REMARKS

Pioneering studies in the latter half of the twentieth century involving stable isotopically labeled carbohydrates revealed substantial shortcomings in the laboratory methods available for isotope labeling. Since then, two core developments have occurred that simplify the isotopic labeling of saccharides significantly: (1) the introduction of new high-yield isotope insertion reactions, CR and MCE, that are simple to perform, reliable, and scalable; and (2) increased commercial availability of carbohydrate-transforming

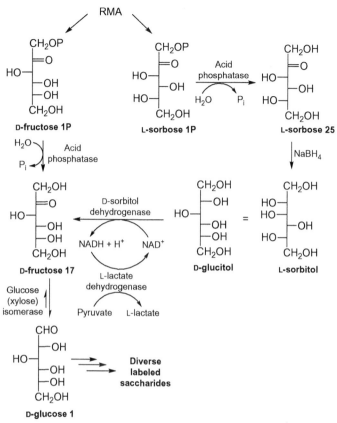

Scheme 27 Enzymic routes to recycle D-fructose 1P and L-sorbose 1P.

enzymes at an affordable cost. These developments have greatly improved access to, and the structural diversity of, isotopically labeled monosaccharides. These innovations have also facilitated the conversion of labeled monosaccharides into an ever-widening range of biologically relevant labeled compounds that either retain their saccharide character (e.g., enzymic assembly into stable isotopically labeled oligosaccharides (Rosevear, Nunez, & Barker, 1982)) or serve as synthetic precursors in the preparation of other types of labeled biomolecules (e.g., ribo- and 2'-deoxyribonucleosides (Kline & Serianni, 1990; Bandyopadhyay, Wu, & Serianni, 1993; Wu & Serianni, 1994); vitamins (Drew et al., 1996)). While notable progress has been made over the past 50 years, new insertion and conversion reactions—chemical, enzymic, and biological—will continue to be discovered and developed in the coming years. At the same time, applications of stable isotopically labeled saccharides and their derivatives to scientific and technical problems are likely to grow, including fundamental structural and mechanistic studies, metabolic pathway elucidation (Meieer, Karlsson, Jensen, Lerche, & Duus, 2011), and noninvasive medical diagnostic testing (e.g., breath tests; Robayo-Torres et al., 2009).

It is not hyperbole to assert that current labeling technology for simple monosaccharides allows access to virtually any ^{13}C, ^{2}H, and/or ^{17,18}O isotopomer that can be imagined, although synthesis costs vary widely with the isotopomeric target. Cost still remains an impediment to more routine use of stable isotopically labeled saccharides in research. The two core insertion reactions described in this report, CR and MCE, have properties and efficiencies often associated with enzyme reactions, and thus are very well suited for the synthesis of labeled compounds. Both reactions are adaptable to many different starting aldoses and, importantly, are scalable to the kilogram level and greater. The most significant shortcoming of CR is its lack of stereospecificity, but methods are available to fully utilize both C2-epimeric aldose products, thus offsetting this shortcoming to a reasonable extent. Commercial production of D-[1-^{13}C]glucose **1**^{1} routinely involves individual CR reactions that are conducted with hundreds of grams of D-arabinose **4** and K^{13}CN, thus reducing the per-gram cost of this compound appreciably. Large-scale isolation and purification of labeled aldoses and other saccharides involve the use of reverse osmosis to rapidly and cost-effectively reduce the volume of aqueous solutions of labeled monosaccharides prior to their purification by chromatography. The CR and MCE reactions have also been used in cGMP syntheses of labeled monosaccharides destined for use in human clinical trials.

While current synthetic methods provide access to virtually any isotopomer of simple monosaccharides, the adage that "availability begets need" fails if cost to the research community is prohibitive. The cost of producing a saccharide isotopomer is largely determined by three factors:

(1) the cost of raw isotope and other reagents required in the synthesis
(2) the properties of the synthetic route (length; reliability; scalability; yield; ease of product purification; product purity requirements; isotopic enrichment requirements; options for recycling labeled by-products into desirable isotopomers)
(3) the cost of labor to perform the synthesis

The above discussion has focused heavily on factor (2) for good reason: Factor (2) is core to determining total cost because it influences the contributions made by factors (1) and (3). Lengthy reaction routes increase labor cost. Unreliable reactions introduce uncertainty, which is reflected in increased cost. Unscalable reactions increase per-gram cost. It is clear that suboptimal reaction routes are tightly coupled to increased cost. However, independent contributions to cost made by factor (1) are not insignificant, and future improvements in enrichment methods to produce, for example, K^{13}CN, will help reduce isotope costs. Enzymic transformations can be limited by the availability and cost of enzymes, but improvements in protein expression will reduce costs in the long run. It can be anticipated that "synthetic" enzymes based on nonpeptide scaffolds may eventually replace amino-acid-based polymeric catalysts. In many labeled syntheses, factor (3) is a major contributor to cost, and in some cases the dominant contributor. Keeping reaction routes short and simple minimizes the cost contribution from (3) by reducing synthesis times and minimizing the level of chemical expertise needed to conduct the synthesis. Likewise, it can be anticipated that further reductions in the contribution of labor to isotope cost will evolve from laboratory automation, and signs already indicate that automation will become more widespread as time evolves (Godfrey, Masquelin & Hemmerle, 2013). This automation will be most easily implemented for reaction routes that employ simple processes. Thus, there is a strong synergy between factors (2) and (3), as there is between factors (2) and (1).

REFERENCES

Angelotti, T., Krisko, M., O'Connor, T., & Serianni, A. S. (1987). [1-^{13}C]Aldono-1,4-lactones: Conformational studies based on ^{1}H-^{1}H, ^{13}C-^{1}H and ^{13}C-^{13}C spin couplings and *ab initio* molecular orbital calculations. *Journal of the American Chemical Society, 109*, 4464–4472.

Angyal, S. J., Bethell, G. S., & Beveridge, R. J. (1979). The separation of sugars and polyols on cation-exchange resin in the calcium form. *Carbohydrate Research, 73*, 9–18.

Bandyopadhyay, T., Wu, J., & Serianni, A. S. (1993). [1'-^{13}C]2'-Deoxyribonucleosides: Structural and conformational insights derived from ^{13}C-^{1}H spin-coupling constants involving C1'. *The Journal of Organic Chemistry, 58*, 5513–5517.

Barker, S. A., Chopra, A. K., Hatt, B. W., & Somers, P. J. (1973). The interaction of areneboronic acids with monosaccharides. *Carbohydrate Research, 26*, 33–40.

Barker, S. A., Hatt, B. W., & Somers, P. J. (1973). The effect of areneboronic acids on the alkaline conversion of D-glucose to D-fructose. *Carbohydrate Research, 26*, 41–53.

Behrens, P. W., Sicotte, V. J., & Delente, J. (1994). Microalgae as a source of stable isotopically labeled compounds. *Journal of Applied Phycology, 6*, 113–121.

Bilik, V., Petrus, L., & Farkas, V. (1975). Reactions of saccharides catalyzed by molybdate ions. XV. Mechanism of the epimerization reaction. *Chemicke Zvesti, 29*, 690–693.

Bilik, V., & Stankovic, L. (1973). Reactions of saccharides catalyzed by molybdate ions. VI. Epimerization of aldotetroses. *Chemicke Zvesti, 27*, 544–546.

Chandrasegaran, S., Kan, L. S., Sillerud, L. O., Skoglund, C., & Bothner-By, A. A. (1985). Isolation and purification of deoxyribonucleosides from 90% ^{13}C-enriched DNA of algal cells and their characterization by ^{1}H and ^{13}C NMR. *Nucleic Acids Research, 13*, 2097–2110.

Clark, E. L., & Barker, R. (1986). General methods for enriching aldoses with oxygen isotopes. *Carbohydrate Research, 153*, 253–261.

Clark, E. L., Hayes, M. L., & Barker, R. (1986). Paramolybdate anion-exchange resin, an improved catalyst for the C1-C2 rearrangement and 2-epimerization of aldoses. *Carbohydrate Research, 153,* 263–270.

Drew, K. N., Church, T. J., Basu, B., Vuorinen, T., & Serianni, A. S. (1996). L-[1-^{13}C]- and [2-^{13}C]ascorbic acid: Synthesis and NMR characterization. *Carbohydrate Research, 284,* 135–143.

Falcicchio, P., Wolterink-Van Loo, S., Franssen, M. C. R., & van der Oost, J. (2014). DHAP-dependent aldolases from (hyper)thermophiles: Biochemistry and applications. *Extremophiles, 18,* 1–13.

Fischer, E. (1889). Reduction von säuren der zuckergruppe. *Berichte der Deutschen Chemischen Gesellschaft, 22,* 2204–2206.

Frandsen, E. K., & Grunnet, N. (1971). Kinetic properties of triokinase from rat liver. *European Journal of Biochemistry, 23,* 588–592.

Freimund, S., Huwig, A., Giffhorn, F., & Köpper, S. (1998). Rare keto-aldoses from enzymatic oxidation: Substrates and oxidation products of pyranose 2-oxidase. *Chemistry—A European Journal, 4,* 2442–2455.

Godfrey, A. G., Masquelin, T., & Hemmerle, H. (2013). A remote-controlled adaptive medchem lab: An innovative approach to enable drug discovery in the 21st century. *Drug Discovery Today, 18,* 795–802.

Hayashi, S., & Lin, E. C. C. (1967). Purification and properties of glycerol kinase from *Escherichia coli. The Journal of Biological Chemistry, 242,* 1030–1035.

Hayes, M. L., Pennings, N. J., Serianni, A. S., & Barker, R. (1982). Epimerization of aldoses by molybdate involving a novel rearrangement of the carbon skeleton. *Journal of the American Chemical Society, 104,* 6764–6769.

Hodge, J. E., & Hofreiter, B. T. (1962). Determination of reducing sugars and carbohydrates. In R. L. Whistler & M. L. Wolfrom (Eds.), *Methods in Carbohydrate Chemistry 1* (pp. 380–394). New York: Academic Press.

Hricoviniova-Bilikova, Z., Hricovini, M., Petrusova, M., Serianni, A. S., & Petrus, L. (1999). Stereospecific molybdic acid-catalyzed isomerization of 2-hexuloses to branched-chain aldoses. *Carbohydrate Research, 319,* 38–46.

Kiliani, H. (1885). Ueber das cyanhydrin der lävulose. *Berichte der Deutschen Chemischen Gesellschaft, 18,* 3066–3072.

King-Morris, M. J., Bondo, P. B., Mrowca, R. A., & Serianni, A. S. (1988). Stable, isotopically substituted carbohydrates: An improved synthesis of (6-^{13}C)aldohexoses. *Carbohydrate Research, 175,* 49–58.

Klepach, T., Carmichael, I., & Serianni, A. S. (2008). ^{13}C-Labeled N-acetyl-neuraminic acid in aqueous solution: Detection and quantification of acyclic *keto, keto* hydrate, and *enol* forms by ^{13}C NMR spectroscopy. *Journal of the American Chemical Society, 130,* 11892–11900.

Klepach, T., Zhang, W., Carmichael, I., & Serianni, A. S. (2008). ^{13}C-^{1}H and ^{13}C-^{13}C NMR J-couplings in ^{13}C-labeled N-acetyl-neuraminic acid: Correlations with molecular structure. *The Journal of Organic Chemistry, 73,* 4376–4387.

Kline, P. C., & Serianni, A. S. (1990). ^{13}C-Enriched ribonucleotides: Synthesis and application of ^{13}C-^{1}H and ^{13}C-^{13}C spin-coupling constants to assess furanose and N-glycoside bond conformations. *Journal of the American Chemical Society, 112,* 7373–7381.

Lin, C.-C., Lin, C.-H., & Wong, C.-H. (1997). Sialic acid aldolse-catalyzed condensation of pyruvate and N-substituted mannosamine: A useful method for the synthesis of N-substituted sialic acids. *Tetrahedron, 38,* 2649–2652.

Mega, T. L., & Vann Etten, R. L. (1990). Oxygen exchange and bond cleavage reactions of carbohydrates studies using the ^{18}O isotope shift in ^{13}C NMR spectroscopy. In J. W. Finley, S. J. Schmidt, & A. S. Serianni (Eds.), *NMR Applications in Biopolymers* (pp. 85–93). New York: Plenum.

Meieer, S., Karlsson, M., Jensen, P. R., Lerche, M. H., & Duus, J. Ø. (2011). Metabolic pathway visualization in living yeast by DNP-NMR. *Molecular BioSystems*, 7, 2834–2836.

Perlin, A. S. (1962). D-, L- and DL-glyceraldehyde. Oxidative degradation of ketohexoses. In R. L. Whistler & M. L. Wolfrom (Eds.), *Methods in Carbohydrate Chemistry 1* (pp. 61–63). New York: Academic Press.

Robayo-Torres, C. C., Opekun, A. R., Quezada-Calvillo, R., Xavier, V., O'Brian Smith, E., Navarrete, M., et al. (2009). ^{13}C-Breath tests for sucrose digestion in congenital sucrase isomaltase deficient and sacrosidase supplemented patients. *Journal of Pediatric Gastroenterology and Nutrition*, 48, 412–418.

Rodrigues, J. R., Couto, A., Cabezas, A., Pinto, R. M., Ribeiro, J. M., Canales, J., et al. (2014). Bifunctional homodimeric triokines/FMC cyclase. *The Journal of Biological Chemistry*, 289, 10620–10636.

Rosevear, P. R., Nunez, H. A., & Barker, R. (1982). Synthesis and solution conformation of the type 2 blood group oligosaccharide αLFuc(1→2)βDGal(1→4)βDGlcNAc. *Biochemistry*, 21, 1421–1431.

Scensny, P., Hirschhorn, S. G., & Rasmussen, J. R. (1983). Convenient synthesis of 3-deoxy-D-*ribo*-hexose and 3-deoxy-D-*arabino*-hexose. *Carbohydrate Research*, 112, 307–312.

Serianni, A. S., & Barker, R. (1979). Isotopically-enriched carbohydrates: The preparation of [^{2}H]-enriched aldoses by catalytic hydrogenolysis of cyanohydrins with ^{2}H$_2$. *Canadian Journal of Chemistry*, 57, 3160–3167.

Serianni, A. S., Cadman, E., Pierce, J., Hayes, M. L., & Barker, R. (1982). Enzymic synthesis of ^{13}C-enriched aldoses, ketoses, and their phosphate esters. *Methods in Enzymology*, 89, 83–92.

Serianni, A. S., Clark, E. L., & Barker, R. (1979). Carbon-13 enriched carbohydrates. Preparation of erythrose, threose, glyceraldehyde and glycolaldehyde with ^{13}C enrichment in various carbon atoms. *Carbohydrate Research*, 72, 79–91.

Serianni, A. S., Nunez, H. A., & Barker, R. (1979). Carbon-13 enriched carbohydrates. Preparation of aldononitriles and their reduction with a palladium catalyst. *Carbohydrate Research*, 72, 71–78.

Serianni, A. S., Nunez, H. A., & Barker, R. (1980). Cyanohydrin synthesis: Studies with [^{13}C]cyanide. *The Journal of Organic Chemistry*, 45, 3329–3341.

Serianni, A. S., Pierce, J., & Barker, R. (1979). Carbon-13 enriched carbohydrates: Preparation of triose, tetrose and pentose phosphates. *Biochemistry*, 18, 1192–1199.

Snyder, J., & Serianni, A. S. (1987a). Synthesis and NMR spectral analysis of unenriched and [1-^{13}C]enriched 5-deoxypentoses and 5-O-methylpentoses. *Carbohydrate Research*, 163, 169–188.

Snyder, J., & Serianni, A. S. (1987b). DL-Apiose substituted with stable isotopes: Synthesis, NMR spectral analysis, and furanose anomerization. *Carbohydrate Research*, 166, 85–99.

Stafford, U., Serianni, A. S., & Varma, A. (1990). Microcomputer-automated reactor for synthesis of ^{13}C-labeled monosaccharides. *AICHE Journal*, 36, 1822–1828.

Taylor, M. E., & Drickamer, K. (2011). *Introduction to Glycobiology*. New York: Oxford University Press.

Vuorinen, T., & Serianni, A. S. (1990a). ^{13}C-Substituted pentos-2-uloses: Synthesis and analysis by ^{1}H and ^{13}C NMR spectroscopy. *Carbohydrate Research*, 207, 185–210.

Vuorinen, T., & Serianni, A. S. (1990b). Synthesis of D-*erythro*-2-pentulose and D-*threo*-2-pentulose and analysis of the ^{13}C and ^{1}H NMR spectra of the (1-^{13}C)- and (2-^{13}C)-substituted sugars. *Carbohydrate Research*, 209, 13–31.

Walker, T. E., London, R. E., Whaley, T. W., Barker, R., & Matwiyoff, N. A. (1976). Carbon-13 nuclear magnetic resonance of [1-^{13}C]-enriched monosaccharides. Signal assignments and orientational dependence of geminal and vicinal carbon-carbon and carbon–hydrogen spin-spin coupling constants. *Journal of the American Chemical Society*, *98*, 5807–5813.

Wolfrom, M. L., & Wood, H. B. (1951). Sodium borohydride as a reducing agent for sugar lactones. *Journal of the American Chemical Society*, *73*, 2933–2934.

Wu, Q., Pan, Q., Zhao, S., Imker, H., & Serianni, A. S. (2007). A disaccharide rearrangement catalyzed by molybdate anion in aqueous solution. *The Journal of Organic Chemistry*, *72*, 3081–3084.

Wu, J., & Serianni, A. S. (1994). ^{13}C-Labeled oligodeoxyribonucleotides: A solution study of a CCAAT-containing sequence at the nuclear factor I recognition site of human adenovirus. *Biopolymers*, *34*, 1175–1186.

Wu, J., Serianni, A. S., & Bondo, P. B. (1992). Multiply-^{13}C-substituted monosaccharides: Synthesis of D-[1,5,6-^{13}C$_3$]glucose and D-(2,5,6-^{13}C$_3$)glucose. *Carbohydrate Research*, *226*, 261–269.

Zhang, W., Carmichael, I., & Serianni, A. S. (2011). Rearrangement of 3-deoxy-D-*erythro*-hexos-2-ulose in aqueous solution: NMR evidence of intramolecular 1,2-hydrogen transfer. *The Journal of Organic Chemistry*, *76*, 8151–8158.

Zhang, W., & Serianni, A. S. (2012). Phosphate-catalyzed degradation of D-glucosone in aqueous solution is accompanied by C1-C2 transposition. *Journal of the American Chemical Society*, *134*, 11511–11524.

Zhao, S., Petrus, L., & Serianni, A. S. (2001). 1-Deoxy-D-xylulose: Synthesis based on molybdate-catalyzed rearrangement of a branched-chain aldotetrose. *Organic Letters*, *3*, 3819–3822.

Zhu, Y., Pan, Q., Thibaudeau, C., Zhao, S., Carmichael, I., & Serianni, A. S. (2006). [^{13}C,^{15}N]2-Acetamido-2-deoxy-D-aldohexoses and their methyl glycosides: Synthesis and NMR investigations of *J*-couplings involving ^{1}H, ^{13}C and ^{15}N. *The Journal of Organic Chemistry*, *71*, 466–479.

Zhu, Y., Zajicek, J., & Serianni, A. S. (2001). Acyclic forms of [1-^{13}C]aldohexoses in aqueous solution: Quantitation by ^{13}C NMR and deuterium isotope effects on tautomeric equilibria. *The Journal of Organic Chemistry*, *66*, 6244–6251.

SECTION IV

RNA Labeling

CHAPTER EIGHTEEN

Stable Isotope-Labeled RNA Phosphoramidites to Facilitate Dynamics by NMR

Christoph H. Wunderlich[*,1], Michael A. Juen[*,1], Regan M. LeBlanc[†,1], Andrew P. Longhini[†,1], T. Kwaku Dayie[†,1,2], Christoph Kreutz[*,1,2]

[*]Institute of Organic Chemistry and Center for Biomolecular Sciences Innsbruck, University of Innsbruck, Innsbruck, Austria
[†]Department of Chemistry and Biochemistry, Center for Biomolecular Structure and Organization, University of Maryland, College Park, Maryland, USA
[1]*Disclosures*: CH Wunderlich none, MA Juen none, C Kreutz, none; RM LeBlanc none, AP Longhini none, TK Dayie, none.
[2]Corresponding authors: e-mail address: dayie@umd.edu; christoph.kreutz@uibk.ac.at

Contents

1. Theory	462
2. Equipment	463
3. Materials	464
3.1 Solutions and Buffers	466
4. Protocol	467
4.1 Duration	467
4.2 Preparation	467
4.3 Caution	467
5. Step 1: Synthesis of 6-^{13}C-Uridine TOM Phosphoramidite	468
5.1 Overview	468
5.2 Duration	468
6. Step 2: Synthesis of 6-^{13}C-Cytidine TOM Phosphoramidite	473
6.1 Overview	473
6.2 Duration	473
7. Step 3: Chemical RNA Synthesis	477
7.1 Overview	477
7.2 Duration	479
8. Step 4: Applications	484
8.1 Overview	484
8.2 Segmental Stable Isotope Labeling Using Chemically Synthesized RNAs	484
8.3 ^{13}C CPMG RD NMR Spectroscopy	485
8.4 ^{13}C ZZ Exchange NMR Spectroscopy	487
8.5 Chemical Exchange Saturation Transfer	488

9. Conclusions	490
Acknowledgments	491
References	492

Abstract

Given that Ribonucleic acids (RNAs) are a central hub of various cellular processes, methods to synthesize these RNAs for biophysical studies are much needed. Here, we showcase the applicability of 6-^{13}C-pyrimidine phosphoramidites to introduce isolated ^{13}C-^{1}H spin pairs into RNAs up to 40 nucleotides long. The method allows the incorporation of 6-^{13}C-uridine and -cytidine residues at any desired position within a target RNA. By site-specific positioning of the ^{13}C-label using RNA solid phase synthesis, these stable isotope-labeling patterns are especially well suited to resolve resonance assignment ambiguities. Of even greater importance, the labeling pattern affords accurate quantification of important functional transitions of biologically relevant RNAs (e.g., riboswitch aptamer domains, viral RNAs, or ribozymes) in the μs- to ms time regime and beyond without complications of one bond carbon scalar couplings. We outline the chemical synthesis of the 6-^{13}C-pyrimidine building blocks and their use in RNA solid phase synthesis and demonstrate their utility in Carr Purcell Meiboom Gill relaxation dispersion, ZZ exchange, and chemical exchange saturation transfer NMR experiments.

1. THEORY

Ribonucleic acid (RNA) is well recognized as a central player in key biological processes such as signaling and gene regulation, catalysis, or viral infections (Blount & Breaker, 2006; Breaker, 2011; Coppins, Hall, & Groisman, 2007; Haller, Souliere, & Micura, 2011; Reining et al., 2013; Serganov & Nudler, 2013). NMR spectroscopy has made significant contributions in revealing molecular details underlying these processes (Al-Hashimi, 2007; Al-Hashimi & Walter, 2008; Campbell, Bouchard, Desjardins, & Legault, 2006; Dayie, 2012; Fürtig, Buck, Richter, & Schwalbe, 2012; Hennig, Williamson, Brodsky, & Battiste, 2001; Johnson & Hoogstraten, 2008; Kloiber, Spitzer, Tollinger, Konrat, & Kreutz, 2011; Latham, Brown, McCallum, & Pardi, 2005; Mittermaier & Kay, 2009; Rinnenthal et al., 2011; Wenter, Reymond, Auweter, Allain, & Pitsch, 2006; Zhang, Sun, Watt, & Al-Hashimi, 2006). The first mandatory step to study both structure and dynamics of biomacromolecules is the introduction of a stable isotope-labeling pattern (Dayie, 2008; Hennig et al., 2001; Lu, Miyazaki, & Summers, 2010). Currently, enzymatic methods to introduce uniformly ^{13}C- and/or ^{15}N-labeled nucleotides into RNA and DNA are state-of-the-art (Hennig et al., 2001; Milligan, Groebe,

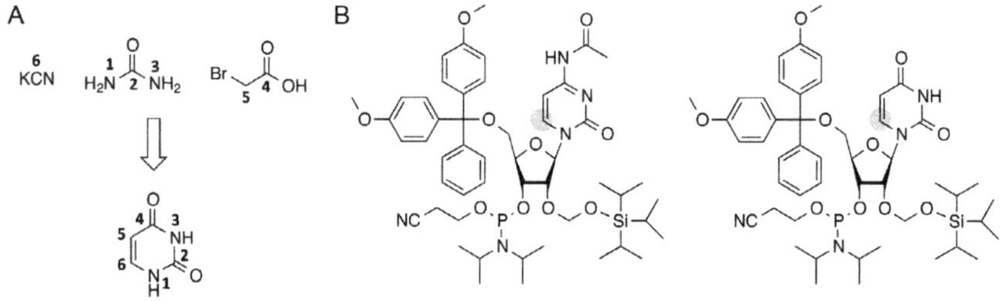

Figure 1 (A) All heteronuclei can be addressed for stable isotope labeling in uracil by choosing the appropriate precursor compounds. (B) 6-^{13}C-cytidine (left) and -uridine (right) triisopropylsilyl-oxmethylprotected RNA phosphoramidites. Orange (gray in the print version) circle: ^{13}C.

Witherell, & Uhlenbeck, 1987; Milligan & Uhlenbeck, 1989). Using these methods, nucleotide-specific-labeling patterns (i.e., only adenosines/uridines or guanosines/cytidines) can be applied in a straightforward manner by mixing labeled and unlabeled (deoxy-) ribonucleotide triphosphates (rNTPs) in polymerase-catalyzed reactions. However, the nature of the stable isotope-labeling pattern can limit the type of NMR experiments that can be run (Wunderlich et al., 2012). An isolated X-^{1}H (X = ^{13}C or ^{15}N) spin topology is particularly important for obtaining accurate parameters from the very powerful relaxation dispersion (RD) and chemical exchange saturation transfer (CEST) NMR experiments that address μs- to ms RNA dynamics. Our labeling strategy eliminates the deleterious strong homonuclear scalar coupling and therefore provides artifact-free dispersion profiles. Here, we outline a robust synthetic access to selectively labeled 6-^{13}C-pyrimidine RNA phosphoramidites using any combination of the following: ^{15}N1, ^{15}N3, and ^{13}C2 from ^{13}C–^{15}N-labeled urea; ^{13}C4 and ^{13}C5 from ^{13}C-labeled bromoacetic acid; and ^{13}C6 from K^{13}CN (Fig. 1A and B).

2. EQUIPMENT

0.22-μm cellulose acetate filters
2-ml Eppendorf tubes
600/800 MHz NMR instrument equipped with ^{1}H/^{13}C/^{31}P probe with gradients
Analytical Dionex DNAPac PA-100 column 4 × 250mm
Balloons with a wall thickness of at least 0.3 mm
Bent adapters with NS-stopcocks
C18 SepPak cartridges (Waters)
Dewar
DNA/RNA synthesizer

Freezer at −20 °C
Glass chromatography columns
High-pressure liquid chromatography system with a column oven (e.g., Thermo Fisher Ultimate 3000)
High vacuum rotary vane pump
Magnetic stirrer with heating and an oil bath or heat block
Magnetic stirring bar
Medium pressure liquid chromatography system (e.g., ÄKTA start)
NMR tubes (5 or 3 mm) suitable for 600 and 800 MHz NMR spectrometer
Parafilm
pH meter and electrode
Reflux condenser
Rotary evaporator with a diaphragm pump
Round-bottom flasks with volumes ranging from 25 to 1000 ml
Semipreparative Dionex DNAPac PA-100 column 9 × 250mm
Separatory funnel
Silica thin layer chromatography plates with fluorescence indicator
Suction filter
Syringes (1, 10, 20, and 50 ml volume)
UV/vis spectrophotometer
UV/vis cuvettes

3. MATERIALS

(4,4′-Dimethoxytriphenyl)methyl chloride (DMT-Cl)
[(Triisopropylsilyl)oxy]methyl chloride (TOM-Cl)
1-O-Acetyl-(2′,3′,5′-O-tribenzoyl)-β-D-ribofuranose (ATBR)
1,2-Dichloroethane
1H-benzylthiotetrazole
2-Cyanoethyl N,N-diisopropylchlorophosphoramidite
2,4,6-Triisopropylbenzenesulfonyl chloride (TPS-Cl)
28% Aqueous ammonia solution
4-(Dimethylamino)pyridine (DMAP)
Acetic acid
Acetic anhydride
Acetonitrile anhydrous
Acetonitrile DNA synthesis grade

Argon
Bis(trimethylsilyl)acetamide
$CDCl_3$
Celite
Citric acid
D_2O
Dibutyltin dichloride (or alternatively: di-*tert*-butyltin dichloride)
Dichloroacetic acid
Dimethylformamide anhydrous
Ethanol
Ethyl acetate
Hexanes (mixture of isomers)
Iodine
Methanol
Methylamine solution in ethanol (8 M)
Methylamine solution in water (30%)
Methylene chloride
Molecular sieve (4 Å)
N-methyl-2-pyrrolidone
N,N-diisopropylethylamine (DIPEA)
N,N-dimethylethylamine (DMEA)
Pyridine anhydrous
RNA support for solid phase synthesis (e.g., Custom primer support from GE healthcare)
Silica gel for column chromatography pore size 60 Å, ≥440 mesh particle size
Sodium acetate trihydrate
Sodium bicarbonate
Sodium perchlorate hydrate
Sodium sulfate anhydrous
Sym-collidine
Tetrabutylammonium fluoride trihydrate (TBAF)
Tetrahydrofuran (containing butylhyroxytoluene (BHT) as stabilizer)
Toluene
Triethylamine
Trimethylsilyl trifluoromethylsulfonate (TMSOTf)
Trizma base
Unlabeled [(triisopropylsilyl)oxy] methyl (TOM) protected RNA phosphoramidites (e.g., from ChemGenes)

Uracils (unlabeled and in various stable isotope-labeled forms by chemical synthesis)
Urea

3.1 Solutions and Buffers

Saturated sodium bicarbonate solution:
    Add 200 g sodium bicarbonate to 1000 ml deionized water.
5% Citric acid
    Dissolve 50 g citric acid in 1000 ml deionized water.
Cap A solution:
    Dissolve 5 g DMAP in 50 ml acetonitrile.
Cap B solution:
    Mix 25 ml acetonitrile with 10 ml acetic anhydride and 15 ml *sym*-collidine.
Oxidation solution:
    Dissolve 250 mg iodine in 25 ml tetrahydrofuran. Then add 10 ml pyridine and 5 ml deionized water.
Detritylation solution:
    Mix 10 ml dichloroacetic acid with 240 ml 1,2-dichloroethane.
100 mM sodium acetate solution:
    Dissolve 13.6 g sodium acetate trihydrate in 1000 ml HPLC grade water.

Dionex column buffer A:

Component	Final concentration	Stock	Amount
Tris–HCl buffer, pH 8.0	25 mM	250 mM	100 ml
urea	6 M	–	360 g

Add water to 1-l. filter through 0.22-µm hydrophilic membrane.

Dionex column buffer B:

Component	Final concentration	Stock	Amount
Tris–HCl buffer, pH 8.0	25 mM	250 mM	100 ml
urea	6 M	–	360 g
Sodium perchlorate hydrate	500 mM	–	70.2 g

Add water to 1-l filter through 0.22-µm hydrophilic membrane.

4. PROTOCOL
4.1 Duration

Preparation	1–3 weeks
Step 1	5–7 days
Step 2	5–7 days
Step 3	5–7 days
Step 4	3 days

4.2 Preparation

Uracils with various stable isotope-labeling patterns are accessible via chemical synthesis as reported earlier (Alvarado, Longhini, et al., 2014). Starting from potassium cyanide and 2-bromoacetic acid, the cyano acetylurea precursor is obtained. In the final step, uracil is formed under reductive reaction conditions using palladium on barium sulfate under a hydrogen atmosphere. Prepare this key intermediate using the previously reported methods to proceed directly with step 1 (Fig. 2).

Purchase the unlabeled TOM protected RNA and DNA phosphoramidites along with the synthesis reagents (e.g., water free (<5 ppm) acetonitrile, 1*H*-benzylthiotetrazole (BTT) from a commercial supplier (e.g., ChemGenes, Glen Research). It is advisable to carry out test oligonucleotide syntheses to optimize the coupling, oxidation, and capping steps. For that purpose, DNA oligonucleotide synthesis should be used before switching to the more expensive RNA phosphoramidite chemistry for fine-tuning.

4.3 Caution

Please check the safety data sheets of the synthesis reagents before starting. Some of these reagents are very hazardous and should be handled by experienced chemists with extreme care in a well-vented fume hood. Phosphoramidites are rather sensitive chemical compounds and are degraded in the presence of acids and water or elevated temperatures. Phosphoramidite solutions (prepared using water-free acetonitrile) should be stored over freshly activated molecular sieve and are stable for several weeks at $-20\ °C$.

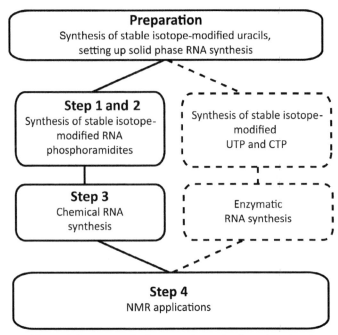

Figure 2 Workflow of the protocol. The part shown in dotted outline is an alternate extension to chemoenzymatic labeling using the selectively stable isotope-modified uracils (Alvarado, LeBlanc, et al., 2014; Alvarado, Longhini, et al., 2014).

5. STEP 1: SYNTHESIS OF 6-^{13}C-URIDINE TOM PHOSPHORAMIDITE

5.1 Overview

Transformations a, b, and c (Fig. 3) were reported earlier (Alvarado, Longhini, et al., 2014). Using the appropriate precursor composition, any heteronucleus of the uracil can be specifically enriched. The stable isotope-modified uracil key intermediate is coupled to a ribose building bock under *Vorbrüggen* nucleosidation conditions (d) (Fig. 3). Subsequent removal of the benzyol protecting groups (e) followed by tritylation (f) gives the precursor for the protection of the 2′-hydroxyl group using the TOM group (g) (Fig. 3). In the last step, the phosphoramidite is obtained (h) (Fig. 3).

5.2 Duration

5–7 days

    1.1. Synthesis of 6-^{13}C-2′,3′,5′-O-tribenzoyluridine: In a 100 ml round-bottom flask, suspend 6-^{13}C-uracil (560 mg, 5 mmol) together with

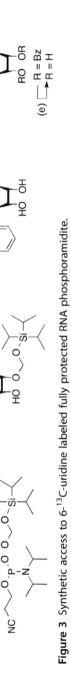

Figure 3 Synthetic access to 6-^{13}C-uridine labeled fully protected RNA phosphoramidite.

ATBR (2.5 g, 5 mmol) in 20 ml dry acetonitrile and evaporated to dryness. Repeat this procedure two times.

1.2. Add 20 ml dry acetonitrile and heat the suspension to 60 °C using a reflux condenser and an oil bath or a heat block.

1.3. Then add bis(trimethylsilyl)acetamide (3.7 ml, 15 mmol) and stir the reaction mixture for 30 min at 60 °C. The suspension should turn into a clear solution.

1.4. Finally, add TMSOTf (3.2 ml, 17.5 mmol) and continue stirring for another 30 min at 60 °C.

1.5. The reaction progress can be monitored using TLC (ethyl acetate/hexanes 1/1, R_f (product) = 0.8).

1.6. After that the reaction mixture is evaporated to yield an orange brown oil which is dissolved in methylene chloride and subsequently washed with saturated sodium bicarbonate solution. The organic phase is dried over sodium sulfate, filtered, and evaporated to dryness to give a yellow solid. The compound can be directly used in the next step.

1.7. Synthesis of 6-^{13}C-uridine: In a 100-ml round-bottom flask crude 6-^{13}C-2′,3′,5′-O-tribenzoyluridine (2.5 g, 4.5 mmol) is dissolved in an ethanolic solution of methylamine (15 ml, 8 M) and an argon atmosphere and stirred at room temperature for 16 h.

1.8. The reaction progress can be monitored using TLC (CH_2Cl_2/methanol 9/1, R_f (product) = 0.1).

1.9. Then, the solution is evaporated to yield a yellow oil, which is dissolved in water and repeatedly extracted with methylene chloride to remove the N-methylbenzamide from the aqueous phase.

1.10. The aqueous phase containing only 6-^{13}C-uridine is then evaporated to dryness resulting in a yellow foam.

1.11. To remove traces of impurities, recrystallization from ethanol is advisable.

1.12. Synthesis of 5′-O-(4,4′-dimethoxytrityl)-6-^{13}C-uridine: The product from the previous step (1.0 g, 4.1 mmol) is coevaporated three times with anhydrous pyridine in a 100-ml round-bottom flask.

1.13. The residual oil is dissolved in 10 ml anhydrous pyridine. The reaction is carried out under an argon atmosphere. Add DMT (1.7 g, 5.0 mmol) in three portions within an hour.

1.14. After 4 h at room temperature, the solvent is evaporated and coevaporated three times with toluene.

1.15. The orange foam is dried in high vacuum for 30 min and then dissolved in methylene chloride and washed with 5% citric acid, water, and saturated sodium bicarbonate solution.

1.16. The organic phase is dried over sodium sulfate, filtered, and evaporated to give the crude product.

1.17. The crude product is further purified by silica column chromatography using a gradient of methylene chloride and methanol (99/1 CH_2Cl_2/methanol to 95/5).

1.18. Synthesis of 5'-O-(4,4'-dimethoxytrityl)-2'-O-[[(triisopropylsilyl)oxy]methyl]-6-^{13}C-uridine: The tritylated uridine (616 mg, 1.13 mmol) is dissolved a mixture of DIPEA(765 μl, 4.5 mmol) and 1,2-dichloroethane (10 ml).

1.19. To the clear solution, dibutyltin dichloride (Bu_2SnCl_2; 616 mg, 2.02 mmol) is added, and the mixture is stirred for 1 h at room temperature.

1.20. Then, the solution is heated to 80 °C using a reflux condenser, and an oil bath or a heat block and TOM-Cl; 276 mg, 1.24 mmol) is added.

1.21. The reaction progress can be monitored using TLC (ethyl acetate/hexanes 1/1, R_f (product) = 0.6).

1.22. After 30 min, the reaction mixture is diluted with methylene chloride and the organic phase is washed with saturated sodium bicarbonate solution.

1.23. The organic phase is dried over sodium sulfate, filtered over Celite, and then evaporated to dryness.

1.24. The crude product is further purified by silica column chromatography using a gradient of hexanes and ethyl acetate (80/20 hexanes/ethyl acetate to 40/60).

1.25. Synthesis of 5'-O-(4,4'-dimethoxytrityl)-2'-O-[[(triisopropylsilyl)oxy]methyl]-6-^{13}C-uridine 3'-O-(2'-cyanoethyl N,N-diisopropylphoshphoramidite): Anhydrous methylene chloride (5 ml) and DMEA (540 μl, 5.0 mmol) are mixed and then added to the product of the previous step (720 mg, 0.98 mmol).

1.26. The clear solution is stirred at room temperature for 15 min.

1.27. Add 2-cyanoethyl N,N-diisopropylchlorophosphoramidite (350 mg, 1.5 mmol) dropwise via a syringe and stir the reaction for 2 h.

1.28. The reaction progress can be monitored using TLC (ethyl acetate/hexanes 7/3 + 1% triethylamine, R_f (product) = 0.5).

1.29. The reaction mixture is then diluted with methylene chloride, and the organic phase is washed with half saturated sodium bicarbonate solution.

1.30. The organic phase is dried over sodium sulfate, filtered, and evaporated to dryness.

1.31. The crude product is further purified by silica column chromatography using a gradient of hexanes and ethyl acetate (50/50 hexanes/ethyl acetate to 40/60 + 1% triethylamine).

1.32. The quality of the final product should be checked by NMR spectroscopy (^{1}H, ^{13}C, and ^{31}P NMR in CDCl$_3$). The ^{31}P NMR spectrum consists of two peaks at 150.5 and 150.1 ppm (referenced to external 85% H$_3$PO$_4$). A ^{31}P peak at approximately 15 ppm indicates a phosphonate impurity, which needs to be removed by silica column chromatography to guarantee high coupling yields in the RNA solid phase synthesis.

5.2.1 Tip

The products of the respective steps should be characterized using NMR spectroscopy (1D-^{1}H and ^{13}C spectra) and mass spectrometry and compared to data from literature before proceeding with the next step.

5.2.2 Tip

In the nucleosidation reaction (step 1.1), the temperature (60 °C) should be checked several times. Overheating will result in the formation of the undesired α-substituted uridine nucleoside.

5.2.3 Tip

In the nucleosidation reaction (step 1.3), let the solution cool to room temperature before adding TMSOTf to avoid a too vigorous reaction.

5.2.4 Tip

The extraction of N-methylbenzamide should be checked using silica TLC (CH$_2$Cl$_2$/MeOH 9/1). After each extraction step, spot the aqueous phase and the organic phase on a silica TLC plate to check the composition of the phases. Normally, after three extraction steps no N-methylbenzamide is found in the aqueous phase.

5.2.5 Tip

In the tomylation reaction (step 1.19), separation of organic and aqueous phase can be retarded especially when carrying out the reaction on a larger scale. In that case, the use of the bis-(*tert*-butyl)tin dichloride is recommended.

5.2.6 Tip

The tomylation reaction gives a distribution of desired 2′-O-regioisomer/undesired 3′-O-regioisomer/starting material of 1/1/1. To improve the yield, collect the 3′-O-regioisomer and remove the TOM protecting group using 1 M tetrabutylammonium fluoride in acetonitrile (1 ml/250 mg) and combine it with the starting material. This recovered material can then be reused in the tomylation reaction.

5.2.7 Tip

When carrying out the silica column chromatography of phosphoramidites, it is absolutely necessary to add 1% triethylamine to the eluents to avoid degradation of the amidites on the column. Triethylamine should also be added to the TLC solvent.

6. STEP 2: SYNTHESIS OF 6-^{13}C-CYTIDINE TOM PHOSPHORAMIDITE

6.1 Overview

Starting from 5′-O-(4,4′-dimethoxytrityl)-2′-O-[[(triisopropylsilyl)oxy]methyl]-6-^{13}C-uridine, the 3′-hydroxyl group is protected using acetic anhydride (a) (Fig. 4). Using TPS-Cl, the O^4 of uridine is activated as a leaving group and can be replaced with an amino group using an ammonium hydroxide solution (b) (Fig. 4). If desired, ^{15}N-labeled ammonium chloride in combination with a base (e.g., K_2CO_3) can be used to introduce a ^{15}N label at the exocyclic amino group at this stage. N-selective acetylation (c) and finally phosphitylation (d) yield the desired 6-^{13}C-cytidine phosphoramidite (Fig. 4).

6.2 Duration

5–7 days

2.1. Synthesis of 3′-O-acetyl-5′-O-(4,4′-dimethoxytrityl)-2′-O-[[(triisopropylsilyl)oxy]-methyl]-6-^{13}C-uridine: In a 50-ml round-

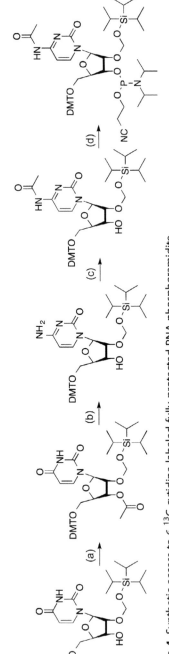

Figure 4 Synthetic access to 6-^{13}C-cytidine labeled fully protected RNA phosphoramidite.

bottom flask, 5'-O-(4,4'-dimethoxytrityl)-2'-O-[[(triisopropylsilyl)oxy]methyl]-6-^{13}C-uridine (350 mg, 0.47 mmol) is dissolved in 1 ml anhydrous pyridine and is stirred at room temperature.

2.2. Add a catalytic amount of DMAP (5 mg, 0.05 mmol).

2.3. Cool the reaction mixture to 0 °C and add acetic anhydride (48 µl, 0.51 mmol) dropwise.

2.4. Remove the ice bath and continue stirring for 2 h.

2.5. The reaction progress can be monitored using TLC (ethyl acetate/hexanes 1/1, R_f (product) = 0.7).

2.6. The solvent is evaporated, and the residue coevaporated twice with toluene.

2.7. The residue is dissolved in methylene chloride, and the organic phase is washed with 5% citric acid, water, and saturated sodium bicarbonate solution.

2.8. Dry the organic phase over sodium sulfate, filter, and evaporate to dryness.

2.9. The crude product can directly be used in the next step.

2.10. Synthesis of 5'-O-(4,4'-dimethoxytrityl)-2'-O-[[(triisopropylsilyl)oxy]-methyl]-6-^{13}C-cytidine: Mix anhydrous methylene chloride (4 ml) and triethylamine (NEt$_3$, 528 µl, 3.8 mmol). To this mixture, add the crude product from the previous step (300 mg, 0.38 mmol) and DMAP (5 mg, 0.05 mmol).

2.11. After 10 min, add TPS-Cl; 176 mg, 0.58 mmol) in small portions over a period of 1 h.

2.12. Stir the reaction for another 4 h at room temperature.

2.13. Dilute the reaction mixture with methylene chloride and wash the organic phase with half saturated sodium bicarbonate solution.

2.14. Dry the organic phase over sodium sulfate, filter, and evaporate to dryness.

2.15. Dry the resulting foam in high vacuum for 30 min.

2.16. Dissolve the crude product from the previous step in 4 ml tetrahydrofuran and add 4 ml 28% aqueous ammonia solution. Continue stirring for 18 h at room temperature.

2.17. After 18 h, evaporate the reaction mixture to yield an oily residue. Take up the residue in 4 ml ethanolic methylamine solution (8 M) and stir for 1 h.

2.18. The reaction progress can be monitored using TLC (CH$_2$Cl$_2$/methanol 96/4, R_f (product) = 0.45).

2.19. Evaporate the solvent and dry the residual oil in high vacuum.
2.20. Dissolve the oil in methylene chloride and wash the organic phase with saturated sodium bicarbonate solution.
2.21. Dry the organic phase over sodium sulfate, filter, and evaporate to dryness.
2.22. Purify the resulting foam from the previous step using silica column chromatography (CH_2Cl_2/methanol 99/1 to 95/5).
2.23. Synthesis of N^4-acetyl-5'-O-(4,4'-dimethoxytrityl)-2'-O-[[(triisopropylsilyl)-oxy]methyl]-6-^{13}C-cytidine: Dissolve the product from the previous step (190 mg, 0.26 mmol) in 1 ml anhydrous dimethylformamide. Then add acetic anhydride (25 µl, 0.26 mmol) dropwise.
2.24. Stir the reaction for 22 h at room temperature.
2.25. The reaction progress can be monitored using TLC (ethyl acetate/hexanes 7/3, R_f (product) = 0.38).
2.26. Quench the reaction by the addition of a few drops of methanol.
2.27. Evaporate the solvent and dissolve the oily residue in methylene chloride.
2.28. Wash the organic phase with saturated sodium bicarbonate solution.
2.29. Dry the organic phase over sodium sulfate, filter, and evaporate to dryness.
2.30. Purify the crude product using silica column chromatography (CH_2Cl_2/methanol 99/1 to 98/2).
2.31. Synthesis of N^4-acetyl-5'-O-(4,4'-dimethoxytrityl)-2'-O-[[(triisopropylsilyl)-oxy]methyl]-6-^{13}C-cytidine 3'-O-(2'-Cyanoethyl N,N-diisopropylphoshphoramidite): Anhydrous methylene chloride (5 ml) and DMEA (250 µl, 2.3 mmol) are mixed and then added to the product of the previous step (130 mg, 0.98 mmol).
2.32. The clear solution is stirred at room temperature for 15 min.
2.33. Add 2-cyanoethyl N,N-diisopropylchlorophosphoramidite (83 mg, 0.35 mmol) dropwise via a syringe and stir the reaction for 2 h.
2.34. The reaction progress can be monitored using TLC (ethyl acetate/hexanes 7/3 + 1% triethylamine, R_f (product) = 0.46).
2.35. The reaction mixture is then diluted with methylene chloride, and the organic phase is washed with half saturated sodium bicarbonate solution.
2.36. The organic phase is dried over sodium sulfate, filtered, and evaporated to dryness.

2.37. The crude product is further purified by silica column chromatography using a gradient of hexanes and ethyl acetate (70/30 hexanes/ethyl acetate + 1% trimethylamine to 30/70 hexanes/ethyl acetate + 1% triethylamine).

2.38. The quality of the final product should be checked by NMR spectroscopy (^{1}H, ^{13}C, and ^{31}P NMR in CDCl$_3$). The ^{31}P NMR spectrum should show two peaks at 150.8 and 150.6 ppm (referenced to external 85% H$_3$PO$_4$). A ^{31}P peak at approximately 15 ppm indicates a phosphonate impurity, which needs to be removed by silica column chromatography to guarantee high coupling yields in the RNA solid phase synthesis.

6.2.1 Tip
The products of the respective steps should be characterized using NMR spectroscopy (1D-^{1}H and ^{13}C spectra) and mass spectrometry and compared to data from literature before proceeding with the next step.

6.2.2 Tip
The crude product from step 2.15 is not stable and should not be stored for a prolonged time at room temperature. It is best used directly after drying in high vacuum.

6.2.3 Tip
When carrying out the silica column chromatography of phosphoramidites, it is absolutely necessary to add 1% triethylamine to the eluents to avoid degradation of the amidites on the column. Triethylamine should also be added to the TLC solvent.

7. STEP 3: CHEMICAL RNA SYNTHESIS
7.1 Overview
Using an automated RNA/DNA synthesizer, the 6-^{13}C-pyrimidine-modified TOM-protected RNA phosphoramidites can be introduced as site-specific stable isotope labels into a target RNA for NMR application. The phosphoramidite chemistry is well established and high-quality synthetic RNAs can be obtained (Fig. 5; Micura, 2002). The main advantage of the chemical oligonucleotide synthesis is that no sequence requirements,

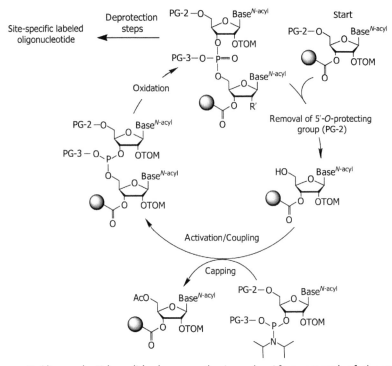

Figure 5 Oligonucleotide solid phase synthesis cycle: After removal of the trityl protecting group, the 5'-OH group is attached to the next building block in the activation/coupling step using a weak acid (e.g., 1H-benzylthiotetrazole) and the respective phosphoramidite. Unreacted 5'-hydroxyls are capped using acetic anhydride. In the final step, the P(III) species is oxidized to the more stable P(V) species before the next synthesis cycle is entered.

as e.g., in the case of T7 RNA polymerase assisted *in vitro* transcription, are encountered. Furthermore, the number and positioning of the stable modified phosphoramidites can be freely chosen. Several DNA/RNA synthesizers are commercially available (e.g., from Applied Biosystems, GE Healthcare, Bioautomation). As aforementioned in Section 4.2, the synthesis cycle should be optimized before using stable isotope-modified RNA phosphoramidites.

Optimized synthesis cycles for an ABI 391 PCRmate are available on request from the authors. The assembled RNA sequence is then deprotected in a two-step procedure: (i) alkaline deprotection: releases the oligonucleotide from the solid support, removes the nucleobase protecting groups and the cyanoethyl groups; (ii) fluoride deprotection: using 1 M

tetrabutylammonium fluoride in tetrahydrofuran the TOM group is cleaved off. Finally, anion exchange chromatography is used to purify the stable isotope-modified RNA.

7.2 Duration
5–7 days

3.1. Freshly prepare phosphoramidite solutions: dissolve 1 g of each TOM protected phoshporamidite in 10 ml water-free acetonitrile (<5 ppm water) and add freshly activated molecular sieve.

3.2. Freshly prepare activator solution: dissolve 1*H*-benzythiotetrazole (1.44 g) in 30 ml water-free acetonitrile and add freshly activated molecular sieve.

3.3. Fill a 1.3-µmol synthesis column with the appropriate amount of solid support (e.g., 16 mg rA Custom Primer support, 80 µmol g^{-1}, GE Healthcare).

3.4. Attach synthesis reagents (Cap A, Cap B, oxidation solution, detritylation solution, and water-free acetonitrile) to the appropriate ports on the RNA synthesizer. Attach phosphoramidite solutions (rA, rG, rC, U, and stable isotope-modified amidites to the additional ports X1, X2, etc.) to the appropriate ports on the RNA synthesizer.

3.5. Fill the reagent lines with the respective solution. Fill the phosphoramidite lines with a minimal volume of respective amidite solution to reduce loss (especially for the stable isotope-labeled phosphoramidites). Attach the synthesis column and wash the column with water-free acetonitrile to check for leaks.

3.6. Enter the target sequence and start the synthesis procedure.

3.7. Check the first detritylation step of the solid support. A deep red color should be observable.

3.8. Check the coupling of the first phosphoramidite by observing the detritylation step. Again a deep red color of the detritylation solution should be observed. Let the synthesis proceed until the sequence assembly is finished. If possible check the final detritylation step. A red color of the solution hints at a successful sequence synthesis.

3.9. Dry the solid support in high vacuum for 30 min. Transfer the solid support into a 1.5-ml Eppendorf© tube with a sealing.

3.10. Ethanolic methylamine/aqueous methylamine deprotection: Add 650 µl aqueous methylamine solution (30%) and 650 µl ethanolic methylamine solution (8 *M*) to the solid support and let stand for

6 h at room temperature. Vortex 1/h. Or alternatively: aqueous methylamine/aqueous ammonia (AMA) deprotection: Add 650 μl aqueous methylamine solution (30%) and 650 μl ammonium hydroxide solution (28%) and heat to 37 °C for 1.5 h.

3.11. Centrifuge to pellet the solid support on the bottom of the tube and transfer the supernatant to a 10-ml round-bottom flask. Wash the solid support three times with 1 ml tetrahydrofuran/water (1/1, v/v) and add the washings to the 10-ml round-bottom flask. Evaporate the solution in the flask to dryness and dry the residue in high vacuum for 30 min to remove the bases.

3.12. Dissolve the residue in the 10-ml round-bottom flask in 1.6 ml 1 M tetrabutylammonium fluoride solution in tetrahydrofuran. If the residue does not readily dissolve, add 100 μl N-methyl-2-pyrrolidone to obtain a clear solution. Stopper the 10-ml round-bottom flask tightly with the aid of Parafilm© and keep the deprotection solution at 33 °C for at least 14 h. Quench the $2'$-O-deprotection reaction with 1.6 ml 1 M triethylammonium acetate solution. Evaporate the deprotection solution to an approximate volume of 1 ml.

3.13. Load the crude RNA on a HiPrep desalting 26/10 column (GE Healthcare) using a MPLC system with UV and conductivity detection. Use HPLC grade water (18 $M\Omega$ cm) to elute the crude RNA. The RNA elutes first and can be identified by its high UV absorbance at 254 nm, whereas the later eluting salts lead to strong increase in conductivity. Collect the RNA in a 50-ml round-bottom flask. Evaporate the RNA containing fraction to dryness and redissolve the oligonucleotide in 1 ml HPLC grade water. The solution of the crude oligonucleotide product can be stored at -20 °C for several weeks without degradation (Fig. 6).

3.14. To check the quality of the crude RNA, anion exchange chromatography using an analytical Dionex DNAPac PA-100 (4 × 250 mm) column and buffer A and B on a HPLC system should be carried out. For RNAs comprising less than 20 nucleotides, a gradient from 0% to 40% B in A in 30 min is sufficient, and for larger RNAs, a gradient from 0% to 60% B in A in 45 min needs to be applied. The chromatographic analysis is carried out at 80 °C to fully denature the RNA (Fig. 7).

3.15. To further purify the crude RNA semipreparative anion exchange chromatography on a Dionex DNAPac PA-100 (9 × 250 mm) column using buffer A and B is carried out. The gradient needs to be

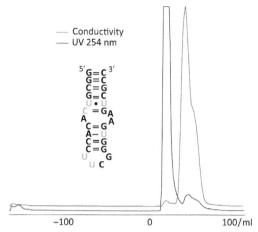

Figure 6 Desalting of crude RNA on a ÄKTA start system (GE Healthcare) using a HiPrep desalting column 26/10. 6-^{13}C-pyrimidine labels are highlighted in orange (gray in the print version).

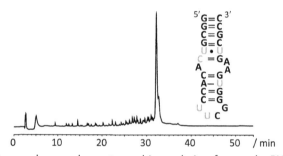

Figure 7 Anion exchange chromatographic analysis of a crude RNA comprising 27 nucleotides. The analysis was carried out on a Dionex DNAPac PA-100 column (4 × 250mm) using a gradient of 0–60% buffer B in buffer A in 45 min at 80 °C. The UV absorbance trace at 260 nm is shown.

optimized for every target RNA and depends on its length and nucleotide composition. The chromatographic purification is carried out at 80 °C to fully denature the RNA. Several runs (typically 5–6) loading approximately 30 nmol crude RNA per run are needed to purify the product of a RNA synthesis on a 1.3 µmol scale (Fig. 8).

3.16. Dilute the fractions containing desired RNA with an equal volume of 100 mM sodium acetate solution.

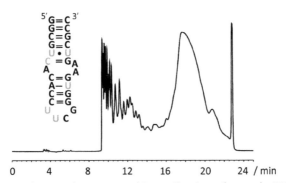

Figure 8 Anion exchange chromatographic purification of a crude RNA comprising 27 nucleotides. A gradient of 30–40% buffer B in buffer A in 15 min at 80 °C was used. The UV absorbance trace at 260 nm is shown.

3.17. Prepare a C18 SepPak cartridge (Waters): (i) 3 column volumes (CV) acetonitrile, (ii) 3 CV acetonitrile/water (1/1, v/v), (iii) 3 CV water, (iv) 3 CV 100 mM sodium acetate.
3.18. Then load the RNA containing fractions on the SepPak cartridge.
3.19. Elute the buffer salts by washing the RNA with 3 CV water.
3.20. Elute the RNA using 3 CV acetonitrile/water (1/1, v/v) into a 50-ml round-bottom flask.
3.21. Evaporate the solution containing the target RNA to dryness and redissolve the RNA in 1 ml HPLC grade water.
3.22. To check the quality of the purified RNA, anion exchange chromatography using an analytical Dionex DNAPac PA-100 (4 × 250 mm) column and buffer A and B on a HPLC system should be carried out. For RNAs comprising less than 20 nucleotides, a gradient from 0% to 40% B in A in 30 min should be used, for larger RNAs a gradient from 0% to 60% B in A needs to be applied. The chromatographic analysis is carried out at 80 °C to fully denature the RNA (Fig. 9).
3.23. The yield of purified target RNA can be calculated by applying the Beer–Lambert law:

$$c = A/\left(\varepsilon^{260\text{nm}} l\right)$$

with c concentration in mol l^{-1}, A absorption at 260 nm, $\varepsilon^{260\text{nm}}$ extinction coefficient at 260 nm in l mol^{-1} cm^{-1} and l length of solution light passes in cm. The extinction coefficient $\varepsilon^{260\text{nm}}$ of the RNA can be estimated by adding the $\varepsilon^{260\text{nm}}$ values of the monomers.

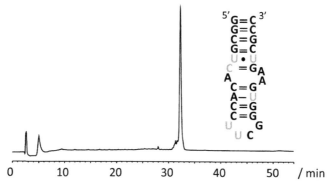

Figure 9 Anion exchange chromatographic analysis of a purified RNA comprising 27 nucleotides. The analysis was carried out on a Dionex DNAPac PA-100 column (4 × 250mm) using a gradient of 0% to 60% buffer B in buffer A in 45 min at 80 °C. The UV absorbance trace at 260 nm is shown. The purity is estimated by peak integration to be >95%.

Nucleotide	ε^{260nm} [l mol^{-1} cm^{-1}]
rA	15,300
rC	7400
rG	11,700
U	9900

3.24. Lyophilize the desired amount of RNA for the NMR spectroscopic application.

7.2.1 Tip
Do not overfill the synthesis column with solid support as the growing RNA chain could clog the column.

7.2.2 Tip
If using the AMA procedure at elevated temperatures, allow cooling to room temperature before opening the tube to avoid spilling of the deprotection solution.

7.2.3 Tip
Use tetrahydrofuran containing the stabilizer BHT for the preparation of the TBAF solution. Stabilizer-free tetrahydrofuran can form radicals leading to irreproducible deprotection results.

7.2.4 Tip

The integrity and homogeneity of the purified RNA (from step 3.21) can further be checked using LC-ESI mass spectrometry and denaturing and native PAGE.

8. STEP 4: APPLICATIONS

8.1 Overview

Using these stable isotope-modified RNA phosphoramidites, an isolated ^{1}H–^{13}C-spin pair is introduced which allows the application of Carr Purcell Meiboom Gill (CPMG) RD NMR experiments in a straightforward manner (Wunderlich et al., 2012). Additionally, a significant side benefit of this methodology is that the site selectively labeled nucleobase can be coupled enzymatically with ribose to make labeled rNTPs (Alvarado, LeBlanc, et al., 2014; Alvarado, Longhini, et al., 2014). Importantly, this very powerful methodology can be exploited to detect and quantify important functional dynamics occurring on the μs- to ms time scale. Thus, structural transitions of RNA important for ligand binding or catalysis can be investigated readily. In this section, we discuss the potential of these chemically synthesized RNAs in a segmental isotope-labeling protocol to address functional dynamics in larger RNAs, as well as its incorporation using chemoenzymatic tricks.

We present ^{13}C CPMG RD data on a 27 nt RNA that mimics the bacterial 16S decoding site (A site), which is involved in discriminating cognate and near-cognate tRNAs (Dethoff, Petzold, Chugh, Casiano-Negroni, & Al-Hashimi, 2012). The labeling pattern is also well suited to study slower RNA secondary structure refolding kinetics occurring at exchange rates at 1 s^{-1}. We illustrate the applicability of the stable isotope-labeled RNAs in a ^{13}C ZZ exchange NMR experiment addressing the refolding kinetics between two hairpin folds, one representing the terminator and the other the antiterminator fold of the *Fsu* preQ$_1$ riboswitch system (Rieder, Kreutz, & Micura, 2010). Finally, we also showcase the usefulness to CEST experiments on a 48 nt RNA labeled with ^{13}C on the base chemically but coupled to the ribose enzymatically (Alvarado, LeBlanc, et al., 2014; Alvarado, Longhini, et al., 2014).

8.2 Segmental Stable Isotope Labeling Using Chemically Synthesized RNAs

Segmental isotope labeling is the key methodology when studying high-molecular weight nucleic acid systems where spectral overlap dominates

(Duss, Maris, von Schroetter, & Allain, 2010; Lebars et al., 2014; Lu, Miyazaki, & Summers, 2010; Nelissen et al., 2008; Tzakos, Easton, & Lukavsky, 2007). In short, a stable isotope-labeled RNA produced either by chemical or enzymatic methods is ligated to another (e.g., unlabeled) RNA giving access to larger constructs with a few NMR active residues to address structural and dynamic features. The procedure makes use of two enzymes, the T4 RNA ligase and T4 DNA ligase, but also deoxyribozymes were used to ligate two RNA strands (Dayie, 2008; Wawrzyniak-Turek & Höbartner, 2014). When using T4 RNA or DNA ligase a 5′-monophosphate is required on the donor strand and a 3′-hydroxyl group at the acceptor fragment. The T4 DNA ligase further needs a DNA splint as the enzyme catalyzes the joining of RNA to another RNA strand in a duplex molecule but will not join single-stranded nucleic acids. In contrast, T4 RNA ligase accepts single-stranded RNAs as substrates. Solid phase RNA synthesis is perfectly suited to craft oligonucleotides with the desired 5′-monophosphate (using a commercially available phosphorylation reagent) and 3′-OH, termini which are not directly accessible via *in vitro* transcription. For *in vitro* synthesis, a ribozyme sequence (e.g., hepatitis delta virus) must be appended at the 5′- and 3′-ends of the donor fragment RNA (Dayie, 2008; Kieft & Batey, 2004), and the 5′-diphosphate is cleaved enzymatically by RNA 5′-polyphosphatase to obtain the desired monophosphorylated group at the 5′-termini (Chen, Zuo, Wang, & Dayie 2012; Dayie, 2008).

We explored the applicability of chemically synthesized RNAs for segmental stable isotope labeling. Here, we showcase the tremendous potential of the approach by the ligation of two RNAs comprising ^{13}C-methyl modifications naturally occurring in transfer RNAs (Fig. 10). Ligation of a 39 nt acceptor and a 37 nt donor strand using T4 RNA ligase yielded a ^{13}C-methyl-modified phenylalanine tRNA. NMR studies on the dynamics of the ^{13}C-methyl-modified residues are currently carried out in our laboratories.

8.3 ^{13}C CPMG RD NMR Spectroscopy

The 27 nt A-site RNA was modified using 6-^{13}C-uridine and 6-^{13}C-cytidine phosphoramidites. The successful incorporation of the six stable isotope labels (five 6-$_{13}$C-uridines and one 6-^{13}C-cytidine) was confirmed by NMR spectroscopy. A standard heteronuclear single quantum correlation (HSQC) spectrum did show the expected six ^{1}H–^{13}C-correlation peaks (Fig. 11A). We then addressed the previously reported conformational dynamics of this RNA using ^{13}C CPMG RD NMR spectroscopy

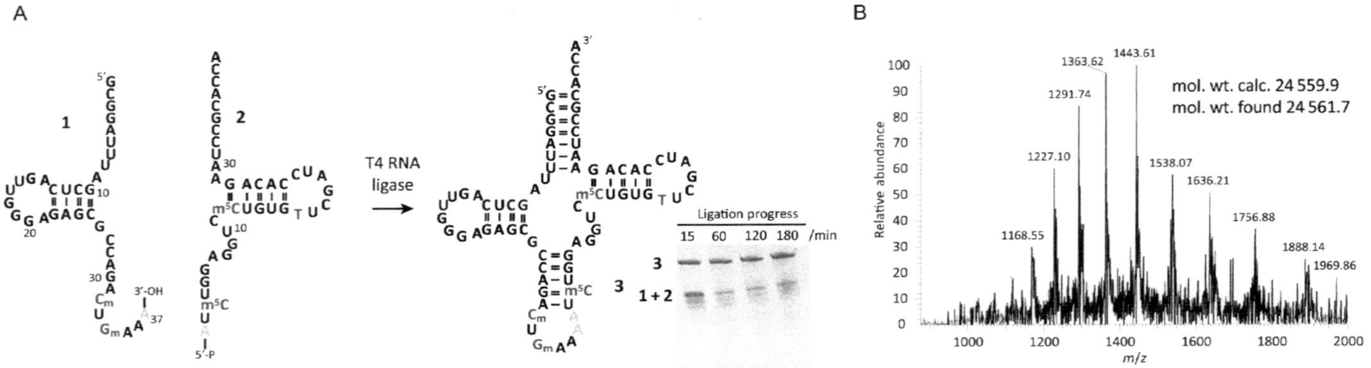

Figure 10 (A) One step T4 RNA ligase catalyzed joining of a 37 and 39 nt RNA comprising ^{13}C-modified methyl residues (highlighted in red (darkgray in the print version), ligation site highlighted in orange (gray in the print version)) yielding a full length 76 nt tRNA^{Phe}. Ligation progress can be readily followed using denaturing polyacrylamide gel electrophoresis (see inset). (B) Electrospray Ionization mass spectrometric analysis of tRNA^{Phe} ligation product.

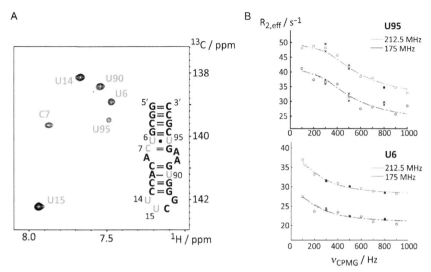

Figure 11 (A) Standard heteronuclear single quantum correlation experiment confirming the successful incorporation of six 6-^{13}C-pyrimidine labels into the 27 nt A-site RNA. (B) Dispersion profiles of U6 and U95 at 212.5 (red) and 175 (blue) MHz carbon larmor frequency. Dots represent the experimental data, and black x represent repeat experiments. The dashed and dotted blue and red lines represent the individual and global fits for a two-state equilibrium in the intermediate exchange regime. (See the color plate.)

(Fig. 11B) (Dethoff et al., 2012). For that purpose, we used a longitudinal optimized transverse relaxation optimized spectroscopy (TROSY) selected dispersion experiment (Weininger, Respondek, & Akke, 2012). This experiment is especially useful for the fast relaxing aromatic RNA nucleobase ^{13}C atoms introduced by the presented labeling pattern.

Two residues could be identified that displayed a nonflat dispersion profile—U6 and U95. Data were recorded at two field strengths (16.4 and 20 T at 288 K), and the dispersion profiles were fitted using an exact solution for a two-site exchange process. The results are summarized (Table 1) and are in good agreement with results obtained from $R_{1\rho}$ dispersion profiles at 298 K (Dethoff et al., 2012).

8.4 ^{13}C ZZ Exchange NMR Spectroscopy

In gene regulatory riboswitch RNAs, slower dynamics occuring at a few milliseconds to seconds can be of special importance for their biological function. This class of noncoding RNAs switches their tertiary and/or secondary structure in response to the presence or absence of a low-molecular

Table 1 Summary of Exchange Data Obtained from ^{13}C-CPMG RD Experiments of the A-Site RNA at 288 K

	k_{ex}/s^{-1}	$p_A/\%^a$	$p_B/\%^a$	$\Delta\delta(^{13}C)/ppm^b$
Global parameters	1204 ± 256	98.1 ± 0.8	1.9 ± 0.8	
U6				2.42 ± 0.4
U95				0.87 ± 0.6

$^a p_A$ refers to the ground state and p_B to the excited state population.
^{b} ^{13}C-chemical shift difference between ground state A and excited state B for the residues exhibiting nonflat dispersion profiles are given.

weight metabolites (Haller et al., 2011). We identified a bistable terminator antiterminator sequence partition (Fig. 12A, dashed box; Rieder et al., 2010). To address the equilibrium refolding kinetics of this naturally occurring bistable RNA element, a single 6-^{13}C-cytidine substitution was introduced, and ^{13}C-longitudinal exchange NMR spectroscopy was applied. Integration of the ^{1}H–^{13}C-correlation peaks in a standard HSQC spectrum yielded the equilibrium constant favoring fold B ($K_{AB}=3$). Subsequently, the forward and backward rates of the secondary structure equilibrium were determined at 33 °C using an approach introduced by analyzing longitudinal relaxation rates with and without exchange contribution (Fig. 12B). The forward rate constant k_{AB} amounts to 3.3 ± 0.1 s^{-1} and the rate k_{BA} for the folding process from state B to A to 1.4 ± 0.3 s^{-1}.

8.5 Chemical Exchange Saturation Transfer

To characterize the nature of riboswitch RNA transitions on timescales slower than CPMG but faster than ZZ exchange, CEST experiments are carried out by varying the position of a very weak ^{13}C radiofrequency (rf) B_1 field and quantified indirectly using the intensities of the major state (NMR-visible) correlations (Bouvignies & Kay, 2012a,2012b; Fawzi, Ying, Ghirlando, Torchia, & Clore, 2011; Forsén & Hoffman, 1963; Longhini et al., 2015; Vallurupalli, Bouvignies, & Kay, 2012; Zhao, Hansen, & Zhang, 2014). As expected, when the B_1 field position coincides with or is close to the resonance frequency of the ^{13}C resonant peak in the ground state, the intensity of this ground state peak decreases because of rf saturation effect (Forsén & Hoffman, 1963). Significantly, when the weak B_1 field is applied close to the position of the corresponding peak from the excited (NMR-invisible) state, the

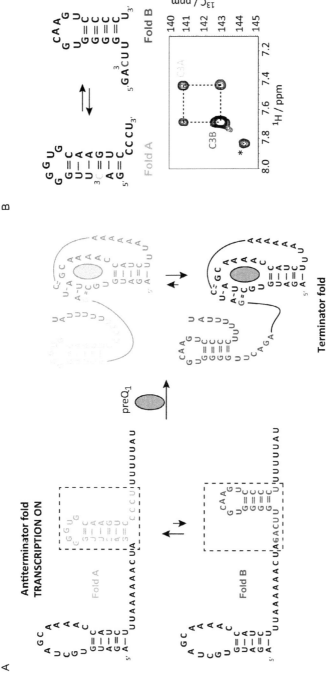

Figure 12 (A) Competing terminator antiterminator RNA folds and the influence of preQ$_1$ binding on the 2° structure equilibrium. (B) ^{13}C longitudinal exchange spectrum at a mixing time of 200 ms and at 33 °C. (See the color plate.)

ground-state peak intensity also decreases due to the transfer of the rf perturbation from the excited state to the ground state by chemical exchange. A plot of the intensity of the ground-state (observed) peak as a function of the position of the B_1 field indicates a pair of dips, with the larger dip at the resonance frequency of the carbon in the ground state and the smaller dip at the frequency of the corresponding ^{13}C resonant peak in the excited state (Fig. 13). Thus, a visual inspection of the plotted data readily reveals that the NMR-invisible state is shifted into a carbon frequency region expected for residues in a Watson–Crick base-pair configuration (Fig. 13). Data were recorded at 16.4 T, and the CEST profiles were fitted using the Bloch–McConnell equations assuming a two-site exchange process (Longhini et al., 2015). Global fitting of CEST profile to a single two-state model resulted in $k_{ex} = 687 \pm 24$ s^{-1} and minor excited population $p_{ES} = 2.3 \pm 0.1\%$, and a ^{13}C-chemical shift difference between ground state and excited state of 3.3 ppm. With our selective labels, the large $^1J_{C1'-C2'}$ (~45 Hz) couplings do not complicate the measurements and need not been taken explicitly into account in the data analysis of sugar (C1′) CEST profiles as required in previous studies using uniformly labeled RNA (Zhao et al., 2014).

9. CONCLUSIONS

Here, we presented a broad and versatile ^{13}C-labeling approach for RNA using 6-^{13}C-pyrimidine phosphoramidite building blocks or chemoenzymatic coupling approach, which can be introduced into RNA by oligonucleotide solid phase or enzymatic synthesis. The 6-^{13}C-pyrimidine labels are generally applicable to study conformational dynamics and the refolding kinetics in RNAs comprising up to 30–40 nucleotides using standard NMR techniques. For larger RNAs, the labeling protocol can be used in combination with enzymatic ligation strategies or chemoenzymatic approaches as outlined. In that case, pulse sequences comprising a TROSY element are particularly useful. The building blocks are currently used to site-specifically modify functional nucleic acids, such as riboswitch aptamer domains or ribozymes, to give fundamental insights into the functional dynamics of these RNAs. We anticipate that this methodology of incorporating modified or labeled nucleotides into specifically designated positions or regions of RNA is very general and will open up new avenues for not only studying the structure and dynamics of large RNAs but will also be useful in biotechnological applications such as RNA biosensors.

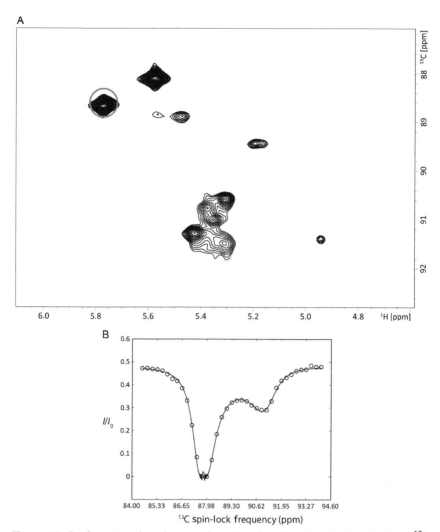

Figure 13 Conformational exchange of a 48 nt fluoride riboswitch probed by ^{13}C CEST NMR spectroscopy (Longhini, LeBlanc et al., 2015). (A) C1′ TROSY HSQC of 6-^{13}C-selectively UTP-labeled fluoride riboswitch shows peak with CEST profile and carbon chemical shift of minor population shift. (B) CEST Profile of C1′ residue from 2D HSQC CEST of fluoride riboswitch at 35 °C with a spin-lock field of 37.9 Hz. Solid lines are the fits to the experimental data to the Bloch–McConnell equation, assuming a two-site exchange process.

ACKNOWLEDGMENTS

C.K. thanks Martin Tollinger and Thomas Moschen for support in fitting the CPMG RD data. C.K. also thanks Ronald Micura for continuous support. Supported partially by NIH P50 GM103297 (T.K.D.), NMR instrumentation by NSF DBI1040158 (T.K.D.), and the Austrian Sciences Fund (I844 and P26550 to C.K.).

REFERENCES

Al-Hashimi, H. M. (2007). Beyond static structures of RNA by NMR: Folding, refolding, and dynamics at atomic resolution. *Biopolymers*, 86(5–6), 345–347.

Al-Hashimi, H. M., & Walter, N. G. (2008). RNA dynamics: It is about time. *Current Opinion in Structural Biology*, 18(3), 321–329.

Alvarado, L. J., LeBlanc, R. M., Longhini, A. P., Keane, S. C., Jain, N., Yildiz, Z. F., et al. (2014). Regio-selective chemical-enzymatic synthesis of pyrimidine nucleotides facilitates RNA structure and dynamics studies. *ChemBioChem*, 15(11), 1573–1577. http://dx.doi.org/10.1002/cbic.201402130.

Alvarado, L. J., Longhini, A. P., LeBlanc, R. M., Chen, B., Kreutz, C., & Dayie, T. K. (2014). Chemo-enzymatic synthesis of selectively 13C/15N-labeled RNA for NMR structural and dynamics studies. In H. B. A. Donald (Ed.), *Methods in enzymology: Vol. 549* (pp. 133–162). Amsterdam/New York: Academic Press, Elsevier. chapter seven.

Blount, K. F., & Breaker, R. R. (2006). Riboswitches as antibacterial drug targets. *Nature Biotechnology*, 24(12), 1558–1564.

Bouvignies, G., & Kay, L. E. (2012a). A 2D 13C-CEST experiment for studying slowly exchanging protein systems using methyl probes: An application to protein folding. *Journal of Biomolecular NMR*, 53(4), 303–310. http://dx.doi.org/10.1007/s10858-012-9640-7.

Bouvignies, G., & Kay, L. E. (2012b). Measurement of proton chemical shifts in invisible states of slowly exchanging protein systems by chemical exchange saturation transfer. *The Journal of Physical Chemistry. B*, 116(49), 14311–14317. http://dx.doi.org/10.1021/jp311109u.

Breaker, R. R. (2011). Prospects for riboswitch discovery and analysis. *Molecular Cell*, 43(6), 867–879. http://dx.doi.org/10.1016/j.molcel.2011.08.024.

Campbell, D. O., Bouchard, P., Desjardins, G., & Legault, P. (2006). NMR structure of varkud satellite ribozyme stem loop V in the presence of magnesium ions and localization of metal-binding sites. *Biochemistry*, 45(35), 10591–10605. http://dx.doi.org/10.1021/bi0607150.

Chen, B., Zuo, X., Wang, Y.-X., & Dayie, T. K. (2012). Multiple conformations of SAM-II riboswitch detected with SAXS and NMR spectroscopy. *Nucleic Acids Research*, 40(7), 3117–3130.

Coppins, R. L., Hall, K. B., & Groisman, E. A. (2007). The intricate world of riboswitches. *Current Opinion in Microbiology*, 10(2), 176–181.

Dayie, K. T. (2008). Key labeling technologies to tackle sizeable problems in RNA structural biology. *International Journal of Molecular Sciences*, 9(7), 1214–1240.

Dethoff, E. A., Petzold, K., Chugh, J., Casiano-Negroni, A., & Al-Hashimi, H. M. (2012). Visualizing transient low-populated structures of RNA. *Nature*, 491(7426), 724–728. http://www.nature.com/nature/journal/v491/n7426/abs/nature11498.html, supplementary-information.

Duss, O., Maris, C., von Schroetter, C., & Allain, F. H. T. (2010). A fast, efficient and sequence-independent method for flexible multiple segmental isotope labeling of RNA using ribozyme and RNase H cleavage. *Nucleic Acids Research*, 38(20), e188.

Fawzi, N. L., Ying, J., Ghirlando, R., Torchia, D. A., & Clore, G. M. (2011). Atomic resolution dynamics on the surface of amyloid β protofibrils probed by solution NMR. *Nature*, 480(7376), 268–272. http://dx.doi.org/10.1038/nature10577.

Forsén, S., & Hoffman, R. A. (1963). Study of moderately rapid chemical exchange reactions by means of nuclear magnetic double resonance. *The Journal of Chemical Physics*, 39(11), 2892–2901. http://dx.doi.org/10.1063/1.1734121.

Fürtig, B., Buck, J., Richter, C., & Schwalbe, H. (2012). Functional dynamics of RNA ribozymes studied by NMR spectroscopy ribozymes. In J. S. Hartig (Ed.), *Methods in molecular biology: Vol. 848* (pp. 185–199). Berlin; Heidelberg, Germany; New York: Humana Press/Springer.

Haller, A., Souliere, M. F., & Micura, R. (2011). The dynamic nature of RNA as key to understanding riboswitch mechanisms. *Accounts of Chemical Research, 44*(12), 1339–1348. http://dx.doi.org/10.1021/ar200035g.

Hennig, M., Williamson, J. R., Brodsky, A. S., & Battiste, J. L. (2001). Recent advances in RNA structure determination by NMR. *Current protocols in nucleic acid chemistry*. Hoboken, New Jersey: John Wiley & Sons, Inc.

Johnson, J. E., & Hoogstraten, C. G. (2008). Extensive backbone dynamics in the GCAA RNA tetraloop analyzed using 13C NMR spin relaxation and specific isotope labeling. *Journal of the American Chemical Society, 130*(49), 16757–16769. http://dx.doi.org/10.1021/ja805759z.

Kieft, J. S., & Batey, R. T. (2004). A general method for rapid and nondenaturing purification of RNAs. *RNA, 10,* 988–995.

Kloiber, K., Spitzer, R., Tollinger, M., Konrat, R., & Kreutz, C. (2011). Probing RNA dynamics via longitudinal exchange and CPMG relaxation dispersion NMR spectroscopy using a sensitive 13C-methyl label. *Nucleic Acids Research, 39,* 4340–4351.

Latham, M. P., Brown, D. J., McCallum, S. A., & Pardi, A. (2005). NMR methods for studying the structure and dynamics of RNA. *ChemBioChem, 6*(9), 1492–1505. http://dx.doi.org/10.1002/cbic.200500123.

Lebars, I., Vileno, B., Bourbigot, S., Turek, P., Wolff, P., & Kieffer, B. (2014). A fully enzymatic method for site-directed spin labeling of long RNA. *Nucleic Acids Research, 42*(15), e117.

Longhini, A. P., Leblanc, R. M., Becette, O., Salguero, C., Wunderlich, C. H., Johnson, B. A., et al. (2015). Chemo-enzymatic synthesis of site-specific isotopically labeled nucleotides for use in NMR resonance assignment, dynamics and structural characterizations, submitted.

Lu, K., Miyazaki, Y., & Summers, M. (2010). Isotope labeling strategies for NMR studies of RNA. *Journal of Biomolecular NMR, 46*(1), 113–125. http://dx.doi.org/10.1007/s10858-009-9375-2.

Micura, R. (2002). Small interfering RNAs and their chemical synthesis. *Angewandte Chemie, International Edition, 41*(13), 2265–2269. http://dx.doi.org/10.1002/1521-3773 (20020703)41:13<2265::AID-ANIE2265>3.0.CO;2-3.

Milligan, J. F., Groebe, D. R., Witherell, G. W., & Uhlenbeck, O. C. (1987). Oligoribonucleotide synthesis using T7 RNA polymerase and synthetic DNA templates. *Nucleic Acids Research, 15*(21), 8783–8798.

Milligan, J. F., & Uhlenbeck, O. C. (1989). Synthesis of small RNAs using T7 RNA polymerase. *Methods in Enzymology, 180,* 51–62.

Mittermaier, A. K., & Kay, L. E. (2009). Observing biological dynamics at atomic resolution using NMR. *Trends in Biochemical Sciences, 34*(12), 601–611.

Nelissen, F. H. T., van Gammeren, A. J., Tessari, M., Girard, F. C., Heus, H A., & Wijmenga, S. S. (2008). Multiple segmental and selective isotope labeling of large RNA for NMR structural studies. *Nucleic Acids Research, 36*(14), e89.

Reining, A., Nozinovic, S., Schlepckow, K., Buhr, F., Furtig, B., & Schwalbe, H. (2013). Three-state mechanism couples ligand and temperature sensing in riboswitches. *Nature, 499*(7458), 355–359. http://dx.doi.org/10.1038/nature12378. http://www.nature.com/nature/journal/v499/n7458/abs/nature12378.html, supplementary-information.

Rieder, U., Kreutz, C., & Micura, R. (2010). Folding of a transcriptionally acting PreQ1 riboswitch. *Proceedings of the National Academy of Sciences of the United States of America, 107*(24), 10804–10809.

Rinnenthal, J., Buck, J., Ferner, J., Wacker, A., Furtig, B., & Schwalbe, H. (2011). Mapping the landscape of RNA dynamics with NMR spectroscopy. *Accounts of Chemical Research*, *44*(12), 1292–1301. http://dx.doi.org/10.1021/ar200137d.

Serganov, A., & Nudler, E. (2013). A decade of riboswitches. *Cell*, *152*(1–2), 17–24. http://dx.doi.org/10.1016/j.cell.2012.12.024.

Tzakos, A. G., Easton, L. E., & Lukavsky, P. J. (2007). Preparation of large RNA oligonucleotides with complementary isotope-labeled segments for NMR structural studies. *Nature Protocols*, *2*(9), 2139–2147.

Vallurupalli, P., Bouvignies, G., & Kay, L. E. (2012). Studying "invisible" excited protein states in slow exchange with a major state conformation. *Journal of the American Chemical Society*, *134*(19), 8148–8161. http://dx.doi.org/10.1021/ja3001419.

Wawrzyniak-Turek, K., & Höbartner, C. (2014). Deoxyribozyme-mediated ligation for incorporating EPR spin labels and reporter groups into RNA. In H. B. A. Donald (Ed.), *Methods in enzymology: Vol. 549* (pp. 85–104). Amsterdam/New York: Academic Press, Elsevier. chapter four.

Weininger, U., Respondek, M., & Akke, M. (2012). Conformational exchange of aromatic side chains characterized by L-optimized TROSY-selected 13C CPMG relaxation dispersion. *Journal of Biomolecular NMR*, *54*(1), 9–14. http://dx.doi.org/10.1007/s10858-012-9656-z.

Wenter, P., Reymond, L., Auweter, S. D., Allain, F. H. T., & Pitsch, S. (2006). Short, synthetic and selectively 13C-labeled RNA sequences for the NMR structure determination of protein-RNA complexes. *Nucleic Acids Research*, *34*(11), e79.

Wunderlich, C. H., Spitzer, R., Santner, T., Fauster, K., Tollinger, M., & Kreutz, C. (2012). Synthesis of (6-13C)pyrimidine nucleotides as spin-labels for RNA dynamics. *Journal of the American Chemical Society*, *134*(17), 7558–7569. http://dx.doi.org/10.1021/ja302148g.

Zhang, Q., Sun, X., Watt, E. D., & Al-Hashimi, H. M. (2006). Resolving the motional modes that code for RNA adaptation. *Science*, *311*(5761), 653–656.

Zhao, B., Hansen, A. L., & Zhang, Q. (2014). Characterizing slow chemical exchange in nucleic acids by carbon CEST and low spin-lock field R1ρ NMR spectroscopy. *Journal of the American Chemical Society*, *136*(1), 20–23. http://dx.doi.org/10.1021/ja409835y.

CHAPTER NINETEEN

In Vivo, Large-Scale Preparation of Uniformly ^{15}N- and Site-Specifically ^{13}C-Labeled Homogeneous, Recombinant RNA for NMR Studies

My T. Le*, Rachel E. Brown[†], Anne E. Simon[†], T. Kwaku Dayie*,[1]
*Department of Chemistry and Biochemistry, Center for Biomolecular Structure and Organization, University of Maryland, College Park, Maryland, USA
[†]Department of Chemistry and Biochemistry, Department of Cellular Biology and Molecular Genetics, Center for Biomolecular Structure and Organization, University of Maryland, College Park, Maryland, USA
[1]Corresponding author: e-mail address: dayie@umd.edu

Contents

1. Theory — 497
2. Equipment — 499
3. Materials — 502
   3.1 Solutions for Bacterial Growth — 504
   3.2 Solutions for Recombinant tRNA-Scaffold Purification — 507
   3.3 Solution for DNAzyme Cleavage — 508
   3.4 Solutions for Denaturing PAGE — 508
4. Protocol — 510
   4.1 Duration — 510
   4.2 Preparation — 511
   4.3 Caution — 511
5. Step 1: Pilot of the Expression of the Recombinant tRNA-Scaffold Plasmid in Wild-Type K12 E. coli — 512
   5.1 Overview — 512
   5.2 Duration — 512
   5.3 Caution — 513
   5.4 Tip — 513
   5.5 Tip — 513
   5.6 Tip — 514
6. Step 2: Double Selection of High-Expressing E. coli Clones — 514
   6.1 Overview — 514
   6.2 Duration — 515
   6.3 Tip — 516
   6.4 Tip — 516
   6.5 Tip — 516

Methods in Enzymology, Volume 565
ISSN 0076-6879
http://dx.doi.org/10.1016/bs.mie.2015.07.020

© 2015 Elsevier Inc.
All rights reserved.

6.6	Tip	516
6.7	Tip	516
6.8	Tip	516
7. Step 3: Large-Scale Expression in Labeled SPG Minimal Media		517
7.1	Overview	517
7.2	Duration	517
7.3	Tip	518
7.4	Tip	518
7.5	Tip	518
7.6	Tip	518
7.7	Tip	518
8. Step 4: Total Cellular RNA Extraction		518
8.1	Overview	518
8.2	Duration	519
8.3	Tip	519
8.4	Tip	519
8.5	Tip	519
8.6	Tip	519
8.7	Tip	520
8.8	Tip	520
9. Step 5a: Purification of the Recombinant tRNA-Scaffold Using Anion-Exchange Chromatography		520
9.1	Overview	520
9.2	Duration	520
9.3	Tip	522
9.4	Tip	522
9.5	Tip	522
9.6	Tip	522
9.7	Tip	522
9.8	Tip	522
9.9	Tip	522
10. Step 5b: Purification of the Recombinant tRNA-Scaffold Using Affinity Chromatography		523
10.1	Overview	523
10.2	Duration	524
11. Step 6: Excision and Purification of the RNA of Interest		524
11.1	Overview	524
11.2	Duration	525
11.3	Tip	526
11.4	Tip	526
11.5	Tip	526
11.6	Tip	527
11.7	Tip	527
11.8	Tip	527
11.9	Tip	527

12. Step 7: NMR Applications	527
12.1 Overview	527
13. Conclusion	530
Acknowledgments	532
References	532

Abstract

Knowledge of how ribonucleic acid (RNA) structures fold to form intricate, three-dimensional structures has provided fundamental insights into understanding the biological functions of RNA. Nuclear magnetic resonance (NMR) spectroscopy is a particularly useful high-resolution technique to investigate the dynamic structure of RNA. Effective study of RNA by NMR requires enrichment with isotopes of ^{13}C or ^{15}N or both. Here, we present a method to produce milligram quantities of uniformly ^{15}N- and site-specifically ^{13}C-labeled RNAs using wild-type K12 and mutant tktA *Escherichia coli* in combination with a tRNA-scaffold approach. The method includes a double selection protocol to obtain an *E. coli* clone with consistently high expression of the recombinant tRNA-scaffold. We also present protocols for the purification of the tRNA-scaffold from a total cellular RNA extract and the excision of the RNA of interest from the tRNA-scaffold using DNAzymes. Finally, we showcase NMR applications to demonstrate the benefit of using *in vivo* site-specifically ^{13}C-labeled RNA.

1. THEORY

Ribonucleic acid (RNA) participates in many cellular processes such as transcription, translation, gene regulation, and immunity (Ahmad, 2006; Breaker, 2009; D'Souza & Summers, 2005; Simon & Miller, 2013). The diversity of biological functions of RNA is due to its ability to assume dynamic, three-dimensional (3D) conformations (Houck-Loomis et al., 2011; Mortimer, Kidwell, & Doudna, 2014; Steitz, 2008). Many biophysical methods have been developed to solve the 3D structures of RNAs (Wang, Zuo, Wang, Yu, & Butcher, 2010; Yang, Parisien, Major, & Roux, 2010). Of these, nuclear magnetic resonance (NMR) spectroscopy, which can provide structural information at atomic resolution, has emerged as a tool of choice to solve RNA structures and probe dynamics (Latham, Brown, McCallum, & Pardi, 2005). However, to fully leverage ultrahigh-field NMR instrumentation for uniformly labeled samples, three bottlenecks in RNA analysis need to be overcome: (i) extensive chemical shift overlap of resonances, (ii) strong ^{13}C–^{13}C dipolar and scalar couplings of adjacent carbon atoms, and (iii) rapid signal loss due to line broadening for RNA molecules longer than 50 nucleotides (nt) (Dayie, 2008).

To overcome these problems, commercial, fully labeled ribonucleotides (rNTPs) can be replaced with in-house prepared, site-specifically labeled ones. For instance, site-specifically labeled rNTPs have been obtained from selective biomass bioproduction and chemoenzymatic synthesis (Dayie & Thakur, 2010; Thakur & Dayie, 2011; Thakur, Luo, Chen, Eldho, & Dayie, 2012; Thakur, Sama, Jackson, Chen, & Dayie, 2010). When these site-specifically labeled rNTPs are used in combination with ultrahigh-field NMR spectroscopy, the spectral resolution increases, the signal-to-noise ratio is enhanced due to the removal of ^{13}C–^{13}C coupling, and the structural assignment of RNAs is simplified (Alvarado, LeBlanc, et al., 2014; Hoogstraten & Johnson, 2008; Johnson, Julien, & Hoogstraten, 2006; Thakur et al., 2012; Thakur, Sama, et al., 2010). However, these methods both rely on in vitro transcription by T7 RNA polymerase, which is known to produce heterogeneous 5' and 3' RNA ends. The presence of these faulty RNA sequences due to transcription errors can create different chemical environments that lead to reduction of signal intensities in NMR and introduce ambiguity and uncertainty in interpretation of NMR data (Dayie, 2008; Furtig, Richter, Wohnert, & Schwalbe, 2003).

Ponchon and Dardel developed an in vivo, recombinant RNA expression method to circumvent the expense of making large quantities of RNAs using in vitro transcription (Ponchon, Beauvais, Nonin-Lecomte, & Dardel, 2009; Ponchon & Dardel, 2007). For this method, they used a transfer RNA (tRNA)-scaffold plasmid to express recombinant RNAs in the anticodon loop of a lysine tRNA (tRNA$_{lys}$)-scaffold in Escherichia coli, disguising the RNAs from degradation by cellular nucleases. Several other groups have modified this method by using the 5S ribosomal RNA to replace the tRNA$_{lys}$ or by using isopropyl β-D-1-thiogalactopyranoside induction to replace continuous expression (Liu et al., 2010; Nelissen et al., 2012; Ponchon & Dardel, 2011). However, with these previous approaches, one can only produce RNAs uniformly labeled with ^{15}N and ^{13}C isotopes. Large, uniformly labeled RNAs generated using these methods still have ^{13}C–^{13}C coupling, leading to signal loss and low resolution.

We proposed a widely applicable method to produce recombinant, site-specifically labeled RNAs by growing wild type and mutant E. coli in defined nitrogen and carbon sources (Le, Brown, Longhini, Simon, & Dayie, 2015). This method is a hybrid of the tRNA-scaffold approach (Ponchon & Dardel, 2007, 2011) and the selective biomass bioproduction method (Hoffman & Holland, 1995; Thakur et al., 2012). Briefly, the process entails cloning, bacterial transformation and growth, RNA extraction and purification, and DNAzyme cleavage followed by a second purification step (Schlosser, Gu, Sule, & Li, 2008; Schlosser & Li, 2010). The RNAs of interest were inserted into a tRNA-scaffold plasmid within the anticodon loop of the tRNA$_{lys}$. Two *E. coli* strains, wild type and mutant tktA, which lacks the transketolase enzyme, were transformed with the plasmid and grown in minimal media supplemented with ^{15}N-labeled ammonium sulfate [(^{15}NH$_4$)$_2$SO$_4$] and a site-specifically labeled ^{13}C carbon source. When grown on minimal media supplemented with 1-^{13}C-acetate, the wild-type K12 *E. coli* incorporates ^{13}C at the C2 and C4 base carbons of the pyrimidines, at the C4 and C6 base carbons of the purines, and at the C3′ position of the ribose ring (Hoffman & Holland, 1995). Similarly, the mutant tktA strain redirects most of the metabolic flux through the oxidative pentose phosphate pathway so that cells grown in minimal media supplemented with 1-^{13}C-glucose produce rNTPs with ^{13}C isotopes highly incorporated at the C1′ and C5′ ribose carbons, the C8 base carbons of purines, and the C5 and C6 base carbons of pyrimidines (Thakur et al., 2012).

Here, we provide detailed steps of our hybrid recombinant tRNA-scaffold and selective biomass bioproduction method (Figs. 1 and 2) that has been used successfully to make RNAs with sizes ranging from 48 to 118 nt. We also present NMR experiments typically used in RNA resonance assignment to demonstrate improvement in spectral quality. Applications of this technology should prove valuable in overcoming the major obstacles of chemical shift overlap, rapid signal loss, and strong ^{13}C-^{13}C dipolar and scalar couplings found in uniformly labeled RNA molecules longer than 50 nt.

2. EQUIPMENT

0.22-μm sterile cellulose acetate filters (GE Healthcare)
0.22-μm sterile syringe filters
1-L Nalgene bottles for centrifugation
1.5-mL microcentrifuge tubes (RNase, DNase, pyrogen-free)
4-L glass culture flasks
15-mL conical tubes (RNase, DNase, pyrogen-free)
50-mL 3 kDa molecular weight cutoff (MWCO) spin columns (Millipore)
50-mL conical tubes (RNase, DNase, pyrogen-free)
50-mL SuperClear Ultra-High Performance centrifuge tubes with plug caps (VWR)

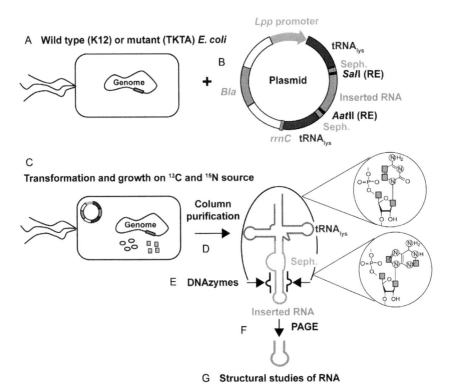

Figure 1 General scheme for the expression and purification of the recombinant tRNA-scaffold in wild type and mutant E. coli strains. (A) Wild type and mutant E. coli strains are made chemically competent using $CaCl_2$ so that they can be transformed using the pBSKrnaSeph plasmid (E. coli tRNA$_{lys}$, Sephadex aptamer) (Ponchon et al., 2009). (B) General diagram of the tRNA-scaffold plasmid: The RNA of interest is inserted at SalI and AatII restriction sites, which surround the anticodon loop of the tRNA$_{lys}$-scaffold. The sephadex tag is included for affinity purification. The RNA transcript is under the control of the lipoprotein promoter (lpp) and terminated with a ribosomal RNA operon transcription terminator (rrnC). (C) Incorporation of NMR-active isotopes into the recombinant tRNA-scaffold: The tRNA-scaffold plasmid is taken up by E. coli using heat-shock transformation. E. coli grown in minimal media supplemented with ^{15}N-ammonium sulfate and/or 1-^{13}C-glucose or 1-^{13}C-acetate incorporates these isotopic nuclei into all cellular RNAs, including the overexpressed recombinant tRNA-scaffold. ^{13}C isotopes are denoted by purple squares and ^{15}N isotopes are denoted by yellow circles. (D) The recombinant tRNA-scaffold is extracted and then purified by anion exchange or affinity chromatography. (E) Two DNAzymes are used to cleave the 5′ and 3′ ends of the inserted RNA. (F) Inserted RNAs are purified using PAGE. (G) The final RNA products are used for structural studies. (See the color plate.)

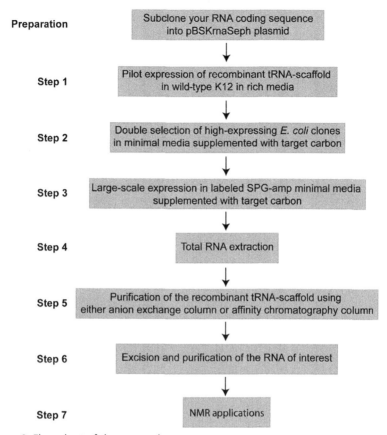

Figure 2 Flow chart of the protocol.

50-mL superloop (GE Healthcare)
50-mL XK 26/60 empty columns (GE Healthcare)
100-mL glass culture flasks
500-mL glass culture flasks
600/800 MHz NMR instrument equipped with at least ^{1}H, ^{15}N, and ^{13}C probes
Aluminum foil
Autoclave
Cuvettes
Electronic pipette controller
Elutrap electroelution system with BT1 and BT2 membranes (Whatman)
Fiberlite F15-8 × 50cy fixed-angle rotor (ThermoScientific)

Fiberlite F8-6 × 1000y fixed-angle large capacity rotor (ThermoScientific)
Freezers (−20 and −80 °C)
Heating block
HiLoad 26/600 Superdex 75 pg column (GE Healthcare)
Ice machine
Incubator with shaking
Liquid chromatography system
Low-volume Shigemi NMR tubes
Micropipettors
Micropipettor tips (RNase, DNase, pyrogen-free)
Nutating tabletop shaker
PCR machine
Petri dishes
pH meter and electrode
Polyacrylamide gel electrophoresis (PAGE) equipment, analytical and preparative size
Razor blade
Refrigerated microcentrifuge
Refrigerator (4 °C)
Serological pipettes (RNase, DNase, pyrogen-free)
Sorvall Legend XTR tabletop centrifuge with swinging bucket rotor
Sorvall RC-5C superspeed centrifuge
Sorvall SS-34 fixed-angle rotor with phenol-resistant tubes
SpeedVac™ Concentrator
Syringes
UV/Vis spectrophotometers (cuvette reader and NanoDrop)
Vacuum filtration system
Vortex mixer
Water bath

3. MATERIALS

Unless stated otherwise, chemicals were obtained from Sigma-Aldrich.

1-^{13}C-acetate (sodium salt) ($CH_3^{13}CO_2Na$) (Isotec)
1-^{13}C-glucose ($^{13}CC_5H_{12}O_6$) (Isotec)
3-(N-morpholino) propanesulfonic acid (MOPS)

5′ and 3′ 8–17 RNA-cleaving synthetic deoxyribozymes (DNAzymes) (Integrated DNA Technologies)
$^{15}N_2$-ammonium sulfate [$(^{15}NH_4)_2SO_4$] (Isotec)
18-nt synthetic DNA oligonucleotide (5′-GCCCGAACAGGGAC TTGAA-3′) (Integrated DNA Technologies)
AatII restriction enzyme (New England Biolabs)
Absolute ethanol (EtOH)
Acid phenol:chloroform (5:1), pH 4.5 (Life Technologies)
Acrylamide
Agar, bacteriological
Ammonium persulfate (APS)
Ammonium sulfate (($NH_4)_2SO_4$)
Ampicillin
Autoclaved water
Bis-acrylamide
BME Vitamins 100× Solution (Sigma)
Boric acid (H_3BO)
Bromophenol blue
Calcium chloride ($CaCl_2$)
Cobalt (II) chloride hexahydrate ($CoCl_2 \cdot 6H_2O$)
Copper (II) chloride dihydrate ($CuCl_2 \cdot 2H_2O$)
Ethylenediaminetetraacetic acid (EDTA), anhydrous
Ethidium bromide (EtBr)
Formamide
Glacial acetic acid (CH_3COOH)
Glucose ($C_6H_{12}O_6$)
Glycerol ($C_3H_8O_3$)
Hydrochloric acid (HCl)
Iron (III) chloride hexahydrate ($FeCl_3 \cdot 6H_2O$)
K12 wild-type E. coli (3722 (CGSC # 4401:F+) from the Coli Genetic Stock Center (CGSC))
Kanamycin
Magnesium chloride ($MgCl_2$)
Magnesium sulfate ($MgSO_4$)
Manganese (II) chloride tetrahydrate ($MnCl_2 \cdot 4H_2O$)
Nickel (II) chloride hexahydrate ($NiCl_2 \cdot 6H_2O$)
pBSKrnaSeph plasmid (kindly provided by Dr. Frédéric Dardel)
Potassium chloride (KCl)
Potassium phosphate, dibasic (K_2HPO_4)

Potassium phosphate, monobasic (KH_2PO_4)
SalI restriction enzyme (New England Biolabs)
Shikimic acid
Sodium acetate (NaOAc)
Sodium chloride (NaCl)
Sodium dodecyl sulfate (SDS)
Sodium hydroxide (NaOH)
Sodium molybdatepentahydrate ($Na_2MoO_4 \cdot 5H_2O$)
Sodium phosphate, dibasic (Na_2HPO_4)
Sodium selenite pentahydrate ($Na_2SeO_3 \cdot 5H_2O$)
Source 15Q anion exchange resin (GE Healthcare)
Spermidine
Tetramethylethylenediamine (TEMED)
tktA mutant *E. coli* (CGSC # 11606, F-Δ(araD-araB)567, ΔlacZ4787(::rrnB-3), λ^-, ΔtktA783::kan, rph-1, Δ(rhaD-rhaB)568, hsdR514)
Tris base
Tryptone
Urea
Xylene cyanol
Yeast extract
Zinc sulfate heptahydrate ($ZnSO_4 \cdot 7H_2O$)

3.1 Solutions for Bacterial Growth

100 mg/mL Ampicillin stock: Dissolve 1 g ampicillin (amp) in 10 mL autoclaved water (final volume). Filter-sterilize and store at $-20\,°C$ in 1-mL aliquots.

100 mg/mL Kanamycin stock: Dissolve 1 g kanamycin in 10 mL autoclaved water (final volume). Filter-sterilize and store at $-20\,°C$ in 1-mL aliquots.

Luria–Bertani–Miller-Amp (LB-Amp)-Rich Media

Component	Final Concentration (w/v)	Stock	Amount (g)
Tryptone	0.10%	–	10
Yeast extract	0.05%	–	5
NaCl	0.10%	–	10

Adjust pH to 7.0 using 10 M NaOH. Add water to 1 L. Autoclave. Add 1 mL ampicillin stock to a final concentration of 100 µg/mL and store at 4 °C.

2× TY-Amp-Rich Media

Component	Final Concentration (w/v)	Stock	Amount (g)
Tryptone	0.16%	–	16
Yeast extract	0.10%	–	10
NaCl	0.05%	–	5

Adjust to pH 7.0 using 10 M NaOH. Add water to 1 L. Autoclave. Add 1 mL ampicillin stock to a final concentration of 100 µg/mL and store at 4 °C.

10% (w/v) Glucose stock: Dissolve 10 g glucose in 100 mL heated autoclaved water (final volume). Filter-sterilize and store at 4 °C. Isotopically labeled 1-^{13}C-glucose is freshly prepared in 20 mL aliquots. Dissolve 2 g 1-^{13}C-glucose in 20 mL heated autoclaved water. Filter-sterilize.

15% (w/v) NaOAc stock: Dissolve 15 g NaOAc in 100 mL autoclaved water (final volume). Adjust pH to 7.4 using glacial acetic acid. Filter-sterilize. Isotopically labeled 1-^{13}C-acetate is freshly prepared in 20 mL aliquots. Dissolve 3 g 1-^{13}C-acetate in 20 mL autoclaved water. Filter-sterilize.

1 mg/mL Shikimic acid stock: Dissolve 50 mg shikimic acid in 50 mL autoclaved water (final volume). Filter-sterilize and store at 4 °C.

Studier Phosphate Buffer (SPG) (Studier, 2005; Thakur, Brown, Sama, Jackson, & Dayie, 2010)

Component	Final Concentration (mM)	Stock	Amount (g)
$(NH_4)_2SO_4$	25	–	3.3
KH_2PO_4	50	–	6.8
Na_2HPO_4	50	–	7.1

Add water to 1 L. Adjust pH to 7.0 using approximately 5 mL 10 M NaOH, autoclave, and store at 4 °C. To prepare ^{15}N-labeled media, replace unlabeled $(NH_4)_2SO_4$ with ^{15}N-$(NH_4)_2SO_4$.

10,000× Trace Metals Solution (Thakur, Brown, et al., 2010; Tyler et al., 2005)

Component	Final Concentration (mM)	Stock (M)	Amount (mL)
$FeCl_3 6H_2O$ in 0.1 M HCl	50	0.1	5
$CaCl_2$	20	1	0.2
$MnCl_2 \cdot 4H_2O$	10	1	0.1
$ZnSO_4 \cdot 7H_2O$	10	1	0.1

$CoCl_2 \cdot 6H_2O$	2	0.2	0.1
$CuCl_2 \cdot 2H_2O$	2	0.1	0.2
$NiCl_2 \cdot 6H_2O$	2	0.2	0.1
$Na_2MoO_4 \cdot 5H_2O$	2	0.1	0.2
$Na_2SeO_3 \cdot 5H_2O$	2	0.1	0.2
H_3BO	2	0.1	0.2

Autoclave all stock solutions of individual metals except acidified $FeCl_3 \cdot 6H_2O$ (filter-sterilize). Add autoclaved water to 10 mL. Wrap in aluminum foil.

Completed SPG-Amp Minimal Media for 1 L Growth with Either Glucose or Acetate (Hoffman & Holland, 1995; Thakur, Brown, et al., 2010; Thakur et al., 2012)

Component	Final Concentration	Stock	Amount
SPG	–	–	960–980 mL
$MgSO_4$	2 mM	2 M	1 mL
Trace metals solution	0.2 ×	10,000 ×	20 μL
BME Vitamins 100 × Solution	0.25 ×	100 ×	2.5 mL
Ampicillin	100 μg/mL	100 mg/mL	1 mL
Choose a carbon source			
Glucose	0.2% (w/v) or 0.4% (w/v)	10% (w/v)	20 mL for 0.2% (w/v) or 40 mL for 0.4% (w/v)
NaOAc	0.3% (w/v)	15% (w/v)	20 mL
When growing tktA mutant *E. coli*			
Shikimic acid	25 μg/mL	1 mg/mL	25 mL
Kanamycin	30 μg/mL	100 mg/mL	300 μL

Store at 4 °C for up to 1 week.

LB–agar plates: Add 7.5 g bacteriological agar to 500 mL LB and autoclave. Allow the media to cool to approximately 50 °C and add 0.5 mL 100 mg/mL ampicillin stock. Pour 10 mL (each) into petri dishes. Let the liquid solidify and store at 4 °C.

SPG–agar plates: Add 7.5 g bacteriological agar to 500 mL SPG and autoclave. Allow the media to cool to approximately 50 °C and then add 0.5 mL 2 M MgSO$_4$ stock, 10 µL trace metals solution, 2.5 mL BME Vitamins 100× Solution, 0.5 mL ampicillin stock, and 10 mL 10% (w/v) glucose stock or 15% (w/v) NaOAc stock. When growing tktA *E. coli*, also add 12.5 mL shikimic acid stock and 150 µL kanamycin stock. Pour 10 mL (each) into petri dishes. Let the liquid solidify and store at 4 °C.

3.2 Solutions for Recombinant tRNA-Scaffold Purification

1 M Tris–HCl, pH 7.0 stock: Dissolve 121 g in 1 L water (final volume). Adjust to pH 7.0 with 35.5% (w/v) HCl. Autoclave.

Lysis Buffer

Component	Final Concentration (mM)	Stock (M)	Amount (mL)
Tris–HCl, pH 7.0	10	1	10
MgCl$_2$	10	1	10

Add autoclaved water to 1 L.

1 M Potassium phosphate buffer, pH 7.0 stock: Dissolve 107.12 g K$_2$HPO$_4$ and 52.39 g KH$_2$PO$_4$ in 1 L water (final volume). Autoclave.

Purification buffer: 40 mM Potassium phosphate buffer, pH 7.0: Add 40 mL 1 M potassium phosphate buffer stock, pH 7.0–960 mL autoclaved water. Filter-sterilize.

Elution Buffer: 40 mM Potassium Phosphate Buffer, pH 7.0 and 1 M NaCl

Component	Final Concentration	Stock	Amount
Potassium phosphate buffer, pH 7.0 stock	40 mM	1 M	40 mL
NaCl	1 M	–	58.44 g

Add autoclaved water to 1 L. Filter-sterilize.

Purification Buffer Containing 100 mM NaCl

Component	Final Concentration (mM)	Stock (M)	Amount (mL)
Potassium phosphate buffer, pH 7.0 stock	40	1	40
NaCl	100	5	20

Add autoclaved water to 1 L. Filter-sterilize.

Purification Buffer Containing 4 M Urea

Component	Final Concentration	Stock	Amount
Potassium phosphate buffer, pH 7.0 stock	40 mM	1 M	40 mL
Urea	4 M	–	240 g

Add autoclaved water to 1 L. Filter-sterilize.

3.3 Solution for DNAzyme Cleavage

2 × DNAzyme Buffer

Component	Final Concentration	Stock (M)	Amount
MOPS, pH 7.2	100 mM	1	5 mL
NaCl	1 M	5	10 mL
KCl	250 mM	1	12.5 mL
MgCl$_2$	15 mM	1	750 µL
MnCl$_2$·4H$_2$O	30 mM	1	1.5 mL

Autoclave all stock solutions. Add autoclaved water to 50 mL and store at 4 °C. Add freshly prepared 100 mM spermidine to each reaction for a final concentration of 1 mM spermidine.

3.4 Solutions for Denaturing PAGE

EDTA, pH 8.0 stock solution: Add 800 mL autoclaved water to 146 g EDTA under magnetic stirring. Slowly add 10 M NaOH to adjust the pH to 8.0 and dissolve the EDTA. Add water to 1 L and filter-sterilize.

10 × Tris-Borate-EDTA (TBE) Buffer Stock

Component	Final Concentration	Stock	Amount
Tris	0.9 M	–	108 g
Boric acid	0.9 M	–	240 g
EDTA, pH 8.0	10 mM	500 mM	20 mL

Add water to 1 L. Sterile-filter.

To make 1 L 1 × TBE running buffer, add 100 mL 10 × TBE to 900 mL water.

1 × Tris–Borate–EDTA (TBE) Buffer, 8 M Urea

Component	Final Concentration	Stock	Amount
TBE	1 ×	10 ×	100 mL
Urea	8 M	–	480.5 g

Add water to 1 L. Sterile-filter.

40% (w/v) 19:1 Acrylamide/bis-acrylamide: Dissolve 1000 g acrylamide and 52.6 g bis-acrylamide in 2631.5 mL autoclaved water (final volume). Wrap in aluminum foil.

20% (w/v) 19:1 Acrylamide/Bis-acrylamide Denaturing PAGE Stock Solution

Component	Final Concentration	Stock	Amount
40% (w/v) 19:1 Acrylamide/bis-acrylamide	10% (w/v)	20% (w/v)	500 mL
TBE	1 ×	10 ×	100 mL
Urea	8 M	–	480.5 g

Add water to 1 L. Wrap in aluminum foil.

Analytical 10% (w/v) 19:1 Acrylamide/Bis-Acrylamide Denaturing PAGE Gel

Component	Final Concentration	Stock	Amount
20% (w/v) 19:1 Acrylamide/bis-acrylamide denaturing PAGE stock solution	10% (w/v)	20% (w/v)	5 mL
1 × TBE, 8 M urea	1 ×	1 ×	5 mL
APS	0.07% (w/v)	10% (w/v)	70 µL
TEMED	0.2% (v/v)	100%	20 µL

Allow the gel to polymerize for 20 min.

Preparative 10% (w/v) 19:1 Acrylamide/Bis-acrylamide Denaturing PAGE Gel

Component	Final Concentration	Stock	Amount
20% (w/v) 19:1 Acrylamide/bis-acrylamide denaturing PAGE stock solution	10% (w/v)	20% (w/v)	200 mL
1× TBE, 8 M urea	1×	1×	200 mL
APS	0.05% (w/v)	10% (w/v)	2 mL
TEMED	0.04% (v/v)	100%	150 µL

Allow the gel to polymerize for 1 h.

2× RNA Loading Buffer

Component	Final Concentration	Stock	Amount
Formamide	95% (v/v)	100%	9.5 mL
SDS	0.025% (w/v)	10% (w/v)	25 µL
EDTA, pH 8.0	0.5 mM	50 mM	100 µL
Bromophenol blue	0.05% (w/v)	–	5 mg
Xylene cyanol	0.05% (w/v)	–	5 mg

Add autoclaved water to 10 mL.

4. PROTOCOL
4.1 Duration

Preparation	2–3 Weeks
Step 1	2 days, O/N
Step 2	3 days
Step 3	2 days, O/N
Step 4	1 day
Step 5a or Step 5b	2 days, O/N
Step 6	1 day
Step 7	2–4 days
Step 8	Depends on experiments

4.2 Preparation

Subclone the RNA of interest into the pBSKrnaSeph plasmid using PCR and "sticky-end" ligation techniques found in Ponchon et al. (2009). In brief, the cDNA insert is generated by PCR (for long RNAs) or by synthetic oligonucleotide assembly (for short RNAs) and then digested along with the plasmid using the restriction enzymes *Aat*II and *Sal*I to create sticky ends. The insert is ligated with T4 DNA ligase (Ponchon et al., 2009). Confirm correct ligation by sequencing.

Note that several different plasmids can be used for cloning to accommodate different RNA insert sequences and purification procedures, which are listed in Ponchon et al. (2009).

Wild-type K12 and mutant tktA *E. coli* are made chemically competent using $CaCl_2$ following a standard protocol (Nakata, Tang, & Yokoyama, 1997). *E. coli* are stored in glycerol at -80 °C. The isotopically labeled carbon and nitrogen substrates can be purchased from commercial sources (Isotec, Cambridge Isotope Laboratories).

4.3 Caution

When designing your vector, there are several important things to consider. First, it is necessary to separate the RNA of interest from other cleavage products of the final DNAzyme reaction, which include the D arm and the T arm of the $tRNA_{lys}$-scaffold, using denaturing PAGE (see Step 5). Therefore, extra nucleotides can be added to the D and T arms of the tRNA-scaffold to create a clear size difference. Second, it is important to consider the first and last nucleotide of the inserted RNA, as they will form the $5'$ and $3'$ dinucleotide cleavage sites of the DNAzymes. According to Schlosser et al. (2008), the dinucleotide sites most efficiently cleaved are $5'$-GG-$3'$, $5'$-GA-$3'$, and $5'$-AG-$3'$ (Schlosser et al., 2008). Note that we have only tested these three dinucleotide cleavage sites. To use other dinucleotide sequences, further optimization may be necessary.

It is important that all microbiology equipment be sterile to avoid contamination of bacterial cultures. Additionally, all solutions must be autoclaved or sterile-filtered before being used for *E. coli* growth. Furthermore, RNAse-free conditions are crucial for maintaining the integrity of the RNA products. Therefore, it is important to keep microbiology equipment separate from RNA handling equipment. Again, equipment should be autoclaved or purchased RNase, DNAse, and pyrogen-free.

5. STEP 1: PILOT OF THE EXPRESSION OF THE RECOMBINANT tRNA-SCAFFOLD PLASMID IN WILD-TYPE K12 *E. COLI*

5.1 Overview

In this step, RNA minipreps are used to assess the expression of the newly constructed vector by K12 wild-type *E. coli* in $2\times$ TY-amp-rich media grown over night (Ponchon et al., 2009). Total cellular RNA is extracted using acid phenol:chloroform (5:1). Proteins, DNA, carbohydrates, and lipids are precipitated or partitioned into the organic phase, while RNA is partitioned into the aqueous phase. The expression of the recombinant tRNA-scaffold is confirmed using analytical denaturing PAGE (Fig. 3).

5.2 Duration

2 days, overnight

(1) *Transformation*: Thaw 10 µL competent K12 cells on ice.
(2) Add 10 ng plasmid to the competent cells and gently tap the tube to mix. Incubate on ice for a minimum of 30 min.

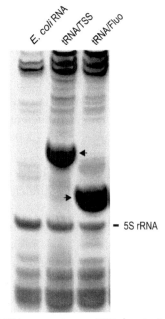

Figure 3 Visualize piloted RNA expression on 10% denaturing PAGE. The RNA PAGE gel was stained with EtBr and visualized under UV light. Arrows indicate the expressed recombinant tRNA-scaffolds.

(3) Heat-shock the cells in a water bath at 42 °C for 40 s. Immediately incubate the tube on ice for a minimum of 2 min.
(4) Add 200 μL LB-amp and spread the bacteria over an LB–agar plate. Incubate at 37 °C for 12–16 h.
(5) Choose several colonies and use each to inoculate 10 mL 2 × TY-amp in 50 mL conical tubes or 100 mL glass culture flasks. Incubate at 37 °C, 300 RPM overnight.
(6) *RNA extraction*: Pellet cells by centrifugation at 4 °C, $1700 \times g$ (Sorvall Legend XTR tabletop centrifuge with swinging bucket rotor) for 1 min.
(7) Resuspend pellets in 500 μL lysis buffer and transfer to 1.5-mL microcentrifuge tubes.
(8) Add an equal volume (500 μL) of acid phenol:chloroform (5:1), pH 4.5, and vortex the mixture.
(9) Incubate on a nutating tabletop shaker at 4 °C for 30 min.
(10) Centrifuge at 4 °C, $17,000 \times g$ for 1 min (refrigerated microcentrifuge). Carefully transfer 400 μL of the aqueous phase to a new tube.
(11) *Denaturing PAGE analysis*: Add 60 μL of 0.5 M EDTA, pH 8.0 to each sample to a final concentration of approximately 80 mM to prevent the interference of salts in the denaturing PAGE analysis.
(12) Mix 2 μL of each sample with 2 μL of 2× RNA loading buffer and analyze using a 10% (v/v) denaturing PAGE.
(13) Visualize the gel by soaking in 0.5 μg/mL EtBr to assess expression of the recombinant tRNA-scaffolds.

5.3 Caution

Phenol is toxic and may also dissolve certain plastics. Use glass or phenol-resistant plastic equipment and handle in a fume hood.

5.4 Tip

A nonrecombinant tRNA-scaffold without an insert can be expressed similarly and used as a control.

5.5 Tip

When growing *E. coli* in 50-mL conical tubes, partially unscrew the cap to allow sufficient aeration of the culture.

5.6 Tip

If the recombinant tRNA-scaffold is poorly expressed in rich media, there may be a mutation in the promoter. In addition, highly structured RNAs tend to show higher levels of expression in E. *coli* due to enhanced protection from cellular nucleases.

6. STEP 2: DOUBLE SELECTION OF HIGH-EXPRESSING E. COLI CLONES

6.1 Overview

Large-scale batch cultures starting from single transformed colonies produced high variability of recombinant tRNA-scaffold expression (Fig. 4). To resolve this issue, a double selection process was adapted for optimizing expression of recombinant proteins (Sivashanmugam et al., 2009). This double selection approach was first applied for the E. *coli* mutant strain tktA in SPG-amp supplemented with 0.2% (w/v) $1\text{-}^{13}C$-glucose. Competent tktA is transformed with the plasmid and plated on SPG–agar plates supplemented with 0.2% (w/v) $1\text{-}^{13}C$-glucose. Five random colonies are chosen and

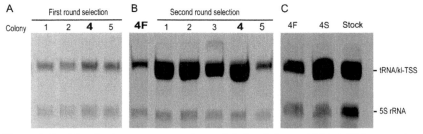

Figure 4 Double selection process to select a tktA E. *coli* clone with high RNA expression in 0.2% (w/v) glucose and D_2O. (A) Five colonies were first grown on 0.4% (w/v) glucose and D_2O-agar plate and then inoculated with selected media. Only four colonies grew. When the absorbance of the culture reached an OD_{600} of 0.5, 10 μL of the liquid culture was spread on new agar plates made with the desired selection media. The total RNA of the remaining culture was collected at an OD_{600} of 1.0, extracted, and visualized by 10% (w/v) denaturing PAGE. The PAGE gel was stained with EtBr. Colony 4 (bold 4F), which yielded slightly higher tRNA/kl-TSS expression, was subjected to a second round of selection. (B) The growth process was repeated using five progeny colonies derived from colony 4F. At an OD_{600} of 0.5, a glycerol cell stock was prepared by mixing equal volumes of liquid culture with 30% (v/v) glycerol. The expression of each colony was compared using 10% (w/v) denaturing PAGE. Progeny colony 4 (bold 4S) exhibited high expression compared to the expression of parental colony 4 (4F). (C) Cells of the glycerol stock of colony 4S were reinoculated. Total cellular RNA was extracted and compared to the total cellular RNA of the parental colonies in each selection round.

grown in rich media. The cells are pelleted and grown further in SPG-amp supplemented with 0.2% (w/v) 1-^{13}C-glucose. The total cellular RNA of each colony is extracted and analyzed using denaturing PAGE to select the colony that best expresses the recombinant tRNA-scaffold (Fig. 4A). Five progeny colonies derived from the initial colony are selected at random and the process is repeated, resulting in a glycerol stock of a single clone that can be used to inoculate all future batch cultures (Fig. 4A and B).

6.2 Duration
3 days
(1) Prepare agar plates made with the final media desired for the large-scale batch culture. When growing tktA, the final media is SPG-amp supplemented with 0.2% (w/v) glucose, 25 μg/mL shikimic acid, and 30 μg/mL kanamycin.
(2) Transform and plate the desired *E. coli* strain (tktA) as described in Step 1.
(3) *First selection*: Randomly select five individual colonies. Use each colony to inoculate 1 mL LB-amp in 1.5-mL microcentrifuge tubes. Incubate at 37 °C, 300 RPM for 3 h.
(4) Pellet the cultures at 4 °C at $1700 \times g$ for 1 min. Remove 900 μL media and resuspend each pellet in the remaining media.
(5) Use each solution to inoculate 10 mL of the desired final media, SPG-amp supplemented with 0.2% (w/v) glucose, 25 μg/mL shikimic acid, and 30 μg/mL kanamycin, in 50-mL conical tubes or 100-mL glass culture flasks.
(6) Measure the absorbance of the culture at 600 nm (OD_{600}) every hour. At an OD_{600} of 0.5, remove 100 μL of each culture and spread it on new agar plates labeled colony 1–5. Incubate as above. Allow the cultures to grow until an OD_{600} of 1.0.
(7) Extract total cellular RNA from each culture and compare the recombinant tRNA-scaffold expression of each culture using 10% (w/v) denaturing PAGE analysis as described in Step 1. Choose the plate of the colony with the highest recombinant RNA expression.
(8) *Second selection*: Randomly select five progeny colonies from the plate of colony from Step 7. Use each progeny colony to inoculate 1 mL of LB-amp in 1.5-mL microcentrifuge tubes.
(9) Measure the OD_{600} of the culture each hour. At an OD_{600} of 0.5, remove 500 μL of each culture and add it to 500 μL 30% (v/v) glycerol

to create a glycerol stock for each colony. Label each of the glycerol stocks and store them at −80 °C. Continue incubating the cultures.

(10) At an OD_{600} of 1.0, extract total cellular RNA from each culture and compare the recombinant tRNA-scaffold expression using 10% (v/v) denaturing PAGE analysis. Use the glycerol stock of the colony with the strongest recombinant RNA expression to inoculate future batch cultures of mutant *E. coli* in the appropriate minimal media.

6.3 Tip

Due to the expense of labeled materials, unlabeled materials should be used first for the double selection process.

6.4 Tip

To save resources, you may use sectioned or mini petri dishes.

6.5 Tip

Pipette the plasmid gently into the competent *E. coli* to avoid damaging the cells.

6.6 Tip

Warm up the agar plates to 37 °C before spreading the bacteria for faster drying.

6.7 Tip

Remove plates from the incubator before colonies overlap or satellite colonies form.

6.8 Tip

To maintain high viability of the glycerol stock when selecting a colony in SPG-amp supplemented with acetate, we recommend pelleting the cells and replacing the media with LB-amp before preparing the glycerol stock in the second selection step.

7. STEP 3: LARGE-SCALE EXPRESSION IN LABELED SPG MINIMAL MEDIA

7.1 Overview

In this step, cells are slowly enriched in 10 mL rich media and 100 mL minimal media cultures before scaling up to a 1-L batch culture. The cell density of each culture is closely monitored throughout the process. To maintain the growth rate, the initial OD_{600} of the culture should be 0.05 (Thakur, Brown, et al., 2010). This is because RNAs are highly expressed in the log phase bacterial growth, so cells should be transferred to the next largest volume when their OD_{600} is 0.5. In general, we found that the maximum OD_{600} correlated to maximum recombinant tRNA-scaffold expression. Therefore, at 1 L growth, cells are collected 2 h into the stationary phase to maximize the recombinant tRNA-scaffold yield.

7.2 Duration

2 days, overnight

(1) *Starter culture*: Inoculate 10 mL LB-amp with 1 μL of the glycerol stock of colony 4S (Fig. 4C) in a 50-mL conical tube or 100-mL glass culture flask. Incubate at 37 °C, 300 RPM.

(2) Measure the OD_{600} of the culture each hour. When it reaches an OD_{600} of 0.5 (approximately 4 h), pellet the culture by centrifugation at 16 °C, $1700 \times g$ for 10 min.

(3) *Mid-scale culture*: Discard the media and resuspend the pellet in 100 mL SPG-amp supplemented with the desired final carbon source in a 500-mL flask. Incubate at 37 °C, 300 RPM.

(4) Again, measure OD_{600} of the culture each hour. When it reaches an OD_{600} of 0.5, pellet the culture by centrifugation at 16 °C, $1700 \times g$ for 10 min.

(5) *Large-scale culture*: Resuspend the pellet in 1 mL SPG-amp and the desired final carbon source. Use this solution to inoculate 2×500 mL of the final culture media in 4 L flasks. Incubate at 37 °C, 300 RPM.

(6) Measure the OD_{600} each hour and collect the cells by centrifugation once they reach the maximum OD_{600} (approximately 2 h into the stationary phase, or 8–16 h depending on the *E. coli* strain and media).

7.3 Tip

Harvesting the recombinant tRNA-scaffold at the correct time is crucial for preventing RNA degradation by cellular nucleases, despite the protective effect of the scaffold. We recommend using unlabeled materials to chart a growth curve and estimate the expected maximum cell density and RNA yield.

7.4 Tip

We use 0.4% (w/v) glucose in the mid-scale culture to increase the *E. coli* growth rate before using 0.2% (w/v) glucose in the large-scale culture. However, due to the expense of isotopically labeled glucose, we use unlabeled glucose for the mid-scale culture. 1-^{13}C-acetate is significantly less expensive, so you should use a final concentration of labeled 0.3% (w/v) 1-^{13}C-acetate in the mid-scale culture as well as in the large-scale culture. ^{15}N-labeled ammonium sulfate may also be used.

7.5 Tip

Note the higher centrifugation temperature (16 °C) in minimal media due to the fragility of the *E. coli* mutants.

7.6 Tip

For large-scale growth in minimal media, we highly recommend growing cells in two 4 L flasks containing 500 mL minimal media.

7.7 Tip

Possible stopping point: Freeze cell pellets at −20 °C.

8. STEP 4: TOTAL CELLULAR RNA EXTRACTION

8.1 Overview

This step is a large-scale RNA prep similar to the RNA minipreps in Step 1. Again, total cellular RNA is separated from proteins, lipids, and carbohydrates using a lysis buffer and acid phenol:chloroform (5:1), pH 4.5. The total cellular RNA containing the recombinant tRNA-scaffold is ethanol precipitated using a high salt concentration. Total cellular RNA is pelleted and air-dried.

8.2 Duration

1 day

(1) Centrifuge the batch culture in two 1 L Nalgene bottles at $4700 \times g$ (Sorvall RC-5C super speed centrifuge with Fiberlite F8-6 × 1000y fixed-angle large capacity rotor) for 30 min at 4 °C.

(2) Weigh the pellet. With the cell pellets in the 50-mL phenol-resistant tubes, add 2.5 mL lysis buffer per g pellet. Vortex vigorously until all the cells are resuspended.

(3) Add 1 × volume of acid phenol:chloroform (5:1), pH 4.5. Vortex vigorously to homogenize.

(4) Incubate on a rotating tabletop shaker at 4 °C for 30 min.

(5) Centrifuge at 4 °C, $12,100 \times g$ (Sorvall RC-5C superspeed centrifuge, Sorvall SS-34 fixed-angle rotor) for 30 min.

(6) Carefully transfer the upper aqueous layer into a new 50-mL Super-Clear Ultra-High Performance centrifuge tube with plug cap. Set aside 200 µL for PAGE analysis of recombinant tRNA-scaffold expression. Add 0.1 × volumes 5 M NaCl and 3 × volumes absolute EtOH. Vortex for 10 s.

(7) Incubate at −20 °C overnight to precipitate the total cellular RNA.

(8) Centrifuge at 4 °C, $12,800 \times g$ (Sorvall Legend XTR tabletop centrifuge, Fiberlite™ F15-8 × 50cy fixed-angle rotor) for 1 h.

(9) Carefully discard the supernatant. Air dry the pellet for 15 min.

8.3 Tip

The average cell pellet from 1 L batch culture weighs 4–8 g.

8.4 Tip

Acidic phenol:chloroform is important because it precipitates DNA, while alkali phenol:chloroform does not (Sambrook & Russell, 2001).

8.5 Tip

tRNAs are preferentially ethanol precipitated in the presence of 0.5 M NaCl rather than 0.3 M NaOAc (also commonly used for RNA precipitation). After ethanol precipitation, RNA should appear as a clear to white precipitate.

8.6 Tip

The total cellular RNA pellet can be resuspended with 70% (v/v) EtOH to remove excess salt before anion-exchange purification.

8.7 Tip
Always use analytical denaturing PAGE to confirm recombinant tRNA-scaffold expression prior to anion-exchange purification.

8.8 Tip
Possible stopping point: Freeze RNA pellet at −20 °C.

9. STEP 5A: PURIFICATION OF THE RECOMBINANT tRNA-SCAFFOLD USING ANION-EXCHANGE CHROMATOGRAPHY

9.1 Overview
Different cellular RNAs have different sizes and shapes, so they possess different electrostatic potentials and charge densities. Due to these RNA characteristics, anion-exchange chromatography can be used to separate the recombinant tRNA-scaffold from other cellular RNAs. Here, we utilize an in-house packed Source 15Q column and an extension of the salt gradients and elution steps used by Nelissen et al. (2012) to purify the recombinant tRNA-scaffold. Collected fractions are confirmed by analytical denaturing PAGE, combined, and solvent exchanged. An example of the anion-exchange purification of tRNA/Fluoride-binding riboswitch (tRNA/Fluo) is presented in Fig. 5 (Baker et al., 2012; Ren, Rajashankar, & Patel, 2012). Alternatively, Section 10 can be utilized depending on the column available and the RNA of interest.

9.2 Duration
2 days, overnight
(1) Resuspend the RNA pellet in 10 mL autoclaved water.
(2) Centrifuge at 4 °C, 12,800 × g (Sorvall Legend XTR tabletop centrifuge, Fiberlite F15-8 × 50cy fixed-angle rotor) for 1 h to pellet any remaining cellular debris. Carefully transfer the supernatant to a new tube.
(3) Load the supernatant into the 50-mL superloop (GE Healthcare) using a 10 mL syringe.
(4) *Preparation of the Source 15Q column*: Pack 50 mL (1 column volume, or 1 CV) Source 15Q resin into a 50-mL XK 26/60 empty column following the manufacturer's instructions. Before running, rinse the column with 2 CV 1× purification buffer and 2 CV 1× elution buffer at 2 mL/min.

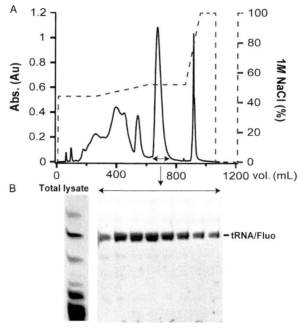

Figure 5 Purification of tRNA/Fluo using the Source 15Q anion-exchange column. (A) The chromatogram was monitored at 254 nm. Cellular RNA, including small RNAs, tRNAs, and 5S rRNA, was eluted using a gradient from 0.44% to 52% (v/v) 1 M NaCl. tRNA/Fluo was eluted at 0.52 M NaCl. An additional step consisting of a high salt concentration, from 52% to 100% (v/v) 1 M NaCl, removed all large RNAs from the column, allowing the Source 15Q column to be reused. (B) The quality of the purified tRNA/Fluo was examined by 10% (w/v) denaturing PAGE. The gel was stained with EtBr and visualized using UV light. Only the fractions containing tRNA/Fluo were loaded onto the gel.

(5) *FPLC purification protocol*:
   A. Equilibrate the Source 15Q column with 5 CV 100% purification buffer.
   B. Inject the sample from the superloop.
   C. Elute the sample with: 5 CV 44% (v/v) elution buffer; 10 CV linear gradient from 44% to 52% (v/v) elution buffer; 5 CV 52% (v/v) elution buffer; 2 CV linear gradient from 52% to 100% (v/v) elution buffer; and 2 CV 100% elution buffer.
   D. Collect elution with a UV absorbance above 0.2 Au at 254 nm in 8 mL fractions.
(6) *Wash and storage of the column*: Wash the column with: 2 CV 0.5 M NaOH, 4 CV 2 M NaCl, and 4 CV autoclaved water. Store the column in 30% (v/v) EtOH by flushing it with 2 CV.

(7) Determine which fractions contain the tRNA-scaffold using 10% (w/v) analytical denaturing PAGE.

(8) Pool the fractions containing the recombinant tRNA-scaffold using a 3 kDa MWCO spin column at 4 °C, $1900 \times g$ (Sorvall Legend XTR tabletop centrifuge with swinging bucket rotor).

(9) Exchange the concentrated fractions $5\times$ with solvent.

9.3 Tip

If the pellet does not dissolve readily, additional autoclaved water can be added. However, some of the precipitate may not dissolve, and any residual debris must be removed by centrifugation before purification.

9.4 Tip

It is advisable to use a superloop to inject your sample to prevent dilution and reduced resolution of the purified recombinant tRNA-scaffold.

9.5 Tip

The FPLC protocol can proceed overnight unattended.

9.6 Tip

The pressure threshold of the column is 0.4 MPa.

9.7 Tip

The recombinant tRNA-scaffold is eluted at 52% (v/v) elution buffer.

9.8 Tip

In the past, we have observed some contaminants in the fractions containing the tRNA-scaffold. You may decide to omit fractions that greatly overlap with other cellular RNAs. We chose to include these fractions due to the second purification step in our protocol, which removes the DNAzymes, 18-nt oligonucleotide, and tRNA-scaffold fragments from the inserted RNA of interest after the DNAzyme digestion of the recombinant tRNA-scaffold.

9.9 Tip

Alternative stopping point: The purified recombinant tRNA-scaffold may be frozen before DNAzyme digestion or exchanged into a buffer as desired for biophysical studies.

10. STEP 5B: PURIFICATION OF THE RECOMBINANT tRNA-SCAFFOLD USING AFFINITY CHROMATOGRAPHY

10.1 Overview

This step is an alternative to Section 9 (Fig. 6). The tRNA-scaffold contains a sephadex tag that binds to dextran resin, so affinity chromatography may also be used depending on the column or resin available and the length of the RNA of interest. Previous protocols used the Sephadex G-100 or G-200 as

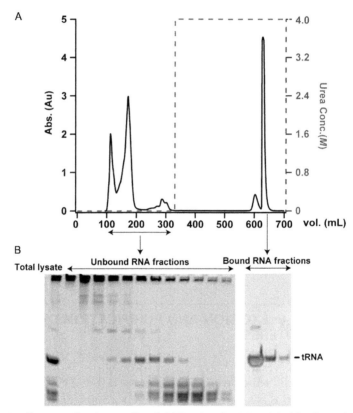

Figure 6 Affinity purification profile of tRNA using the HiLoad 26/600 Superdex 75 pg column. (A) Chromatogram of column fractions: After sample injection, the column was first rinsed with 1 CV 40 mM MOPS pH 7, 150 mM NaCl, and 1.5 CV purification buffer containing 4 M urea. (B) Eluted fractions were analyzed by 10% (w/v) denaturing PAGE. Unbound RNA fractions were loaded from left to right. Large RNAs were eluted first. The bound tRNA was eluted under denaturing conditions in the purification buffer containing urea.

an affinity column (Nelissen et al., 2012; Ponchon et al., 2009). Here, we automate the procedure by using a HiLoad 26/600 Superdex 75 pg column (GE Healthcare). The bound recombinant tRNA-scaffold is eluted with purification buffer containing 4 M urea. However, note that although affinity chromatography can be completed in less time, anion-exchange chromatography yields a purer product.

10.2 Duration

1 day

(1) Prepare the RNA sample and load it into the superloop as described in Step 5a.
(2) *FPLC purification protocol*:
   A. Equilibrate the column with 1 CV purification buffer containing 100 mM NaCl.
   B. Inject the sample from the superloop.
   C. Rinse the column with 1.5 CV purification buffer containing 100 mM NaCl.
   D. Elute the sample with 1 CV purification buffer containing 4 M urea.
   E. Collect elution with a UV absorbance above 0.2 Au in 5 mL fractions.
(3) *Rinse and storage of the column*: Rinse the column with 1 CV of purification buffer and 1 CV water. Store the column by flushing it with 2 CV 30% (v/v) EtOH.
(4) Collect and analyze RNA fraction as described in Step 5a.

11. STEP 6: EXCISION AND PURIFICATION OF THE RNA OF INTEREST

11.1 Overview

In this step, 8–17 RNA-cleaving deoxyribozymes (DNAzymes) are used to excise the inserted RNA of interest from the recombinant tRNA-scaffold (Schlosser et al., 2008). These DNAzymes contain catalytic core sequences specific to dinucleotide RNA cleavage sites flanked by two substrate-recognition domains that hybridize with the RNA on either side of the dinucleotide cleavage site. An additional 18-nt synthetic DNA oligonucleotide (5′-GCCCGAACAGGGACTTGAA-3′), complementary to the 3′ side of the tRNA acceptor stem and T arm of the tRNA-scaffold, is included

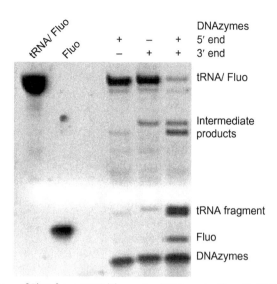

Figure 7 Excision of Fluo from tRNA/Fluo using DNAzymes. Two 8–17 DNAzymes were designed to cleave the 5′ and 3′ ends of the fluoride-binding riboswitch aptamer from tRNA/Fluo. tRNA/Fluo was annealed to either one or both DNAzymes and the 18-nt oligonucleotide by snap-cooling. The mixture was then incubated in DNAzyme buffer for 48 h at RT. The products were analyzed by 10% (w/v) denaturing PAGE.

to melt the tRNA acceptor stem. This allows the DNAzymes to access their specific dinucleotide cleavage sites. The extent of reaction is analyzed using denaturing PAGE (Fig. 7).

11.2 Duration

2–4 days

(1) Design and purchase two DNAzymes to cleave the two ends of the inserted RNA of interest from the recombinant tRNA-scaffold (Integrated DNA Technologies). Select the optimal DNAzyme core sequence based on the dinucleotide pair at the desired cleavage site (Schlosser et al., 2008). Include 12-nt "handle" sequences (complementary to the recombinant tRNA-scaffold sequences on either side of the cleavage site) on both sides of the catalytic core.

(2) Estimate the recombinant tRNA-scaffold concentration of the sample. Add the 5′ and 3′ DNAzymes and 18-nt oligonucleotide that is complementary to the T arm of the tRNA-scaffold in the appropriate ratios. Typical DNAzyme:DNAzyme:oligonucleotide:tRNA-scaffold ratios range from 2:2:2:1 to 4:4:2:1.

(3) Add autoclaved water to achieve approximately 7 μM (7 nmol/mL) recombinant tRNA-scaffold.
(4) Heat the mixture at 90 °C for 3 min. Snap-cool on ice for 10 min.
(5) Add an equal volume of 2× DNAzyme buffer.
(6) Allow the reaction to proceed at room temperature for 48–72 h.
(7) Add 1 μL of 0.5 M EDTA, pH 8.0 and 9 μL of 2× RNA loading buffer to a 10 μL sample of the reaction. Assess the extent of cleavage by 10% (w/v) denaturing PAGE.
(8) If satisfied, quench the reaction by adding 0.5 M EDTA, pH 8.0 to a final concentration of 80 mM.
(9) Exchange the mixture several times into water to remove all salts and then concentrate the solution to less than 1 mL using a 3-kDa MWCO spin column as described above.
(10) Purify the product RNA from DNA oligonucleotides and tRNA-scaffold fragments by 10% (w/v) preparative denaturing PAGE as described previously by Petrov, Wu, Puglisi, and Puglisi (2013). Excise the band corresponding to the product RNA and extract the product RNA from the gel using the Elutrap electroelution system.
(11) Concentrate and exchange the purified RNA into water 5× using a 3 kDa MWCO spin column as described above. Lyophilize and redissolve the sample in the desired NMR buffer.

11.3 Tip

For more accurate determination of RNA concentration, we recommend performing an alkali digestion of a small aliquot of RNA with 1 M NaOH, followed by neutralization with 1 M HCl.

11.4 Tip

We recommend optimization of DNAzyme annealing length, DNAzyme: tRNA-scaffold ratio, and reaction time to achieve the optimal yield of final product. (The ranges provided have been optimized for the RNAs studied in our work.) All optimizations should be carried out using unlabeled materials.

11.5 Tip

It is also useful to run small-scale reactions containing unlabeled recombinant tRNA-scaffold and each DNAzyme separately. The patterns provided on an analytical denaturing PAGE gel will allow you to confirm which band on your preparative PAGE gel is your final RNA product. This is especially recommended if the length of your desired product RNA is similar in molecular weight to one or both arms of the tRNA-scaffold.

11.6 Tip

The annealing step should occur only in water, without salt, to prevent degradation of the RNA. Therefore, it is critical to buffer exchange the recombinant tRNA-scaffold extensively with water after anion-exchange purification.

11.7 Tip

Note that the DNAzymes and 18-nt oligonucleotide may also be recovered during PAGE purification and electroelution to reduce expense. The DNAzymes will not be separated because they are nearly the same molecular weight, but this is inconsequential because they will later be used in the reaction together.

11.8 Tip

If the products of the DNAzyme cleavage reaction are similar in molecular weight, you may want to use 12% (w/v) or 15% (w/v) preparative denaturing PAGE.

11.9 Tip

Alternate stopping point: The quenched reaction or final, solvent exchanged product can be stored at -20 °C.

12. STEP 7: NMR APPLICATIONS
12.1 Overview

Major advances in protein-labeling technologies inspired various new multidimensional heteronuclear NMR techniques that have revolutionized solution NMR structure determination of proteins (Carlomagno, 2014; Dominguez, Schubert, Duss, Ravindranathan, & Allain, 2011; Gardner & Kay, 1998; Gobl, Madl, Simon, & Sattler, 2014; Sattler, Schleucher, & Griesinger, 1999). In a similar vein, comparable advances in RNA labeling technologies are important for moving the field of RNA NMR forward. The *in vivo* labeling technologies described here join a slew of other advances *in vitro* labeling approaches that are beginning to help us along that path

(Alvarado, LeBlanc, et al., 2014; Alvarado, Longhini, et al., 2014). The hybrid method proposed here combines the best of the tRNA-scaffold approach (Ponchon et al., 2009) and the selective biomass bioproduction method (Hoffman & Holland, 1995; Thakur et al., 2012).

To showcase the versatility of our approach, we show 2D ^{1}H–^{15}N and 2D ^{1}H–^{13}C correlation spectra as well as a 3D-edited HNCO experiment (Dingley, Masse, Feigon, & Grzesiek, 2000; Muhandiram & Kay, 1994). First, the imino region of the ^{1}H–^{15}N HSQC correlation spectrum of the 71-nt kl-TSS RNA (a cap-independent translation enhancer located in the 3′ UTR of *Pea enation mosaic virus*) depicts the expected number of cross peaks that corresponds to the A–U and G–C Watson–Crick base pairs in this RNA (Fig. 8A). Indeed, the A–U and G–C Watson–Crick base paired spectral regions of the tRNA/kl-TSS overlay well with the peaks of the tRNA-scaffold alone and those of the kl-TSS RNA excised from the tRNA-scaffold (Fig. 8B).

While this ^{15}N-labeling is useful for deciphering the Watson–Crick base paired regions in an RNA, it has very limited utility compared to ^{13}C-labeling that allows access to more widely distributed sites within different RNA structural elements and provides several benefits over ^{15}N only labeling (Dayie, Brodsky, & Williamson, 2002; Dayie, Tolbert, & Williamson, 1998). *In vivo* labeling of RNA using an *E. coli* strain deficient in the transketolase gene (tktA) enables site-specific placement of ^{13}C isotopes at predictable locations within the ribose and nucleobase moieties, depending on the labeling pattern of the carbon source used (Josephson & Fraenkel, 1969, 1974; Thakur et al., 2012). *In vivo* production of RNA grown within the tktA *E. coli* strain on 1-^{13}C-glucose affords selective isotopic enrichment of the ribose C5′ positions (∼75%) C1′ positions (∼20%) and negligible labeling elsewhere in the sugar ring (<5% for C2′, C3′, and C4′ positions) (Thakur et al., 2012). A clear benefit of this labeling is that unwanted one bond ^{13}C1′–^{13}C2′ or ^{13}C4′–^{13}C5′ scalar and dipolar couplings are completely removed (Fig. 9). Another added benefit of RNAs synthesized in tktA grown in 1-^{13}C-glucose is that purine C2 and C8 carbon atoms and pyrimidines C5 and C6 carbon atoms are selectively enriched with the ^{13}C isotope without any residual carbon–carbon coupling (Fig. 9), as reported previously (Thakur et al., 2012). We anticipate that these labeling patterns will be useful for many structural, dynamic, and functional RNA studies.

A third benefit of this hybrid approach is that, in addition to the protonated carbon sites typically probed in traditional NMR studies, our *in vivo* labeling provides direct access to the nonprotonated carbonyl carbon

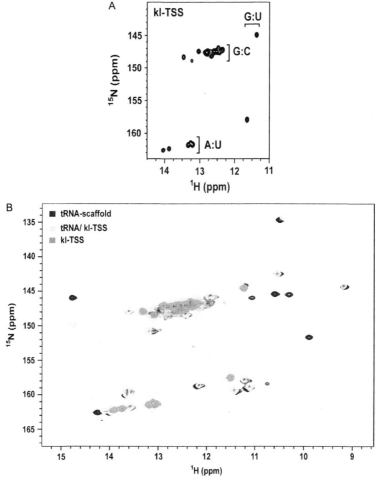

Figure 8 (A) ^{1}H–^{15}N HQSC spectra of the kl-TSS. Tentative base pairs are labeled. (B) Overlay of ^{1}H–^{15}N HQSC spectra of the tRNA-scaffold (blue), tRNA/kl-TSS (cyan), and the kl-TSS (purple). (See the color plate.)

groups, such as the C2 and C4 of uridines and C6 of guanosine. These nonprotonated sites are involved in hydrogen bonding, and their chemical shift appears sensitive to the nature of the hydrogen bond found in A–U, G–U, or U–U base pairs (Furtig et al., 2003). *In vivo* production of tRNA/SAM-II (*S*-adenosylmethionine (SAM) metabolite binding riboswitch) (Chen, Zuo, Wang, & Dayie, 2012) using K12 *E. coli* grown in SPG-amp supplemented with ^{15}N$_2$-ammonium sulfate and 1-^{13}C-acetate as the sole carbon source enabled the following sites to be specifically enriched with ^{13}C isotopes: The C4 carbons of all rNTPs, the C2 carbons

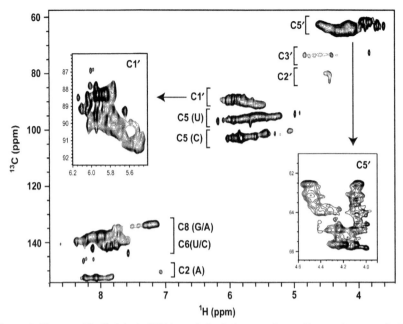

Figure 9 Site-specifically labeled RNA made in tktA grown in media supplemented with 1-^{13}C-glucose. The carbon-labeling pattern of RNAs made in tktA is represented by the 2D ^{1}H–^{13}C spectra of the kl-TSS RNA. The kl-TSS RNA was labeled at C5 and C6 of pyrimidines, C8 of purines, and C1′ (right inset) and C5′ of ribose (bottom inset). The C2′ and C3′ ribose are labeled with low efficiency.

of cytosines and uridines, the C6 carbons of guanosines and adenosines, and the C2′ and C3′ ribose carbons. With these labeling patterns, the imino protons could be readily linked with the carbonyl carbons using 3D HNCO NMR experiments (Figs. 10A). The proton and nitrogen chemical positions within the 3D ^{15}N-edited HNCO spectrum of SAM-II corresponded exactly with those obtained in traditional 2D imino^{1}H–^{15}N HSQC spectrum (Fig. 10B). Using HNCO for signal readout should remove spectra overlap by spreading the ^{1}H–^{15}N correlation map along a third dimension to resolve overlapped proton and nitrogen resonances and allow mapping sites of ligand or drug binding to RNAs of biological interest (Fig. 10A and B).

13. CONCLUSION

We present here a fast, efficient, and cost-effective protocol for obtaining mg quantities of uniformly labeled ^{15}N and site-specifically

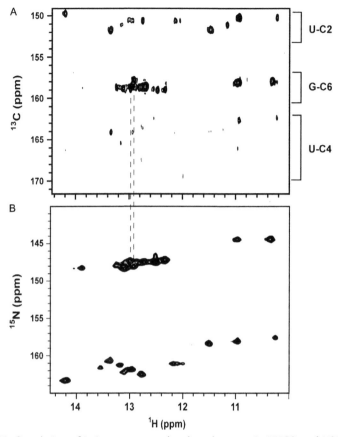

Figure 10 Correlation of imino groups and carbonyl groups in HNCO and HSQC using SAM-II (52 nt). (A) 3D ^{13}C-edited HNCO spectra. The carbonyl groups of U-C2, C4, and G-C6 were labeled. (B) 2D ^{1}H–^{15}N HSQC spectra of SAM-II.

labeled ^{13}C RNAs for NMR structural studies. Our method has many advantages, including: (1) use of inexpensive starting materials to prepare *in vivo*, homogeneous RNAs; (2) labels that overcome chemical shift overlap problems and remove scalar coupling and ^{13}C–^{13}C dipolar coupling of large RNAs due to site-specific labeling; (3) labeled RNAs that can be recycled by nuclease digestion followed by rephosphorylation to produce NTPs for making isotopically labeled RNAs. In addition, a variety of multi-dimensional NMR experiments can be applied using our labeling schemes to illuminate RNA function. Examples of some of those applications (2D ^{1}H–^{15}N/^{1}H–^{13}C correlation spectra as well as 3D HNCO) were presented to illustrate the benefits of this labeling approach. We anticipate that this hybrid selective biomass and *in vivo* tRNA-scaffold approach for synthesizing RNAs will open up new avenues of RNA studies using multidimensional NMR technologies to better characterize the structural and dynamic basis of RNA function.

ACKNOWLEDGMENTS

We would like to thank Dr. Frédéric Dardel for providing the recombinant tRNA-scaffold plasmid. We would also like to thank Dr. Chandar Thakur, Dr. Bin Chen, Dr. Luigi Alvarado, Andrew Longhini, and Regan LeBlanc for their expertise in *E. coli* growth and NMR experiments. Support for this research was provided by a CMB training grant NIGMS T32GM080201 (M.T.L.), NSF grants MCB 1411836 and MCB 1300425 (A.E. S.), NSF CHE1213668 (T.K.D.), and NSF DBI1040158 (T.K.D.). Support for this research was also provided by a grant to the University of Maryland from the Howard Hughes Medical Institute Undergraduate Science Education Program (R.E.B).

REFERENCES

Ahmad, S. (2006). RNA world—Behind the scenes. *Nature Reviews. Genetics, 7*, 242–243. http://dx.doi.org/10.1038/nrg1852.

Alvarado, L. J., LeBlanc, R. M., Longhini, A. P., Keane, S. C., Jain, N., Yildiz, Z. F., et al. (2014). Regio-selective chemical-enzymatic synthesis of pyrimidine nucleotides facilitates RNA structure and dynamics studies. *Chembiochem, 15*, 1573–1577. http://dx.doi.org/10.1002/cbic.201402130.

Alvarado, L. J., Longhini, A. P., LeBlanc, R. M., Chen, B., Kreutz, C., & Dayie, T. K. (2014). Chemo-enzymatic synthesis of selectively $^{13}C/^{15}N$-labeled RNA for NMR structural and dynamics studies. *Methods in Enzymology, 549*, 133–162. http://dx.doi.org/10.1016/B978-0-12-801122-5.00007-6.

Baker, J. L., Sudarsan, N., Weinberg, Z., Roth, A., Stockbridge, R. B., & Breaker, R. R. (2012). Widespread genetic switches and toxicity resistance proteins for fluoride. *Science, 335*, 233–235. http://dx.doi.org/10.1126/science.1215063.

Breaker, R. R. (2009). Riboswitches: From ancient gene-control systems to modern drug targets. *Future Microbiology, 4*, 771–773. http://dx.doi.org/10.2217/fmb.09.46.

Carlomagno, T. (2014). Present and future of NMR for RNA-protein complexes: A perspective of integrated structural biology. *Journal of Magnetic Resonance, 241*, 126–136. http://dx.doi.org/10.1016/j.jmr.2013.10.007.

Chen, B., Zuo, X., Wang, Y. X., & Dayie, T. K. (2012). Multiple conformations of SAM-II riboswitch detected with SAXS and NMR spectroscopy. *Nucleic Acids Research, 40*, 3117–3130. http://dx.doi.org/10.1093/nar/gkr1154.

Dayie, K. T. (2008). Key labeling technologies to tackle sizeable problems in RNA structural biology. *International Journal of Molecular Sciences, 9*, 1214–1240. http://dx.doi.org/10.3390/ijms9071214.

Dayie, K. T., Brodsky, A. S., & Williamson, J. R. (2002). Base flexibility in HIV-2 TAR RNA mapped by solution ^{15}N, ^{13}C NMR relaxation. *Journal of Molecular Biology, 317*, 263–278. http://dx.doi.org/10.1006/jmbi.2001.5424.

Dayie, T. K., & Thakur, C. S. (2010). Site-specific labeling of nucleotides for making RNA for high resolution NMR studies using an *E. coli* strain disabled in the oxidative pentose phosphate pathway. *Journal of Biomolecular NMR, 47*, 19–31. http://dx.doi.org/10.1007/s10858-010-9405-0.

Dayie, K. T., Tolbert, T. J., & Williamson, J. R. (1998). 3D C(CC)H TOCSY experiment for assigning protons and carbons in uniformly ^{13}C- and selectively ^{2}H-labeled RNA. *Journal of Magnetic Resonance, 130*, 97–101. http://dx.doi.org/10.1006/jmre.1997.1286.

Dingley, A. J., Masse, J. E., Feigon, J., & Grzesiek, S. (2000). Characterization of the hydrogen bond network in guanosine quartets by internucleotide 3hJ(NC)' and 2hJ(NN) scalar couplings. *Journal of Biomolecular NMR, 16*, 279–289.

Dominguez, C., Schubert, M., Duss, O., Ravindranathan, S., & Allain, F. H. T. (2011). Structure determination and dynamics of protein-RNA complexes by NMR spectroscopy. *Progress in Nuclear Magnetic Resonance Spectroscopy, 58*, 1–61. http://dx.doi.org/10.1016/j.pnmrs.2010.10.001.

D'Souza, V., & Summers, M. F. (2005). How retroviruses select their genomes. *Nature Reviews. Microbiology, 3*, 643–655. http://dx.doi.org/10.1038/nrmicro1210.

Furtig, B., Richter, C., Wohnert, J., & Schwalbe, H. (2003). NMR spectroscopy of RNA. *Chembiochem, 4*, 936–962. http://dx.doi.org/10.1002/cbic.200300700.

Gardner, K. H., & Kay, L. E. (1998). The use of ^{2}H, ^{13}C, ^{15}N multidimensional NMR to study the structure and dynamics of proteins. *Annual Review of Biophysics and Biomolecular Structure, 27*, 357–406. http://dx.doi.org/10.1146/annurev.biophys.27.1.357.

Gobl, C., Madl, T., Simon, B., & Sattler, M. (2014). NMR approaches for structural analysis of multidomain proteins and complexes in solution. *Progress in Nuclear Magnetic Resonance Spectroscopy, 80*, 26–63. http://dx.doi.org/10.1016/j.pnmrs.2014.05.003.

Hoffman, D. W., & Holland, J. A. (1995). Preparation of carbon-13 labeled ribonucleotides using acetate as an isotope source. *Nucleic Acids Research, 23*, 3361–3362.

Hoogstraten, C. G., & Johnson, J. E. (2008). Metabolic labeling: Taking advantage of bacterial pathways to prepare spectroscopically useful isotope patterns in proteins and nucleic acids. *Concepts in Magnetic Resonance Part A, 32A*, 34–55. http://dx.doi.org/10.1002/cmr.a.20103.

Houck-Loomis, B., Durney, M. A., Salguero, C., Shankar, N., Nagle, J. M., Goff, S. P., et al. (2011). An equilibrium-dependent retroviral mRNA switch regulates translational recoding. *Nature, 480*, 561–564. http://dx.doi.org/10.1038/nature10657.

Johnson, J. E., Jr., Julien, K. R., & Hoogstraten, C. G. (2006). Alternate-site isotopic labeling of ribonucleotides for NMR studies of ribose conformational dynamics in RNA. *Journal of Biomolecular NMR, 35*, 261–274. http://dx.doi.org/10.1007/s10858-006-9041-x.

Josephson, B. L., & Fraenkel, D. G. (1969). Transketolase mutants of *Escherichia coli*. *Journal of Bacteriology, 100*, 1289–1295.

Josephson, B. L., & Fraenkel, D. G. (1974). Sugar metabolism in transketolase mutants of *Escherichia coli*. *Journal of Bacteriology, 118*, 1082–1089.

Latham, M. P., Brown, D. J., McCallum, S. A., & Pardi, A. (2005). NMR methods for studying the structure and dynamics of RNA. *Chembiochem, 6*, 1492–1505. http://dx.doi.org/10.1002/cbic.200500123.

Le, M. T., Brown, R. M., Longhini, A. P., Simon, A. E., & Dayie, T. K. (2015). In vivo, site-specific labeling of homogeneous, recombinant RNA in wild type and mutant *E. coli* for NMR structural and dynamic studies. (In preparation)

Liu, Y., Stepanov, V. G., Strych, U., Willson, R. C., Jackson, G. W., & Fox, G. E. (2010). DNAzyme-mediated recovery of small recombinant RNAs from a 5S rRNA-derived chimera expressed in *Escherichia coli*. *BMC Biotechnology, 10*, 85. http://dx.doi.org/10.1186/1472-6750-10-85.

Mortimer, S. A., Kidwell, M. A., & Doudna, J. A. (2014). Insights into RNA structure and function from genome-wide studies. *Nature Reviews. Genetics, 15*, 469–479. http://dx.doi.org/10.1038/Nrg3681.

Muhandiram, D. R., & Kay, L. E. (1994). Gradient-enhanced triple-resonance 3-dimensional NMR experiments with improved sensitivity. *Journal of Magnetic Resonance. Series B, 103*, 203–216. http://dx.doi.org/10.1006/jmrb.1994.1032.

Nakata, Y., Tang, X., & Yokoyama, K. K. (1997). Preparation of competent cells for high-efficiency plasmid transformation of *Escherichia coli*. *Methods in Molecular Biology (Clifton, NJ)*, *69*, 129–137.

Nelissen, F. H., Leunissen, E. H., van de Laar, L., Tessari, M., Heus, H. A., & Wijmenga, S. S. (2012). Fast production of homogeneous recombinant RNA-towards large-scale production of RNA. *Nucleic Acids Research*, *40*, e102. http://dx.doi.org/10.1093/nar/gks292.

Petrov, A., Wu, T., Puglisi, E. V., & Puglisi, J. D. (2013). RNA purification by preparative polyacrylamide gel electrophoresis. *Methods in Enzymology*, *530*, 315–330. http://dx.doi.org/10.1016/B978-0-12-420037-1.00017-8.

Ponchon, L., Beauvais, G., Nonin-Lecomte, S., & Dardel, F. (2009). A generic protocol for the expression and purification of recombinant RNA in *Escherichia coli* using a tRNA scaffold. *Nature Protocols*, *4*, 947–959. http://dx.doi.org/10.1038/nprot.2009.67.

Ponchon, L., & Dardel, F. (2007). Recombinant RNA technology: The tRNA scaffold. *Nature Methods*, *4*, 571–576. http://dx.doi.org/10.1038/nmeth1058.

Ponchon, L., & Dardel, F. (2011). Large scale expression and purification of recombinant RNA in *Escherichia coli*. *Methods*, *54*, 267–273. http://dx.doi.org/10.1016/j.ymeth.2011.02.007.

Ren, A., Rajashankar, K. R., & Patel, D. J. (2012). Fluoride ion encapsulation by Mg^{2+} ions and phosphates in a fluoride riboswitch. *Nature*, *486*, 85–89. http://dx.doi.org/10.1038/nature11152.

Sambrook, J., & Russell, D. W. (2001). *Preparation of organic reagents. Molecular cloning: A laboratory manual*. Cold Spring Harbor, NY: A. 1.23. Cold Spring Harbor Laboratory Press.

Sattler, M., Schleucher, J., & Griesinger, C. (1999). Heteronuclear multidimensional NMR experiments for the structure determination of proteins in solution employing pulsed field gradients. *Progress in Nuclear Magnetic Resonance Spectroscopy*, *34*, 93–158. http://dx.doi.org/10.1016/S0079-6565(98)00025-9.

Schlosser, K., Gu, J., Sule, L., & Li, Y. (2008). Sequence-function relationships provide new insight into the cleavage site selectivity of the 8-17 RNA-cleaving deoxyribozyme. *Nucleic Acids Research*, *36*, 1472–1481. http://dx.doi.org/10.1093/nar/gkm1175.

Schlosser, K., & Li, Y. (2010). A versatile endoribonuclease mimic made of DNA: Characteristics and applications of the 8-17 RNA-cleaving DNAzyme. *Chembiochem*, *11*, 866–879. http://dx.doi.org/10.1002/cbic.200900786.

Simon, A. E., & Miller, W. A. (2013). 3' Cap-independent translation enhancers of plant viruses. *Annual Review of Microbiology*, *67*, 21–42. http://dx.doi.org/10.1146/annurev-micro-092412-155609.

Sivashanmugam, A., Murray, V., Cui, C., Zhang, Y., Wang, J., & Li, Q. (2009). Practical protocols for production of very high yields of recombinant proteins using *Escherichia coli*. *Protein Science*, *18*, 936–948. http://dx.doi.org/10.1002/pro.102.

Steitz, T. A. (2008). A structural understanding of the dynamic ribosome machine. *Nature Reviews. Molecular Cell Biology*, *9*, 242–253. http://dx.doi.org/10.1038/nrm2352.

Studier, F. W. (2005). Protein production by auto-induction in high-density shaking cultures. *Protein Expression and Purification*, *41*, 207–234. http://dx.doi.org/10.1016/j.pep.2005.01.016.

Thakur, C. S., Brown, M. E., Sama, J. N., Jackson, M. E., & Dayie, T. K. (2010). Growth of wildtype and mutant *E. coli* strains in minimal media for optimal production of nucleic acids for preparing labeled nucleotides. *Applied Microbiology and Biotechnology*, *88*, 771–779. http://dx.doi.org/10.1007/s00253-010-2813-y.

Thakur, C. S., & Dayie, T. K. (2011). Asymmetry of ^{13}C labeled 3-pyruvate affords improved site specific labeling of RNA for NMR spectroscopy. *Journal of Biomolecular NMR*, *52*, 65–77. http://dx.doi.org/10.1007/s10858-011-9582-5.

Thakur, C. S., Luo, Y., Chen, B., Eldho, N. V., & Dayie, T. K. (2012). Biomass production of site selective $^{13}C/^{15}N$ nucleotides using wild type and a transketolase *E. coli* mutant for labeling RNA for high resolution NMR. *Journal of Biomolecular NMR, 52*, 103–114. http://dx.doi.org/10.1007/s10858-011-9586-1.

Thakur, C. S., Sama, J. N., Jackson, M. E., Chen, B., & Dayie, T. K. (2010). Selective ^{13}C labeling of nucleotides for large RNA NMR spectroscopy using an *E. coli* strain disabled in the TCA cycle. *Journal of Biomolecular NMR, 48*, 179–192. http://dx.doi.org/10.1007/s10858-010-9454-4.

Tyler, R. C., Sreenath, H. K., Singh, S., Aceti, D. J., Bingman, C. A., Markley, J. L., et al. (2005). Auto-induction medium for the production of [U-^{15}N]- and [U-^{13}C, U-^{15}N]-labeled proteins for NMR screening and structure determination. *Protein Expression and Purification, 40*, 268–278. http://dx.doi.org/10.1016/j.pep.2004.12.024.

Wang, Y. X., Zuo, X., Wang, J., Yu, P., & Butcher, S. E. (2010). Rapid global structure determination of large RNA and RNA complexes using NMR and small-angle X-ray scattering. *Methods, 52*, 180–191. http://dx.doi.org/10.1016/j.ymeth.2010.06.009.

Yang, S. C., Parisien, M., Major, F., & Roux, B. (2010). RNA structure determination using SAXS data. *Journal of Physical Chemistry B, 114*, 10039–10048. http://dx.doi.org/10.1021/Jp1057308.

CHAPTER TWENTY

Cut and Paste RNA for Nuclear Magnetic Resonance, Paramagnetic Resonance Enhancement, and Electron Paramagnetic Resonance Structural Studies

Olivier Duss[1,2,3], Nana Diarra dit Konté[3], Frédéric H.-T. Allain[1]
Institute for Molecular Biology and Biophysics, ETH Zürich, Zürich, Switzerland
[1]Corresponding authors: e-mail address: olivier.duss@alumni.ethz.ch; allain@mol.biol.ethz.ch

Contents

1. Introduction 538
2. Cut and Paste RNA Approach 540
   2.1 Principle 540
   2.2 Building Blocks 540
   2.3 Combination and Ligation 543
3. Production of Small (<10 nts) Isotopically Labeled RNAs 546
   3.1 Introduction 546
   3.2 Double RNase H Cleavage 549
   3.3 Combined RNase H and VS Ribozyme Cleavage 550
4. Protocol A: Production of Small Spin-Labeled RNA Fragments 552
5. Protocol B: Production of Unlabeled and Isotopically Labeled RNA Fragments 554
   5.1 In Vitro Transcription and Ribozyme Cleavage 554
   5.2 Purification by Denaturing Anion-Exchange HPLC 554
   5.3 Sequence-Specific RNase H Cleavage 555
   5.4 VS Ribozyme Cleavage of Purified RNA Fragments 556
   5.5 Tips 556
6. Protocol C: Ligation 557
   6.1 Protocol 557
   6.2 Tips 558

---

[2] Current address: Department of Integrative Structural and Computational Biology, The Scripps Research Institute, La Jolla, United States.
[3] These authors contributed equally to this work.

7. Summary and Outlook 559
Acknowledgments 560
References 561

Abstract

RNA is a crucial regulator involved in most molecular processes of life. Understanding its function at the molecular level requires high-resolution structural information. However, the dynamic nature of RNA complicates structure determination because crystallization is often not possible or can result in crystal-packing artifacts resulting in nonnative structures. To study RNA and its complexes in solution, we described an approach in which large multi-domain RNA or protein–RNA complex structures can be determined at high resolution from isolated domains determined by nuclear magnetic resonance (NMR) spectroscopy, and then constructing the entire macromolecular structure using electron paramagnetic resonance (EPR) long-range distance constraints. Every step in this structure determination approach requires different types of isotope or spin-labeled RNAs. Here, we present a simple modular RNA cut and paste approach including protocols to generate (1) small isotopically labeled RNAs (<10 nucleotides) for NMR structural studies, which cannot be obtained by standard protocols, (2) large segmentally isotope and/or spin-labeled RNAs for diamagnetic NMR and paramagnetic relaxation enhancement NMR, and (3) large spin-labeled RNAs for pulse EPR spectroscopy.

1. INTRODUCTION

Despite consisting of only four different building blocks, RNA can form complex three-dimensional structures, many with catalytic activity. In the ribosome, RNA constitutes the catalytic center connecting single amino acids into functional polypeptide chains (Steitz, 2008). RNA splicing, a molecular process which is generating from long precursor RNAs different messenger RNAs, thereby contributing to the huge protein diversity in eukaryotes, is also carried out by a set of different RNA molecules (Wahl, Will, & Luhrmann, 2009). In prokaryotes (and also in some eukaryotes), riboswitches are functional RNA structures, which change their structure upon binding small molecules (Breaker, 2011). These structural changes often contribute to regulation of gene expression.

Besides having catalytic activity, RNAs can also function as protein scaffolds, can guide macromolecular machines such as the RISC complex to specific messenger RNAs, and can interfere with translation and mRNA stability (Mercer, Dinger, & Mattick, 2009; Morris & Mattick, 2014). RNAs have been shown to be able to act as microRNA sponges in eukaryotes (Hansen et al., 2013) or protein sponges in prokaryotes (Duss, Michel, Yulikov, et al., 2014).

Understanding how RNA works at the molecular and atomic level helps to understand the origin of diseases providing opportunities for therapeutic intervention or provides the basis for designing novel drugs targeting key molecular processes in pathogenic prokaryotes. However, in stark contrast to proteins, high-resolution structural information on RNA is much more difficult to obtain. Crystallographic approaches are often unsuccessful due to the structural flexibility of RNA and classical nuclear magnetic resonance (NMR) spectroscopy is restricted only to small RNAs with limited biological relevance.

Compared to proteins, RNA resonances have shorter transverse relaxation times due to the generally more extended nature of RNA structures resulting in less favorable tumbling properties (Dominguez, Schubert, Duss, Ravindranathan, & Allain, 2011; Tzakos, Grace, Lukavsky, & Riek, 2006). Faster transverse relaxation results in broader line shapes complicating NMR spectra interpretation. Because RNAs have only four chemically similar nucleosides, RNA spectra show more severe signal overlap compared to proteins that have 20 chemically more distinct amino acids, further hampering solution NMR structure determination.

To reduce signal overlap in RNA, we have presented a simple and flexible approach for multiple segmental isotope labeling of RNA (Duss, Lukavsky, & Allain, 2012; Duss, Maris, von Schroetter, & Allain, 2010). By isotope labeling only a few nucleotides within a large biologically relevant RNA, resonance overlap is drastically reduced and specific atoms can be used as structural reporters in complex macromolecular assemblies and processes.

Furthermore, we have developed a combined NMR-EPR (electron paramagnetic resonance)-based approach for determining high-resolution structures of large RNAs and protein–RNA complexes in solution (Duss, Michel, Yulikov, et al., 2014; Duss, Yulikov, Jeschke, & Allain, 2014, 2015). In this approach, NMR restraints obtained from isolated structural domains are combined with long-range distance constraints obtained by site-directed spin-labeling and measurements with the double electron–electron resonance (DEER) EPR spectroscopy technique.

In this chapter, we review how to generate different types of isotope and/or spin-labeled RNAs for EPR and NMR structural studies using a simple cut and paste approach for RNA. We start by explaining the principles of the cut and paste approach. In the second part, we present original data on how to produce small (<10 nucleotides) isotopically labeled RNAs, which could not be obtained so far using standard protocols. Then, we provide step-by-step protocols for (1) spin-labeling 4-thiouridine-modified RNAs, (2) generating unlabeled and isotopically labeled RNA fragments of any size, and (3) ligation of differently isotope and/or spin-labeled RNA fragments. We conclude this chapter, by summarizing the chapter and providing an outlook on how these segmental isotope- and spin-labeled RNAs help in studying structure, dynamics, and assembly of large RNA and protein–RNA complexes.

2. CUT AND PASTE RNA APPROACH

2.1 Principle

Here, we propose a versatile and flexible approach for generating different types of isotope and/or spin-labeled RNAs for NMR and EPR structural studies (Fig. 1). In a first step, up to three types of RNA fragments (building blocks) are generated, namely, unlabeled (fragment B1 in Fig. 1), isotope-labeled (B2), and spin-labeled (A) RNAs. In a second step, these RNA fragments can be combined at will and then ligated to generate several types of isotope and/ or spin-labeled RNAs.

2.2 Building Blocks

2.2.1 Spin-Labeled RNA Fragments

Small modified RNAs (<30 nucleotides) are commercially available. The most commonly used modified nucleotide that is incorporated into RNA by chemical synthesis is 4-thiouridine. A 3-(2-iodoacetamido)-proxyl (IA-proxyl) spin label can be attached to the sulfur of the 4-thiouridine base using a simple protocol (Ramos & Varani, 1998) (Fig. 2). The attachment efficiency is usually close to 100%. It has been shown that the 4-thiouridine substitution within a double-stranded RNA itself and its subsequent modification with a nitroxyl spin label does not significantly alter the structure of the RNA (Qin, Hideg, Feigon, & Hubbell, 2003). Although upon spin-label attachment one hydrogen bond acceptor is lost, the base remains stacked within the helix. A variety of other spin labels in RNA have also been used and are reviewed elsewhere (Zhang, Cekan, Sigurdsson, & Qin, 2009). A step-by-step protocol is provided in Section 4.

2.2.2 Unlabeled and Isotopically Labeled RNA Fragments

If the unlabeled RNA fragments are small enough (<~20–30 nucleotides), they can be obtained by chemical synthesis and are commercially available.

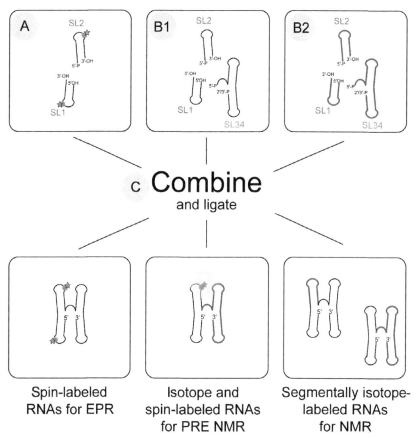

Figure 1 Principle of cut and paste approach to generate RNAs for NMR, PRE NMR, and EPR structural studies. Step-by-step protocols are provided at the end of the chapter for producing small spin-labeled RNA fragments (protocol A), unlabeled, and isotope-labeled RNA fragments (protocol B), and how to ligate the different RNA fragments to obtain differently labeled full-lengths RNAs (protocol C).

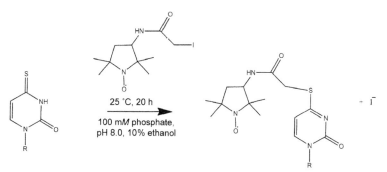

Figure 2 Spin-labeling of 4-thiouridine-modified RNA fragment with 3-(2-iodoacetamido)-proxyl (Ramos & Varani, 1998).

However, larger unlabeled RNAs and isotopically labeled RNAs (small and large) have to be generated differently.

Longer labeled or unlabeled RNA fragments could be produced by *in vitro* transcription. To be compatible for further ligation, the acceptor fragments (fragments 5′ to the site of ligation) require a 3′-hydroxyl group, while the donor fragments (fragments 3′ to the site of ligation) require a 5′-monophosphate (Fig. 3). The 5′-monophosphate can be obtained by priming the transcription reaction with GMP. However, GMP priming is not complete. In addition, the fragment obviously has to start with a guanosine. A 3′-hydroxyl group is generated by *in vitro* run-off transcription. However, run-off transcription also generates untemplated nucleotides at the 3′-end, resulting in inhomogeneous RNA fragments.

A more convenient option is the use of multiple chimera-mediated sequence-specific RNase H cleavage from a precursor RNA obtained by *in vitro* transcription (Duss et al., 2010, 2012). Hybridization of an RNA of interest with a 2′-O-methyl-RNA/DNA chimera in presence of RNase H leads to sequence-specific cleavage (Duss et al., 2010; Lapham &

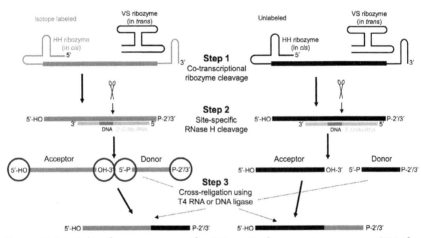

Figure 3 Principle of sequence-specific RNase H cleavage to generate RNA fragments with correctly protected termini for subsequent ligation. The isotope-labeled material is highlighted in red, the unlabeled material in black. In the 2′-O-methyl RNA/DNA chimera, the DNA is in dark blue, and the 2′-O-methyl RNA in light blue. The termini of both acceptor and donor fragments are encircled in green. Scissors indicate RNase H cleavage sites. P-2′/3′ stands for a 2′/3′-cyclic phosphate. *This figure has been taken from Duss et al. (2010) by permission of Oxford University Press.* (See the color plate.)

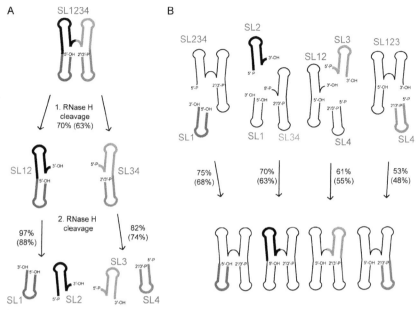

Figure 4 Principle of multiple segmental isotope labeling using sequence-specific RNase H cleavage to generate RNA fragments (A) with subsequent ligation of isotope-labeled and unlabeled RNA fragments using splinted T4 DNA ligation (B). The RNase H cleavage and T4 DNA ligation efficiencies are shown; values in brackets are after denaturing anion-exchange HPLC purification. *This figure has been taken from Duss et al. (2010) by permission of Oxford University Press.* (See the color plate.)

Crothers, 1996) (Figs. 3 and 4). A 5′-hammerhead (HH) ribozyme in *cis* and 3′ *Neurospora* Varkud satellite (VS) ribozyme for cleavage in *trans* can be further introduced into the RNA precursor to obtain homogenous ends and allowing the simultaneous generation of multiple RNA fragments from a single RNA precursor (Duss et al., 2010, 2012) (Figs. 3 and 4). Note that sequence-specific RNase H cleavage generates a 3′-hydroxyl group and a 5′-monophosphate at the cleavage site, which are the termini required for successful ligation (Fig. 3). Thus, using sequence-specific RNase H cleavage, no modification of the termini is required prior to ligation. Protocols for ribozyme and sequence-specific RNase H cleavage are provided in Section 5.

2.3 Combination and Ligation

Once the different types of RNA fragments have been obtained, they are combined and then ligated to generate different types of RNAs for

NMR and EPR structural studies. For successful ligation, the 3′-end of the acceptor fragment has to be a free hydroxyl group while the 5′-end of the donor fragment has to contain a monophosphate (Fig. 3). Ligation is usually performed using T4 DNA ligase and hybridizing a DNA oligonucleotide (DNA splint) to the RNA parts close to the site of ligation. This splinted ligation prevents ligation of several RNA fragments in the wrong sequential order. We typically observed ligation efficiencies between 50% and 75% for 2–3 fragment ligations (Duss et al., 2010).

Nonsplinted T4 RNA ligation can also be used if the acceptor and donor RNA fragments can be brought into proximity by base-pairing. Ligation with T4 RNA ligase has high sequence preference for a purine at the 3′-end of the acceptor fragment and pyrimidine at the 5′-end of the donor fragment (Moore & Query, 2000). With favorable nucleotides at the ligation site, the ligation efficiency is higher than with T4 DNA ligase and can be close to 100% (Duss et al., 2010). T4 RNA ligations are usually only successful for 2-fragment ligations and if the fragments are protected in such a way that ligation in the wrong sequential order becomes impossible (Fig. 3).

2.3.1 Segmentally Labeled RNAs for NMR Structural Studies

By combining several isotope labeled with unlabeled RNA fragments followed by splinted T4 DNA ligation, segmentally isotope-labeled RNAs can be obtained (Figs. 3 and 4) (Duss et al., 2010, 2012).

Segmental isotope labeling is absolutely required when studying RNAs of moderate to large size. For example, it allows verifying if an RNA domain such as a hairpin has the same structure in isolation or in context of an entire biologically functional RNA or protein–RNA complex. Twenty-nine nucleotides at the 3′-end of the intact 230-kDa HIV-1 5′-leader RNA dimer (total length of 712 nucleotides) were isotopically labeled, which helped to detect structures that are important for HIV-1 genome packaging (Lu et al., 2011).

Segmental isotope labeling of RNA can also help to monitor assembly of large protein–RNA complexes. For example, we separately isotope labeled in the noncoding RNA RsmZ, the eight-binding sites for the homodimeric RsmE protein, to monitor the sequential, specific, and cooperative assembly of the RsmZ RNA and five RsmE protein dimers (Duss, Michel, Yulikov, et al., 2014).

Labeling only a small part within a larger RNA significantly reduces the number of resonances in ^{13}C- and ^{15}N-edited correlation spectra. This is crucial for the detection of well-resolved resonances in order to measure a sufficient amount of residual dipolar couplings for successful structure determination of intermediate to large-size RNAs (Tzakos et al., 2006).

2.3.2 Spin-Labeled RNAs for Pulsed EPR

Site-specific attachment of two spin labels into large RNAs allows measuring long-range distance constraints (~20–80 Å) with pulsed EPR (Duss et al., 2015). Chemical synthesis is the method of choice for introducing modified nucleotides into smaller RNAs (<30–50 nucleotides). However, many biologically relevant RNAs are significantly larger and cannot be obtained by the same strategy. While the introduction of spin labels at the 5'- or 3'-ends of larger RNAs have been described and reviewed (Sowa & Qin, 2008), internal labeling of larger RNAs is more challenging.

Two methods have been introduced by the Hobartner lab. In a first study, they performed ligation of small RNA fragments containing spin labels using a deoxyribozyme (Buttner, Seikowski, Wawrzyniak, Ochmann, & Hobartner, 2013). However, successful ligation requires no cytosine to be present at the 3'-end of the acceptor fragment and either an AG or an AA dinucleotide at the 5'-end of the donor fragment. In the second approach, spin-labeled guanosines were posttranscriptionally attached to the 2'-OH groups of several internal adenosines of *in vitro* transcribed RNAs using terbium assisted deoxyribozymes (Buttner, Javadi-Zarnaghi, & Hobartner, 2014). The evolution of new deoxyribozymes with relaxed sequence requirements at the site of ligation will broaden the applicability of these methods for the production of doubly spin-labeled RNAs for DEER measurements.

In contrast, we have obtained spin-labeled RNAs using splinted enzymatic ligation of shorter spin-labeled and unlabeled fragments using T4 DNA ligase (Duss, Yulikov, et al., 2014). This approach has no sequence requirements at the ligation site and virtually no upper RNA size limitation. The labeling efficiency after purification is generally in the range of 85–100% for all doubly spin-labeled RNAs. The ligation of unlabeled and spin-labeled fragments should also allow different types of spin labels to be introduced into the same RNA for orthogonal spin-labeling techniques (Yulikov, 2015).

2.3.3 Isotopically and Spin-Labeled RNAs for PRE NMR

While two spin labels on a single RNA allow the measurement of long-range distances between the spin labels with EPR, introduction of a single spin label on a macromolecule results into increased relaxation of nearby NMR active nuclei providing intermediate-range (~20–25 Å) distance information. This paramagnetic relaxation effect has been well exploited for studying structure and dynamics of proteins, protein–protein, and protein–DNA complexes (Clore & Iwahara, 2009; Gobl, Madl, Simon, & Sattler, 2014). A few studies have also used paramagnetic resonance enhancement (PRE) effects for studying protein–RNA complexes by attaching the spin label on the RNA and observing the PREs on the isotopically labeled protein (Ramos & Varani, 1998). Later, a short chemically synthesized RNA was labeled at the 5′-end with a 2,2,6,6-tetramethylpiperidine 1-oxyl spin label and the paramagnetic relaxation effect observed on ^{13}C-labeled nucleotides on the same RNA (Wunderlich et al., 2013). Although providing valuable information, observing the paramagnetic relaxation enhancement on longer isotope-labeled RNAs has only recently been achieved, yet by annealing noncovalently a short spin-labeled RNA fragment to a longer isotope-labeled RNA fragment (Helmling et al., 2014). However, this approach is limited to specific cases requiring a stable annealing between the two RNA fragments.

Using the cut and paste approach (Fig. 1), large spin-labeled RNAs in which specific segments are isotopically labeled could be obtained in high enough quantities for NMR structural studies without sequence requirements and with minimal restrictions in choosing the spin-labeling sites and parts to be isotope labeled. Such samples could provide important information on RNA structural changes in riboswitches or could help characterizing structural dynamics of RNAs. It should be noted that the intermediate distance range of PRE (<20–25 Å), might be too short-range to observe an effect between different RNA structural domains within a larger RNA. Therefore, prior structural information such as long-range distance information from pulse EPR would be required on such a system.

3. PRODUCTION OF SMALL (<10 nts) Isotopically Labeled RNAs

3.1 Introduction

Uniform or nucleotide-specific isotope labeling of RNA is crucial for the success of NMR structural studies of RNA and protein–RNA complexes, especially when studying systems of molecular weights greater than 15 kDa (Dominguez et al., 2011).

RNAs larger than 18–20 nucleotides can efficiently be transcribed *in vitro* using isotopically labeled NTPs (Dominguez et al., 2011). However, the production of isotope-labeled RNAs smaller than ~10 nucleotides cannot be achieved using the same procedure. *In vitro* transcription of RNAs smaller than 10 nucleotides is inefficient or not possible at all. In addition, during *in vitro* transcription several abortive products of up to 12 nucleotides (Ramirez-Tapia & Martin, 2012) can be formed, which makes the separation of the transcript of interest and the abortive products impossible. Other drawbacks of *in vitro* transcription are the sequence requirements at the 5′-end (efficient *in vitro* transcription using T7 polymerase requires at least one G at the 5′-end) and the inhomogeneity at the 3′-end (usually one to several additional nucleotides are added at the 3′-end), which can be deleterious for NMR spectral quality. Additional nucleotides on the termini of small single-stranded RNAs increase the spectral overlap, can lead to sliding of the RNA on the protein, to binding in more than one register and to changes in binding affinity (potentially changing the exchange regime).

Small RNAs (<15 nts) for NMR structural studies are usually produced by chemical synthesis using phosphoramidites. Unfortunately, isotopically labeled phosphoramidites are not commercially available.

One option to produce a small isotopically labeled RNA is to engineer a ribozyme at the 5′-end. A HH ribozyme placed at the 5′-end cleaves very efficiently co-transcriptionally and has no sequence requirements. However, this approach does not allow to obtain RNAs smaller than 7–10 nucleotides (specially for AU-rich sequences), first, because successful catalysis requires that the 5′-end of the target RNA base-pairs efficiently with the HH ribozyme (which is not the case if the melting temperature of this duplex is too small) and second, because co-transcriptional HH ribozyme cleavage might prevent separation of the cleaved RNA of interest from abortive products of unsuccessful transcription initiation (see above) (Ramirez-Tapia & Martin, 2012).

A more promising option is to exploit the chimera-mediated sequence-specific cleavage of RNA by RNase H (see Section 2.2.2). In the following, we will present two different approaches both supported with biologically relevant example RNAs. In the first approach, the small RNA of interest is liberated from the precursor RNA by two sequence-specific RNase H cleavages (Fig. 5A). In the second approach, the 5′-end of the RNA of interest is removed by sequence-specific RNase H cleavage while the 3′-end is cleaved by a VS ribozyme in *trans* (Fig. 5B). In both approaches, the 3′-end of the RNA of interest is protected by a stem-loop structure during sequence-specific RNase H cleavage. This is crucial for preventing unspecific RNA degradation by other ribonucleases likely present in small amounts due to co-purification with recombinant RNase H (see further below).

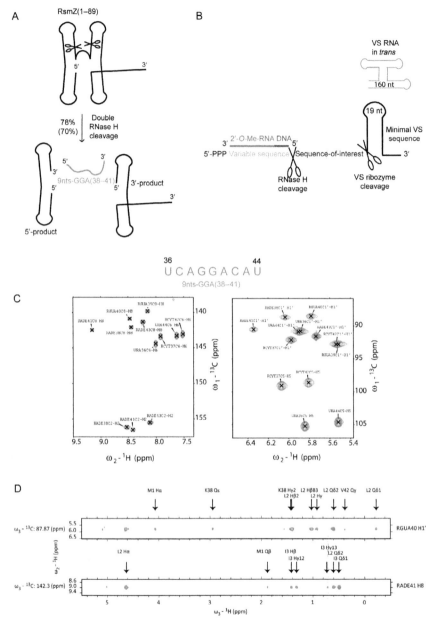

Figure 5 Production of small isotope-labeled RNAs. Small isotope-labeled RNAs can be obtained by double RNase H cleavage (A) or combined RNase H and VS ribozyme cleavage (B) from a larger RNA precursor. (A) Scheme of double RNase H cleavage of the RsmZ (1–89) precursor RNA into a 5′-product, a 3′-product, and the 9 nts-GGA (39–41) RNA of interest (red). This requires two 2′-O-Me-RNA/DNA chimeras specifically hybridizing

3.2 Double RNase H Cleavage

The noncoding RNA RsmZ sequesters several RsmE proteins from the ribosome-binding sites of certain mRNAs thereby activating translation initiation (Duss, Michel, Yulikov, et al., 2014). To study the molecular basis of this sequestration process, we determined the solution structure of the RsmZ RNA bound to three RsmE protein dimers. The first step in structure determination was to obtain the high-resolution NMR structures of RsmE bound to the different sites on the RsmZ RNA in isolation. The third RsmE protein dimer binds to a single-stranded linker region between stem-loop 2 and 3. Therefore, we determined the solution NMR structure of RsmE bound to two 9-nucleotides RNAs comprising the linker between SL2 and SL3 of the RsmZ RNA (Duss, Michel, Diarra dit Konte, et al., 2014).

To isotopically label this 9-nucleotides RNA, we performed from an *in vitro* transcribed RsmZ precursor RNA consisting of SL1–SL4, a double RNase H cleavage using two chimeras (see Fig. 5A). The two chimeras for sequence-specific RNase H cleavage could be added in catalytic amounts and the two cleavages were performed simultaneously. The RNase H cleavage efficiency for this double cleavage was nearly 80% (and nearly 70% after purification). After subsequent purification by high-performance liquid chromatography (HPLC), we could obtain close to 300 nmol of pure isotopically labeled 9-nucleotides RNA from a 10 ml transcription reaction of the precursor RsmZ RNA. The identity and purity of the 9- nucleotides RNA could be confirmed by matrix-assisted laser desorption/ionization mass spectroscopy (MALDI-MS) and acquisition of a ^{1}H-^{13}C heteronuclear

---

Figure 5—Cont'd to the desired cleavage sites on the RsmZ (1–89) RNA (not shown for simplicity). (B) Combined RNase H and VS cleavage: The RNA of interest (green) is flanked by a well-transcribing sequence (gray) at the 5′-end and a minimal VS ribozyme stem-loop (SL) sequence at the 3′-end. The 5′-sequence is cleaved off with an RNase H catalyzed reaction mediated by a 2′-O-Me-RNA (cyan)/DNA (blue) chimera. The VS ribozyme minimal sequence at the 3′-end (black) is cleaved off with the corresponding VS RNA in *trans* (orange). (C) ^{1}H-^{13}C HSQC spectra of the aromatic region (left) and aliphatic H1′ and H5 region (right) of the 21-kDa complex of ^{13}C,^{15}N-labeled 9 nts-GGA(39–41) RNA bound to the RsmE protein (Duss, Michel, Diarra dit Konte, Schubert, & Allain, 2014). (D) G40 H1′ and A41 H8 strips of the 3D filtered-edited-^{1}H-^{13}C HSQC-NOESY spectrum of the same protein–RNA complex as in (C). (See the color plate.)

single-quantum correlation (HSQC) spectrum of the RNA resonances. The ^{1}H-^{13}C HSQC spectra of the isotopically labeled 9-nucleotides RNA in complex with RsmE (21.5 kDa) are shown in Fig. 5C. The possibility of isotopically labeling the RNA component allowed the measurement of 154 intermolecular nuclear Overhauser enhancements (NOEs) between the RNA and the protein (Fig. 5D), which was crucial to solve the solution structure.

3.3 Combined RNase H and VS Ribozyme Cleavage

CUG-BP 2 and DND1 are two human proteins that regulate gene expression through binding uracil-rich sequences in the 3′-UTRs of mRNAs. In tumorous cells, CUG-BP 2 stabilizes COX-2 mRNA and inhibits its translation by binding to AU-rich sequences (Mukhopadhyay, Houchen, Kennedy, Dieckgraefe, & Anant, 2003). In contrast, DND1 prohibits miRNA-mediated gene suppression by sequestering their target sites on mRNAs (Kedde et al., 2007). To elucidate the molecular mechanisms underlying these processes, we aim at solving the solution structures of these two proteins in complex with their target RNA sequences, AUUUAAUU and CUUAUUUG, respectively. Because only a limited number of intermolecular protein–RNA and intramolecular RNA–RNA NOEs could be obtained with unlabeled RNAs, we aimed at labeling the RNAs with ^{13}C and ^{15}N.

To obtain these RNAs, we combined RNase H and VS ribozyme cleavage from a precursor RNA consisting of a well-transcribing 5′-sequence followed by the RNA of interest and a minimal VS ribozyme substrate sequence at the 3′-end (Fig. 5B). There are several advantages to such a design. First, since it is sufficient that the chimera for site-specific RNase H cleavage is complementary to the 5′-region upstream of the cleavage site, the same chimera can be used for any RNA to be cleaved (Lapham & Crothers, 1996) (Fig. 5B). Second, the 3′-VS ribozyme stem-loop protects the RNA from unspecific cleavage, which likely occurs due to co-purification of ribonucleases during RNase H enzyme purification. For the AUUUAAUU and CUUAUUUG precursor RNAs, the presence of a 5′-stem-loop did not protect the 5′-cleavage fragment and did not seem to impact the 3′ fragment, the one of interest. However, the VS stem-loop at the 3′-end was necessary to prevent complete degradation of both cleavage products (Fig. 6B).

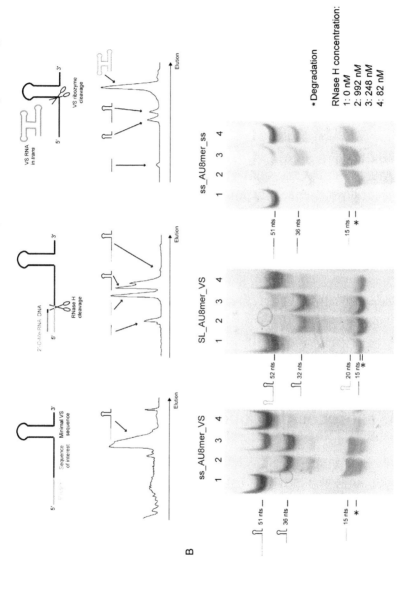

Figure 6 Detailed procedure of AUUUAAUU and CUUAUUUG production. (A) Scheme of all the different steps. The steps are described in the upper panel and the corresponding HPLC run are in the lower panel. (B) Gels of RNase H site-specific cleavage. The 2′-O-methyl RNA/DNA chimera can diffuse out of the gel during the destaining process, hence the absence of the corresponding band in the left and right gel. (See the color plate.)

In principle, the VS ribozyme cleavage step could be done co-transcriptionally as for example, successfully performed for obtaining the structured 72 nucleotides noncoding RNA RsmZ (Fig. 3) (Duss et al., 2010). However, for the constructs used in this study, the co-transcriptional ribozyme cleavage was very inefficient, possibly due to the structural context of the substrate VS stem-loop. Furthermore, the VS ribozyme stem-loop at the 3′-end of the RNA of interest was crucial to protect the unstructured RNA of interest during subsequent RNase H cleavage (see above). Therefore, we performed three steps: first, we transcribed and HPLC-purified the precursor RNA; second, we performed sequence-specific RNase H cleavage and purified the intermediary products; and third, we proceeded with the VS ribozyme cleavage and a final purification (Fig. 6A).

The RNase H cleavage efficiency was nearly 100% but the precursor RNA yielded two intermediary products and only one of them was a functional substrate for the VS ribozyme. VS ribozyme cleavage of the functional product resulted into almost 100% cleavage. After purification by HPLC, we could obtain close to 30 nmol of pure isotopically labeled AUUUAAUU and of unlabeled CUUUAUUG RNAs from 100 nmol of precursor RNAs. The identity and purity of the RNAs were confirmed by MALDI-MS and acquisition of a ^{1}H-^{13}C HSQC spectrum of the RNA resonances (Fig. 7). Note that with this approach, the RNAs obtained will have a 5′-phosphate at the 5′-end and a 2′-3′ cyclic phosphate at the 3′-end. The NMR spectra will therefore differ from chemically synthesized RNAs which often have hydroxyls at both the 5′ and 3′-ends.

4. PROTOCOL A: PRODUCTION OF SMALL SPIN-LABELED RNA FRAGMENTS

1. Small RNAs (10–25 nucleotides) containing a single 4-thiouridine modification at a specific position are commercially available (e.g., Dharmacon). The RNAs are first deprotected according to the manufacturer's instructions.
2. The 3-(2-iodoacetamido)-proxyl (IA) spin label is attached to the modified RNA according to an adapted protocol from Ramos and Varani (1998): The IA-proxyl spin label is first dissolved in 100% methanol at a 100 mM concentration. The modified RNA is resuspended into reaction buffer containing 100 mM potassium phosphate (pH 8.0), 80–110 μM 4-thiouridine-modified RNA, 8–11 mM IA-proxyl spin

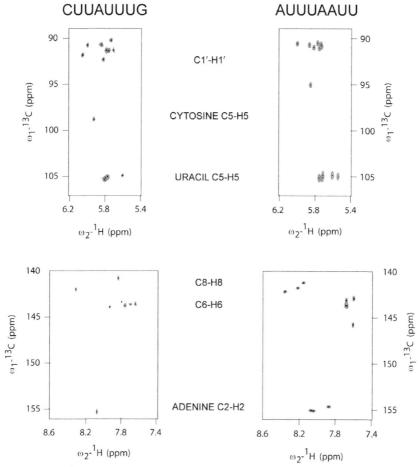

Figure 7 ^{1}H-^{13}C HSQC spectra of the aliphatic (top panels) and aromatic (bottom panels) regions of ^{13}C, ^{15}N-labeled AUUUAAUU (30 µM, 2 scans at 900 MHz), and unlabeled CUUAUUUG (70 µM, 400 scans at 600 MHz) RNAs.

label (100-fold molar excess over modified RNA). The reaction mix is incubated for 20–28 h at 25 °C in the dark under vigorous shaking.

3. The spin-labeled RNA is purified using a buffer exchange NAP-10 column from GE healthcare. All the fractions having an absorption ratio A260/230 greater than 1 are collected. The RNA concentration is determined by UV-spectroscopy. If the RNA is directly used for EPR measurements without further ligation, the purified RNA is subjected to a second NAP-10 column purification step.

5. PROTOCOL B: PRODUCTION OF UNLABELED AND ISOTOPICALLY LABELED RNA FRAGMENTS

5.1 In Vitro Transcription and Ribozyme Cleavage

1. For generating the RNA precursor, transcription yields of the RNA precursor are optimized on 40 µl small-scale reactions with changing concentrations of $MgCl_2$, linearized plasmid DNA, NTPs, and T7 polymerase, and testing the influence of the addition of pyrophosphatase and/or GMP.
2. The best condition is scaled up to a large-scale reaction of 5–20 ml. A typical reaction contains 42.5 mM $MgCl_2$, 4.5 mM of each NTP, 33 ng/µl linearized plasmid (possibly containing a HH ribozyme 5′ to the sequence of interest (Duss et al., 2010; Ferre-D'Amare & Doudna, 1996), 10 µM separately transcribed VS ribozyme RNA (Duss et al., 2010; Guo & Collins, 1995), and 1.7 µM in-house produced T7 polymerase (Price, Oubridge, Varani, & Nagai, 1998) in a transcription buffer containing 40 mM Tris–HCl (pH 8.0), 1 mM spermidine, 0.01% Triton X-100, and 5 mM DTT. After 4–6 h of transcription, the reaction mixture is heated to 65 °C for 15 min to complete the ribozyme cleavage. However, it might be necessary to perform several cycles of thermal cycling if the ribozyme cleavage efficiencies are unsatisfactory (for example, repeating three times 5–15 min at 65 °C and 15–60 min at 37 °C). Finally, the reaction is stopped with the addition of 100 mM EDTA, pH 8.0.

5.2 Purification by Denaturing Anion-Exchange HPLC

1. The transcription mixture is filtrated using a 0.22 µm filter and then purified by anion-exchange chromatography on a preparative Dionex DNAPac PA-100 column (22 × 250 mm) at 85 °C. Flow rate: 20 ml/min; eluent A: 12.5 mM Tris–HCl (pH 8.0), 6 M urea; eluent B: 12.5 mM Tris–HCl (pH 8.0), 0.5 M $NaClO_4$, 6 M urea; detection at 260 nm; 30–75% B gradient within 18 min or 0–65% B gradient within 26 min (Duss et al., 2010).
2. Fractions containing the purified RNA are determined by 6–16% urea acrylamide gels and then liberated from urea and desalted either by dialysis against water or by n-butanol extraction of the aqueous phase until RNA precipitation (Cathala & Brunel, 1990). The RNA precipitate is redissolved into 1 ml of water and precipitated with a 30–50 ml of n-butanol followed by centrifugation. This procedure is repeated three

times in total. After freeze-drying the pellet, the RNA is dissolved in water to obtain a concentration of a few hundred micromolar.

5.3 Sequence-Specific RNase H Cleavage

1. The RNA precursor is then subjected to sequence-specific RNase H cleavage assisted with a 2′-O-methyl-RNA/DNA chimera hybridized to the site of cleavage (Figs. 3 and 8). For successful cleavage, the chimera has to be designed such that four DNA nucleotides hybridize directly 5′ to the cleavage site on the RNA and several additional 2′-O-methyl-RNA nucleotides hybridize further upstream on the RNA to be cleaved (Fig. 5B). Although not required for cleavage, 2′-O-methyl-RNA nucleotides can also be designed to hybridize to the RNA downstream of the cleavage site (Fig. 3). We found that chimeras that are 14–16 nucleotides long and hybridize both upstream and downstream of the cleavage sites are typically yielding the most complete and specific cleavage results (up to 95% specific cleavage) (Duss et al., 2010, 2012).
2. The best conditions for cleavage are obtained by small-scale reactions (typically 500 pmol RNA in 15 μl reaction volume). The most important optimization parameters are the RNase H enzyme concentration (NEB, or in-house produced (Ponchon, Beauvais, Nonin-Lecomte, & Dardel, 2009)) and the ratio between the RNA and the 2′-O-methyl-RNA/DNA chimera (Duss et al., 2010). Although most RNase H cleavage reactions can be performed using only 5% stoichiometric amount of 2′-O-methyl-RNA/DNA chimera compared to RNA to be cleaved, some reactions are less sensitive to unspecific cleavage when using stoichiometric amounts of chimera. Generally, the reactions are incubated for 1 h at 37 °C. Some reactions can also be

ss_RNA_VS

5′-GGGAUCACACAAUACAUUUAAUUAAGGGCGUCGUCGCCCCGAGGAUUGAUA-3′

Chimera

5′-GTATUmGmUmGmUmGmAmUmCmCmCm-3′

Figure 8 Example of a precursor RNA and the corresponding 2′-O-methyl-RNA/DNA chimera required for site-specific cleavage mediated by RNase H. The sequence of interest is in red (light gray in the print version) letters, the 5′-region complementary to the chimera in gray and the VS 3′ stem-loop in black. Four additional nucleotides elongate the VS stem-loop and are dispensable. The 2′-O-methyl-RNA nucleotides are labeled with an "m."

performed at lower temperature (e.g., 4 °C) to minimize unspecific cleavage.
3. The best conditions are scaled up to a large-scale reaction (typically 20–200 nmol). Note that when the RNase H enzyme concentration is scaled up proportionally like the other reaction components, this can lead to unspecific cleavage. Therefore, the RNase H concentration should be reduced around 10-fold in the large-scale reaction. The completion of the reaction should be verified by denaturing polyacrylamide gels. In case of incomplete cleavage, more RNase H can be added and the reaction proceeded by another hour. A typical reaction to cleave 200 nmol of RNA is performed in 6 ml volume containing 33 μM RNA, 1.65 μM chimera, 80 nM in-house produced RNase H in 50 mM Tris–HCl (pH 7.5), 100 mM NaCl, and 10 mM MgCl$_2$.
4. The reactions are directly loaded onto an anion-exchange HPLC followed by n-butanol extraction and lyophilization as mentioned above.

5.4 VS Ribozyme Cleavage of Purified RNA Fragments

1. The concentrations of MgCl$_2$ and separately transcribed VS ribozyme RNA (Duss et al., 2010; Guo & Collins, 1995) are optimized in small-scale reactions (30 μl).
2. A large-scale VS ribozyme cleavage reaction is performed. A typical 1 ml large-scale reaction contains between 10 and 50 μM of RNA and VS ribozyme RNA in 50 mM Tris–HCl (pH 7.5), 100 mM NaCl, and 120 mM MgCl$_2$. The cleavage can be conducted at 37 °C for 45 min followed by 15 min at 65 °C. Depending on the RNA construct, thermal cycling is required to increase VS ribozyme cleavage efficiency. Thereby, the sample is heated to 65 °C for 5–15 min and then cooled down to 37 °C for 15–60 min. The cycling is repeated three to four times. Finally, the reaction is directly loaded onto an anion-exchange HPLC followed by n-butanol extraction and lyophilization as mentioned above.

5.5 Tips

1. To facilitate cleavage product separation during HPLC purification, additional nucleotides can be added 3′ of the VS stem-loop and the 5′-region of the precursor RNA can be designed longer than the complementary 2′-O-methyl-RNA/DNA chimera.

2. Transcription from chemically synthesized oligonucleotides usually results in RNA precursors with less homogenous ends compared to transcription from linearized plasmid DNA templates. We observed that performing RNase H cleavage reactions from RNA precursor obtained from plasmid DNA templates results into much less unspecific RNA degradation and cleaner RNase H cleavage patterns.
3. Performing the RNase H cleavage at 20–25 °C instead of 37 °C can help reducing unspecific degradation.

6. PROTOCOL C: LIGATION

Ligations of different unlabeled, isotope, and/or spin-labeled fragments are performed using either DNA splinted ligation using T4 DNA ligase or nonsplinted ligation with T4 RNA ligase.

6.1 Protocol

1. Ligation reactions are first performed on small-scale reactions:
   a. *For T4 DNA ligase-based ligations*: Typically, reactions with 200 pmol RNA fragments in 20 μl reaction volume are performed by optimizing the T4 DNA enzyme concentration (NEB, fermentas, or in-house), the reaction time, the reaction temperature, and testing the influence of PEG-4000. The DNA splints, which are added in 1–1.2-fold excess in respect to the RNA fragments, are usually annealed to the RNA fragments prior to ligation. However, we found that depending on the secondary structure of the RNA, annealing of the DNA splints to the RNA fragments is not required. The DNA splints should be designed such that they typically hybridize to 15–20 nucleotides on each side of the ligation site (see also tips below). Furthermore, the formation of secondary structures within the DNA splint should be minimized if possible. Alternatively, longer DNA splints can be designed which form longer (more stable) intermolecular base-pairs with the RNA fragments compared to intramolecular DNA–DNA base-pairs. For such cases, slow annealing at high RNA fragment and DNA splint concentrations is crucial to maximize RNA/DNA duplex formation compared to intramolecular structure formation.
   b. *For T4 RNA ligase-based ligations*: Typically reactions with 400 pmol RNA fragments in 10 μl reaction volume are performed by mainly

optimizing the T4 RNA enzyme concentration (NEB) and testing the addition of BSA.
2. The best reaction conditions are scaled up for the large-scale reactions:
   a. *For T4 DNA ligase-based ligations*: A typical large-scale ligation reaction (e.g., 30–50 nmol of RNA fragments) is performed for 2–6 h at 37 °C and is 10 μM in RNA fragments, 10–15 μM in DNA splint oligo, 10% in PEG-4000, 40 mM Tris–HCl (pH 7.8), 0.5 mM ATP, 10 mM MgCl$_2$, 10 mM DTT, 50 U T4 DNA ligase (fermentas) per nmol of RNA to be ligated or 2 μM final concentration of in-house produced T4 DNA ligase.
   b. *For T4 RNA ligase-based ligations*: A typical large-scale ligation reaction (e.g., 100 nmol of RNA fragments) using T4 RNA ligase is 40 μM in both RNA fragments, 1× in NEB ligation buffer (50 mM Tris–HCl (pH 7.8), 1 mM ATP, 10 mM MgCl$_2$, 10 mM DTT), 1× in BSA using 5 U T4 RNA ligase per nmol of RNA to be ligated. The reaction is performed for 2 h at 37 °C.
3. The reactions are purified by anion-exchange HPLC followed by *n*-butanol extraction and lyophilization as described in Section 5. Note that the harsh HPLC purification of spin-labeled RNA does not lead to cleavage of the spin label.
4. The spin-labeling efficiencies are determined by CW EPR and are for most RNAs in the range 85–100%.

6.2 Tips

1. To maximize the ligation efficiency and reduce the degradation of the RNA (by co-purified RNases during purification of T4 DNA ligase), long DNA splints covering almost the entire RNA sequence can be used.
2. For ligation of fragments containing spin labels, to prevent low ligation efficiencies, the spin label should not be attached to nucleotides closer than four nucleotides from the site of ligation. Nevertheless, ligation efficiencies close to 20% were obtained for a 4-piece ligation in which one spin label was attached only three nucleotides away from one ligation site (Duss, Yulikov, et al., 2014).
3. For ligation of fragments containing spin labels, the ligation reactions are performed in the absence of DTT in the ligation buffer to prevent reduction of the spin label. The ligation efficiency (typically 20–40%) is not diminished by omitting DTT.

7. SUMMARY AND OUTLOOK

In this chapter, we present approaches for obtaining different types of isotope- and/ or spin-labeled RNAs for NMR and EPR structural studies.

For NMR structural studies of protein–RNA complexes consisting of a small RNA in complex with a protein, we provide a protocol to generate small (<10 nts) isotope-labeled RNAs by sequence-specific RNase H cleavage or by combining the latter with VS ribozyme cleavage. Both approaches take advantage of a protective stem-loop at the $3'$-end of the RNA of interest, which seems to be important to protect the small (often unstructured) RNA from degradation by nucleases which co-purify with the RNase H enzyme. We obtain pure and high amounts of isotope-labeled RNAs, which are sufficient for acquiring three- or four-dimensional NMR experiments (>1 mM concentrations). Structure determination of such complexes is often rendered difficult because of the difficulty to assign resonances (especially for the strongly overlapped RNA sugar protons) and the lack of a sufficiently high number of intermolecular NOEs. By isotope labeling the RNA, resonance overlap can significantly be reduced and the acquisition of three-dimensional nuclear Overhauser effect spectroscopy (NOESY) RNA-edited spectra increases the number of NOEs that can be acquired. Furthermore, by using different labeling scheme for RNA and protein, a set of filtered and edited spectra can be recorded to increase the number of intermolecular NOEs between protein and RNA (Dominguez et al., 2011).

We also provide a simple cut and paste approach to generate large segmentally isotope- and/or spin-labeled RNAs. First, different types of smaller RNA fragments are produced by sequence-specific RNase H or ribozyme cleavage from larger fragments or the fragments are commercially available. These fragments may be unlabeled, isotopically labeled, or spin-labeled. In a second step, several of these fragments are combined at will and ligated to generate large RNAs for NMR and EPR structural studies.

By combining unlabeled and isotopically labeled fragments, large segmentally labeled RNAs are made. Segmentally labeled RNAs have reduced resonance overlap which is essential to obtain well-resolved resonance pairs in two-dimensional heteronuclear correlation spectra for detecting a sufficient amount of residual dipolar coupling or paramagnetic resonance enhancement data for successful structure determination. In addition, segmental isotope labeling of RNA helps resonance assignment, allows detecting structural elements in large biologically relevant RNAs (Lu et al., 2011), or helps monitoring assembly of multi-component protein–RNA complexes (Duss, Michel, Yulikov, et al., 2014).

By combining unlabeled with spin-labeled RNAs, large doubly spin-labeled RNAs are obtained. Using EPR spectroscopy, long-range distances (20–100 Å) between the two spin-labels can be measured (Duss et al., 2015). If sufficient distances can be measured, solution structures of large RNAs or protein–RNA complexes can be obtained even if present in more than a single conformation in solution (Duss, Michel, Yulikov, et al., 2014; Duss, Yulikov, et al., 2014). In contrast to NMR, there is technically no size limitation in EPR spectroscopy, and RNAs and protein–RNA complexes with megadalton molecular weight should be in principle accessible by EPR (Duss et al., 2015).

Finally, by combining all three types of fragments—unlabeled, isotope-labeled, and spin-labeled—large RNAs could be obtained containing a site-specifically attached spin-label on one part of the RNA with another part of the RNA isotopically labeled. If the spin-label is close in space (<20–25 Å) to the isotope-labeled part, intermediate-range distance information within the RNA could be obtained by PRE.

While the PRE effect is only intermediate-range and only limited structural information on larger multi-domain RNAs may be obtained, pseudo contact shifts (PCS) could have a greater potential to provide rich information on larger RNAs and protein–RNA complexes. PCS arise when the environment of the unpaired electron is not isotropic like the nitroxide radical. In addition to information on long-range distances (up to 40 Å), PCS also provide orientation information (Otting, 2010).

Overall, using a simple cut and paste approach, various types of large segmental isotope- and/or spin-labeled RNAs can be obtained, which provides the basis to study the structure, dynamics, and mechanism of large RNAs and protein–RNA complexes using NMR and EPR spectroscopy.

ACKNOWLEDGMENTS

Research of the Allain lab described in this chapter is/was supported by the Swiss National Science Foundation (SNF) grant Nr. 3100A0-118118, 31003ab-133134, and 31003A-149921, and the SNF-NCCR structural biology Iso-lab.

REFERENCES

Breaker, R. R. (2011). Prospects for riboswitch discovery and analysis. *Molecular Cell*, *43*(6), 867–879.
Buttner, L., Javadi-Zarnaghi, F., & Hobartner, C. (2014). Site-specific labeling of RNA at internal ribose hydroxyl groups: Terbium-assisted deoxyribozymes at work. *Journal of the American Chemical Society*, *136*(22), 8131–8137.
Buttner, L., Seikowski, J., Wawrzyniak, K., Ochmann, A., & Hobartner, C. (2013). Synthesis of spin-labeled riboswitch RNAs using convertible nucleosides and DNA-catalyzed RNA ligation. *Bioorganic and Medicinal Chemistry*, *21*(20), 6171–6180.
Cathala, G., & Brunel, C. (1990). Use of n-butanol for efficient recovery of minute amounts of small RNA fragments and branched nucleotides from dilute solutions. *Nucleic Acids Research*, *18*(1), 201.
Clore, G. M., & Iwahara, J. (2009). Theory, practice, and applications of paramagnetic relaxation enhancement for the characterization of transient low-population states of biological macromolecules and their complexes. *Chemical Reviews*, *109*(9), 4108–4139.
Dominguez, C., Schubert, M., Duss, O., Ravindranathan, S., & Allain, F. H. (2011). Structure determination and dynamics of protein-RNA complexes by NMR spectroscopy. *Progress in Nuclear Magnetic Resonance Spectroscopy*, *58*(1–2), 1–61.
Duss, O., Lukavsky, P. J., & Allain, F. H. (2012). Isotope labeling and segmental labeling of larger RNAs for NMR structural studies. *Advances in Experimental Medicine and Biology*, *992*, 121–144.
Duss, O., Maris, C., von Schroetter, C., & Allain, F. H. (2010). A fast, efficient and sequence-independent method for flexible multiple segmental isotope labeling of RNA using ribozyme and RNase H cleavage. *Nucleic Acids Research*, *38*(20), e188.
Duss, O., Michel, E., Diarra dit Konte, N., Schubert, M., & Allain, F. H. (2014). Molecular basis for the wide range of affinity found in Csr/Rsm protein-RNA recognition. *Nucleic Acids Research*, *42*(8), 5332–5346.
Duss, O., Michel, E., Yulikov, M., Schubert, M., Jeschke, G., & Allain, F. H. (2014). Structural basis of the non-coding RNA RsmZ acting as a protein sponge. *Nature*, *509*(7502), 588–592.
Duss, O., Yulikov, M., Jeschke, G., & Allain, F. H. (2014). EPR-aided approach for solution structure determination of large RNAs or protein-RNA complexes. *Nature Communications*, *5*, 3669.
Duss, O., Yulikov, M., Jeschke, G., & Allain, F. H. (2015). Combining NMR and EPR to determine structures of large RNAs and protein-RNA complexes in solution. *Methods in Enzymology*, *558*, 279–331.
Ferre-D'Amare, A. R., & Doudna, J. A. (1996). Use of cis- and trans-ribozymes to remove 5' and 3' heterogeneities from milligrams of in vitro transcribed RNA. *Nucleic Acids Research*, *24*(5), 977–978.
Gobl, C., Madl, T., Simon, B., & Sattler, M. (2014). NMR approaches for structural analysis of multidomain proteins and complexes in solution. *Progress in Nuclear Magnetic Resonance Spectroscopy*, *80*, 26–63.
Guo, H. C., & Collins, R. A. (1995). Efficient trans-cleavage of a stem-loop RNA substrate by a ribozyme derived from Neurospora VS RNA. *EMBO Journal*, *14*(2), 368–376.
Hansen, T. B., Jensen, T. I., Clausen, B. H., Bramsen, J. B., Finsen, B., Damgaard, C. K., et al. (2013). Natural RNA circles function as efficient microRNA sponges. *Nature*, *495*(7441), 384–388 (Research Support, Non-U.S. Gov't).
Helmling, C., Bessi, I., Wacker, A., Schnorr, K. A., Jonker, H. R., Richter, C., et al. (2014). Noncovalent spin labeling of riboswitch RNAs to obtain long-range structural NMR restraints. *ACS Chemical Biology*, *9*(6), 1330–1339.

Kedde, M., Strasser, M. J., Boldajipour, B., Oude Vrielink, J. A., Slanchev, K., le Sage, C., et al. (2007). RNA-binding protein Dnd1 inhibits microRNA access to target mRNA. *Cell*, *131*(7), 1273–1286.

Lapham, J., & Crothers, D. M. (1996). RNase H cleavage for processing of in vitro transcribed RNA for NMR studies and RNA ligation. *RNA*, *2*(3), 289–296.

Lu, K., Heng, X., Garyu, L., Monti, S., Garcia, E. L., Kharytonchyk, S., et al. (2011). NMR detection of structures in the HIV-1 5′-leader RNA that regulate genome packaging. *Science*, *334*(6053), 242–245.

Mercer, T. R., Dinger, M. E., & Mattick, J. S. (2009). Long non-coding RNAs: Insights into functions. *Nature Reviews Genetics*, *10*(3), 155–159 (Research Support, Non-U.S. Gov't Review).

Moore, M. J., & Query, C. C. (2000). Joining of RNAs by splinted ligation. *Methods in Enzymology*, *317*, 109–123.

Morris, K. V., & Mattick, J. S. (2014). The rise of regulatory RNA. *Nature Reviews Genetics*, *15*(6), 423–437.

Mukhopadhyay, D., Houchen, C. W., Kennedy, S., Dieckgraefe, B. K., & Anant, S. (2003). Coupled mRNA stabilization and translational silencing of cyclooxygenase-2 by a novel RNA binding protein, CUGBP2. *Molecular Cell*, *11*(1), 113–126.

Otting, G. (2010). Protein NMR using paramagnetic ions. *Annual Review of Biophysics*, *39*, 387–405.

Ponchon, L., Beauvais, G., Nonin-Lecomte, S., & Dardel, F. (2009). A generic protocol for the expression and purification of recombinant RNA in Escherichia coli using a tRNA scaffold. *Nature Protocols*, *4*(6), 947–959.

Price, R. P., Oubridge, C., Varani, G., & Nagai, K. (1998). Preparation of RNA: Protein complexes for X-ray crystallography and NMR. In C. W. J. Smith (Ed.), *RNA-protein interactions: A practical approach* (pp. 37–74). Oxford: Oxford University Press.

Qin, P. Z., Hideg, K., Feigon, J., & Hubbell, W. L. (2003). Monitoring RNA base structure and dynamics using site-directed spin labeling. *Biochemistry*, *42*(22), 6772–6783.

Ramirez-Tapia, L. E., & Martin, C. T. (2012). New insights into the mechanism of initial transcription: The T7 RNA polymerase mutant P266L transitions to elongation at longer RNA lengths than wild type. *The Journal of Biological Chemistry*, *287*(44), 37352–37361.

Ramos, A., & Varani, G. (1998). A new method to detect long-range protein-RNA contacts: NMR detection of electron-proton relaxation induced by nitroxide spin-labeled RNA. *Journal of the American Chemical Society*, *120*(42), 10992–10993.

Sowa, G. Z., & Qin, P. Z. (2008). Site-directed spin labeling studies on nucleic acid structure and dynamics. *Progress in Nucleic Acid Research and Molecular Biology*, *82*, 147–197.

Steitz, T. A. (2008). A structural understanding of the dynamic ribosome machine. *Nature Reviews Molecular Cell Biology*, *9*(3), 242–253.

Tzakos, A. G., Grace, C. R. R., Lukavsky, P. J., & Riek, R. (2006). NMR techniques for very large proteins and RNAs in solution. *Annual Review of Biophysics and Biomolecular Structure*, *35*, 319–342.

Wahl, M. C., Will, C. L., & Luhrmann, R. (2009). The spliceosome: Design principles of a dynamic RNP machine. *Cell*, *136*(4), 701–718 (Research Support, Non-U.S. Gov't Review).

Wunderlich, C. H., Huber, R. G., Spitzer, R., Liedl, K. R., Kloiber, K., & Kreutz, C. (2013). A novel paramagnetic relaxation enhancement tag for nucleic acids: A tool to study structure and dynamics of RNA. *ACS Chemical Biology*, *8*(12), 2697–2706.

Yulikov, M. (2015). Spectroscopically orthogonal spin labels and distance measurements in biomolecules. *Electron Paramagnetic Resonance*, *24*, 1–31.

Zhang, X., Cekan, P., Sigurdsson, S. T., & Qin, P. Z. (2009). Studying RNA using site-directed spin-labeling and continuous-wave electron paramagnetic resonance spectroscopy. *Methods in Enzymology*, *469*, 303–328.

AUTHOR INDEX

Note: Page numbers followed by "f" indicate figures and "t" indicate tables.

A

Abdine, A., 375–376
Abdul-Manan, N., 373, 375–376
Abe, H., 103–106
Abraham, E., 111–112
Abraham, S.J., 254–255
Aceti, D.J., 32, 35, 505t
Acharya, P., 292–293
Acharya, R., 372t
Adachi, J., 326–327
Adachi, O., 128–131, 137
Adney, W.S., 127
Adriaensens, P., 254–255
Aebersold, R., 349–350
Agarwal, V., 194
Agemark, M., 197–198
Agre, P., 197–198
Ahmad, S., 497
Ahmed, M.A.M., 169–170, 200–201, 203–204
Ahn, J.H., 312–314
Ahuja, S., 194–195
Airenne, K.J., 248–249
Akaike, T., 128–129
Akasaka, R., 338–339, 339–340f
Akbey, U., 194
Akke, M., 485–487
Al-Abdul-Wahid, M.S., 70–71, 74
Alakhov, Y.B., 312–314
Alarcon, M., 259
Albert, K., 217–221
Alcaide, F., 392–393
Alemany, R., 298
Alexandersson, E., 197–198
Alexandrov, A.A., 126
Alexandrov, K., 374
Alexiev, U., 148, 163–164
Al-Hashimi, H.M., 462–463, 484–487
Al-Ibraheem, J., 157–158
Al-Khateeb, N., 106–107, 109–110
Allain, F.H.T., 318, 390–419, 462–463, 484–485, 527–528, 538–553, 557–560

Aller, S.G., 196
Al-Rubeai, M., 295–296
Alvarado, L.J., 467–468, 468f, 484, 498, 527–528
Amaya, K., 398, 399t, 400–401, 404, 411
Anant, S., 550
Anderson, D., 248–249
Anderson, D.E., 173
Anderson, E.H., 5
Anderson, L., 348
Anderson, S.M., 127–129
Andjus, P.R., 126
Ando, E., 312–314
Andre, I., 207
Andre, N., 196
André, S., 71–73, 90
Andreas, L.B., 194
Andrews, A.T., 325–326, 377t, 406t
Andronesi, O.C., 169–170
Angelotti, T., 428–429
Anglister, J., 300
Angyal, S.J., 429, 435, 437–438, 445–447
Anicuta, S.-G., 107
Anthony-Cahill, S.J., 70–71, 80–81
Antich, P.P., 254–255
Anufrieva, N.V., 377t
Aoki, K., 295–296, 297f, 303
Aoki, M., 312–316, 319, 323, 334
Appenroth, K.-J., 223–224
Arakawa, T., 196
Aramini, J.M., 74–75
Aranguren, M., 111–112
Aranko, A.S., 394–395
Arashida, T., 128–129
Arata, Y., 291–292, 292t, 300
Archer, S.J., 291–292
Ardini, E., 91
Argarana, C.E., 358
Arif, A., 415
Arif, B., 248
Arkin, I.T., 134
Arking, I.T., 218t, 220–221

Arlow, D.H., 252
Arnold, A.A., 374
Arseniev, A.S., 69–70
Arthanari, H., 194–195, 381–382
Arulanandam, A., 291–292
Arya, R., 290–291
Asada, H., 196
Ashby, R., 101–102
Ashurst, J., 148–149
Asimakopoulou, A., 377t
Aslimovska, L., 203–204
Assenberg, R., 246–249, 291–292
Astley, O.M., 112
Astot, C., 225
Atalla, R.H., 112
Atrasz, R.G., 149
Atreya, H.S., 28, 168–186, 250, 295–296, 298–300
Augé, S., 107
Aulton-Jones, M., 375–376
Austin, R.J., 207
Auweter, S.D., 400–401, 462–463
Ayers, B., 391, 393–394
Ayers, D.F., 377t
Ayling, A., 322
Aziz, R., 68–70

B

Baaske, P., 370–371
Babich, A., 295–296
Bachor, R., 126
Backliwal, G., 292–293
Backmark, A., 196–198
Baell, J.B., 86–87
Baenziger, J.E., 203f, 204–205
Bagaria, A., 375–376
Bajaj, V.S., 194
Baker, A., 126
Baker, J.L., 520
Balan, V., 135–136
Baldansuren, A., 46–50, 48t, 51t, 52–54, 58t, 61–64
Baldus, M., 169–170, 194
Bali, G., 112, 116–117, 125–126, 133–134, 216–217, 218t, 220–223, 225–226, 229–234
Ball, G., 290–291
Ball, L.J., 169–170

Ballmer-Hofer, K., 291–292
Bamann, C., 196
Bamberg, E., 196
Bandiera, T., 91
Bandyopadhyay, T., 452–454
Baneyx, F., 18–21, 30–31, 322
Banigan, J.R., 169–170, 185
Bann, J.G., 70, 73–74
Baranov, V.I., 312–314
Barber, J., 6
Barbet-Massin, E., 194
Bardiaux, B., 194
Barker, C., 295–296, 297f, 303
Barker, R., 428–430, 433–437, 439–442, 445–447, 452–454
Barker, S.A., 436
Barnes, E.M., 7
Barnoud, F., 116–117, 129, 135–136
Barnwal, R.P., 168–169
Barry, G.F., 249
Barth, A., 204–205
Barve, M., 169–170
Basu, B., 445, 452–454
Basu, J., 126
Batchelor, W., 111–112
Bates, C.J., 325–326, 377t, 406t
Bates, R.G., 14
Batey, R.T., 484–485
Battiste, J.L., 462–463
Batzoglou, S., 400
Baumgartner, N., 291–292
Bax, A., 5, 168–170, 184
Bayburt, T.H., 31t
Bayrhuber, M., 194–195, 370
Bazzacco, P., 370–371
Beardall, J., 115
Beaucage, G., 100–101
Beaugrand, M., 374
Beauprez, J., 31–32
Beauvais, G., 498, 500f, 511–512, 523–524, 527–528, 555
Becette, O., 488–490
Bechtold, I., 248–249
Beck, K., 321–322
Becker, S., 169–170, 369, 372t
Beckwith, J., 321–322
Behrens, P.W., 427
Beilby, M., 112

Belenky, M., 194
Belikov, V.M., 377t
Bellstedt, P., 46–47, 169–170, 173, 175t, 178, 181, 186
Benner, J., 392–394, 399t, 400–401
Benov, L., 157–158
Benziman, M., 112, 137
Berardi, M.J., 194–195
Berestovsky, G.N., 126
Berg, P., 295–296
Berger, I., 248–249, 291–292
Berger, P., 248–249
Bergey, D.H., 127–128
Berk, A.J., 295–296
Bernardo, E.B., 127–129
Bernhard, F., 314–315, 368–382, 394–396
Berrier, C., 370–371, 375–376
Bertelsen, E.B., 395–396
Bertrand, K., 271–272
Bessi, I., 545–546
Bethell, G.S., 429, 435, 437–438, 445–447
Beveridge, R.J., 429, 435, 437–438, 445–447
Bewley, C.A., 290–303
Beynon, R.J., 349, 360–361, 365
Bezerra, A.G., 196–197, 199
Bezsonova, I., 73–74
Bhate, M.P., 372t
Bhatia, C.R., 217–219, 224–225
Bhattacharya, A., 290–291
Bickford, E.D., 225–226
Bielecki, S., 128–129
Bielik, A., 299
Bieniossek, C., 248–249
Bieszke, J.A., 196
Bilik, V., 436–437
Billon-Denis, E., 370–371
Bingman, C.A., 32, 35, 505t
Birdsall, B., 71–73
Birken, S., 291–292, 292t
Bitsch, F., 250, 252, 256, 262–265, 269, 273
Bittl, R., 196
Bjorkman, P.J., 293–295
Blackwell, J., 99
Blaise, C., 223–224
Blake, M.I., 219, 228–229
Blakeley, M.P., 6–7
Blaschek, W., 127

Blaschke, U.K., 393–394, 400–401, 412–413
Blechner, S.L., 5
Bley, T., 106–107, 109–110
Blissard, G.W., 259–261
Bloch, P.L., 37
Bloor, D., 6
Blount, K.F., 462–463
Bock, C., 375–376
Bogle, I.D.L., 8t
Bohm, M., 137
Bohn, A., 127–128
Boime, I., 70–74
Boisbouvier, J., 379
Boland, C., 370
Boldajipour, B., 550
Bondarenko, V., 372t
Bondo, P.B., 438–439
Bonelli, B., 107
Bonhoeffer, K.F., 215
Bonner, D.M., 103
Booth, I.R., 7
Bordes, F., 207
Bordignon, E., 372t
Borkovich, K.A., 196
Bos, M.P., 194
Bose, S., 219–221, 223–224
Bosman, G.J., 194–195, 256–258
Bothner-By, A.A., 427
Bouchard, P., 462–463
Bouhss, A., 370
Bourbigot, S., 484–485
Bourret, R.B., 74
Boutell, J., 312–314
Bouvignies, G., 488–490
Bovee-Geurts, P.H.M., 194–195, 252, 270
Bowyer, J.R., 225–226
Box, M.E., 254–255
Boze, H., 198
Bracic, G., 57–61
Bracken, C., 6, 36–37
Brady, J.W., 127
Bramsen, J.B., 538
Brandt, H., 100–101
Braun, P., 290–291
Bray, J., 125–126
Breaker, R.R., 462–463, 497, 520, 538
Brem, J., 69–71, 86–87

Brender, J.R., 69–70, 88
Briand, L., 185–186
Bringmann, P., 292–293
Britovsek, G., 127
Brock, A., 326–327
Brockwell, D., 6
Brodsky, A.S., 462–463, 528
Brown, A.J., 127–129
Brown, D.J., 462–463, 497
Brown, G., 325–326
Brown, J.W., 225
Brown, L.S., 169–170, 194–207
Brown, M.E., 505–506t, 517
Brown, P.A., 111–112
Brown, R.E., 497–532
Brown, R.M., 127–128
Brüggert, M., 252
Brunel, C., 554
Brutscher, B., 46–47, 168–169, 175–176, 176t
Buck, J., 462–463
Budhiono, A., 127–128
Buhr, F., 462–463
Bujons, J., 254–255
Bullard, J.M., 377t
Burd, A.M., 149
Burges, H., 248–249
Burgess, J., 217–219
Burgess, R.R., 409–410
Bürgin, M., 292–293
Burkhardt, N., 5
Burrell, M.M., 256–257
Burrows, K.M., 112
Burz, D.S., 271–272
Busche, A.E., 394–395
Butcher, S.E., 497
Buttner, L., 545
Byrd, R.A., 206f
Byrne, B., 196
Byron, O., 138–139
Bystrov, V.F., 69–70
Byung, Y., 74

C

Cabezas, A., 447
Cadalbert, R., 194
Cadman, E., 439–442, 445–447
Cai, J., 254–255
Cai, M.L., 6

Cairney, J., 127
Calvar, I.L., 127–128
Camakaris, H., 48t, 51t, 52–53
Campbell, D.O., 462–463
Campbell, E.B., 196
Campbell, I.D., 70–75, 175–176, 395–397
Campos-Olivas, R., 68–70
Canales, J., 447
Cannon, R.E., 127–129
Cantor, C.R., 358
Capadona, J.R., 111–112
Capel, M.S., 28
Capes, M.D., 163–164
Cappuccio, J.A., 374
Carlomagno, T., 527–528
Carlson, E.D., 312–314
Carmichael, I., 442–444
Carnal, S., 249
Carr, S., 185–186
Carralot, J.-P., 291–292
Casagrande, F., 194–195
Casiano-Negroni, A., 484–487
Catchmark, J.M., 116–117, 127–128
Cathala, G., 554
Catoire, L.J., 370–371, 374
Cavanagh, J., 85, 168, 177
Ceccarelli, E.A., 30–31
Cecil, R., 156
Cekan, P., 540
Celis, V., 377t
Cellitti, S.E., 70, 79–80
Chagas, B., 106–107
Chaiken, I.M., 69–70
Chait, B.T., 291–292, 292t
Chambliss, C.K., 135–136
Chandrasegaran, S., 427
Chang, C., 325–326
Chang, D.K., 207
Chang, P.K., 325–326, 377t, 406t
Chanliaud, E., 112
Chanzy, H., 112, 138–139
Chapman, B.E., 259, 262–265
Charlebois, R.L., 148, 163–164
Charteris, A., 324–325, 406t
Chartrand, É., 374
Chary, K.V.R., 28, 168–170, 176–177, 184–185
Chasapis, C.T., 377t

Chatterjee, A., 169–170
Chauder, B.A., 273–274
Chen, B., 462–463, 467–468, 468f, 484–485, 498–499, 506t, 527–530
Chen, C.A., 292–293
Chen, C.Y., 207
Chen, F., 234
Chen, H., 71–73, 86, 299
Chen, K., 295
Chen, K.-E., 7
Chen, L., 292–293
Chen, M., 293–296
Chen, W.-N., 382
Chen, W.-P., 215–221, 225
Chen, X., 7
Chen, Y.C., 207
Cheng, C.H., 207
Cheng, G., 111–112
Cheng, K.-C., 116–117, 127–128
Chenuet, S., 292–293
Cherouati, N., 196
Cherry, S., 400–401
Chiguru, S., 71–73
Chillon, M., 298
Chittaboina, S., 196
Chiu, E.J.C., 194–195, 370
Cho, H.S., 324–325
Choe, S., 194–195, 372t
Choi, C.H., 312–314
Choi, C.Y., 312–315, 321–322
Choi, J., 291–292
Choi, S.K., 46–64
Chong, S., 399t
Chopra, A.K., 436
Chou, J.J., 194–195
Chou, S.H., 207
Chow, B.Y., 197
Chow, J.Y.H., 7, 20f
Christen, P., 377t
Christie, M.P., 7
Chu, H.S., 312–314
Chu, M., 194–195
Chugh, J., 484–487
Chujo, R., 100–101, 104–106
Chumpitazi-Hermoza, B.F., 116–117, 135–136
Chundawat, S.P.S., 113–114, 124–142, 218t, 220–221

Chung, J., 300
Chuong, A.S., 197
Church, T.J., 445, 452–454
Cirino, G., 377t
Claridge, T.D.W., 69–71, 86–87
Clark, E.L., 428–430, 433–434, 436–437, 447
Clark-Lewis, I., 391–392
Clausen, B.H., 538
Clausen, T., 377t
Clauser, K.R., 5
Cleveland IV, T.E., 148–164
Cline, S.W., 148, 163–164
Clore, G.M., 6, 71–73, 488–490, 545–546
Cocucci, M., 217–219
Cohen, J.D., 215–221, 225
Cohen, J.S., 69–70
Cohen, L.A., 254–255
Cole, P.A., 391–393, 396–397, 399t, 400–401, 418
Colebrook, S.A., 395–397
Coleman, T.A., 292–293
Coletta, C., 377t
Collart, F.R., 46–47, 53–54
Collins, R.A., 554, 556
Colombini, M., 194–195
Condreay, J.P., 248–249
Conley, J.D., 325–326, 376–379, 377t, 406t
Connolly, P.J., 373, 375–376
Constantinescu, A., 254–255
Contzen, J., 6–7
Cooke, R.J., 219–220, 223–224
Cookson, D.J., 115–116
Cooley, N., 312–314
Cooperman, B.S., 315
Cope, B.T., 219–221, 223–224
Coppins, R.L., 462–463
Cordeiro, N., 111–112
Corin, K., 370–371
Cosgrove, T., 107
Cosper, N.J., 54, 62–64
Cotton, G.J., 393–394, 400–401, 412–413
Coulombe, R., 71–73, 90
Couperwhite, I., 127–129
Couto, A., 447
Covington, A.K., 14
Cowan-Jacob, S.W., 252, 262f, 269, 273
Cowburn, D., 318, 328, 391, 393–394

Cowieson, N., 115–116
Cox, M., 254–255
Craggs, T.D., 69–70, 87
Craigie, R., 6
Cramer, S.P., 314–315
Crane, F.A., 219, 228–229
Craven, C.J., 375–376
Crawford, D.J., 223–224
Creemers, A.F.L., 252, 270
Cremer, H., 248–249
Crespi, H.L., 125–126, 148–149, 214–221, 223–226
Crestini, C., 106–107
Croizier, G., 248–249
Crompton, K.E., 107
Crooks, E.T., 292–293
Cross, T.A., 194, 372t
Crothers, D.M., 542–543, 550
Crowe, D., 254–255
Crowley, P.B., 78
Cruz, L., 106–107
Cubeddu, L., 290–291
Cui, C.X., 6, 514–515
Curley, J.M., 103
Curmi, P.G., 290–291
Curry, J., 215–216, 219–220, 230–231
Cutting, B., 252, 273
Czaja, W.K., 127–129
Czeslik, C., 318, 328, 394–395, 418
Czihak, C., 112

D

da Costa Sousa, L., 135–136
DaBoll, H.F., 215
Dabydeen, S.A., 259–261
daCosta, C.J., 203f, 204–205
Dahlquist, F.W., 173
Dai, J., 194, 372t
Daigre, C., 295–296
Dale, B.E., 135–136
Dalmas, F., 100
Dalvit, C., 71–73, 90–91
Damgaard, C.K., 538
Danielson, J.A., 197–198
Danielson, M.A., 68–70, 86
Dannatt, H.R., 194
Danthinne, X., 295–296, 297f, 303

Dardel, F., 498–499, 500f, 511–512, 523–524, 527–528, 555
Darie, C.C., 28
Darnell, J.E., 295–296
Darwish, T.A, 98–117
Das, B.B., 194–195
Das, N., 194, 372t
DasSarma, P., 163–164
DasSarma, S., 148, 163–164
Datta, S., 111–112
Daubresse, N., 234
Daugelat, S., 392–393
Dauvergne, M.T., 6–7
Davalos, R., 130
Davanloo, P., 316–317, 320
Davies, D.D., 219–220, 223–224
Davies, G.J., 137
Davis, H.M., 135–136
Davison, B.H, 124–142
Davoudpour, Y., 111–112
Dawson, P.E., 391–392, 399t, 400
Dayie, K.T., 291–292, 462–463, 484–485, 497–498, 528
Dayie, T.K., 462–490, 497–532
de Ghellinck, A., 107, 109, 125–126
De Jesus, M., 292–293
De Lausnay, S., 31–32
De Meutter, J., 196–197
De Mey, M., 31–32
Dean, F., 377t
Deepa, B., 111–112
DeGrip, W.J., 194–195, 252, 256–258, 270
DeLange, F., 252, 270
Delente, J., 427
Delmer, D.P., 112
Demange, P., 107, 196–197
Demidkina, T.V., 377t
Demirci, A., 116–117, 127–128
Denbow, M.L., 107
Deng, F., 259–261
DeNiro, M.J., 215–216, 220–221
Denisov, I.G., 374–375
Desaive, C., 28
Desjardins, G., 462–463
Desmyter, A., 196
Dethoff, E.A., 484–487
Deville, C., 251, 257–259, 269, 273–274
Dey, P.M., 127, 225–226, 234, 235f

Di Cola, E., 100
Di Renzo, F., 107
Diarra dit Konté, N., 538–553, 557–560
Dick, L.R., 254–255
Didenko, T., 69–70
Dieckgraefe, B.K., 550
Diederich, F., 71–73
Diehl, A., 194
DiIanni, C.L., 377t
Dikanov, S.A., 46–47
Dimitrijev-Dwyer, M., 7
Ding, S.Y., 127
Dinger, M.E., 538
Dingley, A.J., 528
Diprose, J.M., 291–292
Dirheimer, G., 377t
Dixon, N.E., 312–314, 375–379
Dmitriev, O.Y., 47
Do, Q.T., 377t
Doblin, M.S., 112
Dobre, L., 107
Dobry, A.S., 197
Doherty, M.K., 349, 360–361, 365
Dohmae, N., 252, 312–316, 326, 369, 375–376, 396, 417–418
Doi, M., 69–70
Doi, Y., 100–101, 103–106
Dolezal, K., 225
Dolnikowski, G.G., 217–220, 224
Dolnkowski, G.G., 217–221
Dominguez, C., 527–528, 539, 546–547, 559
Dominguez, M.A., 73–74
Donald, A.M., 112
Donald, H.B.A., 467–468, 468f, 484–485
Dondapati, S.K., 369
Dong, S., 259–261, 370–371
Doolittle, W.F., 148, 163–164
Doria-Rose, N.A., 292–293
Dötsch, V., 53, 168–169, 271–272, 314–315, 368–382, 394–396
Doudna, J.A., 497, 554
Doverskog, M., 259, 262–265
Dowhan, D., 296–298
Doyle, S., 400–401
Drake, S.K., 74
Drew, K.N., 445, 452–454
Drews, M., 259, 262–265

Drickamer, K., 424
Dror, R.O., 252
Druz, A., 292–293
D'Silva, P., 168–170, 172, 175–181, 175t, 180f
D'Souza, V., 497
Du, X., 293–296
Dübel, S., 293
Dubendorff, J.W., 18–19
Dubey, A., 168–186
Duff, A.P., 4–23
Duffy, A.M., 298
Dufresne, A., 111–112
Duhr, S., 370–371
Dunn, J.J., 316–317, 320
Dunn, S., 225–226
Dunstall, T.G., 126
Dupureur, C.M., 73–74
Durney, M.A., 497
Durst, F., 369, 373, 375–376
Duss, O., 484–485, 527–528, 538–553, 557–560
Duus, J.Ø., 452–454
Düx, P., 148–149
Dykes, G.A., 113–114
Dyson, H.J., 168, 395–396

E

Easton, L.E., 484–485
Eckert, C.A., 127
Edgar, R.C., 400
Egbringhoff, H., 148–149
Egorova-Zachernyuk, T.A., 194–195, 256–258
Ehrmann, M., 321–322
Eichhorn, S.J., 111–112
Eichmann, C., 194–195, 370, 372t
Eilers, M., 194–195
Eizen, N., 137
Ekvall, M., 197–198
Elaswarapu, R., 248–249
Elbein, A.D., 299
Elberson, M.A., 28–43
Eldho, N.V., 498–499, 506t, 527–528
Eliceiri, K.W., 159f
Elkind, M.M., 126
Emami, S., 194–207

Endo, Y., 266–268, 312–314, 377t, 381–382
Enfors, S.O., 7
Engelhard, C., 196
Engelman, D.M., 28, 30
Englander, S.W., 390
Era, S., 196–197
Erat, M., 400–401
Erata, T., 127–128
Erfani, S., 194–195
Eriani, G., 377t
Eriksson, H.M., 370–371
Eriksson, M., 111–112
Esaki, N., 377t
Esposito, D., 292–293
Esquivies, L., 370, 373
Etezady-Esfarjani, T., 324–325, 373, 380–381
Etzkorn, M., 372t, 374–375
Evanics, F., 73–74
Evans, B.R., 112, 116–117, 124–142, 214–239
Evans, J.N.S., 68–70
Evans, T.C., 392–394, 398, 399t, 400–401, 404, 411
Ezure, T., 312–314

F

Faeh, C., 71–73
Fairbrother, W.J., 85, 168, 177
Falcicchio, P., 452
Faleev, N.G., 377t
Falke, J.J., 68–70, 73–74, 86
Falzone, C.J., 325–326, 376–379, 377t, 406t
Fan, B.X., 112
Fan, Y., 194–207
Farkas, V., 436–437
Farmer, B.T., 5–6
Faucher, A.M., 71–73, 90
Fauster, K., 462–463, 484
Fawzi, N.L., 488–490
Feeney, J., 71–73, 107, 196–197
Feicht, R., 69–70
Feigon, J., 528, 540
Feldman, L.T., 295–296
Fellert, M., 196
Fendrich, G., 250, 252, 256, 262–265, 262f, 269, 273

Fenton, W.A., 390
Fenz, S.F., 369
Férard, J.-F., 223–224
Ferguson, S.J., 70–75, 175–176
Fernández, C., 250–253, 254–255f, 257–259, 262–269, 273–274, 279–281, 300, 379
Ferner, J., 462–463
Ferre-D'Amare, A.R., 554
Fesik, S.W., 186, 256–259, 273–274, 291–292, 292t
Feuerstein, S., 168–169
Fey, M., 194
Fiaux, J., 395–396
Fielding, L., 71–73
Fierke, C.A., 5–6
Fiez-Vandal, C., 196
Figeys, D., 28
Findlay, H., 197–200
Fink, H.P., 127–128
Finkelstein, D.I., 107
Finley, J.W., 433–434
Finnerty, C.M., 259–261
Finsen, B., 538
Fischer, E., 428–429
Fischer, G.A., 325–326, 377t, 406t
Fischer, M., 69–70
Fischer, S., 111–112
Fitzgerald, D.J., 248–249
Fitzpatrick, T.B., 377t
Flanagan, B.M., 113–114
Fleischmann, E.M., 148
Fletcher, J.E., 374
Flocco, M., 91
Fogliatto, G.P., 91
Fojud, Z., 100–101
Forino, M., 91
Forman-Kay, J., 73–74
Forsén, S., 488–490
Forsythe, J.S., 107
Foster, L.J.R., 98–117
Foston, M.B., 116–117, 133–134, 138–139, 214–217, 218t, 221–223, 225–226, 229–234
Fowler, L.J., 324–325, 406t
Fox, B.G., 46, 312–314
Fox, G.E., 498
Fox, P.L., 415

Fraenkel, D.G., 528
Frandsen, E.K., 447
Frank, A.M., 46–47, 53–54
Frank, A.O., 273–274
Franks, W.T., 186, 194
Franssen, M.C.R., 452
Franz, G., 127
Franz, J.E., 377t
Fraser, M., 248–249
Freeborn, B.R., 28
Freedberg, D.I., 295
Freedman, M.H., 69–70
Freimund, S., 444
Freitas, F., 106–107
Frericks Schmidt, H.L., 47–50, 48t, 51t, 52–53, 57–61, 58t, 64
Frericks, H.L., 55–57
Frey, T., 300
Frieden, C., 70, 73–74
Frölich, N., 369
Frueh, D.P., 168
Frutos, S., 91
Fujiwara, T., 326–327
Fukazawa, R., 46–64
Fukushima, K., 234
Fuller, R.C., 103
Fuller, R.W., 262–268
Funamoto, S., 196
Fürtig, B., 462–463, 498, 528–530
Furutani, Y., 197, 199
Fux, C., 292–293

G

Gabius, H.-J., 71–73, 90
Gadea, B., 271–272
Gagnaire, D., 116–117, 128–129, 135–136
Gamble, T.R., 5
Gan, R., 312–314
Gangloff, J., 377t
Gans, P., 379
Gao, D., 135–136
Gaponenko, V., 206f, 254–255
Garau, G., 91
Garbe, D., 318, 328, 394–395, 418
Garces, R.G., 194–195
Garcia, A.J., 248–249
Garcia, E.L., 544, 559–560
Gardner, K.H., 5–6, 324, 381–382, 390, 527–528
Garrone, E., 107
Garvey, C.J., 98–117
Garyu, L., 544, 559–560
Gaskell, S.J., 349, 360–361, 365
Gaskin, M., 315
Gatenholm, P., 130
Gautier, A., 194–195
Gautier, M.F., 198
Gayen, A., 169–170, 185
Geisse, S., 246–247, 292–293
Gelan, J., 254–255
Gelev, V., 194–195, 374, 381–382
Gennis, R.B., 46–64
Gentz, S., 292–293
George, S.J., 314–315
Georgescu, J., 252
Georgiev, I.S., 290–293, 295, 300, 301f, 303
Gerber, S.A., 349
Gerhartz, B., 249
Gerig, J.T., 68–70, 73–74, 85–86
Getmanova, E.V., 300
Ghazi, A., 370–371, 375–376
Ghirlando, R., 488–490
Ghitti, M., 71–73
Ghoshdastider, U., 370, 374
Ghosh-Roy, A., 169–170
Giaretti, W., 28
Gibbons, N.E., 149
Gidley, M.J., 112–114, 214–215
Giffhorn, F., 444
Gilbert, E.P., 214–215
Gilbert, P., 6
Gilbert, R.J., 138–139
Gileadi, O., 125–126, 248–249
Gill, A., 382
Giller, K., 369, 372t
Gillies, R., 126
Giordano, M., 115
Giordano, P., 90
Giovannozzi-Sermanni, G., 106–107
Giralt, E., 91
Girard, F.C., 484–485
Girvin, M.E., 186
Giumarro, C., 217–219, 224–228
Giusti, F., 374
Glaser, L., 127–128

Glaubitz, C., 203–204, 375–376
Glinschert, A., 71–73, 90
Gluzman, Y., 295–296
Gobl, C., 527–528, 545–546
Goddard, A.D., 194–195
Godfrey, A.G., 454–455
Goff, S.P., 497
Goldman, A., 6–7, 148, 163–164
Goncalves, J.A., 194–195
Goncalves-Miskiewicz, M., 128–129
Gooley, A., 290–291
Goren, M.A., 312–314
Gorham, R.P., 225–226
Gor'kov, P.L., 203–204
Gorlach, M., 46–47, 169–170, 173, 175t, 178, 181, 186
Gorman, J., 292–293
Gossert, A.D., 246–284, 300
Goto, M., 314–315, 322–323
Goto, N.K., 186, 324
Goto, Y., 100–101, 104–106
Gottstein, D., 370–371, 375
Goudreau, N., 71–73, 90
Grabe, M., 372t
Graber, R., 377t
Grace, C.R.R., 539, 544–545
Grady-Smith, C.S., 314–315
Graepel, K.W., 293–296
Graff, P., 250, 252, 256, 262–265, 269, 273
Graham, F.L., 298
Grandi, G., 409
Gras, A., 196
Graslund, S., 125–126
Gravel, A., 374
Gray, D.M., 254–255
Gray, W.M., 215–221, 225
Greathouse, G.A., 128–129
Green, L.S., 377t
Grego, S., 219–220, 223–224
Griesinger, C., 527–528
Griffin, J.H., 399t
Griffin, R.G., 194
Griffith, M.C., 70–71, 80–81
Grinkova, Y.V., 31t, 374–375
Groebe, D.R., 462–463
Groeger, G., 106–107, 109–110
Groisman, E.A., 462–463
Gronenborn, A.M., 6, 68–91
Gross, P.R., 126
Gross, R.A., 100–102
Gross, S.S., 54
Grossfield, A., 194–195
Grotzfeld, R.M., 273
Grunnet, N., 447
Grusak, M.A., 217–221, 224
Grzesiek, S., 269, 528
Gu, J., 498–499, 511, 524–525
Guardado-Calvo, P., 251, 257–259, 269, 273–274
Guittet, E., 375–376
Gunasekera, T.S., 5, 30, 36–37
Güntert, P., 53, 324, 381–382, 395–396
Guo, H.C., 554, 556
Guthrie, E., 299
Gutmann, S., 250–253, 254–255f, 273, 279–280, 300
Gygi, S.P., 349

H

Haase, W., 53
Habeck, M., 194
Haberstock, S., 314–315, 369–370, 374
Hackeng, T.M., 399t
Haegeman, G., 293
Haertlein, M., 6–7, 107, 109, 125–126
Hafner, S., 46–47, 169–170, 173, 175t, 178, 181, 186
Häggström, L., 259, 262–265
Hagn, F., 194–195, 372t, 374–375, 381–382
Hahn, G.H., 312–314
Hajduk, P.J., 71–73, 91
Hall, K.B., 462–463
Hall, S.J., 7, 21
Hallac, R.R., 71–73
Hallberg, B.M., 125–126
Haller, A., 462–463, 487–488
Hallgren, K., 197–200
Halpern, L.A., 217–219, 224–226
Hamatsu, J., 262–265, 272, 272f
Hamelin, O., 379
Hames, B.D., 312–314
Hamilton, K., 74–75
Hamilton, W.A., 7, 20f
Hammerton, K.M., 99, 102–106
Hammill, J.T., 79–80, 82
Hamzeh, Y., 111–112

Han, L., 259
Han, X., 197
Hanada, K., 312–314, 319–320
Hanawa-Suetsugu, K., 338–339, 339–340*f*
Handschumacher, R.E., 325–326, 377*t*, 406*t*
Hansen, A.L., 488–490
Hansen, A.P., 186, 256–259, 291–292, 292*t*
Hansen, T.B., 538
Hao, X., 70, 79–80
Haon, S., 107
Haque, M.E., 88–89
Hara, T., 312–314
Harada, Y., 163–164
Harborne, J.B., 225–226, 234, 235*f*
Hare, J.T., 262–265
Hargous, Y., 400–401
Haris, P.I., 46
Harjani, J.R., 86–87
Harner, M.J., 273–274
Harris, M., 128–129
Harrison, S.C., 194–195
Hart, R., 46–50, 64
Hart, S.A., 391, 395
Hartig, J.S., 462–463
Hartley, J.L., 292–293, 299
Hartman, K., 69–70, 88
Hartman, S.C., 325–326, 377*t*, 406*t*
Hasan, T., 126
Hatanaka, K., 128–129
Hato, M., 314–315, 322–323, 370–371
Hatt, B.W., 436
Hatzoglou, M., 415
Hawley, A.M., 115–116
Hayashi, A., 326–327
Hayashi, F., 336–337, 337–338*f*
Hayashi, S., 447
Hayashi, T., 196–197
Hayashi, Y., 54
Hayes, M.L., 436–437, 439–442, 445–447
Hazemann, I., 6–7
Hazen, J.L., 79–80, 82
He, F., 333, 333–334*f*, 336, 336*f*
He, J.H., 113–114, 116–117, 124–142, 218*t*, 220–223
He, L., 7
He, M., 312–314
Headlam, M.J., 376–379

Hebel, D., 254–255
Heberle, J., 196
Hedfalk, K., 196–198
Hefferon, K.L., 259–261
Hefke, F., 375–376
Hegeman, A.D., 215–221, 225
Heibeck, T.H., 382
Heikinheimo, P., 6–7
Hein, C., 314–315, 368–371
Heise, H., 169–170
Heller, R., 292–293
Heller, W.T., 138–140, 214–215
Helmling, C., 545–546
Helms, G.L., 68–70
Hemmerle, H., 454–455
Henderson, B.R., 376–379
Henderson, C.A., 149
Hendra, P.J., 236–239
Hendrickson, W.A., 358
Heng, X., 544, 559–560
Hennig, M., 462–463
Henning, A., 400–401
Henrich, E., 314–315, 368–371
Heraud, P., 115
Herberich, B., 326–327
Herrmann, S., 128–129
Herzfeld, J., 194
Hestrin, S., 112–114, 127–129
Hettrich, K., 111–112
Heublein, B., 127–128
Heus, H.A., 484–485, 498, 520, 523–524
Hideg, K., 540
Higgins, S.J., 312–314
Hildebrandt, A.C., 225–226, 234–235
Hildinger, M., 292–293
Hiller, S., 194–195, 324–325, 373–374, 380–381
Hilty, C., 379
Himmel, M.E., 127
Hinck, A.P., 291–292
Hinniger, A., 250–253, 254–255*f*, 273, 279–280, 300
Hino, M., 349
Hino, N., 326–327
Hino, T., 196
Hinterstoisser, B., 134, 218*t*, 220–221
Hirai, Y., 85–86
Hirao, I., 326–327

Hiratake, J., 324–326, 377t, 406t
Hiroaki, H., 271–272
Hirs, C.H.W., 215
Hirschhorn, S.G., 431–432
Hitchenns, T.K., 85
Hitchman, R.B., 249
Hoang, J., 69–70, 87
Höbartner, C., 484–485, 545
Hochuli, M., 30
Hodge, J.E., 435, 437–438
Hodge, K., 225
Hodgman, C.E., 312–314
Hoeltzli, S.D., 70, 73–74
Hoevels, Y., 369–370
Hoffman, D.W., 498–499, 506t, 527–528
Hoffman, R.A., 488–490
Hofreiter, B.T., 435, 437–438
Hofstetter, K., 134, 218t, 220–221
Hoheisel, S., 254–255
Holak, T.A., 252
Holbrey, J.D., 111–112
Holden, P.J., 4–23, 98–117
Holland, J.A., 498–499, 506t, 527–528
Holt, J.G., 127–128
Holt, S., 111–112
Homer, R.J., 377t
Hong, J.S., 7
Hong, M., 176–177, 194
Hoogstraten, C.G., 462–463, 498
Hoopes, J.T., 28–43
Hopkins, R.F., 292–293
Hore, P.J., 69–70, 87
Hori, H., 54
Hori, T., 314–315, 322–323
Horne, M.K., 107
Horsefield, R., 196
Horst, R., 69–71, 86–87, 254–255, 390
Hortin, G., 70–74
Horwich, A.L., 390, 395–396
Hoshino, K., 127–128
Hosur, R.V., 169–170
Houchen, C.W., 550
Houck-Loomis, B., 497
Hounslow, A.M., 375–376
Hourai, Y., 262–265, 272, 272f
Howard, B.H., 295–296
Howell, S.C., 194–196
Howlett, B.J., 197

Howley, P.M., 295–296
Hoyer, W., 169–170
Hricovini, M., 437
Hricoviniova-Bilikova, Z., 437
Hu, F., 194
Hu, H., 70, 79–80
Hu, S.-H., 7
Hu, Y.-C., 248–249
Hu, Z., 259–261, 391, 394–395
Huang, C.C., 5–6
Huang, J.T., 295–296
Huang, W., 207
Huang, Y., 6, 236–239
Hubbell, W.L., 540
Huber, R., 377t
Huber, R.G., 545–546
Huber, T., 382
Huger, J., 248–249
Huh, B., 234
Hull, W.E., 70
Humpula, J.F., 135–136
Hung, I., 194, 203–204, 372t
Hunter, C.L., 348
Hurst, D.P., 194–195
Hurwitz, J., 35t, 36f
Huster, D., 194–195
Hutchins, H., 194
Huwig, A., 444
Hwang, E., 370, 373
Hwang, G.S., 415–416
Hwang, P.M., 390
Hyberts, S.G., 194–195, 381–382
Hyean-Woo, L., 74
Hytonen, V.P., 358
Hyun-Won, K., 74

I

Ichikawa, O., 196–197, 199t, 207
Idnurm, A., 197
Iguchi, M., 127–128
Ikeda, M., 336, 336f
Ikeda-Suno, C., 196
Ikeya, T., 53
Illarionov, B., 69–70
Imai, T., 54, 62–64
Imanishi, S., 312–314
Imasaki, T., 248–249
Imataka, H., 312–314

Imker, H., 437
Imperiale, M.J., 295–296, 297f, 303
Inagaki, F., 395
Infantino, A.S., 71–73, 90
Inoue, A., 349–350
Inoue, M., 312–316, 319, 323, 334
Inoue, Y., 100–101, 104–106
Iraha, F., 326–327
Ishihara, G., 314–315, 322–323
Ishikawa, T., 127–128
Ishino, T., 128–129
Ishizuka, Y., 323
Issaly, N., 198
Ito, K., 314–315, 322–323
Ito, M., 312–314
Ito, N., 318, 328
Ito, Y., 252, 271–272, 312–316, 326, 369, 375–376, 396, 417–418
Itoh, T.J., 126
Ivanov, V.T., 69–70
Iwahara, J., 545–546
Iwai, H., 393–397, 400–401
Iwanari, H., 196
Iwasaki, N., 312–316, 322–324
Iwasaki, T., 46–64
Iwashita, Y., 324, 381–382, 395–396

J

Jaakola, V.P., 148, 163–164
Jackson, A.M., 312–314
Jackson, G.W., 498
Jackson, J.C., 79–80, 82
Jackson, M.E., 498, 505–506t, 517
Jackson, S.E., 69–70, 87
Jacobs, W.R., 392–393
Jacobsson, U., 259, 262–265
Jacob-Wilk, D., 112
Jacrot, B., 28
Jahic, M., 7
Jahnke, W., 250–253, 254–255f, 262f, 269, 273, 279–280, 300
Jain, N., 468f, 484, 498, 527–528
Jaipuria, G., 168–170, 172, 175–181, 175t, 180f
Jakes, K., 176–177
James, D.C., 299
James, M., 7
Janning, P., 318, 328, 394–395, 418

Jardetzky, O., 256–257, 265
Jaremko, Ł., 369, 372t
Jaremko, M., 369, 372t
Jarrott, R.J., 7
Jarvis, D.L., 248–249
Jaudzems, K., 194, 382
Javadi-Zarnaghi, F., 545
Javeed, F., 300
Jayakumar, A., 7
Jeffers, S.A., 251, 257–259, 269, 273–274
Jeffries, C.M., 7, 20f
Jenkins, N., 299
Jensen, P.R., 452–454
Jensen, T.I., 538
Jeon, Y.H., 194–196
Jeong, E.J., 415–416
Jerabek-Willemsen, M., 370–371
Jergic, S., 375–379
Jeronimidis, G., 112
Jeschke, G., 538–539, 544–545, 549, 558–560
Jestin, J., 100
Jewett, M.C., 312–314
Ji, X., 234
Jia, J., 415
Jiang, S., 298
Jipa, I., 107
João, C., 106–107
Joh, N.H., 372t
Johanson, U., 197–198
John, R.A., 324–325, 406t
Johnsen, M., 112
Johnson, B.A., 488–490
Johnson, D.K., 127
Johnson, J.E., 462–463, 498
Johnson, P.E., 224
Johnson, S., 91
Johnston, W.A., 374
Jones, A.D., 113–114, 124–142, 218t, 220–221
Jones, D.H., 70, 79–80
Jones, D.T., 400, 415–416
Jones, P.P., 292–293
Jong-Whan, C., 74
Jonker, H.R., 545–546
Jonoobi, M., 111–112
Josephson, B.L., 528
Jostock, T., 293

Jouanin, L., 234
Jouault, N., 100
Joudrier, P., 198
Judge, P.J., 194
Julien, K.R., 498
Jun, S.Y., 396–397, 401–402
Juen, M.A., 462–490
Junemann, R., 5
Jung, C., 6–7
Jung, J.S., 197–198
Jung, K.H., 200–201
Jung, S., 74
Junge, F., 53, 194–195, 314–315, 369–370, 373, 375–376
Jurga, S., 100–101

K

Kadokura, H., 321–322
Kahru, A., 7
Kai, A., 128–129
Kai, L., 53, 194–195, 369–370, 374
Kainosho, M., 53, 168–169, 252, 269–271, 312–314, 324, 381–382, 395–396
Kaiser, L., 370–371
Kaitsu, Y., 312–314
Kajikawa, M., 291–292
Kajimoto, K., 349
Kaldenhoff, R., 369
Kamen, D.E., 186
Kamo, N., 314–315, 322–323
Kan, L.S., 427
Kandori, H., 197, 199
Kane Dickson, V., 196
Kanelis, V., 395–396
Kang, C., 372*t*
Kang, D., 374
Kang, S.H., 396–397, 401–402
Kansiz, M., 115
Kappl, R., 57–61
Kaptein, R., 85
Karan, R., 163–164
Karbyshev, M., 53, 194–195, 370, 373, 375–376
Kärkkäinen, H.-R., 248–249
Karlsson, M., 452–454
Karsten, W.E., 325–326, 376–379, 377*t*, 406*t*
Kasai, T., 336–337, 337–338*f*
Kasprzak, A.J., 262–265

Kataev, A.A., 126
Kataoka, M., 349
Kato, H., 324–326, 377*t*, 406*t*
Kato, K., 300
Kato, T., 291–292
Katona, E., 126
Katritch, V., 69–71, 86–87, 196, 254–255
Katsuta, N., 196
Katz, J.J., 125–126, 214–221, 218*t*, 223–226, 228–229
Katzen, F., 321–322, 374
Kaushal, G.P., 299
Kavran, J.M., 292–295
Kawagoe, Y., 112
Kawahara, K., 69–70
Kawamura, I., 194, 200–201, 372*t*
Kawano, S., 127–128
Kawecki, M., 127–128
Kay, L.E., 5–6, 169–170, 176–177, 184–186, 254–255, 324, 381–382, 390, 395–396, 462–463, 488–490, 527–528
Keane, S.C., 468*f*, 484, 498, 527–528
Keatinge-Clay, A.T., 271–272
Keckes, J., 112
Kedde, M., 550
Kefala, G., 370, 373
Keire, D.A., 295
Kelle, R., 252, 270
Kelly, A.E., 271–272
Kelly, M.J.S., 169–170
Kelman, Z., 28–43, 148–164
Kemter, K., 69–70
Kendall, F., 28
Kennedy, S., 550
Kent, S.B., 391–392
Kerfah, R., 379
Keum, J.W., 312–314
Khabibullina, N.F., 374
Khan, F., 69–70, 87
Kharytonchyk, S., 544, 559–560
Kido, J.-I., 349
Kidwell, M.A., 497
Kieffer, B., 484–485
Kieft, J.S., 484–485
Kiga, D., 326–327
Kigawa, T., 252, 312–316, 319–327, 334, 369, 375–379, 377*t*, 396–397, 405–408, 417–418

Kikukawa, T., 314–315, 322–323, 370–371
Kiliani, H., 428–429
Kim, D.M., 312–315, 321–322, 396–397, 401–402
Kim, H.G., 128–131
Kim, H.J., 194–196, 396–397, 401–402
Kim, H.S., 128–131
Kim, H.W., 70–75, 175–176
Kim, J., 111–112
Kim, K.H., 415–416
Kim, K.K., 128–131
Kim, K.S., 415–416
Kim, M.J., 415–416
Kim, M.S., 292–295, 377t
Kim, S., 415–416
Kim, S.-J., 168–170, 184
Kim, S.U., 234
Kim, S.Y., 314–315, 322–323, 370–371
Kim, T.W., 312–315, 321–322
Kim, Y., 325–326
Kim, Y.G., 128–131
Kimata, N., 194–195
Kimura, M., 196–197
Kimura, T., 194–195
Kimura-Someya, T., 314–315, 322–323
King, G.J., 7
King-Morris, M.J., 438–439
Kingston, R.E., 292–293
Kirby, N.M., 115–116
Kirk, K.L., 254–255
Kirpichnikov, M.P., 372t
Kirschner, M.W., 349
Kirshenbaum, I., 216t, 236–239
Kitevski, J.L., 73–74
Kitevski-Leblanc, J.L., 68–71, 74, 85–87
Kiviniemi, A., 71–73
Kjeldgaard, M., 28
Kjellbom, P., 197–198
Klaassen, C.H.W., 252, 270
Klammt, C., 53, 194–195, 370, 373, 395–396
Klein-Seetharaman, J., 292t, 300
Klemm, D., 127–128
Klepach, T., 443
Kline, P.C., 452–454
Kloiber, K., 462–463, 545–546
Klopp, J., 249, 257–259, 262–269, 273–274, 281
Klumpp, M., 291–292

Knecht, R., 250, 256, 262–265, 269
Knoppers, M.H., 291–292
Knott, R.B., 107
Kobashigawa, Y., 169–170, 395
Kobayashi, M., 134
Kobayashi, T., 312–314, 326–327
Koch, H.G., 321–322
Kodama, K., 323, 326–327
Koenig, J.L., 99
Kofuku, Y., 194–195, 252, 257–258, 265–268, 270–271, 271f
Koglin, A., 395–396
Kohl, A., 196
Kohno, T., 266–268, 377t
Koizumi, M., 324–326, 377t, 406t
Kokubo, T., 271–272
Kolb, V., 409
Kolks, M.A., 358
Komives, E.A., 186, 196–197
Kommer, A., 409
Kondo, K., 194–195, 252, 257–258, 265–268, 270–271, 271f
Kong, L., 299
Konrat, R., 168, 462–463
Kopeina, G.S., 374
Köpper, S., 444
Koshiba, S., 312–316, 322–326, 334, 336–337, 337–338f, 376–379, 377t, 396–397, 405–408, 418
Kossiakoff, A.A., 5
Kost, T.A., 248–249
Kotin, R.M., 248–249
Kounosu, A., 46–47, 54, 62–64
Koutsopoulos, S., 370–371
Kovac, B., 106–107
Kozak, M., 293
Krabben, L., 194
Kragl, U., 252, 270
Kragt, A., 107, 196–197
Krause, N., 196
Kreutz, C., 462–490, 527–528, 545–546
Krieger, C., 169–170
Krishna Mohan, P.M., 169–170
Krishna, N.R., 196–197, 200–201
Krishnamoorthy, J., 88
Krishnamurthy, S., 194
Krishnan, C., 135–136
Krishnarjuna, B., 168–170, 172, 175–181, 175t, 180f

Krisko, M., 428–429
Kriz, A., 291–292
Krois, A.S., 88–89
Krucker, M., 217–221
Krystynowicz, A., 128–129
Kubick, S., 369
Kuchel, P.W., 259, 262–265
Kuchenreuther, J.M., 314–315
Kudlicka, K., 127–128
Kuestner, R.E., 207
Kuhlbrandt, W., 196
Kukimoto-Niino, M., 338–339, 339–340f
Kukol, A., 134, 218t, 220–221
Kulkarni, P.V., 254–255
Kulomaa, M.S., 358
Kumasaka, T., 47–50, 48t, 51t, 52–53, 57–61, 58t, 64
Kumeta, H., 395
Kuppan, K.V., 382
Kuprov, I., 69–70, 87
Kuramitsu, S., 315–316
Kurek, I., 112
Kuryatov, A.B., 69–70
Kushner, D.J., 126
Kutscha, H., 46–47, 169–170, 173, 175t, 178, 181, 186
Kuwasako, K., 333, 333–334f
Kuwata, K., 196–197
Kuznetsov, N.A., 377t
Kuznetsova, A.A., 377t
Kwan, A.H., 7, 20f
Kwon, Y.D., 292–293
Kwong, P.D., 290–303
Kyne, C., 78
Kyogoku, Y., 318, 328, 391, 398

L

LaBaer, J., 290–291
Labroo, V.M., 254–255
Ladbury, J.E., 74–75
Ladizhansky, V., 169–170, 194–207
Ladner, J.E., 35t, 36f
Lagpacan, L., 70, 79–80
LaGuerre, A., 368–382
Lahti, R., 6–7
Laitinen, O.H., 358
Lake, V., 4–23
Lam, W.L., 148, 163–164
Lambruschini, C., 91
Lamond, A.I., 225
Lamprecht, J., 126
Langan, P., 112–114, 124–142, 214–215, 218t, 220–221
Lange, S., 194
Langer, J.A., 28
Lapham, J., 542–543, 550
Lapierre, C., 234
Larda, S.T., 69–70, 87
Laroche, Y., 196–197
Latham, M.P., 462–463, 497
Lauwereys, M., 196–197
Laux, V., 107, 109, 125–126
Law, M.-F., 295–296
Lazarides, A.A., 374–375
Le, M.T., 497–532
Leahy, D.J., 292–295
Le Bon, C., 370–371
le Sage, C., 550
Lebars, I., 484–485
Lebaupain, F., 370–371
LeBlanc, R.M., 462–490
Ledwidge, R., 271–272
Lee, G., 112
Lee, I., 127–128
Lee, J.C., 207
Lee, K., 349–350
Lee, K.Y., 396–397, 401–402
Lee, M.D., 382
Lee, S.C., 248–249
Lee, S.J., 128–131
Lee, Y., 254–255
Leegood, R.C., 225–226
Legault, P., 462–463
Lehninger, A.L., 252
Leiting, B., 6
LeMaster, D.M., 5, 256–257, 377t
Lenz, R.W., 100–101, 103
Leonard, P.G., 74–75
Lerche, M.H., 452–454
Les, D.H., 223–224
Lesch, H.P., 248–249
Leung, E.W.W., 86–87
Leung, S., 293–296
Leunissen, E.H., 498, 520, 523–524
Lewandowska, D., 225

Lewis, G.N., 215–216
Li, C., 69–70, 74–75, 88–89, 298
Li, D., 370–371
Li, F., 370–371
Li, H., 336–337, 337–338f
Li, J., 291–293
Li, M., 197, 266–268
Li, S., 348–365
Li, Q.Q., 6, 372t, 514–515
Li, W., 91
Li, X., 382
Li, Y., 498–499, 511, 524–525
Li, Z., 54, 62–64
Liang, B., 368
Liebetanz, J., 252, 273
Liedl, K.R., 545–546
Lin, C.-C., 443
Lin, C.-H., 443
Lin, E.C.C., 447
Lin, M.T., 46–64
Lin, S.-P., 116–117, 127–128
Lindkvist-Petersson, K., 197–200
Lindley, N.D., 107, 196–197
Lindner, P., 116
Linser, R., 7, 194–195, 381–382
Liu, D., 318, 328
Liu, J.J., 69–71, 86–87, 254–255
Liu, J.-R., 116–117, 127–128
Liu, Q., 54
Liu, R., 236–239
Liu, S., 348–365
Liu, X.-M., 252, 270
Liu, X.Q., 391, 394–395
Liu, Y., 498
Liu, Z.L., 111–112
Ljunggren, J., 259, 262–265
Loewen, M.C., 300
Logan, T.M., 262–265
Logusch, E.W., 377t
Loh, C.-T., 324–325, 376–381, 377t, 395–396, 406t
Löhr, F., 53, 368–382, 395–396
Loira Calvar, I., 116–117
London, R.E., 428–429
Long, J., 185–186
Long, S.B., 196
Longhini, A.P., 462–490, 498, 527–528
Longo, P.A., 292–295

Loscha, K.V., 379, 381–382
Lu, K., 462–463, 484–485, 544, 559–560
Lu, M., 6, 36–37
Luchette, P.A., 88–89
Luck, L.A., 69–70, 73–74
Luckow, V.A., 247–249
Ludwig, C., 318, 328, 394–395, 418
Luhrmann, R., 538
Lukavsky, P.J., 484–485, 539, 542–545, 555
Lundstrom, K., 196
Luo, Y., 498–499, 506t, 527–528
Luo, Z., 370–371
Lustbader, J.W., 291–292, 292t
Lutz, E.A., 69–70, 74–75, 88
Lyerla, J.R.J., 69–70
Lynch, D.L., 194–195
Lynch, R.M., 292–293
Lynn, G.W., 139–140
Lyukmanova, E.N., 372t, 374

M

Ma, L., 298
Ma, L.-C., 74–75
Ma, Y., 370, 374
Määttä, A.I., 248–249
MacCoss, M.J., 349–350
Mack, J., 71–73, 91
Mackinnon, R., 196
MacLean, B., 349–350
Madden, K.R., 207
Madin, K., 312–314
Madl, T., 527–528, 545–546
Madlansacay, M.R., 382
Maeda, M., 252, 270–271
Maenaka, K., 291–292
Maertens, J., 31–32
Magnelli, P., 299
Magnin, T., 196
Mahnke, M., 248–249
Mahon, M.M., 377t
Mähönen, A.J., 248–249
Major, F., 497
Majumder, M., 415
Makena, A., 69–71, 86–87
Makinen, M.W., 5–6
Makino, S., 312–314, 377t, 381–382
Mak-Jurkauskas, M.L., 194
Malia, T.J., 194–195

Malthouse, J.P., 377t
Malyvanh, V.P., 127–128
Mamontov, E., 112
Manglik, A., 252
Manley, P.W., 252, 262f, 269
Mann, L.R., 30
Mann, M., 251, 348–349
Mann, S.G., 249
Manni, S., 372t
Manning, J.M., 325–326, 377t, 406t
Manzana, W., 292–293
Mao, H., 391, 395
Maramorosch, K., 252
Marassi, F.M., 368
Marcotte, I., 374
Marcovich, N.E., 111–112
Maréchal, E., 107, 109, 125–126
Marin, M., 196
Maris, C., 484–485, 539, 542–544, 542–543f, 552, 554–556
Markley, J.L., 32, 35, 46, 256–257, 265, 312–314, 505t
Marley, J., 6, 36–37
Marsh, E.N.G., 69–70, 86, 88
Marsilio, F., 6
Martasek, P., 54
Martens, H.I., 254–255
Marti, S., 249
Martin, C.T., 546–547
Martin, D.B., 207
Martin, D.E., 99, 104, 107
Martinez, M., 196
Martínez-Sanz, M., 214–215
Mary, S., 370–371
Maslennikov, I., 194–195, 370, 372t, 373
Mason, R.P., 71–73, 254–255
Masquelin, T., 454–455
Masse, J.E., 528
Massou, S., 107, 196–197
Masters, B.S.S., 54
Masterson, L.R., 46, 53, 194–195
Masuda, K., 291–292, 292t
Masutani, M., 312–314
Matei, E., 71–73, 90
Mathew, A., 111–112
Matsubara, Y., 377t, 381–382
Matsuda, N., 312–316, 320–321, 323

Matsuda, T., 312–316, 319–327, 334, 376–379, 377t, 396–397, 405–408, 418
Matsuo, H., 168–169
Matsuoka, M., 128–131, 137
Matsushita, K., 128–131, 137
Matsushita, S., 46–64
Matsuzaki, K., 128–129
Matthews, D.E., 135–136
Mattick, J.S., 538
Matwiyoff, N.A., 428–429
May, R., 5
Mayer, R., 112
Mayr, L.M., 246–247
Mazar, A.P., 186, 256–259, 291–292, 292t
McBee, J.K., 349–350
McCallum, S.A., 462–463, 497
McCarthy, A.A., 196
McClary, J., 292–293
McConnell, H.M., 300
McDermott, A.E., 168
McDonald, J.F., 377t
McGaughey, J., 138–139, 214–215, 221, 231–232, 234
McIntosh, L.P., 173
McIntosh, T.S., 135–136
McKenney, K., 29
McLaggan, D., 7
McLellan, J.S., 293–296
McNaughton, D., 115
Mcphee, J.R., 156
Meadows, R.P., 291–292
Mega, T.L., 433–434
Mehl, R.A., 70, 79–80, 82
Mehta, V.D., 254–255
Mei Ono, A., 381–382, 395–396
Meieer, S., 452–454
Meier, D., 293
Meilleur, F., 6–7
Meister, A., 325–326, 377t, 406t
Melendez, M.G., 73–74
Melnichenko, Y.B., 138–139
Meneses, P.I., 259–261
Menzel, C., 293
Meola, A., 251, 257–259, 269, 273–274
Mercer, T.R., 538
Mertens, H.D.T., 115–116
Messens, J., 196–197
Messerschmidt, A., 377t

Meyhack, B., 250, 256, 262–265, 269
Michel, E., 318, 369, 390–419, 538–539, 544, 548f, 549, 559–560
Micura, R., 462–463, 477–478, 484, 487–488
Middelberg, A.P.J., 7, 8t, 409–410
Mikami, S., 312–314
Mikawa, T., 262–265, 272, 272f
Mikkelsen, D., 113–114
Mildorf, T.J., 252
Miller, A., 214–215
Miller, J., 223–224
Miller, L.K., 247–249
Miller, S.M., 271–272
Miller, W.A., 497
Milligan, J.F., 462–463
Milne, J.J., 377t
Milon, A., 107, 196
Mineev, K.S., 372t
Minor, F.W., 128–129
Mirza, U.A., 291–292, 292t
Mirzaei, H., 349–350
Mishima-Tsumagari, C., 338–339, 339–340f
Mitchell, M., 299
Mitsuhashi, J., 252
Mitsui, T., 326–327
Mittermaier, A.K., 462–463
Miyajima-Nakano, Y., 46–64
Miyake-Stoner, S., 79–80, 82
Miyata, Y., 312–316, 319
Miyauchi, S., 314–315, 322–323
Miyazaki, Y., 462–463, 484–485
Miyazawa-Onami, M., 196–197, 199t, 207
Mizumura, T., 194–195, 257–258, 265–268, 270–271, 271f
Moeller, M., 148, 163–164
Moffatt, B.A., 18–19
Molina, D.M., 196
Molloy, B.B., 262–268
Mondal, S., 168–169, 172, 178
Mongelli, N., 90–91
Monsma, S.A., 259–261
Monteith, W.B., 78
Monti, A., 232–233
Monti, S., 544, 559–560
Moody, M., 223–224
Mookerjee, A., 126

Moon, C., 197–198
Moore, D., 296–298
Moore, G., 254–255
Moore, M.J., 544
Moore, P.B., 5–6, 30
Moore, S., 325–326, 377t, 406t
Mootz, H.D., 418
Moras, D., 315–316
Morcombe, C.R., 206f
Morgan, W.D., 107, 196–197
Morita, E.H., 266–268, 377t
Moritz, T., 225
Morozova, E.A., 377t
Morrill, K.L., 149
Morris, K.V., 538
Morris, V.K., 7
Morrison, I.M., 236–239
Mortimer, S.A., 497
Moses, V., 30
Moskau, D., 90
Motoda, Y., 312–314, 319–320
Motooka, D., 69–70
Moulin, G., 198
Mowrey, D.D., 372t
Mozetic, M., 111–112
Mrowca, R.A., 438–439
Muchmore, D.C., 173
Mudie, S.T., 115–116
Mueller, G.A., 324
Muhandiram, D.R., 528
Muir, T.W., 391–394, 396–397, 399t, 400–401, 412–413, 418
Mukai, T., 326–327
Mukherjee, S., 169–170, 176–177
Mukhopadhyay, D., 550
Mukhopadhyay, R., 415
Mulcair, M.D., 86–87
Müller, D., 371
Muller, K., 71–73
Muller, M., 112, 196, 321–322
Muller, P., 194–195
Mulligan, R.C., 295–296
Münch, D., 370–371
Munekata, M., 127–128
Munro, R.A., 194–207
Murashige, T., 225–226
Murata, T., 314–315, 322–323
Murphy, G.M., 216t, 236–239

Murphy, J., 28
Murray, C.J., 382
Murray, D.T., 194
Murray, P.D., 292–293
Murray, V., 6, 514–515
Murton, J.K., 111–112
Musco, G., 71–73
Mustafi, D., 5–6
Mustafi, S.M., 169–170, 176–177
Muto, Y., 312–316, 320, 323–324, 336, 336f
Myles, D.A.A., 6–7, 214–215
Myszka, D.G., 293–295
Myyr, M., 223–224

N

Nabel, G.J., 290–291, 295–296, 297f, 300, 301f, 303
Nagai, K., 554
Nagano, M., 291–292
Nagao, S., 85–86
Nagle, J.M., 497
Nair, S.S., 232–233
Nairn, R., 298
Nakabayashi, K., 334–335, 335f
Nakada-Nakura, Y., 196
Nakajima, R., 312–316, 320–321
Nakamura, S., 206f
Nakata, Y., 511
Nakatsu, T., 324–326, 377t, 406t
Nakayama, H., 326–327
Nakayama, T., 377t
Narasimhulu, K.V., 46–50, 64
Nars, G., 207
Navarrete, M., 452–454
Neidhardt, F.C., 37
Neilan, B.A., 127–129
Nelissen, F.H.T., 484–485, 498, 520, 523–524
Nelson, A., 7
Nelson, D.J., 254–255
Nemoto, N., 69–70
Nettesheim, D.G., 291–292
Nettleship, J.E., 291–292
Neuschul, P., 128–129
Neutze, R., 197–198
Nevins, J.R., 295–296
Newkirk, M.M., 254–255

Nguyen, G.T., 377t
Nguyen, N., 292–293
Nicolini, C., 28
Niegowski, D., 196
Nimlos, M.R., 127
Nishi, Y., 69–70
Nishikawa, K., 349–350
Nishimoto, S.-I., 71–73
Nishimura, M., 291–292
Nishino, T., 54
Nishiyama, Y., 112, 138–139, 214–215
Nixon, C.P., 249
Nobutoshi, I., 391
Nokhrin, S., 47
Nomura, S., 163–164
Nonin-Lecomte, S., 498, 500f, 511–512, 523–524, 527–528, 555
Norden, K., 196–198
Nordlund, H.R., 358
Nordlund, P., 125–126
Noren, C.J., 70–71, 80–81
Northcote, D.H., 217–219
Nothnagel, H.J., 194–195
Notley, S.M., 111–112
Nozinovic, S., 462–463
Nozirov, F., 100–101
Nudler, E., 462–463
Numata, K., 319
Nunez, H.A., 428–429, 435–436, 445, 452–454
Nunokawa, E., 312–314, 319–320
Nyblom, M., 197–200
Nygaard, R., 252

O

O'Brian Smith, E., 452–454
O'Connor, T., 428–429
O'Donovan, D., 262–265, 272, 272f
O'Neill, H.M., 124–142, 214–217, 218t, 221–223, 225–226, 229–234
O'Reilly, D.R., 247–249
Oates, J., 194–195
Oberg, F., 197–200
Ober-Reynolds, B., 315
Oberstrass, F.C., 400–401
O'Brien, T., 298
O'Byrne, C., 7
Ochmann, A., 545

Ochsner, U.A., 377t
O'Connell, J.F., 6
Oda, J.I., 324–326, 377t, 406t
Oda, N., 318, 328, 391
Odier, L., 116–117, 129, 135–136
O'Doherty, A.M., 298
Oesterhelt, D., 148–149, 153–154
Ogasawara, T., 266–268, 312–314, 377t
O'Grady, C., 47
Ogura, K., 169–170, 395
Oh, I.S., 312–315, 321–322
Ohki, S.Y., 168–169
Ohlson, E., 196
Öhman, L., 259, 262–265
Ohmori, D., 46–47, 57–61
Ohsawa, N., 336, 336f, 338–339, 339–340f
Ohtake, K., 326–327
Ohtsuki, T., 326–327
Okada, S., 126
Okamura, M., 370, 373
Okano, T., 138–139
Okano, Y., 196–197
Okayama, H., 292–293
Okitsu, T., 194, 372t
Oksman, K., 111–112
Okuda, K., 127–128
Okude, J., 194–195, 252, 257–258, 265–268, 270–271, 271f
Okuni, T., 326–327
Oladi, R., 111–112
Olah, G.A., 5
O'Leary, J.M., 196–197, 199t, 200
Olejniczak, E.T., 71–73, 91, 291–292
Olins, P.O., 249
Oliveira, R., 106–107
Oliver, R.P., 197
Olivier, J., 219–220, 223–224
Ollerenshaw, J.E., 390
Ondruschka, J., 106–107, 109–110
O'Neill, B.K., 8t
O'Neill, H.M., 112, 116–117, 125–129, 133–134, 138–140, 214–217, 218t, 220–223, 229–232, 234
Ong, S.-E., 251, 348–349
Onkelinx, E., 254–255
Ono, C., 248–249
Ono, M.A., 324
Oomens, A.G.P., 259–261

Opefi, C.A., 194–195
Opekun, A.R., 452–454
Opella, S.J., 194, 368
Oppenheimer, N.J., 85
Orbán, E., 368–371
Orekhov, V.Y., 194–195
Osanai, T., 319
Osborne, M., 254–255
Oscarson, S., 71–73, 90
Oschkinat, H., 168–169
Oss, A., 194
Oswald, R.E., 168–169
Otomo, T., 318, 328, 391, 398
Otting, G., 312–314, 324–325, 375–382, 377t, 395–396, 406t, 560
Oubridge, C., 554
Oude Vrielink, J.A., 550
Oue, S., 54
Ounaies, Z., 111–112
Ovchinnikov, Y.A., 69–70
Ovodov, S.Y., 312–314
Owens, R.J., 291–292
Ozawa, K., 312–314, 375–379

P

Paabo, M., 14
Paalme, T., 5, 7
Pack, A., 113–114, 218t, 220–221
Padhy, N., 126
Pahler, A., 358
Paliy, O., 5–6, 30, 36–37
Palmer, A.G., 85, 168, 177
Pan, M., 35t, 36f
Pan, Q., 437, 442–443
Pancera, M., 292–293
Panopoulos, P., 377t
Pantoja-Uceda, D., 168–169
Papeo, G., 90
Paramonov, A.S., 374
Pardi, A., 462–463, 497
Parekh, R.B., 299
Parisien, M., 497
Park, C.G., 312–315, 321–322
Park, E., 291–292
Park, K.-H., 370–371, 375–376
Park, S.H., 194–195
Parker, I.H., 112, 115–116
Parker, M.J., 375–376

Parmelee, D., 292–293
Parris, W.E., 381–382
Passauer, L., 111–112
Patel, D.J., 520
Patzelt, H., 148–149
Paul, W., 106
Paulus, H., 399t
Pautsch, A., 369–370
Paweletz, N., 126
Pawson, T., 381–382
Pear, J.R., 112
Pederson, T.M., 186, 256–259, 291–292, 292t
Pedi, L., 196
Peeler, J.C., 70, 80
Peleg, Y., 291–292
Pell, A.J., 194
Pellé, X., 273
Pelton, J.G., 324–325
Peng, L., 71–73, 86
Pennings, N.J., 436–437
Penttinen, O.-P., 223–224
Penzel, S., 194
Perez, J.A., 70–75, 175–176
Perez, S., 127
Perkins, S.J., 214–215, 221
Perler, F.B., 392–393, 398, 399t, 400–401, 404, 411
Perlin, A.S., 448–449
Pervushin, K., 373, 390
Pessey, O., 379
Peterkofsy, A., 29
Peterson, T.C., 374
Petridis, L., 112, 138–139, 214–215
Petritis, B., 315
Petros, A.M., 186, 256–259, 291–292, 292t
Petrov, A., 526
Petrus, L., 436–437
Petrusova, M., 437
Petzold, K., 484–487
Pfeffer, I., 69–71, 86–87
Phan, J., 273–274
Phillips, R.S., 377t
Pickford, A.R., 196–197, 199t, 200
Pielak, G.J., 69–70, 74–75, 88–89
Pierce, J., 429, 439–442, 445–447
Pijlman, G.P., 249
Pikus, J.D., 46

Pillai, C.K.S., 106
Pingali, S.V., 113–114, 138–140, 214–215, 218t, 220–221
Pingali, S.J., 124–142
Pinnow, M., 111–112
Pinto, R.M., 447
Piomelli, D., 91
Pipich, V., 100
Pistorius, A.M., 256–258
Pitman, M.C., 194–195
Pitsch, S., 462–463
Pittard, J., 48t, 51t, 52–53
Plevin, M.J., 46–47, 168–169, 175–176, 176t, 379
Plückthun, A., 393–394, 400–401
Pollak, S., 291–292, 292t
Pollet, B., 234
Pollok, B.A., 391, 395
Ponchon, L., 498–499, 500f, 511–512, 523–524, 527–528, 555
Popot, J.-L., 370–371, 374
Portman, K.L., 185–186
Pörtner, R., 248–249
Possee, R.D., 249
Post, J.A., 194
Potter, H., 292–293
Pound, A., 291–292, 292t
Prade, L., 377t
Prasanna, C., 168–186
Pratt, J.M., 312–314, 349, 360–361, 365
Pratt, R., 215–216, 219–220, 230–231
Preston, G.M., 197–198
Preston, R.J., 28–43
Price, R.P., 554
Prosser, R.S., 68–71, 73–74, 85–89
Proudfoot, M., 325–326
Proverbio, D., 314–315, 369, 374
Prual, C., 196
Pucci, B., 370–371
Puech, V., 107, 196–197
Puglisi, E.V., 526
Puglisi, J.D., 526
Puskar, L., 99, 104, 107
Put, J., 254–255
Putter, I., 256–257, 265
Putzbach, K., 217–221
Pyzyna, B., 149

Q

Qi, R., 324–325, 376–381, 377t, 395–396, 406t
Qian, S.W., 291–292
Qian, X., 197
Qin, J., 217–220, 224
Qin, P.Z., 540, 545
Qiu, D., 107
Qiu, Z., 259–261
Query, C.C., 544
Quezada-Calvillo, R., 452–454
Quignard, F., 107

R

Raap, J., 252, 270
Ragauskas, A.J., 116–117, 127, 133–134, 216–217, 218t, 220–223, 225–226, 229–234
Raghavan, M., 293–295
Rahman-Huq, N., 291–292
Raina, S., 197–198
Rajagopalan, M.R., 194, 372t
Rajashankar, K.R., 520
Ramachandran, R., 46–47, 169–170, 173, 175t, 178, 181, 186
Ramakrishnan, V., 6, 28
Ramamoorthy, A., 69–70, 88
Ramirez-Tapia, L.E., 546–547
Ramon, A., 196
Ramos, A., 540, 541f, 545–546, 552
Ramqvist, A.-K., 259
Rance, M., 85, 168, 177
Rane, S., 100–101
Rantalainen, A.-L., 223–224
Rasband, W.S., 159f
Raschle, T., 372t, 374–375
Rasia, R.M., 46–47, 175–176, 176t
Rasmussen, J.R., 431–432
Rauf, F., 315
Ravindranathan, S., 527–528, 539, 546–547, 559
Ray Wu, L.G.K.M., 295–296
Ray, P.S., 415
Rayment, I., 254–255
Reckel, S., 53, 194–195, 369–371, 373, 375–376
Recny, M.A., 291–292

Reddy, P.T., 28–43
Reese, M.L., 375
Rekas, A., 4–23
Reeves, D.T., 112, 125–126, 216–217, 218t, 220–223, 229–232, 234
Reeves, P.J., 194–195, 300
Rehbein, K., 194
Rehm, T., 252
Rehn, M., 148, 163–164
Reichl, H., 295–296
Reinhart, C., 196
Reining, A., 462–463
Reis, M.A., 106–107
Reitz, O., 215
Rej, R., 266–268
Religa, T.L., 254–255, 390
Remaud-Simeon, M., 207
Rempe, C.S., 138–139
Rempel, B.L., 47
Ren, A., 520
Renault, M., 194
Rendahl, A.K., 215–221, 225
Respondek, M., 485–487
Retel, J.S., 194
Reverdatto, S., 271–272
Revtovich, S.V., 377t
Reymond, L., 462–463
Ribble, W., 377t
Ribeiro, J.M., 447
Rice, A.J., 374
Rich, R.L., 293–295
Richard, R.B., 292–293
Richards, F.M., 5, 256–257
Richards, K.S., 249
Richardson, C.D., 248
Richmond, T.J., 248–249
Richter, C., 292t, 300, 462–463, 498, 528–530, 545–546
Richter, D., 100
Richter, G., 69–70, 169–170
Riedel, D., 169–170
Rieder, U., 484, 487–488
Rieffel, S., 249
Riek, R., 372t, 373, 390, 539, 544–545
Rienstra, C.M., 55–57, 186
Rinnenthal, J., 462–463
Ritz, E., 203–204
Rivers, J., 349, 360–361, 365

Robayo-Torres, C.C., 452–454
Robb, F.T., 148
Roberts, A.B., 291–292
Roberts, G.C., 71–73
Roberts, K., 219–220, 223–224
Robinson, G., 223–224
Robinson, J.L., 149
Robinson, R.A., 14
Robson, C.L., 299
Rodrigues, J.R., 447
Rodriguez, E., 196–197, 200–201
Roe, A.J., 7
Roest, S., 249, 257–259, 262–269, 273–274, 281
Rogers, R.D., 111–112
Rojas, A., 130
Rokop, S.E., 5
Rolando, C., 234
Roman, L.J., 54
Romeo, E., 91
Roos, C., 314–315, 369–371, 374
Rosano, G.N.L., 30–31
Roschier, M.M., 248–249
Rosell, F., 254–255
Rosen, L.A., 227–228
Rosen, M.K., 381–382
Rosenberg, A.H., 316–317, 320
Rosenblum, G., 315
Rosenzweig, R., 254–255
Rosevear, P.R., 377t, 452–454
Ross, P., 112
Roth, A., 520
Rotsaert, F.J., 46
Rout, A., 168–169
Roux, B., 497
Rowan, S.J., 111–112
Rowe, W.B., 325–326, 377t, 406t
Rowen, R., 157
Rozanov, D.V., 91
Ruderman, J., 271–272
Rueter, A., 186, 256–259, 291–292, 292t
Ruf, R.A.S., 69–70, 74–75, 88
Rule, G.S., 85
Ruotolo, B.T., 88
Ruschak, A.M., 254–255, 324
Russell, C.B., 173
Russell, D.W., 519
Russell, R.A., 98–117
Russell, R.M., 217–220, 224
Russell, W.C., 298
Ruterjans, H., 53
Ruvinov, S.B., 377t
Ryabova, L.A., 312–314
Rydzik, A.M., 69–71, 86–87

S

Sacchi, G.A., 217–219
Sachse, R., 369
Saeki, M., 314–315, 322–323
Sahl, H.-G., 370–371
Sahu, S.C., 377t, 381–382
Said, S., 100
Saini, K.S., 290–291
Saio, T., 169–170
Saito, M., 312–314, 319–320
Sakaguchi, K., 6
Sakai, T., 271–272
Sakai, W.S., 217–219
Sakamoto, A., 326–327
Sakamoto, K., 326–327
Sakharova, I.S., 377t
Sakurai, M., 100–101, 104–106
Salguero, C., 488–490, 497
Salmén, L., 134, 218t, 220–221
Salopek-Sondi, B., 74
Sama, J.N., 498, 505–506t, 517
Samain, D., 127
Sambrook, J., 519
Samoilova, R.I., 46–50, 48t, 51t, 52–53, 57–61, 58t, 64
Sandberg, G.J., 225
Sanders, C.R., 88–89, 168, 194–196
Sanders-Loehr, J., 46
Sano, M., 130
Santner, T., 462–463, 484
Santoro, J., 168–169
Saporovskaya, M.B., 377t
Sarramegna, V., 196
Sarver, N., 295–296
Sasakawa, H., 291–292
Sastry, M., 290–303
Satkowski, M.M., 100–101
Sato, A., 326–327
Sato, H., 126
Sato, S., 326–327
Sattler, M., 527–528, 545–546

Sau, S., 315
Saul, J., 315
Saurel, O., 207
Saves, I., 207
Savinov, A.Y., 91
Sawada, S., 377t
Sawasaki, T., 312–314
Saxena, I.M., 127–128
Scanlon, M.J., 86–87
Scarpelli, R., 91
Scensny, P., 431–432
Schaeffer, P.M., 376–379
Schafer, B., 53
Schafer, U., 321–322
Schaffitzel, C., 248–249
Schalkwyk, L.C., 148, 163–164
Schaller, H., 107, 109, 125–126
Schechtman, L.A., 100–101
Scheidt, H.A., 194–195
Schenk, R.M., 225–226, 234–235
Schink, A., 391, 395
Schirch, V., 377t
Schlepckow, K., 462–463
Schlesinger, R., 196
Schleucher, J., 527–528
Schlichting, I., 6–7
Schlinkert, R., 395–397
Schlosser, K., 498–499, 511, 524–525
Schmid, K., 291–292
Schmidt, P., 194–195
Schmidt, S.J., 433–434
Schmidt, T., 369
Schmidt, T.G.M., 358
Schmidt-Rohr, K., 99
Schmieder, P., 168–170
Schnapp, G., 369–370
Schneider, B., 369, 373, 375–376
Schneider, C.A., 159f
Schneider, R.J., 295–296
Schneider, T., 370–371
Schnorr, K.A., 545–546
Schober, H., 112
Schoenborn, B.P., 6, 28
Schott, A.-K., 69–70
Schramm, M., 112–114, 127–129
Schreckengost, W.E., 112
Schroeter, D., 126

Schubert, M., 168–169, 527–528, 538–539, 544, 546–547, 548f, 549, 559–560
Schultz, P.G., 70–71, 80–81, 326–327
Schulz, R., 112
Schutz, V., 418
Schwahn, D., 100
Schwalbe, H., 292t, 300, 462–463, 498, 528–530
Schwartz, A.M., 128–129
Schwarz, D., 53, 194–195, 369–370, 373, 375–376
Schwarzer, D., 318, 328, 394–395, 418
Schweins, R., 100
Scott, K.L., 196
Scott, R.A., 54, 62–64
Searle, M.S., 185–186
Sears, V.F., 98
Sehgal, S.N., 149
Seiboth, T., 46–47, 169–170, 173, 175t, 178, 181, 186
Seikowski, J., 545
Seki, E., 312–316, 319–324
Sekiguchi, M., 169–170
Sekine, S., 338–339, 339–340f
Selenko, P., 271–272
Serber, Z., 271–272
Serganov, A., 462–463
Serianni, A.S., 424–455
Seyoum, E., 372t
Sferrazza, M., 107, 109, 125–126
Shah, J., 127–128
Shah, R., 112, 124–142, 214–239
Shah, S.T.A., 370
Shahid, S.A., 194
Shankar, N., 497
Shanker, S., 252
Shao, Y., 293
Sharaf, N.G., 68–91
Sharma, C.P., 106
Sharma, K.S., 370–371
Sharma, L.N., 135–136
Shea, C.R., 126
Shekhtman, A., 271–272
Shenkarev, Z.O., 374
Sheppard, N.C., 299
Sherry, A.D., 254–255
Sheta, E.A., 54
Shi, F., 101–102

Shi, J., 324–325
Shi, L., 169–170, 194–205, 203f, 206f, 372t
Shibabe, S., 220
Shibata, T., 312–316, 369
Shih, W.M., 194–195
Shikata, M., 312–314
Shimada, I., 125–126, 194–197, 199t, 207, 291–292, 292t, 300
Shimamura, T., 196
Shimba, S., 126
Shimizu, M., 266–268, 377t
Shimizu, N., 57–61
Shimizu, Y., 349–350
Shimono, K., 314–315, 322–323, 370–371
Shindo, K., 291–292, 292t
Shinohara, Y., 349
Shinya, N., 314–315, 322–323, 370–371
Shirai, T., 262–265, 272, 272f
Shiraishi, Y., 194–195, 252, 257–258, 265–268, 270–271, 271f
Shirk, H.G., 128–129
Shiroishi, M., 196
Shirokov, V.A., 369, 409
Shirouzu, M., 314–315, 322–323, 326–327, 333–335, 333–335f
Shirzad-Wasei, N., 194–195
Shoji, S., 336, 336f
Shokes, J.E., 54, 62–64
Shortle, D., 168–170, 173, 182–184
Shrestha, B., 246–284
Shu, F., 6, 28
Shulepko, M.A., 372t, 374
Siaterli, E., 249
Sicotte, V.J., 427
Siegel, S.M., 217–219, 224–228
Sigurdsson, S.T., 540
Sillerud, L.O., 427
Simon, A.E., 497–532
Simon, B., 148–149, 527–528, 545–546
Simon, G.P., 112, 115–116
Simon, M.I., 74
Simpson, D.M., 349, 360–361, 365
Singarapu, K.K., 377t, 381–382
Singer, A., 325–326
Singh, S., 32, 35, 196, 505t
Sinkkonen, A., 223–224
Sitarska, A., 257–259, 262–269, 273–274, 281

Sivashanmugam, A., 6, 514–515
Sizun, C., 251, 257–259, 269, 273–274
Skarina, T., 325–326
Skelton, N.J., 85, 168, 177
Skerra, A., 358
Skoglund, C., 427
Skoog, F., 225–226
Skora, L., 246–284
Skrisovska, L., 318, 391, 396, 399t, 400–403, 405–410
Slade, K.M., 69–70, 74–75, 88
Slanchev, K., 550
Sligar, S.G., 31t, 374–375
Smalla, M., 168–169
Smee, C., 248–249
Smiley, J., 298
Smirnov, A., 107
Smith, B.L., 197–198
Smith, C.J., 266–268
Smith, C.W.J., 554
Smith, D.F., 37
Smith, H.H., 217–219, 224–225
Smith, R.H., 248–249
Smith, S.O., 194–195
Smith, T.J., 266–268
Smotrina, T., 107
Snoswell, M.A., 8t
Snyder, J., 431–432, 442
Sobhanifar, S., 53, 194–195, 369–370, 373, 375–376
Soda, K., 377t
Soetaert, W., 31–32
Sohn, J.H., 74
Sokaribo, A., 47
Solnick, D., 295–296
Solomon, P., 197
Solsona, O., 198
Soma, A., 326–327
Somers, P.J., 436
Son, H.J., 128–131
Sondhi, D., 391–393, 396–397, 399t, 400–401, 418
Sonenberg, N., 312–314
Sonnichsen, F., 168
Soper, M.T., 88
Sosa-Peinado, A., 5–6
Soto, C., 292–293
Souliere, M.F., 462–463, 487–488

Southern, P., 295–296
Southworth, M.W., 392–393, 398, 399t, 400–401, 404, 411
Sowa, G.Z., 545
Sparks, S.W., 5
Spear, S.K., 111–112
Sperling, L.J., 47–50, 48t, 51t, 52–53, 57–61, 58t, 64
Spicer, L.D., 5–6
Spiess, H.W., 99
Spindel, W., 126
Spirin, A.S., 312–316, 323, 401–402, 409
Spitalcri, A., 71–73
Spitzer, R., 462–463, 484, 545–546
Sporn, M.B., 291–292
Sprangers, R., 390
Spudich, E.N., 196
Spudich, J.L., 196
Sreenath, H.K., 32, 35, 505t
Stafford, U., 432
Stalker, D.M., 112
Stamatoyannopoulos, J.A., 349–350
Stanbury, P.F., 7, 21
Stankovic, L., 436–437
Starkey, M., 248–249
Staunton, D., 395–397
Stefer, S., 314–315, 369
Steffan, T., 196
Stehle, J., 370–371, 375
Steitz, T.A., 497, 538
Stemmann, O., 349
Stepanov, V.G., 498
Stergachis, A.B., 349–350
Sternberg, L., 215–216
Sterne, K.A., 291–292
Stevens, R.C., 69–71, 86–87, 254–255
Stewart, D., 236–239
Stewart, I.I., 28
Stewart-Jones, G.B.E., 299
Stockbridge, R.B., 520
Stoeckenius, W., 148–149, 153–154, 157
Stomp, A.M., 223–224
Storme, V., 196–197
Stover, P., 377t
Strack, D., 234, 235f
Strappe, P.M., 298
Strasser, M.J., 550

Strauss, A., 250–253, 254–255f, 256, 262–265, 262f, 269, 273, 279–280, 300
Stroescu, M., 107
Strych, U., 498
Studier, F.W., 18–19, 29, 80, 316–317, 320, 505t
Stuhrmann, H.B., 214–215
Su, J., 111–112
Su, X.-C., 324–325, 375–381, 377t, 395–396, 406t
Suckale, N., 106–107, 109–110
Sudarsan, N., 520
Sudesh, K., 103–106
Sugiki, T., 196–197, 199t, 207
Sugiura, M., 71–73
Sugiyama, J., 112
Sukumaran, S., 100–101
Sule, L., 498–499, 511, 524–525
Sumii, M., 197, 199
Summers, M., 248–249, 462–463, 484–485
Summers, M.F., 497
Sun, Q., 125–126, 216–217, 218t, 220–223, 229–232, 234
Sun, Q.N., 112
Sun, X., 462–463
Sunde, M., 7
Sussman, J.L., 315–316
Suzuki, A., 85–86
Suzuki, T., 312–314, 349–350
Suzuki, Y., 69–70, 86, 88
Swapna, G.V.T., 74–75
Swarbrick, J., 290–291
Swartz, J.R., 312–316, 321–323
Sweeney, P.J., 256–257
Sykes, B.D., 70
Synowiecki, J., 106–107, 109–110
Szcześniak, E., 100–101
Szekely, K., 194
Szyperski, T., 30, 186

T

Tabor, R.F., 111–112
Tai, H., 85–86
Tajima, K., 127–128
Takagi, Y., 248–249
Takahashi, H., 125–126, 194–197, 199t, 207, 291–292, 292t, 300
Takahashi, M., 326–327, 333, 333–334f

Takai, K., 312–314
Takano, T., 207
Takeda, M., 53, 312–314, 370–371, 375
Takegoshi, K., 206f
Takeuchi, K., 207
Takio, K., 252, 312–316, 326, 369, 375–376, 396, 417–418
Talebzadeh-Farooji, M., 394–395
Talmont, F., 107, 196–197
Tamm, L.K., 204–205, 368
Tamura, K., 54
Tan, R., 298
Tanabe, K., 71–73
Tanaka, A., 312–316, 320–321
Tanaka, H., 377t
Tanaka, R., 266–268, 377t
Tanaka, T., 262–265, 272, 272f, 418
Tang, F., 217–221
Tang, G., 217–220, 224
Tang, M., 47–50, 48t, 51t, 52–53, 57–61, 58t, 64
Tang, P., 372t
Tang, X., 511
Tapaneeyakorn, S., 194–195
Taravel, F.R., 116–117, 128–129, 135–136
Tarrago, T., 91
Tarui, H., 312–314
Tashiro, K., 134
Tatulian, S.A., 204–205
Taylor, G.F., 194
Taylor, J.E., 7, 20f
Taylor, M.E., 424
Telenti, A., 392–393
ten Have, S., 225
Tenno, T., 271–272
Terada, T., 252, 312–339, 369, 375–376, 396, 417–418
Terao, T., 206f
Terauchi, T., 324, 381–382, 395–396
Terpe, K., 30–31, 315
Teruya, K., 398
Terwilliger, T.C., 315–316
Terwisscha van Scheltinga, A.C., 196
Tessari, M., 484–485, 498, 520, 523–524
Testori, E., 194
Thach, W., 69–70, 87
Thakur, A., 168–170, 172, 175–181, 175t, 180f

Thakur, C.S., 498–499, 505–506t, 517, 527–528
Thibaudeau, C., 442–443
Thoma, R., 291–292
Thomas, A.F., 214–217
Thomas, J.G., 322
Thomas, L., 194–195
Thomson, T., 28
Thornton, K.C., 73–74
Thotakura, N.R., 292–293
Tian, Y., 194–195
Tiisma, K., 7
Tillemans, V., 225
Tillman, T.S., 372t
Timasheff, S.N., 215
Tochio, H., 271–272
Tochio, N., 312–316, 322–326, 334–337, 335f, 337–338f, 376–379, 377t, 396–397, 405–408, 418
Toivanen, P.I., 248–249
Tolbert, T.J., 393–394, 528
Tollinger, M., 462–463, 484
Tomari, Y., 349–350
Tomida, M., 196–197
Tomizawa, T., 336–337, 337–338f
Tommassen, J., 194
Tommassen-van Boxtel, R., 194
Tomo, Y., 312–316, 319
Tonelli, M., 377t, 381–382
Tong, K.I., 418
Torchia, D.A., 5, 488–490
Torizawa, T., 324, 381–382, 395–396
Tornroth-Horsefield, S., 196
Torres, J., 134, 218t, 220–221
Tozawa, Y., 368
Traaseth, N.J., 46, 53, 169–170, 185, 194–195
Tranter, D., 194–195
Trbovic, N., 395–396
Trelease, S.F., 215
Trent Franks, W., 194
Trewhella, J., 5
Trolard, R., 5–6
Tropea, J.E., 299, 400–401
Tropis, M., 107, 196–197
Trowitzsch, S., 291–292
Trutnau, M., 106–107, 109–110
Tsetlin, V.I., 69–70

Tshudy, D.W., 100–101
Tsuchida, T., 128–131, 137
Tsuda, K., 333–336, 333–336f
Tsuji, T., 252, 269–271, 395–396
Tsuji, Y., 234
Tsujimoto, H., 196
Tsutsui, M., 312–316, 369
Tsvetikova, M.A., 377t
Tugarinov, V., 169–170, 176–177, 184–185, 390, 395–396
Tuominen, V.U., 6–7
Turek, P., 484–485
Turkiewicz, M., 128–129
Turner, G.J., 148, 163–164
Turner, M.B., 111–112
Turnland, J.R., 224
Tyler, R.C., 32, 35, 505t
Tzakos, A.G., 484–485, 539, 544–545
Tzitzilonis, C., 372t

U

Uchida, K., 291–292
Uchiyama, S., 69–70
Uden, P.C., 100–101
Ueda, T., 194–195, 252, 257–258, 265–268, 270–271, 271f
Uegaki, K., 318, 328, 391, 398
Uemori, Y., 127–128
Uhlenbeck, O.C., 462–463
Ullrich, S.J., 375–376
Ulmer, H.W., 100–101
Ulrich, A.S., 148–149
Umehara, T., 334–335, 335f
Unger, T., 291–292
Unno, K., 126
Uphaus, R.A., 219, 228–229
Urban, V.S., 124–142, 214–215
Urey, H.C., 216t, 236–239
Ushio, M., 169–170
Usmanova, D., 372t
Utsumi, H., 126

V

Vajpai, N., 194–195, 252, 262f, 269, 370
Valentin, R., 107
Vallurupalli, P., 488–490
van Berkel, S.S., 69–71, 86–87
Van Craenenbroeck, K., 293

van de Laar, L., 498, 520, 523–524
van der Oost, J., 452
Van Dorsselaer, A., 6–7
van Gammeren, A.J., 484–485
van Horn, E., 30
Van Horn, W.D., 194–196
van Oers, M.M., 249
van Oostrum, J., 194–195
van Rossum, B.J., 194
Vanatalu, K., 5, 7
Vanderhart, D.L., 112
Vanhoenacker, P., 293
Vann Etten, R.L., 433–434
Varani, G., 540, 541f, 545–546, 552, 554
Varma, A., 432
Vasella, A., 137
Vasilescu, V., 126
Vasiliauskaite, I., 251, 257–259, 269, 273–274
Vasko, P.D., 99
Veglia, G., 46, 53, 194–195
Velyvis, A., 324
Venters, R.A., 5–6
Verardi, R., 46, 53, 194–195
Verhoefen, M.-K., 370–371, 375
Verhoeven, M.A., 375–376
Veronesi, M., 90–91
Vialard, J., 248
Viel, S., 71–73, 86
Vileno, B., 484–485
Villafranca, J.J., 377t
Villalba, C., 324–325, 373, 380–381
Vilu, R., 5, 7
Vincendo, M., 116–117, 129, 135–136
Viola, R.E., 325–326, 376–379, 377t, 406t
Virta, P., 71–73
Vismeh, R., 135–136
Vitali, F., 400–401
Vlak, J.M., 249, 259–261
Vogel, H.J., 103
Vogl, G., 112
Volkert, B., 111–112
Volkmann, G., 394–397
von Schroetter, C., 484–485, 539, 542–544, 542–543f, 552, 554–556
Vostrikov, V.V., 46, 53, 194–195
Voytas, D., 296–298
Vucelic, D., 126

Vuister, G.W., 168–170, 184
Vukoti, K., 194–195
Vulpetti, A., 71–73
Vuorinen, T., 444–445, 452–454

W

Waber, J., 217–219
Wacker, A., 462–463, 545–546
Wada, A., 194, 372t
Wada, T., 314–315, 322–323, 370–371
Waegeman, H., 31–32
Wagberg, L., 111–112
Wagner, G., 168–169, 194–195, 271–272, 291–292, 372t, 374–375, 381–382
Wagner, R., 196
Wahl, M.C., 377t, 538
Wakamori, M., 326–327
Wakiyama, M., 312–314
Walker, D.M., 377t
Walker, J.M., 256–257
Walker, M.T., 266–268
Walker, T.E., 428–429
Wall, V.E., 292–293
Wallace-Williams, S.E., 252, 270
Walter, N.G., 462–463
Walton, W.J., 262–265
Walz, T., 374
Wan, P.T., 246–247
Wang, C., 111–112
Wang, C.L., 5
Wang, G.-F., 69–70, 74–75, 88–89
Wang, H., 7
Wang, J.J., 6, 497, 514–515
Wang, L., 7, 326–327
Wang, M., 259–261
Wang, S., 194, 203–204, 372t
Wang, T., 372t
Wang, X., 370–371
Wang, Y., 107
Wang, Y.-X., 462–463, 484–485, 497, 528–530
Ward, A., 196
Ward, M.E., 194, 203–204
Ware, G.C., 30
Warnock, J.N., 295–296
Warschawski, D.E., 374
Waschuk, S.A., 196–197, 199
Wasilko, D., 248–249
Wasserman, M., 194–195

Watanabe, S., 314–315, 319, 321–322
Watt, E.D., 462–463
Watts, A., 194–195
Watts, J., 349–350
Waugh, D.S., 46–53, 51t, 52f, 173, 175–176, 400–401
Wawrzyniak, K., 545
Wawrzyniak-Turek, K., 484–485
Weatherbee, J.A., 291–292
Weaver, C.M., 224
Weigel, P.H., 299
Weigelt, J., 125–126
Weinberg, Z., 520
Weinberger, C., 295–296
Weininger, U., 485–487
Weiss, K.L., 139–140
Welcher, B., 293
Wemmer, D.E., 324–325
Weng, Y., 196
Wenter, P., 462–463
Werner, K., 292t, 300
Weski, J., 321–322
Wetterholm, A., 196
Weyand, S., 196
Whaley, T.W., 428–429
Whistler, R.L., 435, 437–438, 448–449
Whitaker, A., 7, 21
Whited, G., 169–170, 200–201, 203–204
Whittaker, J.W., 207
Whitten, A.E., 7
Wibert, K.B., 216–217
Wider, G., 373, 379, 390
Wienken, C.J., 370–371
Wignall, G.D., 138–139
Wijmenga, S.S., 484–485, 498, 520, 523–524
Wikstrom, J., 197–200
Wiktorowska-Jezierska, A., 128–129
Wilde, K.L., 4–23, 99, 102–106
Will, C.L., 538
Willbold, D., 168–169
Williams, C.K., 127
Williams, J.S., 377t
Williams, K., 290–291
Williams, K.S., 399t
Williamson, J.R., 462–463, 528
Willis, C.L., 194–195
Willis, K., 291–292
Willis, R.C., 324–326, 381–382

Willson, R.C., 498
Wilson, H.M., 236–239
Winfrey, S., 293
Winter, G., 196
Winther-Jensen, B., 111–112
Winzor, D.J., 185–186
Witherell, G.W., 462–463
Withka, J.M., 291–292
Wohnert, J., 498, 528–530
Wolff, P., 484–485
Wolfrom, M.L., 428–429, 433, 435, 437–438, 448–449
Wolterink-Van Loo, S., 452
Wong, C.-H., 393–394, 443
Wong, D.T., 262–268
Wong, T.Y., 91
Wong, Y.L., 372t
Wood, B., 115
Wood, H.B., 428–429, 433
Wood, M.J., 186, 196–197
Woods, A.G., 28
Woodward, J., 127–128
Woolfolk, C.A., 325–326
Wotkowicz, C., 399t
Wright, P.E., 168, 395–396
Wu, C., 168–170, 184
Wu, H., 391, 394–395
Wu, J., 438–439, 452–454
Wu, L., 293
Wu, P.S.C., 312–314, 375–379
Wu, Q., 437
Wu, T., 526
Wu, X., 292–293
Wu, Y., 372t
Wulhfard, S., 292–293
Wullschleger, S.D., 232–233
Wunderlich, C.H., 462–490, 545–546
Wurm, F.M., 292–293
Wüthrich, K., 30, 69–71, 86–87, 168, 254–255, 318, 324–325, 369, 373, 379–381, 390–391, 395–397, 399t, 400–403, 405–409, 418
Wyss, D.F., 291–292

X

Xavier, V., 452–454
Xia, W., 292–293
Xian, F., 348–365
Xu, B., 7

Xu, L., 290–291, 293, 295, 300, 301f, 303
Xu, M.Q., 392–394, 398, 399t, 400–401, 404, 411
Xu, R., 318, 328, 391, 393–394
Xu, X., 299
Xu, Y., 372t

Y

Yabuki, T., 252, 312–316, 319–324, 326–327, 369, 375–376, 396, 417–418
Yagi, H., 86–87, 291–292
Yakir, D., 215–216, 220–221
Yamada, K., 248–249
Yamada, Y., 127–128
Yamaguchi, Y., 291–292
Yamaguchi-Nunokawa, E., 323
Yamamoto, M., 418
Yamamoto, T., 349
Yamamoto, Y., 85–86
Yamanaka, S., 127–128
Yamasaki, K., 314–315, 334
Yamashita, H., 127–128
Yamazaki, T., 318, 328, 391, 398
Yamskov, I.A., 377t
Yan, L.Z., 400
Yanagisawa, T., 326–327
Yanamala, N.V.K., 300
Yang, J., 48t, 51t, 52–53, 203–204
Yang, J.-P., 374
Yang, S.C., 497
Yang, X.-Y., 215–221, 225
Yang, Y., 293–296
Yang, Y.J., 88–89
Yang, Z.-Y., 293
Yao, P., 415
Yap, L.L., 46–50, 55–57, 64
Yasuda, S., 234
Yik, J.H.N., 299
Yildiz, Z.F., 468f, 484, 498, 527–528
Ying, J., 488–490
Yoda, K., 220
Yokochi, M., 169–170
Yokogawa, T., 326–327, 349–350
Yokoyama, J., 324–326, 376–379, 377t, 396–397, 405–408, 418
Yokoyama, K.K., 511
Yokoyama, S., 312–339
Yoneyama, M., 334–335, 335f
Yoshida, Y., 312–316, 369

Yoshie, N., 100–101, 104–106
Yoshinaga, F., 128–131, 137
Young, D.J., 127–128
Yu, H., 236–239
Yu, J., 196
Yu, J.-X., 71–73
Yu, L., 71–73, 91
Yu, P., 497
Yu, T.-Y., 374
Yu, Y., 169–170
Yulikov, M., 538–539, 544–545, 549, 558–560
Yun, S.I., 100–101, 111–112
Yurugi-Kobayashi, T., 196

Z

Zajicek, J., 434
Zakomyrdina, L.N., 377t
Zanders, E.D., 291–292
Zanetti, G., 395–397
Zeder-Lutz, G., 196
Zegada-Lizarazu, W., 232–233
Zemb, T., 116
Zerbs, S., 46–47, 53–54
Zettler, J., 318, 328, 394–395, 418
Zhang, C., 293
Zhang, Q., 70, 79–80, 186, 462–463, 488–490
Zhang, S., 370–371
Zhang, W., 169–170, 200–201, 203–204, 424–455
Zhang, X., 540
Zhang, Y.H., 6, 194, 514–515
Zhao, B., 488–490
Zhao, S., 424–455
Zheng, J., 370–371
Zheng, X., 370–371
Zhou, D.H., 55–57
Zhou, H.-X., 194, 372t
Zhou, T., 292–293
Zhou, Y., 88
Zhou, Z., 377t
Zhu, Y., 292–293, 434, 442–443
Zhuo, R., 196
Ziarelli, F., 71–73, 86
Ziegler, U., 291–292
Zigoneanu, I.G., 88–89
Zilm, K.W., 206f
Zimmerman, E.S., 382
Zitomer, R., 271–272
Zocher, M., 371
Zoonens, M., 374
Zou, H., 349
Zou, Y., 252
Zscherp, C., 204–205
Zubay, G., 312–314, 316–317
Züger, S., 394–395
Zuo, X., 462–463, 484–485, 497, 528–530
Zweckstetter, M., 369, 372t

SUBJECT INDEX

Note: Page numbers followed by "*f*" indicate figures, "*t*" indicate tables, and "*s*" indicate schemes.

A

Abelson kinase (Abl) spectrum, 259–261, 262*f*
Acid hydrolysis, 256–259
Adaptation protocol, 6
Adenovirus
  adenoviral cosmid (pVRC1194), 296–298
  expression system
    large-scale production, 300
    small-scale production, 299
  generation, 298
  recombinant adenoviral genome, 296–298
  shuttle vector (pVRC1290), 296–298, 297*f*
Affinity chromatography, purification
  duration, 524
  HiLoad 26/600 Superdex 75 pg column, 523–524, 523*f*
Aldoses
  extension, biologically important
    GlcNAc, 442–443, 442*s*
    glycosylamines, 442–443
    Neu5NAc, 443, 443*s*
  internal carbons
    coupling, CR and MCE, 439, 441*s*
    RMA, 439–442, 441*s*
  MCE
    carbon skeletal rearrangement, 436, 437*s*
    C2-epimers, 435–436
    examples, 437, 438*s*
    generalized laboratory protocol, 437–438
    multiple labeling, chain inversion, 438–439
Algal hydrolysate, 149
Alzheimer amyloid precursor protein (APP), 336–337, 337*f*
Amino acid isotope scrambling, 405–408, 406*t*

Amino acid scrambling enzymes and inhibition strategies, 376–379, 377*t*
Amino acid-selective isotope labeling
  E. coli auxotrophs
    challenges, 53
    cofactors, 54
    complication, 52–53
    ideal genotypes, 50, 51*t*
    L-aspartate and L-glutamate, 50–52, 52*f*
    membrane protein complex Cyt bo3, 55–57, 56*t*
    multiple deletions, 54–55
    ^{15}Nε-glutamine-labeled FdxB, 57–61, 58*t*
    ^{14}N (in natural abundance, N/A) lysine-labeled ARF, 61–63
    ^{14}N(N/A) tyrosine-labeled ARF, 63–64
    single-gene knockout, 50
    strains and properties, 47–50, 48*t*
  metabolic scrambling, *E. coli*, 47
  protein–ligand and protein–protein interactions, 46
Amino acid selective unlabeling
  factors affecting, 176–177
  OBP3, 185–186
  optimal combination, 177–180, 178*f*, 180*f*
  phenylalanine (Phe), 184
  polytopic membrane protein EmrE, 185
  protein–staphylococcal nuclease, 182–184
  resonance assignments, 177, 186
  sequential assignments, 177, 180–182, 181*f*, 183*f*
  Val and Leu methyl residues, 184–185
Amino acid-specific labeling, 395–396
Amino acids, stable isotope-labeling
  ^{1}H$_2$O-based reaction, 324–325
  main-chain NH$_3$ group, exchange, 325–326

595

Amino acids, stable isotope-labeling (*Continued*)
  SAIL method, 324
  selective labeling, 323
  Tyr, 323–324
  uniform labeling, 323
Amino acid-type selective isotope labeling, insect cells
  chemical labeling, 254–255
  dropout media, 279–280
  excess labeling, 253, 255f
  medium preparation, 280
  protocol, 280–281
  replacement of, 252, 254f
  scrambling *vs.* label dilution, 252, 253f
Aminooxyacetate (AOA)
  PLP-requiring enzyme, 376–379
  proton exchange inhibition, 381–382
Analytical denaturing PAGE
  total cellular RNA extraction, 520
  wild-type K12 *E. coli*, expression, 512, 512f
Anion-exchange chromatography
  centrifugation, 522
  chemical RNA synthesis, 481–483f
  contaminants removal, 522
  duration, 520–522
  FPLC protocol, 521–522
  Source 15Q column, 520, 521f
Annual ryegrass
  crystalline index (CrI), 231–232
  growth rate, 230–231, 231f
  vernalization/cold-hardening, 231–232
Arabidopsis thaliana development, 225
Asp-Phe-Gly (DFG) motif, 273
Attenuated total reflection Fourier transform infrared spectroscopy (ATR FT-IR) experiments, 110
Autoimmune thyroid disease (AITD), 334–335, 335f
Autoinduction
  lac operon, 29
  vs. manual induction, 29–30
  protein expression, 42–43
Autolysis, 256–259
Auxotrophs, *E. coli*. See *E. coli*

B

Back protonation/back exchange, 380
Bacterial cellulose
  chemical and physical characterization, 133–134
  deuteration
    bacterial strain, choice of, 130
    cell growth, adaptation of, 131–132
    growth media, 130–131, 132t
    purification, 133
  FTIR, 134, 135f
  isotopic redistribution, 128–129
  mass spectrometry
    enzymatic digestion, 137
    HPLC-refractive index (RI) analysis, 135–136
    individual ion abundances, 135–136
    LC-MS analytical results, 137, 138f
  occurrence and characteristics, 127
  physical properties, 127–128
  production, 128–129
  SANS analysis
    disk-like forms, 141–142, 141f
    instrument configurations, 139–140
    neutron scattering, 138–139
    scattering profiles, 140–141, 140f
  structure and morphology, 127–128
Bacterial cellulose, deuteration
  biosynthetic pathways, 116–117
  cellulose chains, 112, 113f
  cellulose crystallite, 117
  characteristics, 111–112
  different deuteration schemes, 113–114, 114t
  growth media, 114, 114t
  raw pellicles, 115
  SANS curves, 116, 116f
  SLD, 113–114, 116
  WAXS, 115–116, 115f
Bacteriorhodopsin
  detergent solubilization and gel filtration, 161–162
  expression, 158
  lysis and membrane harvest, 158–160
  protein calibration standards, 163
  solutions, 158

sucrose gradient ultracentrifugation, 160–161
verification, isotopic labeling
  MALDI-MS, 162–163, 162f
  solutions, 163
Baculovirus expression vector system (BEVS)
  Bac-to-Bac system, 249
  features, 248–249
  life cycle, 248
  transfection and infection, 279
β2-adrenergic receptor (β2AR), 270–271, 271f
Bioexpress 2000, 269
Biopolyesters deuteration
  bacterial cellulose
    biosynthetic pathways, 116–117
    cellulose chains, 112, 113f
    cellulose crystallite, 117
    characteristics, 111–112
    different deuteration schemes, 113–114, 114t
    growth media, 114, 114t
    raw pellicles, 115
    SANS curves, 116, 116f
    SLD, 113–114, 116
    WAXS, 115–116, 115f
  chitosan
    characteristics, 106
    cost-effective source, 111
    DDA, 106
    extraction and purification, 109–110
    FT-IR analysis, 110, 111f
    hyphal and nonhyphal fungi, 106–107
    Pichia pastoris, 107–109
  PHB
    D-PHB, 100–101
    generic PHA monomer, structure, 100–101, 100f
    LB medium, 101
    nitrogen-free minimal medium, 101–102
    polymer extraction and analysis, 102
  PHO
    characterization studies, 103
    production and deuteration, 103–104

Biopolymer deuteration
  applications, biopolymer, 98
  bacterial cellulose
    biosynthetic pathways, 116–117
    cellulose chains, 112, 113f
    cellulose crystallite, 117
    characteristics, 111–112
    different deuteration schemes, 113–114, 114t
    growth media, 114, 114t
    raw pellicles, 115
    SANS curves, 116, 116f
    SLD, 113–114, 116
    WAXS, 115–116, 115f
  chitosan
    characteristics, 106
    cost-effective source, 111
    DDA, 106
    extraction and purification, 109–110
    FT-IR analysis, 110, 111f
    hyphal and nonhyphal fungi, 106–107
    Pichia pastoris, 107–109
  FT-IR microscopy, 99
  microbial synthesis, 99
  neutron scattering techniques, 98
  NMR, 99
  PHB
    D-PHB, 100–101
    generic PHA monomer, structure, 100–101, 100f
    LB medium, 101
    nitrogen-free minimal medium, 101–102
    polymer extraction and analysis, 102
  PHO
    characterization studies, 103
    production and deuteration, 103–104
  selective deuteration
    ^{13}C NMR, 104–106, 105f
    deuterium substitution, 104–106
    GC–MS spectrum, 104
  vibrational spectroscopy, 98
Bioreactors, 7
Biosynthetic amino acid type-specific incorporation
  amino acid combinations, 76

Biosynthetic amino acid type-specific incorporation (*Continued*)
  aminoacyl-tRNA synthetase, 73–74
  biosynthetic microbial protein expression, 73–74
  constructs preparation and transformation, 75
  key steps, 74–75
  limitation, 70–71, 72*f*
  media, 78*t*
  m-fluoro-L-tyrosine-containing protein, 75–76, 76*f*
  mutagenesis, 74
  NMR-based assignments, 74
  resonance assignments, 74
  stock solutions, 77*t*
  tryptophan precursor fluoroindole, 78
Bromodomain, ATAD2, 273–274

C

Capsid protein-nucleocapsid (CA-NC), 90
Carr Purcell Meiboom Gill (CPMG) RD NMR experiments, 484
^{13}C CPMG RD NMR spectroscopy, 485–487, 487*f*
6-^{13}C-cytidine synthesis
  duration, 473–477
  product characterisation, 477
  synthetic access, 473, 474*f*
  temperature, 477
  triethylamine, 477
Cell-free expression
  amino acid isotope scrambling, 405–408, 406*t*
  batch mode (BM), 405–408
  composition, reaction mixture, 405–408, 407*t*
  continuous exchange cell-free (CECF) mode, 405–408
  optimization, 409
  production of, 408–409
  stock solutions, 407*t*, 408
Cell-free generation, MPs
  amino acid scrambling enzymes and inhibition strategies, 376–379, 377*t*
  AOA, 376–379
  CMCs, 370–371
  D-CF, 370–371

E. coli
  labeling, advantage, 379
  lysates, 368
  expression and purification protocols, 371–372
  expression yield, 369
  L-CF, 371
  ^{15}N–^{1}H BEST-TROSY, 369, 370*f*
  NMR spectroscopy
    characteristics, 372–373
    homo-oligomerization, 373
    liquid-state, 374
    NDs, 374–375
  ^{15}N-TROSY spectra, 376–379, 380*f*
  P-CF, 370
  perdeuteration
    advantage, 380
    AOA and L-methionine sulfoximine, 381–382
    back protonation scrambling, 380–381
  production, 368
  protein functionality, 369
  sample quality, 369
  stable isotopes, labeling, 375–376, 376*f*
Cell harvest
  with cell wash, 155
  without cell wash, 155
Cell lysis
  french press and microfluidizer, 157
  general considerations, 156
  low-salt, 157
  sonication, 157–158
Cellulose. *See* Bacterial cellulose
Chemical exchange saturation transfer (CEST), 488–490, 491*f*
Chemical RNA synthesis
  anion exchange chromatographic analysis, 481–483*f*
  duration, 479–484
  integrity and homogeneity, 484
  oligonucleotide solid phase synthesis cycle, 477–479, 478*f*
  temperature, 483
  tetrahydrofuran, 483
Chemical shift anisotropy (CSA), 85–86
Chitosan, deuteration
  characteristics, 106

DDA, 106
  extraction and purification, 109–110
  FT-IR analysis, 110, 111f
  hyphal and nonhyphal fungi, 106–107
  Pichia pastoris, 107–109
^{13}C-labeling
  biochemical reaction, 262–265, 263f
  HNCA pathway, 265
  methyl groups, 262–265, 264f
^{13}C/^{15}N-labeled proteins, large-scale production, 331–332
Combinatorial selective unlabeling, 170–171
Competition-based experiments, 90
Conventional recombinant expression methods, 314–315
Coupled transcription–translation
  batch and dialysis modes, 312–314, 313f
  reaction solution, 316–317
Critical micellar concentrations (CMCs), 370–371
Crystalline index (CrI), 231–232
6-^{13}C-uridine synthesis
  bis-(tert-butyl)tin dichloride, 473
  duration, 468–473
  N-methylbenzamide extraction, 472
  synthetic access, 468, 469f
  temperature, 472
  triethylamine, 473
Cut and paste RNA approach
  catalytic activity, 538
  combination and ligation
    isotopically and spin-labeled RNAs, PRE NMR, 545–546
    segmentally labeled RNAs, 542f, 544–545
    spin-labeled RNAs, pulsed EPR, 542f, 545
  EPR spectroscopy, 560
  functions, 538
  large segmentally labeled RNAs, 559–560
  ligation
    efficiency, 558
    protocol, 557–558
  NMR-EPR, 539
  NOEs, 559
  PCS, 560

principle, 540, 541f
  RNA splicing, 538
  small isotopically labeled RNAs
    combined RNase H and VS ribozyme cleavage, 550–552
    double RNase H cleavage, 549–550
    HH ribozyme, 547
    in vitro transcription, 546–547, 554
    phosphoramidites, 547
    purification, denaturing anion-exchange HPLC, 554–555
    ribozyme cleavage, 554
    VS ribozyme cleavage, 556
  RNA precursor, 557
  RNase H, 547
  sequence-specific RNase H cleavage, 555–556
  spin-labeled, production protocol, 552–553
  spin-labeled RNA fragments, 540, 541f
  structural flexibility, 539
  transverse relaxation, 539
  unlabeled and isotopically labeled RNA fragments
    multiple segmental isotope labeling, 542–543, 543f
    sequence-specific RNase H cleavage, 542, 542f
Cyanohydrin reduction (CR)
  advantages, 429
  cyclization to glycosylamines, 429–430, 430s
  disadvantage, 435–436
  generalized laboratory protocol, 434–435
  hydrogen isotopes, 433
  K^{13}CN, 432
  ketosugars, 429–430, 432s
  KF synthesis, 428–429
  oxygen atom exchange, 433–434, 433s
  permutations, 433s, 434
  skeletal rearrangements, 431, 431s
^{13}C ZZ exchange NMR spectroscopy, 487–488, 488t, 489f

D

Dedicator of cytokinesis 2 (DOCK2), 338–339, 338–339f
Degree of deacetylation (DDA), 106

Degree of polymerization (DP), 221–223
Detergent-based (D-CF) mode, MPs, 370–371
Deuterated expression
 "ModC1" medium composition, 8–9, 8t
 solution preparation, 8
Deuterated PHB (D-PHB), 100–101
Deuterated protein production
 adaptation pathways, 14, 14s
 commercial supercompetent expression cells, 19–20
 90% D_2O, SANS, 14–15
 100% D_2O, unlabeled glycerol, 15–16
 gentle handling, cultures, 21
 level quantification, 16–17
 ^{15}N ammonium hydroxide/sodium hydroxide, base feed, 21–22
 perdeuteration, 100% D_2O, 16
 plasmid choice, 18–19, 20f
 precipitating media, 19
 staged starter cultures, 21
 typical deuteration levels, 22–23
Deuteration
 bacterial cellulose
  bacterial strain, choice of, 130
  cell growth, adaptation of, 131–132
  growth media, 130–131, 132t
  purification, 133
 protein labeling, 29, 31t
Deuterium (^2H) labeling, plants
 annual ryegrass
  CrI, 231–232
  growth rate, 230–231, 231f
  vernalization/cold-hardening, 231–232
 chemical and physical characterization, 221–223
 cultivation methods
  advantage, 228–229
  basic requirements, 225–226
  parameters, 225–226, 227t
  phytotron perfusion chambers, 225–226, 227t
  transpiration and water vapor exchange, 227–228
 D_2O
  chemical and physical properties, 216–217, 216t, 218t
  $vs.$ H_2O, plant growth, 217–219
  inhibitory effects, 215–216
  natural variation, 219–220
  scattering properties, 214–215
 DP, 221–223
 GPC, 221–223
 lignocellulosic biomass, 221–223
 metabolic studies
  Arabidopsis thaliana development, 225
  duckweed *Lemna*, 223–224
  nutritional tracer studies, 224
 multiple-chamber perfusion system, 229–230, 230f
 PDI, 221–223
 SANS, 214–215
 substitution
  analysis, 220–221
  NMR studies, 214–215
 switchgrass
  hemicellulose and glucan contents, 232–233
  long-term hydroponic cultures, 232–233, 233f
 winter grain rye, D_2O and phenylalanine
  FTIR analysis, 236–239, 238f
  growth rate, comparison, 235–236, 236–237f, 238t
  plant growth solutions, 234–235
  shikimate biosynthetic pathway, 234, 235f
Deuterium oxide (D_2O)
 biological toxicity and neutron scattering
  photosynthetic eukaryotes, 125–126
  transcription and translation level, 126
 chemical and physical properties, 216–217, 216t, 218t
 $vs.$ H_2O, plant growth, 217–219
 inhibitory effects, 215–216
 natural variation, 219–220
 scattering properties, 214–215
D-fructose 17 and L-sorbose 25, 452
2D HSQC spectrum, 169, 170f
Dictyostelium discoideum, 290–291
DNA template preparation
 elements, 351–352, 352f
 forward and reverse primers, 352–353, 354t
 PCR amplification, 353–355, 354–355t

Subject Index

peptide length, 355
purification kit, 355
synthetic oligonucleotides, 352–353, 353f
DNAzymes, 524–525, 525f
2D [^{15}N,^{1}H]-HNCO spectrum, 416f, 417
Domain-selective dual ^{13}C/^{15}N-labeling, 417–418
Double selection, high expressing *E. coli* clones
  duration, 515–516
  glycerol stock, 516
  mutant strain tktA, 514–515, 514f
Duckweed *Lemna*, 223–224

E

E. coli cell-free protein synthesis method
  advantages, 312–315
  coding region, 318
  conventional recombinant expression methods, 314–315
  coupled transcription–translation
    batch and dialysis modes, 312–314, 313f
    reaction solution, 316–317
  disulfide bonds, 321–322
  ligand complexes, 322–323
  low-molecular mass components, 321
  mammalian protein production, 315
  molecular chaperones, 322
  PDB-deposited structures, 315–316, 316t
  reaction modes, 317–318, 317t
  reaction solution, 312–314
  S30 extract and tRNAs, 320–321
  stable isotope labeling, 315–316
  tags, 319
  template DNA, 319–320, 319f
  T7 RNA polymerase, 319f, 320
Engulfment and cell motility 1 (ELMO1), 338–339, 338–339f
Enzyme-mediated aldol condensation
  asymmetric labeling, 449–451, 451s
  chemo-enzymic synthesis, 445, 446s
  glycerol kinase, 447, 447s
  2-ketohexoses
    top- and bottom-half labeling, 449
    top-half labeling, 448–449
  RMA, 445–447, 446s, 451–452
  symmetric labeling, 449–451
  TIM, 445–447

triokinase, 447, 448s
Escherichia coli (E. coli)
  auxotrophs, amino acid-selective isotope labeling
    challenges, 53
    cofactors, 54
    complication, 52–53
    ideal genotypes, 50, 51t
    L-aspartate and L-glutamate, 50–52, 52f
    membrane protein complex Cyt bo3, 55–57, 56t
    metabolic scrambling, 47
    multiple deletions, 54–55
    15Nε-glutamine-labeled FdxB, 57–61, 58t
    ^{14}N (in natural abundance, N/A) lysine-labeled ARF, 61–63
    ^{14}N(N/A) tyrosine-labeled ARF, 63–64
    single-gene knockout, 50
    strains and properties, 47–50, 48t
  cell-free generation, MP
    labeling, advantage, 379
    lysates, 368
  expression system, insect cells, 246–247
  in vivo expression, 418
  precursor constructs
    minimal medium, 403–404, 403t
    recombinant precursor expression protocol, 404–405
  protein labeling
    advantages, 28
    plasmid and strain selection, 30–32
    protocols, 28
    stable isotope labeling, types, 28
Eukaryotic membrane proteins, 205–207, 206f
Excision and purification, RNA
  accurate determination, 526
  DNAzymes, 524–525, 525f
  duration, 525–526
  optimizations, 526
  PAGE gel, 526–527
Expressed protein ligation (EPL)
  requirements, 397–398
  segmental isotope labeling, 392–394
Expression system, insect cells
  advantage, 247

Expression system, insect cells (*Continued*)
  BEVS, 248
  E. coli, 246–247
  High Five cells, 247
  limitation, 247
  optimization, 249
  S2 cells, 247
  Sf9 and Sf21 cells, 247
  virus generation protocols, 248–249

F

^{19}F-containing ligands
  biochemical screening, 91
  competition-based experiments, 90
  line broadening, monitoring, 89–90
  magnetization transfer
    experiments, 91
^{19}F-modified aromatic amino acids
  biosynthetic amino acid type-specific
    incorporation
    amino acid combinations, 76
    aminoacyl-tRNA synthetase, 73–74
    biosynthetic microbial protein
      expression, 73–74
    constructs preparation and
      transformation, 75
    key steps, 74–75
    limitation, 70–71, 72*f*
    media, 78*t*
    m-fluoro-L-tyrosine-containing
      protein, 75–76, 76*f*
    mutagenesis, 74
    NMR-based assignments, 74
    resonance assignments, 74
    stock solutions, 77*t*
    tryptophan precursor fluoroindole, 78
  biosynthetic methods, 73
  ligand-observe NMR experiments
    biochemical screening, 91
    competition-based experiments, 90
    line broadening, monitoring, 89–90
    magnetization transfer experiments, 91
  molecular probe, 69–70
  posttranslational covalent modification,
    70–71, 72*f*
  protein–ligand interactions, 71–73
  protein-observe NMR experiments
    protein aggregation, 88
    protein–ligand interactions, 86–87
    protein–lipid interactions, 88–89
    protein un/folding, 87
  site-specific incorporation
    amber tRNA evolution, 80–81
    constructs preparation and
      transformation, 81–82
    C-terminal purification tag, 82
    media, 84*t*
    negative and positive control, 83–84
    nonsense stop codons and nonsense
      suppressor-tRNAs, 70–71
    principles, 79–80
    4-(trifluoromethyl)- phenylalanine-
      containing protein expression, 82–83
    tRNA/tRNA synthetase pair, 80
  solubility and stability, NMR sample, 85
  solution NMR, 68
  spectrometer magnetic field strength,
    85–86
  spin ½ NMR active nuclei, 68–69, 69*t*
Fourier transform infrared spectroscopy
  (FTIR)
  bacterial cellulose, 134, 135*f*
  chitosan, deuteration, 110, 111*f*
  wavenumber assignments, 216–217, 218*t*
  winter grain rye, D$_2$O and phenylalanine,
    236–239, 238*f*
FPLC protocol, 521–522, 524
Fragment-based drug discovery (FBDD),
  273–274

G

Gel permeation chromatography (GPC),
  221–223
G-protein-coupled receptor (GPCR)
  rhodopsin, 270

H

Halobacterium salinarum, isotopic labeling
  algal growth medium, 151
  algal hydrolysate, 149
  buffered salt solution, 150*t*
  characteristics, 148
  culture media, 149
  deuterated buffered salt solution, 151*t*
  deuterated media, 148
  frozen storage, culture, 152

80% glycerol/medium, 151
growth
  agar plates, 152
  isotopically labeled medium, 153
maintenance procedure, 149
protein expression, 149
protein purification
  bacteriorhodopsin, expression and purification, 158–162
  cell growth, 154
  cell harvest, 154–156
  cell lysis, 156–158
  protein expression, factors affecting, 153–154
  verification, MALDI-MS, 162–163
reagents, 150
standard growth medium, 151
starter culture growth, nonlabeled medium, 152
trace metals solution, 150t
HAQP1. *See* Human aquaporin-1 (hAQP1)
Heteronuclear single quantum coherence (HSQC) spectrum
  combined RNase H and VS ribozyme cleavage, 552, 553f
  NMR
    2D, 169, 170f
    hybrid recombinant tRNA scaffold, 528, 529f
    site-directed labeling, 326
HiLoad 26/600 Superdex 75 pg column, 523–524, 523f
Human aquaporin-1 (hAQP1)
  large-scale isotope labeling, 201–202
  purification, 203–205
  sample production, 204–205
  small-scale natural abundance expression, 199–200
Human glutamyl-prolyl tRNA synthetase (EPRS), 415, 416f
Hybrid recombinant tRNAscaffold
  advantages, 530–532
  affinity chromatography, purification
    duration, 524
    HiLoad 26/600 Superdex 75 pg column, 523–524, 523f
  anion-exchange chromatography, purification

  centrifugation, 522
  contaminants removal, 522
  duration, 520–522
  FPLC protocol, 521–522
  Source 15Q column, 520, 521f
bacterial growth solutions, 504–507
denaturing PAGE solutions, 508–510
DNAzyme cleavage solutions, 508
double selection, high expressing *E. coli* clones
  duration, 515–516
  glycerol stock, 516
  mutant strain tktA, 514–515, 514f
equipment, 499–502
excision and purification
  accurate determination, 526
  DNAzymes, 524–525, 525f
  duration, 525–526
  optimization, 526
  PAGE gel, 526–527
expression, wild-type K12 *E. coli*
  analytical denaturing PAGE, 512, 512f
  duration, 512–513
  nonrecombinant tRNA-scaffold, 513
  safety measures, 513
large-scale expression, labeled SPG minimal media
  cell density and RNA yield, 518
  duration, 517
  growth rate, 517
  mid-scale culture, 518
materials, 502–510
NMR applications
  benefits, 528–530
  $^1H-^{15}N$ HSQC correlation spectrum, 528, 529f
  imino groups and carbonyl groups, correlation, 528–530, 531f
  in vivo labeling, 528
  tktA grown, 1-^{13}C-glucose, 528, 530f
protocol
  duration, 510
  flowchart, 499, 501f
  preparation, 511
  safety measures, 511
purification solutions, 507–508
total cellular RNA extraction
  analytical denaturing PAGE, 520

Hybrid recombinant tRNAscaffold
  (*Continued*)
  duration, 519
  RNA precipitation, 519
  seperation, 518

I

In-cell NMR
  protein expression, 271–272
  resonance experiments, 272, 272f
Insect cells
  amino acid type-specific
    chemical labeling, 254–255
    dropout media, 279–280
    excess labeling, 253, 255f
    medium preparation, 280
    protocol, 280–281
    replacement of, 252, 254f
    scrambling *vs.* label dilution, 252, 253f
  applications
    drug discovery, 273–274
    in-cell NMR, 271–272
    membrane proteins, 270–271
    structural studies, 269
  cell culture, 275–276
  cell lines, 274
  commercial media, 274
  consumables, 275
  equipment, 275
  expression system
    advantage, 247
    baculovirus life cycle, 248
    E. coli, 246–247
    High Five cells, 247
    limitation, 247
    optimization, 249
    S2 cells, 247
    Sf9 and Sf21 cells, 247
    virus generation protocols, 248–249
  isotope labeling
    adverse effects, cell metabolism, 251–252
    amino acids sources, 250–251
    carry over from preculture, 251
    unlabeled medium, BV suspension, 251
  reagents, 274–275
  stock cell lines generation, 277
  transfection and infection, BV, 279
  uniform isotope labeling
    ^{13}C-labeling, 262–265
    commercial media, 256
    culture monitor, 284
    economic media, 256–259
    harvest, 284
    ^{2}H-labeling, 265–268
    infection, 282
    medium change, 283
    medium preparation, 282
    ^{15}N-labeling, 259–261
    starter culture preparation, 282
    virus amplification, 279
Intrinsically disordered proteins (IDPs), 168
In vivo expression, *E.coli*, 418
Isotope incorporation, protein labeling, 32–34, 33t
Isotope-labeled RNA phosphoramidites. *See* Stable isotope-labeled RNA phosphoramidites
Isotope labeling
  insect cells
    adverse effects, cell metabolism, 251–252
    amino acids sources, 250–251
    carry over from preculture, 251
    unlabeled medium, BV suspension, 251
  ssNMR, 194–195
Isotope scrambling
  cell-free expression, 405–408, 406t
  metabolic precursor, 176
  ^{14}N, 173, 175t
  reduction methods, 175–176
Isotopomers
  D-glucose, 425–427, 426s
  nomenclature, 425–427, 426s
  synthetic strategies, 427

K

Kilani–Fischer (KF) synthesis, 428–429

L

Labeled SPG minimal media, expression
  cell density and RNA yield, 518
  duration, 517
  growth rate, 517
  mid-scale culture, 518

L-aspartate, 50–53, 52f
Leptosphaeria rhodopsin (LR)
  large-scale isotope labeling, 200–201
  sample production, 204–205
  small-scale natural abundance expression, 199
L-glutamate, 50–53, 52f
Ligand-observe NMR experiments
  biochemical screening, 91
  competition-based experiments, 90
  line broadening, monitoring, 89–90
  magnetization transfer experiments, 91
Ligation
  efficiency, 558
  precursor fragments
    C-terminal precursors, 411–412
    ligation competent state, 409–410
    N-terminal precursors, 410–411
    on-column protocol, 413–415, 414f
    reactive fragments, 412
    sequential ligation, 412–413
  protocol, 557–558
Lignocellulosic biomass, 221–223
Line broadening, 89–90
Lipid-based (L-CF) mode, MPs, 371
Low-molecular-weight creatine phosphate tyrosine (LMCPY), 321
LR. *See* Leptosphaeria rhodopsin (LR)

M

MALDI-MS
  bacteriorhodopsin, 162–163, 162f
  protein purification, 162–163
Mammalian expression system
  Dictyostelium discoideum, 290–291
  eukaryotic systems, 290–291
  functions, 291–292
  isotopically enriched proteins, 291–292, 292t
  large-scale protein production, 291–292
  transient transfection
    adenoviral shuttle vector, 296, 297f
    adenoviruses, generation of, 298
    gene expression, 292–293, 294f
    mammalian viruses, 295–296
    procedure, 293
    protein expression and construct selection, 293–295
    recombinant adenoviral genome, 296–298, 297f
Manual induction
  vs. autoinduction, 29–30
  M9 medium, 36–37
  MTM, 36–37
  phosphate buffer, 37
  protein expression, 41–42
Mass spectrometry
  enzymatic digestion, 137
  HPLC-refractive index (RI) analysis, 135–136
  individual ion abundances, 135–136
  LC-MS analytical results, 137, 138f
Membrane protein complex Cyt bo3, 55–57, 56t
Membrane proteins (MPs)
  CF systems, 368
  D-CF, 370–371
  functional, 368
  functionality assessment, 371–372
  GPCR rhodopsin, 270
  L-CF, 371
  methionine methyl resonances, β2AR, 270–271, 271f
  P-CF, 370
  structures, 372t
Metabolic scrambling, *E. coli*, 47
ML40K1, 61–63
Modified Neidhardt media
  composition, 40
  preparation, 37–38
Modified Tyler media (MTM)
  manual induction, 36–37
  media composition, 40
Molybdate-catalyzed epimerization (MCE), aldoses
  carbon skeletal rearrangement, 436, 437s
  C2-epimers, 435–436
  examples, 437, 438s
  generalized laboratory protocol, 437–438
Monosaccharides, stable isotope labeling aldoses
  extension, biologically important, 442–443
  internal carbons, 439–442
  multiple labeling, chain inversion, 438–439

Monosaccharides, stable isotope labeling (*Continued*)
applications, 452–454
chemical reactions, selection, 427–428
chemical structures, 425s
core developments, 452–454
^{13}CO, role of, 427–428, 428s
CR
  advantages, 429
  cyclization to glycosylamines, 429–430, 430s
  disadvantage, 435–436
  generalized laboratory protocol, 434–435
  hydrogen isotopes, 433
  K^{13}CN, 432
  ketosugars, 429–430, 432s
  KF synthesis, 428–429
  oxygen atom exchange, 433–434, 433s
  permutations, 433s, 434
  skeletal rearrangements, 431, 431s
D-fructose 17 and L-sorbose 25, 452
enzyme-mediated aldol condensation
  asymmetric labeling, 449–451, 451s
  chemo-enzymic synthesis, 445, 446s
  glycerol kinase, 447, 447s
  RMA, 445–447, 446s, 451–452
  symmetric labeling, 449–451
  TIM, 445–447
  top- and bottom-half labeling, 2-ketohexoses, 449
  top-half labeling, 2-ketohexoses, 448–449
  triokinase, 447, 448s
functions, carbohydrates, 424
isotopomers
  D-glucose, 425–427, 426s
  nomenclature, 425–427, 426s
  synthetic strategies, 427
MCE, aldoses
  carbon skeletal rearrangement, 436, 437s
  C2-epimers, 435–436
  examples, 437, 438s
  generalized laboratory protocol, 437–438
relative carbonyl reactivities, osones
  definition, osones, 444
  2-ketoses, 445

selective reactivity, 444s, 445
reverse osmosis, 454
saccharide isotopomer, 454
synthetic enzymes, 454–455
transformative reactions, 424
uniform labeling, 424
Multidimensional NMR experiments, 168
Multiple-chamber perfusion system, 229–230, 230f
Multiple labeling, protein
  ^{13}C, 18
  ^{15}N, 17–18
Multiple reaction monitoring (MRM), 348–349, 363

N

Nanodiscs (NDs)
  cell-free generation, 374–375
  characteristics, 374
  liquid-state NMR, 374–375
Native chemical ligation (NCL), 391–392
Natural polyhistidine tag (NHis-tag), 319
^{15}Nε-glutamine-labeled FdxB, 57–61, 58t
Neutron crystallography, 4–5
Neutron scattering
  biological toxicity, D2O
    photosynthetic eukaryotes, 125–126
    transcription and translation level, 126
  D-labeling techniques, 124–125
  SANS analysis, 138–139
  SLDs, 124–125, 125f
Neutron scattering techniques, 98
^{15}N-^{1}H BEST-TROSY, 369, 370f
^{15}N-labeled proteins, medium-scale production
  apparatus, 327f, 328–331
  protocol, 329–330
  standard composition, reaction solution, 328–331, 329t
^{15}N-labeling
  Abl spectrum, 259–261, 262f
  ammonia-containing media, 259–261
  biochemical reaction, 259, 260f
  cell culturing protocols, 259–261, 261f
^{14}N (in natural abundance, N/A) lysine-labeled ARF, 61–63
N-terminal tags, 319
^{15}N-TROSY spectra, 376–379, 380f

Subject Index

^{14}N(N/A) tyrosine-labeled ARF, 63–64
Nuclear magnetic resonance (NMR)
  advantage, 169–170
  amino acid selective unlabeling
    choice of, 176–180, 178f, 180f
    OBP3, 185–186
    phenylalanine (Phe), 184
    polytopic membrane protein EmrE, 185
    protein–staphylococcal nuclease, 182–184
    resonance assignment, 186
    sequential assignments, 180–182, 181f, 183f
    Val and Leu methyl residues, 184–185
  based assignments, 74
  cell-free generation, MP
    characteristics, 372–373
    homo-oligmerization, 373
    liquid-state, 374
    NDs, 374–375
  combinatorial selective unlabeling, 170–171
  2D HSQC spectrum, 169, 170f
  ^{1}H active nuclei, 4–5
  hybrid recombinant tRNAscaffold
    benefits, 528–530
    ^{1}H—^{15}N HSQC correlation spectrum, 528, 529f
    imino groups and carbonyl groups, correlation, 528–530, 531f
    in vivo labeling, 528
    tktA grown, 1-13C-glucose, 528, 530f
  IDPs, 168
  isotope scrambling
    metabolic precursor, 176
    ^{14}N, 173, 175t
    reduction methods, 175–176
  ligand-observe experiments
    biochemical screening, 91
    competition-based experiments, 90
    line broadening, monitoring, 89–90
    magnetization transfer experiments, 91
  limitation, 390
  multiple labeling, protein
    ^{13}C, 18
    ^{15}N, 17–18
  protein-observe experiments
    protein aggregation, 88

  protein–ligand interactions, 86–87
  protein–lipid interactions, 88–89
  protein un/folding, 87
  resonances/peaks, identification of, 172, 173–174f
  reverse labeling, 169
  sample preparation, 171–172
  selective isotope labeling, 169
  sequence-specific resonance assignments, 168
  solubility and stability, sample, 85
  solution NMR, 68
  specific pulse sequences, 168–169
  spectral overlap/crowding, 168
Nuclear Overhauser effects (NOEs), 91
Nutritional tracer, 224

O

Odorant-binding protein (OBP3), 185–186
Oligonucleotide solid phase synthesis cycle, 477–479, 478f
On-column ligation, 413–415, 414f
Optimal ligation site
  importance, 398–399, 399t
  recommendations, 400

P

Panicum virgatum. See Switchgrass
Peptide complexes
  APP, 336–337, 337f
  DOCK2, 338–339, 338–339f
  ELMO1, 338–339, 338–339f
  PID2, 336–337, 337f
  solution structures, NMR spectroscopy, 336–337, 338f
  tested fusion constructs, 338–339, 340f
Peptide synthesis, PURE systems
  components, 356, 358t
  formylmethionine, 355–356
  translation, 356, 357t
Perdeuteration
  cell-free generation, MP
    advantage, 380
    AOA and L-methionine sulfoximine, 381–382
    back protonation scrambling, 380–381
    ^{13}C labeling, 18
    100% D$_2$O, 16
    protein, 6

Phosphotyrosine interaction domain 2 (PID2), 336–337, 337f
Phytotron perfusion chambers, 225–226, 227t
Pichia pastoris
  characteristics, sample, 205–207
  chitosan, 107–109
  G-protein-coupled receptors (GPCRs), 196
  hAQP1, 203–205
  large-scale isotope labeling
    hAQP1, 201–202
    LR, 200–201
  limitation, 207
  lipid reconstitution optimization, 203–204
  minimize protein losses, 202–203
  protein targets and vectors, 197–198
  protein-to-lipid ratio, 203–205
  sample optimization, 205–207
  sample production
    amide I band, 204–205
    cell breakage, 202–203
    characteristics, 205–207
    eukaryotic membrane proteins, 205–207, 206f
    His-tag affinity protein purification, 202–203
    lipid reconstitution optimization, 203–204
    LR and hAQP1, 204–205
    prescreening, 205–207
    quantification, 203
    static FTIR spectra, 203, 203f
  small-scale natural abundance expression
    best-producing colonies, 198
    biochemical and economic considerations, 198
    hAQP1, 199–200
    LR, 199
    media recipes, 198, 199t
    spectral quality, 205–207, 206f
    static FTIR spectra, 203, 203f
Poly(3-hydroxybutyrate) (PHB)
  D-PHB, 100–101
  generic PHA monomer, structure, 100–101, 100f
  LB medium, 101
  nitrogen-free minimal medium, 101–102
  polymer extraction and analysis, 102
Poly-ADP-ribose polymerases (PARPs), 333, 333f
Polydispersity index (PDI), 221–223
Poly-3-hydroxyoctanoate (PHO)
  characterization studies, 103
  production and deuteration, 103–104
Posttranslational covalent modification, 70–71, 72f
Precipitate (P-CF) mode, MPs, 370
Precursor constructs
  cell-free expression
    amino acid isotope scrambling, 405–408, 406t
    batch mode (BM), 405–408
    composition, reaction mixture, 405–408, 407t
    continuous exchange cell-free (CECF) mode, 405–408
    optimization, 409
    production of, 408–409
    stock solutions, 407t, 408
  cloning, target protein fragments, 402–403, 402f
  E. coli
    minimal medium, 403–404, 403t
    recombinant precursor expression protocol, 404–405
  ligation-competence, 400–401
  recombinant precursor expression, 404–405
  TEV protease, 400–401
  unlabeled/uniformly labeled precursors, 401–402
Protein biosynthesis, 312–314
Protein expression
  adenoviral expression system
    large-scale production, 300
    small-scale production, 299
  glycosylation pattern, 299
  manual induction, 41–42
  $^{15}N/^{13}C$ proline-CGM6750 CUSTOM media, 303
  NMR characterization
    HIV-1 gp120 outer domain (OD), 300–301, 301f
    $^{1}H-^{15}N$ HSQC spectra, 301–302, 302f
    selective labeling, amino acids, 300

Protein labeling
  autoinduction
    lac operon, 29
    vs. manual induction, 29–30
  deuteration, 29, 31t
  E. coli
    advantages, 28
    plasmid and strain selection, 30–32
    protocols, 28
    stable isotope labeling, types, 28
  expression
    autoinduction, 42–43
    manual induction, 41–42
  isotope incorporation, 32–34, 33t
  manual induction
    vs. autoinduction, 29–30
    M9, 36–37
    MTM, 36–37
    phosphate buffer, 37
  media composition
    autoinduction, 40
    modified Neidhardt media, 40
    modified Tyler media, 40
  media preparation
    50x autoinduction carbon source stock, 38
    50x carbon source stock, 38
    20x modified Neidhardt stock, 37–38
    20x nitrogen phosphate stock, 38
    10,000x trace elements stock, 38–39
    10,000x vitamin stock, 39
Protein-observe NMR experiments
  protein aggregation, 88
  protein–ligand interactions, 86–87
  protein–lipid interactions, 88–89
  protein un/folding, 87
Protein perdeuteration, 6
Protein purification, *Halobacterium salinarum* bacteriorhodopsin
  detergent solubilization and gel filtration, 161–162
  expression, 158
  lysis and membrane harvest, 158–160
  MALDI-MS, verification, 162–163, 162f
  protein calibration standards, 163
  solutions, 158, 163
  sucrose gradient ultracentrifugation, 160–161

cell growth, 154
cell harvest
  with cell wash, 155
  without cell wash, 155
cell lysis
  french press and microfluidizer, 157
  general considerations, 156
  low-salt, 157
  sonication, 157–158
protein expression, factors affecting, 153–154
Protein synthesis using recombinant elements (PURE) systems
  E. coli-based reconstituted cell-free protein translation system, 350
  enrichment and digestion, synthesized peptide
    affinity tags, 358
    purification and digestion, 359–360
    Strep-Tactin beads, 360
    Strep-tag, 358–359
  peptide synthesis
    components, 356, 358t
    formylmethionine, 355–356
    translation, 356, 357t
  proof-of-concept experiment, 363–364, 364f
  vs. QconCATs, 365
  quantification, synthesized peptide
    absolute quantification, 361–362, 362t, 363f
    LOQ, 363
    MRM detection, 363
    reference peptide, accuracy, 360–361
    Strep-tag sequence, 361
Protein trans-splicing (PTS), 394–395
Pseudo contact shifts (PCS), 560
Purified precursors, ligation
  C-terminal precursors, 411–412
  N-terminal precursors, 410–411
  reactive fragments, 412
  sequential ligation, 412–413

Q

Quantitative concatemers (QconCATs)
  advantages, 349
  limitations, 365

R

Rabbit muscle aldolase (RMA)
  enzyme-mediated aldol condensation, 445–447, 446s, 451–452
  internal carbons, 439–442, 441s
Recombinant adenoviral genome, 296–298
Recombinant expression methods, 314–315
Recombinant precursor expression, 404–405
Reconstituted cell-free protein translation system, 350
Relative carbonyl reactivities, osones
  definition, osones, 444
  2-ketoses, 445
  selective reactivity, 444s, 445
Relaxation dispersion (RD), 484
Reverse labeling technique
  ML40K1, 61–63
  NMR spectroscopy, 169
  RF4RIL, 63–64
Rhodopsin
  LR, 197
  microbial, 203
  proteorhodopsin, 203–204
Ribonucleic acid (RNA)
  chemically synthesized, 484–485, 486f
  expression and purification, 499, 500f
  functions, 462–463
  limitation, 497
  mutant tktA strain, 498–499
  ribonucleotides (rNTP), 498
  tRNA-scaffold plasmid, 498
Ribonucleotides (rNTP), 498
RNA phosphoramidites. *See* Stable isotope-labeled RNA phosphoramidites
RNA splicing, 538

S

Scattering length density (SLD)
  bacterial cellulose, deuteration, 113–114, 116
  neutron scattering, 124–125, 125f
SDS–PAGE analysis, 13
Seed germination experiments, 219
Segmental amino acid-type isotope labeling
  advantages, 396
  2D [^{15}N,^{1}H]-HNCO spectrum, 416f, 417
  domain-selective dual ^{13}C/^{15}N-labeling, 417–418
  EPRS, 415, 416f
  in vivo expression, E.coli, 418
  ligation, precursor fragments
    ligation competent state, 409–410
    on-column protocol, 413–415
    purified precursors, 410–413
  methodological approach, 396–397
  optimal ligation site
    importance, 398–399, 399t
    recommendations, 400
  precursor constructs
    cell-free expression, 405–409, 406t
    cloning, target protein fragments, 402–403, 402f
    E. coli, 403–405, 403t
    ligation-competence, 400–401
    recombinant precursor expression, 404–405
    TEV protease, 400–401
    unlabeled/uniformly labeled precursors, 401–402
  requirements, EPL, 397–398
  VEGFA, 415, 416f
Segmental isotope labeling
  EPL, 392–394
  NCL, 391–392
  PTS, 394–395
  sortase A-mediated ligation, 395
Segmental stable isotope labeling, 484–485, 486f
Selective unlabeling, amino acid. *See* Amino acid selective unlabeling
Sequence-specific resonance assignments, 168
Shotgun proteomics, 349–350
Site-directed labeling
  aminoacylation, 326
  chemical acylation, 326
  orthogonality, 326–327
Site-specific incorporation
  amber tRNA evolution, 80–81
  constructs preparation and transformation, 81–82
  C-terminal purification tag, 82
  media, 84t
  negative and positive control, 83–84

nonsense stop codons and nonsense
  suppressor-tRNAs, 70–71
 principles, 79–80
 4-(trifluoromethyl)- phenylalanine-
  containing protein expression,
  82–83
 tRNA/tRNA synthetase pair, 80
Small-angle neutron scattering (SANS)
 bacterial cellulose
  disk-like forms, 141–142, 141f
  instrument configurations,
   139–140
  neutron scattering, 138–139
  scattering profiles, 140–141, 140f
 bacterial cellulose, deuteration, 116,
  116f
 deuterated ribosomal materials, 5
 90% D_2O, 14–15
 M9 medium, 5
Small isotopically labeled RNAs
 combined RNase H and VS ribozyme
  cleavage, 550–552
  advantages, 550
  CUG-BP 2 and DND1, 550
  ^1H-^{13}C HSQC spectra, 552, 553f
  double RNase H cleavage, 549–550
  HH ribozyme, 547
  in vitro transcription, 546–547, 554
  phosphoramidites, 547
  purification, denaturing anion-exchange
   HPLC, 554–555
  ribozyme cleavage, 554
  VS ribozyme cleavage, 556
  RNA precursor, 557
  RNase H, 547
  sequence-specific RNase H cleavage,
   555–556
  spin-labeled, production protocol,
   552–553
Small ubiquitin-related modifier (SUMO)
  tag, 319
Solid-phase peptide synthesis (SPPS), 349
Solid-state NMR (ssNMR)
 E. coli expression systems,
  194–195
 isotope labeling, 194–195
 multidimensional studies, 195–196
 Pichia pastoris

characteristics, sample, 205–207
GPCRs, 196
hAQP1, 203–205
large-scale isotope labeling,
  200–202
limitation, 207
lipid reconstitution optimization,
  203–204
minimize protein losses, 202–203
protein targets and vectors, 197–198
protein-to-lipid ratio, 203–205
sample optimization, 205–207
sample production, 202–207
small-scale natural abundance
  expression, 198–200
spectral quality, 205–207, 206f
static FTIR spectra, 203, 203f
technical challenges, 194
technological improvements, 194
Sortase A-mediated ligation, 395
Source 15Q column, 520, 521f
Spectral overlap/crowding, 168
"spy" ^{19}F-containing molecule, 90
Stable isotope-labeled peptides
 DNA template preparation
  elements, 351–352, 352f
  forward and reverse primers, 352–353,
   354t
  PCR amplification, 353–355,
   354–355t
  peptide length, 355
  purification kit, 355
  synthetic oligonucleotides, 352–353,
   353f
 enrichment and digestion, synthesized
  peptide
  affinity tags, 358
  purification and digestion, 359–360
  Strep-Tactin beads, 360
  Strep-tag, 358–359
 equipment, materials, and buffers,
  350–351
 MRM, 348–349
 PURE systems
  E. coli-based reconstituted cell-free
   protein translation system, 350
  enrichment and digestion, 358–360
  peptide synthesis, 355–357

Stable isotope-labeled peptides (*Continued*)
  proof-of-concept experiment,
    363–364, 364*f*
  quantification, 360–363
  QconCATs, 349, 365
  quantification, synthesized peptide
    absolute quantification, 361–362, 362*t*, 363*f*
    LOQ, 363
    MRM detection, 363
    reference peptide, accuracy, 360–361
    Strep-tag sequence, 361
  shotgun proteomics, 349–350
  spike synthetic labeled peptides, 349
  SPPS, 349
  targeted assays, 348
Stable isotope-labeled RNA phosphoramidites
  applications
    ^{13}C CPMG RD NMR spectroscopy, 485–487, 487*f*
    CEST, 488–490, 491*f*
    chemically synthesized RNAs, 484–485, 486*f*
    CPMG RD NMR experiments, 484
    ^{13}C ZZ exchange NMR spectroscopy, 487–488, 488*t*, 489*f*
  6-^{13}C-cytidine synthesis
    duration, 473–477
    product characterisation, 477
    synthetic access, 473, 474*f*
    temperature, 477
    triethylamine, 477
  chemical RNA synthesis
    anion exchange chromatographic analysis, 481–483*f*
    duration, 479–484
    integrity and homogeneity, 484
    oligonucleotide solid phase synthesis cycle, 477–479, 478*f*
    temperature, 483
    tetrahydrofuran, 483
  6-^{13}C-uridine synthesis
    bis-(tert-butyl)tin dichloride, 473
    duration, 468–473
    N-methylbenzamide extraction, 472
    synthetic access, 468, 469*f*
    temperature, 472
    triethylamine, 473
  duration, protocol, 467
  equipment, 463–464
  functions, RNA, 462–463
  materials, 464–466
  preparation, 467, 468*f*
  robust synthetic access, 462–463, 463*f*
  safety measures, 467
  solutions and buffers, 466, 466*t*
Stable isotope labeling, proteins
  amino acids
    ^{1}H$_2$O-based reaction, 324–325
    main-chain NH$_3$ group, exchange, 325–326
    SAIL method, 324
    selective labeling, 323
    Tyr, 323–324
    uniform labeling, 323
  constructs and cell-free reaction conditions, 327–328, 327*f*
  large-scale production, ^{13}C/^{15}N-labeled proteins, 331–332
  medium-scale production, ^{15}N-labeled proteins
    apparatus, 327*f*, 328–331
    protocol, 329–330
    standard composition, reaction solution, 328–331, 329*t*
  peptide complexes
    APP, 336–337, 337*f*
    DOCK2, 338–339, 338–339*f*
    ELMO1, 338–339, 338–339*f*
    PID2, 336–337, 337*f*
    solution structures, NMR spectroscopy, 336–337, 338*f*
    tested fusion constructs, 338–339, 340*f*
  purification, synthesized proteins, 332–333
  site-directed labeling
    aminoacylation, 326
    chemical acylation, 326
    orthogonality, 326–327
  WWE domains, 333, 333*f*
  zinc-binding proteins
    ZBR domain, Emi2, 336, 336*f*
    ZFAT, 334–335, 335*f*

Stereo-array isotope labeling (SAIL)
    method, 324
Sucrose gradient ultracentrifugation,
    160–161
Switchgrass
  hemicellulose and glucan contents,
      232–233
  long-term hydroponic cultures, 232–233,
      233f
Synthesized peptide
  enrichment and digestion
    affinity tags, 358
    purification and digestion, 359–360
    Strep-Tactin beads, 360
    Strep-tag, 358–359
  quantification
    absolute quantification, 361–362, 362t,
        363f
    LOQ, 363
    MRM detection, 363
    reference peptide, accuracy, 360–361
    Strep-tag sequence, 361

T

Targeted assays, 348
Tobacco etch virus (TEV) protease,
    400–401
Total cellular RNA extraction
  analytical denaturing PAGE, 520
  duration, 519
  RNA precipitation, 519
  seperation, 518
Transfer RNA (tRNA)-scaffold plasmid,
    498
Transient transfection
  adenoviral shuttle vector, 296, 297f
  adenoviruses, generation of, 298
  gene expression, 292–293, 294f
  mammalian viruses, 295–296
  procedure, 293
  protein expression and construct
      selection, 293–295
  recombinant adenoviral genome,
      296–298, 297f
Transverse relaxation, 539
Triple labeling, protein. See Multiple
    labeling, protein
Tryptophan precursor fluoroindole, 78

U

Uniform isotope labeling, insect cells
  ^{13}C-labeling
    biochemical reaction, 262–265, 263f
    HNCA pathway, 265
    methyl groups, 262–265, 264f
  commercial media, 256
  culture monitor, 284
  economic media
    advantages, 258–259
    amino acid composition, 256–257,
        257f
    yeast-based extracts, 257–258
  harvest, 284
  ^{2}H-labeling
    alpha-protonation, glycine, 266–268,
        268f
    deuteration, eukaryotic cells,
        265, 267f
    metabolic pathways, 266–268, 268f
    protocols, 265–266
  infection, 282
  medium change, 283
  medium preparation, 282
  ^{15}N-labeling
    Abl spectrum, 259–261, 262f
    ammonia-containing media, 259–261
    biochemical reaction, 259, 260f
    cell culturing protocols, 259–261, 261f
  starter culture preparation, 282
Unlabeled and isotopically labeled RNA
    fragments
  multiple segmental isotope labeling,
      542–543, 543f
  sequence-specific RNase H cleavage,
      542, 542f
Unlabeled protein production
  expression cells and staged culturing
    flask culture 1, 10
    flask culture 2, 10
    flask culture 3, 10
    transformation, 9–10, 9s
  1 L bioreactor culture
    growth rate expectations and
        prediction, 12–13
    induction and harvest, 11–12
    SDS–PAGE analysis, 13
    set-up and inoculation, 10–11

V

Vascular endothelial growth factor-A (VEGFA), 415, 416f

W

Wide angle X-ray scattering (WAXS), 115–116, 115f
Wild-type K12 *E. coli*, expression
  analytical denaturing PAGE, 512, 512f
  duration, 512–513
  nonrecombinant tRNA-scaffold, 513
  safety measures, 513
Winter grain rye, D_2O and phenylalanine FTIR analysis, 236–239, 238f
  growth rate, comparison, 235–236, 236–237f, 238t
  plant growth solutions, 234–235
  shikimate biosynthetic pathway, 234, 235f

Z

Zinc-binding proteins
  ZBR domain, Emi2, 336, 336f
  ZFAT, 334–335, 335f
Zinc-binding region (ZBR) domain, Emi2, 336, 336f
Zinc-finger gene in AITD susceptibility region (ZFAT), 334–335, 335f

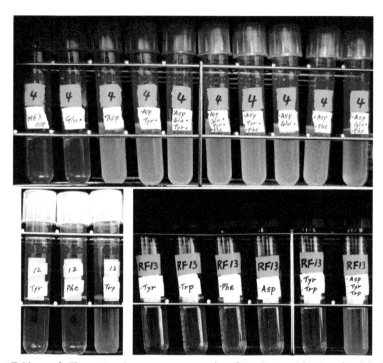

Myat T. Lin et al., Figure 1 L-Aspartate auxotrophy of RF4 (#4, top) having the deletions of *aspC* and *tyrB* and representing an ideal genotype for tyrosine labeling (Table 1). For comparison, RF12 (#12, bottom) displays L-tryptophan auxotrophy, whereas RF13 (#13, bottom), having the knockouts of *aspC*, *tyrB*, *trpA*, and *trpB*, and representing ideal genotypes for tyrosine and tryptophan labeling, does not grow in M63 minimal medium in the presence of either L-tyrosine, L-phenylalanine, or L-tryptophan alone (see also Table 1). These examples illustrate that *E. coli* amino acid metabolic network is more complex than the simple sums of individual amino acid biosynthesis pathway catalogs (Waugh, 1996) (cf. Table 1). Concentration(s) of amino acid(s) added to M63 minimal medium (where exist): 0.4 g/L L-glutamate, 0.4 g/L L-aspartate, 0.17 g/L L-tyrosine, 0.13 g/L L-phenylalanine, and 0.1 g/L L-tryptophan.

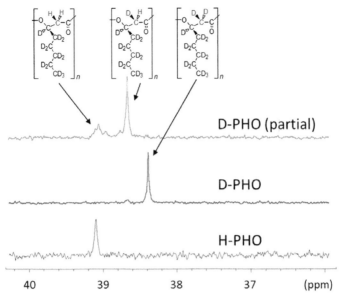

Robert A. Russell et al., Figure 3 Expansion of Fig. 2 in the region of 39 ppm.

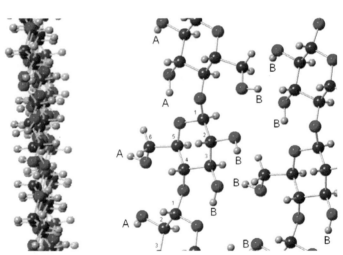

Robert A. Russell et al., Figure 6 Cellulose chains in a single hydrogen-bonded sheet of the monoclinic lattice in bacterial cellulose. The left-hand side shows a number of cellulose chains edge on. The pink axial atoms are those nonexchangeable hydrogens (attached to carbon atoms), and the blue atoms are those equatorial hydrogens which participate in hydrogen bonding. The right-hand side shows a region of the hydrogen-bonded sheet with an edge occupying chain. The –OH groups at the edge the sheet, A, are exchangeable in the solvent. Those –OH groups labeled B are in the interior of the cellulose crystallite and do not normally exchange.

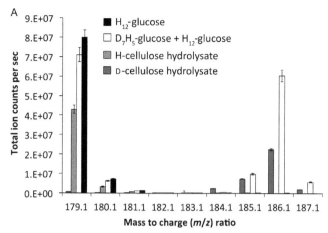
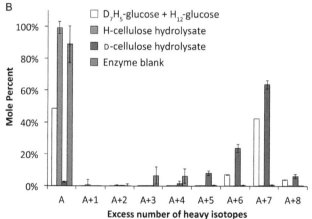

Hugh O'Neill et al., Figure 3 (A) Relative ion abundance and (B) calculated mole percent of major molecular species in hydrolysis products of deuterated and unlabeled cellulose. LC–MS was carried out under Q1 multiple ion monitoring using electrospray ionization in negative-ion mode to detect pseudomolecular ion ($[M-H]^-$) species in the range of m/z 179–187. Note that D_7H_5-glucose+H_{12}-glucose is an equimolar mixture of each. Error bars represent standard deviations for duplicate hydrolysate reaction mixtures. Percent mole of heavy isotopes relative to the unlabeled species (m/z 179) was calculated based on the procedure described by K. Biemann in *Mass spectrometry: Organic chemical applications*, McGraw-Hill, NY, 1962. Reproduced from He et al. (2014) with kind permission from Springer Science and Business Media.

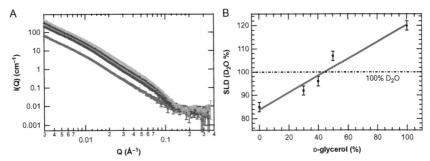

Hugh O'Neill et al., Figure 4 SANS analysis of purified bacterial cellulose with different levels of deuterium incorporation. (A) SANS profiles of celluloses grown using 100%, 50%, 40%, 30%, and 0% deuterated glycerol (orange, purple, green, blue, and red) measured in 100% H_2O. (B) Scattering length densities (SLDs) of different cellulose samples related to the fraction of deuterated glycerol present in the growth medium. *Reproduced from He et al. (2014) with kind permission from Springer Science and Business Media.*

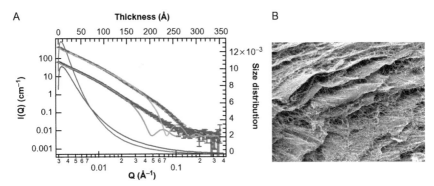

Hugh O'Neill et al., Figure 5 SANS analysis of purified bacterial cellulose with 0% and 100% deuterium incorporation. (A) SANS profiles of celluloses grown using 100% (blue open circles) and 0% (red dots) deuterated glycerol measured in 100% H_2O. Large disks (closest approximation to sheet-like forms) with monodisperse thickness (solid lines) and Schulz's polydisperse thickness (dashed lines) were used as models to fit data. The thickness distribution (100% and 0% deuterated glycerol—blue and red lines, respectively) is plotted against the thickness (top) versus size distribution (right). (B) SEM image of protiated bacterial cellulose (scale bar 50 μm). *Reproduced from He et al. (2014) with kind permission from Springer Science and Business Media.*

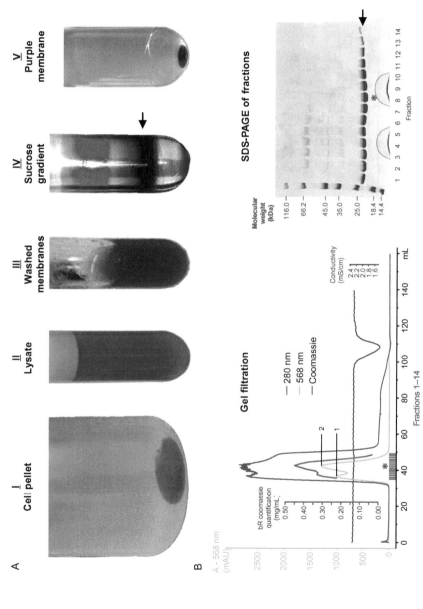

Thomas E. Cleveland IV and Zvi Kelman, Figure 1 See legend on next page.

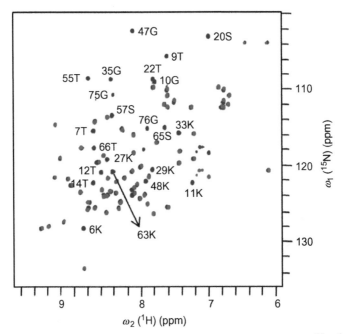

Chinmayi Prasanna et al., Figure 2 An overlay of selected region of 2D [^{15}N–^{1}H] HSQC spectrum of uniformly ^{15}N labeled (blue) and lysine, threonine (K, T) selectively unlabeled (red) samples of ubiquitin. Note that in addition to Lys and Thr, the resonances of Gly and Ser are also absent in the selectively unlabeled sample due to isotope scrambling as discussed in the text.

Thomas E. Cleveland IV and Zvi Kelman, Figure 1 Purification of bR. (A) The isolation of the "purple membrane" (PM) is illustrated. The following stages of the protein purification are shown: cell harvest (I), lysis in low-osmotic buffer (II), resuspended membranes after several wash cycles with low-osmotic buffer (III), the band of PM after passage through a sucrose step gradient (IV), and the final PM pellet after washing to remove sucrose (V). The two-colored compounds visualized in the sucrose gradient step are bacterioruberin (red, upper band) and bR (purple, marked with an arrow). The preparation changes from reddish to purple as the bacterioruberin is increasingly eliminated. (B) A final gel filtration chromatography step is used to isolate soluble and monodisperse bR. The first peak (labeled "1") is aggregated bR in the void volume and is discarded, while peak "2" is retained. The peak bR fraction is labeled with an asterisk (*) on the gel filtration chromatogram (left), and on the corresponding SDS-PAGE gel of its fractions (right). UV absorbance during gel filtration was measured at 568 nm, which gives a signal specific to bR; and at 280 nm, which detects total protein as well as Triton X-100. Due to the high absorbance of Triton X-100, the detector saturates at this wavelength. In addition, the Coomassie-stained bR band (marked with arrow) on the SDS-PAGE was quantified using ImageJ (Schneider, Rasband, & Eliceiri, 2012), and the quantification overlaid on the UV trace. Finally, the conductivity trace is shown; bR is generally purified in low ionic strength buffers.

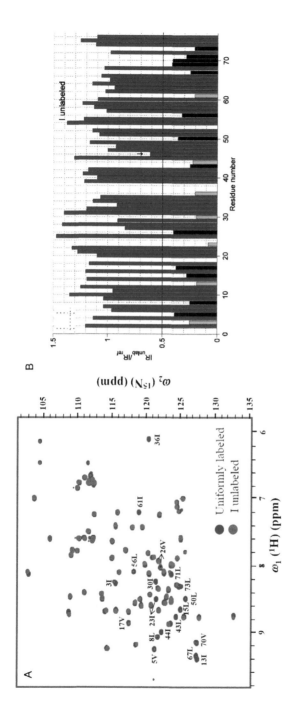

Chinmayi Prasanna et al., Figure 3 (A) An overlay of selected region of 2D [^{15}N–^{1}H] HSQC spectrum of uniformly ^{15}N labeled (blue) and isoleucine selectively unlabeled (red) samples of ubiquitin. Assignments for the unlabeled residues are indicated by the residue number. (B) Normalized intensity plot. Plots of $IR_{unlab,i}/IR_{ref,i}$: $IR_{unlab,i} = I_i^{unlab}/I_{control}^{unlab}$ and $IR_{ref,i} = I_i^{ref}/I_{control}^{ref}$ where i denotes the residue number. The colored bars indicate the following: black: $IR_{unlab,i}/IR_{ref,i} <0.5$, i.e., undergoing strong isotope scrambling (undesired residues corresponding to Leu and Val), green: desired selectively unlabeled residues (Ile), and blue: uniformly labeled residues. I denotes volume of the peak and "control" cenotes reference residue which does not undergo any effect of unlabeling in both selectively unlabeled and the reference sample. Residues E24 and G53 were not assigned and hence are absent along with Pro. The control residue chosen was G47.

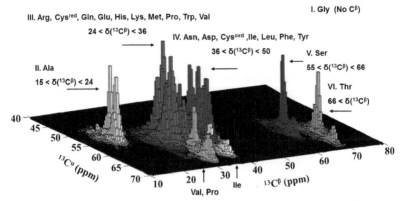

Chinmayi Prasanna et al., Figure 4 Amino acid types and their $^{13}C^{\alpha}$ and $^{13}C^{\beta}$ chemical shifts range (in ppm).

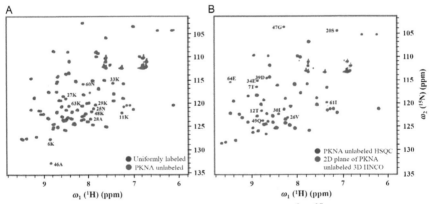

Chinmayi Prasanna et al., Figure 6 (A) An overlay of a 2D ^{1}H–^{15}N HSQC spectrum of PKNA selectively unlabeled sample (red) with that of a uniformly labeled sample of ubiquitin. (B) The overlay of a 2D ^{1}H–^{15}N HSQC of PKNA unlabeled sample (blue) with the 2D ^{1}H–^{15}N plane of a 3D HNCO. The residues which are $i+1$ to the unlabeled PKNA residues are additionally in the 2D plane of 3D HNCO when compared to the HSQC of PKNA sample.

Barbara R. Evans and Riddhi Shah, Figure 6 Compared to winter grain rye seedlings grown for 15 days in H_2O (A) and in 2 mM deuterated phenylalanine-$d8$ (B), seedlings grown for 18 days in 50% D_2O (C), and in 2 mM deuterated phenylalanine-$d8$ in 50% D_2O (D) are shorter and have stunted roots. Size bars = 10 cm.

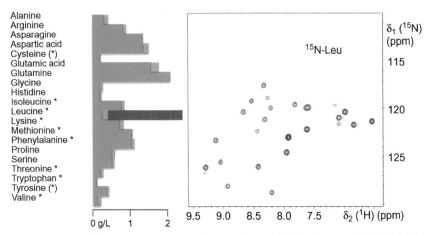

Lukasz Skora et al., Figure 3 *Amino acid-type selective labeling by excess.* In the left diagram, the amino acid composition of a customized medium for Leu labeling is shown. Amino acid amounts derived from pure sources and yeast extract are shown with green and brown bars, respectively. Labeled amino acids are indicated by magenta bars. Abl kinase was expressed in full SF-4 medium where 2 g/L of ^{15}N-Leucine was added. The final incorporation level of ^{15}N-Leu was determined to be 84% by mass spectrometry. *Figure reproduced from Gossert et al. (2011) with kind permission from Springer Science and Business Media.*

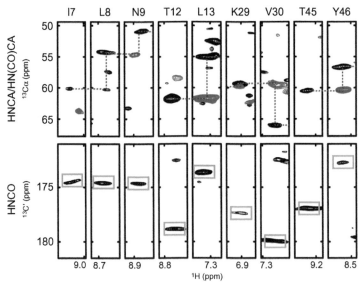

Lukasz Skora et al., Figure 15 *In-cell triple resonance experiments obtained in Sf9 insect cells.* ^{13}C,^{1}H^{N} strips of a 3D HNCA (black) and HN(CO)CA (red) spectra (overlaid, upper panel) and the 3D HNCO spectrum (lower panel) of Sf9 cells expressing the protein GB1. Cells were grown in ^{13}C,^{15}N-labeled Bioexpress 2000 medium (Cambridge isotope labs) for protein expression. *Reprinted with permission from Hamatsu et al. (2013). Copyright 2013 American Chemical Society.*

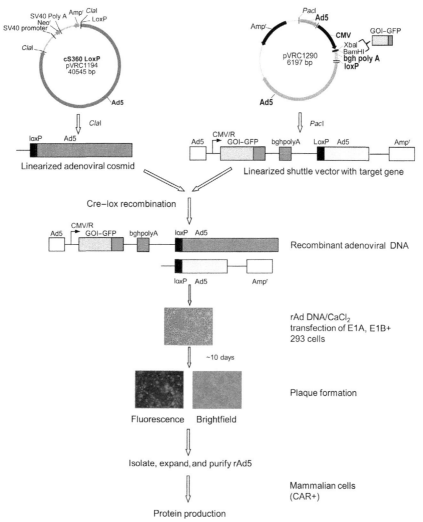

Mallika Sastry et al., Figure 2 Recombinant adenovirus as a tool to obtain isotopically labeled proteins. The target gene is cloned into a shuttle vector (pVRC1290) using the restriction sites (e.g., Xba1, BamH1) in the multiple cloning site; adenoviral cosmid DNA (pVRC1194) and shuttle vector are each linearized with ClaI and PacI, respectively, then recombined *in vitro* with Cre–Lox recombinase to obtain recombinant adenoviral genome (Aoki et al., 1999). The recombined adenoviral type 5 DNA is transfected into 293 adherent helper mammalian cells and recombinant adenovirus is isolated and purified using well-established methods. Target protein production is achieved by infecting CAR+ mammalian cells, such as A549 or CHO(CAR+).

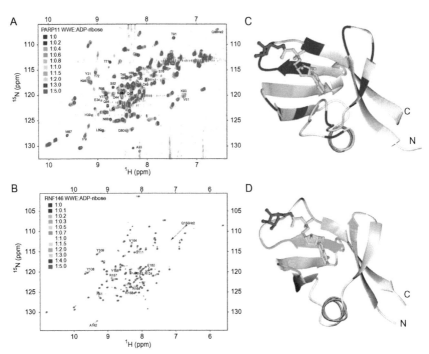

Takaho Terada and Shigeyuki Yokoyama, Figure 5 NMR chemical shift perturbations by ADP-ribose on the ^{1}H–^{15}N HSQC spectra of the WWE domains of PARP11 (A) and RNF146 (B), and color-coded mapping of the degrees of chemical shift changes of the perturbed residues on the model structures ((C) and (D), respectively). *Adopted with modifications from He et al. (2012).*

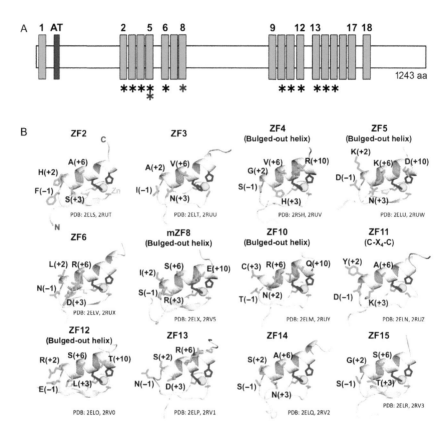

Takaho Terada and Shigeyuki Yokoyama, Figure 6 Zinc-finger motifs of ZFAT. (A) The positions of the 18 C_2H_2-type zinc-finger motifs (ZF1–ZF18) and the AT-hook motif (AT) are indicated by the green and blue boxes, respectively, along the primary structure of ZFAT. The black and violet asterisks indicate the zinc-finger motifs with solution structures determined for the human and mouse ZFAT proteins, respectively. (B) Ribbon diagrams of the solution structures of the isolated zinc-fingers, ZF2–ZF6, ZF8, ZF10–ZF15, of ZFAT. The zinc ion and the zinc-coordinating His and Cys residues of each zinc-finger motif are shown as a yellow ball and magenta and cyan stick models, respectively. The residues potentially involved in DNA recognition are shown as green stick models. *Adopted with modifications from Tochio et al. (2015).*

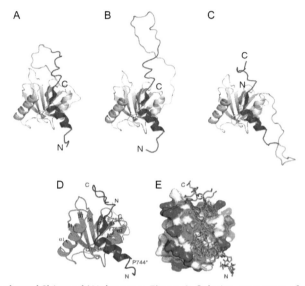

Takaho Terada and Shigeyuki Yokoyama, Figure 9 Solution structures of chimeras I–III and the complex between the Fe65L1 PID2 domain and the APP 32-mer peptide. (A–C) Ribbon representations of chimeras I–III, respectively. The Fe65L1 PID2 domain, the APP 32-mer peptide, and the linker are shown in gray, magenta, and yellow, respectively. (D) Ribbon representation of the complex between the Fe65L1 PID2 domain and the APP 32-mer peptide. The PID2 domain and the APP 32-mer peptide are shown in green and magenta, respectively. (E) Surface presentation of the Fe65L1 PID2 domain in complex with the APP 32-mer peptide, shown as a stick model. The PID2 surface is colored with the hydrophobic residues (Ala, Val, Leu, Ile, Met, Phe, and Trp) in green, the positively charged residues (Arg and Lys) in blue, the negatively charged residues (Asp and Glu) in red, and the other residues in gray. *Adopted with modifications from Li et al. (2008).*

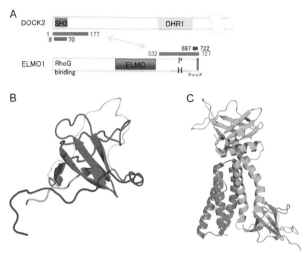

Takaho Terada and Shigeyuki Yokoyama, Figure 10 The complex of the SH3 domain of DOCK2 and the Pro-rich region of ELMO1. (A) Domain organizations of the N-terminal parts of DOCK2 and ELMO1. The regions involved in the association between DOCK2 and ELMO1 are residues 8–70 (red) and residues 697–722 (blue), respectively, for NMR spectroscopy and residues 1–177 (orange) and residues 532–727 (green) for crystallography. (B) The solution structure of the ELMO1 (697–722)–DOCK2 (8–70) fusion protein. The vector-derived peptide, the ELMO1 peptide, the linker, and the DOCK2 SH3 domain are colored light green, blue, gray, and red, respectively. (C) The crystal structure of the DOCK2 (1–177)·ELMOl (532–727) complex. DOCK2 and ELMO1 are shown in orange and green, respectively. *Adopted with modifications from Hanawa-Suetsugu et al. (2012).*

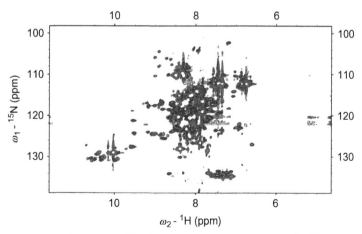

Aisha LaGuerre et al., Figure 1 ^{15}N–^{1}H BEST-TROSY of a uniformly ^{2}H,^{15}N-labeled four TMS integral membrane protein with (blue) and without (red) binding partner. TROSY spectra were measured on MP samples of 300 μM in 20 mM sodium acetate buffer pH 4 at 45° on an 800 MHz spectrometer.

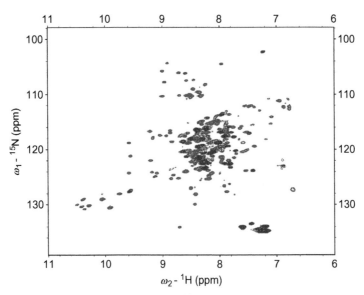

Aisha LaGuerre et al., Figure 2 Overlay of ^{15}N–^{1}H BEST-TROSY spectra of a uniformly and selectively labeled four TMS integral MP. The protein was uniformly labeled by CF expression in presence of either a complete set of ^{2}H,^{15}N-labeled amino acids (red) or a combination of the ^{2}H,^{15}N-labeled amino acids arginine, lysine, threonine, and tyrosine (blue). Amino acids were supplied at 4 mM final concentrations. TROSY spectra were measured on MP samples of 200 μM in 20 mM sodium acetate buffer pH 4 at 45° on an 800 MHz spectrometer.

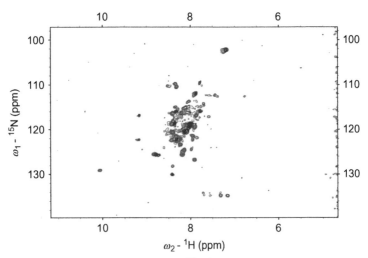

Aisha LaGuerre et al., Figure 3 Overlay of ^{15}N–^{1}H BEST-TROSY spectra of a four TMS integral membrane protein selectively labeled with ^{13}C, ^{15}N, ^{2}H aspartate and threonine. Blue: expression with scrambling inhibitors, aminooxyacetate, D-malic acid, and D-cycloserine; red: expression without scrambling inhibitors. TROSY spectra were measured on MP samples of 200 μM in 20 mM sodium acetate buffer pH 4 at 45° on a 600 MHz spectrometer.

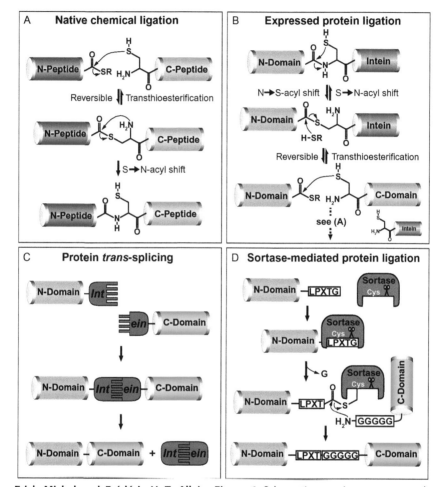

Erich Michel and Frédéric H.-T. Allain, Figure 1 Schematic overview on commonly applied methods for chemo-selective protein ligation: (A) native chemical ligation (NCL), (B) expressed protein ligation (EPL), (C) protein *trans*-splicing (PTS), and (D) sortase-mediated protein ligation.

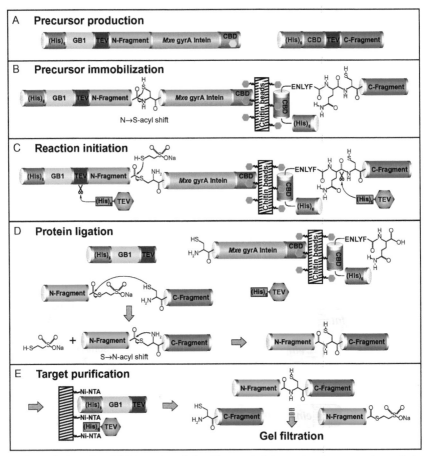

Erich Michel and Frédéric H.-T. Allain, Figure 4 Schematic representation of the on-column protein ligation protocol. (A) Crude solutions containing ligation-competent N- and C-terminal precursor molecules prepared by cell-free and/or recombinant expression in *E. coli* are mixed and passed over chitin beads. (B) The encoded chitin-binding domain retains the precursor molecules on the column. (C) The ligation reaction is initiated by addition of MESNA and TEV protease, which induces intein-mediated α-thioester generation and liberation of the N-terminal cysteine, respectively. (D) After ligation for at least 16 h at 25 °C, the reaction mixture is eluted (E) and is passed over Ni-NTA beads to remove the $(His)_6$-tagged protein constructs. The ligated protein is finally separated from unreacted fragments by gel filtration.

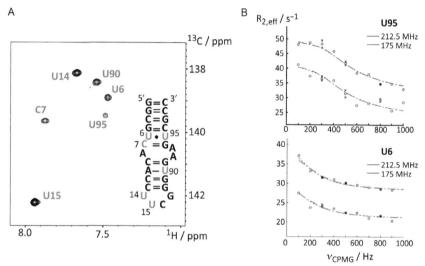

Christoph H. Wunderlich et al., Figure 11 (A) Standard heteronuclear single quantum correlation experiment confirming the successful incorporation of six 6-^{13}C-pyrimidine labels into the 27 nt A-site RNA. (B) Dispersion profiles of U6 and U95 at 212.5 (red) and 175 (blue) MHz carbon larmor frequency. Dots represent the experimental data, and black x represent repeat experiments. The dashed and dotted blue and red lines represent the individual and global fits for a two-state equilibrium in the intermediate exchange regime.

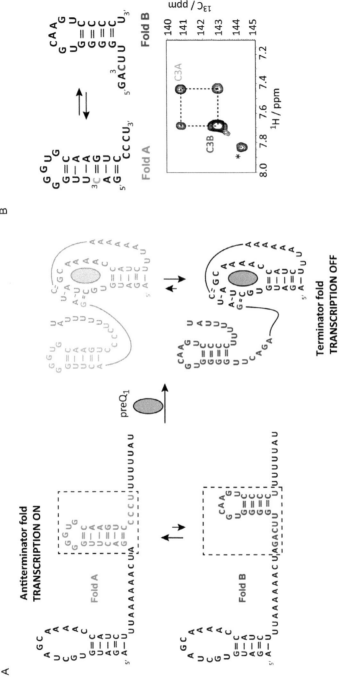

Christoph H. Wunderlich et al., Figure 12 (A) Competing terminator antiterminator RNA folds and the influence of preQ$_1$ binding on the 2° structure equilibrium. (B) ^{13}C longitudinal exchange spectrum at a mixing time of 200 ms and at 33 °C.

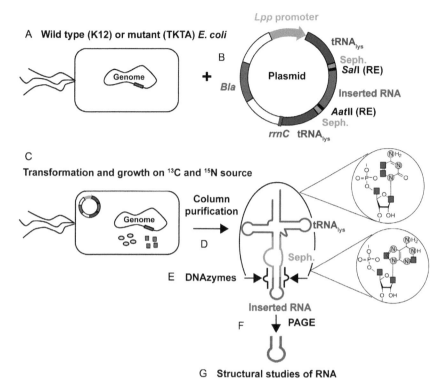

My T. Le et al., Figure 1 General scheme for the expression and purification of the recombinant tRNA-scaffold in wild type and mutant *E. coli* strains. (A) Wild type and mutant *E. coli* strains are made chemically competent using $CaCl_2$ so that they can be transformed using the pBSKrnaSeph plasmid (*E. coli* tRNA$_{lys}$, Sephadex aptamer) (Ponchon et al., 2009). (B) General diagram of the tRNA-scaffold plasmid: The RNA of interest is inserted at *Sal*I and *Aat*II restriction sites, which surround the anticodon loop of the tRNA$_{lys}$-scaffold. The sephadex tag is included for affinity purification. The RNA transcript is under the control of the lipoprotein promoter (*lpp*) and terminated with a ribosomal RNA operon transcription terminator (*rrnC*). (C) Incorporation of NMR-active isotopes into the recombinant tRNA-scaffold: The tRNA-scaffold plasmid is taken up by *E. coli* using heat-shock transformation. *E. coli* grown in minimal media supplemented with ^{15}N-ammonium sulfate and/or 1-^{13}C-glucose or 1-^{13}C-acetate incorporates these isotopic nuclei into all cellular RNAs, including the overexpressed recombinant tRNA-scaffold. ^{13}C isotopes are denoted by purple squares and ^{15}N isotopes are denoted by yellow circles. (D) The recombinant tRNA-scaffold is extracted and then purified by anion exchange or affinity chromatography. (E) Two DNAzymes are used to cleave the 5' and 3' ends of the inserted RNA. (F) Inserted RNAs are purified using PAGE. (G) The final RNA products are used for structural studies.

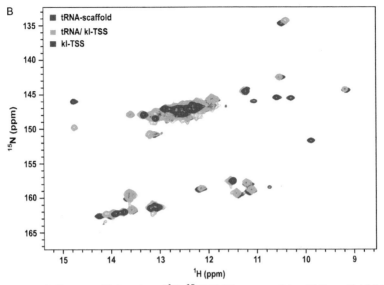

My T. Le et al., Figure 8 (B) Overlay of ^{1}H–^{15}N HQSC spectra of the tRNA-scaffold (blue), tRNA/kl-TSS (cyan), and the kl-TSS (purple).

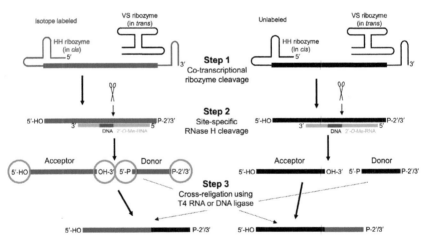

Olivier Duss et al., Figure 3 Principle of sequence-specific RNase H cleavage to generate RNA fragments with correctly protected termini for subsequent ligation. The isotope-labeled material is highlighted in red, the unlabeled material in black. In the 2′-O-methyl RNA/DNA chimera, the DNA is in dark blue, and the 2′-O-methyl RNA in light blue. The termini of both acceptor and donor fragments are encircled in green. Scissors indicate RNase H cleavage sites. P-2′/3′ stands for a 2′/3′-cyclic phosphate. *This figure has been taken from Duss et al. (2010) by permission of Oxford University Press.*

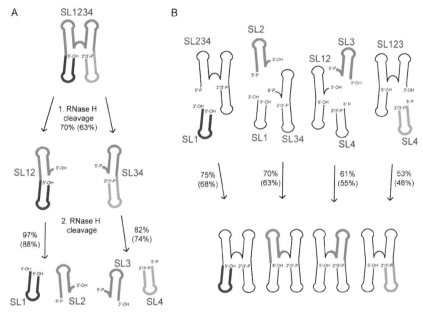

Olivier Duss *et al.*, Figure 4 Principle of multiple segmental isotope labeling using sequence-specific RNase H cleavage to generate RNA fragments (A) with subsequent ligation of isotope-labeled and unlabeled RNA fragments using splinted T4 DNA ligation (B). The RNase H cleavage and T4 DNA ligation efficiencies are shown; values in brackets are after denaturing anion-exchange HPLC purification. *This figure has been taken from Duss et al. (2010) by permission of Oxford University Press.*

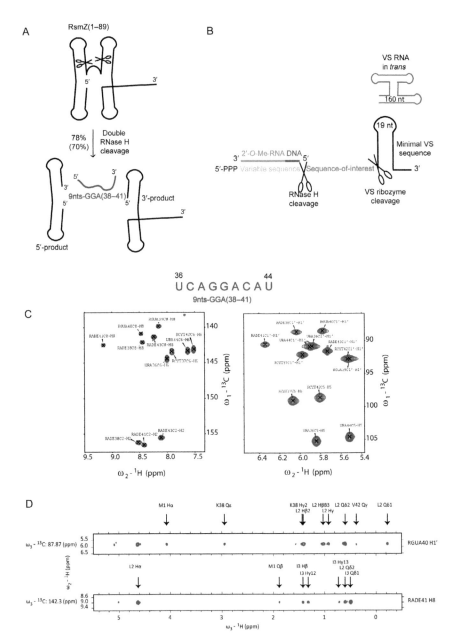

Olivier Duss et al., Figure 5 Production of small isotope-labeled RNAs. Small isotope-labeled RNAs can be obtained by double RNase H cleavage (A) or combined RNase H and VS ribozyme cleavage (B) from a larger RNA precursor. (A) Scheme of double RNase H cleavage of the RsmZ (1–89) precursor RNA into a 5′-product, a 3′-product, and the 9 nts-GGA (39–41) RNA of interest (red). This requires two 2′-O-Me-RNA/DNA chimeras specifically hybridizing to the desired cleavage sites on the RsmZ (1–89) RNA (not shown for simplicity). (B) Combined RNase H and VS cleavage: The RNA of interest (green) is flanked by a well-transcribing sequence (gray) at the 5′-end and a minimal VS ribozyme stem-loop (SL) sequence at the 3′-end. The 5′-sequence is cleaved off with an RNase H catalyzed reaction mediated by a 2′-O-Me-RNA (cyan)/DNA (blue) chimera. The VS ribozyme minimal sequence at the 3′-end (black) is cleaved off with the corresponding VS RNA in *trans* (orange). (C) ^{1}H-^{13}C HSQC spectra of the aromatic region (left) and aliphatic H1′ and H5 region (right) of the 21-kDa complex of ^{13}C,^{15}N-labeled 9 nts-GGA(39–41) RNA bound to the RsmE protein (Duss, Michel, Diarra dit Konte, Schubert, & Allain, 2014). (D) G40 H1′ and A41 H8 strips of the 3D filtered-edited-^{1}H-^{13}C HSQC-NOESY spectrum of the same protein–RNA complex as in (C).

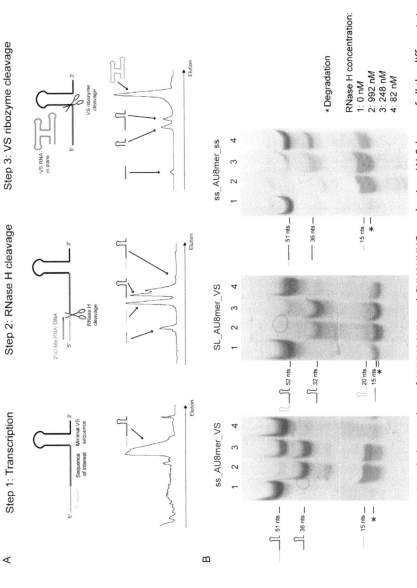

Olivier Duss et al., Figure 6 Detailed procedure of AUUUAAUU and CUUAUUUG production. (A) Scheme of all the different steps. The steps are described in the upper panel and the corresponding HPLC run are in the lower panel. (B) Gels of RNase H site-specific cleavage. The 2'-O-methyl RNA/DNA chimera can diffuse out of the gel during the destaining process, hence the absence of the corresponding band in the left and right gel.

Edwards Brothers Malloy
Ann Arbor MI. USA
November 4, 2015